Y0-BYB-153

Methods in Cell Biology

VOLUME 82

Laser Manipulation of Cells and Tissues

Series Editors

Leslie Wilson

Department of Molecular, Cellular and Developmental Biology
University of California
Santa Barbara, California

Paul Matsudaira

Whitehead Institute for Biomedical Research
Department of Biology
Division of Biological Engineering
Massachusetts Institute of Technology
Cambridge, Massachusetts

Methods in Cell Biology

VOLUME 82

Laser Manipulation of Cells and Tissues

Edited by

Michael W. Berns

Beckman Laser Institute
University of California
Irvine, California

Karl Otto Greulich

Leibniz Institute for Age Research
Abteilung Einzelzell und Einzelmolekueltechniken
Jena, Germany

AMSTERDAM • BOSTON • HEIDELBERG • LONDON
NEW YORK • OXFORD • PARIS • SAN DIEGO
SAN FRANCISCO • SINGAPORE • SYDNEY • TOKYO
Academic Press is an imprint of Elsevier

Cover Photo Credit: Pictures depicting different aspects of light interaction with cells. *Top Right* from Berns, Chapter 1 and Berns *et al.*, 1981, Science 213: 505–513. Image reproduced with permission of the copyright owner the American Association for the Advancement of Science. *Top Left* from Botvinick and Wang, Chapter 18. *Bottom Right* from Vogel *et al.*, Chapter 5. *Bottom Left* from Schütze *et al.* Chapter 23.

Academic Press is an imprint of Elsevier
525 B Street, Suite 1900, San Diego, California 92101-4495, USA
84 Theobald's Road, London WC1X 8RR, UK

This book is printed on acid-free paper. ♾

For information on all Academic Press publications
visit our Web site at www.books.elsevier.com

ISBN-13: 978-0-12-370648-5
ISBN-10: 0-12-370648-3

PRINTED IN THE UNITED STATES OF AMERICA
07 08 09 10 9 8 7 6 5 4 3 2 1

CONTENTS

PART III Laser Scissors/Ablation

CONTRIBUTORS

Numbers in parentheses indicate the pages on which the authors' contributions begin.

Nancy L. Allbritton (709), Department of Physiology and Biophysics, School of Health Sciences, University of California, Irvine, California 92697; Departments of Biomedical Engineering, Chemistry, and Chemical Engineering and Materials Science, University of California, Irvine, California 92697

Alexander R. Ball, Jr. (377), Department of Biological Chemistry, School of Medicine, University of California, Irvine, California 92697

Judith Baumgart (293), Laser Zentrum Hannover, Hollerithallee 8, D-30419 Hannover, Germany

Michael W. Berns (3, 457, 725), Departments of Biomedical Engineering, Developmental and Cell Biology, and Surgery, Beckman Laser Institute, University of California, Irvine, California 92612; Departments of Bioengineering, Whitaker Institute for Biomedical Engineering, University of California, San Diego, La Jolla, California 92093

Elliot L. Botvinick (81, 497, 601), Beckman Laser Institute, University of California, Irvine, California 92612

A. Buchstaller (649), Institute of Pathology, Ludwig Maximillians University, Munich, Germany

Tomáš Čižmár (467), Institute of Scientific Instruments ASCR, v.v.i., Academy of Sciences of the Czech Republic, 61264 Brno, Czech Republic

Imran Clark (309), Cyntellect, Inc., San Diego, California 92121

Julien Colombelli (267), Light Microscopy Group, Cell Biology and Biophysics Unit, European Molecular Biology Laboratory (EMBL), D-69117 Heidelberg, Germany

Brian R. Daniels (111), Department of Chemical and Biomolecular Engineering, The Johns Hopkins University, Baltimore, Maryland 21218

Benayahu Dafna (675), Department of Cell and Developmental Biology, Sackler School of Medicine, Tel-Aviv University, Israel

Kishan Dholakia (467), SUPA, School of Physics and Astronomy, University of St. Andrews, Fife, KY16 9SS Scotland

Thomas J. Diefenbach (335), Department of Physiology, Tufts University School of Medicine, Boston, Massachusetts 02111

Diarmaid H. Douglas-Hamilton (409), Hamilton Thorne Biosciences, Beverly, Massachusetts 01915

Emma Eriksson (629), Department of Physics, Göteborg University, SE-412 96 Göteborg, Sweden

Brenda K. Eustace (335), Vertex Pharmaceuticals Inc., Cambridge, Massachusetts 02140

Trisha Eustaquio (309), Cyntellect, Inc., San Diego, California 92121

Andreas Gebert (153), Institute of Anatomy, University of Lübeck, Ratzeburger Allee 160, D-23538 Lübeck, Germany

Mattias Goksör (629), Department of Physics, Göteborg University, SE-412 96 Göteborg, Sweden

Karl Otto Greulich (59, 725), Leibniz Institute of Age Research/Fritz Lipmann Institute, D-07745 Jena, Germany, e-mail: Kog@fli-leibniz.de.

P. K. Gupta (563), Laser Biomedical Applications and Instrumentation Division, Raja Ramanna Centre for Advanced Technology, Indore 452013, India

Naomi S. Hachiya (355) Department of Neurophysiology, Tokyo Medical University, 6-1-1 Shinjuku, Shinjuku-ku, Tokyo 160-8402, Japan

Jason T. Heale (377), Department of Biological Chemistry, School of Medicine, University of California, Irvine, California 92697

Norman R. Heckenberg (207, 525), Centre for Biophotonics and Laser Science, School of Physical Sciences, The University of Queensland, Brisbane QLD 4072, Australia

Alexander Heisterkamp (293), Laser Zentrum Hannover, Hollerithallee 8, D-30419 Hannover, Germany

Polla Hergert (239), Division of Molecular Medicine, Wadsworth Center, Albany, New York 12201

Roman Hobza (433), Laboratory of Plant Developmental Genetics, Institute of Biophysics, Academy of Sciences of the Czech Republic, CZ-612 65 Brno, Czech Republic

Diane Hoffman-Kim (335), Department of Molecular Pharmacology, Physiology, and Biotechnology, Center for Biomedical Engineering, Brown University, Providence, Rhode Island 02912

Verena Horneffer (153), Institute of Biomedical Optics, University of Lübeck, Peter-Monnik Weg 4, D-23562 Lübeck, Germany

Kwame L. Hoyte (309), Cyntellect, Inc., San Diego, California 92121

Gereon Hüttmann (153), Institute of Biomedical Optics, University of Lübeck, Peter-Monnik Weg 4, D-23562 Lübeck, Germany

Shur Irena (675), Department of Cell and Developmental Biology, Sackler School of Medicine, Tel-Aviv University, Israel

Daniel G. Jay (335), Department of Physiology, Tufts University School of Medicine, Boston, Massachusetts 02111

Kiyotoshi Kaneko (355), Department of Neurophysiology, Tokyo Medical University, 6-1-1 Shinjuku, Shinjuku-ku, Tokyo 160-8402, Japan

Alexey Khodjakov (239, 725), Division of Molecular Medicine, Wadsworth Center, Albany, New York 12201; Department of Biomedical Sciences, SUNY, Albany, New York 12222; Marine Biological Laboratory, Woods Hole, Massachusetts 02543

Jong-Soo Kim (377), Department of Biological Chemistry, School of Medicine, University of California, Irvine, California 92697

Gregor Knöner (207, 525), Centre for Biophotonics and Laser Science, School of Physical Sciences, The University of Queensland, Brisbane QLD 4072, Australia

Manfred R. Koller (309), Cyntellect, Inc., San Diego, California 92121

Xiangduo Kong (377), Department of Biological Chemistry, School of Medicine, University of California, Irvine, California 92697

Tatiana B. Krasieva (377), Beckman Laser Institute, Department of Surgery, Laser Microbeam and Medical Program, University of California, Irvine, California 92697

Norbert Linz (153), Institute of Biomedical Optics, University of Lübeck, Peter-Monnik Weg 4, D-23562 Lübeck, Germany

Jadranka Lončarek (239), Division of Molecular Medicine, Wadsworth Center, Albany, New York 12201

Kathrin Lorenz (153), Institute of Biomedical Optics, University of Lübeck, Peter-Monnik Weg 4, D-23562 Lübeck, Germany

Holger Lubatschowski (293), Laser Zentrum Hannover, Hollerithallee 8, D-30419 Hannover, Germany

Michael P. MacDonald (467), SUPA, School of Physics and Astronomy, University of St. Andrews, Fife, KY16 9SS Scotland

Valentin Magidson (239), Division of Molecular Medicine, Wadsworth Center, Albany, New York 12201

Iva Z. Maxwell (293), Department of Engineering and Applied Science, Harvard University, Cambridge, Massachusetts 02138

Eric Mazur (293), Department of Engineering and Applied Science, Harvard University, Cambridge, Massachusetts 02138

S. K. Mohanty (563), Laser Biomedical Applications and Instrumentation Division, Raja Ramanna Centre for Advanced Technology, Indore 452013, India

Jaclyn M. Nascimento (601), Department of Electrical and Computer Engineering, University of California, San Diego, La Jolla, California 92093

Allison R. Nelson (709), Department of Physiology and Biophysics, School of Health Sciences, University of California, Irvine, California 92697

Anaclet Ngezahayo (293), Institute of Biophysics, Leibniz University Hannover, D-30419 Hannover, Germany

Timo A. Nieminen (207, 525), Centre for Biophotonics and Laser Science, School of Physical Sciences, The University of Queensland, Brisbane QLD 4072, Australia

Yilmaz Niyaz (649), PALM Microlaser Technologies GmbH, Am Neuland 9 + 12, 82347 Bernried, Germany

Simon Parkin (525), Centre for Biophotonics and Laser Science, School of Physical Sciences, The University of Queensland, Queensland 4072, Australia

Pedro A. Quinto-Su (113), Department of Chemical Engineering & Materials Science, University of California, Irvine, California 92697; Laser Microbeam and Medical Program, Beckman Laser Institute, University of California, Irvine, California 92612

Emmanuel G. Reynaud (267), Light Microscopy Group, Cell Biology and Biophysics Unit, European Molecular Biology Laboratory (EMBL), D-69117 Heidelberg, Germany

Kate Rhodes (309), Cyntellect, Inc., San Diego, California 92121

Conly L. Rieder (239), Division of Molecular Medicine, Wadsworth Center, Albany, New York 12201; Department of Biomedical Sciences, SUNY, Albany, New York 12222; Marine Biological Laboratory, Woods Hole, Massachusetts 02543

Socher Rina (675), Department of Cell and Developmental Biology, Sackler School of Medicine, Tel-Aviv University, Israel

Halina Rubinsztein-Dunlop (207, 525), Centre for Biophotonics and Laser Science, School of Physical Sciences, The University of Queensland, Brisbane QLD 4072, Australia

K. Schütze (649), PALM Microlaser Technologies GmbH, Am Neuland 9 + 12, 82347 Bernried, Germany

Jan Scrimgeour (629), Department of Physics, Göteborg University, SE-412 96 Göteborg, Sweden

Jagesh V. Shah (81), Laboratory for Cellular Systems Biology and Molecular Imaging, Department of System's Biology, Harvard Medical School, Boston, Massachusetts 02115

Bing Shao (601), Department of Electrical and Computer Engineering, University of California, San Diego, La Jolla, California 92093

Linda Z. Shi (601), Department of Bioengineering, University of California, San Diego, La Jolla, California 92093

Christopher E. Sims (709), Department of Physiology and Biophysics, School of Health Sciences, University of California, Irvine, California 92697

Wolfgang Singer (525), Centre for Biophotonics and Laser Science, School of Physical Sciences, The University of Queensland, Queensland 4072, Australia

Ernst H. K. Stelzer (267), Light Microscopy Group, Cell Biology and Biophysics Unit, European Molecular Biology Laboratory (EMBL), D-69117 Heidelberg, Germany

M. Stich (649), PALM Microlaser Technologies GmbH, Am Neuland 9 + 12, 82347 Bernried, Germany

Yona Tadir (409), Beckman Laser Institute, University of California, Irvine, California 92612; Department of Obstetrics and Gynecology, Rabin Medical Center, Tel Aviv University, Israel

Vasan Venugopalan (113), Department of Chemical Engineering & Materials Science, University of California, Irvine, California 92697; Laser Microbeam and Medical Program, Beckman Laser Institute, University of California, Irvine, California 92612

Alfred Vogel (153), Institute of Biomedical Optics, University of Lübeck, Peter-Monnik Weg 4, D-23562 Lübeck, Germany

Annette Vogt (725), Max Planck Institute for History of Science, 14195 Berlin, Germany

Boris Vyskot (433), Laboratory of Plant Developmental Genetics, Institute of Biophysics, Academy of Sciences of the Czech Republic, CZ-612 65 Brno, Czech Republic

Hong-Yang Wang (689), International Cooperation Laboratory on Signal Transduction, Eastern Hepatobiliary Surgery Institute, The Second Military Medical University, Shanghai 200438, People's Republic of China

Yingxiao Wang (497), Department of Bioengineering, Molecular and Integrative Physiology, Neuroscience Program, Beckman Institute for Advanced Science and Technology, University of Illinois, Urbana-Champaign, Illinois 61801

Kyoko Yokomori (377), Department of Biological Chemistry, School of Medicine, University of California, Irvine, California 92697

Michelle Zatcoff (309), Cyntellect, Inc., San Diego, California 92121

Pavel Zemánek (467), Institute of Scientific Instruments ASCR, v.v.i., Academy of Sciences of the Czech Republic, 61264 Brno, Czech Republic

Weihua Zeng (377), Department of Biological Chemistry, School of Medicine, University of California, Irvine, California 92697

PREFACE

Both of us are so excited about the combination of optical and molecular tools that are now available that we wish we were just starting our research careers! We feel that this excitement *must* be conveyed to both young and "seasoned" researchers who are looking for ways to analyze and solve problems that, perhaps, the more conventional methods alone are unable to do.

With the above in mind, we undertook an assembly of chapters that deal with the long history of the "microbeam" approach (see Chapters 1, 2, and 7), and the rest of the book which is a group of chapters that primarily focus on the use of light as (1) a microablation tool (scissors), Chapters 4, 5, and 7–15; (2) a noninvasive force-generator (tweezers), Chapters 6 and 16–22; and (3) a mechanism to catapult molecules and fragments of cells into other devices for molecular analysis (catapulting and capture), Chapters 23–26.

But before embarking on an exploration of this book, we would like to recount for the reader/browser some of the key people and our own *moments* in a field that both of us have devoted our careers to.

One of us (M.W.B.) was introduced to the "microbeam" by one of my Ph.D. committee professors, Adrian Srb, a geneticist at Cornell University, who took me aside one day in 1965 and said, "There's a new thing called a laser. We have one down in the basement of Stimson Hall [now an historical building at Cornell University in Ithaca New York] that is coupled to a microscope. No one knows quite what to do with it, so why don't you see if it will be useful in your experiments to 'micro-dissect' the *proliferation zone* in the anal segment of the millipedes that you are studying." My Ph.D. project was to try to understand how the millipede regulates the number of body segments and legs that it makes during its postembryonic development. Ironically, the system did not work for my application because the ruby laser wavelength 694.3 nm [see diagram and photo, Figures 2–7 and 2–8 in Berns (1974)] would not pass though the hard pigmented exoskeleton of the millipedes. However in doing my background research for these experiments, I learned that lasers were not the first radiation sources attached to microscopes.

In fact, we both discovered that there was a long history of "microbeams" going all the way back to 1912 when Sergei Tschachotin (also spelled *Tschakotine* and *Chakhotin*) published his first description of an ultraviolet light source attached to a microscope for the purpose of partial irradiation of biological objects (Tschachotin, 1912). In fact, the trials and tribulations of Sergei Tschachotin throughout his prolific scientific and sociopolitical life is an intriguing story in its own right. In many ways, Tschachotin was a scientific and political "nomad" (see his biography in Chapter 27). He was an outspoken critic of both the Nazi and

Stalin regimes, and he literally roamed throughout Europe as a "man without a country" at times being a scientist and at times being an author and political critic. This made him persona non grata in many countries from the 1920s until his return to the Soviet Union in the 1950s. He went to Germany for his graduate studies ultimately attaining the Ph.D. from the University of Heidelberg in 1908. His scientific tradition was continued in the 1970s and 1980s by the brothers Drs. Christoph and Thomas Cremer. They introduced one of us (K.O.G.) into the field, who had his scientific cradle in Heidelberg until he changed, after the reunification of Germany, to Jena in the former German Democratic Republic. This revelation now explains the long successful tradition of microbeam irradiation at the University of Heidelberg (see References in Chapter 1 to the 1970s work of Christoph and Thomas Cremer, and Jurgen Bereiter-Hahn).

Certainly, Sergei Tschachotin must be considered one of the twentieth century pioneers of experimental cytology and embryology. Without his seminal work, this book simply would not have been written. He also was a brave man who was not afraid to express his political views against totalitarian and immoral political regimes. For that we are all in his debt.

When the laser came on the scene in the early 1960's, it was immediately recognized as a potential, powerful, and revolutionizing force for the field of microbeam irradiation. Through the work of Frenchmen, Marcel Bessis and G. Nomarski in 1963, and subsequent studies by the French "School" of microbeams (see Chapter 1) the laser was shown to have unique advantages attributable to its monochromaticity and intensity. By 1967, when the first NATO Advanced Study Institute on "Microbeam and Partial Cell Irradiation" was held in Cannes, France (September 20–29), it was clear that the laser was the "light wave" of the future for microbeams. When the second NATO Advanced Study Institute on the same subject was held in June of 1970, in Stresa, Italy, it was recognized that the laser was, indeed, the radiation of choice for selective cell manipulation through the microscope. With the invention of optical tweezers and its adaptation for nondestructive cell manipulation by Arthur Ashkin in 1987, the laser microbeam "toolbox" became complete in such a way as could never be imagined by the two true generally unsung pioneers, Sergei Tchakhotine and Marcel Bessis.

Now as we gaze over the past and present landscape of a field that started almost 100 years ago by a man who was a scientist and political critic (at a time when political critics somehow disappeared), *microbeams* are no longer a technology primarily of the wealthier countries such as the United States, Japan, and Europe. This field is now "global" as evidenced by the list of contributors from all over the World.

Michael W. Berns
Karl Otto Greulich

References

Berns, M. (1974). "Biological Microirradiation: Classical and Laser Sources." Prentice Hall, NY.

Tschachotin, S. (1912). Die mikroskopische Strahlenstich Methode, eine Zelloperationsmethode. *Biol. Zentralbl.* **32,** 623.

PART I

Reviews

CHAPTER 1

A History of Laser Scissors (Microbeams)

Michael W. Berns

Departments of Biomedical Engineering, Developmental and
Cell Biology, and Surgery
Beckman Laser Institute, University of California
Irvine, California 92612

Department of Bioengineering
Whitaker Institute for Biomedical Engineering
University of California, San Diego
La Jolla, California 92093

This introductory chapter reviews the history of microbeams starting with the original UV microbeam work of Tchakhotine in 1912 and covers the progress and application of microbeams through 2006. The main focus of the chapter is on

0091-679X/07 $35.00
DOI: 10.1016/S0091-679X(06)82001-7

laser "scissors" starting with Marcel Bessis' and colleagues work with the ruby laser microbeam in Paris in 1962. Following this introduction, a section is devoted to describing the different laser microbeam systems and then the rest of the chapter is devoted to applications in cell and developmental biology. The approach is to focus on the organelle/structure and describe how the laser microbeam has been applied to studying its structure and/or function. Since considerable work has been done on chromosomes and the mitotic spindle (Section V.A and C), these topics have been divided in distinct subsections. Other topics discussed are injection of foreign DNA through the cell membrane (optoporation/optoinjection), cell migration, the nucleolus, mitochondria, cytoplasmic filaments, and embryos fate-mapping. A final technology section is devoted to discussing the pros and cons of building/buying your own laser microbeam system and the option of using the Internet-based RoboLase system. Throughout the chapter, reference is made to other chapters in the book that go into more detail on the subjects briefly mentioned.

I. Introduction

For the purposes of this chapter, "laser scissors" (or the more historical term "laser microbeam") is defined as "the use of a laser for microscopic alteration and/or ablation of subcellular, cellular, or tissue samples." Laser scissors may be used to ablate or alter an organelle inside a cell in order to better understand the function of that organelle such as the centriole in cell division (La Terra *et al.*, 2005; see Chapter 7 by Magidson *et al.*, this volume) and cell movement (Koonce *et al.*, 1984). Or it may be used to destroy the entire cell itself in order to free the internal constituents for analysis (Sims *et al.*, 1998; see Chapter 26 by Nelson *et al.*, this volume). Or it may be used to cut out small segments of chromosomes in order to create a genetic library (Monajembashi *et al.*, 1986; see Chapter 2 by Greulich and Chapter 15 by Hobza and Vyskot, this volume). Or it may be used to "catapult" and capture either whole cells or fragments of cells from a diseased tissue in order to analyze the genetic makeup of that cell/tissue (see Chapter 23 by Schütze *et al.* and Chapter 24 by Dafna *et al.*, this volume). It may also be used to destroy single cells or groups of cells in developing organisms in order to understand the developmental fate of those cells—such as in the formation of the nervous system (Chalfie *et al.*, 1985; Hall and Russel, 1991). Clinically it may be used to carve a trench in the *zona pellucida* layer surrounding human eggs in order to improve the fertility (Neev *et al.*, 1992; Tadir *et al.*, 1992; see Chapter 14 by Tadir and Douglas-Hamilton, this volume). These are just a few examples of the multitude of studies and practical applications of laser scissors, most of which will be discussed to varying degrees in this book.

II. Prehistory of Laser Scissors

In 1912, the Russian, Sergej Tschachotin, built an ultraviolet (UV) light microirradiation device for the purposes of destroying regions of single cells, embryos, and organisms (Tschachotin, 1912). This was the beginning of a field which has been variably referred to as partial cell irradiation (PCI) in the earlier literature, and in the later literature as microirradiation, microbeams, and now laser scissors and/or laser microbeams (Berns, 1974a, Berns *et al.*, 1998b).

For most of his studies, Tschachotin used a magnesium spark to generate 280-nm UV light that was focused by a quartz objective from below while the specimen was observed with a standard glass objective from above (Fig. 1; see also excellent discussion on Tschachotin's contributions in Chapter 7 by Magidson *et al.*, this volume). Optical limitations at that time allowed for a focused UV spot of a diameter of 5 μm (Tschachotin, 1912, 1921, 1937). Even with these limitations, between the period of 1912–1955 Tschachotin used the UV microbeam to study cells from a wide variety of cells and organisms: eggs from sea urchins (Tschachotin, 1920, 1929) and mollusks (Tschachotin, 1938), numerous protozoans such as *Paramecium* (Tschachotin, 1935a, b), *Vorticella* (Tschachotin, 1936a), and *Euglena* (Tschachotin, 1936b). He published

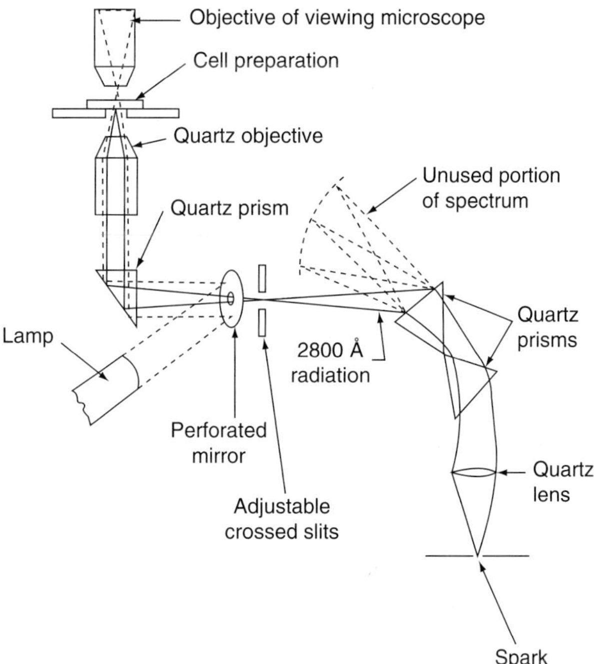

Fig. 1 Tschachotin's 280-nm microbeam device. Adapted from Zirkle (1957) and Berns (1974a).

papers in Russian, German, and French journals, and a final short review "Problems and Methods in Experimental Cytology," in Italian in 1955 (Tschachotin, 1955).

In one series of studies on the biology of the ciliate *Paramecium,* he irradiated the contractile vacuoles, the organelles that regulate osmotic balance between the cell and its surrounding water environment (Tschachotin, 1935c,d). He was able to show in *Paramecium* with two contractile vacuoles that if one was damaged with the UV microbeam it swelled and stopped its contractile activity. However, in response to this disruption, the other one increased its contractile rate in order to compensate for the damaged one. This study demonstrated that there was some kind of internal regulatory mechanism(s) between the two contractile vacuoles in order to maintain physiological equilibrium between the cell and its environment. In fact, when both contractile vacuoles were destroyed with the UV microbeam, either the *Paramecium* died or in some cases, new contractile vacuoles were formed to replace the damaged ones. The control and regulation of the synthesis of an organelle in response to an experimentally induced physiological deficit is still one of the few examples of this phenomenon in biology.

Although much of the early groundbreaking work with UV microbeams was done by Tschachotin in the first half of the twentieth century, parallel efforts to expose parts of cells to ionizing X-rays were being undertaken (Vintemberger, 1929a,b,c). Subsequent studies also employed PCI with gamma rays and alpha particles (Zirkle, 1932). However, PCI with ionizing radiation posed a far more complex set of problems with respect to constructing either partial cell shielding and/or aperture devices that limited a selected region in the cell to radiation exposure. This was an extremely difficult task fraught with numerous technical hurdles (see discussion in Berns, 1974a, pp. 21–26; Zirkle, 1957).

By and large, most microbeam irradiation was with UV light. Major advances were made with the development of low-resolution-reflecting quartz objectives (Fig. 2; Forer, 1966; Zirkle,1957), and ultimately with the development of achromatically corrected UV transmitting quartz objectives (Fig. 3; Bessis and Normarski, 1959). The latter development was significant because, as a result, it became possible to focus the UV light to near-diffraction limits yielding focal spots of 0.2–5 μm in diameter (Bessis and Nomarski, 1960). This approach had a significant advantage over the use of reflecting objectives which could not be used in a phase-contrast mode, and due to the low numerical aperture could not focus the beam to less than 2 μm in diameter (Zirkle, 1957). Notwithstanding the limitations of reflecting and quartz objectives, numerous studies were conducted on subcellular irradiation of chromosomes (Zirkle and Uretz, 1963), disruption of the mitotic spindle, and cytoplasm (Forer, 1965; Zirkle, 1970). In fact, studies using nonlaser UV microbeams have continued well into the laser microbeam era by several laboratories (Spurck *et al.*, 1990; Walker *et al.*, 1989; Waterman-Storer and Salmon, 1998). Clearly, PCI using UV from nonlaser light sources has had a long history, starting in 1912 and continuing to the present day.

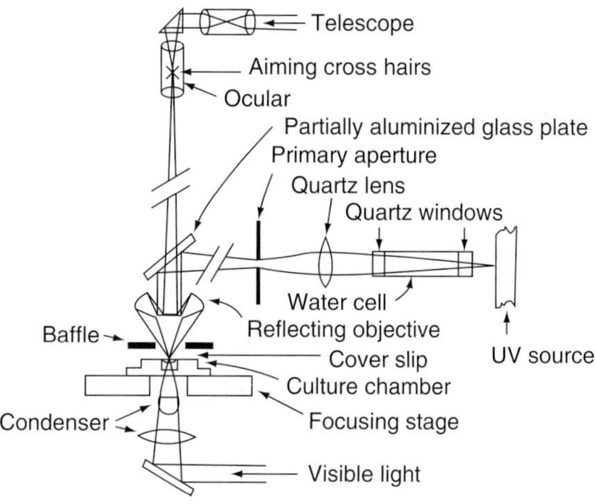

Fig. 2 Reflecting objective UV microbeam. Adapted from Zirkle (1957) and Berns (1974a).

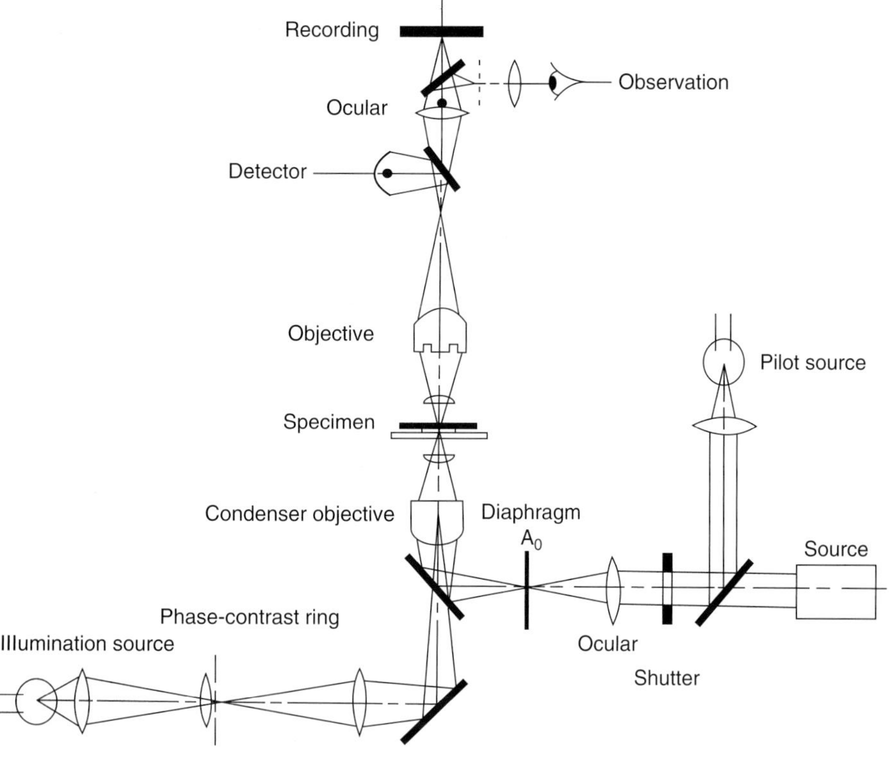

Fig. 3 The Bessis–Nomarski condenser-objective UV microbeam. Adapted from Bessis and Nomarski (1959).

III. The Ruby Laser Microbeam: The Beginning

The first laser was built in 1960 at Hughes Aircraft Company, California by Theodore Maiman (1960). In 1962, Marcel Bessis and his colleagues in Paris built the first laser microbeam using the ruby laser (Bessis *et al.*, 1962). Because the radiation was from the visible (694.3 nm) portion of the spectrum, it could be delivered to the specimen through a standard 100× magnification glass objective with a numerical aperture of 1.25. The focus spot diameter of this system was 2.5 μm (Fig. 4). The system employed a light source in an epi-illumination mode, two dielectric filters, two polarizers, and the laser. The system was very simple in design and did not use either expensive quartz optics or low-resolution-reflecting objectives.

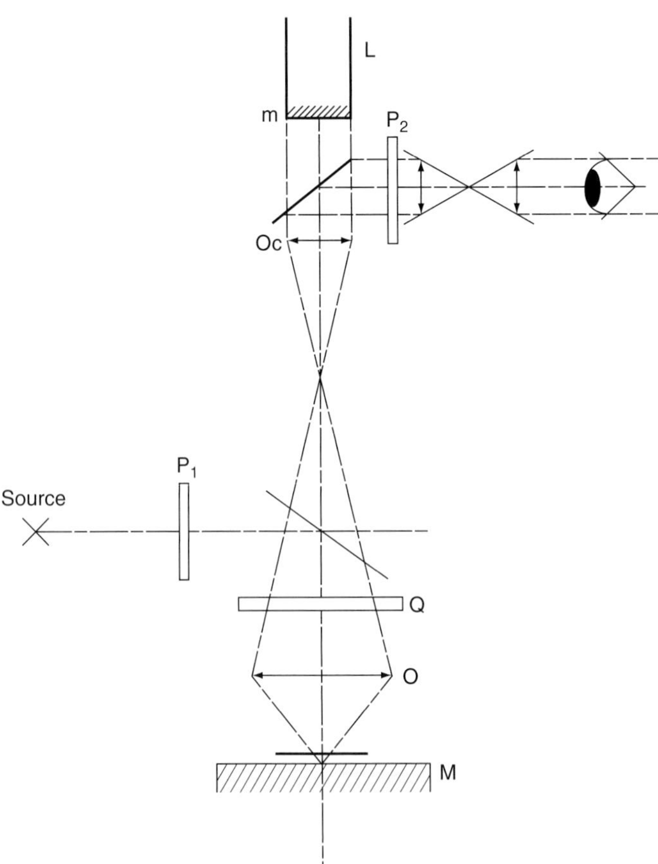

Fig. 4 The Bessis *et al.*, Ruby laser microbeam: L, laser; m, laser emission surface; M, dielectric mirror; Oc, ocular; O, objective; Q, quartz quarter wave plate; P1 and P2, polarizers. Adapted from Bessis *et al.* (1962).

The ruby laser microbeam was used to alter cells and tissue in a precise way (see early reviews by Bereiter-Hahn, 1972a; Berns and Salet, 1972; Moreno and Salet, 1969). These studies focused on the effects of the laser microbeam on the mitochondria in the cytoplasm of vertebrate tissues culture cells and involved mitochondrial-specific cytochemical staining (Amy and Storb, 1965), detailed transmission electron microscopy of the laser-induced structural effects (Storb *et al.*, 1966), and cell viability studies following irradiation (Amy *et al.*, 1967; Wertz *et al.*, 1967). These studies laid the groundwork for future studies that established the laser microbeam as an important tool in cell and developmental biology (Berns *et al.*, 1981).

The ruby laser microbeam also was used on single-cell organisms, gametes, and embryos. Soon after Marcel Bessis and his colleagues described ruby laser micro-irradiation of cells, Norman Saks and his colleagues in New York City demonstrated the following after ruby laser irradiation to subcellular regions of the cytoplasm in the single-cell organism, *Amoeba*: (1) a reduction in cytoplasmic streaming, (2) ejection of the damaged cytoplasm, and (3) a reduction in growth rate (Saks and Roth, 1963; Saks *et al.*, 1965). In gametes, ruby laser microirradiation also was used to localize function along the length of the flagellum of sperm from sea urchins (*Strongylocentrous*) and starfish (*Pisaster*; Goldstein, 1969). These studies looked at the relationships between length of flagellum fragments and continued wave propagation, as well as beat patterns distal and proximal to the laser-induced damage site.

Eggs and embryos were studied using the ruby laser microbeam. For example, in nuclear transplantation studies, the laser was used to destroy the haploid female nucleus followed by the implantation of a diploid nucleus into an enucleated egg in frogs (McKinnell *et al.*, 1969). It was also used for selective ablation of the germinal crescent in chick embryos. This resulted in chickens without the ability to make gametes (Mims and Mckinnell, 1971). These studies as well as the early studies by Bessis and his colleagues in Paris were the prelude to a wide range of subcellular, cellular, and embryo ablation ("fate-mapping") studies using other laser microbeam systems which contributed greatly to a better understanding of cell biology and the development of a wide variety of organisms (Berns, 1974a; Berns *et al.*, 1981 reprinted with the permission of *Science* as appendix at the end of this chapter).

IV. Laser Systems: Beyond the Ruby Laser

It would be fair to say that each time a new laser was developed, it was interfaced with a microscope and used for microbeam irradiation. The key laser parameters of interest were wavelength and pulse duration. Wavelength was important because of the ability to match it with either natural absorbance or absorbance facilitated through an applied chromophore (Berns and Rounds, 1970a,b; Berns and Salet, 1972). Pulse duration became important with the advent of the Q-switched nanosecond lasers because of the ability to generate nonlinear multiphoton ablation (Berns *et al.*, 1981; Calmettes and Berns, 1983). With the development of the picosecond and femtosecond lasers it has become possible to combine both multiphoton ablation with multiphoton-excited fluorescence in the same cell and even the same organelle (Berns *et al.*, 2000; Sacconi *et al.*, 2005). In addition, the femtosecond laser has been used to produce 100-nm size cuts in fixed and dried human chromosomes (Konig *et al.*, 2001). Excellent reviews on the mechanisms of short pulse laser ablation can be found in Vogel *et al.* (2005), Vogel and Venugopalan (2003), and Chapter 4 by Quinto-Su and Venugopalan, this volume.

Following the ruby laser, the blue-green (488 and 514 nm; 50 μs) argon ion laser was the next laser to become available, and it was rapidly adopted as a "workhorse" for cellular micromanipulation/ablation studies. The first such study was published in *Nature* in 1969 and was done in the laboratory of Donald E. Rounds, Pasadena, California (Berns *et al.*, 1969a). The system was actually designed and built by a brilliant self-trained engineer Robert E. Olson. It utilized a rotating front-surfaced mirror with a 5-mm hole drilled-through so that the 3-mm-diameter laser could pass into the microscope (Fig. 5). The mirror rotated at 60 Hz with virtually no vibration. This resulted in a stable image projected onto to the video camera. Also, because of

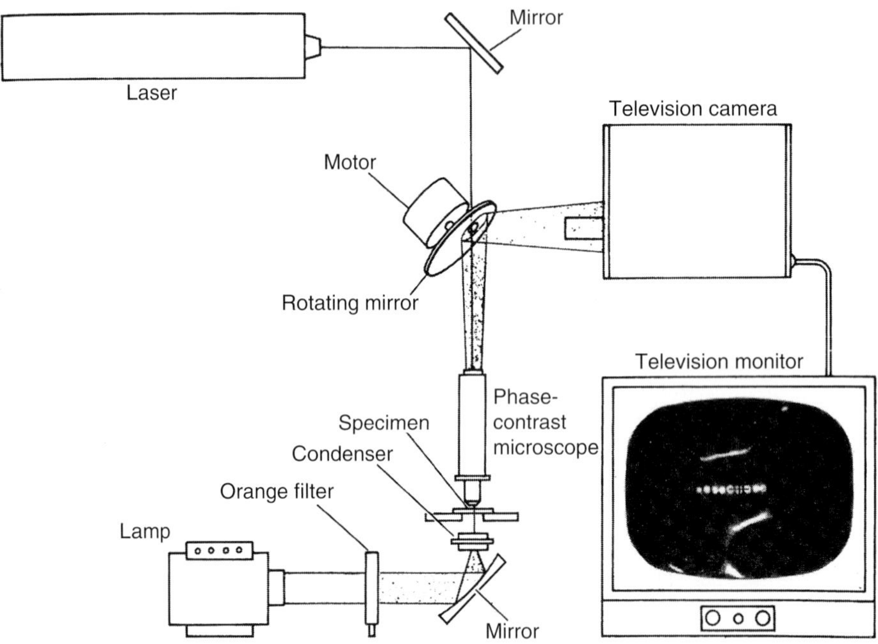

Fig. 5 Argon ion laser microbeam at the Pasadena Foundation for Medical Research, Pasadena, California, 1968. The laser emitted a 50-μs mixed wavelength (primary wavelengths were 488 and 514 nm) beam that had a diameter of 3 mm and a pulse repetition rate of 60 Hz. The laser had peak power of 1 W and an average power of 3 mW. Single or multiple pulses of the laser passed through a 5-mm-diameter hole a synchronized rotating (60 Hz) aluminum-coated front-surfaced mirror. Images of the target cell (organelle) were projected off the front surface-spinning mirror into the television camera and displayed on the monitor. A target crosshair was drawn on the television monitor.

the fast rotation rate, the hole through which the laser passed was not visible in the video image. The laser and the rotating mirror were electronically synchronized so that the 50-μs duration laser pulse passed through the hole when it was positioned directly above the microscope optical path. This system was used in the first studies on chromosome surgery in live cells (Berns *et al.*, 1969a,b, 1970a), microdissection of the nucleolus in live cells (Berns *et al.*, 1969c), and microirradiation of mitochondria in contracting cardiac cells *in vitro* (Berns *et al.*, 1970b).

In addition to the early French (ruby laser) and American (argon laser) laser microbeam studies, it is also important to recognize that about this same time Jurgen Bereiter-Hahn, at the Institute für Kinematische Zellforschung der Universitat Frankfurt, Germany, was using a microscope-focused helium–cadmium laser at 441.6 nm to alter mitochondria in epithelial cells of the South African clawed toad, *Xenopus laevis* (Bereiter-Hahn, 1972a). The effect on the mitochondria was attributed to natural absorbance by the mitochondrial respiratory pigments. In addition, ruby laser cell microsurgery on fish melanocytes was performed at about this same time (Egner and Bereiter-Hahn, 1970). Bereiter-Hahn's early work on melanocytes and mitochondria and his two review articles on the laser "micromanipulator" in medicine and biology (Bereiter-Hahn, 1972a,b) set the stage for a long line of excellent studies by German scientists. See subsequent studies by Christoph and Thomas Cremer with the UV laser microbeam (Cremer *et al.*, 1974), followed by Lohs-Schardin's work with the Cremers' and Christiane Nusslein-Volhard (Lohs-Schardin *et al.*, 1979a,b) using the UV laser to study developing drosophila embryos by selective microablation, and more recent UV studies using the pulsed third harmonic wavelength (355 nm) of the Nd:YAG laser to cut single microtubules in live cells in order to study their polymerization/depolymerization dynamics (Colombelli *et al.*, 2005; see Chapter 8 by Colombelli *et al.*, this volume). In 1986, the groups of professors Karl Otto Greulich and Jürgen Wolfrum used an excimer-pumped dye laser to microdissect chromosomes (Monajembashi *et al.*, 1986). These investigators subsequently perfected chromosome microdissection for the generation of DNA libraries using the much simpler pulsed nitrogen laser (see Section V.A.4 and Chapter 2 by Greulich, this volume). More recently, the near-infrared femtosecond laser has been used for delicate subcellular microsurgery on fixed chromosomes with damage measured as small as 100 nm (Konig *et al.*, 2001).

An important early technical advance in the 1970s was the development and perfection of a wide range of dichroic interference filters (Ifs) with optical coatings resistant to laser-induced damage. A dichroic If-based reflection of the argon ion laser beam into the microscope (Fig. 6A and B; Berns, 1971) led to the use of a dye laser microbeam that was tunable from the UV to the near-infrared region of the spectrum (Berns, 1972). The use of a series of dichroic filters in combination with tunable dye lasers and multiple harmonic generation allowed virtually any combination of wavelength reflection and transmission. Using an array of dichroic optics and 10 ns and 100 ps Nd:YAG/dye lasers, a laser microbeam system was built that could be used to ablate portions of chromosomes and a large number of other subcellular organelles (see Fig. 1 in Appendix, this chapter). A subsequent iteration of this system (Fig. 7) included the integration of two trapping laser beams such that a chromosome could be held in the traps, cut, and the two separate chromosome pieces moved independently using two laser traps (see "Chromosome Micromanipulation" at www.robolase.ucsd.edu/movies.html). This system, the confocal ablation trapping system (CATS; Berns *et al.*, 1998b), became the core microbeam system of the NIH LAMMP Biotechnology Resource at the Beckman Laser Institute at the University of California, Irvine, established in 1980; http://www.bli.uci.edu/lammp/.

As mentioned previously, the 257-nm-pulsed UV laser was first described in 1974 (Cremer *et al.*, 1974) and it was rapidly adopted in developmental biology to study drosophila embryos (Lohs-Schardin *et al.*, 1979a,b). These and other embryonic and developmental studies will be discussed in Section VI. Additionally, the 337-nm-pulsed nitrogen laser as well as other lasers, currently are being used to

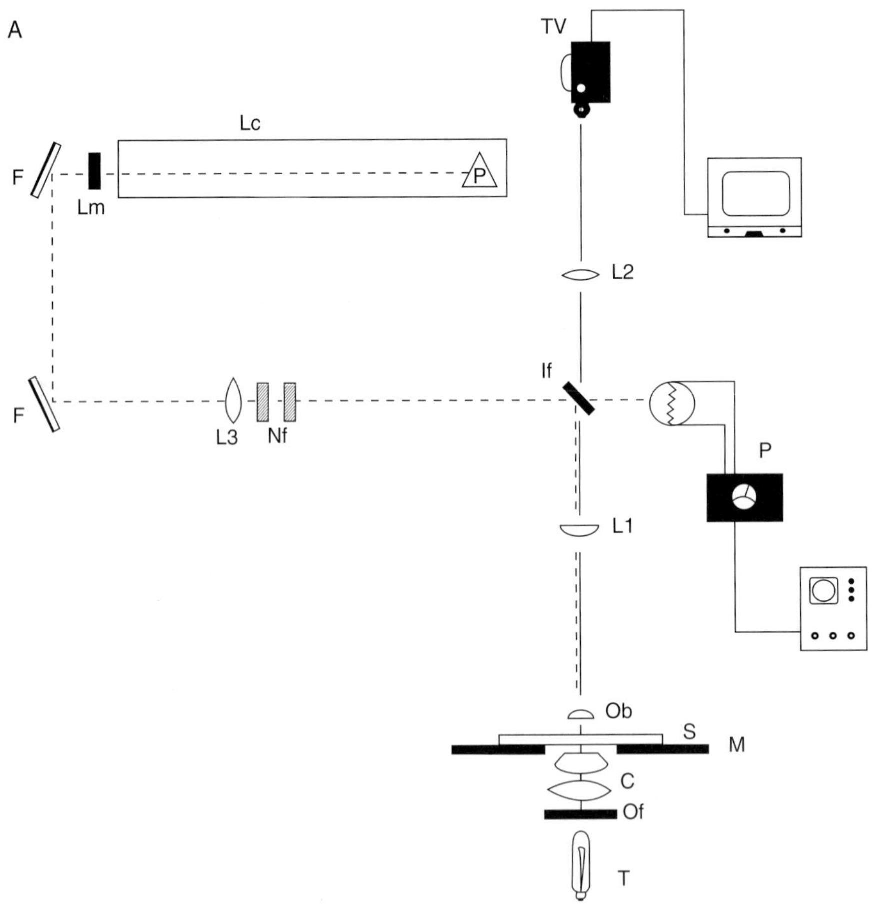

Fig. 6 (A) Diagram of laser microbeam: C, condenser; F, front-surfaced mirrors; If, interference filter; L_1, 60-mm focal length lens; L_2, 10-cm focal length lens; L_3, 100-cm focal length lens; Lc, laser cavity; Lm, laser output mirror, M, microscope stage; Nf, calibrated neutral density filters; Ob, 100× Zeiss neofluar objective; Of, orange filter; P, photodiode with attached photometer and oscilloscope; Δ, wavelength selector prism; S, specimen chamber; TV, television camera and monitor; T, tungsten light source; -----, laser beam; ——, substage illumination. (B) Argon ion laser microbeam at the University of Michigan, Ann Arbor, 1970–1972, and the University of California, Irvine, 1972–1980. The laser was variable pulsed (1–200 pulses/s) with a 35-W peak power operating at 60 Hz. The 3-mm beam was reflected into the microscope by a dichroic interference filter [If in (A)] that reflects 90% of the light below 520 nm into the microscope and transmits 90% of the light above 520 nm to the video camera mounted above the microscope [see (A)].

induce double-strand breaks (DSBs) in DNA in order to study the recruitment and function of DNA repair molecules (Kim *et al.*, 2005; Rogakou *et al.*, 1999; also see Chapter 11 by Hoffman-Kim *et al.*, this volume).

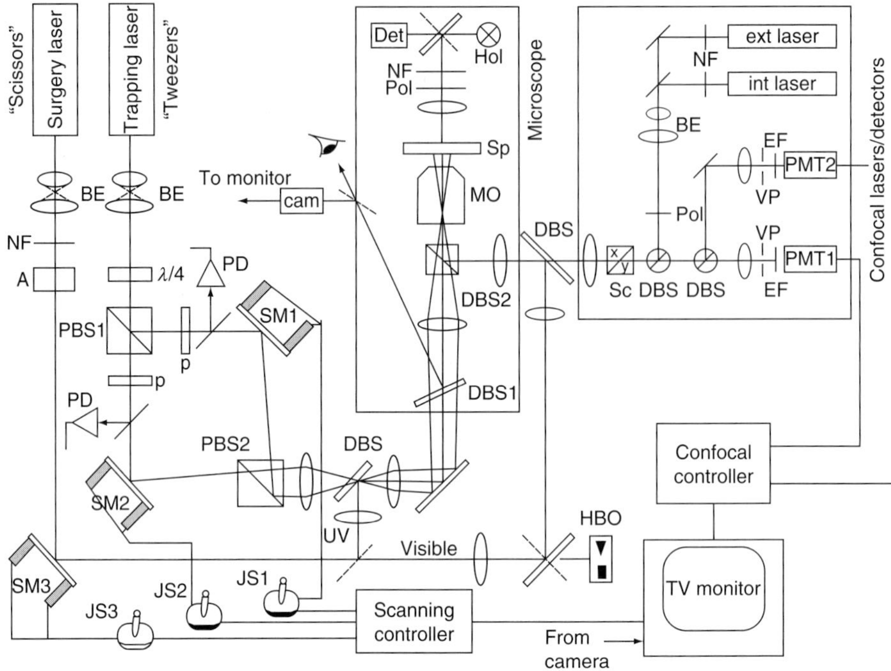

Fig. 7 Confocal ablation trapping system (CATS). Two external lasers are brought into the confocal microscope. The "tweezers" laser is a Ti:sapphire laser and is divided into two beams by prism beam splitter (PBS1) and recombined into two coaxial beams by PBS2. Movement of each of the two tweezer beams is controlled by scanning mirrors (SM1 and SM2) which are controlled by two joystick controllers (JS1 and JS2). The "scissors" laser is an Nd:YAG laser that can operate at the fundamental 1.06 μm and the second (532 nm), third (355 nm), and fourth (266 nm) harmonic wavelengths. The scissors beam reflects off a joystick-controlled (JS3) scanning mirror (SM3) and enters the microscope by two different paths depending on the wavelength used. Other elements in the diagram are as follows: BE, beam expander; A, attenuator; PD, photodiode detector; DBS, dichroic beam splitter; UV, ultraviolet reflector; CAM, video camera; MO, microscope objective; Sp, specimen; Pol, polarizer; NF, neutral density filter; Sc, x-y scanner; PMT, photomultiplier tube; HBO, mercury lamp.

Laser ablation (scissors) also has been combined with laser tweezers (optical traps; Greulich *et al.*, 1989). In one of the first published studies combining these two laser manipulation tools, individual chromosomes were manipulated using combined laser scissors and tweezers (Greulich, 1999, p. 194; Seeger *et al.*, 1991). Chromosomes removed from cells and placed in solution on either microscope slides or cover glasses were cut with a nitrogen laser (337 nm) and the laser-dissected pieces were moved around with optical tweezers. Using this method, over 100 chromosomes segments could be generated and collected without mechanical contact and in a sterile manner. In another study, combined laser scissors and tweezers were used to facilitate the fusion of two myeloma cells (Wiegand Steubing *et al.*, 1991; see video clip titled

"Cell Fusion" online at www.robolase.ucsd.edu/movies.html). In this study, a 1.06-μm Nd:YAG laser (220-mW average power in the focal spot) was use to trap a myeloma cell in culture and bring it in contact with a second myeloma cell. When the cell membranes were in contact with each other, a UV (337 nm) nitrogen laser beam was focused on to the two apposed cell membranes. The result was the creation of a fusion of the two cell membranes resulting in the formation of a single fused cell. Although the initial frequency of successful fusion and survival was only 1%, the addition of polyethylene glycol (PEG) increased cell survival to 11%, thus making this method a potential viable approach to producing selected cell hybrids in situations where only a few cells are available for hybridization. Additionally, chromosomes within live cells in mitosis have been cut with laser scissors and the cut segments moved around within the living cell (Liang *et al.*, 1994). Despite these early studies, remarkably few studies have taken advantage of the combination of laser scissors and tweezers for cell manipulation. However, the development of fluorescent fusion proteins (Khodjakov *et al.*, 1997) and advances in microscopic imaging systems have generated renewed interest in combining these two optical technologies to manipulate and study cells.

V. Cells and Organelles

A. Chromosomes

1. Structural: Surgery and Damage Analysis

As described in the previous section, the first experiment using a laser microbeam to alter chromosomes in live dividing cells was published in *Nature* magazine in 1969 (Berns *et al.*, 1969a). This study used a 1-W average power argon ion laser emitting at the primary wavelengths of 488 and 514 nm (50-μs pulse width; microjoules per pulse). Because both the energy and power in the focal spot were not high enough to cause a visible alteration in irradiated chromosomes, a vital stain (acridine orange) was used to sensitize the chromosomes to the laser beam. Since acridine orange intercalates directly between the base pairs of the DNA double helix, it was possible to sensitize the DNA and the chromosome to the laser light. The result was the production of micron-size lesions in the chromosome that appeared as actual "nicks" or full breaks in the chromosomes (Fig. 8A and B). A subsequent study demonstrated that by careful manipulation of the acridine orange dye concentration and laser parameters it was possible to selectively vary the degree of the chromosome damage from a slight phase "paling" (dye concentration of 0.02 μg/ml) to a complete chromosome break (dye concentration of 25 μg/ml; Berns *et al.*, 1969c). Interestingly, the phase "paling," which represented a change in refractive index as seen with the phase-contrast microscope but distinct structural damage when observed with the transmission electron microscope (Fig. 8C; Rattner and Berns, 1974), appeared similar to the paling seen in subsequent studies with nanosecond and picosecond 532-nm Nd:YAG and recently 200-fs 800-nm lasers (Fig. 9). In addition, detailed

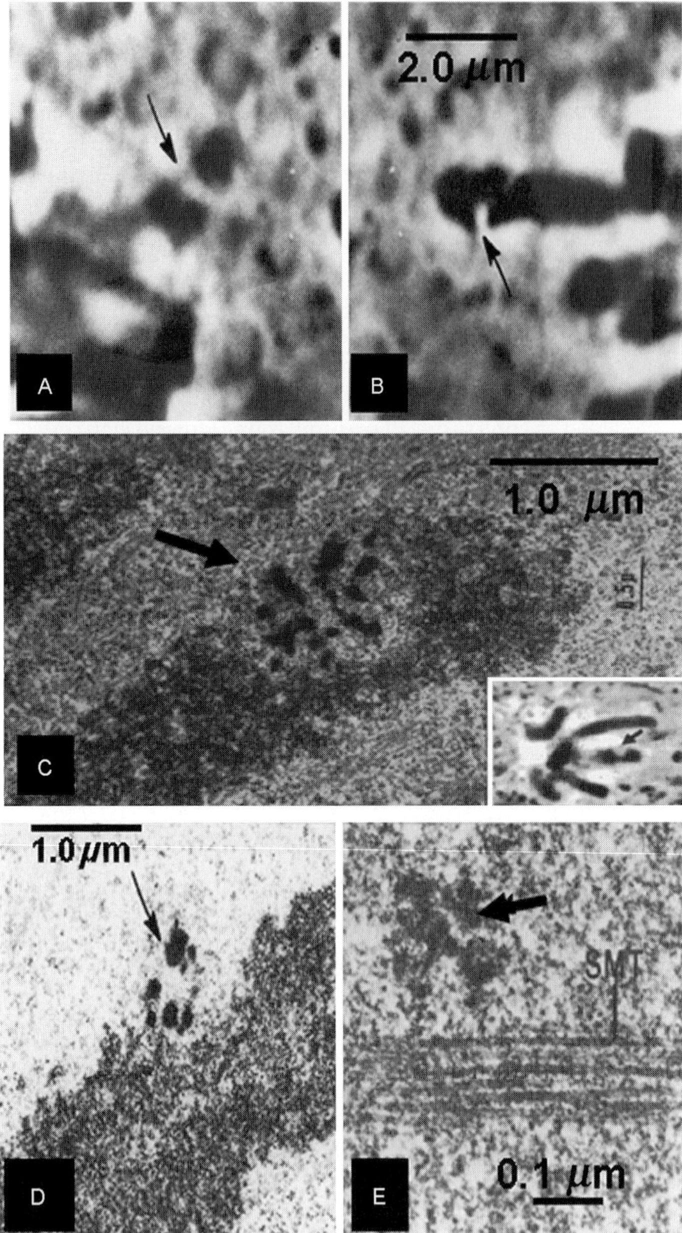

Fig. 8 Chromosome lesions produced with an argon ion laser (514 and 488 nm) in anaphase mitotic chromosomes of salamander lung cell treated with 0.0025% of acridine orange. (A) Several laser exposures across chromosome resulted in the tip being severed. (B) A piece cut out of second chromosome with a single laser pulse (Berns *et al.*, 1969a). (C) Transmission electron micrograph of chromosome lesion as in (B) produced with high fluence (1,000 $\mu J/\mu m^3$) without acridine orange (arrow); scale

time-sequence analysis of the femtosecond lesions showed phase-dark material aggregating in the region of the paling spot within 45 s of the laser exposure. This material gradually increased over the next 30 min as the chromosomes progressed through anaphase separation. It is possible that the phase-dark material contains proteins associated with the repair of the massive amounts of DNA damage caused by the laser exposure (see discussion in the next paragraph). Additionally it should be noted that after laser exposure to multiple chromosome sites, the cell continued through cell division normally. Thus, the complex biochemical machinery regulating chromosome movement, pole separation, and cytokinesis into two separate daughter cells was not severely affected by the laser exposure.

The initial 1969 (Berns *et al.*, 1969a,c) dye-sensitization studies of chromosomes were followed by other studies employing additional photosensitizing agents such as bromodeoxyuridine (Berns *et al.*, 1976), psoralens (Cremer *et al.*, 1981; Peterson and Berns, 1978), and more recently ethidium bromide (Berns *et al.*, 2000). In all of these studies, the photosensitizing agents were used to confer light-absorbing properties to the chromosome so that selective targeting and ablation could be achieved in order to study some aspect of chromosome structure and function. These early studies established the framework for current studies by many investigators around the world who use photosensitizing agents such as Hoechst dye and bromodeoxyuridine in combination with laser microbeam irradiation to produce DNA DSBs. Using fluorescent antibody probes and green fluorescent protein (GFP) gene fusion technology, the recruitment of DNA repair factors into the damage sites has been observed (Kim *et al.*, 2005). The ability to detect DNA repair proteins being recruited into the damage site has resulted in significant progress in understanding the molecular basis of DNA repair (see Chapter 11 by Hoffman-Kim *et al.*, this volume). These studies are possible because of the marriage of the laser microscope technology with the vast array of fluorescent molecular tools such as gene fusion proteins (Khodjakov *et al.*, 1997) and monoclonal and polyclonal antibodies.

A key accomplishment of the early chromosome laser microbeam studies was the demonstration that the cells could survive the irradiation for hours and even days. Initially it was possible to ablate part of a single mitotic chromosome and follow the cell until the next division cycle (Berns *et al.*, 1971a). Subsequently it was possible to isolate and clone the irradiated cells (Basehoar and Berns, 1973). Achievement of that goal ultimately resulted in the ability to inactivate specific genetic sites on selected chromosomes and to clone the individual irradiated cells into stable populations that

---

bar = 0.5 μm. Inset is phase-contrast picture of chromosomes in live cell. Arrow indicates the "phase paling" at the irradiated point. (D) Transmission electron micrograph of a lesion (arrow) place on the side of one chromatid resulting in the destruction of the kinetochore using the 532-nm frequency doubled Nd:YAG with a pulse duration of 180 ns and a fluence of @500–1000 μJ/cm^2 (see McNeill and Berns, 1981). (E) Transmission electron micrograph of a lesion (arrow) placed on the side of the band of spindle microtubules in dividing cell of the fungus *N. haematococca* (Aist and Berns, 1981) using laser parameters similar to (E). Figures C, D, and E reproduced from the *J. Cell Biol.* **62**, **88**, and **91**.

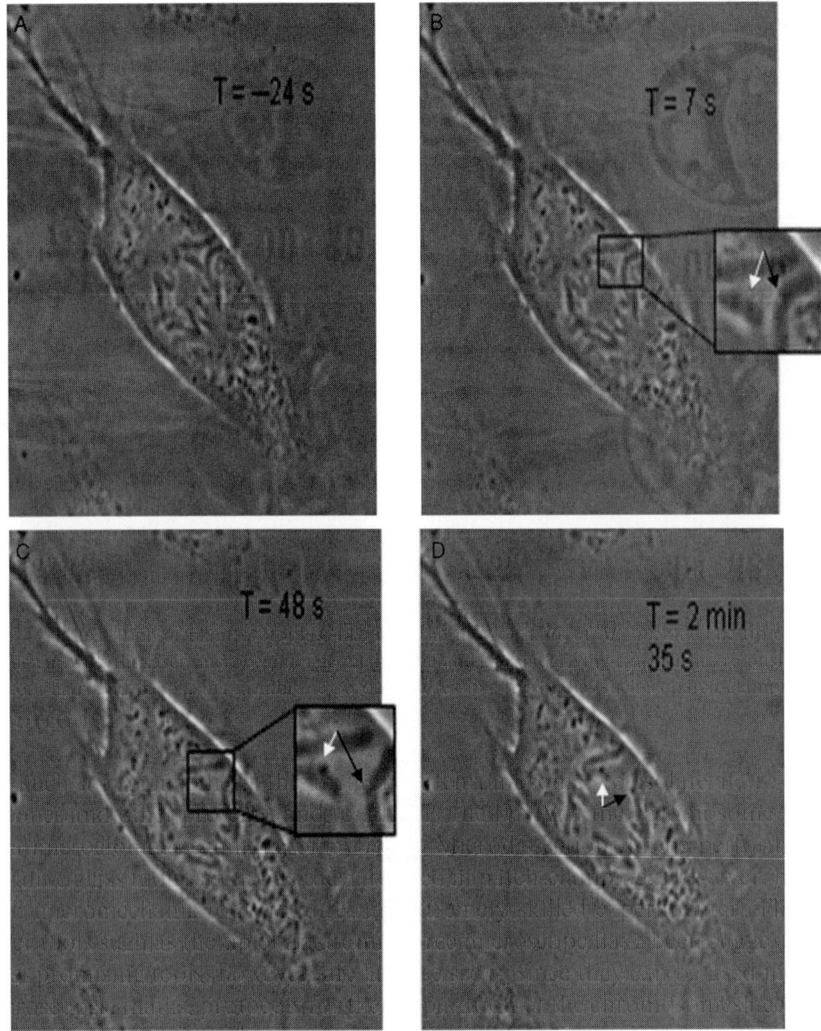

Fig. 9 Mitotic PtK2 rat kangaroo cell with two lesions placed on chromosomes in anaphase using the 800 nm; 76 MHz; 200-fs laser. Laser fluence and irradiance in the focused laser spot were 762 J/cm^2 and 5.1×10^4 W/cm^2, respectively. The laser was focused to a diffraction-limited spot using a 63× 1.4 NA oil immersion objective. (A) Early anaphase 24 s prior to laser exposure. (B) Postexposure to two lesions. The first which is 17-s postlaser exposure (white arrow) is more visible as a paling spot than the second (black arrow) which is 7-s postexposure. (C) The first lesion (white arrow) exhibits distinct phase-dark material accumulated in the lesion zone and the second (black arrow) is just beginning to show accumulation of phase-dark material in the lesion area. (D) After 2 min 35 s both lesions now show strong accumulation of phase-dark material in lesion area. Cell is progressing through anaphase and eventually divided normally into two daughter cells. Images courtesy of Christopher Lee, Veronica Gomez, Alexander Dvornikov, and Nicole Wakida, Beckman Laser Institute, University of California, Irvine.

maintained the laser-inactivated genetic deficiency (Berns *et al.*, 1979; see Figs. 5–8 in Appendix, this chapter). These early studies were the forerunners of the gene cloning and DNA sequencing studies (Monajembashi *et al.*, 1986; Ponelies *et al.*, 1997), which eventually led to the development of the commercial instruments for molecular analysis and diagnostics (see Chapter 2 by Greulich and Chapter 23 by Schütze *et al.*, this volume).

Of considerable interest was characterization of the structural and chemical nature of the laser-induced alteration in the chromosome. Serial section electron microscopy and cytochemical staining analysis of chromosomes irradiated with the argon, dye, and later the Q-switched nanosecond Nd:YAG lasers led to several conclusions: (1) micron and submicron lesions could be produced in chromosomes, (2) the lesions observed under phase-contrast microscopy could be relocated and characterized using the electron microscope and there was no extended damage outside of the region seen with the light microscope (Rattner and Berns, 1974), (3) it appeared that careful manipulation of laser parameters with and without the use of photosensitizing agents permitted selective alteration of either the DNA or protein (histone) backbone of the chromosome (Berns and Floyd, 1971; Rattner and Berns, 1974), and (4) cells could survive the irradiation, undergo subsequent cell division, and eventually be cloned into viable populations (see Section V.A.2).

2. Genetic Manipulation

The studies on selective inactivation of the ribosomal genes (also referred to as the rDNA, the nucleolar organizer, and the secondary constriction) were the first to demonstrate that the laser microbeam could be used to inactivate a selected genetic site in a live cell. Initially this was achieved by laser microbeam irradiation of the secondary constriction region on chromosomes of primary lung tissue culture explants of the salamander using DNA-specific acridine orange photosensitization (*Taricha granulosa*; Berns *et al.*, 1970a). In this study, the secondary constriction on the tips of the large chromosomes were highly visible when the chromosomes were aligned either on the metaphase plate or when they were moving toward the mitotic poles during anaphase. It was possible to irradiate the secondary constrictions with the argon ion laser beam in cells treated with nontoxic levels of acridine orange and show that the irradiated site was not able to organize a nucleolus when the cell entered G1 interphase. Presumably, the rDNA genes located at the secondary constriction region of the chromosome had been inactivated or damaged by the laser light interaction with the acridine that was intercalated into the DNA. Since these cells had several nucleolar organizer rDNA sites on different chromosomes, it was possible to reduce the number of nucleoli formed proportional to the number of nucleolar organizer sites irradiated.

Although the previous study demonstrated that the laser microbeam had the precision and specificity to inactivate active genetic regions on a targeted chromosome, the salamander cells were from primary lung cultures and they did not have the capacity to divide and form viable clones. One of two key developments at this time was the introduction of the Tasmanian rat kangaroo (*Potorous tridactylis*) cell lines, PtK1 (female) and PtK2 (male) (Walen and Brown, 1962). These cells remain flat throughout mitosis and they have only 12–13 chromosomes which remain relatively stable from one generation to the next. Thus, it was possible to follow an irradiated cell through a cell cycle to the next mitosis (Berns *et al.*, 1971a). Because of the flat nature of these cells, it was also possible to recognize the chromosomes during mitosis and focus the laser on to the secondary constriction region of the X chromosome in either the female rat kangaroo cells (PtK1) or the male cells (PtK2)—the region containing the nucleolar, ribosomal genes (Berns *et al.*, 1972a). It was possible to inactivate these genes and subsequently reduce the number of nucleoli formed in the daughter cells down to one for the female XX cells and none for the male XY cells. The male cells subsequently formed numerous "micronucleoli" which possibly were aggregates of ribosomal protein sequestered in the cell during mitosis and reaggregated following mitosis. However, it was also possible that these small nucleoli were true nucleoli formed from ribosomal genes scattered throughout the genome that were activated by the cell following destruction of the major rDNA sites.

The rDNA irradiation experiments led to experiments in which an rDNA-irradiated cell was cloned into a viable population that was deficient in the deleted genetic site (Berns *et al.*, 1979). The second key development was the demonstration that the intensity of photons in the focused laser spot was high enough to produce damage without the use of the photosensitizing agent acridine orange (Berns *et al.*, 1971b). This observation led to the first suggestion of a two-photon effect in a living cell (Berns, 1976; Berns and Floyd, 1971; Berns *et al.*, 1971b), and it opened the way for non-dye treatment of cells and their subsequent long-term survival and clonability. Several key conclusions were made from these studies: (1) an irradiated cell could divide again, (2) the cell replicated the irradiated chromosome as evidenced by the presence of the replicate in subsequent daughter cells and in the clonal population, and (3) the contiguous DNA molecule that extends through the irradiated secondary constriction does not contain DNA sequences complementary to ribosomal DNA as evidenced by the lack of *in situ* DNA–RNA hybridization (see Appendix, this chapter). Also, it was observed that the unirradiated rDNA-containing chromosome appeared to hybridize considerably more rDNA complement than normal, suggesting a possible gene amplification process following laser ablation. In the context of studies on DNA repair following laser ablation (Kim *et al.*, 2005; see Chapter 11 by Hoffman-Kim *et al.*, this volume), it is likely that substantial DNA repair occurred following irradiation of the rDNA. This provides a reasonable explanation for the replication of the irradiated chromosome region.

3. Change in Ploidy (Chromosome Number)

Manipulation of chromosome complement involved the removal of individual chromosomes from single cells by destroying the connection (the kinetochore) of the chromosome to the mitotic spindle (Berns, 1974c; Fig. 10). As a result, the

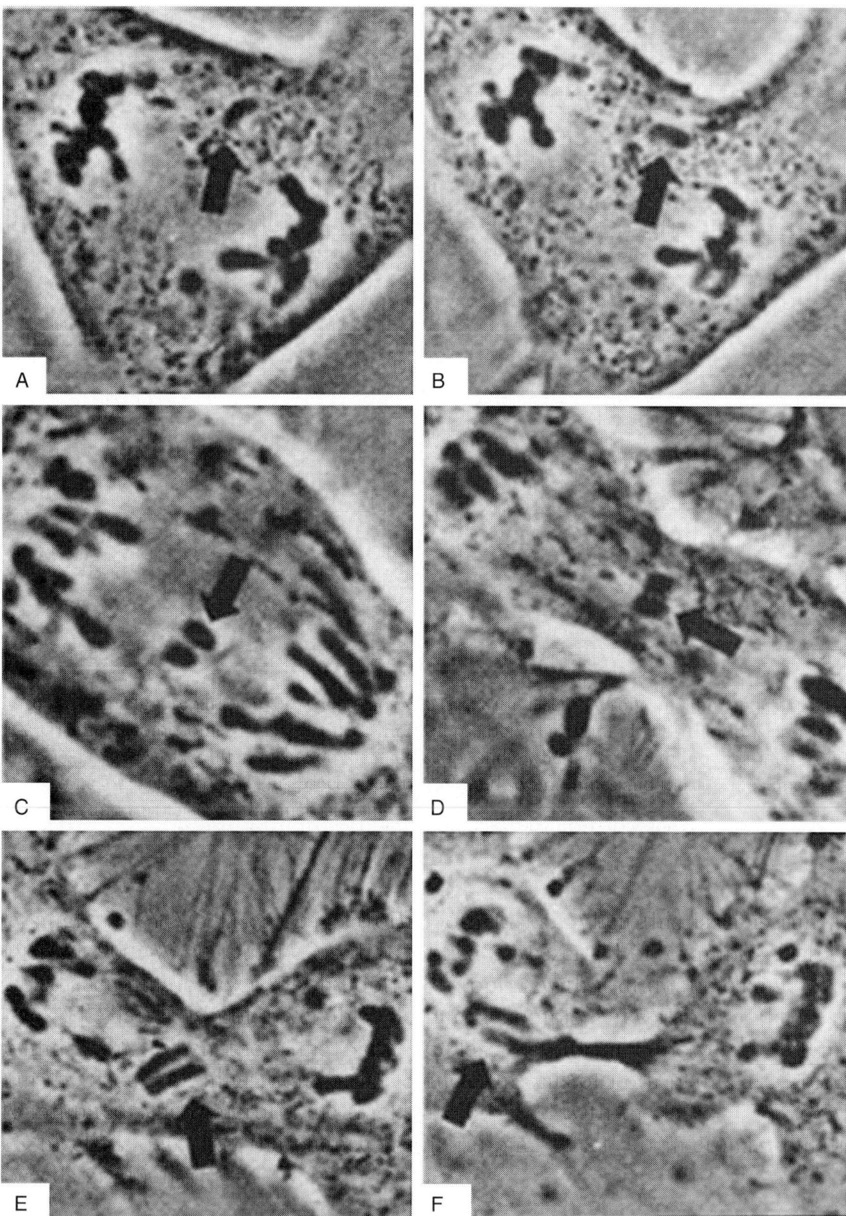

Fig. 10 (A and B) Anaphase cell demonstrating the loss of an irradiated chromosome from the nucleus. The irradiated chromosome can be seen in the cytoplasm: the two light micrographs show slightly different focal planes in the cell; the two chromatids are indicated by the arrows. (C and D) Irradiated chromosomes lost during mitosis as a result of being caught in the stem body; (C) anaphase cell showing the two chromosomes remaining behind at the metaphase plate; (D) the same cell in telophase showing the two chromosomes being caught in the stem body; neither daughter cell received one of the chromosomes. (E and F) Both microirradiated chromosomes are incorporated into one daughter cell (E); late telophase showing both chromosomes being incorporated into one nucleus (F). (See Berns, 1974b.)

irradiated chromosome became detached from the mitotic spindle and was either excluded from the cell in the constriction body when the cell divided or it randomly drifted into one of the daughter cells creating either an aneuploid or trisomic cell (Berns, 1974b,c). A surprising finding of these studies was that the clonal population of cells appeared to have replaced the lost chromosome, thus suggesting that the cell had a mechanism for replacing lost chromosomes by making an extra replica of the remaining homologue. In addition, cells with 0.5-μm ablation regions also were clonable and led to the suggestion that DNA repair might occur and that "it might also be possible to study the repair of chromosomal damage caused by laser irradiation" (Berns, 1974c). Thirty years later, laser microbeam damage to nuclei and chromosomes has uncovered a host of proteins that are recruited to the damage DNA sites and are involved specifically in the repair of DNA DSBs (Kim *et al.*, 2005; Kruhlak *et al.*, 2006; see Chapter 11 by Hoffman-Kim *et al.*, this volume).

It was also possible to sever chromosomes such that fragments of the chromosomes drifted in the cell unattached to the mitotic spindle. The fragments were either excluded from cells during cytokinesis or they were randomly incorporated into daughter cells, often being encapsulated in a micronucleus. The possibility to create a wide variety of partial and whole chromosome abnormalities by this method was suggested (Berns, 1974b).

4. Microdissection for Gene Mapping and DNA Sequencing

In 1986, a paper was published describing the laser microdissection of chromosomes (Monajembashi *et al.*, 1986). This groundbreaking paper described the use of a 20-ns excimer-pumped dye laser to cut human lymphocyte chromosomes into slices of less than 0.5 μm (corresponding to about 30 Mb of DNA). They suggest that the laser-generated chromosome fragment can be used for the construction of specific gene libraries—sets of DNA probes specific to the genes on the dissected chromosome segment. The ability to more easily generate DNA libraries and eventually sequence the genes in the laser-cut chromosome regions of specific interest would be of great value in generating DNA libraries and probes from specific regions of chromosomes. Such a laser dissection method would have distinct advantages over the time-consuming and tedious process of microneedle dissection of chromosomes.

Laser microdissection has been used successfully for the generation of DNA libraries in human (Djabali *et al.*, 1991; Hadano *et al.*, 1991; He *et al.*, 1997; Ludecke *et al.*, 1989; Upadhyaya *et al.*, 1995), insect (Ponelies *et al.*, 1989), and plant cells (Fukui *et al.*, 1992; Greulich, 1999). Although the early studies employed the expensive and technically challenging excimer-pumped dye laser (Monajembashi *et al.*, 1986), subsequent studies have used much simpler and cheaper lasers such as the UV nitrogen laser at 337 nm (Ponelies *et al.*, 1989), the argon ion laser at 488 and 514 nm, and the green 532-nm second harmonic of the Q-switched Nd:YAG laser (He *et al.*, 1997). For a thorough and complete discussion of the use of the laser to cut and clone DNA sequences from chromosomes see Greulich (1999, pp. 202–203) and Chapter 2 by Greulich, this volume. Additionally, the use of the laser to cut and catapult nuclear (DNA) fragments onto substrates and into receptacles for subsequent PCR and DNA analysis has evolved from these early chromosome cutting experiments and is discussed in Chapter 24 by Dafna *et al.* and Chapter 2 by Greulich, this volume. The laser "catapulting" and collection techniques have been widely adopted in the field of molecular pathology and have become a regular commercial products of Microlaser Technologies (Zeiss-P.A.L.M., Bernried, Germany), Leica Microsystems (Glattbrugg, Switzerland), and Arcturus Biosciences, Inc. (Mountain View, California).

Finally, the use of the laser to cut chromosomes and then trap and move the fragments with a laser tweezers has been described both for laser-microdissected chromosomes outside of cells (Berns *et al.*, 1994, 1998a,b, 2006; Greulich 1999; Seeger *et al.*, 1991) and for the study of the intracellular movement of cut chromosomes (Liang *et al.*, 1994). The use of a combined laser scissors and trapping approach to manipulate and study the cell and organelle function is an area that shows promise for future research and commercial development.

B. Through the Membrane: "Optoporation/Optoinjection"

Laser perforation of the cell membrane in order to allow foreign DNA into the cell and ultimately into the host cell genome was first demonstrated in 1984 and 1986 by the Japanese (Kurata *et al.*, 1986; Tsukakoshi *et al.*, 1984). They used the third harmonic 355 nm of the Nd:YAG laser and successfully transfected normal rat kidney cells with foreign DNA with a 0.6% transfection frequency. Shortly thereafter, human fibrosarcoma cells were successfully transfected with the *neo* gene that conferred resistance to the antibiotic aminoglycoside G 418 (Tao *et al.*, 1987). Although success rate of stably transfected tarnsformants was only 0.3% in these studies, this still was 10–100 times more efficient than standard chemical-mediated DNA transfer. It was suggested that the laser perforation of the cell membrane was sealed very quickly thus facilitating cell survival. It was also suggested that the transformation frequency could be improved on by further optimization of the laser parameters and the cell-handling conditions. Although the transformation frequencies were about the same as when a microcapillary was used to inject DNA directly into individual cells, the speed, the lack of need of a highly trained person to do microinjection, and the potential for automated optical injection are all potential advantages of laser-mediated cellular transfection.

Since the initial studies of laser-facilitated injection of foreign molecules, numerous studies have been reported using other laser systems: (1) the CW argon laser operating at 488 nm (Palumbo *et al.*, 1996; Schneckenburger *et al.*, 2002), (2) the third harmonic of the pulsed Nd:YAG laser at 355 nm (Shirahata *et al.*, 2001), (3) the 532-nm second harmonic of the Nd:YAG laser (Krasieva *et al.*, 1998; Soughayer *et al.*, 2000), (4) the 1064-nm fundamental wavelength of the Nd:YAG laser (Mohanty *et al.*, 2003), (5) the femtosecond, Ti:Sapphire laser (Tirlapur and Konig, 2002a,b; Zeira *et al.*, 2003), and (6) a compact 405-nm diode laser (Paterson *et al.*, 2005).

Laser-induced gene transfer in plant cells would appear to be even more advantageous, as the strong cellulose-containing cell wall is a major barrier to both chemical and microcapillary-facilitated gene transfer. Several studies by Weber and colleagues have established the feasibility of injecting DNA into plant cells and chloroplasts using a pulsed UV laser (Weber *et al.*, 1988a,b, 1989a,b, 1990a,b). An integrated discussion of these studies can be found in Greulich (1999), and in Chapter 2 by Greulich, this volume.

The pioneering work of these investigators has improved the likelihood of commercial agricultural application of laser-facilitated gene transfer for the improvement of crop production. Key experiments in both rapeseed and tobacco demonstrated a successful stable transformation frequency of 0.5%. The DNA isolated from leaves of mature second-generation plants successfully hybridized against the probe DNA, thus clearly demonstrating that the incorporated genes were passed on to successive generations (Greulich, 1999, pp. 220–221). Similarly, foreign genes were inserted into cells from the growing callus of rice plants, and complete plantlets were generated that contained the inserted genes in all cells of the plant. Since these plants were not propagated through successive generations, it is not known if the transfection was stable and if it occurred in the plant gametes (Guo *et al.*, 1995). However, this was the first known case of an entire organism being grown from a single cell that had specific genes inserted by this method. In 2002, a femtosecond near-infrared laser was described as a noninvasive tool for the nanoprocessing and targeted transfection in plants without compromising viability (Tirlapur and Konig, 2002a,b).

From all the studies described and referenced in this section, it is clear that laser-facilitated introduction of genes and other foreign molecules into cells does work. It is unclear if there is an optimal laser for this, and it is also not clear whether focusing the laser onto single cells is the optimal method for getting foreign material into cells. For example, it is possible to achieve efficient injection of foreign material into a large number of cells irradiated simultaneously by using a low-power, low-numerical aperture objective (Clark *et al.*, 2006; see Chapter 10 by Rhodes *et al.*, this volume). This group has termed this process "laserfection," and has developed two commercial instruments based on this concept (http://www.cyntellect.com). The mechanism by which the cell membrane is affected by the laser irradiation is not known.

C. Cell Division

1. Background

Cell division (mitosis and meiosis) has been studied using classic nonlaser UV microbeams for decades and continues to the present day (see Section I). UV, visible, and infrared laser beams have been used as well. It is not clear if either the laser or nonlaser sources are any better, though it can be argued that the confinement of cell damage only to the focused spot is much easier with laser sources than the nonlaser UV focused spots. With the high peak power of the nanosecond and picosecond laser beams, it has been shown that due to nonlinear processes, including multiphoton absorption (Calmettes and Berns, 1983), the laser damage can be confined not only to a diffraction-limited spot, but also to a subdiffraction-limited spot due to the high photon density at the peak of the Gaussian center of a TEMoo mode laser beam (Berns, 1974a; Botvinick and Berns, 2005; Appendix, this chapter). With the femtosecond lasers, damage zones as small as 100 nm have been reported in chromosomes (Konig *et al.*, 2001) and individual stress fibers (Kumar *et al.*, 2006). However, when the 50-μs-pulsed argon ion laser (488 and 514 nm) was first used in 1969–1972, nontoxic vital dyes (chromophores) had to be used to sensitize the target organelle to the light (Berns and Floyd, 1971; Berns *et al.*, 1969a,b, and c).

2. Centrosomes: "To Dye For"

Dye-facilitated "photomolecular dissection" (PMD) has been valuable in helping to discern the structure and function of the centrosomes. These structures play key roles in the orderly and equal separation of chromosomes at cell division. However, the precise role(s) of the centrosome and its organelle subcomponents, the centrioles and the pericentriolar material (PCM), in this process was (and is still) not well understood.

In 1977 in a landmark paper, R. R. Gould and G. Borisy at the University of Wisconsin, Madison, Wisconsin (Gould and Borisy, 1977) demonstrated that PCM isolated from centrosomes could nucleate microtubules. Thus, the idea that the PCM was the microtubule-organizing center (MTOC) for the mitotic spindle became generally accepted. At about that same time two studies were published in which (1) PCM was selectively microirradiated (and destroyed) by argon ion laser microbeam exposure following cellular treatment with the nucleic acid-binding dye acridine orange (Berns *et al.*, 1977) and (2) the centrioles were selectively destroyed followed by dye sensitization and irradiation with a 100-ns Nd:YAG laser at 473 nm (Berns and Richardson, 1977). In the first case, following the dye + laser treatment, the cells were not able to organize a normal mitotic spindle and undergo chromosome separation (see experimental summary, Figs. 9–13 in Appendix, this chapter). Although microtubules were seen at the chromosome kinetochores (the microtubule attachment sites), none were seen at the cell poles and especially in the PCM. In the case of the laser-ablated centrioles, the cells divided normally and electron microscopic analysis revealed microtubules associated with the PCM in the absence of structurally intact centrioles (which had been damaged by the dye + laser treatment). This series of PMD studies came to the same conclusions about the PCM as the biochemical studies of Gould and Borisy (1977): the PCM is the MTOC and is critical to spindle organization and function. A role for the centriole in cell division remained elusive.

A subsequent laser-dye PMD study using different types of photoinduced nucleic acid-binding psoralen molecules was able to confirm the presence of RNA and not DNA in the centrosome (Peterson and Berns, 1978). In 2006,

RNA from centrosomes were isolated, purified, and partially sequenced (Alliegro *et al.*, 2006). These authors showed by biochemical means and subcellular *in situ* hybridization that the initial transcripts were associated with centrosomes and not the cell nucleus. This centrosome RNA is called cnRNA, and its specific role(s) in cell division and centrosome duplications have yet to be determined. It has not been determined if cnRNA is associated with either the PCM or the centriole.

The combination of GFP technology and laser microsurgery has created a paradigm shift because it is now possible to target cell structures not previously (or readily) visible in living cells (Khodjakov *et al.*, 1997). Individual GFP-fluorescing spindle fibers and microtubules can be cut and their dynamics studied (Colombelli *et al.*, 2005), and individual centrioles within a centrosome can be destroyed and subsequent cell division and centriole replication studied (La Terra *et al.*, 2005; see Chapter 7 by Magidson *et al.*, this volume).

3. Chromosome Movements

The ability to remove chromosomes from the mitotic spindle by destroying the kinetochore attachment sites to the spindle microtubules (Berns, 1974b,c) evolved into a study designed to better understand the factors that control chromosome movement during mitosis (McNeill and Berns, 1981). The pulsed nanosecond 532-nm Nd:YAG laser was used to destroy a single kinetochore on one side of a double-chromatid chromosome in early metaphase prior to the chromosome reaching the metaphase plate. This study had several key findings. First, destruction of a kinetochore (Fig. 8D) on one side of a double-chromatid chromosome resulted in the entire double-chromatid chromosome moving through the metaphase plate to the spindle pole to which the unirradiated kinetochore was still attached (Fig. 11, adapted from McNeill and Berns, 1981). This result demonstrated that spindle pole forces are exerted on the chromosome prior to the initiation of anaphase, and that these forces are sufficient to move a chromosome with twice the normal mass toward that pole at a velocity equal to the normal rate of anaphase chromosome movement. However, there was a brief reduction in the velocity of the irradiated chromosome as it passed through the mass of chromosomes at the metaphase plate (see diagram in section "Mitotic Organelles," Appendix, this chapter). In more recent studies using a similar 532-nm nanosecond Nd:YAG laser microbeam (Rieder *et al.*, 1986), it was observed that laser-severed arms of chromosomes were pushed from the spindle by "ejection forces" mediated by nonkinetochore-bound microtubules. This result could explain where the additional force came from to move the chromosomes at the normal rate as seen in the earlier experiment (McNeill and Berns, 1981). Second, the remaining chromosomes that were aligned on the metaphase plate underwent normal anaphase separation even though the irradiated chromosome never moved to the metaphase plate. This demonstrated that cells will undergo anaphase even if all the chromosomes do not move to the metaphase plate. In this experiment, the destruction

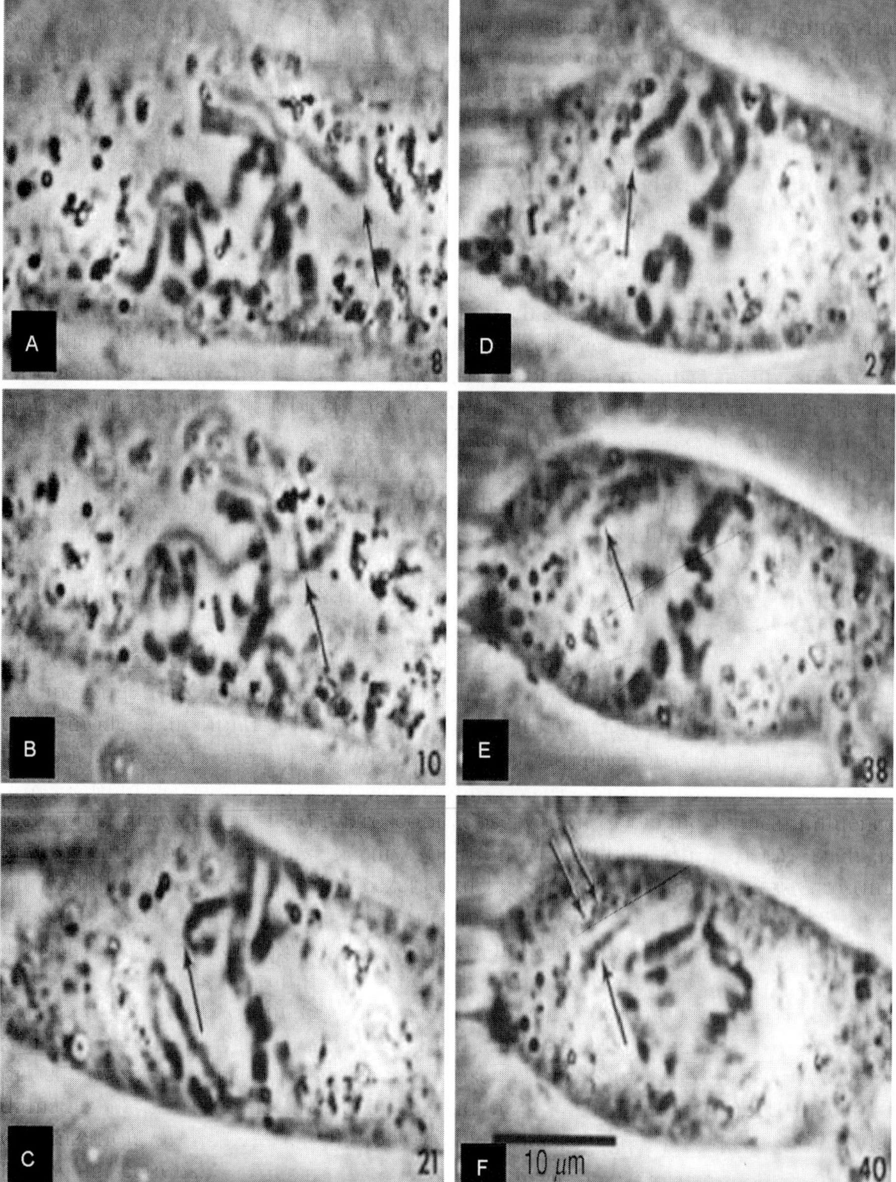

Fig. 11 Series of phase-contrast micrographs of a PtK2 cell in which one kinetochore was irradiated during prometaphase. Time (in minutes) from irradiation is shown in bottom right-hand corner of frames. (A) Prometaphase. The irradiated chromosome (arrow) is close to the right-hand pole. (B and C) The irradiated chromosome is moving across the spindle toward the opposite pole. (D) Metaphase. The irradiated chromosome remains near the left-hand pole while the other chromosomes are aligned on the metaphase plate. (E) Very early anaphase. The chromatids of the irradiated chromosome have separated by ~1 μm. (F) Mid-anaphase. The chromatids of the irradiated chromosome remain parallel and slightly separated from each other. The irradiated chromatid is indicated by a double arrow and the unirradiated chromatid by a single arrow (McNeill and Berns, 1981). Reproduced from the *J. Cell Biol.* 1981, **88,** 543–553. Copyright 1981 The Rockefeller Press.

of the kinetochore likely resulted in elimination of the kinetochore "wait" signal so the cell proceeded through the metaphase checkpoint unaware that one chromosome had not moved to the metaphase plate. Third, the two chromatids of the irradiated chromosome separated (but did not undergo any poleward movements) at the same time that the rest of the metaphase chromosomes separated. This observation supports current proof that the chemical signals for separation of chromatids are distinct from the signals and reactions connected to the poleward movement of the chromosomes.

In addition to the early studies just discussed, the laser microbeam has evolved into an important tool to study chromosome and spindle dynamics in a broad group of organisms and cells. In particular, Conly Rieder and Alexey Khodjakov of the Wadsworth VA Medical Center in Albany, New York have used the laser microbeam to study mitosis in cells from humans, salamanders, rat kangaroos, *Haemanthus* (water lily), fruit flies, and yeast (see Chapter 7 by Magidson *et al.*, this volume). Their studies, more than any others, have used the laser microbeam to unravel the mysteries of an important cellular process.

4. Fungi, Yeast, Fruit Flies, and Round Worms

A separate section is being devoted to studies on four additional organisms because in all four cases the laser microbeam has been a significant tool in elucidating how these cells divide.

The initial laser microbeam studies on the ascomycete fungus *Nectria haematococca* (see *Fusarium, Fungi imperfecti*) were of interest because these cells are particularly small, and the mechanical properties of their mitotic spindle were poorly understood. A prominent structural element of the spindle is a dense band of microtubules that extends between two spindle pole bodies (SPBs; see Fig. 8E). It was suspected that this band of microtubules served to push the SPBs apart during anaphase. Surprisingly, when the band was severed with the laser rather than collapsing toward the center of the spindle, the two SPBs moved apart at an accelerated rate. This suggested that the microtubule band served as a "governor, or "rate-brake," of spindle pole (and chromosome) separation (Aist and Berns, 1981). Further studies showed that these forces were localized in extranuclear cytoplasmic asters, were mediated by astral microtubules, and were capable of affecting the rate of spindle elongation (Aist *et al.*, 1991). These studies showed that the forces for SPB separation were located external to nucleus and the spindle, pulling the SPBs toward cell membrane attachment points. In another study on a basidiomycete fungus that causes root rot (*Helicobasidium mompa*), the laser was used to sever the central spindle during anaphase B (when the chromosomes had already reached the spindle poles). The results were similar to the laser microbeam study in the ascomycete fungus *N. haematococca*: when the central spindle was severed, the two SPBs separated at an accelerated rate (Bayles *et al.*, 1993). On the basis of the results of these two studies and those of several genetic studies (Berlin *et al.*, 1990; Sullivan and Huffaker, 1992; Winey *et al.*, 1991), it was concluded that cytoplasmic astral pulling forces and spindle pushing forces may be common to all fungi. Additionally, a study on animal cells in which the laser microbeam was used to irradiate the spindle pole resulted in a reduction in the astral microtubules and an acceleration of separation at anaphase B (Aist *et al.*, 1993). The results in biological systems as diverse as fungi, yeast, and vertebrate cells suggest that structures external to the spindle proper may play key roles in the final stages of cell division.

The laser microbeam also has played a significant role in understanding cell division in fission yeast (*Schizosaccharomyces pombe*), also a relatively small organism with a small mitotic spindle. Using a 532-nm Nd:YAG nanosecond laser, Khodjakov *et al.* (2004) were able to demonstrate that the laser microbeam could cut intracellular structures with a high degree of specificity. They were able to cut midzone microtubules at various stages of anaphase B and demonstrate that the midzone region is necessary for stabilization of microtubule ends and for spindle elongation. In addition, by cutting the spindle near the SPB, they were able to demonstrate that the SPB is not necessary for spindle elongation, but rather contributed to the attachment of the chromosomes and the nuclear envelope to the spindle. In a more recent study in the same organism, the role of the SPB in cytokinesis was studied using the same 532-nm Nd:YAG laser (Magidson *et al.*, 2006). In this organism, cytokinesis is thought to be controlled by the daughter SPB through a regulatory pathway called the septation initiation network (SIN). The authors demonstrated that laser ablation of both, but not a single SPB, resulted in failure of cytokinesis. Ablation of only the daughter SPB often led to activation of the SIN on the mother SPB and successful cytokinesis occurred. It was concluded that either SPB could drive cytokinesis.

Studies on the fruit fly, *Drosophila melanogaster*, have contributed perhaps more than any organism to our general understanding of genetics and development (see Section VI for developmental studies and for an extensive discussion of laser microbeam studies on mitosis). In these studies, the 532-nm nanosecond Nd:YAG laser microbeam was used in combination with other methods to study kinetochore microtubule dynamics. Specifically, *D. melanogaster* S2 cells expressing α-tubulin tagged with GFP, RNA interference, laser microsurgery, and photobleaching were used to show that the kinetochore protein MAST/Orbit—the single CLASP orthologue in *Drosophila*—is an essential component for microtubule subunit incorporation into fluxing kinetochore fibers (Maiato *et al.*, 2004, 2005; see Chapter 7 by Magidson *et al.*, this volume).

Extensive laser microbeam studies on mitosis in the round worm *Caenorhabditis elegans* have been conducted by Anthony Hyman and his group in Germany. In particular, they studied the forces during cell division that controlled the formation of daughter cells of different sizes by a process of asymmetric division (Grill *et al.*, 2001). In these studies, they used a nitrogen (337 nm) laser to ablate the central spindle in order to reveal the forces acting on each spindle pole in the developing embryo. They discovered that (1) forces acting on the spindle pole were external to the spindle itself and (2) these forces were unequal—resulting in the posterior daughter cell being larger than the anterior one. The different sizes of the daughter cells in the developing *C. elegans* embryo are essential to the generation of cell diversity during animal development. In a subsequent study, this same group used the 337-nm nitrogen laser microbeam to fragment the centrosomes at the spindle poles and found that by analyzing the speed of the centrosome fragments it was possible to determine that there was a force imbalance that could be attributed to "a number of force generators" pulling on the astral microtubules that were external to the mitotic spindle (Grill *et al.*, 2003). In a 2004 study (Cowan and Hyman, 2004), laser ablation of the centrosome revealed that it was required for the initiation of the unequal division cell polarity axis established in early development, and that this role of the centrosome might be independent of the well-established role of the centrosome as an MTOC. Thus, one role of the centrosome in the development of *C. elegans* is to initiate the polarity but not to maintain it.

D. Cell Migration

Related to the previous discussion on the role of the centrosome in the establishment of cell polarity (and its major role as a cellular MTOC), there are other possible roles for this ubiquitous cell structure. One of these is as a central controller of cell migration. Such a role was first explored with a UV laser microbeam in 1984 (Koonce *et al.*, 1984). In these studies the UV 266-nm fourth harmonic wavelength of a pulsed nanosecond Nd:YAG laser was used to selectively ablate the centrosome–centriole complex of migrating salamander eosinophils in culture. Remarkably, when this region was destroyed the cells lost their directional movement, and their migration speed was reduced by half (Fig. 12). This study suggested that the centriole was necessary for the controlled migration of these cells. This suggested yet another role for the centriole—other than just in mitosis. A recent study demonstrated that when the centrosome regions of chemoattractant-stimulated migrating mammalian neutrophils were ablated with a femtosecond 800-nm laser beam, the cells appeared to lose their ability to polarize their microtubule network and migrate in response to the chemotactic gradient (Henry Bourne, Jingsong Xu, and Nicole wakida, personal communication). Thus, the laser microbeam has contributed to further our understanding of cell migration and, in particular, the role of the centrosome in that process.

E. Nucleolus

In the early 1960s, it was demonstrated that classic UV-microbeam irradiation of nucleoli in HeLa tissue culture cells resulted in a two-third reduction in cytoplasmic RNA (Perry *et al.*, 1961). At that time, the role of the nucleolus in the production and/or transport of ribosomal RNA was not known.

In the first laser microbeam experiment on the nucleolus, an argon ion laser was used to irradiate the nucleolus. This resulted in a marked reduction of radioactive RNA label in the cytoplasm (Berns *et al.*, 1969b). This confirmed the previous

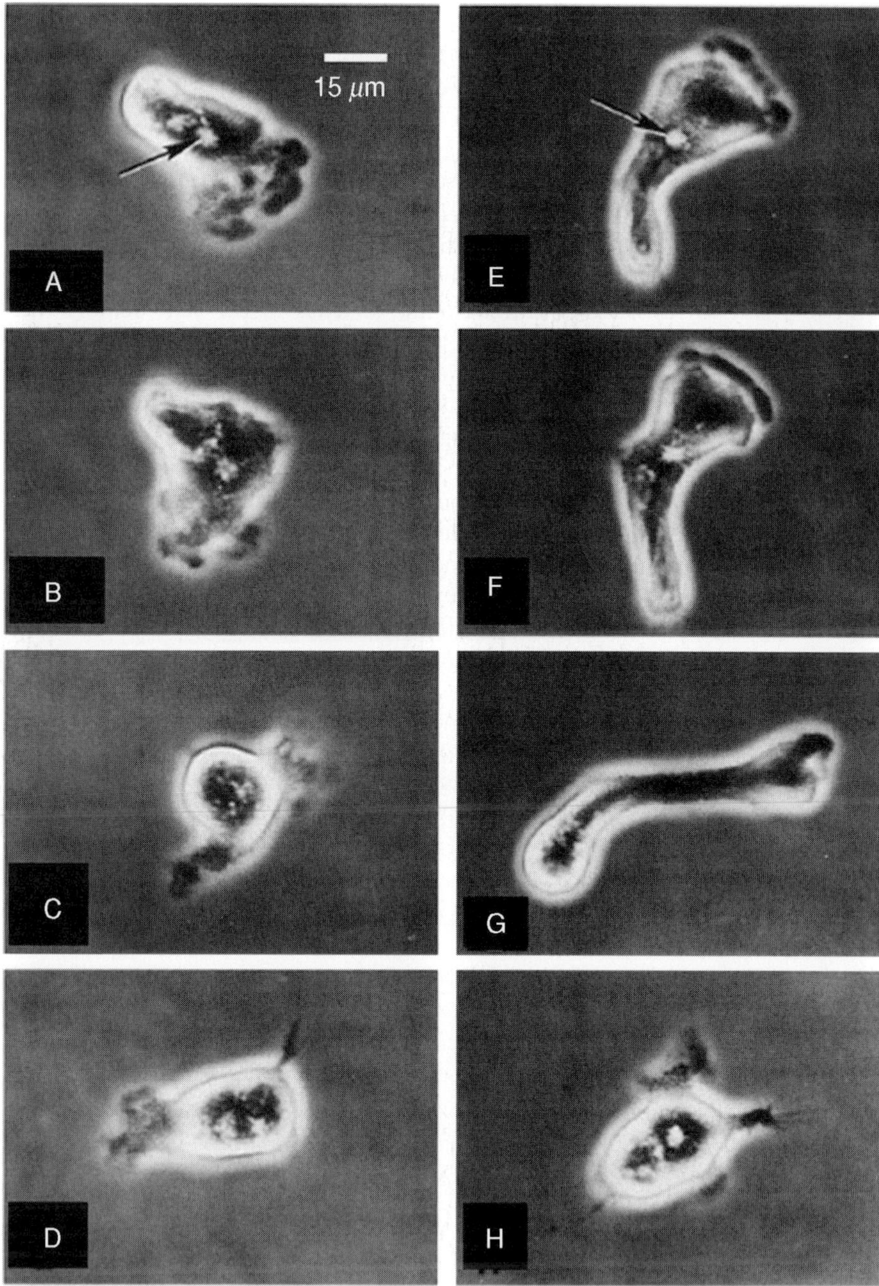

Fig. 12 Laser irradiation of centrosomes in two different migrating salamander eosinophils [cell 1 (A–D) left-hand column; cell 2 (E–H) right-hand column]. Preirradiation cell 1 (A and E), with arrow marking the centrosome region to be irradiated. Immediately after irradiation (B and F). In cell 1 (C), 5.0-min post-laser, cell has rounded and has two small pseudopods. (D) 10.0 min post-laser, cell is still

results of Perry, using a form of radiation that was less generally damaging to the cell than the UV light. Following this study, the nucleolus was further microdissected after segregation of its submicroscopic components by the drug actinomycin D (Brinkley and Berns, 1974). Actinomycin D treatment of cultured cells caused separation of the fibrillar from the granular subcomponents of the nucleolus. The cells were next treated with quinacrine hydrochloride an antimalarial drug that sensitized the nucleolar components to light. An argon ion laser microbeam was then used to destroy either the fibrillar or the granular components. The cells were next incubated in H^3-uridine, an RNA precursor, in order to assay for RNA synthesis. Autoradiographic analysis of the cells demonstrated that damage to the nucleolar fibrillar elements affected RNA synthesis three times greater than damage to the granular component. The results supported the idea at the time that the ribosomal DNA was more closely associated with the nucleolar fibrillar element where it was thought the DNA was initially transcribed to produce the large preribosomal particles. This is now the currently accepted pathway for ribosomal RNA synthesis and ribosome formation. These studies were the first that used a drug to alter the structure of an organelle so that the laser, in combination with a second sensitizing compound, could be used to microdissect its subcomponents in order to determine their function. Subsequent electron microscopy studies of laser-irradiated cells demonstrated that the laser microirradiation and resultant damage had been very specific to the region targeted under the light microscope (Meredith and Berns, 1976).

Studies on the role of the chromosome secondary constriction in nucleolus formation and the effect of laser-induced reduction of nucleoli in daughter cells have been discussed in Section II.A. Irradiation of the nucleolar organizer rDNA regions of chromosomes, and of the nucleolus itself, was a major focus of laser microbeam studies from 1969 to 1981.

F. Mitochondria

As described in Section III, mitochondria were some of the first organelles irradiated with the ruby laser (Amy and Storb, 1965; Amy *et al.*, 1967; Moreno *et al.*, 1969; Storb *et al.*, 1966). These studies involved irradiation of many mitochondria in a focal volume as opposed to focusing the laser into a single mitochondrion.

Argon ion laser (488 and 514 nm) microirradiation of a single mitochondrion was carried out in neonatal rat myocytes primarily because the mitochondria were several microns in diameter and they were clearly visible under phase-contrast microscopy (Berns *et al.*, 1970b). In addition, the respiratory enzymes (particularly the reduced cytochromes *c* and *c*-1) had strong absorption bands in the blue-green

---

partially rounded but moving. This cell was not tracked. For cell 2, 1-min post-laser (G). In this cell, the overall rate of movement decreased from 24.0 to 7.1 μm/min as a result of irradiation. Both cells lost their ability to migrate in a straight path (Koonce *et al.*, 1984). Reproduced from the *J. Cell Biol.* 1984, **98,** 1991–2010. Copyright 1984 The Rockefeller University Press.

region of the spectrum, thus eliminating the need for an external photosensitizing agent. Lesions of varying sizes and degree of severity could be produced by modulating the laser output and by matching laser intensity with the phase-contrast density of the mitochondrion. Lesions in single mitochondria were categorized as either Type I (small phase-contrast dark spot on a lightened background), Type II (perforation/light spot), or Type III (complete destruction). It was further possible to correlate the lesion type with contractile response of a single cell (Berns *et al.*, 1972a,b) or a multicellular network (Berns *et al.*, 1972c). These studies examined the relationships between damage to a single mitochondrion and a resulting change in contractility pattern of the cell, as well as changes in the contractility of other cells in the network. Cells could be "laser-stimulated" to enter a state of uncoordinated contractility in which either the different myofibrillar regions within the cell were contacting out of synchrony or the entire cell network entered an uncoordinated state of contraction. The asynchronous uncoordinated state of contraction, whether within a single cell or an entire network of cells, was termed "fibrillation" because of its resemblance to that condition in the heart. This initial study was followed with a second study specifically to examine the laser and cellular parameters related to fibrillation (Waymire *et al.*, 1976). It was demonstrated that contractile myocytes from auricular tissue were more easily stimulated to fibrillate than cells from the ventricles. This appeared to be regardless of the number of cells in the cell network for the auricular cells as opposed to the situation for ventricular cells. In the latter case, the more cells in the cell network, the more difficult it was to induce fibrillation. The observations indicated that there may be a difference between contractile myocytes from different regions of the heart, and that these *in vitro* differences might be exploited in studies on cardiac pharmacology and physiology.

Another interesting observation of the mitochondrial irradiation studies was that a single mitochondrion could be structurally altered with no apparent visible alteration to the mitochondria in close proximity to it. This observation, which was observed many times, would suggest that the cell is not composed of one large interconnected mitochondrion. (De Giorgi *et al.*, 2000). A recent femtosecond laser microbeam ablation study of mitochondria made similar observations and came to the same conclusion. (Maxwell *et al.*, 2005). Electron microscope studies have demonstrated that within one large mitochondrion a submicron lesion could be produced without damaging the ultrastructure of the cristae membranes and the outer mitochondrial membrane that lie outside the beam focal volume (Adkisson *et al.*, 1973). One of the possible explanations for the observed change in contraction was the release of calcium from the irradiated mitochondrion (Rattner *et al.*, 1976).

Several other studies have examined the induced change in contractility following laser exposure of a single mitochondrion. In a pioneering early study, Salet in Paris suggested that the absorbed energy of the 532-nm laser was actually converted into chemical energy which contributed to the increase in cardiac cell beating frequency (Salet, 1972). In a subsequent study, laser-induced acceleration of heart cell contractility was inhibited when the cells were incubated in the presence of either KCN or ATP (Salet *et al.*, 1979). This led to the conclusion that the laser-mediated change in contractility was chemically mediated, most likely through the light-induced generation of ATP, similar to that found in nature in certain bacteria. However, in another series of experiments, calcium release from the mitochondrion was suggested as one of the possible causes of the observed changes in contraction. In one of these studies, cells were impaled with microelectrodes so that their electrical activity could be monitored during and after the laser exposure (Kitzes *et al.*, 1977). Although it was technically challenging to impale a cell with an electrode and record from the cell while it underwent laser microbeam irradiation, the following were demonstrated: (1) irradiation of a single mitochondrion altered the electrical activity of the cell as evidenced by depolarization of the outer cell membrane, (2) the membrane potential change was accompanied by a change in cell contractility, (3) both the membrane potentials and the cell contraction pattern returned to the preirradiation condition, (4) a class of cells called "pacemakers" was identified that appeared to be easily stimulated to depolarize by laser exposure, and (5) these cells could undergo fibrillation for a period as long as 5 min.

The reason why irradiation of a single mitochondrion in a cardiomyocyte resulted in membrane depolarization and an induced state of altered contraction, including "fibrillation," was not known. The suggestions were either (1) thermal damage to the outer cell membrane caused by absorption of the laser light by the mitochondrion or (2) a release of calcium from the mitochondrion. The thermal theory seemed to receive support from a study in which laser irradiation of carbon particles attached to the surface of cardiac cells was very efficient in inducing contractility changes similar to those observed when either a mitochondrion or even a single myofibril was irradiated (Strahs *et al.*, 1978). Whether the release of calcium from the mitochondrion played a role in the change in contractility has been neither proved nor disproved. However when calcium was removed from the cell culture medium and the calcium chelating agent EGTA added, all cell contractility stopped. Irradiation of mitochondria, myofibrils, and carbon particles on the cell surface under these conditions caused no contractile response. These results suggested that release of calcium alone from a single mitochondrion was probably not sufficient to cause an alteration in either membrane electrical activity or contractility.

It is interesting that the contractile responses of the cardiac cells were similar with the microsecond argon ion laser (Adkisson *et al.*, 1973; Berns *et al.*, 1970a, 1972a,b,c; Kitzes *et al.*, 1977; Waymire *et al.*, 1976) and the Q-switched nanosecond Nd:YAG laser (Salet *et al.*, 1980; Strahs *et al.*, 1978) even though different ablation and contractile mechanisms have been hypothesized. In addition, scanning electron microscope images of the Q-switched Nd:YAG laser lesions in single mitochondria (Fig. 13) and in single blood cells (Fig. 14) revealed focal damage diameters as small as 170–260 nm for the nanosecond systems (Burt *et al.*, 1979). These are similar in size to those recently reported with the femtosecond laser (see Chapter 9 by Heisterkamp *et al.*, this volume; Konig *et al.*, 1999, 2001; Kumar *et al.*, 2006).

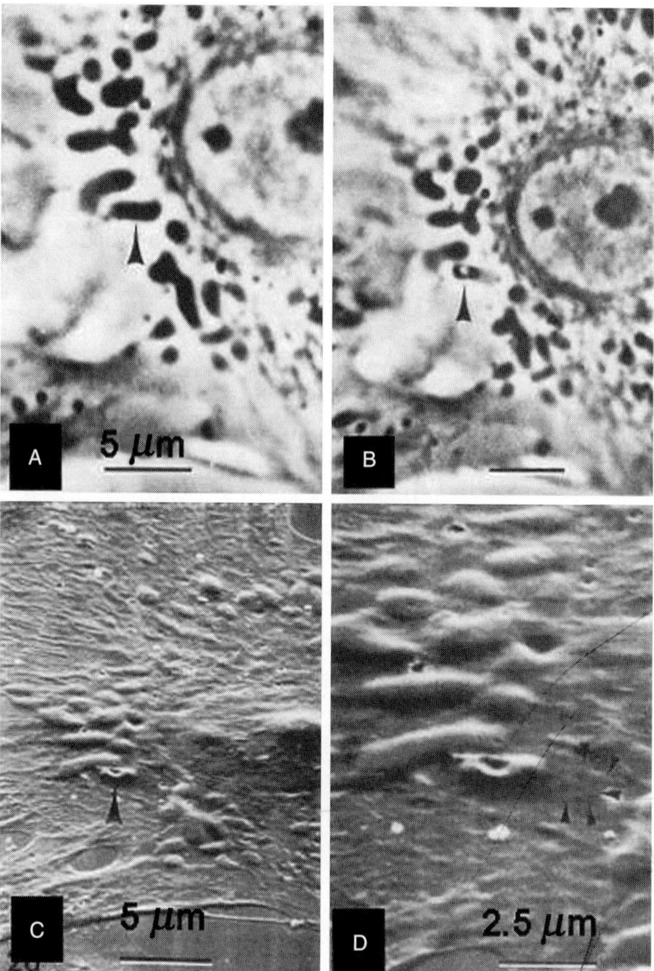

Fig. 13 (A) Phase-contrast micrograph of a heart cell prior to and (B) post-laser irradiation of a single mitochondrion (arrow). This lesion type, classified as moderate, is characterized by a central phase light area surrounded by a phase-dark ring. Note that the remainder of the mitochondrion is paled. Panel (C) is a low magnification scanning electron micrograph (SEM) of the same cell as in (A) and (B). Match-up of organelles visible in the phase-contrast micrographs is possible. Panel (D) is a high magnification SEM image of the target mitochondrion (small arrows correspond to the paled portion of the mitochondrion in the phase-contrast image (B). Also, the raised ring and central dimple of the lesion in the SEM images correspond to the phase-dark ring and central light zone in the post-laser phase-contrast images (B). (See Burt *et al.*, 1979.)

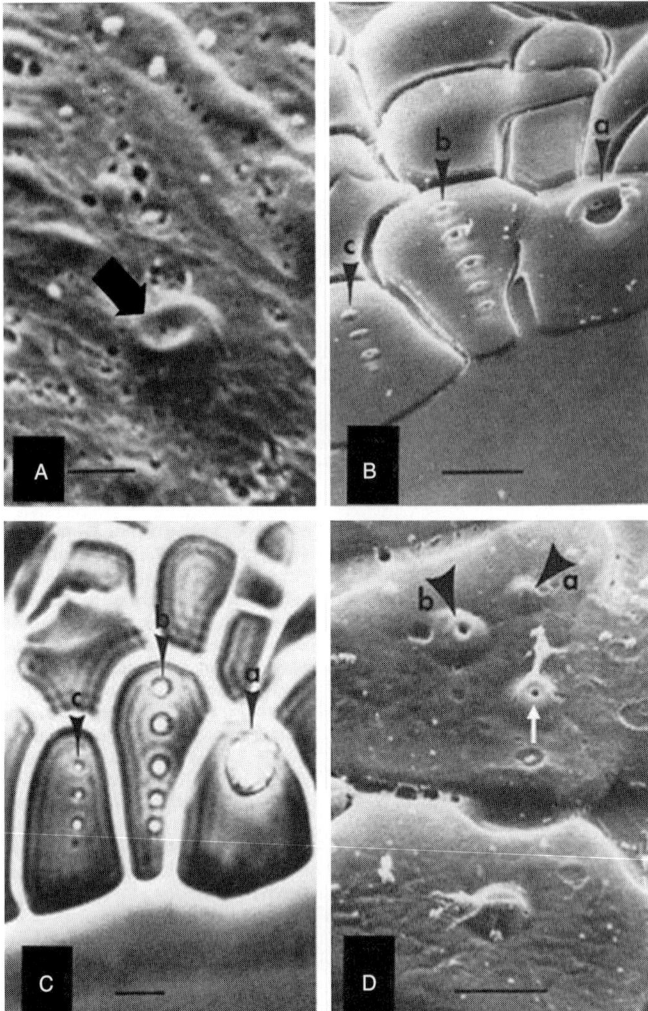

Fig. 14 (A) Scanning electron micrograph (SEM) of a cell irradiated with a 40× objective. Note that the lesion appears very similar to the lesion seen in Fig. 13C and D with a central dimple surrounded by a raised ring. Scale bar = 1 μm. (B) SEM of red blood cells with several lesions made by laser irradiation through the glass: gross perforation (a), raised blister with central dimple and perforation (b), and raised blister with no dimple or perforation (c). Scale bar = 5 μm. (C) Phase-contrast micrograph of the same red blood cells seen in (B): gross perforation (a), raised blister with central dimple and perforation (b), and raised blister with no perforation or dimple (c). Scale bar = 5 μm. (D) SEM of red blood cells with several lesions made by laser irradiation. Note the lesion consisting of a raised blister without central dimple or perforation (a), and another lesion type consisting of a raised blister with central dimple and perforation (b). The central small lesion point is 150–200 nm in diameter. Scale bar = 2 μm. (See Burt *et al.*, 1979.)

G. Cytoplasmic Filaments

Three types of cytoplasmic filaments were microdissected using the Q-switched Nd:YAG laser operating at either 532 or 537 nm: (1) muscle fibers (myofibrils in individual sarcomeres) in cardiomyocytes, (2) stress fibers in nonmuscle cardiac cells, and (3) bands of 100-Å filaments in nonmuscle cardiac cells.

In cardiac muscle fibers, the optical resolution of individual *A* and *Z* bands permitted ultrafine nanodissection of these structures (Strahs *et al.*, 1978; see Figs. 17 and 18 in Appendix, this chapter). Electron microscopy analysis of laser-irradiated *A* bands revealed a discrete electron-dense damage zone within the *A* band of ~0.5 μm in diameter. There was no detectible damage to the rest of the irradiated *A* band. However, damage to a single *Z* band resulted in disorganization of the *Z* band including its lateral expansion such that the distance between the irradiated *Z* band and the next *Z* band almost doubled. This result would suggest that the *Z* band of the muscle fiber is maintained under considerable tension, and when its structural integrity is compromised the tension it is under literally pulls it apart.

Laser nanoscissors were first applied to stress fibers (bundles of actin filaments) in 1979, and demonstrated for the first time that cytoskeletal elements could be studied with laser nanoscissors (Strahs and Berns, 1979). At the time of these early studies, the role of stress fibers in cell shape cytoskeletal organization and cell mechanics was not known. The fact that stress fiber recovery and regrowth was observed following laser cutting suggested that there was a dynamic system of physiological regulation occurring. A conclusion of these early laser nanoscissors experiments was that there are two types of stress fibers, retracting and nonretracting, and that "some fibers are under tension." This suggestion was confirmed 27 years later in an elegant study using femtosecond laser nanoscissors in combination with traction force microscopy and fluorescence photobleaching methods. This study concluded that individual stress fibers are under tension (Kumar *et al.*, 2006). Neither traction force microscopy, fluorescent fusion protein expression, nor photobleaching recovery techniques existed 27 years ago when the first stress fiber nanoscissors experiments were conducted. It is in some way comforting for the scientists who conducted those first experiments to read that "... the technologies described here offer a novel approach to spatially map the cytoskeletal mechanics of living cells on the nanoscale" (Kumar *et al.*, 2006).

Another group of cytoplasmic filaments studied by laser nanosurgery were bands of 100-Å intermediate filaments. They occur in large bands in cells following treatment with colcemid, a microtubule-disrupting agent. Since it had been shown that bands of 100-Å filaments can move within a cell, it was of interest to determine if there was relative movement of the intermediate filaments within the band itself. To test this hypothesis several 500-nm-diameter laser lesions were made within a band and their relative position to each other was tracked over time. No change in relative position was observed, thus leading to the conclusion that there was no movement within the band of filaments. Electron microscopy of the 500-nm lesion revealed substantial alteration outside of the laser focal point. A microplasma lesion mechanism could produce damage outside the focal point via a shock effect on nearby structures.

VI. Embryos and Fate-Mapping

The laser has become a very useful tool for the ablation of selective cells during and after development in many organisms. The first extensive series of studies on fate-mapping (the determination of the developmental fate of individual cells or groups of cells) were performed in Germany on *D. melanogaster* embryos employing the 257-nm UV laser microbeam with a diameter of 10–30 μm (Cremer *et al.*, 1974). This system was used to expose groups of five to seven nuclei in the early embryo and then allow the embryo to develop all the way to the adult where defects were correlated to the regions irradiated in the embryo (Lohs-Schardin *et al.*, 1979a). In this study, over 10,000 eggs were irradiated and scored with respect to defects in the adult fly. The authors suggested that the laser fate-mapping approach compared to other methods would provide real rather than relative locations of adult progenitor cells in the early embryo. Additionally, another study by the same group focused on the fate-map of localized cuticle damage in the larval epidermis (Lohs-Schardin *et al.*, 1979b). Again, compared to previous attempts at fate-mapping the authors concluded that the UV laser method yielded much more accurate and detailed information. Following this early work on fate mapping, a 266-nm fourth harmonic from the Nd:YAG laser was used to kill patterns of cells in leg imaginal disks that had been removed from drosophila larvae and then reimplanted into to abdomens of adult females (Girton and Berns, 1982). These disks underwent pattern regulation either regenerating, duplicating, or triplicating depending on the precise region of the disk that had been ablated. These results were similar to those obtained by genetic mutations and by other means of cell removal, thus allowing the conclusion that wound healing is controlled by the same pattern-formation system. More recently, in *Drosophila*, it has been possible to model morphogenetic forces during development by using laser microsurgery to disrupt dorsal closure in the embryo (Hutson *et al.*, 2003).

In addition to the work on *Drosophila*, the laser microbeam has been applied to study several developmental systems in other insects. Fate-mapping of the silkworm *Bombyx mori* has been conducted using the 355-nm third harmonic of the Nd:YAG laser (Myohara, 1994). Initial studies involved irradiation of the egg at fertilization followed by determination of localized defects in the cuticle. On the basis of the correlation of specific defects with irradiated regions of the egg it was possible to fate-map each larval segment back to specific regions of the egg. In a more recent study, it was possible to actually fate-map the silk gland to a limited region of the egg that also gives rise to the labium (Myohara, 2003).

One of the most productive areas of laser microbeam application has been in developmental neurobiology. An early study in crickets was designed to test the hypothesis that so-called "pioneer fibers," which develop relatively early in insect differentiation, serve to organize the peripheral sensory nerves (Edwards *et al.*, 1981). This hypothesis was tested by using the fourth harmonic 266-nm Nd:YAG laser to ablate a specific region of the developing embryo before and after the formation of pioneer fibers. It was found that pioneer fibers were necessary for the development of the cercal nerve which gives rise to the peripheral sensory nerves in the cricket appendages. But perhaps the most extensive series of studies in developmental neurobiology have involved the selective ablation of cells in the *C. elegans* (an organism that already has been discussed in Section V.C.4). The foremost series of investigations were initiated by Sydney Brenner of Cambridge University (UK) in 1985 (Chalfie *et al.*, 1985). They used the 337-nm nitrogen laser to kill precursor cells of the neural circuit for touch sensitivity. Their analysis revealed that there are two pathways for touch-mediated movement and a single pathway for posterior touch. Further, they were able to discern how the touch circuitry changes as the animal matures. Following these early studies, a whole host of studies on various aspects of *C. elegans* have been conducted by many other investigators. For example, Wakabayashi *et al.* (2004) pursued the neurobiology of locomotion showing that by laser ablation of precursor cells, the regulation of locomotion was controlled by antagonistic neuronal pathways consisting of nine classes of sensory neurons and four classes of interneurons. In a study on the role of cellular interactions in the determination of early *C. elegans* development, early blastomeres were inactivated by laser microsurgery and the cell lineages of irradiated embryos were compared to those of intact (unirradiated) embryos (Schnabel, 1991). It was shown that one specific blastomere was needed for specification of the mesodermal pharyngeal fates (structures) and another blastomere was necessary for the formation of the hypodermal structures, but the formation of the nervous system required both blastomeres. In addition to the studies just mentioned, development of *C. elegans* has been studied using laser microbeams by several other groups including Hall and Russel (1991), Bergmann and Horwitz (1991), and Schierenberg (1984). A good summary of these studies can be found in Greulich (1999, pp. 101–103).

VII. Technology and the Future

The combination of available laser, optical, image analysis/computing, and molecular tools solidifies laser microbeams (including laser tweezers) as major tools for cell study in the twenty-first century. As described by Magidson *et al.*, in Section II.B of Chapter 7, this volume, it is possible to build your own "versatile low-cost microsurgery workstation." They champion the idea of having an open tabletop system, as opposed to a closed more expensive "turnkey" commercial system available by Zeiss-P.A.L.M (http://www.palm-microlaser.com), Cell Robotics, Inc. (http://www.cellrobotics.com), and Leica, Inc. (http://www.leica-microsystems.com/LMD6000). In addition to a major reduction in cost, the open tabletop system can be adjusted and modified as needed. Notwithstanding these advantages, the choice of which laser to use is still of considerable debate, especially with the advent of the near-infrared femtosecond laser and the suggestion that it may be "the laser of choice." However, as can be seen from the many studies described in this chapter and the rest of this volume, researchers have (and still are) successfully illuminating important questions in cell and developmental biology using picosecond, nanosecond, and microsecond

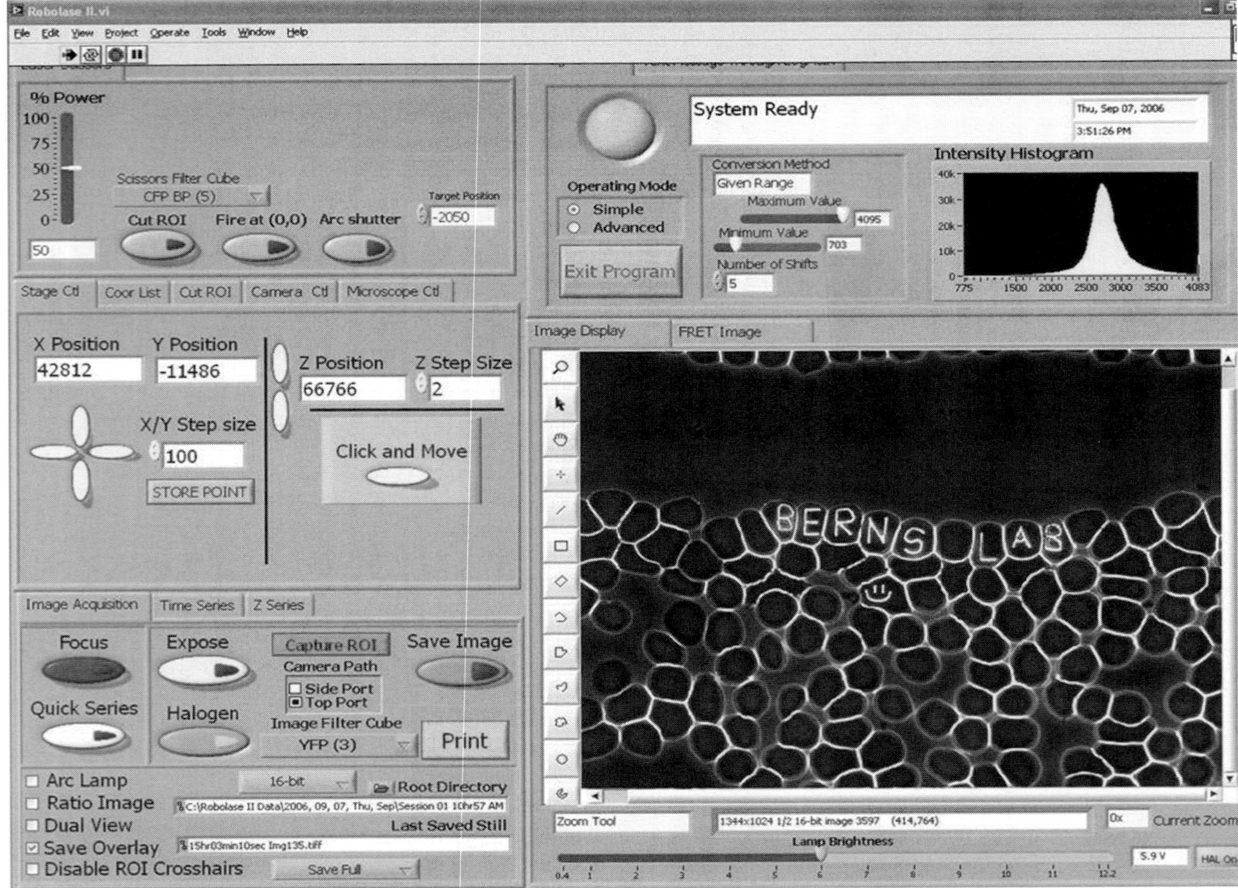

Fig. 15 RoboLase control panel. The control panel as viewed during a *logmein.com* session using the laser scissors. The top left box controls the following: laser power, five different dichroic filter cubes, and various cut-and-fire controls. The row of rectangles immediately below controls drop-down menus for: stage controls, coordinate listing, cut outlined region of interest, and microscope controls. The two large rectangles below this are the drop-down menus for stage controls, providing control of the x/y/z movement of the stage, and the amount of movement of the stage per command, which is adjustable. Below this is a rectangle that provides control over the image: "focus" allows for continual streaming of images and "expose" updates a single image. The large rectangle in the right side contains an image of the cells under the microscope. In this case, they are dried red blood cells on a microscope cover glass being viewed with a 63× 1.4 NA

lasers ranging in wavelength from the UV to the infrared. So which laser to place in your "open-table" laser microbeam is quite a knotty choice.

Another option to having "your own" laser microbeam system, whether it be a self-built or commercial system, is to utilize the system at the LAMMP Biotechnology Resource Center at the University of California Irvine Beckman Laser Institute http://www.bli.uci.edu/lammp/. LAMMP was specifically developed to provide a sophisticated laser microbeam system that includes a confocal microscope with an ablation laser and two trapping lasers—thus the name CATS (see Fig. 7).

A third option is to use the newly developed and evolving RoboLase systems, which when fully operational, will be accessible through the Internet (Botvinick and Berns, 2005; see Chapter 18 by Botvinick and Wang, this volume). RoboLase is actually a linked array of four different laser microscopes (and growing) based at the University of California Irvine and San Diego campuses that provides a variety of ablation, trapping, and imaging capabilities accessible through the Internet. Optimal operation is obtained with gigabit per second bandwidth, but for cells and organelles that are not moving above microns per minute, 10–100 Mb/s speed is adequate for most laser microsurgical studies. With high-speed bandwidth (gigabit per second) swimming sperm on the microscope in California have been visualized, trapped, and measured from the University of Queensland in Australia (see digital video sequence "Trans-Pacific Real Time Sperm Tracking" accessible at www.robolase.ucsd.edu/movies.html). Laser ablation experiments have been conducted remotely from New York, Georgia, and Florida in the United States, and from Australia and Germany internationally.

In future studies, either an individual investigator or a group of collaborators will be able to log on to the system and communicate during their experiments by text messaging directly on the system control panel (Fig. 15) or by phone communication while operating the laser microscope. The downside of this type of experimentation is that any special biological preparation would have to be sent ahead of time to one of the RoboLase laboratories in California, thus making certain types of experiments problematic. On the other hand, once the details of handling the material are worked out, the experiments can be conducted remotely saving considerable time and expense. In addition, the feature of having more than one collaborator log on for the experiment greatly expands the ability to do collaborative research and truly brings laser scissors (and tweezers) to a new level of "global science."

---

microscope objective. The "BERNS LAB" design has been produced by first outlining each letter in the blood cell using the tenth tool down the column on the left side of the image. When this is done the ROI (region of interest) button in the top right panel is activated using the mouse. This activates the laser and the scanning mirror that moves the laser precisely in the patter outlined with the tool. Other controls on the right side of the control panel are: (1) control of microscope lamp brightness (at the bottom), and (2) control of the grayscale range of the imaging camera, which is a Hamamatsu Orca-AG deep-cooled 1344 × 1024 pixel 12-bit digital CCD camera with digital (fire wire) output. See Botvinick and Berns (2005) for more specific details on this system.

VIII. Appendix

Laser Microsurgery in Cell and Developmental Biology

Michael W. Berns, J. Aist, J. Edwards, K. Strahs, J. Girton
P. McNeill, J. B. Rattner, M. Kitzes, M. Hammer-Wilson
L.-H. Liaw, A. Siemens, M. Koonce, S. Peterson, S. Brenner
J. Burt, R. Walter, P. J. Bryant, D. van Dyk, J. Coulombe
T. Cahill, G. S. Berns

The laser microbeam permits selective alteration of part of a subcellular organelle in a single living cell. This alteration can be in a specific class of molecules confined to an area of less than 0.25 micrometer. This capability has been developed over the last 15 years and is now generally available for studies in cell and developmental biology.

Summary. New applications of laser microbeam irradiation to cell and developmental biology include a new instrument with a tunable wavelength (217- to 800-nanometer) laser microbeam and a wide range of energies and exposure durations (down to 25×10^{-12} second). Laser microbeams can be used for microirradiation of selected nucleolar genetic regions and for laser microdissection of mitotic and cytoplasmic organelles. They are also used to disrupt the developing neurosensory appendages of the cricket and the imaginal discs of *Drosophila*.

Just over 12 years ago, the blue-green argon ion laser microbeam was introduced as a potential tool for subcellular microsurgery (*1, 2*). There had been limited success earlier with the red ruby laser (*3*) and the classical ultraviolet microbeam (*4*). The work with the blue-green argon laser led to development of a tunable wavelength flash-lamp-pumped dye laser microbeam (*5*) and later to a dye laser that was pumped by the green wavelength of a low-power neodymium-YAG (yttrium-aluminum-garnet) laser (*6*). The recent establishment of a National Institutes of Health Biotechnology "user" resource has permitted the development of a dye laser microbeam completely tunable from 217 to 800 nanometers, by employing the second (532 nm), third (355 nm), and fourth (265 nm) harmonic wavelengths of a high-power 10-nanosecond pulsed neodymium-YAG laser (Fig. 1). In addition, a separate high-power 25-picosecond neodymium-YAG laser has been integrated into the system to permit exposure of living cells to ultrashort pulses of light. This dual laser system is interfaced with an inverted Zeiss Axiomat microscope and an image array processing computer, which permits the exposure of groups of cells, single cells, or individual organelles within single cells to a variety of wavelengths at various power densities, with time exposures as short as 25 psec. In addition, the use of the sophisticated Zeiss Axiomat microscope and the image processing computer permits a state-of-the-art optical and photometric examination of the biological material.

Since this user system is now available to the scientific community (*7*), we will review some of the earlier key experiments and several recent unpublished experiments that demonstrate the use and versatility of the system in cell and developmental biology.

Principles of Selective Damage

Laser light is intense, coherent, monochromatic electromagnetic radiation. The damage produced by a focused laser beam may be due to classical absorption by natural or applied chromophores and the subsequent generation of heat (*8*), or it may be caused by a photochemical process such as the production of monoadducts or of diadduct cross-linking in the case of laser light–stimulated binding of psoralen to nucleic acids (*9*). A third possibility is the generation of damage by an uncommon physical effect that occurs when ultrahigh photon densities are achieved in very short periods (a few nanoseconds or picoseconds). The resulting nonlinear optical effects, such as multiphoton absorption, dielectric breakdown, and pressure phenomena, occur when the classic law of reciprocity does not hold; these effects may be responsible for some of the disruption observed in biological material (*10*). Whichever of the above damage-producing mechanisms is operating, whether "classical" or "uncommon," the damage often can be confined to a specific cellular or subcellular target in a consistent and controllable way. In addition, once the biophysical mechanism of laser interaction with the molecules is ascertained, the investigator has a method for precise disruption of a specific class of molecules within a strictly delimited region of the living cell. The size of this region may be considerably smaller than the size of the focused laser beam because of the distribution of the target molecules in the target zone. However, the size of the focused laser spot also is of paramount importance, because it defines the maximum volume of biological material that will be available for direct interaction with the laser photons. Though the diameter of the focused laser spot is a direct function of the wavelength, the magnification of the focusing objective, and the numerical aperture of the objective, the actual diameter of the "effective" lesion area may be considerably less than the theoretical limit of the focused laser beam, which is half the wavelength. This is because a high-quality laser beam can be generted in the transverse electromagnetic (TEM_{00}) mode, which results in a beam with a Gaussian energy profile across it. The profile is carried over to the focused spot, producing a "hot spot" of energy in the center. It has been demonstrated consistently (*11*) that by careful attenuation of the raw laser beam, the damage-producing portion in

Michael W. Berns is in the Department of Developmental and Cell Biology, University of California, Irvine 92717; J. Aist is in the Department of Plant Pathology, Cornell University, Ithaca, New York 14853; J. Edwards is in the Department of Zoology, University of Washington, Seattle 98195; K. Strahs is affiliated with Beckman Instruments, Inc., Fullerton, California 92632; J. Girton is a member of the School of Life Sciences, University of Nebraska, Lincoln 68588; P. McNeill, J. B. Rattner, M. Kitzes, M. Hammer-Wilson, L.-H. Liaw, A. Siemens, M. Koonce, S. Peterson, S. Brenner, J. Burt, R. Walter, P. J. Bryant, D. van Dyk, J. Coulombe, T. Cahill, and G. S. Berns are affiliated with the Department of Developmental and Cell Biology, University of California, Irvine 92717.

0036-8075/81/0731-0505$02.00/0

the focused spot can be confined to the central hot spot (that is, the only region within the focused spot that is above the threshold for damage production). As a result, lesions can be routinely produced that are less than 0.25 μm in diameter and frequently 0.1 μm in diameter (see cover).

Monitoring, attenuation, filtering devices
532, 355, 266 nm
10^9 W
Nd-YAG laser
25×10^{-12} sec. (picoseconds)
532, 355, 266 nm
10^6 W
Nd-YAG laser
15×10^{-9} sec. (nanoseconds)
217 - 700 nm
10^5 W
Dye laser
Series of Dichroic Filters
0.5 μm Digital X-Y Stage
Newvicon TV
Digitizer
LSI 11/2 Minicomputor, Dual Floppy Discs
Deanza Image Array Processor (Gray Scale)
Inverted Zeiss Axiomat DIC, Pol., Phase, Quartz Microscope
Monitor
Video Time Lapse Tape Recorder
CRT
Color Monitor

Fig. 1. Diagram of laser microbeam system. The three basic components of the system are the lasers (Quantel YAG 400 and 481/TDL III), the microscope (Zeiss inverted Axiomat equipped for phase contrast, bright field, polarization, and differential interference contrast), and the television computer system (DeAnza IP 5000 image array processor, Sierra LST-1 television camera, and GYYR DA 5300 MKIII videotape system). An LSI-11 minicomputer is used to drive the image array processor. In addition, the image processor–LSI combination is interfaced to the X-Y digital microscope stage in order to provide cell tracking capabilities.

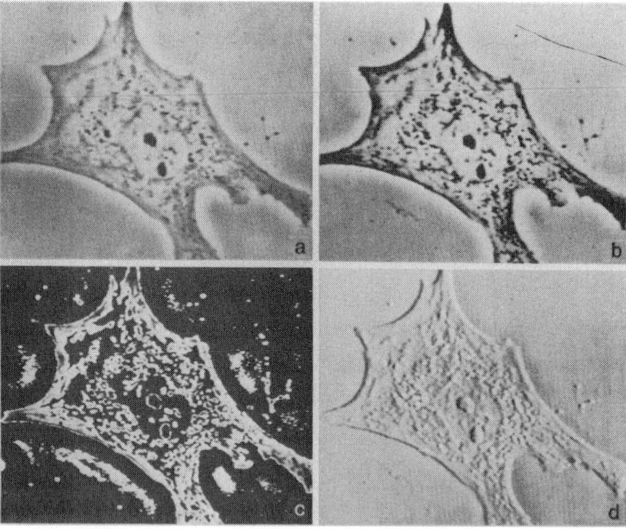

Fig. 2. (a) Live phase-contrast image of PTK_2 kangaroo kidney cell; image photographed directly from television monitor; (b) same cell after computer contrast enhancement by an intensity transformation that resulted in the reassignment of gray values and display of the image in real time; the result is an increase in contrast that enhances specific cellular structures; (c) same cell displayed after contrast enhancement of image boundaries; (d) real-time image of same cell after boundary enhancement by subtracting a slightly offset image from an original unshifted image. This image is virtually identical to a differential interference contrast image. It is generated by the computer by using a standard phase-contrast image with lower light level illumination.

Video Computer Microscopy

Since the laser beam can be focused to produce a damage spot less than the diffraction limits of the light microscope, one of the limiting factors becomes the quality of the optical image itself. It is difficult to focus the laser beam onto a target that is not readily visible.

We have been able to improve upon the optical image of the highest quality light microscope (the Zeiss Axiomat) by using a low light level television camera (Newvicon tube) whose signal is digitized and enhanced by a fast (real time) image-processing computer. Recent advances in computer technology have led to the development of small, relatively inexpensive image array processors that are capable of performing sophisticatd image-processing routines on video images in real time. Real-time processing allows sophisticated routines, such as contrast enhancement, edge detection, background subtraction, multiple image averaging, and pseudocolor enhancement; all of these can be performed on the microscope image during the time of the actual experiment. Figure 2 and the cover photograph are examples of computer-enhanced images. The processor can also be used for more analytical tasks such as calculation of object areas, boundary lengths, and intracellar distances. The integration of the computer into the laser microbeam system has greatly extended the capabilities of this system and the scope of the experiments to be described.

Chromosome Microsurgery

In 1969, a low-power pulsed argon ion laser was focused on chromosomes of living mitotic salamander cells that had been photosensitized with the vital dye acridine orange. The result was the production of a 0.5-μm lesion in the irradiated region of the chromosome (Fig. 3). Subsequent studies on salamander and rat kangaroo cells (PTK_1 and PTK_2) demonstrated that the laser microbeam could be used to selectively inactivate a specific genetic site, the nucleolar genes (*1, 12*). Three different laser microbeam systems were used in these studies: the low-power argon laser, with acridine orange sensitization; a high-power argon laser, without dye photosensitization (most likely a multiphoton process mechanism); and the fourth harmonic (265 nm) of a neodymium-YAG laser. Not only can the nucleolar genes be selectively deleted, causing a loss of nucleoli in the subsequent cell generations, but a corresponding lack of one light-staining Giemsa band in the nucleolar organizer region of the chromosome can be demonstrated in cells cloned from the single irradiated cell (*13*) (Fig. 4). Experiments that include in situ hybridization with ^{3}H-labeled RNA and selective silver staining for the nucleolar organizer have demonstrated the loss of one group of ribosomal genes in the clonal population of cells (*14*) (Figs. 4 to 8). The use of the laser to destroy selected chromosome regions with the subsequent maintenance of this genetic loss is clearly feasible. In addition, it is now relatively easy to manipulate the ribosomal genes in vitro in order to study their regulation and function—a problem of considerable interest in light of the classic genetic studies on the bobbed mutant in *Drosophila* (*15*) and gene amplification

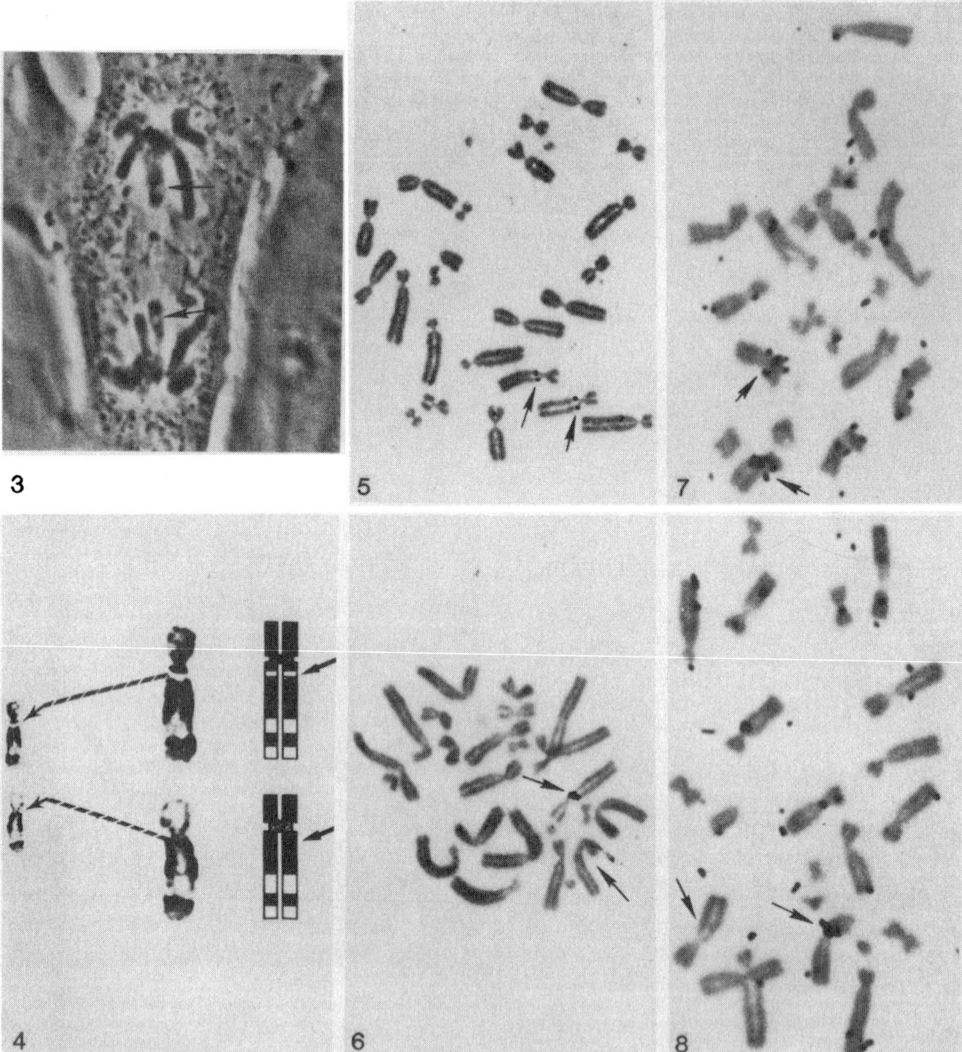

Fig. 3. Phase-contrast photomicrograph of anaphase PTK_2 chromosomes after placement of two lesions, 1 μm in diameter, on the chromosome arms (arrows). Lesions were produced by irradiation with the 514-nm beam of an argon laser with an energy density of 1000 microjoules per square micrometer without dye sensitization. Fig. 4. Giemsa-trypsin–banded chromosomes from PTK_2 clone in which originating cell had one nucleolar organizer secondary constriction irradiated with a 265-nm beam of a YAG laser. Note the deletion of one light-staining chromosome region (arrows). Fig. 5. Silver-stained chromosomes from control nonirradiated PTK_2 cells. Note two clearly stained nucleolar organizer regions. Arrows point to two homologs of a pair. Fig. 6. Silver-stained chromosomes from irradiated clone. Note one heavily stained nucleolar organizer. Fig. 7. In situ hybridization of ^{3}H-labeled ribosomal RNA to control nonirradiated PTK_2 cell. Note two chromosomes with selective hybridization to the nucleolar organizer. Fig. 8. In situ hybridization to cell cloned from the irradiated cell. Note only one chromosome with hybridization to the nucleolar organizer.

in amphibian oocytes (*16*). Our preliminary experiments with silver staining and in situ hybridization suggest the possibility of ribosomal gene "magnification" in vitro after the laser deletion of one group of ribosomal genes (Figs. 4 to 8). This suggestion is based on the finding that the one remaining nucleolar organizer region (after deletion of one) appears to stain twice as deeply as those in cells with the normal two nucleolar organizers. Similarly, there appears to be roughly twice the normal amount of in situ hybridization to the one nucleolar organizer region in the cells cloned from the irradiated cell. Final conclusions on this point, however, must await quantitative ribosomal DNA determinations in vitro.

Whereas the studies discussed above dealt with selective removal of portions of chromosomes, other studies have been done in which entire chromosomes have been removed from mitotic cells (*17*). This can be accomplished by irradiation of the centromere region at metaphase of mitosis. When a centromere with its microtubule attachment site (the kinetochore) is destroyed, the chromatid no longer remains attached to the mitotic spindle. Frequently the chromatid remains behind at the metaphase plate and is caught within the constriction ring at cytokinesis. The chromosome may be incorporated into the cytoplasm of one of the daughter cells. The genetic result is the frequent production of one daughter cell that is missing an entire chromosome and one daughter cell that has an extra chromosome enclosed within a micronucleus. These daughter cells have been observed through the subsequent mitosis (*18*), and the irradiated chromosome duplicates itself without a functional kinetochore. At the next mitosis, the duplicated irradiated chromosome cannot attach to the spindle, and once again a micronucleus is formed. The capability of directed whole chromosome removal permits a class of cytogenetic studies in which investigators can selectively delete chromosomes and thus have a method to complement the already well-developed methods of somatic cell fusion. In addition, the ability to damage a restricted region of a chromosome (such as the centromere) and to observe the cell

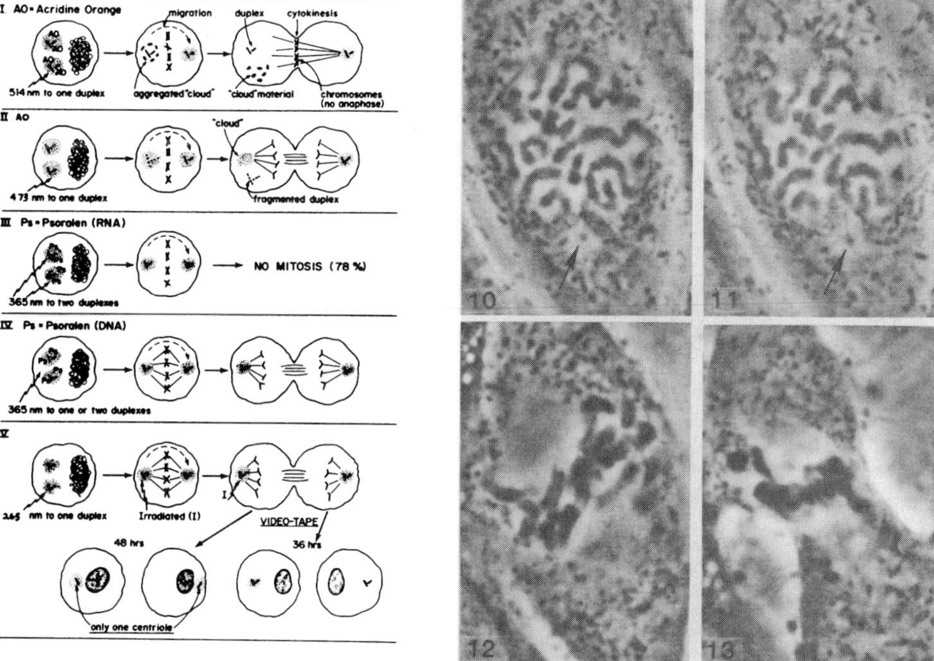

Fig. 9. Summary of experiments in which the centriolar region was irradiated under conditions to produce selective disruption to specific components. (*I*) Acridine orange sensitized the pericentriolar cloud to 488 and 514 nm. (*II*) Irradiation with 473 nm selectively disrupted the centriole proper and left the pericentriolar cloud unaffected; acridine orange was also used as a sensitizer. (*III*) Psoralen (4′-aminomethyl-4,5′,8-trimethylpsoralen) selectively sensitized the pericentriolar cloud by binding to RNA. (*IV*) Psoralen specific for DNA (4′-hydroxymethyl-4, 5′, 8-trimethylpsoralen, 4′-methoxymethyl-4,5′,8-trimethylpsoralen, and 4,5′,8-trimethylpsoralen) had no effect on the immediate process of cell division. (*V*) Irradiation with 265-nm laser light was effective in preventing centriole replication but did not inhibit the immediate cell division process. Furthermore, the fact that the irradiated cell went through a subsequent division without duplicating centrioles demonstrates that centriole replication is not needed for mitosis to occur. This result also implicates the nucleic acid in the process of centriole duplication. Fig. 10. Prophase PTK_2 cell treated with acridine orange to sensitize the centriolar region to the argon ion laser beam of 514 nm. The arrow indicates the centriolar region (dark spot) in the perinuclear clear zone. This is before irradiation. Fig. 11. Immediately after irradiation of the centriolar region. Note the slight increase in the extent of darkening (arrow). Fig. 12. The irradiated cell about 15 minutes after irradiation. Note that the chromosomes have continued to condense and align in a metaphase-like configuration. Fig. 13. At 30 minutes after irradiation, the cell undergoes cytokinesis without any anaphase movement of chromosomes. Ultrastructural examination of this and similarly irradiated cells demonstrated that the pericentriolar material had been selectively damaged.

through its cell cycle to a subsequent mitosis permits studies on chromosome damage and repair from a new perspective.

Mitotic Organelles

Centriolar Zone. Extensive work has been devoted to the use of the laser to selectively disrupt three mitotic structures (centrioles, kinetochores, and microtubules) in order to elucidate their organization and function in the process of cell division. These studies may be the most demanding in terms of understanding and applying the principles of selective damage discussed in the first section of this article. Figure 9 summarizes the centriole experiments.

Centrioles are just within the resolution of the light microsocope. In the PTK_2 cell line, the centriolar duplex is frequently visible in prophase as a phase dark dot 0.25 μm in diameter within a perinuclear clear zone (*19*). Ultrastructurally, the centriolar complex is composed of the centriole proper and a surrounding cloud of material called the pericentriolar cloud. Treatment of prophase cells with nontoxic levels of acridine orange selectively sensitized the pericentriolar cloud to the green beam of either the argon or YAG laser (*20*). Irradiation of the centriolar complex after acridine orange treatment resulted in selective disruption of the cloud without apparently affecting the centriole. The cells progressed toward metaphase, but no anaphase movement of chromosomes occurred, even though the cells went through cytokinesis (Figs. 10 to 13). Since acridine orange binds selectively to nucleic acid, this study supports the earlier finding (*21*) that some nucleic acid is located in the pellicle of *Paramecium*. In our studies, the high degree of sensitivity of the pericentriolar cloud implicated this region as a major site of nucleic acid localization. In addition, the lack of microtubule organization after disruption of the cloud suggests that this region is a microtubular organizing center in vivo, a fact confirmed by Gould and Borisy (*22*) using isolated pericentriolar material. In later laser microbeam studies of the centriolar region, a psoralen compound that is photochemically bound to DNA did not inhibit mitosis after exposure of the centriolar region to the appropriate cross-linking wavelength (365 nm) of laser light (*23*). However, another psoralen compound that, upon exposure to long-wavelength ultraviolet, binds to both DNA and RNA effectively inhibited mitosis after laser microirradiation of the centriolar region. These results suggest that an RNA with secondary structure in the pericentriolar region has a major role in the organization and function of the mitotic spindle. Recent ribonuclease digestion studies support this finding (*24*).

A final series of laser microbeam studies on the centriolar region involved selective destruction of the centriole proper without damage to the pericentriolar cloud (*25*). In these studies, the blue second harmonic wavelength of the YAG laser was used with acridine orange. The biophysical mechanism of damage production was most likely an uncommon physical effect because of the short exposure time (nanoseconds) and high power. It was demonstrated that cells with destroyed centrioles but intact pericentriolar material were capable of proceeding through mitosis in a normal fashion. The role of the centriole in mitosis remains debatable. Selective alteration of the centriolar region and computerized tracking of cells should elucidate centriolar replication and its relation to control of mitosis.

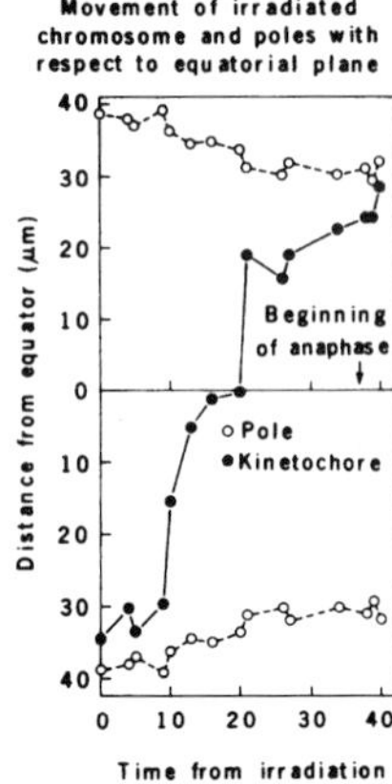

Fig. 14. Graphic depiction of the movement of a double-chromatid chromosome after laser irradiation of one kinetochore. In the diagram, the black circle represents the centromere with two kinetochores, one on each side. The kinetochore closest to the bottom pole was irradiated at time 0. The entire double-chromatid chromosome with only one functional kinetochore subsequently went through the movements depicted by the black circle in the figure. The rate of movement was equivalent to the normal rate of anaphase movement even though the chromosome mass was twice that of a single chromatid.

Kinetochores. The other major mitotic structure that participates in the organization of microtubules is the kinetochore. Using a very finely focused green laser beam, we have destroyed this region of the chromosome and then investigated the dynamics of chromosome movement. When both kinetochores of a metaphase double-chromatid chromosome are destroyed, the chromosome drifts about in the cell and the chromatids separate slightly from each other at the exact time that the rest of the chromosomes initiate their anaphase movements. This observation illustrates that the initial separation of chromatids at anaphase is not mediated by a microtubule force (*18*).

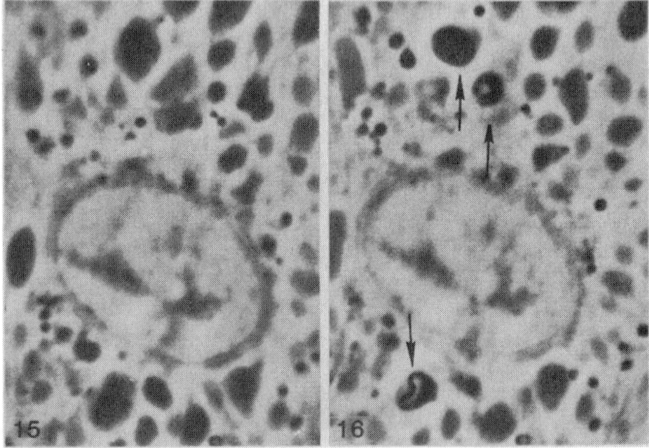

Fig. 15. Phase-contrast micrograph of myocardial cell with large mitochondria just before microirradiation with 514 nm of an argon laser. Fig. 16. The same cell after irradiation. The lesions have different degrees of severity (arrows indicate irradiated mitochondria).

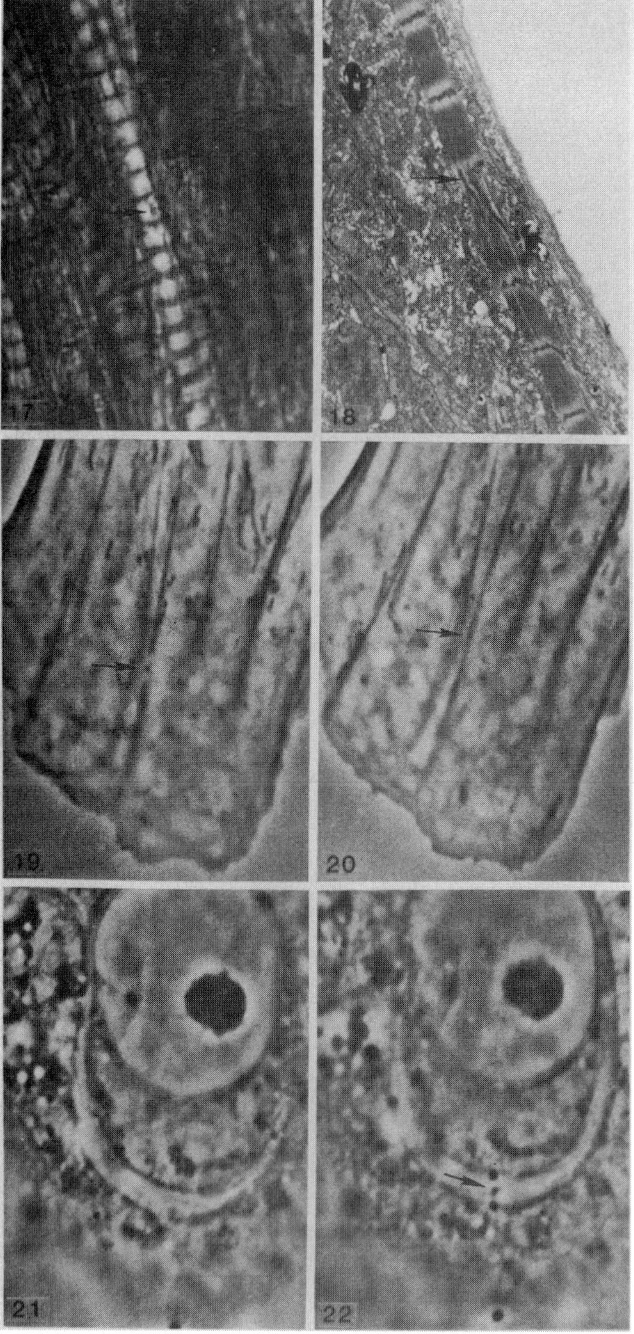

In other studies, only one kinetochore was destroyed, and the chromosome with both chromatids and only one functional kinetochore was tracked (*26*) (Fig. 14). The results are quite dramatic and demonstrate that (i) two functional kinetochores are necessry for the alignment of a chromosome on the metaphase plate and for normal anaphase movement; (ii) bipolar tension on the kinetochore is necessary to stabilize the orientation of the chromosome on the metaphase plate; (iii) irradiation and inactivation of one kinetochore lead to nondisjunction of the irradiated chromosome; (iv) chromatids with irradiated kinetochores retain their ability to replicate but are unable to repair the damaged kinetochore region; and (v) within limits, the velocity with which a kinetochore moves is independent of the mass associated with it.

Microtubules. The microtubules were among the first mitotic structures successfully irradiated with the classical ultraviolet microbeam instruments (*27*). In our studies we have initiated microtubule studies with the laser microbeam on the highly visible dense band of microtubules in dividing fungal cells (*28*).

Of particular interest is the function of the dense microtubular band that extends between the two separating nuclei at the end of fungal mitosis (*29*). Earlier observations led to the hypothesis that this band of microtubules served to push the two nuclei apart. However, laser disruption of the bundle resulted in a threefold increase in rate of nuclear separation; 22.4 μm/min as opposed to 7.6 μm/min in control unirradiated cells (*28*). In addition, damage to the outside of the nucleus (distal to the bundle) resulted in a significant decrease (6.1 μm/min) in the rate of nuclear separation. These experiments indicate that the intranuclear band of microtubules is rate-limiting (slowing down the movement of the nuclei). Furthermore, it appears that the forces for nuclear separation may be coming from

Fig. 17. Polarization photomicrograph of myofibrillar region of contracting myocardial cell in vitro. A single A band (arrow) has been irradiated with the 532-nm beam of the YAG laser. Damage is localized within one A band. **Fig. 18.** Electron micrograph of myofibrillar network that has had one Z line (arrow) irradiated while the cell was in the living state. The damage to one Z line resulted in considerable disarray of the myofilaments in proximity to the irradiated Z line. **Fig. 19.** Phase-contrast micrograph of a single stress fiber of a rat endothelium that has been cut by one pulse of 532-nm light from the YAG laser microbeam. **Fig. 20.** Same cell 1 hour after irradiation. Note that there has been regeneration of the cut stress fiber. **Fig. 21.** Band of 100-Å filaments in a nonmuscle cell from a tertiary culture of neonatal rat heart. **Fig. 22.** Same cell with three laser lesions placed across the band of 100-Å filaments by the 532-nm beam of the YAG laser. This type of system has been used to study relative movement within the band.

the other side of the nuclei, where electron microscopy has revealed a substantial array of astral microtubules.

Cytoplasm

Mitochondria. The laser microbeam has been extensively applied to the subcellular disruption of single mitochondria (*30*). Much of this work has been conducted in contracting mammalian cardiac cells in culture and has had as its major aim elucidation of the factors regulating cardiac cell contractility. Morphologically distinct lesions can be placed in individual mitochondria and the subsequent contractile, electrical, and morphological responses of the cell can be analyzed (Figs. 15 and 16). Salet *et al.* (*31*) appear to have demonstrated that the laser light energy can be trapped and converted directly to adenosine triphosphate by the irradiated organelle; the irradiated cells also undergo a transient increase in beat rate. In other studies, cells have been impaled with microelectrodes prior to selective irradiation, and a distinct depolarization of the cell membrane has been demonstrated after irradiation of one mitochondrion. Only those cells with the classic "pacemaker" action potential (*32*) can be shown to enter a fibrillatory state after irradiation. The nonpacemaker cells exhibit a laser-induced depolarization but maintain normal electrical and contractile activity. In all of the irradiated cells, the cell membrane eventually returns to its normal membrane resting potential, thus suggesting that the laser effect on the cell membrane is transient, probably resulting in a temporary alteration of membrane permeability to specific ions. This kind of investigation permits precise alteration in cardiac cell contractility by producing a well-defined lesion at a predetermined subcellular site. Subsequent repair, recovery, and pharmacologic control of beat arrhythmia is thus studied in a precisely controlled situation.

Myofilaments, stress fibers, and 100-angstrom filaments. The laser microbeam can be used to study other motility-related cytoplasmic cell structures. For example, individual myofibers can be microirradiated at specific subfilament points. It is possible to damage a single Z line or A band in an actively contracting cell (*33*) and then analyze the changes in both contractile pattern and myofilament structure (Figs. 17 and 18).

The cytoplasmic stress fibers of cultured endothelial cells are amenable to selective microirradiation (*34*). It is possible to sever a single stress fiber and observe its repair and regeneration (Figs. 19 and 20). Selective alteration of a specific number of stress fibers at specific locations within the cytoplasm permit detailed studies on the role of these cytoskeletal elements in cell migration and cell shape changes.

Intracellular motility patterns have been studied by placing multiple 0.25-μm lesions in preselected regions of bands of 100-Å filaments and then examining the relative movement of the lesion sites with respect to each other (*33*) (Figs. 21 and 22).

Plant Cell Development—Chloroplast Irradiation

Though no detailed microbeam studies have been conducted on the chloroplasts in plant cells, the potential for such studies is great. Cells with large chloroplasts or multiple distinct chloroplasts would be particularly amenable to study. Entire chloroplasts, parts of a chloroplast, or specific ultrastructural elements of a chloroplast could be selectively damaged by appropriate matching of laser wavelength and chloroplast pigment.

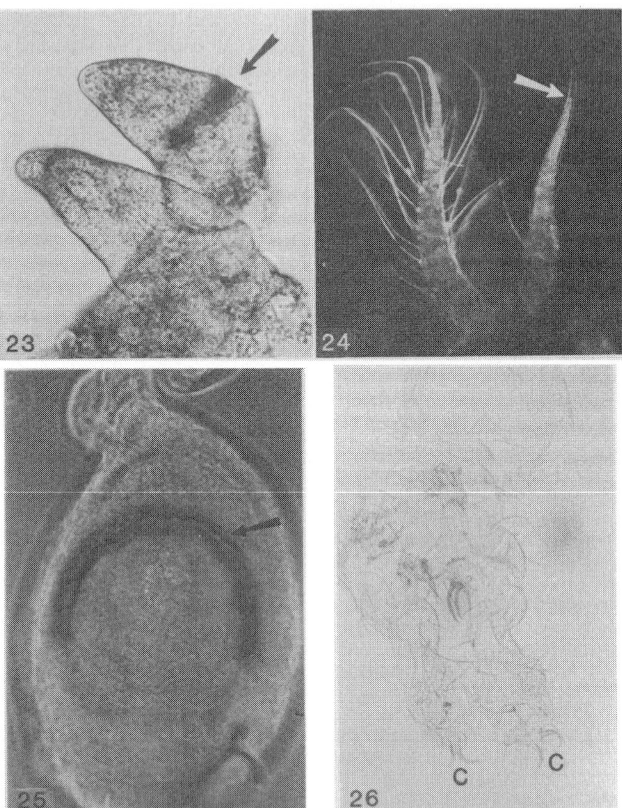

Fig. 23. Cricket embryo with embryonic rudiments of the cerci clearly visible at the rear of the animal. A band of cells approximately 25 μm in width has been killed with the 265-nm fourth harmonic of the YAG laser. The dead cells have been stained with trypan blue to illustrate the region of irradiation (arrow). (×170) Fig. 24. Cricket carried through to hatching with one cercus formed from a cercal rudiment that had been irradiated as indicated in Fig. 23. Note that the cercus developed from the irradiated embryonic rudiment is shorter and lacking in the fine hairs when compared to the cercus developed from the unirradiated cercal rudiment. Various degrees of final cercal differentiation occur depending upon the time, extent, and position of the irradiation. (×63) Fig. 25. Mesothoracic *Drosophila* imaginal disc after laser microirradiation at 265 nm in the lower tibia–upper tarsus region. The disc was stained by trypan blue immediately after irradiation to demonstrate the zone of dead cells corresponding to the lesion site (arrow). Fig. 26. The result of a prothoracic leg imaginal disc irradiated as in Fig. 25, implanted within the abdomen of an adult female for 10 days at 25°C, and then injected into a late third-instar larva for metamorphosis. Note the duplication of extra distal leg structures (claws), labeled *C*.

The efficient absorbance of the argon laser wavelengths (488 and 514 nm) by chloroplasts has been demonstrated in the green alga *Coleochaete* (*35*). In this alga, single cells in the developing multicellular thallus were destroyed by selective irradiation of the one large chloroplast in the cell.

A series of developmental studies was conducted in which a specific number of non-seta- (flagellum) bearing cells in the thallus were destroyed, and the subsequent mitotic and differentiative pattern of the thallus was studied (*35*). These studies revealed that mitosis could be stimulated in the thallus by merely reducing the number of cells in a given region. Mitosis apparently was stimulated when thallus cells were no longer contacted on all sides by other cells. The selective destruction of seta-bearing cells consistently resulted in new seta-bearing cells differentiating from non-seta-bearing cells, so that the number of these cells was always maintained. These studies demonstrated built-in self-regulatory developmental mechanisms for both seta cell differentiation and vegetative cell growth. It was possible to induce a differentiative process by selective removal of a specialized cell type (the seta cells).

Developmental Neurobiology

The laser microbeam is useful in developmental studies when precise destruction of specific cells or groups of cells in the embryo or larva is necessary. In an early study, the ruby laser microbeam was used to destroy the supraesophageal ganglion in spiders in order to analyze the altered web-building behavior (*36*). However, considerable microbeam work has been done on the nervous sytem of the nematode *Caenorhabditis elegans* (*37*). In these studies, specific cells in the embryonic or juvenile nervous systems were destroyed by laser microirradiation, and the subsequent nervous system development and behavior of the organism were analyzed.

In studies by Lors-Schardin *et al.* (*38*), the 257-nm wavelength of a frequency-doubled argon laser has been used to destroy selected regions of the developing *Drosophila* germ band and blastoderm. Up to 45 nuclei were destroyed with a 10- to 30-μm focused laser beam, and the subsequent defects were used to derive "defect maps." According to these investigators, the laser microbeam approach provides a more detailed and accurate developmental fate map than the earlier methods of lesion production did, perhaps because of the ability to selectively destroy a smaller group of cells in a specific target area. The ultraviolet laser system has also been used for studies of chromatin damage and repair (*39, 40*).

We have used the 265-nm fourth harmonic of the YAG laser and the 280-nm second harmonic of the YAG-pumped dye laser to study the development of a neurosensory system in the cricket (*41*). The hypothesis that pioneer fibers, which develop relatively early in the differentiation of insect appendages, serve to organize the peripheral sensory nerves was tested by ablating apical regions of the cercal rudiments in embryos of *Acheta domesticus*. Multiple nerve bundles, rather than the normal middorsal and midventral pair of nerves, were formed within the cercus after laser ablation of the cercal tip before pioneer fiber differentiation, but the cercal nerve was normal when lesions were made after formation of the pioneer fiber tracts and associated glia. These results indicate a necessary morphogenetic role for the pioneer fibers (Figs. 23 and 24).

Pattern Formation

The laser microbeam has been used to induce specific pattern abnormalities by the production of small areas of localized cell death in individual imaginal discs of *Drosophila* larvae. Specific regions of dissected discs were treated with the 265-nm fourth harmonic wavelength of a YAG laser to induce cell death. The effects were then analyzed by culture in vivo and induced metamorphosis to detect pattern duplications and triplications. The key feature of this system was the ability to confine effects of the laser treatment to selected regions of the discs. The irradiated discs were incubated in vitro for a short time after irradiation and then transplanted into the abdomens of host larvae, which were then observed through metamorphosis. This method has made it possible to determine the potential for pattern regulation of a small group of cells in situ, with a resolution much greater than in previous studies (Figs. 25 and 26).

Conclusion

Laser microbeam irradiation has already contributed to the resolution of specific problems in cell and developmental biology. In other investigations, laser microsurgery is just beginning to be applied, and the ultimate contributions of this approach have yet to be realized. A new approach to optical microscopy in which a high-sensitivity television system is combined with an image array processing computer appears to extend greatly the capabilities of laser microbeams and optical microscopy in general. A diffraction-limited focused laser beam could be used to stimulate spectral emissions (such as fluorescence, Raman spectra, resonance Raman spectra) in restricted regions of living cells. This should yield precise physical-chemical data on the structure and organization of the living cell.

References and Notes

1. M. W. Berns, R. S. Olson, D. E. Rounds, *Nature (London)* **221**, 74 (1969).
2. M. W. Berns and D. E. Rounds, *Sci. Am.* **222** (No. 2), 98 (1970).
3. M. Bessis, F. Gires, G. Nomarski, *C.R. Acad. Sci.* **225**, 1010 (1962).
4. G. Moreno, M. Lutz, M. Bessis, *Int. Rev. Exp. Pathol.* **7**, 99 (1969).
5. M. W. Berns, *Nature (London)* **240**, 483 (1972).
6. ——, *Lasers in Physical Chemistry and Biophysics*, J. Joussot-Dubien, Ed. (Elsevier, Amsterdam, 1975), pp. 389–401.
7. The Laser Microbeam Program (LAMP) has been established under the National Institutes of Health, Biotechnology Resource Program of the Division of Research Resources. This facility is available for outside use, and application forms can be obtained by contacting the facility director (Michael W. Berns). All applications will be reviewed by an external advisory committee.
8. M. W. Berns and C. Salet, *Int. Rev. Cytol.* **33**, 131 (1972).
9. S. P. Peterson and M. W. Berns, *Photochem. Photobiol.* **27**, 367 (1978); *J. Cell Sci.* **32**, 197 (1978); *ibid.* **34**, 289 (1978).
10. M. W. Berns, *Biophys. J.* **16**, 973 (1976).
11. ——, *Biological Microirradiation* (Prentice-Hall, Englewood Cliffs, N.J., 1974).
12. ——, D. E. Rounds, R. S. Olson, *Exp. Cell Res.* **56**, 292 (1969); M. W. Berns, Y. Ohnuki, D. E. Rounds, R. S. Olson, *ibid.* **60**, 133 (1970); M. W. Berns, W. K. Cheng, A. D. Floyd, Y. Ohnuki, *Science* **171**, 903 (1971); M. W. Berns and A. D. Floyd, *Exp. Cell Res.* **67**, 305 (1971); M. W. Berns and W. K. Cheng, *ibid.* **69**, 185 (1971); Y. Ohnuki, R. S. Olson, D. E. Rounds, M. W. Berns, *ibid.* **71**, 132 (1972).
13. M. W. Berns, L. K. Chong, M. Hammer-Wilson, K. Miller, A. Siemens, *Chromosoma* **73**, 1 (1979).
14. M. W. Berns, unpublished results.
15. K. D. Tartof, *Proc. Natl. Acad. Sci. U.S.A.* **71**, 1272 (1974).
16. E. H. Davidson, *Gene Activity in Early Development* (Academic Press, New York, 1968).
17. M. W. Berns, *Science* **186**, 700 (1974).
18. S. L. Brenner, L.-H. Liaw, M. W. Berns, *Cell Biophys.*, in press.
19. J. B. Rattner and M. W. Berns, *Chromosoma* **54**, 387 (1976); *Cytobios* **15**, 37 (1976).
20. M. W. Berns, J. B. Rattner, S. Brenner, S. Meredith, *J. Cell Biol.* **72**, 351 (1977).
21. J. Smith-Sonneborn and W. Plaut, *J. Cell Sci.* **2**, 225 (1967); S. R. Heidemann, G. Sander, M. W. Kirschner, *Cell* **10**, 337 (1977).
22. R. R. Gould and G. G. Borisy, *J. Cell Biol.* **73**, 601 (1977).
23. S. P. Peterson and M. W. Berns, *J. Cell Sci.* **34**, 289 (1978).
24. D. Pepper and B. R. Brinkley, *Cell Motil.* **1**, 1 (1980).
25. M. W. Berns and S. M. Richardson, *J. Cell Biol.* **75**, 977 (1977).
26. P. A. McNeill and M. W. Berns, *ibid.*, in press.
27. A. Forer, *ibid.* **25**, 95 (1965); R. E. Zirkle, *Radiat. Res.* **41**, 516 (1970).
28. J. R. Aist and M. W. Berns, *J. Cell Biol.* **87**, 234a (1980).
29. C. L. Wilson and J. R. Aist, *Phytopathology* **57**, 769 (1967); J. R. Aist and P. H. Williams, *J. Cell Biol.* **55**, 368 (1972).
30. M. W. Berns, N. Gamaleja, C. Duffy, R. Olson,

D. E. Rounds, *J. Cell. Physiol.* **76**, 207 (1970); M. W. Berns, D. C. L. Gross, W. K. Cheng, D. Woodring, *J. Mol. Cell. Cardiol.* **4**, 71 (1972); M. W. Berns, D. C. L. Gross, W. K. Cheng, *ibid.*, p. 427; K. P. Adkisson *et al.*, *ibid.* **5**, 559 (1973); J. Rattner, J. Lifsics, S. Meredith, M. W. Berns, *ibid.* **8**, 239 (1976); C. Salet, *C. R. Acad. Sci.* **272**, 2584 (1971); *Exp. Cell Res.* **73**, 360 (1972).
31. C. Salet, G. Moreno, F. A. Vinzens, *Exp. Cell Res.* **120**, 25 (1979).
32. M. Kitzes, G. Twiggs, M. W. Berns, *J. Cell. Physiol.* **93**, 99 (1977).
33. K. R. Strahs, J. M. Burt, M. W. Berns, *Exp. Cell Res.* **113**, 75 (1978).
34. K. R. Strahs and M. W. Berns, *ibid.* **119**, 31 (1979).
35. G. McBride, J. LaBounty, J. Adams, M. Berns, *Dev. Biol.* **37**, 90 (1974).
36. P. N. Witt, *Am. Zool.* **9**, 121 (1969).
37. R. Russel and G. White, personal communication.
38. M. Lors-Schardin, K. Sander, C. Cremer, T. Cremer, C. Zorn, *Dev. Biol.* **68**, 533 (1979).
39. C. Cremer, T. Cremer, C. Zorn, J. Zimmer, *Clin. Genet.* **14**, 286 (1978).
40. C. Zorn, C. Cremer, T. Cremer, J. Zimmer, *Exp. Cell Res.* **124**, 111 (1979).
41. J. S. Edwards, S.-W. Chen, M. W. Berns, *J. Neurosci.*, in press.
42. Supported by NIH grants HL 15740, GM 23445, RRO 1192, and NB07778, and by USAF grant OSR 80-0062.

Acknowledgments

I would like to acknowledge the numerous students and colleagues who over the years contributed to many of the studies described in this chapter. I also thank the following US agencies and institutions who have provided funds for these studies: the National Institutes of Health, the National Science Foundation, the Office of Naval Research, the Air Force Office for Scientific Research, the American Cancer Society, the American Heart Association, the California Cancer Coordinating Committee, and the Arnold and Mabel Beckman Foundation. In addition, I am particularly grateful for the permission granted by Science magazine to reprint, in its entirely, the article "Laser Microsurgery in Cell and Developmental Biology" (Science 213:505–513, 1981). The copyright owner of that article is the American Association for the Advancement of Science, Washington, DC.

References

Adkisson, K. P., Baic, D., Burgott, S., Cheng, W. K., and Berns, M. W. (1973). Argon laser micro-irradiation of mitochondria in rat myocardial cells in tissue culture. IV. Ultrastructural and cytochemical analysis of minimal lesions. *J. Mol. Cell. Cardiol.* **5,** 559–564.

Aist, J. R., and Berns, M. W. (1981). Mechanics of chromosome separation during mitosis in *Fusarium* (Fungi imperci): New evidence from ultrastructural and laser microbeam experiments. *J. Cell Biol.* **91,** 446–458.

Aist, J. R., Tao, W., Bayles, C. J., and Berns, M. W. (1991). Direct experimental evidence for the existence, structural basis and function of astral force during anaphase B *in vivo*. *J. Cell Sci.* **100,** 279–288.

Aist, J. R., Liang, H., and Berns, M. W. (1993). Astral and spindle forces in PtK_2 cells during anaphase B: A laser microbeam study. *J. Cell Sci.* **104,** 1207–1216.

Alliegro, M. C., Alliego, M. A., and Palazzo, R. E. (2006). Centrosome-assisted RNA in surf clam oocytes. *Proc. Natl. Acad. Sci. USA* **103,** 9034–9038.

Amy, R. L., and Storb, R. (1965). Selective mitochondrial damage by a ruby laser microbeam: An electron microscopic study. *Science* **150,** 756–757.

Amy, R. L., Storb, R., Fauconnier, B., and Wertz, R. K. (1967). Ruby laser microirradiation of single tissue culture cells vitally stained with Janus green B. I. Effects observed with the phase contrast microscope. *Exp. Cell Res.* **45,** 361–373.

Basehoar, G., and Berns, M. W. (1973). Cloning of rat kangaroo (PTK_2) cells following laser micro-irradiation of selected mitotic chromosomes. *Science* **179,** 1333–1334.

Bayles, C. J., Aist, J. R., and Berns, M. W. (1993). The mechanics of anaphase B in a basidiomycete as revealed by laser microbeam microsurgery. *Exp. Mycol.* **17,** 191–199.

Bereiter-Hahn, J. (1972a). Laser micro-irradiation as a tool in biology and medicine. Part 2: Method and applications. *Microsc. Acta* **72,** 1–33.

Bereiter-Hahn, J. (1972b). Laser micro-irradiation as a tool in biology and medicine. Part 1: Fundamentals of laser light and of its interactions with biological material. *Microsc. Acta* **71,** 225–241.

Bergmann, C. I., and Horwitz, H. R. (1991). Chemosensory neurons with overlapping functions direct chemotaxis to multiple chemicals in *C. elegans*. *Neuron* **7,** 729–742.

Berlin, V., Styles, C. A., and Fink, G. R. (1990). BIKI, a protein required for microtubule function during mating and mitosis in *Saccharomyces cerevisiae* colocalizes with tubulin. *J. Cell Biol.* **111,** 2573–2586.

Berns, M. W. (1971). A simple and versatile argon laser microbeam. *Exp. Cell Res.* **65,** 470–473.

Berns, M. W. (1972). Partial cell irradiation with a tunable organic dye laser. *Nature* **240,** 483–485.

Berns, M. W. (1974a). "Biological Microirradiation." Prentice-Hall, Englewood Cliffs, NJ.

Berns, M. W. (1974b). Directed chromosome loss by laser microirradiation. *Science* **186,** 700–705.

Berns, M. W. (1974c). Laser microirradiation of chromosomes. *Cold Spring Harbor Symp.* **38,** 165–174.

Berns, M. W. (1976). A possible two-photon effect *in vitro* using a focused laser beam. *Biophys. J.* **16,** 973–977.

Berns, M. W., Aist, J., Edwards, J., Strahs, K., Girton, J., McNeill, P., Rattner, J. B., Kitzes, M., Hammer-Wilson, M, Liaw, L.-H., Siemens, A., Koonce, M., *et al.* (1981). Laser microsurgery in cell and developmental biology. *Science* **213,** 505–513.

Berns, M. W., Botvinick, E., Liaw, L.-H., Sun, C.-H., and Shah, J. (2006). Micromanipulation of chromosomes and the mitotic spindle using laser microsurgery (laser scissors) and laser-induced optical forces (laser tweezers). *In* "Cell Biology: A Laboratory Handbook" (J. E. Celis, ed.), 3rd edn., Vol. 3, pp. 351–363. Elsevier Academic Press, San Diego.

Berns, M. W., Cheng, W. K., Floyd, A. D., and Ohnuki, Y. (1971b). Chromosome lesions produced with an argon laser microbeam without dye sensitization. *Science* **171,** 903–905.

Berns, M. W., Cheng, W. K., and Hoover, G. (1971a). Cell division after laser microirradiation of mitotic chromosomes. *Nature* **233,** 122–123.

Berns, M. W., Chong, L. K., Hammer-Wilson, M., Miller, K., and Siemens, A. (1979). Genetic microsurgery by laser: Establishment of a clonal population of rat kangaroo cells (PTK_2) with a directed deficiency in a chromosomal nucleolar organizer. *Chromosoma* **73,** 1–8.

Berns, M. W., and Floyd, A. D. (1971). Chromosome dissection by laser: A cytochemical and functional analysis. *Exp. Cell Res.* **67,** 305–310.

Berns, M. W., Floyd, A. D., Adkisson, K., Cheng, W. K., Moore, L., Hoover, G., Ustick, K., Burgott, S., and Osial, T. (1972a). Laser microirradiation of the nucleolar organizer in cells of rat kangaroo (*Potorous tridactylis*): Reduction of the nucleolar number and the production of micronucleoli. *Exp. Cell Res.* **75,** 424–432.

Berns, M. W., Gamaleja, N., Olson, R., Duffy, C., and Rounds, D. E. (1970b). Argon laser microirradiation of mitochondria in rat myocardial cells in tissue culture. *J. Cell. Physiol.* **76,** 207–214.

Berns, M. W., Gross, D. C. L., and Cheng, W. K. (1972c). Argon laser microirradiation of mitochondria in rat myocardial cells in tissue culture. III. Irradiation of multicellular groups. *J. Molec. Cell. Cardiol.* **4,** 427–433.

Berns, M. W., Gross, D. C. L., Cheng, W. K., and Woodring, D. (1972b). Argon laser microirradiation of mitochondria in rat myocardial cells in tissue culture. II. Correlation of morphology and function in irradiated single cells. *J. Mol. Cell. Cardiol.* **4,** 71–83.

Berns, M. W., Leonardson, K., and Witter, M. (1976). Laser microbeam irradiation of rat kangaroo cells (PTK_2) following selective sensitization with bromodeoxyuridine and ethidium bromide. *J. Morph.* **149,** 327–337.

Berns, M. W., Liang, H., Sonek, G. J., and Liu, Y. (1994). Micromanipulation of chromosomes using laser microsurgery (optical scissors) and laser-induced optical forces (optical tweezers). *In* "Cell Biology: A Laboratory Handbook" (J. E. Celis, ed.), pp. 217–227. Academic Press, Orlando.

Berns, M. W., Liang, H., Sonek, G. J., and Liu, Y. (1998a). Micromanipulation of chromosomes using laser microsurgery (optical scissors) and laser-induced optical forces (optical tweezers). *In* "Cell Biology: A Laboratory Handbook. Update 1997" (J. E. Celis, ed.), 2nd edn., Vol. 2, pp. 193–202. Academic Press, Orlando.

Berns, M. W., Ohnuki, Y., Rounds, D. E., and Olson, R. S. (1970a). Modification of nucleolar expression following laser microirradiation of chromosomes. *Exp. Cell Res.* **60,** 133–138.

Berns, M. W., Olson, R. S., and Rounds, D. E. (1969a). *In vitro* production of chromosomal lesions using an argon laser microbeam. *Nature* **221,** 74–75.

Berns, M. W., Olson, R. S., and Rounds, D. E. (1969c). Argon laser microirradiation of nucleoli. *J. Cell Biol.* **43,** 621–626.

Berns, M. W., Rattner, J. B., Brenner, S., and Meredith, S. (1977). The role of the centriolar region in animal cell mitosis: A laser microbeam study. *J. Cell Biol.* **72,** 351–367.

Berns, M. W., and Richardson, S. M. (1977). Continuation of mitosis after selective laser microbeam destruction of the centriolar region. *J. Cell Biol.* **75,** 977–982.

Berns, M. W., and Rounds, D. E. (1970a). Cell surgery by laser. *Sci. Am.* **222,** 98–110.

Berns, M. W., and Rounds, D. E. (1970b). Laser microbeam studies on tissue culture cells. *Ann. NY Acad. Sci.* **168,** 550–563.

Berns, M. W., Rounds, D. E., and Olson, R. S. (1969b). Effects of laser microirradiation on chromosomes. *Exp. Cell Res.* **56,** 292–298.

Berns, M. W., and Salet, C. (1972). Laser microbeams for partial cell irradiation. *Int. Rev. Cytol.* **33,** 131–156.

Berns, M. W., Tadir, Y., Liang, H., and Tromberg, B. (1998b). Laser scissors and tweezers. *In* "Methods in Cell Biology, Vol. 55" (M. P. Sheetz, ed.), pp. 71–97. Academic Press, San Diego.

Berns, M. W., Wang, Z., Dunn, A, Wallace, V., and Venugopalan, V. (2000). Gene inactivation by multiphoton-targeted photochemistry. *Proc. Natl. Acad. Sci. USA* **97,** 9504–9507.

Bessis, M., Gires, F., Mayer, G., and Nomarski, G. (1962). Irradiation des organites cellulaires à l'aide d'un laser à rubis. *C. R. Acad. Sci.* **225,** 1010–1012.

Bessis, M., and Nomarski, G. (1959). Conditions de l'irradiation ultra-violette des organites cellulaires. *C. R. Acad. Sci.* **249,** 768–776.

Bessis, M., and Nomarski, G. (1960). Irradiation ultra-violette des organites cellulaires AVEC observation continue en contraste de phase. *J. Biophys. Biochem. Cytol.* **8,** 77.

Botvinick, E., and Berns, M. W. (2005). Internet-based robotic laser scissors and tweezers microscopy. *Microsc. Res. Tech.* **68,** 65–74.

Brinkley, L., and Berns, M. W. (1974). Laser microdissection of actinomycin D segregated nucleoli. *Exp. Cell Res.* **87,** 417–422.

Burt, J. M., Strahs, K. R., and Berns, M. W. (1979). Correlation of cell surface alterations with contractile response in laser microbeam irradiated myocardial cells: A scanning electron microscope study. *Exp. Cell Res.* **118,** 341–351.

Calmettes, P. P., and Berns, M. W. (1983). Laser-induced multiphoton processes in living cells. *Proc. Natl. Acad. Sci. USA* **80,** 7197–7199.

Chalfie, M., Sulston, J. E., White, J. G., Southgate, E., Thomson, J. N., and Brenner, S. (1985). The neural circuit for touch sensitivity in *Caenorhabditis elegans*. *J. Neurosci.* **5,** 956–964.

Clark, I. B., Hanania, E. G., Stevens, J., Gallina, M., Fieck, A., Brandes, R., Palsson, B. O., and Koller, M. R. (2006). Optoinjection for efficient targeted delivery of a broad range of compounds and macromolecules into diverse cell types. *J. Biomed. Opt.* **11,** 014034-1–014034-8.

Colombelli, J., Reynaud, E. G., Rietdorf, J., Pepperkok, R., and Stelzer, E. H. K. (2005). *In vivo* selective cytoskeleton dynamics quantification in interphase cells induced by pulsed ultraviolet laser nanosurgery. *Traffic* **6,** 1093–1102.

Cowan, C. R., and Hyman, A. A. (2004). Centrosomes direct cell polarity independently of microtubule assembly in *C. elegans* embryos. *Nature* **431,** 92–96.

Cremer, C., Zorn, C., and Cremer, T. (1974). An ultraviolet laser microbeam for 257 nm. *Microsc. Acta* **75,** 331–337.

Cremer, T., Peterson, S. P., Cremer, C., and Berns, M. W. (1981). Laser microirradiation of Chinese hamster cells at wavelength 365 nm. Effects of psoralen and caffeine. *Radiat. Res.* **85,** 529–543.

De Giorgi, F., Lartigue, L., and Ichas, F. (2000). Electrical coupling and plasticity of the mitochondrial network. *Cell Calcium* **28,** 365–370.

Djabali, M., Nguyen, C., Biunno, I., Oostra, B. A., Mattei, M. G., Ikeda, J. E., and Jordan, B. R. (1991). Laser microdissection of the fragile X region: Identification of cosmid clones and of conserved sequences in this region. *Genomics* **10,** 1053–1060.

Edwards, J. S., Chen, S.-W., and Berns, M. W. (1981). Cercal sensory development following laser microlesions of embryonic apical cells in *Acheta domesticus*. *J. Neurosci.* **1,** 250–258.

Egner, O., and Bereiter-Hahn, J. (1970). Laser-Strahlenstichversuche an Fish-Melanophoren. *Z. Wiss. Mikrosk.* **70,** 17–22.

Forer, A. (1965). Local reduction of spindle fiber birefringence in living *Nephrotoma suturalis* (Loew) spermatocytes induced by ultraviolet microbeam irradiation. *J. Cell Biol.* **25,** 95–117.

Forer, A. (1966). Simple conversion of reflecting lenses into phase-contrast condensers for ultraviolet light irradiations (ultraviolet microbeam equipment, ultraviolet microscopes). *Exp. Cell Res.* **43,** 688–691.

Fukui, K., Minezawa, M., Kamisugi, Y., Ishikawa, M., Ohmido, N., Yanagisawa, T., Fugishita, M., and Sakai, F. (1992). Microdissection of plant chromosomes by an argon ion laser beam. *Theor. Appl. Genetics* **84**(7–8), 787–794.

Girton, J. R., and Berns, M. W. (1982). Pattern abnormalities induced in *Drosophila* imaginal discs by a UV laser microbeam. *Dev. Biol.* **91,** 73–77.

Goldstein, S. F. (1969). Irradiation of sperm tails by laser microbeam. *J. Exp. Biol.* **51,** 431–441.

Gould, R. R., and Borisy, G. G. (1977). The pericentriolar material in Chinese hamster ovary cells nucleates microtubule formation. *J. Cell Biol.* **73,** 601–615.

Greulich, K. O. (1999). "Micromanipulation by Light in Biology and Medicine." Birkhäuser Verlag, Basel.

Greulich, K. O., Bauder, U., Monajembashi, S., Ponelies, N., Seeger, S., and Wolfrum, J. (1989). UV Laser Mikrostrahl und optische Pinzette (UV laser microbeam and optical tweezers). *LaborPraxis/ Labor 2000*, 36–42.

Grill, S. W., Gonczy, P., Stelzer, E. H., and Hyman, A. A. (2001). Polarity controls forces governing asymmetric spindle positioning in the *Caenorhabditis elegans* embryo. *Nature* **409,** 630–633.

Grill, S. W., Howard, J., Schaffer, E., Stelzer, E. H., and Hyman, A. A. (2003). The distribution of active force generators controls mitotic spindle positions. *Science* **301,** 518–521.

Guo, Y., Liang, H., and Berns, M. W. (1995). Laser-mediated gene transfer in rice. *Physiologia Plantarum* **93,** 19–24.

Hadano, S., Watanabe, M., Yokiu, H., Kogi, M., Kondo, I., Tsuchiya, H., Kanazawa, I., Wakasa, K., and Ikeda, J. (1991). Laser microdissection and single unique primer PCR allow generation of regional chromosome DNA clones from a single human chromosome. *Genomics* **11,** 364–373.

Hall, D. H., and Russel, R. L. (1991). The posterior nervous system of the nematode *Caenorhabditis elegans*: Serial reconstruction of identified neurons and complete pattern of synaptic interactions. *J. Neurosci.* **11,** 1–22.

He, W., Liu, Y., Smith, M., and Berns, M. W. (1997). Laser microdissection for generation of a human chromosome region specific library. *Microsc. Microanal.* **3,** 47–52.

Hutson, M. S., Tokutake, Y., Chang, M. S., Bloor, J. W., Venakides, S., Kiehart, D. P., and Edwards, G. S. (2003). Forces for morphogenesis investigated with laser microsurgery and quantitative modeling. *Science* **300,** 145–149.

Khodjakov, A., Cole, R. W., and Rieder, C. (1997). A synergy of technologies: Combining laser microsurgery with green fluorescent protein tagging. *Cell Motil. Cytoskeleton* **38,** 311–317.

Khodjakov, A., La Terra, S., and Chang, F. (2004). Laser microsurgery in fission yeast: Role of the mitotic spindle midzone in anaphase B. *Curr. Biol.* **14,** 1330–1340.

Kim, J.-S., Krasieva, T. B., Kurumizaka, H., Chen, D. J., Taylor, A. M. R., and Yokomori, K. (2005). Independent and sequential recruitment of NHEJ and HR factors to DNA damage sites in mammalian cells. *J. Cell Biol.* **170,** 341–347.

Kitzes, M., Twiggs, G., and Berns, M. W. (1977). Alteration of membrane electrical activity in rat myocardial cells following selective laser microbeam irradiation. *J. Cell. Physiol.* **93,** 99–104.

Konig, K., Riemann, I., Fischer, P., and Halbhuber, K. (1999). Intracellular nanosurgery with near infrared femtosecond laser pulses. *Cell Mol. Biol.* **45,** 192–201.

Konig, K., Riemann, I., and Fritzsche, W. (2001). Nanodissection of human chromosomes with near-infrared femtosecond laser pulses. *Optics Lett.* **26,** 819–821.

Koonce, M. P., Cloney, R. A., and Berns, M. W. (1984). Laser irradiation of centrosomes in newt eosinophils: Evidence of centriole role in motility. *J. Cell Biol.* **98,** 1999–2010.

Krasieva, T. B., Chapman, C. F., LaMorte, V. J., Venugopalan, V., and Tromberg, B.J (1998). Mechanisms of cell permeabilization by laser microirradiation. *In* "Optical Investigations of Cells *In Vitro* and *In Vivo*," The International Society for Optical Engineering. *Proc. SPIE* **3260,** 38–44.

Kruhlak, M. J., Celeste, A, Dellaire, G., Fernandez-Capetillo, O., Muller, W. G., McNally, J. G., Bazett-Jones, D. P., and Nussenzweig, A. (2006). Changes in chromatin structure and mobility in living cells at sites of DNA double-strand breaks. *J. Cell Biol.* **172,** 823–834.

Kumar, S., Maxwell, I. Z., Heisterkamp, A., Polte, T. R., Lele, T. P., Salanga, M., Mazur, E., and Ingber, D. E. (2006). Viscoelastic retraction of single living stress fibers and its impact on cell shape, cytoskeletal organization, and extracellular matrix mechanics. *Biophys. J.* **90,** 3762–3773.

Kurata, S. I., Tsukakoshi, M., Kasuya, T., and Ikawa, Y. (1986). Laser method for efficient introduction of foreign DNA into cultured tissue cells. *Exp. Cell Res.* **162,** 372–378.

La Terra, S., English, C. N., Hergert, P., McEwen, B. F., Sluder, G., and Khodjakov, A. (2005). The *de novo* centriole assembly pathway in HeLa cells: Cell cycle progression and centriole assembly/maturation. *J. Cell Biol.* **168,** 713–722.

Liang, H., Wright, W. H., Rieder, C. L., Salmon, E. D., Profeta, G., Andrews, J., Liu, Y., Sonek, G. J., and Berns, M. W. (1994). Directed movement of chromosome arms and fragments in mitotic newt lung cells using optical scissors and optical tweezers. *Exp. Cell Res.* **213,** 308–312.

Lohs-Schardin, M., Sander, K., Cremer, C., Cremer, T., and Zorn, C. (1979a). Localized ultraviolet laser microbeam irradiation of early *Drosophila* embryos: Fate maps based on location and frequency of adult defects. *Dev. Biol.* **68,** 533–545.

Lohs-Schardin, M., Cremer, C., and Nusslein-Volhard, C. (1979b). A fate map for the larval epidermis of *Drosophila melanogaster*: Localized cuticle defects following irradiation of the blastoderm with an ultraviolet laser microbeam. *Dev. Biol.* **73,** 239–255.

Ludecke, H. J., Senger, G., Claussen, U., and Horsthemke, B. (1989). Cloning defined regions of the human genome by microdissection of banded chromosomes and enzymatic amplification. *Nature* **338,** 348–350.

Magidson, V., Chang, F., and Khodjakov, A. (2006). Regulation of cytokinesis by spindle-pole bodies. *Nat. Cell Biol.* **8,** 891–893.

Maiato, H., Rieder, C. L., and Khodjakov, A. (2004). Kinetochore-driven formation of kinetochore fibers contributes to spindle assembly during animal mitosis. *J. Cell Biol.* **167,** 831–840.

Maiato, H., Khodjakov, A., and Rieder, C. L. (2005). *Drosophila* CLASP is required for the incorporation of microtubules subunits into fluxing kinetochore fibres. *Nat. Cell Biol.* **7,** 42–47.

Maiman, T. (1960). Stimulated optical radiation in ruby masers. *Nature* **187,** 493–494.

Maxwell, I., Cheung, S., and Mazur, E. (2005). Nanoprocessing of subcellular targets using femtosecond laser pulses. *Med. Laser Appl.* **20,** 193–200.

McKinnell, R., Mims, M. F., and Reed, L. A. (1969). Laser ablation of maternal chromosomes in eggs of *Rana pipiens. Z. Zell.* **93,** 30–35.

McNeill, P. A., and Berns, M. W. (1981). Chromosome behavior after laser microirradiation of a single kinetochore in mitotic PTK_2 cells. *J. Cell Biol.* **88,** 543–553.

Meredith, S., and Berns, M. W. (1976). Light and electron microscopy of laser microirradiated nucleoli and nucleoplasm in tissue culture cells. *J. Morph.* **150,** 785–803.

Mims, M. F., and McKinnell, R. (1971). Laser irradiation of the chick embryo germinal crescent. *J. Embryol. Exp. Morph.* **26,** 31–36.

Mohanty, S. K., Sharma, M., and Gupta, P. K. (2003). Laser-assisted microinjection into targeted animal cells. *Biotechnol. Lett.* **25,** 895–899.

Monajembashi, S., Cremer, C., Cremer, T., Wolfrum, J., and Greulich, K. O. (1986). Microdissection of human chromosomes by a laser microbeam. *Exp. Cell Res.* **167,** 262–265.

Moreno, G., and Salet, C. (1969). Partial cell irradiation by visible and ultraviolet light. *Int. Rev. Exptl. Pathol.* **7,** 99–137.

Moreno, G., Lutz, M., and Bessis, M. (1969). Partial cell irradiation by ultraviolet and visible light: Conventional and laser sources. *Int. Rev. Exp. Pathol.* **7,** 99–137.

Myohara, M. (1994). Fate mapping of the silkworm, *Bombyx mori*, using localized UV irradiation of the egg at fertilization. *Development* **120,** 2869–2877.

Myohara, M. (2003). Fate mapping of the larval silk glands of *Bombyx mori* by UV laser irradiation of the egg at fertilization. *Dev. Genes Evol.* **213,** 178–181.

Neev, J., Tadir, Y., Ho, P., Berns, M. W., Asch, R. H., and Ord, T. (1992). Microscope-delivered ultraviolet laser zona dissection: Principles and practices. *J. Assist. Reprod. Genet.* **9,** 513–523.

Palumbo, G., Caruso, M., Crescenzi, E., Tecce, M. F., Roberti, G., and Colasanti, A. (1996). Targeted gene transfer in eukaryotic cells by dye-assisted laser optoporation. *J. Photochem. Photobiol. B* **36,** 41–46.

Paterson, L., Agate, B., Comrie, M., Ferguson, R., Lake, T. K., Morris, J. E., Carruthers, A. E., Brown, C. T. A., Sibbett, W., Bryant, P. E., Bryant, P. E., and Gunn-Moore, F. (2005). Photoporation and cell transfection using a violet diode laser. *Optics Exp.* **13,** 595–600.

Perry, R. P., Hell, A., and Errera, M. (1961). The role of the nucleolus in ribonucleic acid and protein sythesis. I. Incorporation of cytidine into normal and nucleolar inactivated HeLa cells. *Biochim. Biophys. Acta* **49,** 47–57.

Peterson, S. P., and Berns, M. W. (1978). Evidence for centriolar region RNA functioning in spindle formation in dividing PTK_2 cells. *J. Cell Sci.* **34,** 289–301.

Ponelies, N., Bautz, E. K. F., Monajembashi, S., Wolfrum, J., and Greulich, K. O. (1989). Telomeric sequences derived from laser-microdissected polytene chromosomes. *Chromosoma* **98,** 351–357.

Ponelies, N., Stein, N., and Weber, G. (1997). Microamplification of specific chromosome sequences: An improved method for genome analysis. *Nucl. Acids Res.* **25,** 3555–3557.

Rattner, J. B., and Berns, M. W. (1974). Light and electron microscopy of laser microirradiated chromosomes. *J. Cell Biol.* **62,** 526–533.

Rattner, J. B., Lifsics, M., Meredith, S., and Berns, M. W. (1976). Argon laser microirradiation of mitochondria in rat myocardial cells in tissue culture. VI. Correlation of contractility and ultrastructure. *J. Mol. Cell. Cardiol.* **8,** 239–248.

Rieder, C. L., Davison, E. A., Jensen, L. C., Cassimeris, L., and Salmon, E. D. (1986). Oscillatory movements of monooriented chromosomes and their position relative to the spindle pole result from the ejection properties of the aster and half-spindle. *J. Cell Biol.* **103,** 581–591.

Rogakou, E. P., Boon, C., Redon, C., and Bonner, W. M. (1999). Megabase chromatin domains involved in DNA double-strand breaks *in vivo*. *J. Cell Biol.* **146,** 905–915.

Sacconi, L., Tolic-Norrelykke, I. M., Antolini, R., and Pavone, F. S. (2005). Combined intracellular three-dimensional imaging and selective nanosurgery by a nonlinear microscope. *J. Biomed. Optics* **10,** 014002-1–5.

Saks, N. M., and Roth, C. A. (1963). Ruby laser as a microsurgical instrument. *Science* **141,** 46–47.

Saks, N. M., Zuzolo, R., and Kopac, M. J. (1965). Microsurgery of living cells by ruby laser irradiation. *Ann. NY Acad. Sci.* **122,** 695–712.

Salet, C. (1972). A study of beating frequency of a single myocardial cell. I. Q-switched laser microirradiation of mitochondria. *Exp. Cell Res.* **73,** 360.

Salet, C., Moreno, G., and Lampidis, T. J. (1980). Effects of microirradiation of heart cells in culture at the mitochondrial level. *Biol. Cell.* **37,** 195–198.

Salet, C., Moreno, G., and Vinzens, F. (1979). A study of beating frequency of a single myocardial cell. III. Laser micro-irradiation of mitochondria in the presence of KCN and ATP. *Exp. Cell Res.* **120,** 25–29.

Schierenberg, E. (1984). Altered cell division rates after laser induced cell fusion in nematode embryos. *Dev. Biol.* **101,** 240–245.

Schnabel, R. (1991). Cellular interactions involved in the determination of the early *C. elegans* embryo. *Mech. Dev.* **34,** 85–99.

Schneckenburger, H., Hendinger, A., Sailer, R., Strauss, W. S. L., and Schmitt, M. (2002). Laser-assisted optoporation of single cells. *J. Biomed. Opt.* **7,** 410–416.

Seeger, S., Monajembashi, S., Hutter, K.-J., Futterman, G., Wolfrom, J., and Greulich, K. O. (1991). Application of laser optical tweezers in immunology and molecular genetics. *Cytometry* **12,** 497–504.

Shirahata, Y., Ohkohchi, N., Itagak, H., and Satomi, S. (2001). New technique for gene transfection using laser irradiation. *J. Invest. Med.* **49,** 184–190.

Sims, C. E., Meredith, G. D., Krasieva, T. B., Berns, M. W., Tromberg, B. J., and Allbritton, N. L. (1998). Laser-micropipet combination for single-cell analysis. *Anal. Chem.* **70,** 4570–4577.

Soughayer, J. S., Krasieva, T., Jacobson, S. C., Ramsey, J. M., Tromberg, B. J., and Allbritton, N. L. (2000). Characterization of cellular optoporation with distance. *Anal. Chem.* **72,** 1342–1347.

Spurck, T. P., Stonington, O. G., Snyder, J. A., Pickett-Heaps, J. D., Bajer, A., and Mole-Bajer, J. (1990). UV microbeam irradiations of the mitotic spindle. II. Spindle fiber dynamics and force production. *J. Cell Biol.* **111,** 1505–1518.

Storb, R., Amy, R. L., Wertz, R. K., Fauconnier, B., and Bessis, M. (1966). An electron microscope study of vitally stained single cells irradiated with a ruby laser microbeam. *J. Cell Biol.* **31,** 11–29.

Strahs, K. R., and Berns, M. W. (1979). Laser microirradiation of stress fibers and intermediate filaments in non-muscle cells from cultured rat heart. *Exp. Cell Res.* **119,** 31–45.

Strahs, K. R., Burt, J. M., and Berns, M. W. (1978). Contractility changes in cultured cardiac cells following laser microirradiation of myofibrils and the cell surface. *Exp. Cell Res.* **113,** 75–83.

Sullivan, D. S., and Huffaker, T. C. (1992). Astral microtubules are not required for anaphase B in *Saccharomyces cerevisiae*. *J. Cell. Biol.* **119,** 379–388.

Tadir, Y., Neev, J., and Berns, M. W. (1992). Laser in assisted reproduction and genetics. *J. Assist. Reprod. Genet.* **9,** 303–305.

Tao, W., Wilkinson, J., Stanbridge, E. J., and Berns, M. W. (1987). Direct gene transfer into human cultured cells facilitated by laser micropuncture of the cell membrane. *Proc. Natl. Acad. Sci. USA* **84,** 4180–4184.

Tirlapur, U. K., and Konig, K. (2002a). Femtosecond near-infrared laser pulses as a versatile non-invasive tool for intra-tissue nanoprocessing in plants without compromising viability. *Plant J.* **31,** 365–374.

Tirlapur, U. K., and Konig, K. (2002b). Targeted transfection by femtosecond laser. *Nature* **418,** 290–291.

Tschachotin, S. (1912). Die mikrikopische Strahlenstrich methode, eine Zelloperations methode. *Biol. Zentralbl.* **32,** 623.

Tschachotin, S. (1920). Action localisée des rayons ultraviolets sur le noyau de l'oeuf d'Oursin, par radiopuncture microscopie. *Compt. Rend. Soc. Biol.* **83,** 1593.

Tschachotin, S. (1921). Les changements de la permeabilité de l'oeuf d'Oursin, localizes experimentalement. *Compt. Rend. Soc. Biol.* **84,** 464.

Tschachotin, S. (1929). Attivazione dell'uovo di riccio di mare per mezzo della microraggiopuntura. *Boll. Soc. Ital. Biol. Sper.* **4,** 475–479.

Tschachotin, S. (1935a). Die Mikrostrahlstichmethode und andere Methoden des zytologischen Experimentes. *Handbuch biol. Arbeitsmethoden Abt.* **V 10,** 877.

Tschachotin, S. (1935b). L'effet d'arret de la fonction de la vacuole pulsatile de la Paramecie par micropuncture ultraviolette. *Compt. Rend. Soc. Biol.* **120,** 782–784.

Tschachotin, S. (1935c). Floculation localisée des colloides dans le cellule par la micropuncture ultraviolette. *Compt. Rend.* **200,** 2036–2038.

Tschachotin, S. (1935d). Recherches physiologiques sur les Protozoaires, faites au moyen de la micropuncture ultraviolette. *Compt. Rend.* **200,** 2217–2219.

Tschachotin, S. (1936a). Irradiation localisée du myonème du pedoncule des Vorticelles par micropuncture ultraviolette. *Compt. Rend.* **202,** 1114.

Tschachotin, S. (1936b). La fonction du stigma chez le Flagelle Euglena, etudiée au moyen de la micropuncture ultraviolette. *Compt. Rend. Soc. Biol.* **121,** 1162–1165.

Tschachotin, S. (1937). Das zytologische Mikroexperiment. (Untersuchungen an isolierten Zellen mit der Mikrostrahlstichmethode.). *Arch. Exptl. Zellforsch. Gewebezucht.* **19,** 498–506.

Tschachotin, S. (1938). Parthenogenese experimentale de l'oeuf de la Pholade par micropuncture ultraviolette, aboutissant à une larve vivante. *Compt. Rend.* **206,** 377–379.

Tschachotin, S. (1955). Problems and methods in experimental cytology. (in Italien). *Boll. Soc. Ital. Sper.* **31**(6), 661–663.

Tsukakoshi, M., Kurata, S., Nomiya, Y., Ikawa, Y., and Kasuya, T. (1984). A new method of DNA transfection by laser microbeam cell surgery. *Appl. Phys. B* **35,** 135–140.

Upadhyaya, M., Osborn, M., Maynard, J., Altherr, M., Ikeda, J., and Harper, P. S. (1995). Towards the finer mapping of faciocapulohumoral muscular dystrophy at 4q35: Construction of a laser microdissection library. *J. Med. Gen.* **60,** 244–251.

Vintemberger, P. (1929a). Sur une technique permettant d'irradier, dans des conditions de grande precision, une fraction déterminée d'une cellule volumineuse comme l'oeuf de Grenouille rousse. *Compt. Rend. Soc. Biol.* **102,** 1050–1052.

Vintemberger, P. (1929b). Sur les résultants de l'application d'une très forte dose de rayons X à l'hemispher inférieur, anuclée de l'oeuf de Grenouille rousse. *Compt. Rend. Soc. Biol.* **102,** 1053–1055.

Vintemberger, P. (1929c). Sur les résultants de l'application d'une très forte dose de rayons X à diverses régions de l'oeuf de Grenouille rousse. *Compt. Rend. Soc. Biol.* **102,** 1055–1057.

Vogel, A., Noack, J., Huttman, G., and Paltauf, G. (2005). Mechanisms of femtosecond laser nanosurgery of cells and tissues. *Appl. Phys. B* **81,** 1015–1047.

Vogel, A., and Venugopalan, V. (2003). Mechanisms of pulsed laser ablation in biological tissues. *Chem. Rev.* **107,** 577–644.

Wakabayashi, T., Kitagawa, I., and Shingai, R. (2004). Neurons regulating the duration of forward locomotion in *Caenorhabditis elegans*. *Neurosci. Res.* **50,** 103–111.

Walen, K. H., and Brown, S. W. (1962). Chromosomes in a marsupial (*Potorous tridactylis*) tissue culture. *Nature (Lond.)* **194,** 406.

Walker, R. A., Inoue, S., and Salmon, E. D. (1989). Asymmetric behavior of severed microtubule ends after ultraviolet-microbeam irradiation of individual microtubules *in vitro*. *J. Cell Biol.* **108,** 931–937.

Waterman-Storer, C. M., and Salmon, E. D. (1998). How microtubules get fluorescent speckles. *Biophys. J.* **75,** 2059–2069.

Waymire, K., Kitzes, M., Meredith, S., Twiggs, G., and Berns, M. W. (1976). Argon laser microirradiation of mitochondria in rat myocardial cells in tissue culture. VII. Fibrillation in ventricle and auricle cells. *J. Cell. Physiol.* **89,** 345–353.

Weber, G., Monajembashi, S., Greulich, K. O., and Wolfrum, J. (1988a). Injection of DNA into plant cells using a UV laser microbeam. *Naturwissenschaften* **75,** 36.

Weber, G., Monajembashi, S., Greulich, K. O., and Wolfrum, J. (1988b). Genetic manipulation of plant cells and organelles with a microfocusedlaser beam. *Plant Cell Tiss. Org.* **12,** 219.

Weber, G., Monajembashi, S., Greulich, K. O., and Wolfrum, J. (1990b). Genetic changes induced in higher plants by a UV laser microbeam. *Israel J. Botany* **40,** 115–122.

Weber, G., Monajembashi, S., Wolfrum, J., and Greulich, K. O. (1989a). Uptake of DNA in chloroplasts of *Brassica napus* (L) facilitated by a UV laser microbeam. *Eur. J. Cell Biol.* **49,** 73.

Weber, G., Monajembashi, S., Wolfrum, J., and Greulich, K. O. (1989b). A laser microbeam as a tool to introduce genes into cells and organelles of higher plants. *Ber. Bunsenges. Phys. Chem.* **93,** 252.

Weber, G., Monajembashi, S., Wolfrum, J., and Greulich, K. O. (1990a). Genetic changes induced in higher plant cells by a laser microbeam. *Physiologia Plantarum* **79,** 190–193.

Wertz, R. K., Storb, R., and Amy, R. L. (1967). Laser micro-irradiation of single tissue culture cells vitally stained with Janus green B. III. Effects on the incorporation of an RNA pyrimidine precursor. *Exp. Cell Res.* **45,** 61–71.

Wiegand Steubing, R., Cheng, S., Wright, W. H., Numajiri, Y., and Berns, M. W. (1991). Laser induced cell fusion in combination with optical tweezers: The laser cell fusion trap. *Cytometry* **12,** 505–510.

Winey, M., Goetsch, L., Baum, P., and Byers, B. (1991). MPS1 and MPS2: Novel yeast genes defining distinct steps of spindle pole body duplication. *J. Cell. Biol.* **114,** 745–754.

Zeira, E., Manevitch, A., Khatchatouriants, A., Pappo, O., Hyam, E., Darash-Yahana, M., Tavor, E., Honigman, A., Lewis, A., and Galun, E. (2003). Femtosecond infrared laser—an efficient and safe *in vivo* gene delivery system for prolonged expression. *Mol. Therapy* **8,** 342–350.

Zirkle, R. E. (1932). Some effects of alpha radiation upon plant cells. *J. Cell. Comp. Physiol.* **2,** 251–274.

Zirkle, R. E. (1957). Partial cell irradiation. *Adv. Biol. Med. Phys.* **5,** 103–146.

Zirkle, R. E. (1970). Ultraviolet-microbeam irradiation of newt-cell cytoplasm: Spindle destruction, false anaphase, and delay of true anaphase. *Radiat. Res.* **41,** 516–537.

Zirkle, R. E., and Uretz, R. B. (1963). Action spectrum for paling (decrease in refractive index) of ultraviolet-irradiated chromosome segments. *Proc. Natl. Acad. Sci.* USA **49,** 45–53.

CHAPTER 2

Selected Applications of Laser Scissors and Tweezers and New Applications in Heart Research

Karl Otto Greulich

Leibniz Institute for Age Research/Fritz Lipmann Institute, D-07745 Jena, Germany

METHODS IN CELL BIOLOGY, VOL. 82

0091-679X/07 $35.00
DOI: 10.1016/S0091-679X(06)82002-9

This contribution bridges the gap from early European contributions via laser micromanipulation to recent work on the use of laser microbeams and optical tweezers in studies of basic aspects of heart infarction. Laser transfection, particularly of plant cells and their chloroplasts, and laser microdissection of chromosomes with subsequent generation of chromosome segment-specific DNA libraries and laser-induced cell fusion are reported. With optical tweezers, microgravity can be simulated in roots of the alga *Chara*. Surprisingly, microgravity reduces growth. In some plant cells, CW lasers, in principle suited primarily for optical tweezers, can be used as microbeam. Also, it is shown that natural killer cells mount an attack on leukemia cells even in the absence of specificity, just induced by exerting force with optical tweezers. Finally, with the help of a laser microbeam, lesions can be induced to study wound healing after heart infarction. A modification of optical tweezers, the erythrocyte-mediated force application (EMFA) technique can be used to induce calcium waves not only in tissue reconstituted from excitable heart muscle cells but also from nonexcitable fibroblasts.

I. Laser Micromanipulation—Early European Contributions

The early story of micromanipulation by light, starting with conventional light and being completed using lasers in the 1960s, has been discussed in Chapter 1 by Berns, this volume. He can be seen as the father of laser scissors, laser microbeams, or laser scalpels, as they are termed almost synonymously, although others have published a few earlier papers (Bessis *et al.*, 1962), but did not continue to develop laser irradiation into a workable tool for the cell biologist. In Michael Berns' first publications on laser micromanipulation, he reported on the effects of laser microirradiation on chromosomes, among them in *Nature*, in 1969 (Berns *et al.*, 1969a,b,c), and as early as 1974 he published a first book on the "Biological Microirradiation" (Berns, 1974). A short version of this early story of laser microirradiation is well accessible to the interested reader in an article which appeared in *Science* in 1981 (Berns *et al.*, 1981; full-text copy reprinted in this book). Some publications appeared very recently, so his work covers more than a third of a century (Botvinick and Berns, 2005; Nascimento *et al.*, 2006).

In the 1970s, the technique reached Europe, particularly with the help of Thomas and Christoph Cremer. They used the technique to confirm the classical concept of chromosome territories (for a well accessible, though not earliest paper, see Cremer *et al.*, 1982), long before the chromosome territories were visualized by chromosome-specific DNA-painting probes. One of their collaborative partners who used the laser microbeam to study *Drosophila* development (Lohs-Schardin *et al.*, 1979) was Christine Nüsslein Vollhard, and she later won the 1999 Nobel prize in Biology and Medicine. For some years, also Jürgen Bereiter-Hahn pioneered laser microbeams in cell biology (Bereiter-Hahn, 1972) but then dedicated his work to other subjects.

In 1986, the second tool for laser micromanipulation by light was invented by Arthur Ashkin (Ashkin *et al.*, 1986): the single-beam optical trap or optical tweezers or laser tweezers. Notably, this story also has a Nobel prize winner involved: Steve Chu who won the 1997 physics prize for high-precision optical trapping of atoms and he was one of the authors of the seminal paper mentioned above. The present contribution does not primarily focus on the isolated use of optical tweezers, but rather in the context of complete micromanipulation by light. Nevertheless, the pioneering work of Steve Block's group (Block *et al.*, 1989) should be mentioned. A book devoted completely to laser tweezers appeared in 1998 as Volume 55 of the series "Methods in Cell Biology" (Sheetz, 1998) of which the present book is Volume 82.

The marriage of the two techniques and thus the possibility to perform complete microscope micromanipulation by light was first published in 1989 (Greulich *et al.*, 1989). Probably, since it appeared in the German language, little notice was taken of this work particularly in the United States for a number of years. Around this time, quite a number of techniques useful for the cell biologist have been developed in Europe. The first book on complete laser manipulation by light appeared at the end of 1990s (Greulich, 1999). There and in other contributions of the present book, experimental details on the construction of laser microbeams and optical tweezers can be found. The following description in the present chapter will summarize only applications reported in part of early European work (if not reported elsewhere in this book) and will finally bridge the gap between recent work on the use of the laser microbeam and optical tweezers to study some basic aspects of heart infarction.

II. Laser Microinjection of Genetic Material into Plant Cells

A. Laser Transfection of Animal and Plant Cells

With the many techniques available for introducing foreign genes into cells, it would not appear necessary to introduce an additional technique. However, introduction of genes into selected individual cells in suspension is still difficult. Additionally, plant cells are particularly difficult to transfect because they are surrounded by a rigid cellulose wall. Although walls may be removed enzymatically, not all the resulting protoplasts can be regenerated into plants. Therefore, many different approaches were taken to introduce cloned genes into whole plants or cells. With the small focus of a laser microbeam (diameter less than 1 μm), the cell wall can be opened, minimizing damage to the remainder of the cell (Weber *et al.*, 1988, 1989).

Interestingly, the first experiments using a laser microbeam to transfect animal cells were published in a physics journal (Tsukakoshi *et al.*, 1984). These authors (Kurata *et al.*, 1986) and Michael Berns' group (Tao *et al.*, 1987) were successful in transfecting a number of different cell types with different materials (see also Chapter 1 by Berns, this volume). Transfection of plant cells was subsequently published (Guo *et al.*, 1995; Weber *et al.*, 1988). Laser holes in membranes of cells of *Brassica napus* or *Nicotiana tabacum* were recorded on video. A fluorescent dye was taken up through laser holes into plasmolyzed cells for less than 5 s. Within ca. 5 s after irradiation they were closed (Weber *et al.*, 1989). However, lowered membrane fluidity at temperatures below 11°C prevented the selfhealing of laser holes. Obviously, the opening time of a laser hole can be manipulated to a far extent by a suitable choice of temperature.

B. Survival of Laser-Treated Rapeseed Cells and Transient Laser Transfection of the Glucuronidase Gene

Survival of cells was studied in *B. napus*. After laser microperforation, 80% of single cells and 30% of immature pollen grains survived. One day later, 40% of cells and 25–30% of pollen grains were alive and continued to grow.

The protoplasm of plant cells shrink in hypertonic solutions. When the cell membrane of *B. napus* is microperforated by a laser microbeam, the cells incorporate extracellular medium until they reach their original volume. This effect can be exploited to transfer material, for example foreign DNA, from the environment into the cytoplasm.

A number of constructs have become available to monitor expression of a gene after DNA transfer into a cell. Bacterial glucuronidase (GUS) with the 35 s promoter of *Cauliflower mosaic virus* lends itself to measure the efficiency of DNA uptake into higher plant cells. When it is expressed, the transfected cell adopts a blue (indigo) color, which allows direct visualization of transient expression. In developing embryos of *B. napus*, the expression of GUS and its time course was followed for 24 h after laser treatment. The enzyme activity reached a maximum after 96 h (Fig. 1).

C. Stable Incorporation into Rapeseed Cells and Recovery of Modified Tobacco Plants

Single cells of *B. napus* were used to study laser transfection of plasmid DNA into their genome. Cells were plasmolyzed and shrunken to 80% of their volume in the presence of DNA conveying resistance to the antibiotic hygromycin (pRT 102 hph). During this time, each cell received one laser pulse into its cell wall and plasma membrane. Through the temporary opening in the membrane, DNA and buffer entered the cytoplasm of the cells swelling them to their original volume.

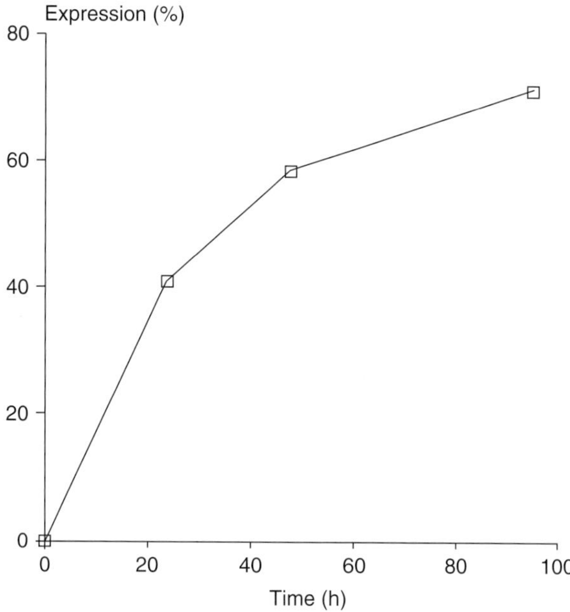

Fig. 1 Transient expression of GUS in embryos of *B. napus*. Plasmid DNA was incorporated into cells by laser irradiation. Expression of the reporter gene was followed over 4 days by histochemical staining. Twenty-four hours after the uptake of DNA, a total of 44% of the embryos expressed the marker gene. Their number increased to 57% and 71%, respectively, after 2 and 4 days of culture.

Approximately 1×10^6 molecules were incorporated into each cell at a DNA concentration of 1 μg/μl. In 1 h, 1000 cells could be irradiated. After irradiation of individual cells ~20% of the cells were resistant to the antibiotic hygromycin and retained their resistance over many cell generations in the absence of selection.

A particularly attractive application of laser transfection is the genetic modification of only a few cells in a whole tissue. This has been achieved with rapeseed embryos. The embryo is in a solution containing the GUS gene described above. Individual cells of the embryo are targeted and microperforated with the laser microbeam. They take up and express the gene. This can be seen by the dark color they adopt (Fig. 2).

Since the frequency of plant regeneration from single cells of *N. tabacum* is known to be better than in most other plant species, individual tobacco cells were used in the following experiment to compare frequencies of transformation obtained with a laser microbeam to those obtained by other methods. Cells were irradiated during plasmolysis. After 7 days of culture, laser-treated and control cells were subjected to a lethal dose of the antibiotic. From a total of 472 irradiated cells, 2 fertile plants were regenerated which had the entire hph gene incorporated into their genome. A frequency of 4.2×10^{-3} based on cells treated by the laser microbeam was found. This is comparable to other efficient methods

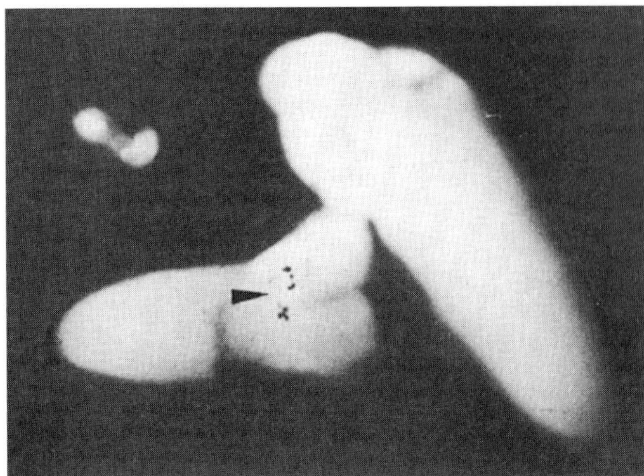

Fig. 2 Microinjection of the GUS gene into selected cells of a rapeseed plant embryo. Cells that express the gene appear dark (indigo), as indicated by the arrow.

for transformation of protoplasts (Weber *et al.*, 1990). These tobacco plants conveyed the hygromycin gene through meiosis into whole plants of the following generation.

D. Transfection of Isolated Chloroplasts and Perforation of Chloroplasts Inside Living Cells

When chloroplasts, with a diameter of only a few micrometer, have to be transfected, laser microinjection is particularly useful. Typically, a small number of DNA molecules are present in one chloroplast. Proteins coded by the chloroplasts are synthesized using a prokaryotic mechanism. In the present experiment, the membrane of isolated chloroplasts was opened with single laser shots. However, compared to irradiating cell walls, the energy of a laser pulse had to be attenuated to prevent bursting of the organelles. Furthermore, the integrity of the entire chloroplast was critically dependent on exact focusing of the laser. For scoring a hit of the membrane, a visible effect on the bright field image of the chloroplast was recorded.

Particularly interesting was the transfer of material from the cytoplasm into a chloroplast inside a living cell (Fig. 3). For that purpose, the laser microbeam was focused onto such a single chloroplast. The membrane of the chloroplast was perforated without damaging the plasma membrane of the cell and material entered the chloroplast.

In order to transfect foreign genes into the chloroplast, it was essential that the laser-induced hole in the chloroplast membrane be sealed quickly after laser treatment. The small diameter of the laser focus (less than 1 μm) and its small depth of field make it feasible to aim the laser exactly at the membrane of chloroplasts inside of cells. Thus, the lifetime of the laser holes could be studied. From irradiation of chloroplasts inside of cells of *B. napus*, a sequence of events

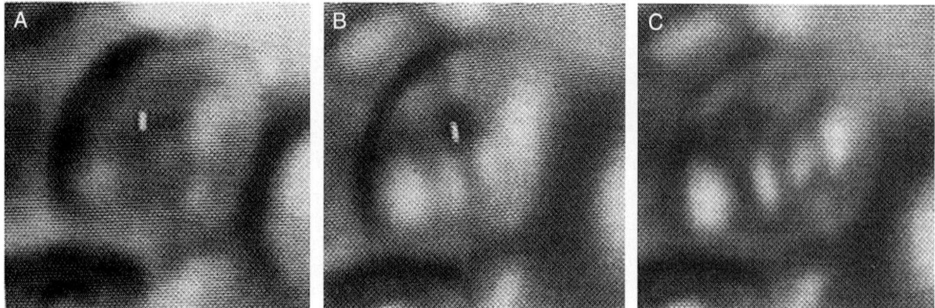

Fig. 3 Microinjection of material from the cytoplasm into a chloroplast in the interior of a rapeseed cell. The laser is directed onto such a chloroplast (A). In (B), a hole is visible in the chloroplast membrane. The chloroplast takes up some of the cytoplasmic material thereby slightly increasing its volume (C).

was recorded on video at 4000× magnification. The data indicated that a laser hole in the chloroplast membrane lasted 1.2 s or less (Weber *et al.*, 1989).

III. Preparation of Plant Protoplast Membrane and Chromosome Segments

A. Preparation of the Protoplast Membrane in Root Tips

Unlike the cell membrane in animal cells, the plasma membrane of plant cells is not easily accessible to experimental manipulation, since it is protected by a rigid cell wall. The usual strategy to get access to the plasma membrane is by enzymatic digestion. This means, however, the cell will be subjected to considerable stress. Also, local removal of the cell wall is not possible. Here, the laser microbeam can be used literally as a microscalpel. This microscalpel can, for example, be used to prepare protoplasts free of cell walls in *Medicago sativa* (Kurkdjian *et al.*, 1993). After dissecting the tips of root hairs, the protoplasm is expelled through the small hole in the cell wall. When a sufficient amount of protoplasm is expelled, it assumes a spherical shape and can separate itself from the root hair. By staining with the cellulose-specific dye Tinopal, it can be shown that the newly formed protoplast is virtually free of cell wall components. Such protoplast material may be better suited for plant membrane studies than protoplasts obtained by enzymatic digestion of the cell wall, since the latter always contains residual cell wall material and since the digestion itself significantly modifies the membrane properties (Fig. 4).

B. Preparation of Chromosome Segment-Specific DNA Probes Using Laser Scissors

For a large number of genetic diseases, it is known in which cytogenetic region of the metaphase chromosomes they are located. For basic studies as well as for diagnostic purposes, DNA probes of those regions are needed. One approach is to

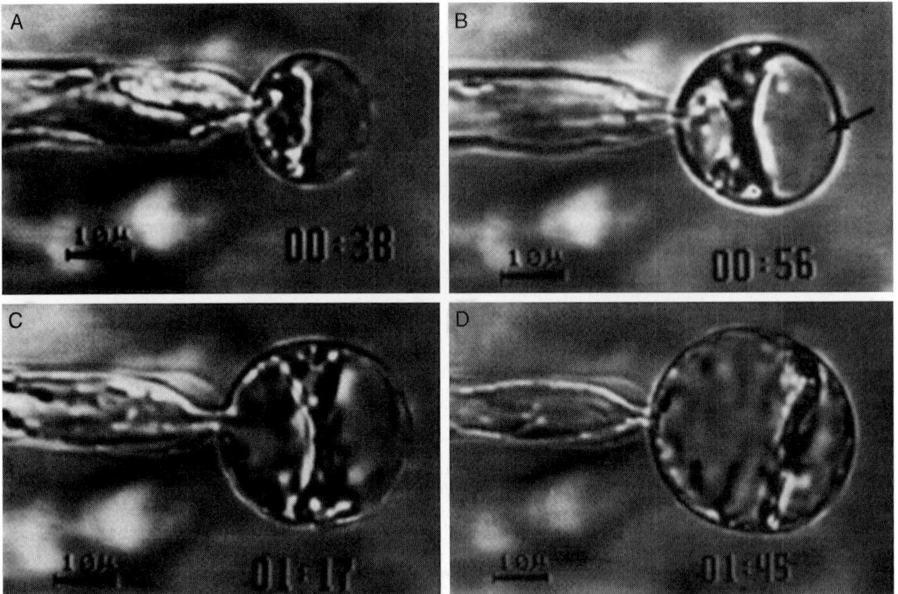

Fig. 4 Preparation of plant protoplast-like material in root hairs of *M. sativae*. The tip of the root hair is cut with the laser microbeam. The membrane and the underlying cytoplasm are pushed out of the root hair, and the plasma membrane becomes accessible for studies such as patch-clamping.

cut such a region out of the metaphase chromosome, to isolate it, to remove histones and other proteins, and to amplify the DNA of the chromosome segment by a microcloning technique or by PCR. Microdissection can be performed with ultrafine glass tools, similar as they are used in patch-clamping. However, handling such micromechanic tools requires an extremely skilled experimenter. Therefore, other tools such as the tip of an atomic force microscope have been suggested. The most promising tools, however, are laser scissors, since they can be used in normal microscopes, that is, protocols for the preparation of the chromosomes need not to be modified as compared to conventional techniques for microdissection.

As mentioned above, one of several techniques such as the polymerase chain reaction (PCR)-based approaches yielding a pool of DNA probes with a size of a few hundred bases in length can be used for DNA amplification. Others such as microcloning into phage vectors generate kilobase-sized DNA probes. The techniques described by Hadano *et al.* (1991) and Djabali *et al.* (1991) combine laser microdissection and different PCR techniques.

An electron microscopic study shows that with unstained chromosomes, an N_2-laser microbeam works with good accuracy in the submicrometer range (Ponelies *et al.*, 1989). For better localization, one would like to work with banded chromosomes. However, the absorption of UV light is highly increased and laser pulses of comparable power density damage the chromosomes.

The generation of chromosome segment-specific DNA libraries after laser microdissection was presented by Ponelies *et al.* (1989) for *Drosophila* chromosomes; Eckelt *et al.* (1989) for human chromosome No. 7; and Lengauer *et al.* (1991) for the use of the library for chromosome panting. With microcloning, only picograms or even femtograms of DNA in nanoliter volumes are available as starting material. On the average, 100–500 recombinant clones (1–3 kb in length) are produced in phage insertion vectors such as Lambda Zap (Stratagene) with ~100 microdissected chromosome segments as starting material (Eckelt *et al.*, 1989).

IV. Laser-Induced Cell and Protoplast Fusion

Cell fusion can combine the genomes of different cells and thus, under favorable circumstances, combine properties of the original cells. Essentially, the biological purpose of sex is to achieve such a combination of properties of the two parents. *In vitro* cell fusion is the basis for the generation of hybridoma cells. Here, the aim is to combine the potential to produce specific antibodies of B lymphocytes, which *in vitro* have only very short survival times, with the longevity of myeloma cells. In almost all cases, the fusion is achieved in a bulk experiment, that is, millions of cells with different specificities are fused and the wanted specificity is subsequently screened for in procedure taking several months. If the fusion could be achieved under complete microscopic control, one might first select the single B lymphocyte with the wanted specificity and fuse the latter with a myeloma cell under visual control. Screening would then no longer be required.

At moderate power density, such a laser fusion can be induced. If two cells are in contact with each other, they may be fused by a short series of laser pulses (Wiegand *et al.*, 1987). The contact may be established either by adhesive forces or due to high cell density on a microscope slide or by specific coupling via a bridging molecule system such as avidin/biotin. So far, plant protoplasts and different types of immune cells have been fused with each other. In the latter case, the contact was enhanced using the optical tweezers, and thereby the fusion yield could be significantly improved. The mechanism of laser-induced cell fusion is not yet clear, although it appears reasonable to assume that surface tension is important.

The fusion of plant cell protoplasts is more difficult since, for each type of plant cell, different physicochemical conditions are required and it takes some time to optimize them. Particularly critical is the adjustment of osmotic pressure of the environment which is quite often achieved by the use of mannitol. Usually, protoplasts are prepared from plant leaves by digestion of the cell wall with a suitable enzyme cocktail. The protoplasts are transferred onto a coverslide at high concentration. Individual pairs of protoplasts stick together due to the hydrophobicity of their membranes. If this is not the case, contact can be established by pushing protoplaststs toward each other with the optical tweezers. In the latter case, protoplasts can be selected, for example according to their size or the number of chloroplasts. In order to

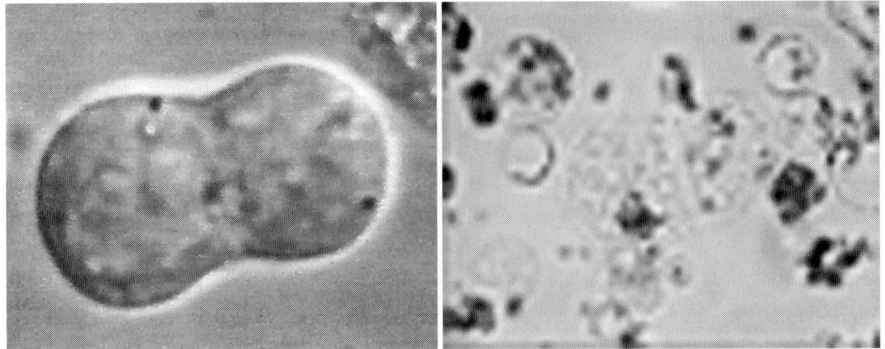

Fig. 5 Left: fusion of two rapeseed protoplasts after a few laser pulses (337 nm) have been directed onto the contact area. The fusion is completed after ~2 min. A third and fourth protoplast can be also fused to the product of the first fusion process. Right: fusion of two mouse myeloma cells. Since these are cells, not protoplasts whose cell wall was removed, the fusion process takes 10 min or more. Cells of very different size can be fused as well.

induce protoplast fusion, the laser microbeam is slightly defocused and directed to the contact area between two protoplasts. Usually, single pulses or short series of 10–20 pulses are sufficient to induce the fusion which is completed after a few seconds up to half a minute, depending on temperature and mannitol concentration (Fig. 5).

The efficiency of laser-induced cell fusion can be significantly increased when, in addition to the laser microbeam, optical tweezers are used to push the cells to be fused toward each other (Steubing *et al.*, 1991).

V. Optical Tweezers Complement the Laser Microbeam

A. Combined Use of Laser Scissors and Tweezers to Isolate Chromosome Segments

In 1989, the laser microbeam was combined with optical tweezers (Greulich *et al.*, 1989). Technically, this meant optimization of all optical elements with respect to the UV ablation wavelength of the 337-nm pulsed nitrogen laser and the 1064-nm Nd:YAG-trapping laser. One of the first applications of complete micromanipulation by light was to cut tips of chromosomes with the laser microbeam (optical scissors) and subsequently to transport the chromosome segments into a glass capillary where they could be transferred for further use in, for example, cloning experiments (Seeger *et al.*, 1991) (Fig. 6).

B. Trapping Lasers Can be Used as Scissors in Selected Plant Cells

Not always are two lasers required to combine the effects of laser microbeams (scissors) and laser tweezers (Holzinger *et al.*, 2002). In plants such as *Microasterias denticularum* or *Pleurebterium tumidivium*, which have large and sensitive nuclei,

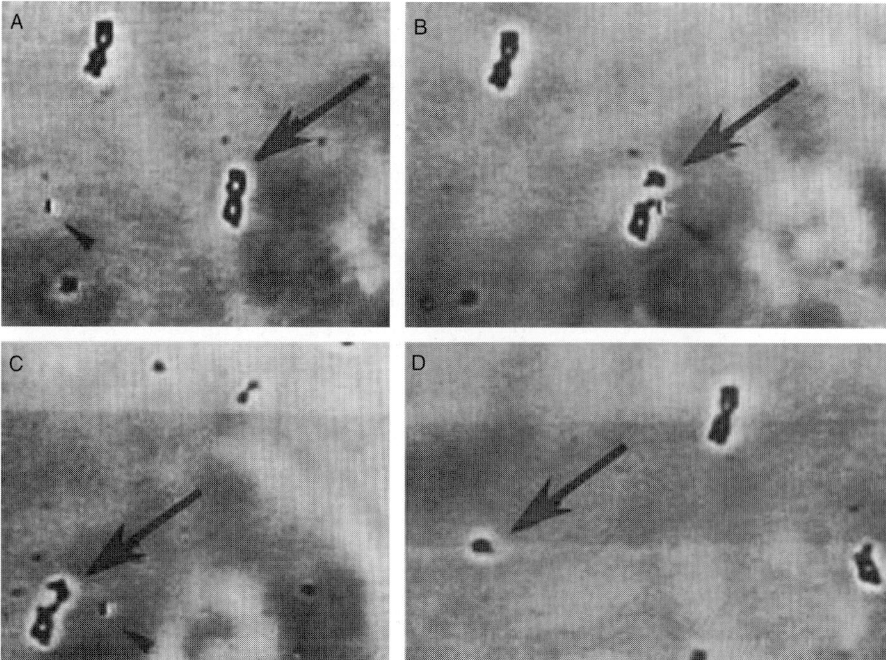

Fig. 6 Complete preparation of a chromosome tip with laser microbeam and optical tweezers. In (A–C), the chromosome tip is cut off in three steps. Between (C) and (D), the tip is transported with optical tweezers to a different site on the microscope slide, where it can be found separated from the main chromosome (D, arrow). This process can be repeated several times and the chromosome tips can be collected and further used for microcloning.

one single Nd:YAG laser can perform both tasks. When this laser is run at a power of 90 mW, it can be used as optical tweezers. For example, secretory vesicles accumulate reversibly by optical trapping with a dramatic change of cell shape. The latter is not reversible, despite a complete reversion of the vesicle accumulation. When the same Nd:YAG laser is run at a power of 180 mW, naturally occurring migrations of large nuclei in cells of both plants are irreversibly inhibited when microtubule tracks are targeted with the laser at this power. This nicely shows that the change from trapping effects to micromanipulation proceeds gradually.

C. Simulating Microgravity in the Alga "*Chara*"

The alga *Chara* has specific tube-like cells, the rhizoids, which it uses to sense gravity. Under normal conditions, the rhizoids always grow downward, that is, into the direction of higher gravity. While the detailed mechanism for this "gravitropism"

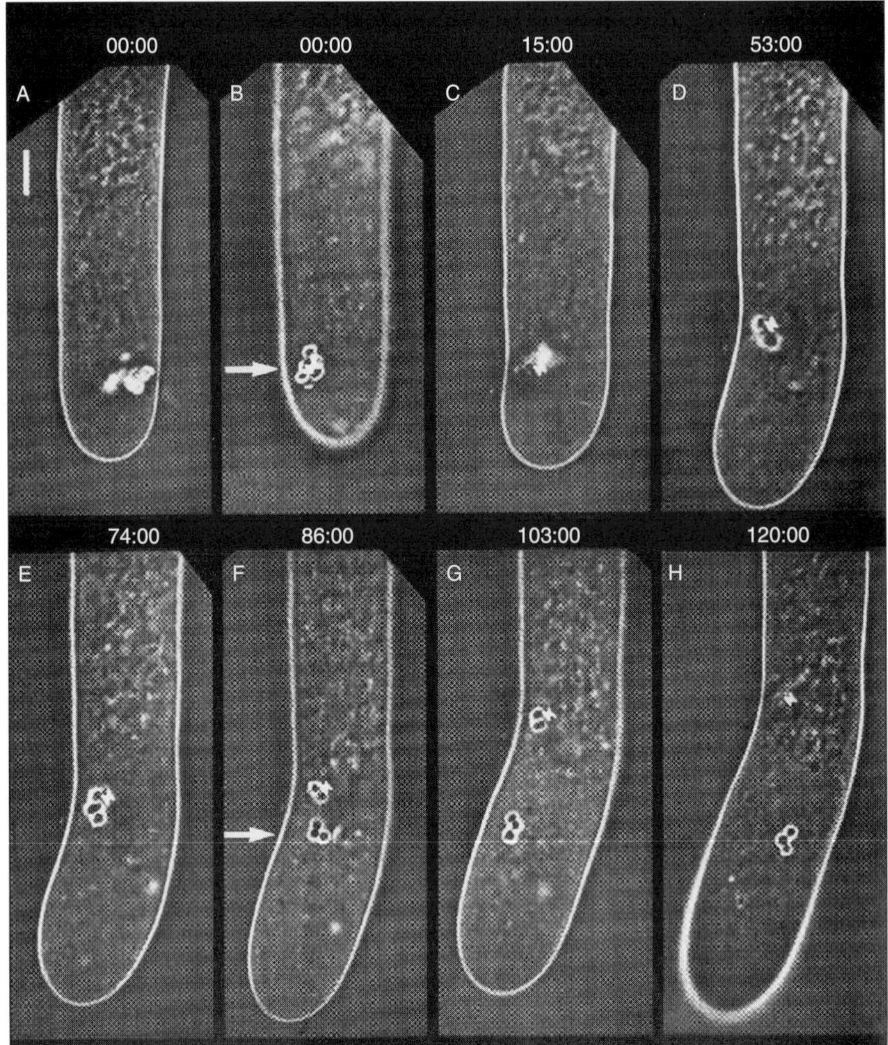

Fig. 7 (A–H) Series of micrographs showing a change in the direction of rhizoid growth induced by continuous statolith displacement with optical tweezers (the exact position of the optical trap is marked by the small H-shaped cursor inside the cells). (A) Tip region of the rhizoid just before the optical-trapping experiment. (B) Lateral displacement of the statoliths into a stable position close to the sidewall. The position of the optical trap remains unchanged during the experiment. The *arrow* points to the trapped statoliths in the laser focus. (C–H) The cell reacts by an extended growth of the opposite cell wall flank, resulting in a change in growth direction. (F–H) Three statoliths (*arrow*) have escaped the optical trap after having been held for a certain period. Note the two statoliths that still remain in the optical trap (just above the *arrow*). Time is given in minutes and seconds, starting after final repositioning of the statoliths, 550×; scale bar = 10 μm.

is not known, there are at least some hints: close to the tips of the rhizoids, dense structures, 1–2 μm in diameter, can be recognized. These structures, called statoliths, are barium sulfate shows the microcrystals enveloped by membranes. Obviously, they sediment under the influence of gravity and control the direction of growth.

They have a higher refractive index than the environment and thus can be easily caught and moved by optical tweezers in the interior of living rhizoids (Leitz *et al.*, 1995). It is not only possible to move one statolith but up to five of such structures can be collected in the focus of the optical tweezers and held permanently. The force required for displacement depends on the direction of displacement, that is it is relevant if the displacement occurs in the direction of the main cell body (basipetal), in the opposite direction (acropetal), or perpendicular to the axis of the tube-like rhizoid (lateral).

Figure 7 shows the microcrystals in the tip region of the alga and also shows the change of growth direction induced by the optical tweezers. Control experiments illuminating the same region but not removing the crystals do not change the growth direction, that is in the present experiment growth is not simply affected by a thermal effect. At time zero, the statoliths are in their original position, determined by gravity. At all other times, up to more than 3 h, the statoliths are displaced. The rhizoids no longer grow toward gravity. That is approximately what one had expected. In other words, the hypothesis that the statoliths mediate gravity sensing is correct. More surprising is the fact that not only the direction but also the speed of growth is changed and that this effect is almost an order of magnitude. Figure 8 shows in the upper part the apical displacement of the statoliths. The lower part, on the same time scale, gives the growth rate of the tip. Note that in this representation the relaxed position of the statoliths is at ~13 μm.

Knowledge on the effects of microgravity on the growth of algae has a quite practical consequence. Since algae may become major source of nutrition during long-term space missions, it is essential to know in advance how they behave under microgravity. Certainly, crucial experiments can only be performed in true microgravity, for example on a space station. However, preliminary experiments can be performed earthbound, using optical tweezers at a fraction of costs, and thus expensive experiments in space can be scheduled much more efficiently.

VI. The Attack of a Natural Killer Cell on a Leukemia Cell

The following experiment (Seeger *et al.*, 1991) describes how, with the help of optical tweezers, the attack of a cell from the human immune system on a cancer cell can be observed from the very first seconds after contact has been established. Immune cells such as cytotoxic T lymphocytes (CTLs) and natural killer (NK) cells circulate through the body and check if other cells show, for example, signs of cancer. Usually, cells are presenting peptides bound to a molecule class called MHC I. If such a peptide is unfamiliar to the CTL, the latter attacks the presenting cell.

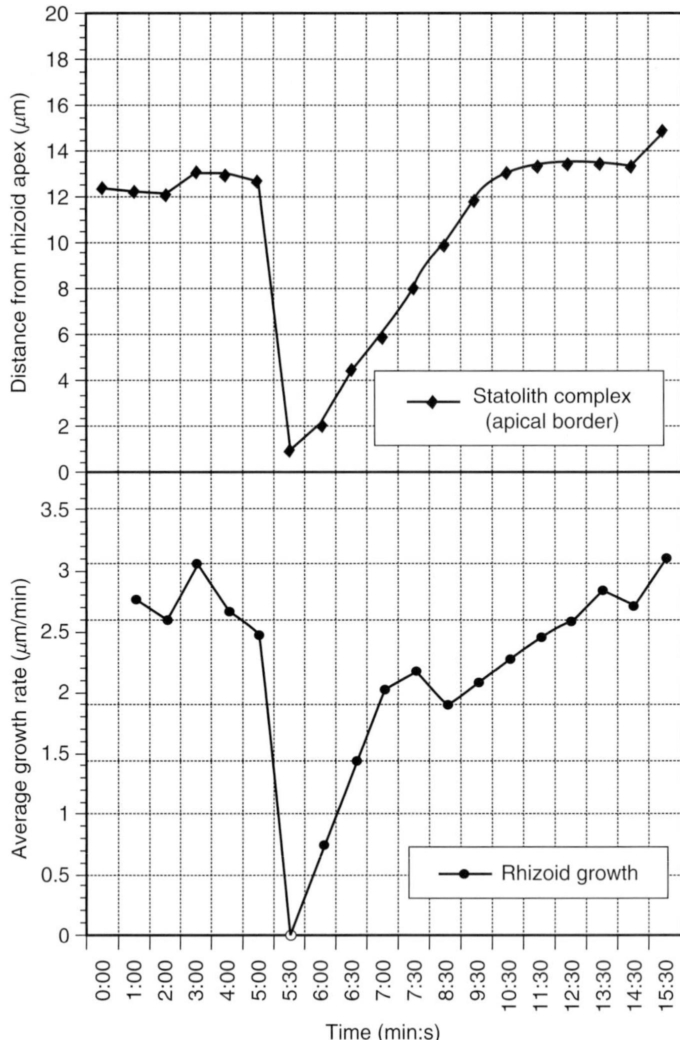

Fig. 8 Acropetal displacement of the statolith complex and effect on tip growth. All statoliths were displaced to the extreme apes of the rhizoid. Optical trapping occurred between 5:00 and 5:30 min; laser power was 1000 mW (about 300 mW in the object plane). The average growth rate and the mean distance of the apical border of the statolith complex from the rhizoid apex before, during, and after statolith displacement by optical tweezers are given. The standard error for the average growth rate is smaller than ± 0.3 μm/min. For technical reasons, the actual growth rate at 5:30 min can only be given as a range between 0 and 0.75 μm/min. Number of rhizoids, $n = 10$.

If for some reason the presenting cell has no MHC I molecule at all, it cannot present the peptide which might indicate that something is wrong. In that case, the NK cell takes over the task of immune surveillance and thus helps to avoid a catastrophic outcome for the whole organism. In order to understand the process of recognition and attack, *in vitro* experiments would be helpful, where contact is established artificially and the process can be observed. Particularly interesting is the contact between cytolytic effector cells such as CTLs or NK cells with target cells. The CTL–target contact is highly specific and governed by complex supramolecular associations between T cell receptor and CD 8 surface molecules on the effector cell and major histocompatibility class I surface molecules complexed with peptides on the target cell. With NK cells, similar mechanisms have been suggested. In order to test such working hypotheses, it is helpful to have other techniques for establishing contact between NK cell and its target cell.

The standard approach is to mix the two types of cells, centrifuge them at low speed and observe what happens after the cells have been resuspended and prepared either for microscopy or for other experiments. This approach is blind for the first minutes after contact between the two reaction partners and therefore it is difficult to study the most important phase of the attack. Alternatively, contact can be established by the use of micromechanical tools, but this may damage the NK cells and thus blur the outcome of the experiment. Here the optical trap can establish contact in an easy and gentle way. With the optical tweezers, the NK cell can be moved toward its target cell and even the natural contact between both cells can be increased. This may help to trigger the activity of the target cell. During this type of experiment, the yield of successful attacks of NK cells on erythroleukemia cells was higher than expected from literature values. An obvious explanation is that the NK attack on the target cells has already started whereas the affinity of recognizing molecules is not sufficiently high to trigger activation. In these cases, facilitating the contact by the optical tweezers may support activation even in those cases where specific contact is not yet fully established. Figure 9 shows the NK cell (small dark cell) attacking the target cell, which changes its morphology by membrane blebbing.

A detailed inspection of membrane blebbing shows that for the first 40–50 s after contact between the NK and cancer cells, no change in morphology is observed. This can be explained by biochemical activation processes that take some time before the cancer cell really is attacked. Interestingly, the further process is not continuous but oscillates. At present, this oscillation is unexplained. A speculative interpretation would be that the attacked cell reacts by repairing damage and that the attacking cell increases its attempts to kill the cancer cell. If this interpretation turns out to be realistic, it would have significant consequences for the understanding of immune surveillance of cancers and might give interesting hints for developing cancer therapies (this chapter has been reproduced from Greulich, 1999).

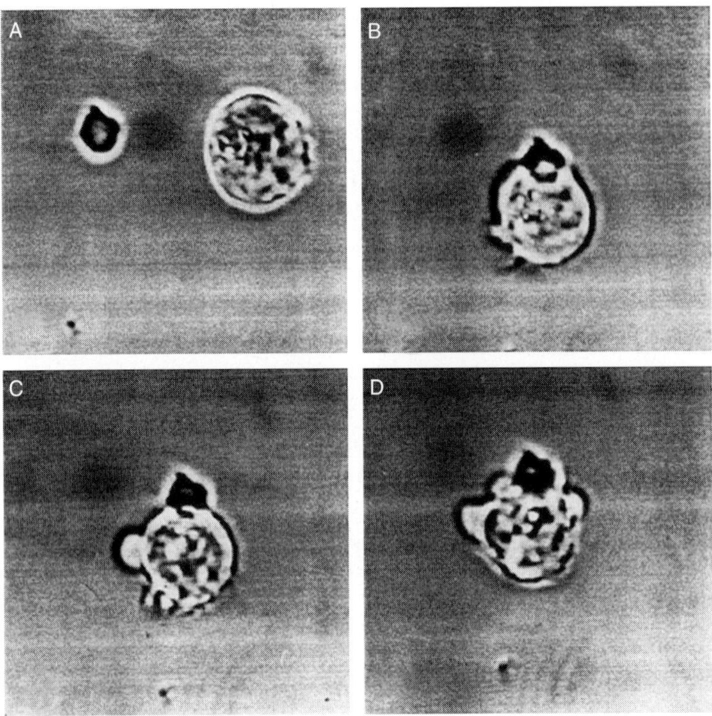

Fig. 9 Attack of an NK cell on an erythroleukemia (K562) cell. The contact is generated by pushing the small NK cells toward the large K562 cell. In the first minute, no effect is visible. The membrane blebbing indicates increasing damage to the K562 cell. This is not an effect of the laser, since similar treatment with the laser alone does not induce membrane blebbing.

VII. Laser Microbeams and Optical Tweezers for Heart Research

A. The Role of Fibroblasts in Heart Mechanics and in Complications after Heart Infarction

Here the use of a laser microbeam and optical tweezers for micromanipulation of heart cells and tissues is introduced (Monajembashi, 2005; Perner, 2004). Whereas more than 80% of the myocardial mass consists of heart muscle cells (cardiomyocytes), they represent only 30% of the cell number. Among the noncardiomyocytes, fibroblasts constitute a predominant fraction. By producing extracellular matrix components, fibroblasts are involved in maintaining the mechanical framework for cardiomyocytes. Furthermore, they play an important role in the complex process of wound healing of the necrotic myocardium following myocardial infarction. This process includes migration and proliferation of fibroblasts into the infarct zone and synthesis of new extracellular matrix proteins. Also this process results in transformation of fibroblasts into actin-expressing myofibroblasts, a process that is responsible for wound contraction and mediation of scar formation.

Such a fibroblast scar is a barrier for incoming calcium waves that organize the coordinated contraction of the heart. In addition, abnormal proliferation of fibroblasts and deposition of the extracellular matrix protein collagen (i.e., fibrosis) alter the normal cardiac architecture and the mechanical behavior of the heart. It is generally accepted that increased extracellular matrix deposition contributes to diastolic stiffness and that fibrosis promotes ventricular dysfunction. Thus, the composition and the mechanical parameter of the scar and the myocardium may, to a large extent, determine the outcome of myocardial infarction.

Another cause of myocardial dysfunction is related to increased pressure-stretch of cardiac myocytes as a consequence of mechanical stress derived for instance from hemodynamic overload (hypertension). Mechanical stress is known to induce cardiac hypertrophy and release of growth-promoting factors. Recent investigations suggest that cardiac fibroblasts act as mechanoelectric transducers in the heart. Moreover, fibroblasts are considered to modulate the contractile activity of the myocardium in response to mechanical changes. These findings emphasize the importance of mechanical influence on myocardial activity. A multitude of methods for mechanical force application exists to produce loads of different magnitude, frequency, and duration either to a cell population or to an individual cell. In bulk approaches, a global mechanical stress is applied to a large number of cells, for instance by stretching of cells grown on a flexible substrate. However, only single-cell assays allow a distinction between locally restricted and more global effects. For most single-cell investigations, methods were used such as microelastic mapping with the force curve mode of an atomic force microscope or with magnetic field devices.

Optical tweezers are particularly suitable for variable force application on the single-cell level or even with subcellular accuracy under microscopic control. This technique allows bright field microscopic observation of the mechanical influence and simultaneously permits the investigation of additional parameters such as intracellular calcium. Optical tweezers can be employed to clarify an interesting aspect of heart infarction. Here it is pertinent to recall that the heart represents a tissue consisting, among others, of heart muscle cells. Each single cell is able to contract, that is to pump independently. In a heart, it is necessary that these contractions are coordinated. This is achieved by calcium waves. How this works in detail, particularly after external stimulation, can be investigated using optical tweezers. On the other hand, a laser microbeam is useful to study how damages in heart tissue are repaired and how such a repair might affect the organization of calcium regulation of heart beating. This is discussed in the following sections.

B. Chicken Embryo Heart Tissue as a Model to Study Wound Healing and Coordination of Heart Beating by Calcium Waves

Here, experiments with chicken cardiomyocytes are described, since the latter can be easily obtained from eggs, which have been bred for a few days. Heart-like tissue can be relatively easily reconstituted from semibreed chicken eggs.

Certainly, it would be better to use human heart tissue, which is however not available live for such experiments, or tissue from mouse hearts which is also more difficult to obtain. The described experiments are therefore a compromise, with the advantage of easy availability and the disadvantage that its results and conclusions can be used only in a limited manner as a model for human heart attack.

C. Optomechanical Stimulation of a Single Cell Starts a Calcium Wave in the Whole Tissue

The mechanical stimulation is induced through optical tweezers, however, not directly. An erythrocyte is used as a handle to transfer the force. Note that the erythrocyte has no biological function—it had just turned out that suitably pre-treated erythrocytes are good force—transmitters. This technique was termed by us *E*rythrocyte-*M*ediated *F*orce *A*pplication (EMFA). For more details of the EMFA technique see Perner *et al.* (2004) or, in German, Monajembashi *et al.* (2005).

If cardiomyocytes of chicken embryos are reconstituted into a tissue-like cell layer and a single cell in this layer is optomechanically stimulated by EMFA, first a calcium burst is set free from intracellular calcium deposits of the directly treated cell. This can be made visible through suitable calcium-sensitive fluorescent dyes (Orange Green 488, BAPTA-1, AM). The increase of calcium concentration in the directly treated cell affects its neighbors so that within a few seconds a calcium wave spreads over the entire tissue. This regular wave stimulates a coordinated beating of all cells that are reached by the calcium wave.

The experiment is especially interesting since it works with fibroblasts. This cell type is generally assumed not to be excitable for calcium waves, in contrast with cardiomyocytes. In Fig. 10, however, it can be clearly observed that through EMFA stimulation, a calcium burst is induced in the directly treated cell. Subsequently, a calcium wave runs across the entire tissue. Apparently, there exists a physicochemical condition in which fibroblast tissue, and thus potential scar tissue, can be made transparent for calcium waves. Making fibroblast scars transparent for calcium waves gives an interesting perspective for developing drugs that reduce the risk for deadly complications after a first heart infarction. Experiments of this type may help in the high-throughput search for "lead substances," which can subsequently be used in a mouse model and can be developed later into drugs for clinical tests. At the end, there would be a medication that would reduce the risk of deadly fibrillations.

D. Injuries Through Laser Microbeam Heal Within 18 h, However, Form Scars

Not only the optical tweezers and EMFA can help to understand problems connected with heart attacks. The laser microbeam also can be used as an extremely fine scalpel to inflict selective lesions in heart tissue and to study the healing process.

Figure 11A (left) shows the tissue directly after injury. Figure 11B (right) shows the situation after 5 h. It is clearly visible how the first cells grow into the

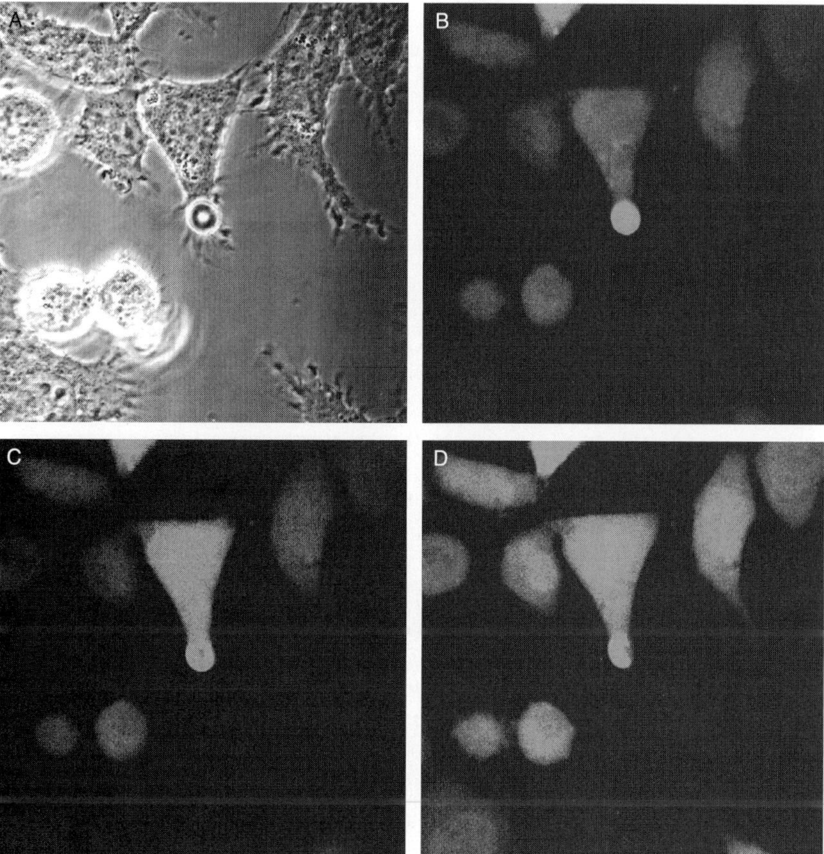

Fig. 10 Induction of a calcium wave in a layer of fibroblasts. An erythrocyte (center, A) is pushed with optical tweezers onto a fibroblasts. The erythrocyte solely acts as a handle, that is has no biological function. This process starts the liberation of calcium in the fibroblast (B), which spreads over the whole cell (C), and finally, after 2–3 s, couples over to neighboring cells and spreads over a wide area in the tissue. The remarkable aspect of this experiment, which works even better with excitable heart muscle cells, is that it indeed does work with nonexcitable fibroblasts. (See Plate 1 in the color insert section.)

empty area and start to fill the gap. After 18 h, the gap is closed (Fig. 11C). The replaced cells are however predominantly dark, spindle-shaped fibroblasts. It shows one of the feared scars has developed. It would be better if the portion of cardiomyocytes observed in the lesion area would be much higher, comparable to the composition of healthy heart tissue. The gap would be closed completely with heart-like tissue, that is a scar would not have developed at all. Since our model is again, as above, suitable for high throughput, it accommodates the search for substances that boost the growth of cardiomyocytes. Surprisingly, such a molecule exists: the insulin-like growth factor, IGF 1. Mice or pigs, which had the gene for IGF 1 implanted into their genome, do not develop scars after heart injury and

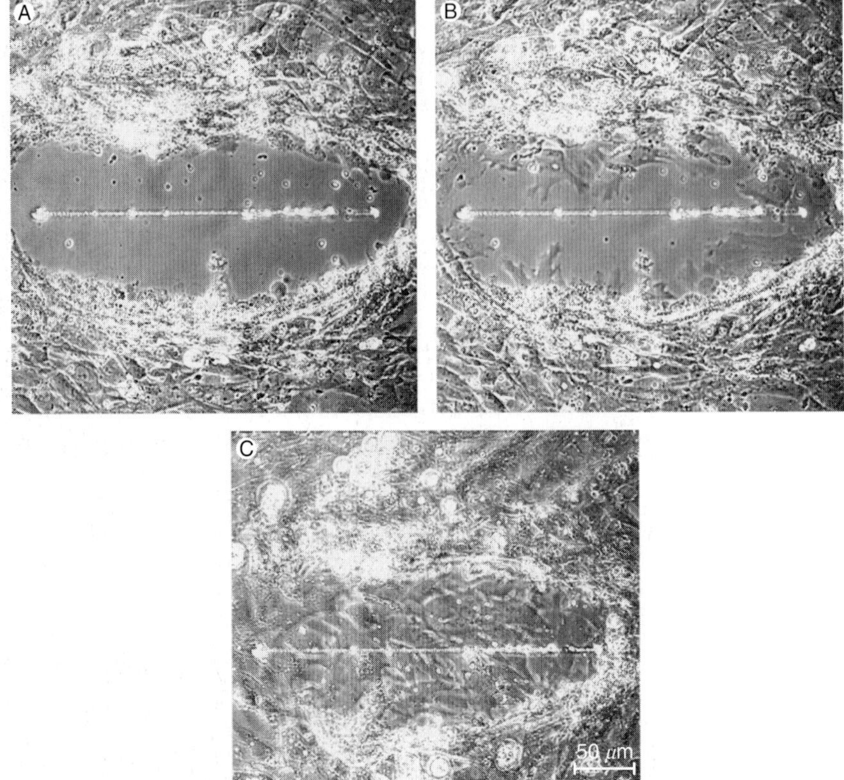

Fig. 11 Injury induced by a laser microbeam to heart tissue. Left (A): severing of the tissue is just completed. The straight line in the center of the lesion is damage of the coverslide, that is, an unwanted side effect. Right (B) dark, spindle-like fibroblasts grow into the lesion. Bottom (C): closing the lesion is complete. In addition to the fibroblasts, a few bright heart muscle cells are also found. But as a whole, a fibroblast scar remains at the site of the lesion.

also live much longer than control animals. Certainly, such a strategy is not applicable in humans. Therefore, a long systematic search for suitable substances lies ahead.

Altogether, in this chapter it has been demonstrated that a combination of laser microbeam and optical tweezers is a microscopy instrument that is multifunctional and versatile for wide variety of studies and applications in biology and biomedicine.

Acknowledgments

I thank all colleagues who have contributed with their experiments to this work and whom I hopefully have cited properly.

References

Ashkin, A., Dziedzic, J. M., Bjorkholm, J. E., and Chu, S. (1986). Observation of a single beam gradient trap for dielectric particles. *Opt. Lett.* **11,** 288–290.

Bereiter-Hahn, J. (1972). Laser als Mikromanipulator in Biologie und Medizin. *Microsc. Acta* **71,** 225–241 and **72,** 1–33.

Berns, M. W. (1974). Biological Microirradiation. Prentice Hall Series on Biological Techniques Englewood Cliffs, New Jersey.

Berns, M. W., Aist, J., Edwards, J., Strahs, K., Girton, J., McNeill, P., Rattner, J. B., Kitzes, M., Hammer-Wilson, M., Liaw, L. H., Siemens, A., Koonce, M., *et al.* (1981). Laser microsurgery in cell and developmental biology. *Science* **213,** 505–513.

Berns, M. W., Olson, R. S., and Rounds, D. E. (1969a). Argon ion laser microirradiation of nucleoli. *J. Cell Biol.* **43,** 1821–1840.

Berns, M. W., Olson, R. S., and Rounds, D. E. (1969b). *In vitro* production of chromosomal lesions using an argon laser microbeam. *Nature (London)* **221,** 74–75.

Berns, M. W., Rounds, D. E., and Olson, R. S. (1969c). Effects of laser microirradiation on chromosomes. *Exp. Cell. Res.* **56,** 292–298.

Bessis, M., Gires, F., Mayer, G., and Nomarski, G. (1962). Irradiation des organites cellulaires a l'aide d'un laser a rubis (Irradiation of cellular organisms with a ruby laser). *C. R. Acad. Sci.* **255,** 1010–1012.

Block, S. M., Blair, D. F., and Berg, H. C. (1989). Compliance of bacterial flagella measured with optical tweezers. *Nature* **338,** 514–518.

Botvinick, E., and Berns, M. W. (2005). Internet based robotic laser scissors and tweezers microscopy. *Microsc. Res. Tech.* **68,** 65–74.

Cremer, T., Cremer, C., Baumann, H., Luedtke, E.-K., Sperling, K., Teuber, V., and Zorn, C. (1982). Rabl's model of the interphase chromosome arrangement tested in Chinese hamster cells by premature chromosome condensation and laser UV microbeam experiments. *Hum. Genet.* **60,** 46–56.

Djabali, M., Nguyen, C., Biunno, I., Oostra, B. A., Mattei, M. G., Ikeda, J. E., and Jordan, B. R. (1991). Laser microdissection of the fragile X region: Identification of cosmid clones and of conserved sequences in this region. *Genomics* **10,** 1053–1060.

Eckelt, A., Ponelies, N., Bautz, E. K. F., Miller, K., Heuer, T., Tümmler, B., Grzeschik, K. H., Wolfrum, J., and Greulich, K. O. (1989). Microdissection in the search for the molecular basis of disease: A chromosome segment specific molecular library. *Ber. Bunsenges. Phys. Chem.* **93,** 1446–1453.

Greulich, K. O. (1999). Micromanipulation by light in biology and medicine. The laser microbeam and optical tweezers. *Birkhäuser Basel* 1–300.

Greulich, K. O., Bauder, U., Monajembashi, S., Ponelies, N., Seeger, S., and Wolfrum, J. (1989). UV laser mikrostrahl und optische pinzette (UV laser microbeam and optical tweezers). *Labor Prax./Labor* **2000,** 36–42.

Guo, Y., Liang, H., and Berns, M. W. (1995). Laser-mediated gene transfer in rice. *Phys. Plant.* 19–24.

Hadano, S., Watanabe, M., Yokiu, H., Kogi, M., Kondo, I., Tsuchiya, H., Kanazawa, I., Wakasa, K., and Ikeda, J. (1991). Laser microdissection and single unique primer PCR allow generation of regional chromosome DNA clones from a single human chromosome. *Genomics* **11,** 364–373.

Holzinger, A., Monajembashi, S., Greulich, K. O., and Lütz-Meindl, U. (2002). Impairment of cytoskeleton dependent vesicle and organelle translocation in green algae: Combined use of a microfocused infrared laser as microbeam and optical tweezers. *J. Microsc.* **208,** 77–83.

Kurata, S. I., Tsukakoshi, M., Kasuya, T., and Ikawa, Y. (1986). Laser method for efficient introduction of foreign DNA into cultured tissue cells. *Exp. Cell Res.* **162,** 372–378.

Kurkdjian, A., Leitz, G., Manigault, P., Harim, A., and Greulich, K. O. (1993). Non-enzymatic access to the plasma membrane of Medicago root hairs by laser microsurgery. *J. Cell Sci.* **105,** 263–268.

Leitz, G., Schnepf, E., and Greulich, K. O. (1995). Micromanipulation of statoliths in gravity sensing Chara rhizoids by optical tweezers. *Planta* **197**(2), 278–288.

Lengauer, C., Eckelt, A., Weith, A., Endlich, N., Miller, K., Lichter, P., Greulich, K. O., and Cremer, T. (1991). Selective staining of a defined chromosomal region by *in situ* hybridization of libraries from laser microdissected chromosomes. *Cytogenet. Cell Genet.* **56,** 27–30.

Lohs-Schardin, M., Cremer, C., and Nüsslein-Volhard, C. (1979). A fate map for the larval epidermis of *Drosophila melanogaster*: Localized cuticle effects following irradiation of the blastoderm with an ultraviolet laser microbeam. *Dev. Biol.* **73,** 239–255.

Monajembashi, S., Perner, B., and Greulich, K. O. (2005). Licht als Werkzeug in der Zell- und Molekularbiologie: Anwendungen in der Herzforschung (light as tool in cell and molecular biology: Applications in heart research). *BIOforum* **9,** 40–41.

Nascimento, J., Botvinick, E. L., Shi, L. Z., Durrant, B., and Berns, M. W. (2006). Analysis of sperm motility using optical tweezers. *J. Biomed. Opt.* **11**(4), 044001.

Perner, B., Monajembashi, S., Rapp, A., Wollweber, L., and Greulich, K. O. (2004). Simulation of heart infarction by laser microbeam and induction of arrythmias by optical tweezers. *SPIE Proc.* **5514,** 179–188.

Ponelies, N., Bautz, E. K. F., Monajembashi, S., Wolfrum, J., and Greulich, K. O. (1989). Telomeric sequences derived from laser microdissected polytene chromosomes. *Chromosoma* **98,** 351–357.

Seeger, S., Monajembashi, S., Hutter, K. J., Futtermann, G., Wolfrum, J., and Greulich, K. O. (1991). Application of laser optical tweezers in immunology and molecular genetics. *Cytometry* **12,** 497–504.

Sheetz, M. P. (ed.) (1998). Laser tweezers in cell biology. *In* "Methods in Cell Biology," Vol. 55. Academic Press, San Diego.

Steubing, R. W., Cheng, S., Wright, W. H., Numajiri, Y., and Berns, M. W. (1991). Laser induced cell fusion in combination with optical tweezers: The laser cell fusion trap. *Cytometry* **12**(6), 505–510.

Tao, W., Wilkinson, J., Stanbridge, E. J., and Berns, M. W. (1987). Direct gene transfer into human cultured cells facilitated by laser micropuncture of the cell membranes. *Proc. Natl. Acad. Sci. USA* **84,** 4180–4184.

Tsukakoshi, M., Kurata, S., Nomiya, Y., Ikawa, Y., and Kasuya, T. (1984). A new method of DNA transfection by laser microbeam cell surgery. *Appl. Phys. B* **35,** 135–140.

Weber, G., Monajembashi, S., Greulich, K. O., and Wolfrum, J. (1988). Injection of DNA into plant cells using a UV laser microbeam. *Naturwissenschaften* **75,** 36–41.

Weber, G., Monajembashi, S., Greulich, K. O., and Wolfrum, J. (1990). Genetic changes induced in higher plants by a UV laser microbeam. *Israel J. Botany* **40,** 115–122.

Weber, G., Monajembashi, S., Wolfrum, J., and Greulich, K. O. (1989). Uptake of DNA in chloroplasts of *Brassica napus* (L) facilitated by a UV laser microbeam. *Eur. J. Cell Biol.* **49,** 73–81.

Wiegand, R., Weber, G., Zimmermann, K., Monajembashi, S., Wolfrum, J., and Greulich, K. O. (1987). Laser induced fusion of mammalian cells and plant protoplasts. *J. Cell Sci.* **88,** 145–150.

CHAPTER 3

Laser-Based Measurements in Cell Biology

Elliot L. Botvinick* and Jagesh V. Shah†

*Beckman Laser Institute, Department of Biomedical Engineering
University of California, Irvine, California 92612

†Laboratory for Cellular Systems Biology and Molecular Imaging, Department of System's Biology
Harvard Medical School, Boston, Massachusetts 02115

In this chapter, we review the imaging techniques and methods of molecular interrogation made possible by integrating laser light sources with microscopy. We discuss the advantages of exciting fluorescence by laser illumination and review commonly used laser-based imaging techniques such as confocal, multiphoton, and total internal reflection microcopy. We also discuss emerging imaging modalities based on intrinsic properties of biological macromolecules such as second harmonic generation imaging and coherent anti-Raman resonance spectroscopy. Super resolution techniques are presented that exceed the theoretical diffraction-limited resolution of a microscope objective. This chapter also focuses on laser-based techniques that can

0091-679X/07 $35.00
DOI: 10.1016/S0091-679X(06)82003-0

report biophysical parameters of fluorescently labeled molecules within living cells. Photobleaching techniques, fluorescence lifetime imaging, and fluorescence correlation methods can measure kinetic rates, molecular diffusion, protein–protein interactions, and concentration of a fluorophore-bound molecule. This chapter provides an introduction to the field of laser-based microscopy enabling readers to determine how best to match their research questions to the current suite of techniques.

I. Introduction

The use of laser sources in microscopy long predate their routine use today in modern cellular-imaging applications such as scanning confocal or multiphoton microscopy (MPM). As outlined in the chapters included in this volume, the first uses of lasers were as sources of high energy light for localized photodamage or microsurgery. Many of the same properties that make them excellent tools for spatially localized damage also provide for high-resolution microscopic imaging and biophysical measurement of molecular parameters within living cells. In part, the wide use of lasers in microscopic imaging, particularly in living cells, has been enabled through the revolution in fluorescent dyes and genetically encoded fluorescent proteins that provide molecular specificity (Shaner *et al.*, 2005; Zhang *et al.*, 2002). In this chapter, we discuss the properties of lasers that make them a unique light source for microscopic imaging and measurement. These properties, such as coherence, monochromaticity, and ultrashort pulses have enabled higher spatial resolution for cellular imaging. More conventional methodologies such as confocal or multiphoton imaging will be discussed alongside some of the recent developments in imaging technologies. Many of the characteristics of lasers also provide the ability to measure molecular parameters within living cells such as protein–protein interactions, diffusion constants, and reaction rates. Many of these measurements have long been considered the domain of *in vitro* systems but now can be interrogated in living cells. The application of lasers to these *in vivo* biophysical methods will also be discussed.

A. The Laser

The laser, originally based on its microwave counterpart the maser of the 1950s, has revolutionized the field of biological measurement. The name laser is an acronym for Light Amplification by Stimulated Emission of Radiation. In a laser, an active laser medium, also referred to as a gain medium, is placed within a resonant optical cavity (for review see Svelto, 1998). An external energy source such as a flash lamp, arc lamp, or light-emitting diode array pumps the electrons of the gain medium into an excited state. If the electrons were allowed to relax to the ground state, spontaneous emission would occur resulting in emitted photons of narrow spectral bandwidth, but of random phase. Instead, a laser uses the principle of stimulated emission to force the emitted photons into a tightly bound range of phase, frequency, and polarization. Photons traveling back and forth due to reflection from the mirrored walls of the optical cavity bombard the excited atoms of the gain medium. Instead of absorbing the photons, the atoms of the gain media release their potential energy by emitting a second photon similar to the first that travels along the same trajectory. The resonant cavity is designed so that a percentage of light passes through one of the mirrors. The escaped light is the laser beam and is typically collimated, coherent, monochromatic, of constant power, and polarized. This is in stark contrast to light emitted from conventional wide-field illumination sources (e.g., mercury or xenon arc lamps) that are incoherent, of wide spectral content, nonpolarized, and exhibit intensity fluctuations.

B. Coherency and Illumination Volume

Laser light is well suited for fluorescence microscopy because it is spatially and temporally coherent, properties that allow a very tight and bright focus in the specimen plane. Temporal coherency means that the wave is nearly perfectly correlated to itself in time at each point (Svelto, 1998). Spatial coherency means that all points within the wavefront are perfectly cross-correlated, or have similar phase independent of time. Consider, for example, a collimated laser beam incident on the back aperture of a microscope objective lens. The light travels as a coherent plane wave. The microscope objective transforms the plane wave into a converging spherical wave. Because of the coherence of the laser, the resulting focus is nearly diffraction-limited. The laser is ideal in that it minimizes the excitation volume, it provides a "bright" excitation field, and it excites fluorescent molecules with a time-invariant intensity (for the case of continuous wave lasers). Furthermore, using the techniques of Q-switching, mode-locking, or gain-switching, laser output can be transformed from continuous wave to pulsed output. Using pulsed lasers much higher peak irradiances can be achieved in the focal spot. The higher peak irradiances allow for nonlinear interactions, such as multiphoton absorption, between the sample and the light as described in sections below.

II. Laser-Based Imaging Methods

A. Laser Scanning Confocal Microscopy

The most common imaging technology based on laser illumination is confocal microscopy. In 1957, Marvin Minsky proposed a "white light" confocal microscope in which the specimen is scanned by the stage (Inoué, 1995). Later, stage scanning confocal microscopes using laser light sources and nonimaging (photomultiplier tube, PMT) detectors were developed (Wilson, 1980; Wilson *et al.*, 1980) for scanning electronic devices. In the 1980s, laser scanning confocal microscopes (LSCMs) (Carlsson *et al.*, 1985, 1987) came into the forefront in which confocal fluorescent images of biological specimens were acquired by steering the laser across a stationary specimen. It was apparent that the out-of-focus light rejection and optical-sectioning capability of LSCM over conventional image collection permitted the visualization of individual biological structures never before visualized in fluorescence (White *et al.*, 1987). In conventional laser-based epi-fluorescence, a steep cone of laser light converges to focus, and then diverges out of the specimen. As the cone narrows, irradiance increases with maximum intensity at the focus spot. In order to maximize quantum yield of the fluorescent molecules in the focus spot, illumination power is typically well above the fluorescence threshold (Tsien and Waggoner, 1995). As a result, fluorescence is emitted and subsequently collected by the objective lens from the true focal plane, as well as from out-of-focus objects. In LSCM, laser light is focused into the specimen and excites fluorescence, which passes through a dichroic filter and is reimaged onto the detection pinhole which is placed "confocal" to the laser focus spot (Carlsson and Aslund, 1987; Carlsson *et al.*, 1985). As a result, fluorescence originating from the laser focus passes through the detector pinhole, while that originating from out-of-focus regions does not. Light passing through the pinhole is detected, and by raster scanning the focus spot through the specimen, an image can be constructed. Although the LSCM is rate limited and may adversely effect live cells, it has been used to provide unique insights into neurobiology (Barger *et al.*, 1995; Goodman *et al.*, 1996; Nitatori *et al.*, 1995), HIV (Amara *et al.*, 1997), intracellular signaling (Bae *et al.*, 1997; Hakansson *et al.*, 1998), hepatocyte function (Buchler *et al.*, 1996), intracellular transport (Lin *et al.*, 1995), roles of intracellular calcium (Cheng *et al.*, 1993), carcinogenesis (Aragane *et al.*, 1998), and RNA processing (Spector *et al.*, 1991), just to name a few.

B. Multipoint Confocal Laser Microscopy

One major concern with LSCM is that the high irradiances required to achieve high frame rates and sufficient signal-to-noise ratios adversely effects cell viability. Additionally, because the quantum yield has a finite maximum value, LSCM is inherently rate limited (Gräf *et al.*, 2005). As a solution, multiple confocal pinholes can be implemented and projected into the specimen simultaneously. As the number of pinholes increases, the dwell time per pixel increases proportionally for the same frame rate, and irradiance can be reduced. While attempts have been made to construct and commercialize multipinhole laser microscopes (MPLMs) using technologies such as liquid crystal spatial light modulators (Smith *et al.*, 2000) or micromirror arrays (Fainman *et al.*, 2001; Verveer *et al.*, 1998), the Nipkow spinning disk has become the technology of choice (Conchello and Lichtman, 2005). The Nipkow disk is a spiral array of equally spaced pinholes and was first applied to microscopy by Petrán *et al.* (1968). An expanded laser beam illuminates the disk, while the tube lens together with microscope objective image the illuminated pinholes into the specimen plane. With each rotation of the disk, the entire field is scanned. The Nipkow disk suffers from poor excitation light efficiency, but has nonetheless been successfully commercialized using an arc lamp white light source (BD Biosciences, California) and has been used in a diverse range of studies including intracellular protein trafficking (Braunagel *et al.*, 2004), nerve myelination (Maria, 2003), and calcium conduction in cerebral endothelia cells (Marrelli *et al.*, 2003). The light efficiency of spinning disk confocal microscopes has been increased to that of LSCM by rotating a second disk containing microlenses coincident with each pinhole of the Nipkow disk. Currently, the CSU10 (confocal scanning unit 10) and the CSU22 from Yokogawa (Yokogawa, Japan) represent the state of the art and scan 20,000 microlenses and pinholes to achieve frame rates of up to 1000 frames per second (Nakano, 2002; Tanaami *et al.*, 2002). For most applications, images are integrated over longer durations and the real benefit of the spinning disk technology over LSCM is cell viability achieved because the necessary irradiance, per pinhole, decreases as the number of pinholes increase for a given frame rate. The Yokogawa spinning disk system is widely used in cell science and has been critical in studies of the cytoskeletal role during cell migration (Adams *et al.*, 2003; Grego *et al.*, 2001; Salmon *et al.*, 2002), the interplay of kinetochores and microtubules during mitosis (Tirnauer *et al.*, 2002), mitotic spindle formation (Ovechkina *et al.*, 2003; Fig. 1), and mitochondrial positioning in yeast (Yaffe *et al.*, 2003).

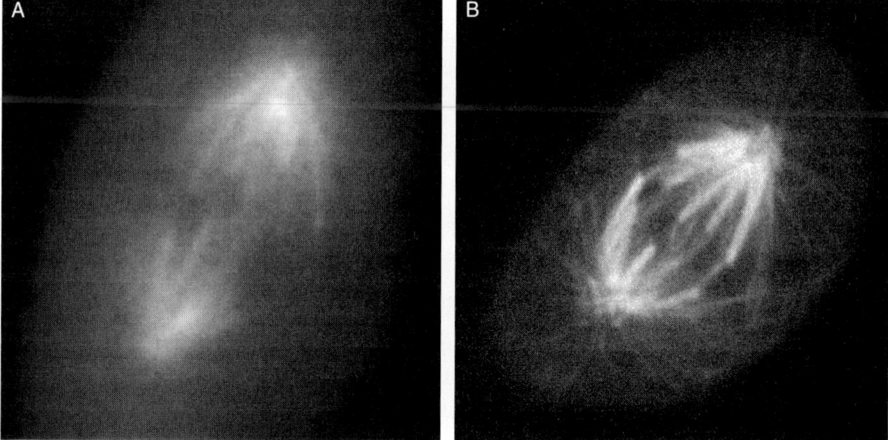

Fig. 1 Confocal microscopy improves spatial resolution within living cells. (A) Conventional epifluorescence image of stable PtK2 cells expressing EYFP-Tubulin. Blurring in the transverse and axial dimensions produces a low resolution and contrast representation of the mitotic spindle. Individual bundles are of poor contrast and single microtubules emanating from the poles are not distinguishable from the background. (B) Image of same cell line acquired by CSU10-based spinning disk confocal head mounted on a Nikon TE2000E2 microscope and illuminated by a mixed gas laser (RM-2018, Spectra-Physics). The improved resolution, particularly in the axial dimension, reveals individual bundles within the mitotic spindle as well as individual microtubules emanating from the spindle poles. Photos courtesy of Jagesh V. Shah and Paul S. Maddox.

C. Total Internal Reflection Microscopy

Total internal reflection microscopy (TIRFM) is a method for imaging cell adhesions, cytoskeletal elements, and membrane components at the glass–cell interface. TIRFM induces total internal reflection between the glass and cell culture media interface sending an evanescent wave a few hundred nanometers deep into the sample with enough irradiance to excite fluorescence. The theory of TIRFM is well described by Reichert and Truskey (1990) and by Gingell *et al.* (1987). The first reported TIRFM system for cells (Axelrod, 1981) implemented a cell culture chamber consisting of a glass coverslip sandwich created by inverting a coverslip containing adhered cells above a second coverslip spaced by a thin membrane. A fused quartz cube was placed on the top coverslip coupled by glycerol. The cube serves to further increase the angle (measured from the optical axis of the objective lens) of laser light from a source sufficiently juxtaposed to the optical axis so that the angle of light between the top coverslip and the cell culture media is greater than the critical angle for total internal reflection. Fluorescence excited by the evanescent wave is collected by the microscope objective and imaged onto a CCD. It is also common to perform TIRFM using only a high numerical aperture objective lens without the relatively complicated setup of the quartz cube-based system (Stout and Axelrod, 1989). Laser light is focused at the back focal plane of the objective, off axis, and at an angle of convergence large enough to illuminate a reasonably large field (Axelrod, 2001a). There are several other schemes for creating TIRFM, reviewed by Axelrod (2001b). Because TIRFM only excites a very thin section of fluorescence at the cell adhesion level, it has enabled several key observations regarding the extent and density of focal adhesions (Mathur *et al.*, 2003), guidance of microtubules to focal adhesions (Krylyshkina *et al.*, 2003), heterogeneity of microtubule-based transport in motile cells (Wadsworth, 1999; Yvon and Wadsworth, 2000), and dynamics of secretory granules (Oheim *et al.*, 1998; Rohrbach, 2000).

D. Nonlinear Imaging

The aforementioned imaging technologies rely on linear absorption of light by molecules to yield fluorescence. In this case, the probability of a fluorescent event is linearly dependent on the beam intensity incident on a fluorescent molecule. A number of imaging modalities are gaining acceptance in cell biology that rely on nonlinear absorption in which the probability of an event (e.g., fluorescence) is not simply proportional to laser intensity (Zipfel *et al.*, 2003). The emergence of nonlinear imaging techniques is in no small part due to the increased availability of turnkey pulsed laser light sources. One such technology is MPM that was first reported by Denk *et al.* (1990). Two-photon absorption was first postulated by Göppert-Meyer (1931) in the 1930s and was first demonstrated shortly after the invention of the laser using a relatively long pulsed laser (ruby laser, $\sim$1 ms pulse duration) to excite a $CaF_2:Eu^{2+}$ crystal (Kaiser and Garrett, 1961). The principle of two-photon absorption is that two photons of near equal energy can interact with a single molecule as if they were a single photon of twice the energy, if they arrive nearly simultaneously. MPM is an umbrella term that includes not only two-photon absorption, but higher modes of nonlinear absorption as well. In the field of biology, MPM has been demonstrated with continuous wave lasers (Booth and Hell, 1998; Kirsch *et al.*, 1998), but in general ultrashort pulse lasers (100 fs or less) in the near infrared (NIR) are used (Curley *et al.*, 1992). MPM provides the advantages of confocal optical sectioning, without the need of the detection pinhole. MPM excites with NIR light, which has the advantage over visible light that endogenous absorption by the cells/tissues and water is relatively low compared to the visible spectrum (Konig *et al.*, 1995; Liang *et al.*, 1996) allowing deeper penetration into tissue samples, multilayer cell cultures, and living tissue. MPM has been used to image pyramidal cells deep within the neocortex (Helmchen and Denk, 2005; Fig. 2), cancerous cell migration in intact tumors (Condeelis and Segall, 2003) and in tissue models (Wang *et al.*, 2002; Wolf *et al.*, 2003), angiogenesis in the developing brain of zebrafish embryos (Lawson and Weinstein, 2002), dynamic translation rates in living dendrites (Job and Eberwine, 2001), calcium dynamics in living tissues (Denk and Detwiler, 1999; Svoboda *et al.*, 1997), and cell fusions in living *Caenorhabditis elegans* embryos (Mohler *et al.*, 1998). Single-beam MPM systems suffer the same rate limitations as LSCM. Accordingly, several groups have developed multipoint MPM (MMM) to gain the speed and viability advantages found in spinning disk confocal microscopes (Egner *et al.*, 2002). MMM have been developed based on beam splitting (Nielsen *et al.*, 2001), a microlens array disk (Bewersdorf *et al.*, 1998; Straub and Hell, 1998), and Yokogawa type spinning disk systems (Fujita *et al.*, 1999).

Two nonlinear imaging modalities: second harmonic generation (SHG and third THG) and coherent anti-Stokes Raman scattering (CARS) are imaging methods that act on endogenous molecules, rather than through the introduction of fluorescent reporters. SHG microscopy, based on a 1978 publication first suggesting the imaging modality (Sheppard and Kompfner, 1978), was first applied to biological tissue by Freund and Deutsch (Freund and Deutsch, 1986; Freund *et al.*, 1986) who examined collagen structure in the rat tail tendon. SHG, or frequency doubling, is a nonlinear process requiring high irradiance illumination through a highly polarizable material with a noncentrosymmetric molecular organization (Campagnola and Loew, 2003). SHG emits light at half the wavelength of the excitation illumination so that two NIR photons from a femtosecond pulsed laser typically used in two-photon microscopy would create a single photon in the visible spectrum at twice the energy. Since SHG is dependent on the scale and the form of molecular ordering within the tissue (Freund and Deutsch, 1986; Zipfel *et al.*, 2003) and collagen is highly organized in the tendon, the method yields high contrast signals in that tissue. SHG has been shown to produce optically sectioned images of the plasma membrane (Campagnola *et al.*, 2002; Peleg *et al.*, 1999), tumor development in hamster cheek pouch mucosa (Guo *et al.*, 1999), collagen in RAFT tissue models (Zoumi *et al.*, 2002), and sarcomeres with the *C. elegans* nematode (Campagnola and Loew, 2003; Fig. 3). CARS is a type of multiphoton "vibrational imaging"

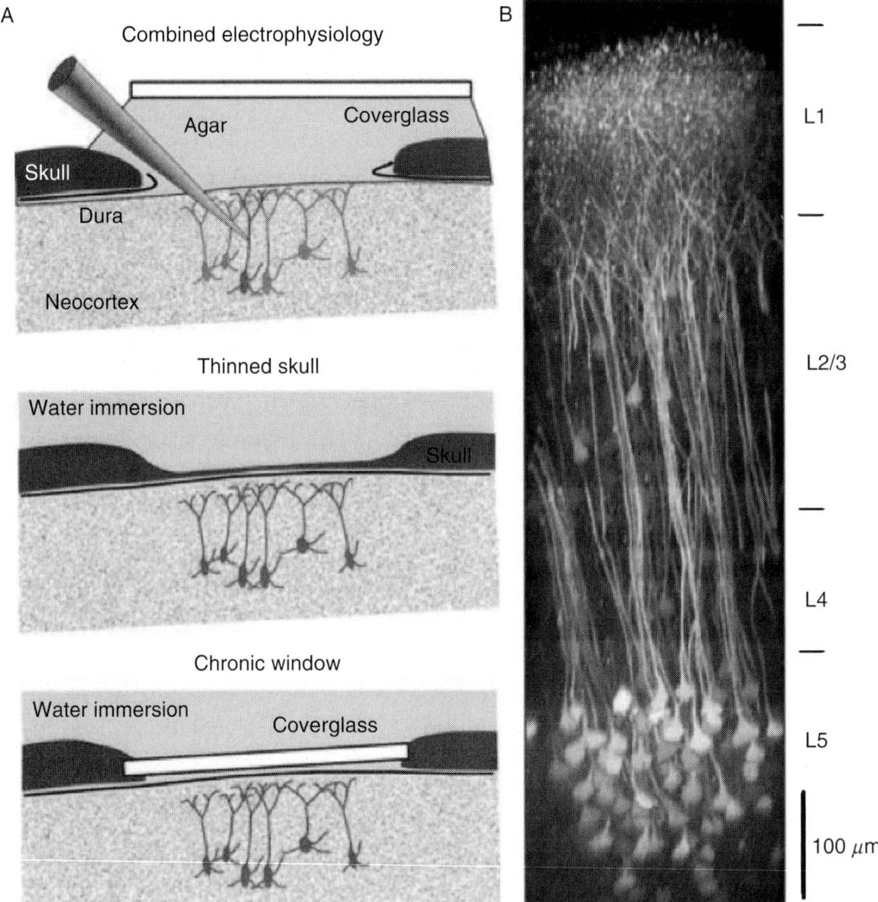

Fig. 2 *In vivo* two-photon imaging in the intact neocortex. (A) Different types of brain access. Open cranial window with the dura mater removed so that micropipettes for cell labeling and electrophysiological recordings can be inserted (top). Pulsation of the exposed brain is reduced by covering the craniotomy with agar and a coverglass. Thinned skull (20- to 4-μm thickness) preparation (middle). Cellular structures are either prelabeled (e.g., with fluorescent proteins in transgenic mice) or stained through a tiny hole lateral to the thinned area. Chronically implanted glass window replacing the skull (bottom). Agar is used underneath the window for stabilization. (B) Example of deep two-photon imaging in mouse neocortex. Maximum intensity side projection of a fluorescence image stack, obtained in a transgenic mouse expressing Clomeleon, a genetically encoded chloride indicator, under the control of the Thy1-promoter, preferentially in deep layer 5 (L5) pyramidal cells. Data were taken with a 10-W pumped Ti:Sapphire oscillator using a 40, NA 0.8 water immersion lens (Zeiss). Note that nearly the entire depth of the neocortex can be imaged. Reprinted by permission from Macmillan Publishers Ltd., Nature (Helmchen and Denk, 2005), copyright (2006).

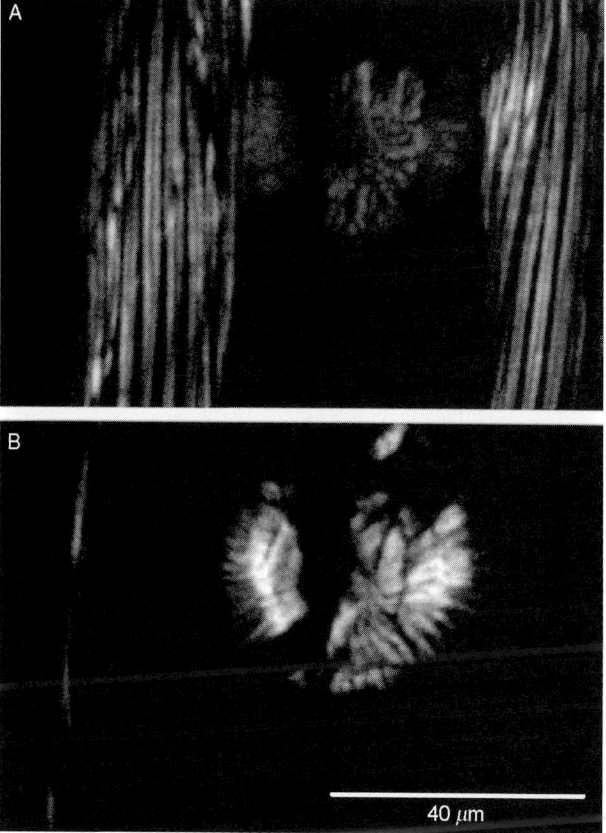

Fig. 3 SHG provides optical sectioning from endogenous molecules. Endogenous SHG imaging of a living adult *C. elegans* nematode, showing two distinct axial slices. (A) The sarcomeres in the body wall muscles are seen at the edges of the animal, as well as in a portion of the chewing mechanism. (B) An optical section further into the same animal, where only the chewing mechanism is observed with substantial SHG intensity. Reprinted by permission from Macmillan Publishers Ltd., Nature (Campagnola and Loew, 2003), copyright (2003).

based on coherent anti-Stokes Raman scattering in which three electric fields: the pump, Stokes, and probe interact with molecules in a sample to generate the anti-Stokes signal (Cheng and Xie, 2004; Duncan *et al.*, 1982; Tolles *et al.*, 1977). Molecular specificity is achieved from characteristic molecular vibrations of endogenous species. CARS is still an experimental technology and has not yet moved into mainstream imaging. It has been used to image lipids in live cells (Chen *et al.*, 2002), to determine the thermodynamic state (liquid crystalline or gel phase) of lipid membranes (Muller and Schins, 2002), to image neutral lipid droplets (LDs) in unstained live fibroblasts (Nan *et al.*, 2003), and to measure real-time intracellular water concentrations in live cells (Potma *et al.*, 2001). CARS has been used to image intact tissues including lipid domains in the mouse ear *in vivo* (Evans *et al.*, 2005), and lipid phase within axonal myelin in live spinal tissue explants (Wang *et al.*, 2005).

E. Super Resolution Microscopy

Lasers can be used to stretch the resolution of microscopy beyond the diffraction limit of light. For example, TIRFM improves axial resolution beyond the diffraction limit because the evanescent field extent is smaller than the axial extent of the point spread function (PSF) of a focused laser. Using acoustic optical modulators to vary the angle of laser incidence in TIRFM with subsequent Laplace transform-based reconstruction, topographical resolutions of tens of nanometers have been achieved (Loerke *et al.*, 2000).

In standing wave fluorescence microscopy two coherent plane waves, either from two apposing objective lenses, or from a single objective opposing a piezo-controlled mirror, interfere in the specimen plane to yield an axial resolution better than 0.05 μm (Bailey *et al.*, 1993). The 4Pi confocal fluorescence microscope presented by Hell and Stelzer (1992b) uses two apposing 1.4 NA objectives to optically compensate for axial spreading of the PSF to achieve 110 nm axial resolution (vs the diffraction limit of $\sim$1 μm) in Nile Blue dye. In type A 4Pi confocal microscopy, both objectives focus a laser beam to a common focal point where constructive interference occurs and the resulting illumination PSF takes on a more spherical geometry thereby reducing axial stretching. In type C 4Pi confocal microscopy, resolution is further enhanced by interference of detected light on a common point detector. The 4Pi design has also been shown to improve two-photon microscopy (Hell and Stelzer, 1992a).

III. Measurement of Cellular State: Molecular Parameters

A central feature in fluorescent imaging is the ability to determine the position and relative levels of a specific molecule. Lasers, as described above, act to provide better spatial resolution and multiplexed detection than conventional light sources due to their coherence and monochromaticity. The ability to provide a high flux of photons to a small diffraction-limited volume can also be used to modulate fluorescent molecules at a specific location or measure the properties of fluorescent molecules. Below we survey a number of biophysical measurements techniques used in living cells that have been enabled or improved by the use of lasers.

A. Photobleaching Methods

1. General Principle

In this family of methods, a protein (or other cellular component) is labeled with a fluorophore and permitted to equilibrate in the cell. Through the use of a laser, a region of these fluorescent molecules can be excited sufficiently so that they are no longer fluorescent, that is undergo photobleaching. These bleached molecules now represent a new species that can be distinguished from the existing fluorescent pool. The kinetics of these molecules can provide a quantitative insight into various physical processes such as diffusion, reaction, and flow. The experimental measurement described above is often referred to as fluorescence recovery (or redistribution) after photobleaching (FRAP).

2. Microscopy Configurations

Practical photobleaching is accomplished via high-intensity light source, such as a laser, that excites the fluorophore with a sufficient number of photons to make the probability of losing its fluorescence significantly large. Many LSCMs can be used as FRAP setups, by increasing the laser intensity in a small region for photobleaching and then lowering the power for imaging the subsequent recovery dynamics. Other setups also use a laser to achieve a diffraction-limited spot of high illumination that is used for single-point FRAP and imaging is then usually achieved by another light source such as an arc lamp. Scanning confocals can photobleach regions of arbitrary shape and size, via the use of the scanhead, permitting the bleaching of a subregion of a fluorescently labeled organelle such as the nucleus (Daigle *et al.*, 2001). Single-point FRAP configurations can accomplish spatial-patterning FRAP through the use of galvanometers that move the laser or lenses that can change the focused beam into lines or more complex shapes.

Some experimental setups utilize lower intensity sources (e.g., mercury or xenon arc lamps) and arise to bleach molecules but require an extended time period to achieve a detectable bleached pool. These longer exposure times compete with intrinsic diffusive mechanisms that permit mixing of the "bright" and "dark" species during the photobleaching period and thus the FRAP dynamics become a complex mix of photobleaching and recovery kinetics. To measure dynamic processes with fast timescales (on the order of tens of seconds to many minutes), a laser is essential to provide a high-intensity flux of photons that can rapidly bleach, and therefore mark, the fluorescent molecules of interest.

3. Measurement Data and Interpretation

In a typical FRAP experiment, the recovery at the point of photobleaching is monitored. This measurement is often made via quantitative imaging but can also be carried out using low-intensity laser excitation to measure local fluorescence. The basic FRAP recovery curve consists of the initial fluorescence level, the bleaching event which reduces the amount of fluorescence and the recovery to a new steady state (Fig. 4). The key parameters in the recovery curve are the rate and extent of recovery. These parameters are obtained by fitting the recovery curve to an exponential recovery equation [Eq. (1)]

$$\bar{F}(t) = A(1 - \mathrm{e}^{-1/\tau}) \tag{1}$$

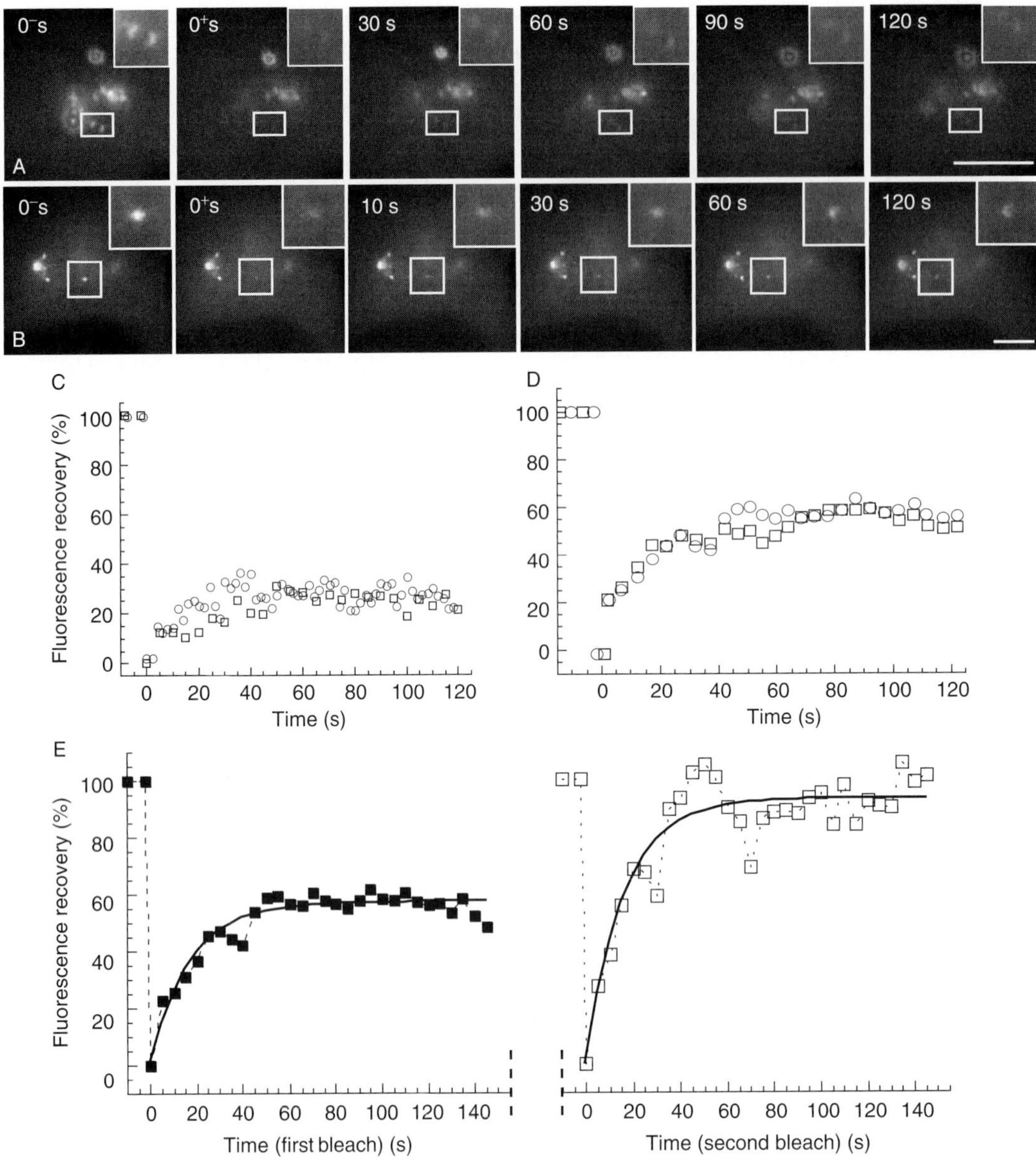

Fig. 4 Fluorescence recovery after photobleaching measurements in living cells. Photobleaching of a mitotic kinetochore protein Mad1 (A) and Mad2 (B) each fused to yellow fluorescent protein (EYFP) stably expressed in PtK2 cells was accomplished with the 532-nm line of an Nd:YAG laser. Quantification of fluorescence recovery at mitotic kinetochores for both Mad1 (C) and Mad2 (D and E) demonstrates that Mad1 recovers to ~25% its original fluorescence over the 5-min time period. However, Mad2 recovers to a level of ~50% (D) which upon serial photobleaching of the same kinetochore recovers to 100% of the prebleach value (E). These data reveal a relatively stable pool of Mad1 but a complex mixture of a stable and transient pool of Mad2 at the unattached kinetochore. Reprinted by permission from Elsevier (Shah *et al.*, 2004), copyright (2004).

where $\bar{F}(t)$ represents the normalized fluorescence of the region that incorporates the initial region fluorescence, fluctuations of the light source, and bleaching due to observation. Recovery curves often exhibit single exponential recovery kinetics with a single recovery time (τ) and extent (A). In more complex kinetics, recovery curve fitting requires modeling the existence of multiple components of recovery [Eq. (2)]

$$\bar{F}(t) = \sum_i A_i(1 - e^{-1/\tau_i}) \tag{2}$$

where each component is parameterized by its time of recovery (the exponential decay term, τ_i) and contribution to extent of recovery (A_i). The extent of recovery is a direct readout of the mobility of the fluorescent moiety in the region. A high extent of recovery indicates a transient interaction at some subcellular scaffold or complete diffusive recovery in a membrane or cytoplasmic compartment. Low extent of recovery indicates a significant immobile fraction ($1 - A$ or $\sum_i A$) that may turnover on timescales longer than that of the experimental measurement (Fig. 4). The time of recovery indicates the timescale of diffusion, reaction kinetics or flow, whichever is the dominant process at work. The interpretation of the recovery time depends greatly on the underlying biology under study. Two simple model systems are the *in vitro* fluid membrane (or *in vitro* solution) and the insoluble scaffold. Within simple *in vitro* soluble systems, a single exponential recovery rate can be readily converted into a diffusion constant, whereas at insoluble scaffolds, the recovery rate is directly related to the off-rate ($1/\tau$) of the reaction between the fluorescent marker and the scaffold (Axelrod *et al.*, 1976; Braeckmans *et al.*, 2003; Bulinski *et al.*, 2001; Lele and Ingber, 2006; Sprague *et al.*, 2004). These two extremes represent the simplest systems in which we can find direct interpretation of the recovery time. *In vivo* cellular measurements can be complicated by the complex structure of membranes and the cytoplasm. However, in combination with a model of the underlying biology, FRAP experiments can reveal kinetic phenomena otherwise inaccessible via fluorescence imaging alone and provide a quantitative measurement of these kinetics.

4. Cellular Applications

Early work using FRAP measured the diffusion of fluorescent reporters within the plasma membrane and laid the groundwork for studies of kinetics and diffusion in many other systems (Axelrod *et al.*, 1976; Jacobson *et al.*, 1976). Since that time, FRAP has been widely employed within many areas of cell biology. Work by Luby-Phelps and Taylor (1988) provided the first measurements of the heterogeneity of diffusion in the cellular cytoplasm using fluorescently labeled dextrans introduce by microinjection. Membranous organelles, such as the nuclear envelope (Daigle *et al.*, 2001), have been studied extensively to dissect the ability of proteins to freely diffuse within the membrane (Axelrod *et al.*, 1976; Jacobson *et al.*, 1976) or membranous compartments (Nehls *et al.*, 2000).

FRAP has also found tremendous utility in the study of nuclear and chromosome dynamics, in part due to the stable nature of some protein scaffolds found near or on chromatin such as chromosomes and kinetochores (Gerlich *et al.*, 2006; Howell *et al.*, 2000; Shah *et al.*, 2004), DNA damage machinery (Bekker-Jensen *et al.*, 2005) and nucleoli (Chen and Huang, 2001), and other more transiently bound nuclear components (Misteli *et al.*, 2000). Similarly, studies related to the cytoskeleton (Bulinski *et al.*, 2001; Pearson *et al.*, 2003) and cytoskeletal-associated structures (Khodjakov and Rieder, 1999; Kisurina-Evgenieva *et al.*, 2004; von Wichert *et al.*, 2003) use FRAP as a central methodology to investigate dynamics of protein association and turnover.

5. Related Methodologies

Modified experimental setups can also monitor the loss in fluorescence that occurs far from a position undergoing constant photobleaching (Delon *et al.*, 2006; Wachsmuth *et al.*, 2003). Through constant photobleaching of fluorescent molecules in one compartment and simultaneously monitoring the loss in fluorescence in another, the dynamic exchange between the compartments can be established. These fluorescence loss in photobleaching or FLIP experiments can provide insight into diffusion and transport that occur between different cellular compartments, for example cytoplasm and nucleus (Belaya *et al.*, 2006; Shimi *et al.*, 2004) or nuclear and subnuclear compartments (Chen and Huang, 2001).

Following the trajectory of bleached molecules by FRAP can also be accomplished by the use of photoactivation of fluorescence or photoconversion. FLAP or fluorescence localization after photoactivation (or photoconversion) follows the emergence of a new fluorescent species that is produced through the action of a laser or high-intensity light source. Fluorescence activation was originally performed by a high-intensity ultraviolet source that "uncaged" or cleaved a chemical moiety that prevented fluorescence (Mitchison, 1989). Such methods found widespread utilization in dynamical cell processes such as cell division (Mitchison, 1989) and cell motility (Theriot and Mitchison, 1991), although the "caged" fluorophores had to be introduced through cellular microinjection. The development of a photoactivatable green fluorescent protein mutant (PA-GFP) (Patterson and Lippincott-Schwartz, 2002) and the photoswitchable fluorescent protein Dronpa (Habuchi *et al.*, 2005) have enabled genetic tagging of proteins and their subsequent activation within living cells without microinjection. Such proteins have been used in dynamic cellular processes such as cell division (Salic *et al.*, 2004). In addition to the activation of fluorescence, a number of proteins have been developed that dramatically change their fluorescent spectra on exposure to ultraviolet wavelengths. Kaede (Ando *et al.*, 2002), EosFP (Wiedenmann *et al.*, 2004), and KikGR (Tsutsui *et al.*, 2005) exhibit dramatic red-shifted spectral shifts in emission providing a spectrally distinct fluorophore after photoconversion. A simple photoconvertible fluorescent protein tag has also been developed by utilizing the efficient energy transfer between concatenated cyan and yellow fluorescent proteins. Without perturbation, excitation of the cyan fluorescent protein produces yellow emission, but after bleaching of the yellow fluorescent protein undergoes a dramatic donor-dequenching producing primarily cyan emission (Shimozono *et al.*, 2006). These proteins represent the state-of-the-art in fluorescence dynamics reporters and should enjoy great success in cell biological applications where dynamic measurements are required.

B. Fluorescence Lifetime Imaging

1. General Principles

While FRAP-based techniques developed from imaging-based modalities, fluorescence spectroscopy techniques have proved to be a driving force for the innovation in microscopy-based biophysical measurements in living cells (Lakowicz, 1999). One such example is fluorescence lifetime imaging (FLIM). The lifetime of a fluorophore is a result of stochastic nature of spontaneous emission of photons from a population of excited fluorophores. The excited fluorophores, when excited nearly simultaneously, do not synchronously emit singlet transition photons, but instead have a characteristic average timescale of transition. Rapid time-resolved measurements after short-pulsed excitation of an ideal fluorophore reveal a single exponential decay in emission intensity. This decay constant is the lifetime of the fluorophore. The lifetime of a fluorophore can be exquisitely sensitive to its surroundings providing a unique measure of local state. Moreover, the measurements of anisotropy (another form of lifetime measurement) as a result of polarized excitation and emission can provide quantitative measurements of rotational diffusion. FLIM techniques are still actively being developed as is the analysis, but its utility in the realm of energy transfer [Fluorescent (or Förster) resonant energy transfer, FRET] measurements, has already provided significant momentum in its use in cell biology.

2. Microscopy Configurations

As with cuvette-based fluorescence spectroscopy measurements, laser-based microscopy provides two methods for the measurement of fluorescence lifetimes. Frequency-domain measurements drive fluorophores with a signal rapidly modulated intensity and detect the modulated response at the detector. Driving the fluorophores with a modulated intensity with varying periods near the fluorescence lifetime results in an emission signal shifted in phase and modulated in amplitude due to the lifetime decay (Harpur *et al.*, 2001). Frequency-domain methods can be implemented on LSCMs, but equivalently can also be done in wide-field microscopy by illuminating the entire microscope field with a modulated excitation. The ability to image by wide-field provides a rapid method of measuring lifetimes on a pixel-by-pixel basis, at a cost of requiring expensive cameras for sensitive detection. The frequency-domain methods, however, can require complex excitation protocols and calculations in the presence of multiple lifetimes (Bastiaens and Squire, 1999).

To accurately resolve multiple lifetimes, time-domain measurements use a pulsed laser to rapidly excite the fluorophores in a small region of the microscope field and detect the emission profile in time. The time-resolved measurements of the fluorescence decay can be evaluated at each scanned position of the laser and the fluorescence lifetime image is thus built pixel-by-pixel. Such time-resolved measurements can also be made on an LSCM equipped with high-time resolution detectors. Here the confocal laser source would not be a continuous wave laser, but a pulsed source of femtosecond to picosecond duration. Many multiphoton scanning microscopes already have much of the hardware for fluorescence lifetime imaging through the use of the pulsed infrared source and nonimaging detectors (Becker *et al.*, 2004).

3. Measurement Data and Interpretation

Frequency-domain measurements use the change in amplitude and phase that occurs as the input signal passes through the fluorescent samples. The calculations result in two lifetimes, one based on the change in amplitude and one based on the change in phase (outlined in French *et al.*, 1998; Lakowicz, 1999). For modulation input periods close to the single fluorescence lifetime in the sample, these values will be similar, but when multiple species are present or if single fluorescent species have multiple lifetimes, these values differ and computation of the lifetimes requires more involved calculations (Harpur *et al.*, 2001). Time-resolved measurements involve fitting the exponential decay captured after the pulsed excitation of the sample. This decay can be directly fit with single or multiple exponentials to obtain the fluorescence lifetimes present in the sample (Lakowicz, 1999).

4. Cellular Applications

Lifetime measurements have enjoyed great success in the area of energy transfer measurements in living cells. FRET is a phenomenon by which excitation of a donor fluorophore results in the emission of acceptor fluorophore through a nonradiative mechanism. The length scale for this transfer must be in the range of 2–7 nm and decays with the sixth power of the increasing distance (Bastiaens and Squire, 1999). This strong distance dependence provides exquisite sensitivity for detecting protein–protein (or other molecular) interactions within living systems. Intensity-based measurements of FRET can be problematic due to the relative concentrations of acceptor and donor fluorophores as well as cross-talk between fluorescent spectra (Jares-Erijman and Jovin, 2003), particularly when using fluorescent protein fusions (Pelet *et al.*, 2006). One interesting result of energy transfer is the reduction in fluorescence lifetime of the donor fluorophore when undergoing energy transfer. Using a variety of measurement modalities, FRET detection via fluorescence lifetimes have become increasingly popular in cell biology applications (Caudron *et al.*, 2005; Delbarre *et al.*, 2006; Kalab *et al.*, 2006; Peter *et al.*, 2005; Ramdya *et al.*, 2003; Fig. 5). Moreover, the possibility of multiphoton excitation provides for measurements made deep within tissues to detect protein–protein interactions within living organisms (Chen and Periasamy, 2004).

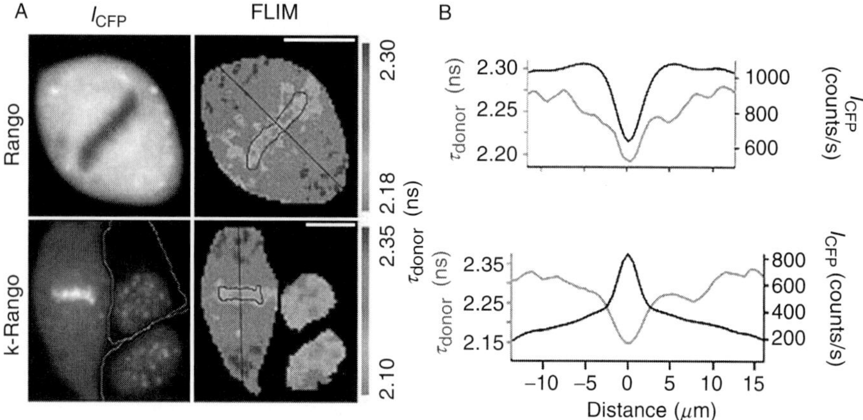

Fig. 5 Fluorescence lifetime/energy transfer imaging within living cells. (A) An FRET probe composed of an ECFP–EYFP fusion (Rango) separated by an Importin-β (protein)-binding domain reports on unbound probe near chromosomes in mitotic cells. The measurement of a control fusion protein (k-Rango) demonstrates a distinct localization but similar lifetime image indicating the specificity of the probe for reporting Importin-β binding. (B) Linescan measurements of fluorescence lifetime and donor fluorescence display similar lifetime profiles bur distinct donor intensities demonstrating the difference between the simple image intensity and the energy transfer reporting an underlying biochemical interaction. Reprinted by permission from Macmillan Publishers Ltd., Nature (Kalab *et al.*, 2006), copyright (2006). (See Plate 2 in the color insert section.)

Lifetime imaging has also been applied to measurements of protein mobility through the use of polarization/anisotropy methods (Clayton *et al.*, 2002; Dix and Verkman, 1990). These techniques provide measures of rotational mobility and are very sensitive to changes in protein size. Rotational correlation times of proteins are generally in the area of tens of nanoseconds. Unfortunately, GFP mutants with lifetimes in the range of ~1–3 ns (Tsien, 1998; Volkmer *et al.*, 2000) cannot provide the temporal resolution for such measurements. As a result, live cell anisotropy measurements for molecular mobility studies require nongenetically encoded fluorophores with long fluorescent lifetimes that need to be covalently coupled to the molecule of interest and directly introduced into the cellular environment. It should be noted, however, that GFP mutants do exhibit significant differences in fluorescence lifetimes, a property that can be exploited for imaging fluorescent proteins with large spectral overlap in emission (Harpur *et al.*, 2001).

C. Fluorescence Correlation Spectroscopy

1. General Principles

Fluorescence correlation spectroscopy (FCS) has become an increasingly popular tool in cell biology due to its ability to discern changes in biological complex size, concentration, and composition in living cells (Medina and Schwille, 2002).

FCS techniques use a laser to produce a small excitation volume (~0.1 to tens of femtoliters) from which the fluctuations in fluorescence are measured at high time resolution (approximately tens of nanoseconds). The origin of the fluctuations can be a result of photophysical processes, reaction kinetics, diffusion, and flow (Magde *et al.*, 1972). Photophysical processes and reaction kinetics have timescales shorter than the average residence time of the fluorophore within the excitation volume, whereas fluctuations due to diffusion and flow are a result of fluorescent molecules leaving and entering the excitation volume. The time average of these fluctuations can also provide the absolute concentrations of fluorescent molecules. The ability to measure such a wide variety of biophysical processes and make absolute measurements of concentration has fueled interest in the use of FCS-based techniques in cell biology.

2. Microscopy Configurations

FCS measurements can be made using a variety of microscope configurations. Basic requirements are a laser to produce the small excitation volume, a high-sensitivity or high time resolution detector [e.g., avalanche photodiode (APD) or PMT], and hardware to convert the measured detection signal into a digital form (Schwille *et al.*, 1999).

Confocal excitation with a continuous wave laser requires a pinhole at the detector to exclude out-of-focus excitation, similar to those already present in laser scanning confocals. Multiphoton excitation does not require the pinhole since only the volume of interest is excited making detection hardware simpler, at the expense of more complex and expensive ultrafast lasers (e.g., Ti:Sapphire femtosecond sources) (Berland *et al.*, 1995; Wang *et al.*, 2006).

Detection hardware can vary depending on the application. High-sensitivity applications, such as measuring low concentrations of fluorescent molecules (~10 pM), require the use of APDs, whereas less sensitive applications can use PMTs. High-time resolution are features of both detectors; however, many single-color FCS configurations use two detectors at half the intensity and cross-correlation, to reduce noise and after-pulsing that may be present in the detectors. This can greatly increase the effective time resolution of the instrument.

Correlation measurements to determine molecular photophysics or mobility can be accomplished online through fast hardware autocorrelators. Alternatively, photon-counting cards with high bandwidth can be used to record fluctuation time series directly and correlation can be done offline. Direct fluctuation recordings can be used in related fluctuation spectroscopy methods such as photon-counting histogram analysis (Chen *et al.*, 1999).

3. Measurement Data and Interpretation

A central feature of FCS measurements is the underlying model of fluorescence fluctuations. Within living cells dynamics of reaction kinetics are relatively slow, particularly when compared to average residence times within the subfemtoliter excitation volume. As a result, most fluctuations are due to the entry and exit of fluorescent species permitting the measurement of a diffusion constant (or equivalently the size of a complex) and concentration. The diffusion constant of the complex of interest can be determined both in the 3D space of the cytoplasm or 2D plane of the membrane. In both cases, the diffusion constant can be derived from fits to the autocorrelation function of the fluorescence fluctuations (Schwille *et al.*, 1999). The equations to fit the autocorrelation decay curve, a measure of the residence time of the fluorescent molecules also contain the average number of molecules contained within the volume. Thus, the fits to the decay curve, even for a single fluorescent species, require two free parameters (Fig. 6). Another parameter derived from the microscope configuration is the excitation volume. This volume can be calibrated by using molecules with a known diffusion constant and at known concentrations and is a sensitive parameter in the fitting of the autocorrelation function. The ability to evaluate absolute concentrations within the cellular environment can be challenging, but relative concentrations can be evaluated with high confidence to follow the dynamic changes in complex formation and dissociation

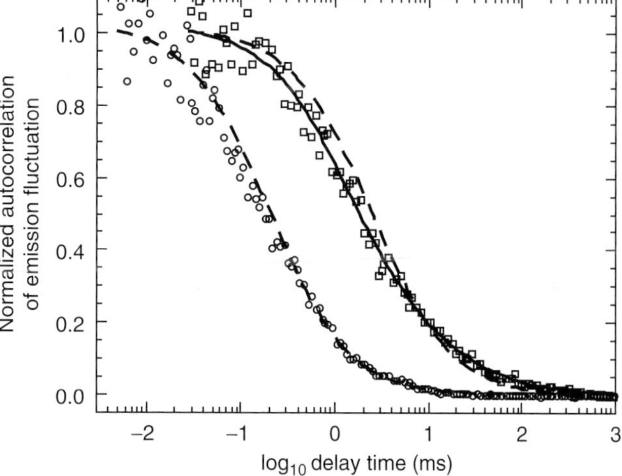

Fig. 6 Fluorescence correlation spectroscopy *in vitro* and within living cells. Fluorescence emission autocorrelation function resulting from Multiphoton excitation of recombinantly expressed cyan fluorescent protein (ECFP) fused to the mitotic checkpoint protein Mad2 (left curve and points) and within PtK2 cells stably expressing ECFP fused to Mad2 (right curves and points). Excitation was accomplished with the 800-nm line of a Ti:Sapphire femtosecond pulsed laser. The *in vitro* measurements demonstrate the fast residence times of *in vitro* proteins (diffusion constant of ~80 $\mu m^2/s$) versus the intracellular measurements resulting in a effective increase of about cytoplasmic viscosity by a factor of four (small complex ~21 $\mu m^2/s$, large complex ~1 $\mu m^2/s$, relative concentration ratio of 6:1, small to large). The intracellular measurements indicate that the Mad2 protein exists in two major complexes in mitotic cells (one component fit — — dashed line, two component fit — solid line) seen by the poor one component fit at long residence times. Only the small complex is present in nonmitotic cells, indicating a mitosis-specific assembly process.

(Wang *et al.*, 2006). In addition, a number of elements such as excitation power, shape of the excitation volume as well as the geometry and obstacles within the cell must all be considered in evaluating the parameters that result from the fit of the autocorrelation function (Berland *et al.*, 2003; Hess *et al.*, 2002; Nagy *et al.*, 2005). With these factors taken into account carefully, FCS can be a unique tool for measurements of cellular biochemistry.

4. Cellular Applications

FCS measurements in living cells are becoming increasingly popular, particularly with the introduction of commercial systems available as additions to scanning confocal microscopes. There has been a great deal of detailed measurement of GFP and its variants to understand the nature of the fluorophore both *in vitro* (Chen *et al.*, 2002) in the cellular environment (Schwille *et al.*, 1999; Wang *et al.*, 2004; Fig. 6). Measurements of intracellular protein dynamics have made important insights into retroviral assembly (Larson *et al.*, 2003), intracellular motility (Kohler *et al.*, 2000), cytoplasmic structure (Weiss *et al.*, 2003), and cell cycle control (Wang *et al.*, 2006). On-going developments in fluorescent proteins and fluorophores (Kogure *et al.*, 2006) and data analysis methods (Chen *et al.*, 1999, 2005; Muller, 2004) promise to provide new insights into cellular dynamics.

5. Related Methodologies

FCS methodologies are being actively developed and the list of related methodologies is changing rapidly (Breusegem *et al.*, 2006). One technique that is particularly well suited to cellular studies is fluorescence cross-correlation spectroscopy (FCCS). Here two spectrally distinct fluorescent species are monitored in the cytoplasm or within the cellular membrane. Autocorrelation analysis of each color separately provides the concentration and diffusion constant parameters previously described. However, cross-correlation between the channels provides concentration and diffusion parameters on complexes containing both fluorophores. Much like FRET, the result is the ability to monitor protein–protein interactions but without the requirement for close apposition of the fluorophores. FCCS-based protein–protein interactions are detected on the basis of the correlated motions into and out of the excitation volume (Bacia *et al.*, 2006). FCCS measurements provide the composition of the complexes, an element not available through single color FCS. Recent work has used FCCS to monitor receptor–ligand interactions (Larson *et al.*, 2005), calcium-signaling (Kim *et al.*, 2005), and *in vivo* protease activity (Kogure *et al.*, 2006).

Many FCS/FCCS instruments are configured for single-point measurements making spatial diffusion maps and spatial cross-correlation measurements impossible. Developments have incorporated position information into FCS measurements providing spatial and temporal correlation profiles (Digman *et al.*, 2005; Sisan *et al.*, 2006; Skinner *et al.*, 2005; Ries and Schwille, 2006). Cross-correlation in time and space will provide novel methods of analyzing cellular complex dynamics and reaction kinetics all within the living cell.

New developments in FCS and FCCS through novel fluorescent proteins that permit single color excitation dual-color emission FCCS or optical configurations for interleaved dual-color excitation (Thews *et al.*, 2005) as well as position scanning are moving the field toward simpler instruments to extract molecular dynamics and protein–protein interactions as they occur in living cells.

IV. Summary

We have presented an overview of the leading and emerging technologies in laser-based imaging and molecular state measurement. While several of these techniques had been proposed in the years preceding laser development, the widespread commercialization and distribution of continuous wave and pulsed laser systems has brought molecular characterization into the biology research laboratory. The transformation of laser systems from the intimidating research platform architecture into the "friendly," closed system "turnkey" architecture is a clear indication that lasers are becoming ubiquitous tools in the biological laboratory. The availability of these turnkey laser systems has enabled the cell biologist to bring a host of techniques that permit imaging and molecular measurements to the study of the biochemical basis of cellular function.

References

Adams, M. C., Salmon, W. C., Gupton, S. L., Cohan, C. S., Wittmann, T., Prigozhina, N., and Waterman-Storer, C. M. (2003). A high-speed multispectral spinning-disk confocal microscope system for fluorescent speckle microscopy of living cells. *Methods* **29**(1), 29–41.

Amara, A., Legall, S., Schwartz, O., Salamero, J., Montes, M., Loetscher, P., Baggiolini, M., Virelizier, J. L., and Arenzanaseisdedos, F. (1997). HIV coreceptor downregulation as antiviral principle: SDF-1 alpha-dependent internalization of the chemokine receptor CXCR4 contributes to inhibition of HIV replication. *J. Exp. Med.* **186**(1), 139–146.

Ando, R., Hama, H., Yamamoto-Hino, M., Mizuno, H., and Miyawaki, A. (2002). An optical marker based on the UV-induced green-to-red photoconversion of a fluorescent protein. *Proc. Natl. Acad. Sci. USA* **99**(20), 12651–12656.

Aragane, Y., Kulms, D., Metze, D., Wilkes, G., Poppelmann, B., Luger, T. A., and Schwarz, T. (1998). Ultraviolet light induces apoptosis via direct activation of CD95 (Fas/APO-1) independently of its ligand CD95L. *J. Cell Biol.* **140**(1), 171–182.

Axelrod, D. (1981). Cell-substrate contacts illuminated by total internal reflection flourescence. *J. Cell. Biol.* **89,** 141–145.

Axelrod, D. (2001a). Selective imaging of surface fluorescence with very high aperture microscope objectives. *J. Biomed. Opt.* **6**(1), 6–13.

Axelrod, D. (2001b). Total internal reflection fluorescence microscopy in cell biology. *Traffic* **2**(11), 764–774.

Axelrod, D., Koppel, D. E., Schlessinger, J., Elson, E., and Webb, W. W. (1976). Mobility measurement by analysis of fluorescence photobleaching recovery kinetics. *Biophys. J.* **16**(9), 1055–1069.

Bacia, K., Kim, S. A., and Schwille, P. (2006). Fluorescence cross-correlation spectroscopy in living cells. *Nat. Methods* **3**(2), 83–89.

Bae, Y. S., Kang, S. W., Seo, M. S., Baines, I. C., Tekle, E., Chock, P. B., and Rhee, S. G. (1997). Epidermal growth factor (EGF)-induced generation of hydrogen peroxide—role in EGF receptor-mediated tyrosine phosphorylation. *J. Biol. Chem.* **272**(1), 217–221.

Bailey, B., Farkas, D. L., Taylor, D. L., and Lanni, F. (1993). Enhancement of axial resolution in fluorescence microscopy by standing-wave excitation. *Nature* **366**(6450), 44–48.

Barger, S. W., Horster, D., Furukawa, K., Goodman, Y., Krieglstein, J., and Mattson, M. P. (1995). Tumor-necrosis-factor-alpha and tumor-necrosis-factor-beta protect neurons against amyloid beta-peptide toxicity—evidence for involvement of a kappa-B-binding factor and attenuation of peroxide and Ca2+ accumulation. *Proc. Natl. Acad. Sci. USA* **92**(20), 9328–9332.

Bastiaens, P. I., and Squire, A. (1999). Fluorescence lifetime imaging microscopy: Spatial resolution of biochemical processes in the cell. *Trends Cell Biol.* **9**(2), 48–52.

Becker, W., Bergmann, A., Hink, M. A., Konig, K., Benndorf, K., and Biskup, C. (2004). Fluorescence lifetime imaging by time-correlated single-photon counting. *Microsc. Res. Tech.* **63**(1), 58–66.

Bekker-Jensen, S., Lukas, C., Melander, F., Bartek, J., and Lukas, J. (2005). Dynamic assembly and sustained retention of 53BP1 at the sites of DNA damage are controlled by Mdc1/NFBD1. *J. Cell Biol.* **170**(2), 201–211.

Belaya, K., Tollervey, D., and Kos, M. (2006). FLIPing heterokaryons to analyze nucleo-cytoplasmic shuttling of yeast proteins. *RNA* **12**(5), 921–930.

Berland, K., and Shen, G. (2003). Excitation saturation in two-photon fluorescence correlation spectroscopy. *Appl. Opt.* **42**(27), 5566–5576.

Berland, K. M., So, P. T., and Gratton, E. (1995). Two-photon fluorescence correlation spectroscopy: Method and application to the intracellular environment. *Biophys. J.* **68**(2), 694–701.

Bewersdorf, J., Pick, R., and Hell, S. W. (1998). Multifocal multiphoton microscopy. *Opt. Lett.* **23**(9), 655–657.

Booth, M. J., and Hell, S. W. (1998). Continuous wave excitation two-photon fluorescence microscopy exemplified with the 647-nm ArKr laser line. *J. Microsc.* **190,** 298–304.

Braeckmans, K., Peeters, L., Sanders, N. N., De Smedt, S. C., and Demeester, J. (2003). Three-dimensional fluorescence recovery after photobleaching with the confocal scanning laser microscope. *Biophys. J.* **85**(4), 2240–2252.

Braunagel, S. C., Williamson, S. T., Saksena, S., Zhong, Z., Russell, W. K., Russell, D. H., and Summers, M. D. (2004). Trafficking of ODV-E66 is mediated via a sorting motif and other viral proteins: Facilitated trafficking to the inner nuclear membrane. *Proc. Natl. Acad. Sci. USA* **101,** 8372–8377.

Breusegem, S. Y., Levi, M., and Barry, N. P. (2006). Fluorescence correlation spectroscopy and fluorescence lifetime imaging microscopy. *Nephron Exp. Nephrol.* **103**(2), e41–e49.

Buchler, M., Konig, J., Brom, R., Kartenbeck, J., Spring, H., Horie, T., and Keppler, D. (1996). cDNA cloning of the hepatocyte canalicular isoform of the multidrug resistance protein, cMrp, reveals a novel conjugate export pump deficient in hyperbilirubinemic mutant rats. *J. Biol. Chem.* **271**(25), 15091–15098.

Bulinski, J. C., Odde, D. J., Howell, B. J., Salmon, T. D., and Waterman-Storer, C. M. (2001). Rapid dynamics of the microtubule binding of ensconsin *in vivo*. *J. Cell. Sci.* **114**(Pt. 21), 3885–3897.

Campagnola, P. J., and Loew, L. M. (2003). Second-harmonic imaging microscopy for visualizing biomolecular arrays in cells, tissues and organisms. *Nature Biotech.* **21**(11), 1356–1360.

Campagnola, P. J., Millard, A. C., Terasaki, M., Hoppe, P. E., Malone, C. J., and Mohler, W. A. (2002). Three-dimensional high-resolution second-harmonic generation imaging of endogenous structural proteins in biological tissues. *Biophys. J.* **82,** 493–508.

Carlsson, K., and Aslund, N. (1987). Confocal imaging for 3-D digital microscopy. *Appl. Opt.* **26**(16), 3232–3238.

Carlsson, K., Danielsson, P. E., Lenz, R., Liljeborg, A., Majlof, L., and Aslund, N. (1985). Three-dimensional microscopy using a confocal laser scanning microscope. *Opt. Lett.* **10**(2), 53–55.

Caudron, M., Bunt, G., Bastiaens, P., and Karsenti, E. (2005). Spatial coordination of spindle assembly by chromosome-mediated signaling gradients. *Science* **309**(5739), 1373–1376.

Chen, D., and Huang, S. (2001). Nucleolar components involved in ribosome biogenesis cycle between the nucleolus and nucleoplasm in interphase cells. *J. Cell Biol.* **153**(1), 169–176.

Chen, J. X., Volkmer, A., Book, L. D., and Xie, X. S. (2002). Multiplex coherent anti-stokes Raman scattering microspectroscopy and study of lipid vesicles. *J. Phys. Chem. B* **106**(34), 8493–8498.

Chen, Y., Muller, J. D., Ruan, Q., and Gratton, E. (2002). Molecular brightness characterization of EGFP *in vivo* by fluorescence fluctuation spectroscopy. *Biophys. J.* **82**(1, Pt. 1), 133–144.

Chen, Y., Muller, J. D., So, P. T., and Gratton, E. (1999). The photon counting histogram in fluorescence fluctuation spectroscopy. *Biophys. J.* **77**(1), 553–567.

Chen, Y., and Periasamy, A. (2004). Characterization of two-photon excitation fluorescence lifetime imaging microscopy for protein localization. *Microsc. Res. Tech.* **63**(1), 72–80.

Chen, Y., Tekmen, M., Hillesheim, L., Skinner, J., Wu, B., and Muller, J. D. (2005). Dual-color photon-counting histogram. *Biophys. J.* **88**(3), 2177–2192.

Cheng, H., Lederer, W. J., and Cannell, M. B. (1993). Calcium sparks—elementary events underlying excitation-contraction coupling in heart-muscle. *Science* **262**(5134), 740–744.

Cheng, J. X., and Xie, X. S. (2004). Coherent anti-Stokes Raman scattering microscopy: Instrumentation, theory, and applications. *J. Phys. Chem. B* **108,** 827–840.

Clayton, A. H., Hanley, Q. S., Arndt-Jovin, D. J., Subramaniam, V., and Jovin, T. M. (2002). Dynamic fluorescence anisotropy imaging microscopy in the frequency domain (rFLIM). *Biophys. J.* **83**(3), 1631–1649.

Conchello, J.-A., and Lichtman, J. W. (2005). Optical sectioning microscopy. *Nat. Methods* **2**(12), 920–931.

Condeelis, J., and Segall, J. E. (2003). Intravital imaging of cell movement in tumours. *Nature Rev. Cancer* **3**(12), 921–930.

Curley, P. F., Ferguson, A. I., White, J. G., and Amos, W. B. (1992). Application of a femtosecond self-sustaining mode-locked Ti:sapphire laser to the field of laser scanning confocal microscopy. *Opt. Quantum Electron.* **24**(8), 851–859.

Daigle, N., Beaudouin, J., Hartnell, L., Imreh, G., Hallberg, E., Lippincott-Schwartz, J., and Ellenberg, J. (2001). Nuclear pore complexes form immobile networks and have a very low turnover in live mammalian cells. *J. Cell Biol.* **154**(1), 71–84.

Delbarre, E., Tramier, M., Coppey-Moisan, M., Gaillard, C., Courvalin, J. C., and Buendia, B. (2006). The truncated prelamin A in Hutchinson-Gilford progeria syndrome alters segregation of A-type and B-type lamin homopolymers. *Hum. Mol. Genet.* **15**(7), 1113–1122.

Delon, A., Usson, Y., Derouard, J., Biben, T., and Souchier, C. (2006). Continuous photobleaching in vesicles and living cells: A measure of diffusion and compartmentation. *Biophys. J.* **90**(7), 2548–2562.

Denk, W., and Detwiler, P. B. (1999). Optical recording of light-evoked calcium signals in the functionally intact retina. *Proc. Natl. Acad. Sci. USA* **96,** 7035–7040.

Denk, W., Strickler, J. H., and Webb, W. W. (1990). Two-photon laser scanning fluorescence microscopy. *Science* **248,** 73–76.

Digman, M. A., Brown, C. M., Sengupta, P., Wiseman, P. W., Horwitz, A. R., and Gratton, E. (2005). Measuring fast dynamics in solutions and cells with a laser scanning microscope. *Biophys. J.* **89**(2), 1317–1327.

Dix, J. A., and Verkman, A. S. (1990). Mapping of fluorescence anisotropy in living cells by ratio imaging. Application to cytoplasmic viscosity. *Biophys. J.* **57**(2), 231–240.

Duncan, M. D., Reintjes, J., and Manuccia, T. J. (1982). Scanning coherent anti-Stokes Raman microscope. *Opt. Lett.* **7,** 350–352.

Egner, A., Andresen, V., and Hell, S. (2002). Comparison of the axial resolution of practical Nipkow-disk confocal fluorescence microscopy with that of multifocal multiphoton microscopy: Theory and experiment. *J. Microsc.* **206**(1), 24–32.

Evans, C. L., Potma, E. O., Puoris'haag, M., Cote, D., Lin, C. P., and Xie, X. S. (2005). Chemical imaging of tissue *in vivo* with video-rate coherent anti-Stokes Raman scattering microscopy. *Proc. Natl. Acad. Sci. USA* **102**(46), 16807–16812.

Fainman, Y., Botvinick, E. L., Price, J. H., and Gough, D. A. (2001). 3D quantitative imaging of the microvasculature with the Texas Instruments digital micromirror device. *In* "Spatial Light Modulators: Technology and Applications." San Deigo, CA.

French, T., So, P. T., Dong, C. Y., Berland, K. M., and Gratton, E. (1998). Fluorescence lifetime imaging techniques for microscopy. *Methods Cell Biol.* **56,** 277–304.

Freund, I., and Deutsch, M. (1986). Second-harmonic microscopy of biological tissue. *Opt. Lett.* **11**(2), 94–96.

Freund, I., Deutsch, M., and Sprecher, A. (1986). Connective tissue polarity. Optical second-harmonic microscopy, crossed-beam summation, and small-angle scattering in rat-tail tendon. *Biophys. J.* **50,** 693–712.

Fujita, K., Nakamura, O., Kaneko, T., Kawata, S., Oyamada, M., and Takamatsu, T. (1999). Real-time imaging of two-photon-induced fluorescence with a microlens-array scanner and a regenerative amplifier. *J. Microsc.* **194**(2/3), 528–531.

Gerlich, D., Koch, B., Dupeux, F., Peters, J. M., and Ellenberg, J. (2006). Live-cell imaging reveals a stable cohesin-chromatin interaction after but not before DNA replication. *Curr. Biol.* **16**(15), 1571–1578.

Gingell, D., Heavens, O. S., and Mellor, J. S. (1987). General electromagnetic theory of total internal reflection fluorescence: The quantitative basis for mapping cell-substratum topography. *J. Cell Sci.* **87** (Pt. 5), 677–693.

Goodman, Y. D., Bruce, A. J., Cheng, B., and Mattson, M. P. (1996). Estrogens attenuate and corticosterone exacerbates excitotoxicity, oxidative injury, and amyloid beta-peptide toxicity in hippocampal neurons. *J. Neurochem.* **66**(5), 1836–1844.

Göppert-Meyer, M. (1931). Über Elementarakte mit zwei Quantensprüngen. Göttinger Dissertation. *Ann. Phys.* **9,** 273–294.

Gräf, R., Rietdorf, J., and Zimmermann, T. (2005). "Live Cell Spinning Disk Microscopy." Springer Berlin, Heidelburg.

Grego, S., Cantillana, V., and Salmon, E. D. (2001). Microtubule treadmilling *in vitro* investigated by fluorescence speckle and confocal microscopy. *Biophys. J.* **81**(1), 66–78.

Guo, Y., Savage, H. E., Liu, F., Schantz, S. P., Ho, P. P., and Alfano, R. R. (1999). Subsurface tumor progression investigated by noninvasive optical second harmonic tomography. *Proc. Natl. Acad. Sci. USA* **96,** 10854–10856.

Habuchi, S., Ando, R., Dedecker, P., Verheijen, W., Mizuno, H., Miyawaki, A., and Hofkens, J. (2005). Reversible single-molecule photoswitching in the GFP-like fluorescent protein Dronpa. *Proc. Natl. Acad. Sci. USA* **102**(27), 9511–9516.

Hakansson, M. L., Brown, H., Ghilardi, N., Skoda, R. C., and Meister, B. (1998). Leptin receptor immunoreactivity in chemically defined target neurons of the hypothalamus. *J. Neurosci.* **18**(1), 559–572.

Harpur, A. G., Wouters, F. S., and Bastiaens, P. I. (2001). Imaging FRET between spectrally similar GFP molecules in single cells. *Nat. Biotechnol.* **19**(2), 167–169.

Helmchen, F., and Denk, W. (2005). Deep tissue two-photon microscopy. *Nat. Methods* **2**(12), 932–940.

Hell, S., and Stelzer, E. H. K. (1992a). Fundamental improvement of resolution with a 4Pi-confocal fluorescence microscope using two-photon excitation. *Opt. Commun.* **93,** 277–282.

Hell, S., and Stelzer, E. H. K. (1992b). Properties of a 4Pi confocal fluorescence microscope. *J. Opt. Soc. Am. A* (*Opt. Image Sci.*) **9**(12), 2159–2166.

Hess, S. T., and Webb, W. W. (2002). Focal volume optics and experimental artifacts in confocal fluorescence correlation spectroscopy. *Biophys. J.* **83**(4), 2300–2317.

Howell, B. J., Hoffman, D. B., Fang, G., Murray, A. W., and Salmon, E. D. (2000). Visualization of Mad2 dynamics at kinetochores, along spindle fibers, and at spindle poles in living cells. *J. Cell Biol.* **150**(6), 1233–1250.

Inoué, S. (1995). Foundations of confocal scanned imaging in light microscopy. *In* "Handbook of Biological Confocal Microscopy" (J. Pawley, ed.), pp. 1–17. Plenum Press, New York.

Jacobson, K., Derzko, Z., Wu, E. S., Hou, Y., and Poste, G. (1976). Measurement of the lateral mobility of cell surface components in single, living cells by fluorescence recovery after photobleaching. *J. Supramol. Struct.* **5**(4), 565(417)-576(428).

Jares-Erijman, E. A., and Jovin, T. M. (2003). FRET imaging. *Nat. Biotechnol.* **21**(11), 1387–1395.

Job, C., and Eberwine, J. (2001). Identification of sites for exponential translation in living dendrites. *Proc. Natl. Acad. Sci. USA* **98**(23), 13037–13042.

Kaiser, W., and Garrett, C. G. B. (1961). Two-photon excitation in CaF_2: Eu^{2+}. *Phys. Rev. Lett.* **7**(6), 229.

Kalab, P., Pralle, A., Isacoff, E. Y., Heald, R., and Weis, K. (2006). Analysis of a RanGTP-regulated gradient in mitotic somatic cells. *Nature* **440**(7084), 697–701.

Khodjakov, A., and Rieder, C. L. (1999). The sudden recruitment of gamma-tubulin to the centrosome at the onset of mitosis and its dynamic exchange throughout the cell cycle, do not require microtubules. *J. Cell Biol.* **146**(3), 585–596.

Kim, S. A., Heinze, K. G., Bacia, K., Waxham, M. N., and Schwille, P. (2005). Two-photon cross-correlation analysis of intracellular reactions with variable stoichiometry. *Biophys. J.* **88**(6), 4319–4336.

Kirsch, A. K., Subramaniam, V., Striker, G., Schnetter, C., Arndt-Jovin, D. J., and Jovin, T. M. (1998). Continuous wave two-photon scanning near-field optical microscopy. *Biophys. J.* **75,** 1513–1521.

Kisurina-Evgenieva, O., Mack, G., Du, Q., Macara, I., Khodjakov, A., and Compton, D. A. (2004). Multiple mechanisms regulate NuMA dynamics at spindle poles. *J. Cell Sci.* **117**(Pt. 26), 6391–6400.

Kogure, T., Karasawa, S., Araki, T., Saito, K., Kinjo, M., and Miyawaki, A. (2006). A fluorescent variant of a protein from the stony coral Montipora facilitates dual-color single-laser fluorescence cross-correlation spectroscopy. *Nat. Biotechnol.* **24**(5), 577–581.

Kohler, R. H., Schwille, P., Webb, W. W., and Hanson, M. R. (2000). Active protein transport through plastid tubules: Velocity quantified by fluorescence correlation spectroscopy. *J. Cell Sci.* **113**(Pt. 22), 3921–3930.

Konig, K., Liang, H., Berns, M. W., and Tromberg, B. J. (1995). Cell damage by near-IR microbeams. *Nature* **377**(6544), 20–21.

Krylyshkina, O., Anderson, K. I., Kaverina, I., Upmann, I., Manstein, D. J., Small, J. V., and Toomre, D. K. (2003). Nanometer targeting of microtubules to focal adhesions. *J. Cell Biol.* **161**(5), 853–859.

Lakowicz, J. R. (1999). "Principles of Fluorescence Spectroscopy." Plenum Press, New York.

Larson, D. R., Gosse, J. A., Holowka, D. A., Baird, B. A., and Webb, W. W. (2005). Temporally resolved interactions between antigen-stimulated IgE receptors and Lyn kinase on living cells. *J. Cell Biol.* **171**(3), 527–536.

Larson, D. R., Ma, Y. M., Vogt, V. M., and Webb, W. W. (2003). Direct measurement of Gag-Gag interaction during retrovirus assembly with FRET and fluorescence correlation spectroscopy. *J. Cell Biol.* **162**(7), 1233–1244.

Lawson, N. D., and Weinstein, B. M. (2002). *In vivo* imaging of embryonic vascular development using transgenic zebrafish. *Develop. Biol.* **248**(2), 307–318.

Lele, T. P., and Ingber, D. E. (2006). A mathematical model to determine molecular kinetic rate constants under non-steady state conditions using fluorescence recovery after photobleaching (FRAP). *Biophys. Chem.* **120**(1), 32–35.

Liang, H., Vu, K. T., Krishnan, P., Trang, T. C., Shin, D., Kimel, S., and Berns, M. W. (1996). Wavelength dependence of cell cloning efficiency after optical trapping. *Biophys. J.* **70**(3), 1529–1533.

Lin, Y. Z., Yao, S. Y., Veach, R. A., Torgerson, T. R., and Hawiger, J. (1995). Inhibition of nuclear translocation of transcription factor Nf-kappa-B by a synthetic peptide-containing a cell membrane-permeable motif and nuclear-localization sequence. *J. Biol. Chem.* **270**(24), 14255–14258.

Loerke, D., Preitz, B., Stuhmer, W., and Oheim, M. (2000). Super-resolution measurements with evanescent-wave fluorescence excitation using variable beam incidence. *J. Biomed Opt.* **5**(1), 23–30.

Luby-Phelps, K., and Taylor, D. L. (1988). Subcellular compartmentalization by local differentiation of cytoplasmic structure. *Cell Motil. Cytoskeleton* **10**(1–2), 28–37.

Magde, D., Elson, E., and Webb, W. W. (1972). Thermodynamic fluctuations in a reacting system—measurement by fluorescence correlation spectroscopy. *Phys. Rev. Lett.* **29**(11), 705.

Maria, N., Provitera, V., Crisci, C., Stancancelli, A., Wendelschafer-Crabb, G., Kennedy, W. R., and Santoro, L. (2003). Quantification of myelinated endings and mechanoreceptors in human digital skin. *Ann. Neurol.* **54,** 197–205.

Marrelli, S. P., Eckmann, M. S., and Hunte, M. S. (2003). Role of endothelial intermediate conductance KCa channels in cerebral EDHF-mediated dilations. *Am. J. Physiol. Heart Circ. Physiol.* **285,** H1590–H1599.

Mathur, A. B., Chan, B. P., Truskey, G. A., and Reichert, W. M. (2003). High-affinity augmentation of endothelial cell attachment: Long-term effects on focal contact and actin filament formation. *J. Biomed. Mater. Res. Part A* **66A**(4), 729–737.

Medina, M. A., and Schwille, P. (2002). Fluorescence correlation spectroscopy for the detection and study of single molecules in biology. *Bioessays* **24**(8), 758–764.

Misteli, T., Gunjan, A., Hock, R., Bustin, M., and Brown, D. T. (2000). Dynamic binding of histone H1 to chromatin in living cells. *Nature* **408**(6814), 877–881.

Mitchison, T. J. (1989). Polewards microtubule flux in the mitotic spindle: Evidence from photoactivation of fluorescence. *J. Cell Biol.* **109**(2), 637–652.

Mohler, W. A., Simske, J. S., Williams-Masson, E. M., Hardin, J. D., and White, J. G. (1998). Dynamics and ultrastructure of developmental cell fusions in the *Caenorhabditis elegans* hypodermis. *Curr. Biol.* **8**(19), 1087–1090.

Muller, J. D. (2004). Cumulant analysis in fluorescence fluctuation spectroscopy. *Biophys. J.* **86**(6), 3981–3992.

Muller, M., and Schins, J. M. (2002). Imaging the thermodynamic state of lipid membranes with multiplex CARS microscopy. *J. Phys. Chem. B* **106**(14), 3715–3723.

Nagy, A., Wu, J., and Berland, K. M. (2005). Characterizing observation volumes and the role of excitation saturation in one-photon fluorescence fluctuation spectroscopy. *J. Biomed. Opt.* **10**(4), 44015.

Nakano, A. (2002). Spinning-disk confocal microscopy—a cutting-edge tool for imaging of membrane traffic. *Cell Struct. Funct.* **27**(5), 349–355.

Nan, X. L., Cheng, J. X., and Xie, X. S. (2003). Vibrational imaging of lipid droplets in live fibroblast cells with coherent anti-Stokes Raman scattering microscopy. *J. Lipid Res.* **44**(11), 2202–2208.

Nehls, S., Snapp, E. L., Cole, N. B., Zaal, K. J., Kenworthy, A. K., Roberts, T. H., Ellenberg, J., Presley, J. F., Siggia, E., and Lippincott-Schwartz, J. (2000). Dynamics and retention of misfolded proteins in native ER membranes. *Nat. Cell Biol.* **2**(5), 288–295.

Nielsen, T., Fricke, M., Hellweg, D., and Andresen, P. (2001). High efficiency beam splitter for multifocal multiphoton microscopy. *J. Microsc.* **201**(3), 368–376.

Nitatori, T., Sato, N., Waguri, S., Karasawa, Y., Araki, H., Shibanai, K., Kominami, E., and Uchiyama, Y. (1995). Delayed neuronal death in the Ca1 pyramidal cell layer of the Gerbil Hippocampus following transient ischemia is apoptosis. *J. Neurosci.* **15**(2), 1001–1011.

Oheim, M., Loerke, D., Stuhmer, W., and Chow, R. H. (1998). The last few milliseconds in the life of a secretory granule—docking, dynamics and fusion visualized by total internal reflection fluorescence microscopy (TIRFM). *Eur. Biophys. J. Biophys. Lett.* **27**(2), 83–98.

Ovechkina, Y., Maddox, P., Oakley, C. E., Xiang, X., Osmani, S. A., Salmon, E. D., and Oakley, B. R. (2003). Spindle formation in Aspergillus is coupled to tubulin movement into the nucleus. *Mol. Biol. Cell.* **14**(5), 2192–2200.

Patterson, G. H., and Lippincott-Schwartz, J. (2002). A photoactivatable GFP for selective photolabeling of proteins and cells. *Science* **297**(5588), 1873–1877.

Pearson, C. G., Maddox, P. S., Zarzar, T. R., Salmon, E. D., and Bloom, K. (2003). Yeast kinetochores do not stabilize Stu2p-dependent spindle microtubule dynamics. *Mol. Biol. Cell.* **14**(10), 4181–4195.

Peleg, G., Lewis, A., Linial, M., and Loew, L. M. (1999). Nonlinear optical measurement of membrane potential around single molecules at selected cellular sites. *Proc. Natl. Acad. Sci. USA* **96,** 6700–6704.

Pelet, S., Previte, M. J., and So, P. T. (2006). Comparing the quantification of Forster resonance energy transfer measurement accuracies based on intensity, spectral, and lifetime imaging. *J. Biomed. Opt.* **11**(3), 34017-1–34017-11.

Peter, M., Ameer-Beg, S. M., Hughes, M. K., Keppler, M. D., Prag, S., Marsh, M., Vojnovic, B., and Ng, T. (2005). Multiphoton-FLIM quantification of the EGFP-mRFP1 FRET pair for localization of membrane receptor-kinase interactions. *Biophys. J.* **88**(2), 1224–1237.

Petrán, M., Hadravský, M., Egger, M. D., and Galambos, R. (1968). Tandem-scanning reflected-light microscope. *J. Opt. Soc. Am.* **58,** 661–664.

Potma, E. O., De Boeij, W. P., Van Haastert, P. J. M., and Wiersma, D. A. (2001). Real-time visualization of intracellular hydrodynamics in single living cells. *Proc. Natl. Acad. Sci. USA* **98**(4), 1577–1582.

Ramdya, P., Skoch, J., Bacskai, B. J., Hyman, B. T., and Berezovska, O. (2003). Activated Notch1 associates with a presenilin-1/gamma-secretase docking site. *J. Neurochem.* **87**(4), 843–850.

Reichert, W. M., and Truskey, G. A. (1990). Total internal reflection fluorescence (TIRF) microscopy. I. Modelling cell contact region fluorescence. *J. Cell Sci.* **96,** 219–230.

Ries, J., and Schwille, P. (2006). Studying slow membrane dynamics with continuous wave scanning fluorescence correlation spectroscopy. *Biophys. J.* **91**(5), 1915–1924.

Rohrbach, A. (2000). Observing secretory granules with a multiangle evanescent wave microscope. *Biophys. J.* **78**(5), 2641–2654.

Salic, A., Waters, J. C., and Mitchison, T. J. (2004). Vertebrate shugoshin links sister centromere cohesion and kinetochore microtubule stability in mitosis. *Cell* **118**(5), 567–578.

Salmon, W. C., Adams, M. C., and Waterman-Storer, C. M. (2002). Dual-wavelength fluorescent speckle microscopy reveals coupling of microtubule and actin movements in migrating cells. *J. Cell Biol.* **158,** 31–37.

Schwille, P., Haupts, U., Maiti, S., and Webb, W. W. (1999). Molecular dynamics in living cells observed by fluorescence correlation spectroscopy with one- and two-photon excitation. *Biophys. J.* **77**(4), 2251–2265.

Shah, J. V., Botvinick, E., Bonday, Z., Furnari, F., Berns, M., and Cleveland, D. W. (2004). Dynamics of centromere and kinetochore proteins; implications for checkpoint signaling and silencing. *Curr. Biol.* **14**(11), 942–952.

Shaner, N. C., Steinbach, P. A., and Tsien, R. Y. (2005). A guide to choosing fluorescent proteins. *Nat. Methods* **2**(12), 905–909.

Sheppard, C. J. R., and Kompfner, R. (1978). Resonant scanning optical microscope. *Appl. Opt.* **17**(18), 2879–2882.

Shimi, T., Koujin, T., Segura-Totten, M., Wilson, K. L., Haraguchi, T., and Hiraoka, Y. (2004). Dynamic interaction between BAF and emerin revealed by FRAP, FLIP, and FRET analyses in living HeLa cells. *J. Struct. Biol.* **147**(1), 31–41.

Shimozono, S., Hosoi, H., Mizuno, H., Fukano, T., Tahara, T., and Miyawaki, A. (2006). Concatenation of cyan and yellow fluorescent proteins for efficient resonance energy transfer. *Biochemistry* **45**(20), 6267–6271.

Sisan, D. R., Arevalo, R., Graves, C., Mcallister, R., and Urbach, J. S. (2006). Spatially-resolved fluorescence correlation spectroscopy using a spinning disk confocal microscope. *Biophys. J.* **106,** 4241–4252.

Skinner, J. P., Chen, Y., and Muller, J. D. (2005). Position-sensitive scanning fluorescence correlation spectroscopy. *Biophys. J.* **89**(2), 1288–1301.

Smith, P. J., Taylov, C. M., Shaw, A. J., and Mccabe, E. M. (2000). Programmable array microscopy with a ferroelectric liquid-crystal spatial light modulator. *Appl. Opt.* **39**(16), 2664–2669.

Spector, D. L., Fu, X. D., and Maniatis, T. (1991). Associations between distinct pre-messenger-RNA splicing components and the cell-nucleus. *EMBO J.* **10**(11), 3467–3481.

Sprague, B. L., Pego, R. L., Stavreva, D. A., and Mcnally, J. G. (2004). Analysis of binding reactions by fluorescence recovery after photobleaching. *Biophys. J.* **86**(6), 3473–3495.

Stout, A. L., and Axelrod, D. (1989). Evanescent field excitation of fluorescence by epi-illumination microscopy. *Appl. Opt.* **28**(24), 5237–5242.

Straub, M., and Hell, S. W. (1998). Multifocal multiphoton microscopy: A fast and efficient tool for 3-D fluorescence imaging. *Bioimaging* **6**(4), 177–184.

Svelto, O. (1998). "Principles of Lasers." Plenum Press, New York.

Svoboda, K., Denk, W., Kleinfeld, D., and Tank, D. W. (1997). *In vivo* dendritic calcium dynamics in neocortical pyramidal neurons. *Nature* **385**(6612), 161–165.

Tanaami, T., Otsuki, S., Tomosada, N., Kosugi, Y., Shimizu, M., and Ishida, H. (2002). High-speed 1-frame/ms scanning confocal microscope with a microlens and Nipkow disks. *Appl. Opt.* **41**(22), 4704–4708.

Theriot, J. A., and Mitchison, T. J. (1991). Actin microfilament dynamics in locomoting cells. *Nature* **352**(6331), 126–131.

Thews, E., Gerken, M., Eckert, R., Zapfel, J., Tietz, C., and Wrachtrup, J. (2005). Cross talk free fluorescence cross correlation spectroscopy in live cells. *Biophys. J.* **89**(3), 2069–2076.

Tirnauer, J. S., Canman, J. C., Salmon, E. D., and Mitchison, T. J. (2002). EB1 targets to kinetochores with attached, polymerizing microtubules. *Mol. Biol. Cell* **13,** 4308–4316.

Tolles, W. M., Nibler, J. W., Mcdonald, J. R., and Harvey, A. B. (1977). Review of theory and application of coherent anti-Stokes Raman-spectroscopy (Cars). *Appl. Spectrosc.* **31**(4), 253–271.

Tsien, R., and Waggoner, A. (1995). Fluorophores for confocal microscopy: Photophysics and photochemistry. *In* "Handbook of Biological Confocal Microscopy" (J. Pawley, ed.), pp. 267–279. Plenum Press, New York.

Tsien, R. Y. (1998). The green fluorescent protein. *Annu. Rev. Biochem.* **67,** 509–544.

Tsutsui, H., Karasawa, S., Shimizu, H., Nukina, N., and Miyawaki, A. (2005). Semi-rational engineering of a coral fluorescent protein into an efficient highlighter. *EMBO Rep.* **6**(3), 233–238.

Verveer, P. J., Hanley, Q. S., Verbeek, P. W., Van Vliet, L. J., and Jovin, T. M. (1998). Theory of confocal fluorescence imaging in the programmable array microscope (PAM). *J. Microsc.* **189,** 192–198.

Volkmer, A., Subramaniam, V., Birch, D. J., and Jovin, T. M. (2000). One- and two-photon excited fluorescence lifetimes and anisotropy decays of green fluorescent proteins. *Biophys. J.* **78**(3), 1589–1598.

Von Wichert, G., Haimovich, B., Feng, G. S., and Sheetz, M. P. (2003). Force-dependent integrin-cytoskeleton linkage formation requires downregulation of focal complex dynamics by Shp2. *EMBO J.* **22**(19), 5023–5035.

Wachsmuth, M., Weidemann, T., Muller, G., Hoffmann-Rohrer, U. W., Knoch, T. A., Waldeck, W., and Langowski, J. (2003). Analyzing intracellular binding and diffusion with continuous fluorescence photobleaching. *Biophys. J.* **84**(5), 3353–3363.

Wadsworth, P. (1999). Regional regulation of microtubule dynamics in polarized, motile cells. *Cell Motil. Cytoskeleton* **42**(1), 48–59.

Wang, H. F., Fu, Y., Zickmund, P., Shi, R. Y., and Cheng, J. X. (2005). Coherent anti-stokes Raman scattering imaging of axonal myelin in live spinal tissues. *Biophys. J.* **89**(1), 581–591.

Wang, W., Wyckoff, J. B., Frohlich, V. C., Oleynikov, Y., Hüttelmaier, S., Zavadil, J., Cermak, L., Bottinger, E. P., Singer, R. H., White, J. G., Segall, J. E., and Condeelis, J. S. (2002). Single cell behavior in metastatic primary mammary tumors correlated with gene expression patterns revealed by molecular profiling. *Cancer Res.* **62**(21), 6278–6288.

Wang, Z., Shah, J. V., Berns, M. W., and Cleveland, D. W. (2006). *In vivo* quantitative studies of dynamic intracellular processes using fluorescence correlation spectroscopy. *Biophys. J.* **91**(1), 343–351.

Wang, Z., Shah, J. V., Chen, Z., Sun, C. H., and Berns, M. W. (2004). Fluorescence correlation spectroscopy investigation of a GFP mutant-enhanced cyan fluorescent protein and its tubulin fusion in living cells with two-photon excitation. *J. Biomed. Opt.* **9**(2), 395–403.

Weiss, M., Hashimoto, H., and Nilsson, T. (2003). Anomalous protein diffusion in living cells as seen by fluorescence correlation spectroscopy. *Biophys. J.* **84**(6), 4043–4052.

White, J. G., Amos, W. B., and Fordham, M. (1987). An evaluation of confocal versus conventional imaging of biological structures by fluorescence light microscopy. *J. Cell Biol.* **105,** 41–48.

Wiedenmann, J., Ivanchenko, S., Oswald, F., Schmitt, F., Rocker, C., Salih, A., Spindler, K.-D., and Nienhaus, G. U. (2004). EosFP, a fluorescent marker protein with UV-inducible green-to-red fluorescence conversion. *PNAS* **101**(45), 15905–15910.

Wilson, T. (1980). Imaging properties and applications of scanning optical microscopes. *Appl. Phys. A: Mater. Sci. Process.* **22**(2), 119–128.

Wilson, T., Gannaway, J. N., and Johnson, P. (1980). A scanning optical microscope for the inspection of semiconductor materials and devices. *J. Microsc.* **118**(3), 309–314.

Wolf, K., Mazo, I., Leung, H., Engelke, K., Von Andrian, U. H., Deryugina, E. I., Strongin, A. Y., Brocker, E. B., and Friedl, P. (2003). Compensation mechanism in tumor cell migration: Mesenchymal-amoeboid transition after blocking of pericellular proteolysis. *J. Cell Biol.* **160**(2), 267–277.

Yaffe, M. P., Stuurman, N., and Vale, R. D. (2003). Mitochondrial positioning in fission yeast is driven by association with dynamic microtubules and mitotic spindle poles. *Proc. Natl. Acad. Sci. USA* **100**(20), 11424–11428.

Yvon, A. M. C., and Wadsworth, P. (2000). Region-specific microtubule transport in motile cells. *J. Cell Biol.* **151**(5), 1003–1012.

Zhang, J., Campbell, R. E., Ting, A. Y., and Tsien, R. Y. (2002). Creating new fluorescent probes for cell biology. *Nat. Rev. Mol. Cell Biol.* **3**(12), 906–918.

Zipfel, W. R., Williams, R. M., and Webb, W. W. (2003). Nonlinear magic: Multiphoton microscopy in the biosciences. *Nat. Biotechnol.* **21**(11), 1369–1377.

Zoumi, A., Yeh, A., and Tromberg, B. J. (2002). Imaging cells and extracellular matrix *in vivo* by using second-harmonic generation and two-photon excited fluorescence. *Proc. Natl. Acad. Sci. USA* **99,** 11014–11019.

PART II

Mechanisms of Laser Interactions

CHAPTER 4

Mechanisms of Laser Cellular Microsurgery

Pedro A. Quinto-Su[*,†] **and Vasan Venugopalan**[*,†]

[*]Department of Chemical Engineering & Materials Science
University of California, Irvine, California 92697

[†]Laser Microbeam and Medical Program
Beckman Laser Institute, University of California, Irvine, California 92612

0091-679X/07 $35.00
DOI: 10.1016/S0091-679X(06)82004-2

This chapter reviews the optics of pulsed laser microbeams and the use of basic instrumentation to provide pulsed laser microbeam capabilities within a microscope platform. Moreover, we review the principal mechanisms by which laser microbeams produce microsurgical effects in cellular targets. We discuss the principal photothermal, photomechanical, and photochemical damage mechanisms as well as their relationship to critical laser microbeam parameters, including wavelength, pulse duration, and numerical aperture. We relate this understanding of damage mechanisms to laser microbeam applications reported in the literature.

I. Introduction

The invention of the laser in 1959 made available a light source with a high degree of collimation, coherence, brilliance, and monochromaticity. Soon thereafter, researchers began to examine the potential use of lasers in biology and medicine. The first studies involving laser–cell interaction used the laser to inactivate cells or cellular organelles (Amy and Storb, 1965; Bessis *et al.*, 1962; Storb *et al.*, 1966) and as a tool to cut and make incisions within the cell using highly focused beams (Berns *et al.*, 1969, 1971, 1981). Other important applications of focused lasers followed, including laser tweezers (Ashkin, 1970; Ashkin *et al.*, 1986) where a focused laser beam can exert precise forces to trap and manipulate small objects. Optical tweezers have found broad applications in cell biology, including their use to understand the mechanical properties of cytoskeletal filaments and DNA as well as the function of motor proteins (Grier, 2003).

As recounted in a comprehensive review of laser nanosurgery (Vogel *et al.*, 2005), optical inactivation of cells and cellular structures was first attempted by Tschachotin (1912), who used a microscope objective to focus the 280-nm emission of a magnesium spark to a 5-μm region on a cell. This work was later refined by Bessis and Nomarski (1960) who were successful in reducing the focused spot diameter to less than a micrometer. However, conventional light sources could only provide low intensities. As a result, long exposure times were necessary to achieve the desired cellular effect. The laser enabled doses of optical radiation sufficient to damage cellular structures to be delivered on time scales of less than a millisecond with better spatial precision.

The improved collimation and spatial coherence of laser radiation made possible, through the use of conventional microscope optics, the production of extremely high photon fluxes localized to a three-dimensional volume with characteristic spatial scales of less than a micrometer. This enabled the targeting of specific structures within the cell, including chromosomes (Berns *et al.*, 1969, 1981), centrioles/centrosomes (Khodjakov *et al.*, 2002; La Terra *et al.*, 2005; Magidson *et al.*, 2006), mitochondria (Amy and Storb, 1965; Storb *et al.*, 1966), kinetochores (Khodjakov *et al.*, 1996), and microtubules (Botvinick *et al.*, 2004; Khodjakov *et al.*, 1997b). The production of focused laser radiation in this manner is simply referred to as a "laser microbeam," and the use of laser microbeams to modify cells and intracellular structures is known as "laser cellular microsurgery."

In this chapter, our aim is to provide basic information regarding the optics of laser microbeams and review our understanding of the principal mechanisms by which laser microbeams produce microsurgical effects in cellular targets. With respect to this latter topic, we limit our discussion to situations where the optical radiation is delivered in relatively short time, on the order of tens of microseconds or less. This chapter is meant to serve as a relatively brief overview of the optics and processes involved in laser cellular microsurgery. Readers interested in a more detailed treatment would be well served to consult the review paper by Vogel *et al.* (2005). In this paper, the current literature on laser cellular microsurgery is comprehensively reviewed, and extensive computational simulations are performed to model and understand the processes involved with an emphasis on femtosecond laser microbeams.

The outline of the remainder of this chapter is as follows. In Section II, we present the optics of laser microbeams by summarizing the basic characteristics of laser radiation and a focused laser microbeam. These principles will enable a determination of the characteristic radiant exposures and irradiances present in the focal volume of the laser microbeam. In Section III, we address various practical considerations involved with the introduction of a laser beam into a microscope platform with the intent of producing a laser microbeam for cellular microsurgery. With these theoretical and practical considerations related to laser microbeam optics dealt with, we turn our attention to the processes that enable pulsed laser microbeams to produce cellular effects. In Section IV, we discuss linear mechanisms of energy deposition within the focal volume of the laser microbeam and introduce key thermal, mechanical, and chemical processes that result. This examination reveals that for the pulse energies typically used in cellular microsurgery of endogenous targets, linear absorption alone is most likely not capable to provide the energy densities necessary for cell microsurgery. This will lead us to Section V where we consider nonlinear absorption processes that result in the generation of free electrons and optical breakdown. In Section VI, we describe the mechanical, thermal, and chemical processes that result from optical breakdown and their role in laser cellular microsurgery. In Section VII, we examine some published studies that achieve laser cellular microsurgery utilizing laser parameters in different regimes of pulse duration. We end with some final thoughts in Section VIII.

II. Lasers and Microbeam Optics

When considering the interaction of pulsed laser microbeams with cells, it is essential to understand the characteristics of laser radiation as well as the spatial distribution of light established within cellular samples. In this section, we provide a simplified description of a laser and characterize the radiation that it produces. We then describe various spatial modes of the laser output and how simple optics can be used to provide a laser microbeam with desired characteristics.

In cellular microsurgery, confinement of the energy deposition is crucial. To achieve this it is helpful to focus the laser beam to very small spot sizes. To illustrate the interplay of the various laser and optical parameters involved, we consider the fundamental or Gaussian mode of the laser and estimate the size that such a beam can be focused by a lens or a microscope objective. Consideration of the Gaussian mode is instructive because it is the only spatial mode that can be focused to the diffraction limit and provides the minimum spot size achievable by an optical system (Silfvast, 1996).

A. Light and the Laser

To describe the functional principles of the laser, we invoke both wave and particle descriptions of light. The wave description is helpful when considering the propagation and focusing of light, while the particle description is helpful when considering light–matter interactions. Light propagation can be represented as a transversely propagating oscillation of the electric field $\vec{E} = \vec{E}_0 \,\mathrm{Re}[\exp(-i\vec{k}x - i\omega t)]$. In this expression, the vector $\vec{E}_0$ gives the amplitude and polarization of the oscillation, $\vec{k} = (2\pi/\lambda)\vec{e}_k$ is the wave vector and defines the direction of propagation which we take to be the z-axis as indicated by the unit vector $\vec{e}_k$, and $\omega = 2\pi c/\lambda$ is the angular oscillation frequency of the wave where λ is the wavelength and $c = 2.9979 \times 10^8$ m/s is the speed of light in vacuum. In the particle description, light consists of quanta known as photons with energy $E = hc/\lambda$ where $h = 6.626\times10^{-34}$ J s is Planck's constant.

As shown in Fig. 1, a basic laser consists of only three components, a gain medium that is placed within an optical resonator (usually composed of a pair of mirrors) and an energy source to excite the molecules within the gain medium. For simplicity, consider an idealized system where the gain medium consists of atoms with only two energy levels: a ground state and an excited state as depicted in Fig. 2. The release of energy by an appropriate pump source (e.g., a flash lamp or spark) can promote the atoms of the gain medium into an excited state. After

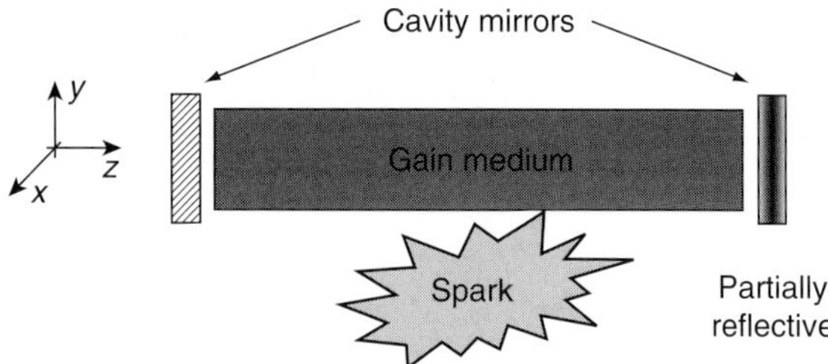

Fig. 1 Basic components of a laser: An optical resonator formed by a couple of mirrors (one of which is only partially reflective), a gain medium, and an pump energy source (e.g., a spark, high-intensity flashlamp, radio frequency current).

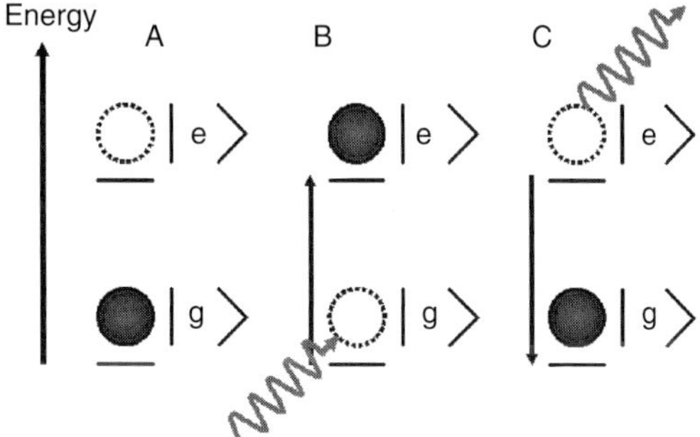

Fig. 2 Energy diagram of a two level atom with ground state |g> and excited state |e>. (A) depicts the initial state of the atom in the ground state. In (B) the atom has absorbed a photon and is promoted to the excited state. In (C) the atom decays back into the ground state by emitting a photon.

some time, these atoms will decay spontaneously back to the ground state. In doing so, they emit photons each having an energy that corresponds to the energy difference between the excited and ground state. These photons are emitted in a random direction and the few that are emitted along the axis of the resonator will be "trapped." On traveling again through the gain medium, the trapped photons can stimulate those atoms still in the excited state to emit a photon. Photons that are emitted in this manner are highly likely to possess the same direction or "mode" as the photon that stimulated the emission. As these light waves oscillate, the simulated emission that occurs in the gain medium is amplified by the added photons. The extraction of light out of the laser cavity is made possible by making one of the mirrors partially transmitting, thereby allowing a subpopulation of the photons that oscillates in the cavity to escape. This results in a collimated beam of light composed of photons of the same wavelength and phase.

Given that the mirrors and gain medium have a finite lateral (x and y) dimension, the wave that oscillates within the resonator can possess many different spatial modes. Denoting z as the direction of light propagation which is also collinear with the longitudinal axis of the resonator, the transverse modes of the electric field within the resonator are proportional to the product of two Hermite polynomials with the Gaussian distribution function (Silfvast, 1996):

$$\mathrm{TEM}_{\mathrm{pq}} = H_{\mathrm{p}}\left(\frac{\sqrt{2}x}{w}\right) H_{\mathrm{q}}\left(\frac{\sqrt{2}y}{w}\right) \exp\left[-(x^2+y^2)/w^2\right] \tag{1}$$

where TEM stands for transverse electric mode, with x and y the transverse directions and w is the distance at which the electric field amplitude decreases by $1/e$ with respect to the maximum amplitude at $r = 0$. TEM_{00} is the "fundamental" mode and has a Gaussian shape because $H_0(u) = 1$. In this case, Eq. (1) becomes $\mathrm{TEM}_{00} = \exp[-(x^2+y^2)/w^2]$.

B. Properties of Gaussian Beams

The TEM_{00} (Gaussian beam) has important properties and often desired for laser microbeam applications. These properties are highly relevant for determining the characteristic focal volume of the laser microbeam and the radiant exposures and irradiances generated therein. Here we shall give a brief summary of the properties of Gaussian laser beams and how they can be altered using simple optical elements.

The Gaussian beam profile is provided by the irradiance which has units of power per unit area and is proportional to the square of the electric field magnitude $I \propto |E|^2$:

$$I(r) = I_0 \exp\left(\frac{-2r^2}{w^2}\right) \tag{2}$$

where I_0 is the peak irradiance, r is the transverse radius $r = (x^2+y^2)^{1/2}$, and w is the Gaussian beam radius. At $r = w$, the irradiance has fallen to $1/e^2$ of its peak value, and 86.5% of the total power of the Gaussian beam is contained at radial locations smaller than w. Figure 3 shows the intensity profile of a Gaussian beam. In general, the Gaussian beam radius w is not constant but varies as a function of the propagation distance z as depicted in Fig. 4. This variation in Gaussian beam radius is given by:

$$w(z) = w_0\left[1 + \left(\frac{\lambda z}{\pi w_0^2}\right)^2\right]^{1/2} \tag{3}$$

At $z = 0$, the beam radius attains a minimum value of w_0 which is known as the beam waist, and $2w_0$ is often referred to as the laser "spot size."

Also of interest is the characteristic (axial) distance in z where the intensity of the Gaussian beam remains substantial. The Rayleigh range z_0 is defined as the z distance from the focal plane at which the peak beam intensity has fallen by a factor of $\sqrt{2}$ and is given by $z_0 = (\pi w_0^2/\lambda)$.

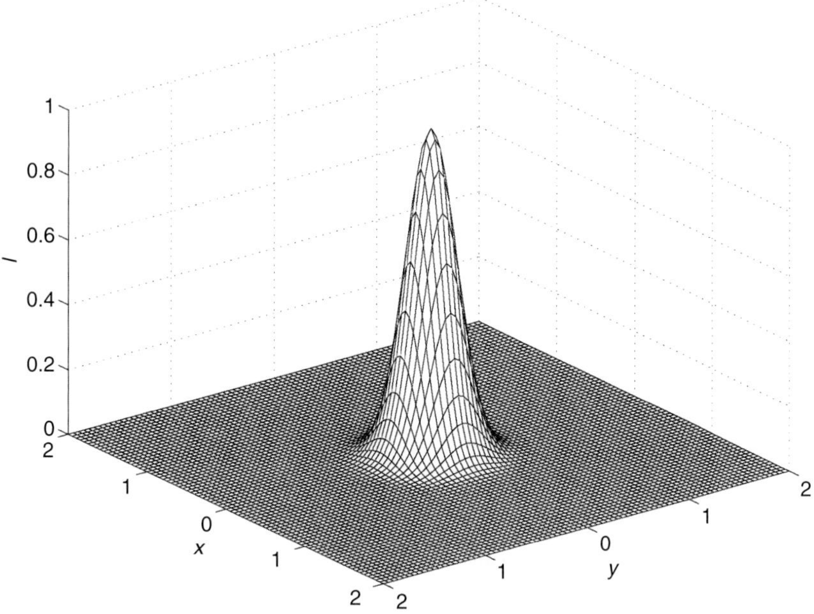

Fig. 3 Gaussian beam intensity in the transverse dimensions x and y.

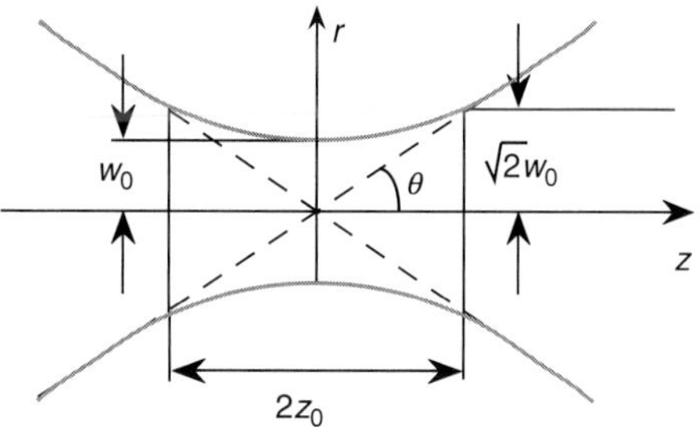

Fig. 4 Width of the Gaussian beam as it propagates in the z direction. w_0 is the waist of the beam, z_0 the Rayleigh range, and θ the divergence angle.

The rate at which this beam diverges is also of importance. As shown in Fig. 4, near the beam waist z_0 (at $z = 0$) the divergence of the beam is actually quite small. However, for large z the beam width $w(z) \rightarrow (\lambda z/\pi w_0)$ and the divergence angle of the laser beam $\theta \rightarrow (\lambda/\pi w_0)$. This divergence angle is related to the numerical aperture of the laser beam by $\mathrm{NA} = n \sin \theta$, where n is the refractive index of the medium.

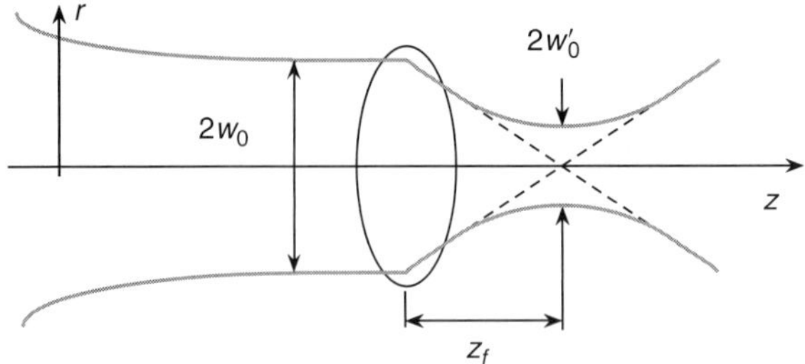

Fig. 5 Focusing a Gaussian beam with waist w_0 by placing a converging lens at its waist. The beam emerging from the lens will have a new waist w'_0 at a distance z_f from the lens.

C. Focusing the Gaussian Beam

Consider focusing a Gaussian beam by placing a thin positive lens with focal length f at the beam waist as shown in Fig. 5. The beam incident on the lens has a Rayleigh range z_0, spot size w_0, and wavelength λ. The new waist is denoted as w'_0 and related to the characteristics of the incident beam by (Milonni and Eberly, 1988):

$$w'_0 = \frac{\lambda f}{\pi w_0}\left[1 + \left(\frac{f^2}{z_0^2}\right)\right]^{-1/2} \tag{4}$$

The location of the new waist is $z_f = f/(1 + f^2/z_0^2)$. In the limit of large z_0, that is, when the incident beam is nearly collimated, these two expressions reduce to $w'_0 = \lambda f/\pi w_0$ and $z_f = f$. Thus as w_0 increases, w'_0 decreases.

It should be noted that other parameters are often used to describe the focusing characteristics of optical elements used to focus the laser beam. One property is the f number $(f/\#)$ that is defined as the ratio of the lens focal length to the effective diameter of the lens, that is, $f/\# = f/d$. The effective diameter of the lens is simply the size of the beam incident on the lens surface. A second important property is the numerical aperture (NA) of a lens or microscope objective. The numerical aperture is simply the radius of the lens/objective divided by the focal length. Thus, $\mathrm{NA} = d/2f$ or $\mathrm{NA} = 1/(2f/\#)$. These expressions are equivalent to the earlier expression in terms of the focusing angle $\mathrm{NA} = n \sin\theta$.

D. Diffraction-Limited Beam Sizes and the *M* Factor

The Gaussian beam is the lowest order spatial mode produced by a laser and the only mode that can be focused to a diffraction-limited spot. The diffraction-limited spot size is the minimum diameter that can be reached by an optical system with a circular aperture of a given f number and is given by $2w_0 = 2.44\lambda(f/\#)$, where w_0 is the radius of the diffraction-limited spot. As a function of the numerical aperture, the diffraction-limited spot size is given by

$$2w_0 = 2.44\lambda/(2\text{NA}) = 1.22\lambda/\text{NA} \tag{5}$$

Thus, a larger numerical aperture focuses the beam to smaller spot sizes. Note that the diffraction-limited spot size is *independent* of the objective magnification but instead depends only on the numerical aperture and wavelength.

Grill and Stelzer (1999) examined the intensity distribution for beams focused by a circular aperture at large numerical apertures. The intensity distribution in three dimensions takes the form of an ellipsoid with its longer axis oriented in the direction of beam propagation. The ratio of this axial dimension of the focal volume to the transverse dimension is

$$\frac{\Delta z}{\Delta r} = \frac{(3 - 2\cos\theta - \cos 2\theta)^{1/2}}{(1 - \cos\theta)} \tag{6}$$

where θ is the focusing angle. Using the expression for the diffraction-limited spot radius w_0, it is possible to calculate the semi-axial dimension of the focal volume $w_0\,\Delta z/\Delta r$. In this ellipsoidal model, intensity distribution can be approximated by

$$I(r,z) = I_0 \exp\left[-2\left(\frac{r^2}{a^2} + \frac{z^2}{b^2}\right)\right] \tag{7}$$

where a and b are the semiaxes of the ellipsoid along the transverse and axial directions, respectively. The ellipsoidal focal volume is given by the expression $V = 4\pi a^2 b/3$. Table I gives the diffraction-limited spot radius, axial dimension calculated from Eq. (6), the Rayleigh range, and the ellipsoidal volume for a beam with $\lambda = 532$ nm.

Unfortunately, most laser beams are not perfectly Gaussian as they contain contributions from several spatial modes. Moreover, the quality of a "perfect" TEM_{00} beam can be adversely affected by aberrations (described in next section) produced by the optical elements in the beam path. The "*M* factor" is used to measure the deviation in the properties of a real laser beam from an ideal Gaussian beam (Silfvast, 1996) and many laser manufacturers specify an *M* factor to describe the beam quality of laser output. To see how this *M* factor is used, consider the following product for an ideal diffraction-limited beam:

Table I
Listing of the Semiaxes and Volume of the Diffraction-Limited Spot Generated by the Focusing of a $\lambda = 532$-nm Laser Microbeam at Various Numerical Apertures

NA	a (μm)	b (μm)	z_0 (μm)	V (μm^3)
0.8	0.40	1.13	0.97	0.78
1.0	0.32	0.86	0.62	0.38
1.3	0.25	0.55	0.37	0.14

Results are based on Eq. (6).

$$w(z)\theta(z) = w(z)\frac{\lambda}{\pi w_0} \tag{8}$$

This product has the minimum value when $w(z) = w_0$, where $w_0\theta = \lambda/\pi$.

A nonideal laser beam can be defined relative to an ideal Gaussian beam. For example, the divergence of a real beam Θ can be expressed as

$$\Theta = M\theta \tag{9}$$

with a real beam waist w_0

$$W_0 = Mw_0 \tag{10}$$

where θ and w_0 are the divergence and the waist of the Gaussian beam, and W_0 is the waist of the real beam. For the real beam, the product $w(z)\theta$ is $\Theta W_0 = \lambda M^2/\pi$, which is M^2 times the value for a Gaussian beam. Thus in the absence of additional optical aberrations, the beam of a real laser can be focused to a spot size M times larger than the ideal diffraction-limited spot size. Additional insights regarding the measurement of beam quality is given by an interesting tutorial by Siegman (1997).

E. Effect of Optical Aberrations on Spot Size

Beyond the multimode emission of a laser, optical aberrations that prevent a laser beam from being focused to a diffraction-limited spot can be produced by lens systems that focus the light along rays that deviate from the idealized paths of Gaussian optics (Hecht, 1974). Such deviations degrade the performance of the optical system and preclude the beam from being focused to the smallest possible spot size.

The presence of optical aberrations in laser microbeam setup can have a dramatically adverse effect because they greatly increase the focal volume of the laser microbeam. As a result, a much larger laser pulse energy is needed to produce a desired irradiance or radiant exposure in the focal volume. Vogel *et al.* (1999a) examined the effect of optical aberrations by replacing achromatic lenses in an optical setup with plano-convex lenses. Waveform distortions of 5.5λ and 18.5λ were produced through the replacement of one and two lenses, respectively. As a result, the measured spot sizes increased from 7.6 to 96.6 μm after the replacement of one lens and then to 130.2 μm after the replacement of both lenses. The diffraction-limited spot sizes in the three cases were 3.5, 2.7, and 3.2 μm, respectively. This shows quite dramatically that if care is not taken to reduce optical aberrations, much larger pulse energies will be required to produce the desired cellular damage and adversely affect the precision of the laser microbeam procedure.

F. Self-Focusing

The use of the smallest possible spot size is often desired to achieve high precision in laser cellular microbeam applications. As a result, high numerical aperture objectives are typically used for both visualization of the sample and delivery of the laser radiation. As a result, the laser microbeam possesses extremely high irradiances in the focal volume. In such cases, one may have to consider the irradiance dependence of the refractive index of the medium. In cases where the refractive index increases with irradiance, the medium itself can act as a positive lens when high irradiances are used. This effect is called self-focusing. In general, self-focusing is undesirable for laser microbeam applications because it can lead to filamentation of the beam which has an adverse effect on the localization of the energy deposition (Vogel *et al.*, 2005).

The polarization of a medium due to an applied electric field can be expressed as a power series in the electric field $P = \sum \chi_n E^n$. This results in the refractive index being expressed as a function of the irradiance $I(I \propto |\vec{E}|^2)$

$$n = n_0 + n_2 I + n_4 I^2 + \ldots \tag{11}$$

If we only consider the first two terms in this expansion, a positive value of n_2 results in an increase in the refractive index with increasing intensity. For a Gaussian beam, the largest refractive index will be produced at the center of the beam. As a result the medium acts as a converging lens, with the tendency to focus the beam to a small spot size. In order for this self-focusing process to be effective in further constraining the beam size, it must overcome the intrinsic diffractive spreading of the electric field. Several expressions have been derived to investigate self-focusing (Boyd, 1992; Milonni and Eberly, 1988; Shen, 1984). Here we follow the treatment of Milonni and Eberly who consider the relative magnitudes of the diffractive spreading and the self-focusing terms in the paraxial approximation to the electromagnetic wave equation (Milonni and Eberly, 1988).

The term in the wave equation that describes this diffractive spreading is proportional to $a_0^{-2}|\vec{E}|$ while the self-focusing term scales with $n_2|\vec{E}|^2|\vec{E}|/n_0$. This latter term simply represents the nonlinear part of the refractive index $(n_2|\vec{E}|^2)$ multiplied by the electric field divided by the linear refractive index. Self-focusing must be considered when these two terms (self-focusing and diffraction) have a comparable magnitude, that is,

$$\frac{k^2 n_2}{n_0}|\vec{E}|^2\vec{E} \sim \frac{\vec{E}}{a_0^2} \tag{12}$$

This reduces to

$$a_0^2|\vec{E}|^2 \sim \frac{n_0}{k^2 n_2} \tag{13}$$

We see that the left-hand side of this equation is equal to the product of the beam area and the square of the electric field that is, the total laser beam power.

Thus, Eq. (13) implies that self-focusing becomes important when a critical power, as opposed to a critical irradiance, of the laser beam is exceeded. By substituting the expression for the beam irradiance $I = (n_0 c\varepsilon_0/2)|E|^2$, we find that the critical power necessary for self-focusing is on the order of

$$P_{\mathrm{cr}} \sim (\pi a_0^2)I = \frac{\pi n_0 c\varepsilon_0}{2} a_0^2 |E|^2 = \frac{\pi n_0 c\varepsilon_0}{2} \frac{n_0}{k^2 n_2} = \frac{c\varepsilon_0 \lambda^2}{8\pi n_2} \tag{14}$$

Experimental results reveal the P_{cr} for water lies in the range of 1–4 MW (Brodeur and Chin, 1998; Feng *et al.*, 1997). Given that the power necessary to achieve a microsurgical effect goes down as the numerical aperture increases, self-focusing can typically be ignored for laser pulses with duration longer than 10 ps when using microscope objectives of NA $\gtrsim$ 0.8. However, we shall see that the irradiance necessary to produce microsurgical effect rises notably as the pulse duration is reduced further. As a result, when using femtosecond laser pulses one should generally use objectives with NA > 0.95 to avoid self-focusing effects (Vogel *et al.*, 2005).

III. Incorporating a Laser Microbeam Within a Microscope Platform

In this section, we describe an experimental setup where a microscope objective is used to both visualize a sample and form a laser microbeam in a fixed location within the sample. We first describe how one can condition the laser beam prior to its introduction into the microscope to best achieve a diffraction-limited laser microbeam within the cellular sample. Once the beam is properly conditioned, we describe the "coupling" of laser radiation into the microscope for delivery through the microscope objective.

A. Beam Preparation

Because most lasers emit radiation with a combination of spatial modes, that is, a combination of TEM's, we would ideally like to eliminate the higher order modes of the beam so that the beam can be focused to its diffraction-limited size by the objective. The simplest way to eliminate the higher order mode components is to employ a spatial filter where the laser beam is focused onto a circular aperture whose size will only allow the TEM_{00} to pass, blocking all the higher order modes.

A sample arrangement is depicted in Fig. 6. The first lens focuses the beam on the aperture, and the second lens recollimates it. To best eliminate the higher order spatial modes of the beam, the size of aperture diameter should be similar to the diameter of the diffraction-limited spot size for the focusing lens or microscope objective, which in terms of the NA and laser wavelength is

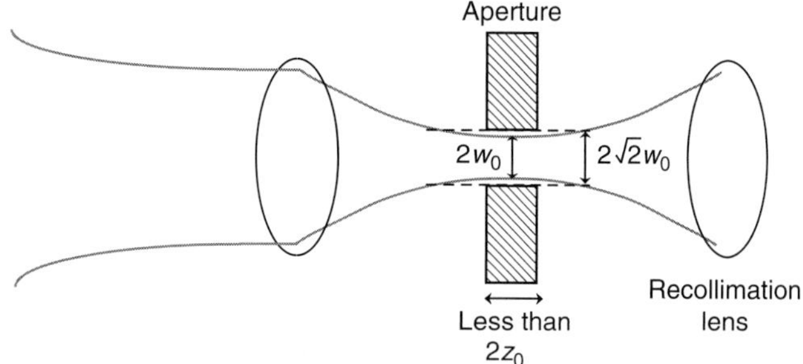

Fig. 6 Spatial filter. The incoming laser beam is expanded to utilize the full diameter of the first lens. The aperture used has a thickness less than double the Rayleigh range ($2z_0$) and a diameter roughly equal to the size of the diffraction-limited spot placed at a distance equivalent to the $z = z_0$. Light rays corresponding to the TEM_{00} mode are shown as passing through the aperture. A lens is necessary after the spatial filter to recollimate the beam.

$$2w_0 = 1.22\lambda/\mathrm{NA} \tag{15}$$

where w_0 is the radius of the diffraction-limited spot. It is important to expand the size of the laser beam, if necessary, to utilize the maximum NA offered by the lens/objective to be used in the spatial filter.

Another parameter to consider is the thickness of the aperture which should be no more than twice the Rayleigh range $z_0 = \pi w_0^2/\lambda$. This is important because the radius of the diffraction-limited beam at $z = z_0$ increases to $w(z_0) = \sqrt{2}w_0$. There is an inverse relationship between diameter of the aperture used in the filter and the resultant beam quality. Specifically, a smaller aperture diameter is more effective in filtering the non-TEM_{00} mode content and improves overall beam quality. However, this comes at the expense of a decrease in transmitted laser energy through the aperture. A good compromise is attained using a pinhole diameter of $1.5w_0 \sim \sqrt{2}w_0$.

B. Beam Delivery in a Microscope Platform

There are several options to couple the laser into the microscope as the beam has only to be guided into the microscope objective which will focus it within the sample. Ideally, the laser beam entering the rear aperture of the microscope objective should be perfectly collimated in order to be focused in the visual plane.

Consider an inverted microscope depicted in Fig. 7. The microscope ports commonly used for introducing a lamp source [i.e., fluorescence (13) or bright field (14) illumination] or a camera (ports 1, 3, or 4) can all be used to introduce the laser beam so long as appropriate care is taken. While the internal microscope optics will direct this beam to the sample, these optics are designed for a different purpose, for example, to condition the illumination from a lamp source (i.e., the fluorescence or condenser optics). As a result, these optics may be detrimental

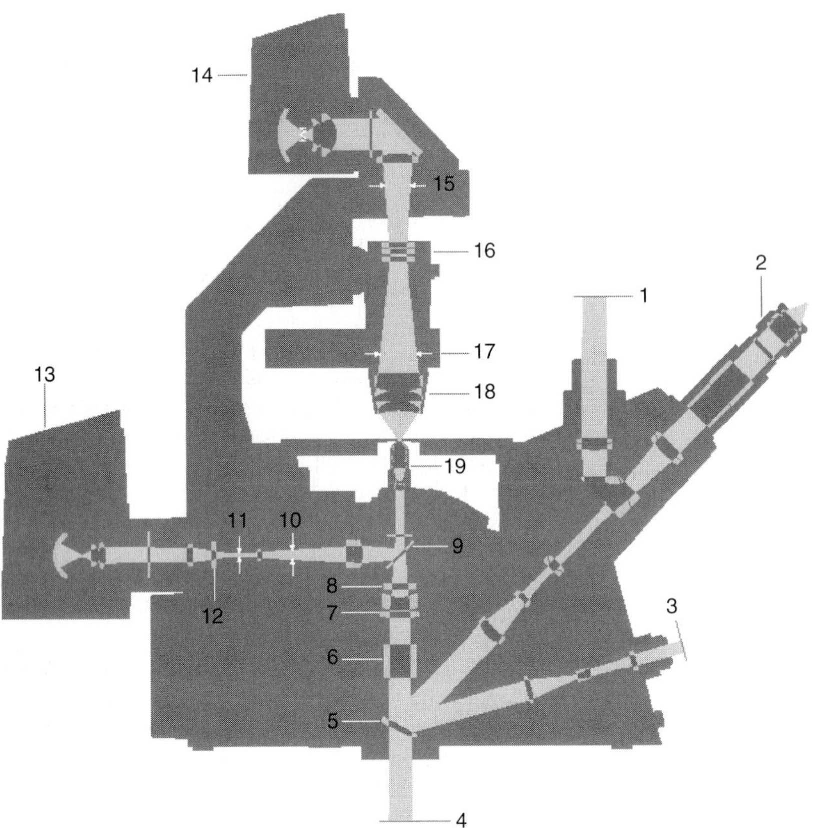

Fig. 7 Schematic of Zeiss 200 Axiovert. Courtesy of Rick Marolt, Zeiss Corporation.

with respect to preserving the collimation of the laser beam. Moreover, care should be taken to introduce appropriate dichroic, band pass, and/or notch filters in the filter cube when a specific port is used to couple the laser light. This is essential to ensure that the combination of the dichroic and the laser wavelength are appropriate to direct the particular laser wavelength to the sample while also blocking the laser wavelength from reaching the ocular or an imaging device and allowing the desired range of wavelengths to be visualized in the ocular (port 2) and/or by a camera (ports 1, 3, or 4).

As a simple example, consider the introduction of a $\lambda = 532$-nm laser beam into the epifluorescence port after removal of the fluorescence lamp (13). In this case, it is generally advisable to remove the fluorescence optics (elements 10, 11, and 12). To direct this beam into the microscope objective, a dichroic is needed in the cube holder (9) that will reflect $\lambda = 532$-nm radiation into the microscope objective but is transparent at other wavelengths to allow imaging of the sample. The transmission

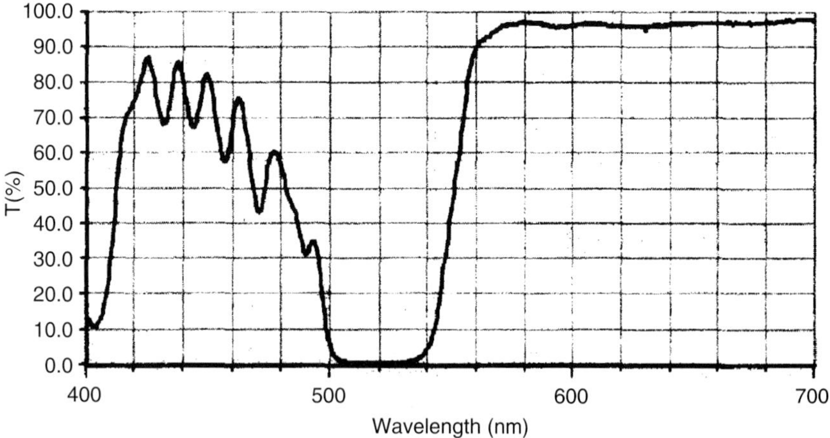

Fig. 8 Reflectance spectra of dichroic placed in a microscope filter cube to direct a $\lambda = 532$-nm microbeam into a microscope objective while transmitting other wavelengths to allow for sample visualization.

spectrum of such a dichroic mirror (Z532RDC, Chroma Technologies, Inc., Rockingham, VT) is shown in Fig. 8. One other consideration is the fact that the dichroic may transmit a small amount of $\lambda = 532$-nm radiation. As such, it is wise to put a 532-nm line filter immediately before the ocular or a camera to minimize stray laser radiation.

Alignment of the laser beam into the microscope objective is very important because a misaligned beam will compromise the ability of the objective to focus the beam to a diffraction-limited spot. This will result in a reduction in the spatial precision with which a biological target can be modified and will also increase the minimum pulse energy required to produce cellular damage. For many applications, control of the focal plane of the laser microbeam relative to the image plane of the microscope objective is desired. This control can be achieved through the use of an assay where the laser microbeam is used to make 'holes in a sample of dried red blood cells (Cole *et al.*, 1995). If the holes produced in the blood cells are in focus, then the focal plane of the laser microbeam is coincident with the image plane. If not, the distance that the microscope objective must be translated to bring the holes into focus represents the distance between the image plane and the focal plane of the laser microbeam. In our system, we simply translate the recollimation lens used after our spatial filter in order to adjust the relative distance between the focal plane of the laser microbeam and the image plane. This translation of the recollimation lens slightly modifies the convergence/divergence of the laser beam prior to entering the microscope objective and is an effective strategy to precisely adjust the separation distance, the image plane, and the focal plane of the laser microbeam.

IV. Linear Absorption and Photothermal, Photomechanical, and Photochemical Processes

In Section II, we discussed the properties of the spatial distribution of light formed by the delivery of an ideal Gaussian laser beam through a microscope objective. In this section, we discuss the optical properties of biological constituents to assess the amount of linear absorption provided by endogenous biological chromophores. This examination will reveal that, for the laser parameters used in many laser microbeam applications, linear absorption alone is unlikely to provide a sufficient energy density within the focal volume for cutting or inactivation of intracellular structures. Nevertheless, an understanding of linear absorption, as well as the thermal, mechanical, and chemical processes that result, gives important insights into laser microbeam processes. In Section V, we will consider the nonlinear processes that likely initiate cellular modification.

A. Linear Absorption

Figure 9 provides the optical absorption spectrum of several important biological chromophores. The magnitude of optical absorption is described in terms of the absorption coefficient μ_a, the reciprocal distance that a photon travels, on average, before experiencing an absorption event. One important observation is that in the wavelength region between the far ultraviolet (UV) and near-infrared (IR)

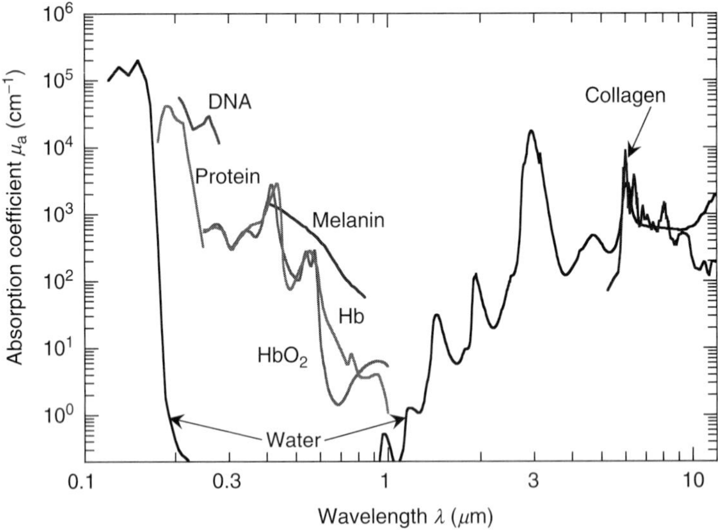

Fig. 9 Optical absorption coefficients of important biological chromophores in the 0.1- to 12-μm spectral region.

($\lambda \approx$ 350–1100 nm), there are no strong endogenous biological absorbers except for melanin and hemoglobin. If we expect linear absorption to play a critical role, this fact stands against the many studies that have employed pulsed visible and near-IR laser sources (e.g., Nd:YAG at λ = 532 and 1064 nm and Ti:Sapphire at $\lambda \sim$ 800 nm) to successfully produce subcellular surgical effects without administration of exogenous dyes.

We can take a quantitative approach to obtain an order of magnitude estimate for the linear absorption necessary in the focal volume to produce a microsurgical effect. Let us assume that the energy density necessary in the focal volume is equivalent to that necessary to vaporize water, that is, $\varepsilon_{\mathrm{vap}} = 2.58$ J/mm^3. Assuming that the laser microbeam is focused to a diffraction-limited spot, the absorbed energy density in the focal plane is given by $\mu_a E_p / \pi w_0^2$, where E_p is the pulse energy of the laser microbeam and w_0 is the radius of the diffraction-limited spot. By equating this expression with $\varepsilon_{\mathrm{vap}}$, solving for μ_a, and expressing the diffraction-limited spot size in terms of the laser wavelength and numerical aperture [see Eq. (5)] we get the following expression for the absorption coefficient necessary to achieve vaporization within the focus of the microbeam:

$$\mu_a = \frac{1.169\lambda^2 \varepsilon_{\mathrm{vap}}}{E_p(\mathrm{NA})^2} \tag{16}$$

Using typical values of NA = 1.3, $E_p = 1$ μJ for laser microsurgery, we find that the required values for μ_a are in the range of ~2–20 cm^{-1} for a wavelength range of λ = 337 nm (N_2 laser) to λ = 1064 nm (Nd:YAG laser). Hemoglobin and melanin aside, no cellular chromophores are present in this wavelength region that can absorb the energy necessary to produce a microsurgical effect. Specifically, the absorption of water, protein, and nucleic acids in this spectral region are far below the minimum given by the above analysis.

B. Photothermal Effects

The absorption of photons results in the excitation of molecules from a ground energy state to an excited state. The energy difference between these two states is given by the photon energy $E_\lambda = hc/\lambda$. Absorption of photons in the UV and visible portion of the spectrum results in the excitation of the electronic energy levels of the molecule, while absorption of IR photons results in the excitation of the vibrational energy levels. If we consider the simple case where photon absorption does not result in a chemical reaction or subsequent photon reemission (e.g., via fluorescence or luminescence), a temperature rise is produced in the molecule. This temperature rise is not always instantaneous. For example, when photon absorption results in the excitation of electronic energy levels within the molecule, it is only once the molecule "relaxes" back to its ground electronic state, and in doing so "shares" the photon energy with the vibrational modes of a molecule, does the molecular temperature actually increase. This process of electronic relaxation followed by a temperature rise is known as an "electron–phonon" interaction and typically occurs on the picosecond time scale. This implies that when ultrashort pulsed laser microbeams are employed, the increase of temperature in the cellular target is fully realized only well after the laser pulse. Nevertheless, if we assume this heating occurs adiabatically, the associated temperature increase is given by:

$$\Delta T(r) = \frac{\varepsilon(r)}{\rho c_v} \tag{17}$$

where $\varepsilon(r)$ is the spatial distribution of the volumetric energy density in the target, ρ the target density, and c_v the specific heat capacity at constant volume. If one wishes to confine the temperature rise to the volume in which the photons are absorbed, it is important to deliver pulsed laser microbeam over a time shorter than the characteristic thermal diffusion time away from the heated volume. To estimate the characteristic thermal diffusion time, we assume that the shape of the heated volume is similar to a cylinder whose length is significantly larger than its diameter. We, in turn, assume that the diameter d of the heated volume is equivalent to that of a diffraction-limited spot. This results in the following expression for the characteristic thermal diffusion time in terms of the laser microbeam wavelength and numerical aperture (Dierickx *et al.*, 1995):

$$t_{\mathrm{d}} = \frac{d^2}{16\kappa} = \frac{0.124\lambda^2}{\kappa(\mathrm{NA})^2} \tag{18}$$

where κ is the thermal diffusivity of the sample. The condition for "thermal confinement" is achieved when the ratio of the laser pulse duration t_{L} to the thermal diffusion time fulfills the condition $(t_{\mathrm{L}}/t_{\mathrm{d}}) \lesssim 1$. Using a value for κ equivalent to that of water (1.44×10^{-3} cm^2/s), calculation of t_{d} over a range of numerical apertures of 0.8–1.3 and laser microbeam wavelengths of 337–1064 nm reveals characteristic thermal diffusion times in the range of 60 ns–1.5 μs. Thus when using laser sources with pulse durations of $\lesssim$60 ns, "thermal confinement" is achieved since the laser pulse duration is significantly less than the characteristic time for thermal diffusion away from the volume where the energy is deposited. Thermal confinement is important in microsurgery to limit the thermal injury of structures outside the focal volume as well as to minimize the energy dose necessary to produce the desired cellular effect.

C. Thermoelastic Stress Generation and Propagation

The temperature increase produced by the energy deposition in the focal volume also results in thermal expansion of the target. If the energy is delivered on a time scale faster than the sample can expand, significant transient stresses can be developed. The generation and propagation of these thermoelastic stresses are governed by the longitudinal speed of sound in the medium c_{a}, the laser pulse duration t_{p}, the characteristic size of the heated volume, and an intrinsic thermophysical property known as the Grüneisen coefficient Γ (Paltauf and Dyer, 2003; Vogel and Venugopalan, 2003). The Grüneisen coefficient is defined as the change of internal stress within a material per unit increase of energy density under constant volume (isochoric) conditions (Vogel and Venugopalan, 2003):

$$\Gamma = \left(\frac{\partial \sigma}{\partial \varepsilon}\right)_v = \frac{\alpha}{\rho c_v \kappa_T} \tag{19}$$

where σ is the internal stress, ε the volumetric energy density, v the specific volume, α the coefficient of thermal expansion, ρ the mass density, c_v the specific heat capacity at constant volume, and κ_T the isothermal compressibility. For a liquid in which energy deposition occurs under constant volume conditions, the initial spatial distribution of the thermoelastic pressure is given by (Vogel *et al.*, 2002a)

$$p(t = 0, \vec{r}) = \Gamma \varepsilon(\vec{r}) \tag{20}$$

where $\vec{r}$ is used to indicate the position within the medium. While the initial spatial distribution of the thermoelastic stress (pressure) is confined only to the heated volume and strictly compressive in nature, the subsequent stress wave propagation will expose regions outside of the heated volume to these stresses. Moreover, because the absorption of photons imparts no significant momentum to the sample, the transient stresses will contain both compressive and tensile components that can produce mechanical damage (Paltauf and Dyer, 2003; Vogel and Venugopalan, 2003; Vogel *et al.*, 2005).

To reduce the magnitude of the thermoelastic stresses, one must deposit energy on a time scale longer than that required for an acoustic wave to propagate out of the heated volume. Again considering the focal microbeam volume as a long cylinder with a diameter equivalent to the diffraction-limited spot size, the characteristic time for stress wave propagation out of the focal volume of the microbeam is given by

$$t_{\mathrm{m}} = \frac{d}{2c_{\mathrm{a}}} = \frac{0.61\lambda}{c_{\mathrm{a}}\mathrm{NA}} \tag{21}$$

Similar to the concept of thermal confinement, "stress confinement" is achieved when the ratio of the laser pulse duration to the stress propagation time fulfills the condition $(t_{\mathrm{L}}/t_{\mathrm{m}}) \lesssim 1$. Using a value of c_{a} equivalent to that of water (1483 m/s), calculation of t_{m} over a range of numerical apertures of 0.8–1.3 and laser microbeam wavelengths of 337–1064 nm reveals characteristic stress propagation times in the range of 100–550 ps. Thus when using pulse durations $\lesssim$100 ps, "thermoelastic stress confinement" is achieved and may be significant toward the generation of mechanical damage. Of particular importance are the tensile thermoelastic stresses as they may lead to the fracture of cellular components or catalyze bubble formation (Paltauf and Dyer, 2003; Vogel and Venugopalan, 2003; Vogel *et al.*, 2005). Analysis by Vogel *et al.* (2005) has revealed that thermoelastic stresses may facilitate precise injury of cellular structures in laser microsurgery when using femtosecond pulses at repetition rates in the kilohertz range.

D. Photochemical Decomposition

Absorption of UV light can also promote direct molecular scission as the photon energy from common UV laser sources exceeds the dissociation energy of many molecular bonds (Vogel and Venugopalan, 2003). However, consideration of the reaction pathways following the excitation by a UV photon reveals that photochemical bond scission competes with the excitation of internal molecular vibrational states that results in heating and forms the basis for possible photothermal decomposition. In Fig. 10, two types of chemical dissociation pathways for a diatomic molecule $A - B$ are shown. The first possibility, shown in Fig. 10A, is that electronic excitation promotes the molecule directly into an electronic excited state with no net bonding and results in a direct dissociation of the molecule into its constituent atoms $A + B$. The second possibility (Fig. 10B) represents the case in which the electronic excitation promotes the molecule into a bound excited state matching that of a second, dissociative electronic state $A + B^*$. Thus, while absorption of the UV photon promotes the molecule into a bound vibrational level within the excited electronic state, as the bond extends, the electronic configuration $A + B^*$ can acquire the repulsive character of a dissociative state or remain bound. While such processes are often discussed in the tissue ablation literature, we are unaware of laser cellular surgery applications where photochemical decomposition has been demonstrated to be the dominant mechanism of action.

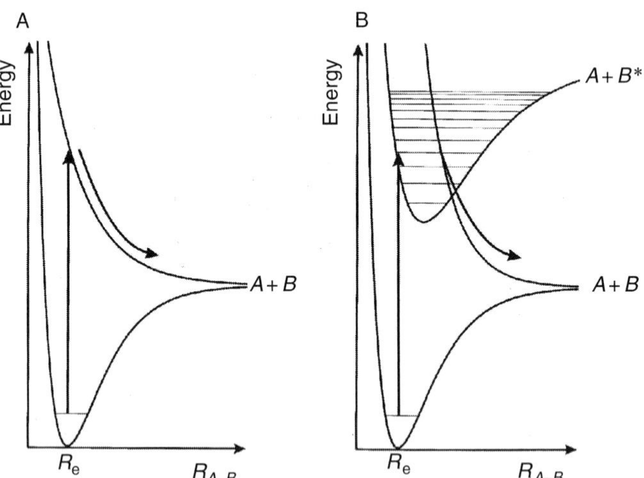

Fig. 10 Energy level diagram illustrating various pathways for photochemical bond breaking. Reprinted with permission from Ashfold and Cook (2002). Copyright 2002, Academic Press.

V. Nonlinear Processes

A. Multiphoton Absorption

Multiphoton absorption refers to a process where k photons interact simultaneously with a molecule to promote it from a ground state to an excited state. The simultaneous absorption of these k photons is equivalent, in terms of energy, to the absorption of a single photon with a shorter wavelength $\lambda_{sp} = \lambda_{mp}/k$, where the subscripts "sp" and "mp" refer to "single-photon" and "multiphoton" absorption, respectively. However, because multiphoton ionization requires the simultaneous interaction of k photons, the rate of absorption $\propto I^k$ as opposed to the rate of linear absorption which is simply $\propto I$. This raises the possibility that materials that may not absorb at a given wavelength λ_0 at low intensities may display absorption of the same radiation at high intensities because k photons are combining to promote an energy level in the molecule that can normally only be populated via absorption of a photon at $\lambda = \lambda_0 /k$.

In the case of pulsed laser microbeams, the maximum intensity is achieved in an ellipsoidal volume as defined by Grill and Stelzer (see Section II.D) by $I \propto \exp[-(r^2/a^2) - (z^2/b^2)]$. Given that the rate of multiphoton absorption is extremely nonlinear αI^k the volume in which multiphoton processes are operative is smaller because $I^k \propto \exp[-(kr^2/a^2) - (kz^2/b^2)]$, so that the lateral and axial dimension becomes narrower by a factor $\propto \sqrt{k}$. As a result, a laser beam can be delivered into a transparent medium and produce a very specific interaction only in the focal volume where the intensities are high. This explains, for example, the ability for a pulsed laser microbeam to modify organelles within a cell without disturbing the membrane because there is no absorption outside the focal volume.

B. Nonlinear Absorption via Optical Breakdown

Nonlinear absorption refers to changes in the intrinsic absorption coefficient (μ_a) of a material at high irradiances. This is different than multiphoton absorption where the intrinsic absorption coefficient of the material is unchanged. Nonlinear absorption is typically observed at irradiances of $\sim 10^{10}$–10^{13} W/cm^2 even in nominally transparent media such as water or cytoplasm. The high electric field strengths lead to ionization of the medium and the generation of quasi-free electrons. In the absence of impurities, this ionization process is initiated by multiphoton absorption. These quasi-free electrons are potent absorbers of the incident radiation that provide a mechanism by which additional energy can be absorbed. The generation of a critical density of quasi-free electrons $\approx 10^{21}$ cm^{-3} is referred to as "plasma formation" and the process in which plasma formation is induced by a laser beam is called "optical breakdown." Laser microbeams of longer (ns) duration have more time to heat the plasma and tend to generate more violent effects than optical breakdown formed by shorter (ps, fs) laser pulses. In the following sections, we consider the processes of multiphoton ionization, quantum tunneling, and avalanche ionization that are involved in the optical breakdown of liquids. We also describe how changes in laser parameters, most notably pulse duration and wavelength, affect the mechanism and dynamics of the optical breakdown process.

C. Dynamics of Optical Breakdown in Liquids

In this section, we examine the process of optical breakdown in liquids. We will consider water in particular as its behavior with respect to optical breakdown is essentially identical to that of biological fluids (Docchio *et al.*, 1986, 1988). To describe the creation of free electrons we treat water as an amorphous semiconductor (Kennedy, 1995; Sacchi, 1991; Williams *et al.*, 1976). This analogy is used because the bound electron states are reminiscent of an insulator while the quasi-free electrons represent a conductive state. More precisely, electrons that have been promoted to a conduction band are termed quasi-free electrons. However, for simplicity, we will use the term "free electron" to refer to a quasi-free electron and "ionization" to refer to the promotion of electrons into the conduction band.

An overview of the optical breakdown process is provided in Fig. 11. The generation of free electrons can occur via one of four mechanisms: (1) multiphoton ionization, (2) quantum tunneling, (3) impact ionization, and (4) thermionic emission. For a particular set of laser parameters, usually only one or two of these mechanisms are important. In general, thermionic emission is not important in laser microbeam applications and will not be given detailed consideration. In what follows we describe the physics of these processes and, in doing so, we describe the details of Fig. 11 as well as the conditions for which each of these mechanisms is important.

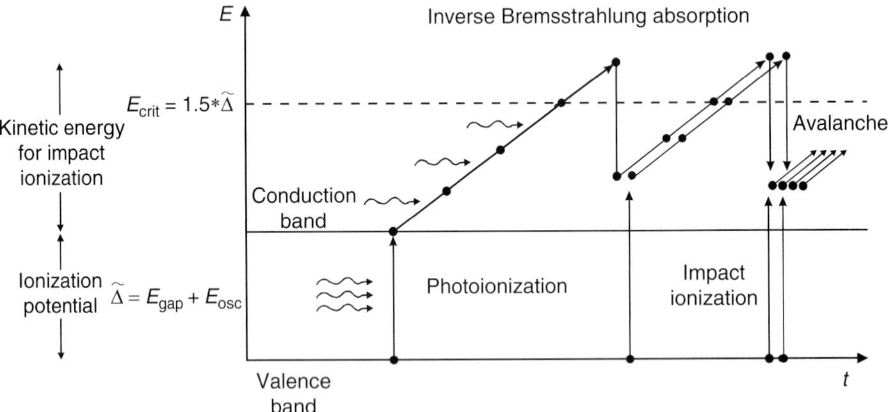

Fig. 11 Schematic diagram of the interplay between photoionization, inverse Bremsstrahlung absorption and impact ionization during plasma formation. See text for further details. Reprinted with permission from Vogel *et al.* (2005). Copyright 2005, Springer.

1. Ionization Potential

The ionization potential must be surmounted in order to produce free electrons. In water, the energy necessary to promote an electron into the conduction band is $\Delta = 6.5$ eV. Photons corresponding to the wavelengths 1064, 800, 532, and 355 nm possess single-photon energies of 1.17, 1.56, 2.34, and 3.51 eV, respectively. Thus the energy of 6, 5, 3, or 2 photons are required at these respective wavelengths to overcome the energy gap between the valence and conduction bands. Moreover, the electrons oscillate as they interact with the electric field of the laser microbeam that demands an additional contribution of energy in order for the electrons to be ionized. This results in the following expression for the ionization potential (Keldysh, 1965; Vogel *et al.*, 2005):

$$\tilde{\Delta} = \Delta + \frac{e^2 E^2}{4m\omega^2} \tag{22}$$

where e is the electron charge, $m = m_c m_v/(m_c + m_v)$ is the reduced mass of the quasi-free electron m_c and the hole in the valence band m_v, E is the electric field amplitude, and $\omega = 2\pi c/\lambda$ is its angular frequency. When the intensities necessary for optical breakdown are moderate, for example, when using nanosecond pulses, the contribution from the electron oscillation is negligible and $\tilde{\Delta} \sim \Delta$. This is no longer true for ultrashort (fs) pulsed laser microbeams when the electric field strengths get very large and the contribution of electron oscillation to the ionization potential becomes significant.

2. Multiphoton Ionization and Quantum Tunneling

Both multiphoton ionization and quantum tunneling are photoionization processes. Unlike impact ionization, these processes do not rely on the existence of free electrons in the focal volume. Given that free electrons are not typically present within biological samples, either multiphoton ionization or quantum tunneling provides the seed electron which is the initiating event for plasma formation in laser microbeam applications.

Multiphoton ionization is the near-simultaneous absorption of several photons. As discussed in the beginning of this section, the rate of this process is proportional to I^k, where k is the number photons required to achieve ionization. Figure 12 shows the energy diagram for water and the effective wavelengths for multiphoton absorption using 1, 2, 3, or 5 photons. In this figure, we assume for simplicity that the contribution of electron oscillation to the ionization potential is negligible so that $\tilde{\Delta} \sim \Delta$.

In quantum tunneling the electron spontaneously appears in the excited (free) state. As indicated by its name, quantum tunneling is a quantum effect and cannot be explained classically. Quantum mechanics provides a finite probability for electrons to spontaneously cross a potential energy barrier into an excited state. The irradiance and frequency of the laser microbeam determine whether multiphoton

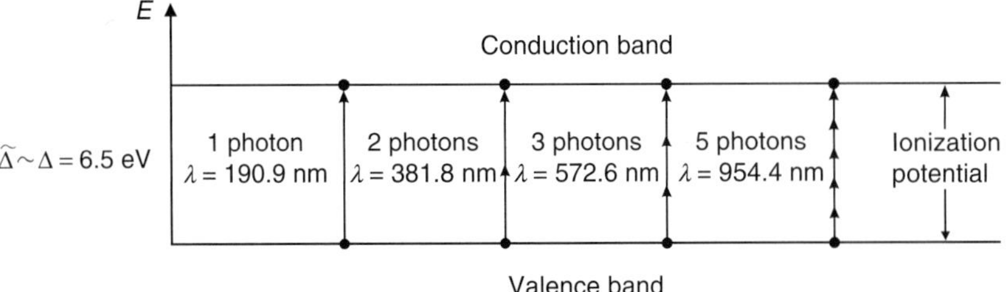

Fig. 12 Schematic diagram of energy levels involved in multiphoton ionization. The band gap energy for water corresponds to a photon with wavelength $\lambda = 190$ nm. However, if simultaneous absorption of 2, 3, or 5 photons occurs, the corresponding maximum wavelength required to overcome the ionization potential are increased to $\lambda = 381$, 572, and 954 nm, respectively.

ionization or quantum tunneling dominates the photoionization process. The Keldysh parameter $\gamma = \omega/\omega_t$ can be used to delimit the different regimes of photoionization (Keldysh, 1965; Vogel *et al.*, 2005), where $1/\omega_t$ is the tunneling time through the atomic potential barrier and ω is the angular frequency of the electric field. Thus γ is simply the ratio of the quantum tunneling time to the oscillation time of the electromagnetic wave. The tunneling time $1/\omega_t$ is inversely proportional to the field strength E^2. Thus as the electric field strength increases, the tunneling time decreases and the probability for electrons to tunnel into the excited state increases. $\gamma \ll 1$ is achieved when using low frequencies and/or large field strengths and leads to a photoionization process dominated by quantum tunneling. By contrast $\gamma \gg 1$ is achieved using high frequencies and/or low to moderate field and leads to a multiphoton ionization dominated process. Typically large field strengths are involved only when using ultrashort pulsed laser microbeams and thus quantum tunneling must be considered only in this case. For all other cases, multiphoton ionization can be considered the dominant mechanism for photoionization. Further details can be found in the review paper by Vogel *et al.* (2005).

Once, photoionization provides the seed electron necessary for the plasma formation, other processes are sometimes available to increase the number of free electrons in the focal volume of the laser microbeam. The principal processes of interest are impact and avalanche ionization that we describe next.

3. Avalanche Ionization

Avalanche (aka cascade) ionization requires the presence of at least one electron in the conduction band that is typically provided by multiphoton ionization or quantum tunneling. Once the free electron is generated within the focal volume, it absorbs photons from the laser microbeam (provided the laser pulse is still present) via inverse Bremsstrahlung absorption. Inverse Bremsstrahlung absorption is a process where a free electron interacts with the potential of an ion/molecule and absorbs a photon. The presence of the ion/molecule is necessary for momentum conservation. The photon energy is converted into kinetic energy by accelerating the electron. In this way an electron can absorb several photons until its kinetic energy is sufficient to generate another free electron through collision with another molecule. This can only occur if the kinetic energy of the electron exceeds a critical value E_{crit} that is related to the effective ionization potential $\tilde{\Delta}$ as

$$E_{\text{crit}} = \left(\frac{1+2\mu}{1+\mu}\right)\tilde{\Delta} \tag{23}$$

where $\mu = m_c/m_v$ is the ratio between the masses of the quasi-free electron and the "hole" left in the valence band. Assuming a value of $\mu = 1$ for water (Vogel *et al.*, 2005), we find that the minimum energy for impact ionization to occur is $E_{\text{crit}} = 1.5\tilde{\Delta}$. The excess energy that remains, $0.5\tilde{\Delta}$, is distributed among the electrons after the collision as shown in Fig. 11 (Ridley, 1999).

This process of an electron colliding with another molecule to generate an additional free electron is called impact ionization. These two electrons can in turn undergo a series of inverse Bremsstrahlung absorption events and gain a sufficient amount of kinetic energy to ionize another two electrons through impact ionization and so on. In this way, the number of free electrons increases geometrically. Moreover, the plasma itself acquires a high energy density, that is, it gets hotter, because inverse Bremsstrahlung absorption provides energy to the electrons in excess of the ionization potential. However, since this process takes time, it can only occur if the laser pulse is sufficiently long. The rate of the avalanche ionization is proportional to the rate of inverse Bremsstrahlung events which, in turn, is $\propto I$.

D. Rate Equation for Free Electrons

The above qualitative descriptions of the various physical mechanisms by which a plasma is generated can be codified within a mathematical framework. This framework describes the temporal evolution of the free electron density through a simple rate equation. The rate equation has two terms that express the generation of free electrons through multiphoton and avalanche ionization processes and two terms that describe the loss of free electrons through diffusion out of the focal volume and electron–ion recombination. This results in the following rate equation for the electron density ρ (Noack and Vogel, 1999; Shen, 1984; Vogel *et al.*, 2005).

$$\frac{d\rho}{dt} = \eta_{\text{mp}} + \eta_{\text{casc}}\rho - g\rho - \eta_{\text{rec}}\rho^2 \tag{24}$$

where η_{mp} and η_{casc} are the rates of multiphoton and cascade ionization respectively, g is the rate of electron loss through diffusion, and η_{rec} is the recombination rate.

Computational results that solve Eq. (24) for two different wavelengths ($\lambda = 580$ and 1064 nm) and three pulse durations (6 ns, 30 ps, and 100 fs) are shown in Fig. 13 and taken from an earlier study by Noack and Vogel (1999). These plots show the evolution of the free electron density during the laser pulse. The results are plotted using a normalized time coordinate in which time is divided by the laser pulse duration (t/τ_{L}). The solid curves represent the total free electron density, while the dotted curve is the contribution that multiphoton absorption provides to the total generation of free electrons. The simulations show that a reduction in the laser pulse duration leads to an increase in the contribution of multiphoton ionization to the total free electron density. This effect is more prominent at $\lambda = 580$ nm where the higher single-photon energy increases substantially the rate of multiphoton ionization. This also makes the time evolution of the free electron density smoother as there is less of a "disconnect" between the rates of multiphoton ionization and avalanche ionization. The strength of the multiphoton and avalanche ionization processes in this rate equation depends on the irradiance of the laser pulse. At low irradiance, avalanche ionization tends to be more important as its rate is $\propto I$, whereas multiphoton ionization (or tunneling depending on γ)

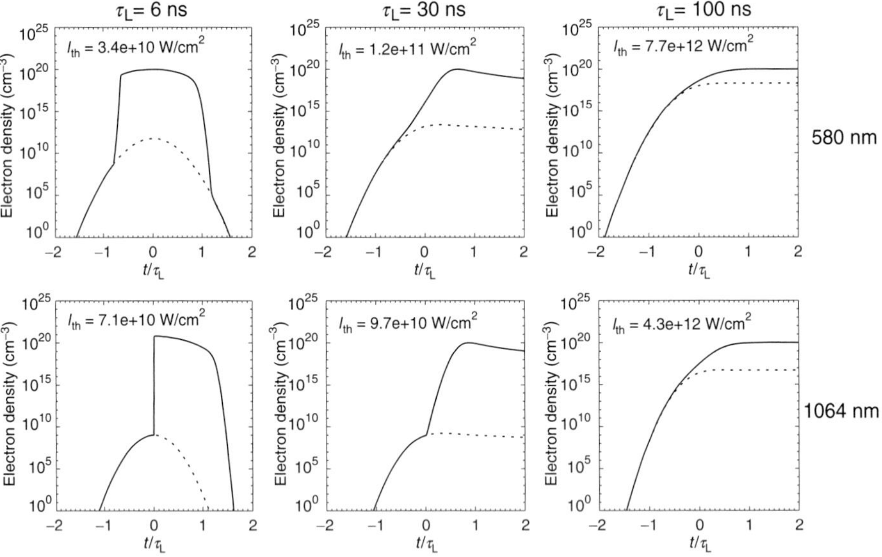

Fig. 13 Evolution of the free electron density at the optical breakdown threshold for different pulse durations and different wavelengths. The solid curves give the total free electron density, while the dotted curves provide the contribution of multiphoton ionization to the total free electron density. Time axis normalized to pulse duration τ_{L}. Reprinted with permission from Noack and Vogel (1999). Copyright 1999, IEEE.

is more important at high irradiances (short pulse durations) as its rate is $\propto I^k$. This allows for the following generalizations regarding the dominant mechanisms involved in plasma formation.

- Long pulses (nanosecond)—Seed electron provided by multiphoton ionization. Plasma formed through avalanche ionization.
- Short pulses (picosecond)—Avalanche and multiphoton ionization processes contribute equally to the generation of free electrons.
- Ultrashort pulses (femtosecond)—Plasma formation is dominated by multiphoton ionization and/or quantum tunneling depending on Keldysh parameter γ. The process is not affected by the linear absorption.

Note that these generalizations are valid in the absence of linear absorption in the medium. If a pathway for linear absorption is provided, for example by the administration of exogenous dyes or gene-expressed fluorescent proteins, seed electrons can be provided through thermionic emission and can lower significantly the minimum radiant exposure/intensity necessary for optical breakdown (Oraevsky *et al.*, 1996).

E. Dependence of Optical Breakdown on Pulse Width and Wavelength

Also of interest is the variation of the threshold irradiance and radiant exposure for optical breakdown with both pulse duration and wavelength. In Fig. 14, the threshold irradiance I_{th} and radiant exposure $\Phi_{th} = I_{th} \times \tau_L$ (Fig. 14B) necessary to reach a critical electron density of $\rho = 10^{21}/cm^3$ is calculated using Eq. (24). Given that there is less time available at shorter pulse durations, we expect that cascade ionization would become less effective in generating free electrons. Thus as the laser pulse duration decreases, plasma formation relies increasingly on the generation of free electrons from multiphoton ionization and quantum tunneling. As a result, the threshold irradiance for plasma formation should increase markedly

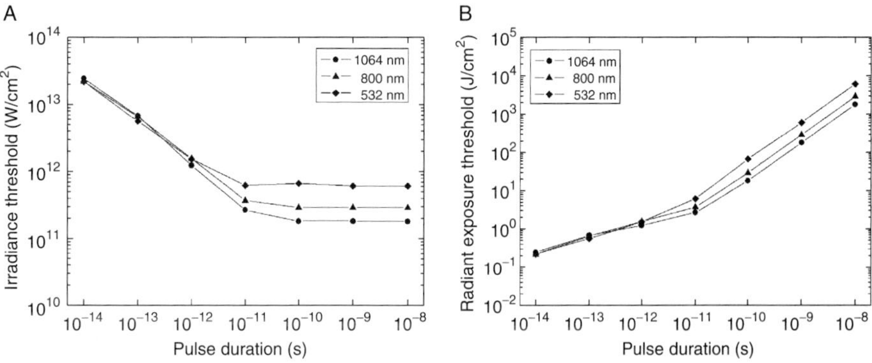

Fig. 14 Calculated threshold irradiances and radiant exposures for optical breakdown as a function of pulse duration at selected wavelengths. Reprinted with permission from Vogel *et al.* (2005). Copyright 2005, Springer.

with decreasing pulse duration. This general feature is evident in computer simulation results shown in Fig. 14A (Vogel *et al.*, 2005). Figure 14B shows the threshold radiant exposure Φ_{th} necessary for plasma formation as a function of pulse duration. This plot has the opposite trend when compared to the threshold irradiance. Specifically, shorter pulse durations result in a smaller energy dose necessary to form a plasma.

As described by Vogel *et al.* (2005), it is possible to identify two regimes in the plots shown in Fig. 14. In the first regime, for pulse durations shorter than 10 ps, electron–ion recombination and electron diffusion are relatively unimportant due to the short irradiation time. As a result, we can consider that only a single set of free electrons is created during irradiation. This means that plasmas formed in this regime all have roughly the same energy density. This makes the threshold radiant exposure Φ_{th} weakly dependent on pulse duration. By contrast, the irradiance will increase considerably with decreasing pulse duration due to the increased reliance on multiphoton absorption or quantum tunneling to generate the free electrons. This regime where the threshold radiant exposure is only weakly dependent on pulse duration for $\tau_L \lesssim 10$ ps has been confirmed by work done by Loesel *et al.* (1996).

In second regime, for pulses longer than 10 ps, free electrons are continually being formed (and are also recombining) during the laser pulse. In this situation, the multiphoton absorption process that provides the seed electrons to trigger the avalanche is the rate-limiting step. This implies that, a critical irradiance must be exceeded to provide the seed electrons. As a result, the threshold radiant exposure for optical breakdown must increase with increasing pulse duration in order to maintain this critical irradiance. This is confirmed by the results shown in Fig. 14A.

F. Effects of Linear Absorption on Threshold Energy

Up to this point, we have not discussed what relevance, if any, linear absorption has to optical breakdown. In fact, linear absorption can be quite important as it can provide a temperature increase that can create free electrons through a process of thermionic emission. For nanosecond pulses, linear absorption can lower dramatically the threshold radiant exposure required for plasma formation because thermionic emission can provide free electrons to initiate the avalanche process (Oraevsky *et al.*, 1996). For picosecond pulses the effect is less pronounced because the initial free electron density is created by multiphoton absorption. For femtosecond pulses linear absorption provides no enhancement as plasma formation is dominated by multiphoton ionization or quantum tunneling.

VI. Thermal, Mechanical, and Chemical Effects Generated by Laser-Induced Plasma Formation

In Section V, we described the processes by which pulsed laser microbeams can ionize biological media and lead to plasma formation. Because free electrons are generated through nonlinear processes, and are avid absorbers of light, they offer a means to deposit energy highly localized to the focal volume of the laser microbeam. We discussed the interplay between the direct generation of (quasi-) free electrons through multiphoton ionization and quantum tunneling mechanisms. These processes of direct photoionization can, in turn, lead to an "amplification" of the free electron density through a succession of inverse Bremsstrahlung absorption events and impact ionization, collectively known as avalanche or cascade ionization. The interplay of these processes is strongly affected by laser pulse duration, leading to a significant reduction in the radiant exposure (and thus pulse energy) required for optical breakdown with decreasing pulse duration.

In this section, we describe the resulting thermal, mechanical, and chemical processes that result from laser-induced plasma formation. The energy deposition provided by the elevated free electron density results in a localized elevation of temperature and pressure. What results is an interesting complex evolution of thermomechanical processes that can only be fully understood by a detailed consideration of the thermal diffusion, thermoelastic, and phase transition processes involved (Vogel and Venugopalan, 2003; Vogel *et al.*, 2005). For the vast majority of applications, it is these thermomechanical processes that are responsible for the resulting cellular injury. However, reductions in the duration of the laser microbeam also reduce the energy density of the free electrons. This occurs not only because the threshold radiant exposure is lower but also because there is less time for the free electrons to be "heated" through inverse Bremsstrahlung absorption events. This reduces the violence of the thermomechanical damage processes and leads to more precise laser-microbeam effects. In fact, comparison of laser parameters used experimentally with numerical simulations has indicated that when using femtosecond pulses, optical breakdown is not a necessary condition for cellular injury. Instead chemical effects resulting from the generation of free electrons below the optical breakdown threshold can be used to produce targeted photodamage and/or the direct fragmentation of biomolecules through the production of reactive oxygen species and the capture of free electrons into antibonding molecular orbitals (Vogel *et al.*, 2002a,b, 2005).

A. Thermomechanical Processes Resulting from Plasma Formation

For pulse durations ($\gtrsim$10 ps, the temperature rise associated with plasma formation leads to direct vaporization of the cellular material, and the high temperatures and pressures generated in the focal volume lead to plasma expansion. This expansion results in the compression of the surrounding material and radiation of a shock wave. The expansion also results in cooling of the plasma, leading to electron–ion recombination and the formation of a cavitation bubble that expands until the internal pressure inside the bubble reaches the vapor pressure of the liquid. At this point, the hydrostatic pressure of the surrounding liquid collapses the expanded gas bubble. As the pulse duration is reduced further, the energy density of the plasma is reduced and the temperature rise alone is not sufficient to vaporize the material. Instead the temperature elevation combines with *tensile* stresses generated by the thermoelastic response of the material to result in bubble formation, that is, fracture. The size of the cavitation bubble and the duration of the expansion and collapse process generally increase with both the pulse duration and pulse energy and range on the order of microseconds for nanosecond laser pulses to tens of nanoseconds for femtosecond laser pulses.

1. Shock Wave and Cavitation Bubble Dynamics

While the pressure magnitudes associated with the shock wave resulting from plasma formation are substantial (Rau *et al.*, 2004; Venugopalan *et al.*, 2002; Vogel *et al.*, 1996), they have not been implicated in the permanent mechanical damage of cellular structures. However, the fluid shear stresses generated by the cavitation bubble expansion and collapse can be quite disruptive (Rau *et al.*, 2006). At large pulse energies, the shear stresses generated by the cavitation bubble can result in the permeabilization of cell membranes and the lysis of entire cells in a region around the focal volume (Krasieva *et al.*, 1998; Rau *et al.*, 2004, 2006; Soughayer *et al.*, 2000). Figure 15 provides a visual display of the time evolution of the plasma, shock wave, and cavitation bubble resulting from the delivery of a 30-μJ, 6-ns pulsed laser microbeam at $\lambda = 532$ nm to a confluent monolayer of PtK2 (rat kangaroo kidney epithelial) cells. In this times series, the plasma formation (A and B), shock wave emission (C–E), cavitation bubble formation (C), expansion (D–I), and collapse (J and K) are clearly visualized. Rau *et al.* (2004, 2006) have provided detailed description and quantitation of the dynamics, as well as an analysis of the hydrodynamic stresses responsible for cellular injury.

Detailed studies have examined the dynamics of shock wave emission and bubble formation to gain an understanding of the conversion of the optical energy of the pulsed laser microbeam into mechanical energy (Rau *et al.*, 2004, 2006; Venugopalan *et al.*, 2002; Vogel *et al.*, 1996, 1999b). These studies reveal important features regarding the relationship between the mechanism of damage and the pulse duration of the microbeam ranging across the nanosecond, picosecond, and femtosecond regimes. The key characteristics of the damage mechanisms that are operative in these three regimes are summarized below.

2. Nanosecond Pulses

While optical breakdown for nanosecond pulses relies on the generation of seed electrons, most often through multiphoton ionization, the dynamics of plasma formation is governed by avalanche ionization. Avalanche ionization provides for significant heating of the plasma. Moreover, the nanosecond pulse duration is much longer than the characteristic time for electron–phonon interactions and thus there is ample time available during irradiation for equilibration between the electronic and vibrational modes in the medium. When using pulse energies significantly in excess of the optical breakdown threshold, this results in a plasma with a large volumetric energy density and temperatures on the order of several thousand

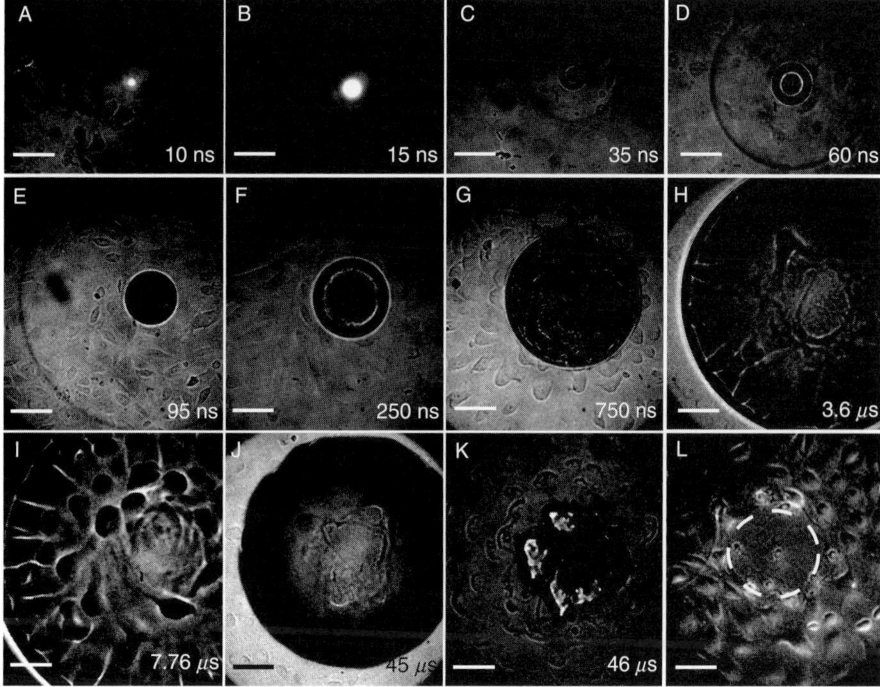

Fig. 15 Time-resolved series of images depicting the time evolution of pulsed laser microbeam cell lysis when delivering a 30-μJ, 6-ns pulsed laser microbeam at $\lambda = 532$ nm via a 40×, 0.8 NA microscope objective to a confluent monolayer of PtK2 cells. Scale bar = 50 μm. See text for further details.

degrees Kelvin (Vogel *et al.*, 1999b). This typically results in the violent effects shown in Fig. 15. Nevertheless, pulsed laser microbeams in the nanosecond regime are used successfully for precise laser microsurgery. Typically, the energies used are not much higher than the threshold for optical breakdown (Hutson *et al.*, 2003; Khodjakov *et al.*, 1997a,b, 2000; Krasieva *et al.*, 1998; Venugopalan *et al.*, 2002).

3. Picosecond Pulses

In the picosecond regime, multiphoton ionization begins to assert itself more strongly in the optical breakdown process. This is because less time is available for the avalanche ionization process, which requires a succession of inverse Bremsstrahlung absorption and impact ionization events, to get established. This has important practical implications for the picosecond microsurgical process. First, because one must rely more on multiphoton ionization for the generation of free electrons, the threshold irradiance for optical breakdown is significantly higher than in the nanosecond regime as shown in Fig. 14A. Second, the smaller contribution from avalanche ionization and, to a lesser degree, incomplete equilibration between the electrons and phonons result in a plasma with a smaller volumetric energy density (and temperature) as compared to optical breakdown using nanosecond pulses. This is true even when considering plasmas of equivalent free electron density. Third, the lower energy density of the plasma leads to a significant reduction in the severity and extent of the damage produced by direct vaporization by the plasma as well as mechanical damage produced by the cavitation bubble dynamics (Vogel *et al.*, 1994).

As compared to nanosecond laser microbeams, these characteristics result in much better precision and wider range of usable pulse energy for picosecond pulses. In contrast to nanosecond laser microbeams, where the minimum pulse energies on the order of $\sim$1 μJ and larger must be used to produce a microsurgical effect, laser microbeams with durations on the order of 30 ps can produce a microsurgical effects using pulse energies in the range of 10–100 nJ. However, while these characteristics provide for much more confined damage, the principal damage mechanisms remain essentially identical to those operative when using nanosecond laser microbeams.

4. Femtosecond Pulses

In the femtosecond regime, avalanche ionization is no longer operative, and multiphoton ionization and quantum tunneling alone generate free electrons. The multiphoton nature of the free electron generation process has two benefits. First, it allows for precise "tuning" of the density of free electrons. This is in contrast to optical breakdown when using nanosecond and, to a lesser degree, picosecond pulses where the interplay between multiphoton and cascade ionization during the laser pulse makes it difficult to precisely adjust the free electron density generated in the focal volume. This allows for the consistent production of free electron densities in the focal volume that are *below* the threshold of optical breakdown. This is commonly referred to as the generation of a "low-density" plasma that often carries with it a low volumetric energy density. The high reactivity of these free electrons enables, with multiple pulses, the creation of subtle photochemical damage of biological media from the generation of reactive oxygen species as well as photodissociation due to the capture of electrons in antibonding molecular orbitals (Vogel *et al.*, 2002a,b, 2005).

Second, the threshold radiant exposure for optical breakdown is minimized when using the shortest available pulses. This means that when using pulse energies that achieve optical breakdown there is very little energy available for disruptive mechanical effects. As a result, femtosecond pulses potentially offer the best option for precise localized microsurgery. However, given the fact that the threshold radiant exposure does not change very much for pulse durations lower than $\approx$10 ps, it is not yet clear how much precision is gained when moving to femtosecond laser pulses. We will discuss particular applications of femtosecond microsurgery in Section VII.

VII. Examination of Laser Parameters Used in Laser Microbeam Applications

We will conclude this chapter by examining the mechanisms that likely mediate the use of laser microbeams in published studies using pulses from the microsecond to femtosecond regimes. A comprehensive analysis of experimental reports utilizing femtosecond laser microbeams is available in Vogel *et al.* (2005).

A. Chromosome Microsurgery Using Microsecond Argon Laser Pulses

Berns *et al.* (1969) reported the use of single 25-μs duration pulses from a multiline argon ion laser with 1.5-W peak power and emission predominately at $\lambda = 514$ and 488 nm through a 100×, 1.3 NA microscope objective to *acridine orange-stained* chromosomes in living mitotic salamander cells. Single pulses produced lesions of submicrometer dimension in anaphase chromosomes that appeared 10–15 min following irradiation. In a later study, a more powerful Argon laser with 12-W peak power and 30-μs pulse duration was used to produce lesions in *unstained* chromosomes that appeared immediately following irradiation (Berns *et al.*, 1971). Figure 16 shows a typical phase-contrast image of a PtK2 cell during anaphase after the placement of two lesions in an unstrained chromosome by pulsed argon ion irradiation (Berns *et al.*, 1981). Berns and coworkers performed further assays to suggest that the mechanism of chromosome damage was different in these two instances. The lesions produced with the assistance of acridine orange staining seemed to damage the DNA within the chromosome whereas the lesions produced without staining seemed to be devoid of histone protein content (Berns *et al.*, 1971).

For the production of lesions in stained chromosomes, Berns *et al.* (1971) estimated that a pulse energy of 10–15 μJ was available in the focal plane. This results in an estimated peak irradiance of 2.18×10^8 W/cm^3 and radiant exposure of 5.46×10^3 J/cm^3. In the second study, 190 μJ was available in the focal volume resulting in a peak irradiance of 3.45×10^9 W/cm^2 and radiant exposure of 1.04×10^5 J/cm^2.

It is quite reasonable to assume that the acridine orange stain provided sufficient linear absorption to provide a thermal mechanism for creation of the lesion. Published absorption spectra of the DNA–acridine orange complex show a broad absorption peak centered at approximately $\lambda = 500$ nm (Suzuki *et al.*, 1972). Although the pulse duration of 25-μs duration is much larger than the characteristic thermal diffusion time ($\approx$200 ns), the high numerical aperture of microbeam radiation produces an extremely small heated volume and provides a highly localized thermal distribution even with thermal diffusion (Vogel *et al.*, 2005). For the unstained chromosomes, the irradiance is 3.45×10^9 W/cm^2 and much too low for multiphoton absorption to be operative. Although the prominent absorption peaks of DNA and histone protein both lie at , $\lambda \lesssim 300$ nm fluorescence spectroscopy studies of proteins such as collagen and elastin reveal weak fluorescence excitation

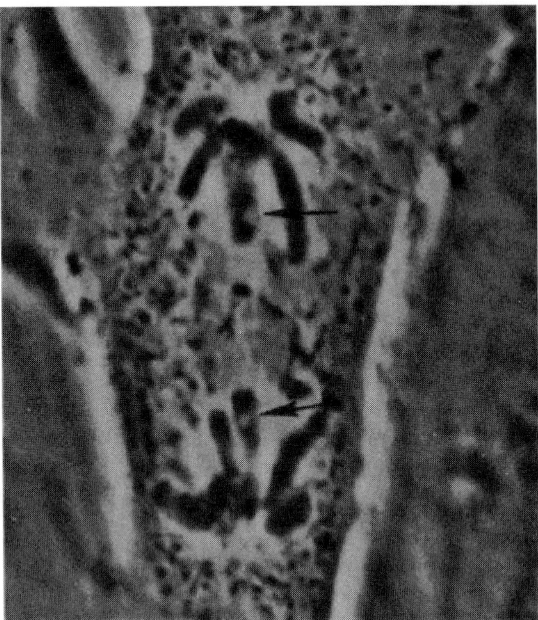

Fig. 16 Phase-contrast photomicrograph of anaphase PtK2 chromosomes after placement of two lesions, 1 μm in diameter on the chromosome arms as indicated by the arrows. Lesions wre produced by irradiation at 514-nm with an Argon laser with a radiant exposure of 10^5 J/cm^2. Reprinted with permission from Berns *et al.* (1981). Copyright 1981, American Association for the Advancement of Science.

peaks at $\lambda \lesssim 410$ and 450 nm (Richards-Kortum and Sevick-Muraca, 1996). Thus it is likely that when using the multiline argon laser which contains emission lines at $\lambda = 457.9$, 488, and 496.5 nm (Berns *et al.*, 1971) combined with the substantial radiant exposure of 1.04×10^5 J/cm^2, sufficient linear absorption is present in the histone proteins to generate the observed lesions.

B. Centrosome Ablation Using Nanosecond 532-nm Laser Microbeams

Centrosomes are associated with the function of microtubule organization during cell division. During prophase the centrosomes align at the equator of the cell making the poles for microtubules. To examine these processes Khodjakov and Rieder have performed centrosome ablation in a mammalian cell line that expresses γ-tubulin green fluorescent protein (GFP) at the centrosome (Khodjakov *et al.*, 2000).

In these studies, complete inactivation of the centrosome was achieved through the delivery of ~30–40 laser pulses at $\lambda = 532$ nm delivered at a repetition rate at 10 Hz with 500-nJ pulse energy and 7-ns pulse duration. The microbeam was formed by a 60×, 1.4 NA microscope objective. This produced a diffraction-limited spot size of 464 nm resulting in an estimated radiant exposure and irradiance in the focal plane of 4.23×10^{10} W/cm^2 and 296 J/cm^2, respectively. Similar laser parameters have also been used by this group to inactivate kinetochores and microtubules (Khodjakov *et al.*, 1996, 1997b). The peak irradiance of 4.23×10^{10} W/cm^2 is a little more than half the threshold irradiance for optical breakdown of water using Q-switched nanosecond laser pulses at $\lambda = 532$ nm for a numerical aperture of 0.9 (Venugopalan *et al.*, 2002). This proximity to the breakdown threshold combined with the linear absorption provided by the GFP-stained centrosomes was likely sufficient to provide for optical breakdown for this set of laser parameters (Oraevsky *et al.*, 1996).

C. Stimulation of Intracellular Ca^{2+} Waves Using Picosecond 775-nm Laser Microbeams

Using a Ti:Sapphire laser with a regenerative amplifier, Iwanaga *et al.* (2005) reported on the delivery of 1.7-ps laser pulses at $\lambda = 775$ nm through a 60×, 0.9 NA objective to trigger intracellular Ca^{2+} waves in HeLa cells. They found that when delivering eight pulses at a 1-kHz repetition rate, the minimum single-pulse radiant exposure to trigger an intracellular Ca^{2+} wave was 0.78 J/cm^2 resulting in a peak irradiance of 4.58×10^{11} W/cm^2. This corresponds very well to the threshold doses for optical breakdown as predicted by the calculations shown in Fig. 14.

D. Use of Femtosecond Laser Microbeams at 800-nm for Transfection and Nanodissection

As described in Section VI, the use of femtosecond lasers provides for the precise adjustment of the free electron density within the focal volume below the threshold for optical breakdown. These highly reactive free electrons can then provide damage through a photochemical mechanism. Alternatively, one can employ larger pulse energies to produce optical breakdown and rely on the cavitation bubble dynamics to contribute mechanical processes to the damage mechanisms.

In the first category, the group of König has used $\lambda = 800$-nm femtosecond pulses delivered at a 80-MHz repetition rate with single pulse energies of ~0.6 nJ for cellular applications (König *et al.*, 2001; Tirlapur and König, 2002). Numerical simulations have estimated this optical dose to provide less than 7% of the energy necessary for optical breakdown and a free electron density of only 10^{15} e^-/cm^3 (Vogel *et al.*, 2005). Nevertheless, the delivery of ~10^6 of these low-energy femtosecond pulses have been shown to achieve cell membrane permeabilization and DNA transfection in live rat kangaroo kidney epithelial (PtK2) and Chinese hamster ovary (CHO) cells (Tirlapur and König, 2002). Presumably it is the accumulation of photochemical effects mediated by the low density of free electrons that are generated which are responsible for the observed damage (Vogel *et al.*, 2005).

In the second category of interactions, the Mazur group has examined the repetitive delivery 200–250 fs duration pulses with significantly higher single-pulse energies (~1.5–5 nJ) delivered at repetition rates of 1 kHz to ablate cellular structures such as F-actin and mitochondria in fixed and live epithelial cells as well as dendrites in living *Caenorhabditis elegans* (Heisterkamp *et al.*, 2005; Maxwell *et al.*, 2005; Shen *et al.*, 2005). Typically 20–500 pulses were delivered to a given location. Unlike the study of Tirlapur and König discussed above, the higher pulse energies are in the neighborhood of the threshold for optical breakdown. The higher pulse energies result in cellular damage that is of a mechanical nature due to cavitation bubble dynamics. However, unlike the cavitation bubble dynamics seen in the nanosecond and picosecond regime, here cavitation bubble formation is achieved through a combination of a significant temperature rise associated with plasma formation and significant tensile thermoelastic stresses due to the achievement of stress confinement when using femtosecond pulses (Vogel *et al.*, 2005). Moreover, the small pulse energies and extreme spatial confinement provide for bubbles with diameter on the order of a few hundred nanometers with lifetimes of tens of nanoseconds (Vogel *et al.*, 2005). These characteristics offer a remarkably precise means for intracellular dissection/modification.

VIII. Concluding Thoughts

Theoretical and experimental work has provided tremendous insight into the operative mechanisms of cellular damage and manipulation in a broad variety of laser microbeam applications. This is an extremely challenging problem as the dynamics are often confined to the nanoscale in both space and time. Clearly, much work remains both experimentally and theoretically to examine these complex phenomena. One very important area is to develop a better optical and physical characterization of biological structures and molecules so that more realistic properties can be incorporated into our models. Equally important is the use of biological assays that are sensitive to, and capable of differentiating between, photothermal, photomechanical, and photochemical processes that are operative in these procedures. Clearly, there remains a wide-open field for more challenging and exciting work!

Acknowledgments

We thank our colleagues that have been involved in this work over the past several years: Nancy Allbritton, Michael Berns, Elliot Botvinick, Arnold Guerra, Amy Hellman, Tatiana Krasieva, Kester Nahen, Kaustubh Rau, Chris Sims, Chung-Ho Sun, Bruce Tromberg, and Alfred Vogel. This work has been supported by the National Institutes of Health through the Laser Microbeam and Medical Program, Grant No. P41-RR-01192, as well as through Grant Nos. R01-RR-14892 and R01-EB-04436.

References

Amy, R. L., and Storb, R. (1965). Selective mitochondrial damage by a ruby laser microbeam: An electron microscopic study. *Science* **150**(3697), 756–758.

Ashfold, M. N. R., and Cook, P. A. (2002). Photochemistry with VUV photons. *In* "Encyclopedia of Physical Science and Technology" (R. A. Meyers, ed.) Academic Press, San Diego.

Ashkin, A. (1970). Acceleration and trapping of particles by radiation pressure. *Phy. Rev. Lett.* **24**(4), 156–159.

Ashkin, A., Dziedzic, J. M., Bjorkholm, J. E., and Chu, S. (1986). Observation of a single-beam gradient force optical trap for dielectric particles. *Opt. Lett.* **11**(5), 288–290.

Berns, M. W., Aist, J., Edwards, J., Strahs, K., Girton, J., McNeill, P., Rattner, J. B., Kitzes, M., Hammer-Wilson, M., Liaw, L.-H., Siemens, A., Koonce, M., *et al.* (1981). Laser microsurgery in cell and developmental biology. *Science* **213**(4507), 505–513.

Berns, M. W., Cheng, W. K., Floyd, A. D., and Ohnuki, Y. (1971). Chromosome lesions produced withan argon laser microbeam without dye sensitization. *Science* **171**(3974), 903–905.

Berns, M. W., Olson, R. S., and Rounds, D. E. (1969). *In vitro* production of chromosomal lesions using an argon laser microbeam. *Nature* **221**(5175), 74–75.

Bessis, M., Gires, F., Mayer, G., and Nomarski, G. (1962). Irradiation des organites cellulaires à l'aide d'un laser à ruby. *C. R. Hebd. Seances Acad. Sci.* **225**(5), 1010–1015.

Bessis, M., and Nomarski, G. (1960). Irradiation ultra-violette des organites cellulaires avec observation continue en contraste de phase. *J. Biophys. Biochem. Cytol.* **8**(3), 777–791.

Botvinick, E. L., Venugopalan, V., Shah, J. V., Liaw, L. H., and Berns, M. W. (2004). Controlled ablation of microtubules using a picosecond laser. *Biophys. J.* **87**(6), 4203–4212.

Boyd, R. W. (1992). "Nonlinear Optics." Academic Press, San Diego, CA.

Brodeur, A., and Chin, S. L. (1998). Band-gap dependence of the ultrafast white-light continuum. *Phys. Rev. Lett.* **80**(20), 4406–4409.

Cole, R. W., Khodjakov, A., Wright, W. H., and Rieder, C. L. (1995). A differential interference contrast-based light microscopic system for laser microsurgery and optical trapping of selected chormosomes during mitosis *in vivo*. *J. Microsc. Soc. Am.* **1**, 203–215.

Dierickx, C. C., Casparian, J. M., Venugopalan, V., Farinelli, W. A., and Anderson, R. R. (1995). Thermal relaxation of port-wine stain vessels probed *in vivo*: The need for 1–10-millisecond laser pulse treatment. *J. Invest. Dermatol.* **105**(5), 709–714.

Docchio, F., Regondi, P., Capon, M. R. C., and Mellerio, J. (1988). Study of the temporal and spatial dynamics of plasmas induced in liquids by nanosecond Nd-YAG laser pulses. 1. Analysis of the plasma starting times. *Appl. Opt.* **27**(17), 3661–3668.

Docchio, F., Sacchi, C. A., and Marshall, J. (1986). Experimental investigation of optical breakdown thresholds in ocular media under single pulse irradiation with different pulse durations. *Lasers Ophthalmol.* **1**, 83–93.

Feng, Q., Moloney, J. V., Newell, A. C., Wright, E. M., Cook, K., Kennedy, P. K., Hammer, D. X., Rockwell, B. A., and Thompson, C. R. (1997). Theory and simulation on the thresholdofwater breakdown induced by focused ultrashort laser pulses. *IEEE J. Quantum Electron.* **33**(2), 127–137.

Grier, D. G. (2003). A revolution in optical manipulation. *Nature* **424**(6950), 810–816.

Grill, S., and Stelzer, E. H. K. (1999). Method to calculate lateral and axial gain factors of optical setups witha large solidangle. *J. Opt. Soc. Am. A* **16**(11), 2658–2664.

Hecht, E. (1974). "Optics." Addison-Wesley, Reading, MA.

Heisterkamp, A., Maxwell, I. Z., Mazur, E., Underwood, J. M., Nickerson, J. A., Kumar, S., and Ingber, D. E. (2005). Pulse energy dependence of subcellular dissection by femtosecond laser pulses. *Opt. Express* **13**(10), 3690–3696.

Hutson, M. S., Tokutake, Y., Chang, M. S., Bloor, J. W., Venakides, S., Kiehart, D. P., and Edwards, G. S. (2003). Forces for morphogenesis investigated with laser microsurgery and quantitative modeling. *Science* **300**(5616), 145–149.

Iwanaga, S., Smith, N., Fujita, K., Kawata, S., and Nakamura, O. (2005). Single-pulse cell stimulation with a near-infrared picosecond laser. *Appl. Phys. Lett.* **87**(24), 243901.

Keldysh, L. V. (1965). Ionization in the field of a strong electromagnetic wave. *Sov. Phy. JETP* **20**(5), 1307–1314.

Kennedy, P. K. (1995). A first-order model for computation of laser-induced breakdown thresholds in ocular and aqueous media: Part I–Theory. *IEEE J. Quantum Electron* **31**(12), 2241–2249.

Khodjakov, A., Cole, R. W., McEwen, B. F., Buttle, K. F., and Rieder, C. L. (1996). Kinetochore moving away from their associated pole do not exert a significant pushing force on the chromosome. *J. Cell Biol.* **135**(3), 315–327.

Khodjakov, A., Cole, R. W., McEwen, B. F., Buttle, K. F., and Rieder, C. L. (1997a). Chromosome fragments possessing only one kinetochore can congress to the spindle equator. *J. Cell Biol.* **136**(2), 229–240.

Khodjakov, A., Cole, R. W., Oakley, B. R., and Rieder, C. L. (2000). Chromosome-independent mitotic spindle formation in vertabrates. *Curr. Biol.* **10**(2), 59–67.

Khodjakov, A., Cole, R. W., and Rieder, C. L. (1997b). A synergy of technologies: Combining laser microsurgery with green fluorescent protein tagging. *Cell Motil. Cytoskeleton* **38**(4), 311–317.

Khodjakov, A., Rieder, C. L., Sluder, G., Cassels, G., Siban, O., and Wang, C. L. (2002). *De novo* formation of centrsomes in vertebrate cells arrested during S phase. *J. Cell Biol.* **158**(7), 1171–1181.

Konig, K., Reimann, I., and Fritzsche, W. (2001). Nanodissection of human chromosomes with near-infrared femtosecond laser pulses. *Opt. Lett.* **26**(11), 819–821.

Krasieva, T. B., Chapman, C. F., Lamorte, V. J., Venugopalan, V., Berns, M. W., and Tromberg, B. J. (1998). Cell permeabilization and molecular transport by laser microirradiation. *Proc. SPIE* **3260,** 38–44.

La Terra, S., English, C. N., Hergert, P., McEwen, B. F., Sluder, G., and Khodjakov, A. (2005). The *de novo* centriole assembly pathway in HeLa cells: Cell cycle progession and centriole assembly/maturation. *J. Cell Biol.* **168**(5), 713–722.

Loesel, F. H., Niemz, M. H., Bille, J. F., and Juhasz, T. (1996). Laser-induced optical breakdown on hard and soft tissues and its dependence on the pulse duration: Experiments and model. *IEEE J. Quantum Electron.* **32**(10), 1717–1722.

Magidson, V., Chang, F., and Khodjakov, A. (2006). Regulation of cytokinesis by spindle-pole bodies. *Nat. Cell Biol.* **8**(8), 891–893.

Maxwell, I., Chung, S., and Mazur, E. (2005). Nanoprocessing of subcellular targets using femtosecond laser pulses. *Med. Laser Appl.* **20**(3), 193–200.

Milonni, P. W., and Eberly, J. H. (1988). "Lasers." John Wiley and Sons, New York, NY.

Noack, J., and Vogel, A. (1999). Laser-induced plasma formation in water at nanosecond to femtosecond time scales: Calculation of thresholds, absorption coefficients, and energy density. *IEEE J. Quantum Electron.* **35**(8), 1156–1167.

Oraevsky, A. A., Silva, L. B. D., Rubenchik, A. M., Feit, M. D., Glinsky, M. E., Perry, M. D., Mam-mini, B. M., Small, W., IV, and Stuart, B. C. (1996). Plasma mediated ablation of biological tissues with nanosecond-to-femtosecond laser pulses: Relative role of linear and nonlinear absorption. *IEEE J. Sel. Topics Quantum Electron.* **2**(4), 801–809.

Paltauf, G., and Dyer, P. E. (2003). Photomechanical processes and effects in ablation. *Chem. Rev.* **103**(2), 487–518.

Rau, K. R., Guerra, A. G., III, Vogel, A., and Venugopalan, V. (2004). Investigation of laser-induced cell lysis using time-resolved imaging. *Appl. Phys. Lett.* **84**(15), 2940–2942.

Rau, K. R., Quinto-Su, P. A., Hellman, A. N., and Venugopalan, V. (2006). Pulsed laser microbeam induced cell lysis: Time-resolved imaging and analysis of hydrodynamic effects. *Biophys. J.* **91**(15), 317–329.

Richards-Kortum, R., and Sevick-Muraca, E. M. (1996). Quantitative optical spectroscopy for tissue diagnosis. *Annu. Rev. Phys. Chem.* **47,** 555–606.

Ridley, B. K. (1999). "Quantum Processes in Semiconductors." Oxford University Press, Oxford, UK.

Sacchi, C. A. (1991). Laser-induced electric breakdown in water. *J. Opt. Soc. Am. B* **8**(2), 337–345.

Shen, N., Datta, D., Schaffer, C. B., LeDuc, P., Ingber, D. E., and Mazur, E. (2005). Ablation of cytoskeletal filaments and mitochondria in live cells using a femtosecond laser nanoscissors. *Mech. Chem. Biosystems* **2**(1), 17–25.

Shen, Y. R. (1984). "Principles of Nonlinear Optics." Wiley-Interscience, New York, NY.

Siegman, A. E. (1997). How to (maybe) measure laser beam quality. Available at: www.stanford.edu/~siegman/beamqualitytutorialosa.pdf.

Silfvast, W. T. (1996). "Laser Fundamentals." Cambridge, New York, NY.

Soughayer, J. E., Krasieva, T. B., Jacobson, S. C., Ramsey, J. M., Tromberg, B. J., and Allbritton, N. L. (2000). Characterization of cellular optoporation with distance. *Anal. Chem.* **72**(6), 1342–1347.

Storb, R., Amy, R. L., Wertz, R. K., Fauconnier, B., and Bessis, M. (1966). An electron microscope study of vitally stained single cells irradiated with a ruby laser microbeam. *J. Cell Biol.* **31**(1), 11–29.

Suzuki, K., Taniguchi, Y., and Miyosawa, Y. (1972). The effect of pressure on the absorption spectra of DNA and DNA-dye complex. *J. Biochem. (Tokyo)* **72**(5), 1087–1091.

Tirlapur, U. K., and König, K. (2002). Targeted transfection by femtosecond laser. *Nature* **418**(6895), 290–291.

Tschachotin, S. (1912). Die mikroskopische strahlenstichmethode, eine zelloperationsmethode. *Biol. Zentralbl.* **32,** 623–630.

Venugopalan, V., Guerra, A., III, Nahen, K., and Vogel, A. (2002). Role of laser-induced plasma formation in pulsed cellular microsurgery and micromanipulation. *Phys. Rev. Lett.* **88**(7), 078103.

Vogel, A., and Venugopalan, V. (2003). Mechanisms of pulsed laser ablation of biological tissues. *Chem. Rev.* **103**(2), 577–644.

Vogel, A., Busch, S., and Parlitz, U. (1996). Shockwave emission and cavitation bubble generation by picosecond and nanosecond optical breakdown in water. *J. Acoust. Soc. Am.* **100**(1), 148–165.

Vogel, A., Capon, M. R. C., Asiyo-Vogel, M. N., and Birngruber, R. (1994). Intraocular photo disruption with picosecond and nanosecond laser pulses: Tissue effects in cornea, lens, and retina. *Investig. Ophthalmol. Vis. Sci.* **35**(7), 3032–3044.

Vogel, A., Nahen, K., Theisen, D., Birngruber, R., Thomas, R. J., and Rockwell, B. A. (1999a). Influence of optical aberrations on laser-induced plasma formation in water and their consequences for intraocular photodisruption. *Appl. Opt.* **38**(16), 3636–3643.

Vogel, A., Noack, J., Nahen, K., Theisen, D., Busch, S., Parlitz, U., Hammer, D. X., Noojin, G. D., Rockwell, B. A., and Birngruber, R. (1999b). Energy balance of optical breakdown in water at nanosecond to femtosecond time scales. *Appl. Phys. B* **68**(2), 271–280.

Vogel, A., Noack, J., Huttman, G., and Paltauf, G. (2002a). Femtosecond-laser-produced low density plasmas in transparent biological media: A tool for the creation of chemical, thermal, and thermomechanical events below the optical breakdown threshold. *Proc. SPIE* **4633A,** 23–37.

Vogel, A., Noack, J., Huttman, G., and Paltauf, G. (2002b). Low-density plasmas below the optical breakdown threshold—Potential hazard for multiphoton microscopy and a tool for the manipulation of intracellular events. *Proc. SPIE* **4620,** 202–216.

Vogel, A., Noack, J., Huttman, G., and Paltauf, G. (2005). Mechanisms of femtosecond laser nanosurgery of cells and tissues. *Appl. Phys. B* **81**(8), 1015–1047.

Williams, F., Varma, S. P., and Hillenius, S. (1976). Liquid water as a lone-pair amorphous-semiconductor. *J. Chem. Phys.* **64**(4), 1549–1554.

CHAPTER 5

Principles of Laser Microdissection and Catapulting of Histologic Specimens and Live Cells

Alfred Vogel,* Verena Horneffer,* Kathrin Lorenz,* Norbert Linz,* Gereon Hüttmann,* and Andreas Gebert[†]

*Institute of Biomedical Optics, University of Lübeck
Peter-Monnik Weg 4, D-23562 Lübeck, Germany

[†]Institute of Anatomy, University of Lübeck
Ratzeburger Allee 160, D-23538 Lübeck, Germany

Rapid contact- and contamination-free procurement of specific samples of histologic material for proteomic and genomic analysis as well as separation and transport of living cells can be achieved by laser microdissection (LMD) of the sample of interest followed by a laser-induced forward transport process [laser pressure "catapulting," (LPC)] of the dissected material.

0091-679X/07 $35.00
DOI: 10.1016/S0091-679X(06)82005-4

We investigated the dynamics of LMD and LPC with focused and defocused laser pulses by means of time-resolved photography. The working mechanism of microdissection was found to be plasma-mediated ablation. Catapulting is driven by plasma formation, when tightly focused pulses are used, and by ablation at the bottom of the sample for moderate and strong defocusing. Driving pressures of several hundred megapascals accelerate the specimen to initial velocities of 100–300 m/s before it is rapidly slowed down by air friction. With strong defocusing, driving pressure and initial flight velocity decrease considerably.

On the basis of a characterization of the thermal and optical properties of the histologic specimens and supporting materials used, we calculated the temporal evolution of the heat distribution in the sample. After laser microdissection and laser pressure catapulting (LMPC), the samples were inspected by scanning electron microscopy. Catapulting with tightly focused or strongly defocused pulses results in very little collateral damage, while slight defocusing involves significant heat and UV exposure of up to about 10% of the specimen volume, especially if samples are catapulted directly from a glass slide.

Time-resolved photography of live-cell catapulting revealed that in defocused catapulting strong shear forces originate from the flow of the thin layer of culture medium covering the cells. By contrast, pulses focused at the periphery of the specimen cause a fast rotational movement that makes the specimen wind its way out of the culture medium, thereby undergoing much less shear stresses. Therefore, the recultivation rate of catapulted cells was much higher when focused pulses were used.

I. Introduction

Procurement of specific samples of histologic material for proteomic and genomic analysis has become important with the increasing refinement of analytic techniques. Moreover, separation and transport of living cells is of interest for stem cell research, organ culture, or tissue engineering. Mechanical separation techniques are tedious, time consuming, and bear the risk of contamination. Therefore, faster laser-based processes have been developed (Eltoum *et al.*, 2002; Kehr, 2003; Thalhammer *et al.*, 2003), and several companies are active in this field. A widespread, rapid, contact- and contamination-free separation method consists in laser microdissection (LMD) of the sample of interest followed by a laser-induced forward transport process, which has been coined laser pressure "catapulting" (LPC) of the dissected material into the vial used for further analysis (Schütze and Lahr, 1998; Schütze *et al.*, 1998). This two-step procedure, consisting of laser microdissection and laser pressure catapulting (LMPC), is illustrated in Fig. 1. The scheme in Fig. 1 shows a setup with an inverted microscope but the same principle, with opposite direction of material transport, has been applied also for upright microscopes (Elvers *et al.*, 2005).

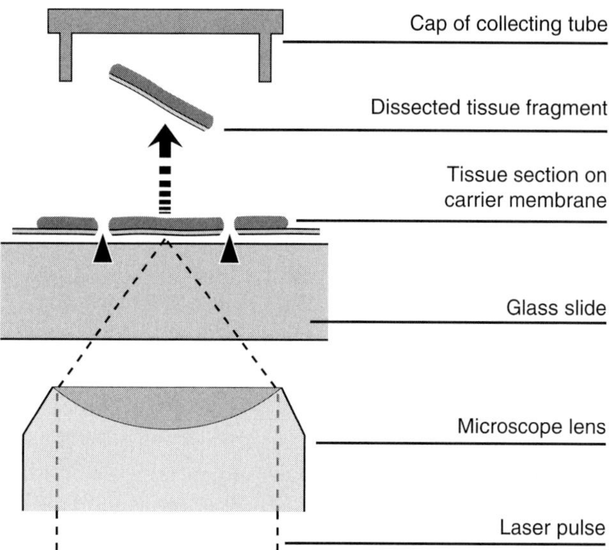

Fig. 1 Principle of separation of small biological objects, exemplified by the isolation of parts of a histologic section. The section is placed on a thin, UV-absorbing polymer foil that is mounted on a routine microscope glass slide. A region of interest is dissected from the section using a series of focused UV-A laser pulses (LMD) and subsequently catapulted (LPC) into the cap of a microfuge tube by a final, often more energetic, laser pulse. The catapulting pulse can be directed either in the center or in the periphery of the dissected specimen.

Historical roots of the individual steps of the process date about 30–40 years back. Soon after the first textbook on materials processing by high-power laser radiation appeared (Ready, 1971), LMD of histologic material using a UV-A laser was introduced in 1976, but the "harvesting" still had to be done with mechanical tools (Isenberg *et al.*, 1976; Meier-Ruge *et al.*, 1976). Ten years later, high-energy laser pulses were used to accelerate metallic flyers to velocities greater than 2.5 km/s for impact studies (Sheffield *et al.*, 1986), and present suggestions to exploit the potential of laser "catapulting" reach from the transfer of semiconductor devices in electronic circuit board manufacture (Mathews *et al.*, 2006) as far as to laser launching of objects into low Earth orbit (Phipps *et al.*, 2000). The underlying physical processes of laser-induced material transport were analyzed, among others, by Anderholm (1970), Fairand and Clauer (1979), Phipps *et al.* (1988, 2000), Romain and Darquey (1990), Fabbro *et al.* (1990), and Bäuerle (2000). Biologists became aware of the potential of laser catapulting in the late 1990s when Schütze and coworkers discovered that a combination of LMD and LPC with focused low-energy pulses can be used to isolate minute amounts of biologic material (Lahr, 2000; Schütze and Lahr, 1998; Schütze *et al.*, 1998). At the same time, a UV-absorbing polymer carrier foil was introduced to enhance the laser light absorption in the specimen and to maintain its mechanical integrity during LPC (Schütze and Lahr, 1998). These advancements, together with an increased demand for separation techniques due to the refinements of proteomic and genomic analysis, and the proof that LMPC is compatible with DNA and RNA recovery (Schütze and Lahr, 1998) paved the way of LMPC into the market (www.PALM-Microlaser.com).

The next step ahead consisted of the development of a protocol for live-cell catapulting and recultivation (Hopp *et al.*, 2005; Mayer *et al.*, 2002; Stich *et al.*, 2003). In parallel, the analysis of the dissection and catapulting mechanisms of biological material progressed. Cell surgery (Amy and Storb, 1965; Berns *et al.*, 1969, 1971; Bessis *et al.*, 1962; König *et al.*, 1999, and references in Vogel *et al.*, 2005a) and LMPC had been developed largely on an empirical basis. After it had been conjectured for a while that microdissection with pulse durations of nanoseconds or shorter is plasma-mediated (Greulich and Pilarczyk, 1998; Meier-Ruge *et al.*, 1976), this hypothesis was experimentally proven in 2002 (Venugopalan *et al.*, 2002), and a detailed theory of plasma-mediated nanosurgery of cells and tissues with femtosecond (fs) pulses was presented in 2005 (Vogel *et al.*, 2005a). In the same year, it was shown that LMD and LPC of histologic material using focused UV nanosecond (ns) laser pulses are also plasma-mediated (Vogel *et al.*, 2005b), and demonstrated that LPC can be performed using both focused and strongly defocused beams (Vogel *et al.*, 2005b). With defocused beams, no plasma is formed, and the catapulting relies on ablation at the bottom of the catapulted sample. Elvers *et al.* (2005) proved that the LPC mechanism applies not only for the separation technique based on an inverted microscope but also for schemes using an upright microscope (www.leica-microsystems.com).

Besides LMPC, a number of related laser-based separation and transport techniques have been developed throughout the past 10 years. Emmert-Buck *et al.* (1996) introduced a laser capture technique in which a thin clear membrane is melted by an IR microbeam onto a small region of the histologic sample. The melted membrane sticks to the cells to be isolated, which can then be lifted and collected in a microfuge tube. This method was later refined by UV laser dissection of the region of interest before the tissue sample is harvested via IR laser-induced adhesion to a membrane (www.arctur.com) or via a specific membrane adhesion cap (www.molecular-machines.com). Laser-induced cell lysis combined with rapid transport of the lysed cell constituents for time-resolved capillary electrophoresis has been established by Allbritton and coworkers (Meredith *et al.*, 2000; Sims *et al.*, 1998), and the mechanisms of this technique were explored by Rau *et al.* (2004, 2006). Deposition of biomaterial films by a matrix-assisted pulsed laser evaporation method (MAPLE direct-write) was demonstrated by Chrisey *et al.* (2003), Wu *et al.* (2003), and Christescu *et al.* (2004). Serra *et al.* (2004) demonstrated the preparation of functional DNA micoarrays through laser-induced forward transfer by means of a titanium dynamic release layer and analyzed its mechanisms (Colina *et al.*, 2006). MAPLE direct-write was used also to create three-dimensional (3D) heterogeneous cell patterns that are of interest for tissue engineering (Barron *et al.*, 2004; Ringeisen *et al.*, 2004; Wu *et al.*, 2003). While the above work was performed utilizing ns laser pulses (mostly at UV wavelengths), several researchers employed fs pulses for biofilm and DNA printing (Karaiskou *et al.*, 2003; Zergioti *et al.*, 2005) and shock wave induced cell isolation, (Hosokawa *et al.*, 2004). Laser-induced forward transfer with fs pulses was found to be more directional than with ns pulses (Zergioti *et al.*, 2003).

The present study focuses on the investigation of LMD and LPC using UV ns laser pulses because this technique is widely used in many laboratories. It is the goal to elucidate the mechanisms and potential side effects of LMPC in both dry and liquid environments (i.e., for histologic specimens and living cells). On the basis of this analysis, strategies for an improvement of the two techniques will be discussed.

II. Features of the Microbeam System and Materials Used

We used a PALM microbeam system equipped with an N_2 laser ($\lambda = 337$nm) emitting pulses of 3-ns (FWHM) duration. The laser beam is coupled into an inverted microscope (Zeiss Axiovert 200) with a motorized, computer-controlled stage. The microscope objectives used in this study were Zeiss LD Plan-Neofluar 40×/0.6 corr, Plan Neofluar 20×/0.5, Fluar 10×/0.5, and Fluar 5×/0.25. The UV-absorbing polymer foil mounted on the glass slides carrying the histologic sections (Fig. 1) consists of polyethylene naphthalate (PEN) and is 1.35-μm thick.

An energy calibration (Fig. 2) showed that the relation between the setting at the control box and the actual energy transmitted through the objectives is logarithmic. The transmitted energy is largest for objectives with small magnification because

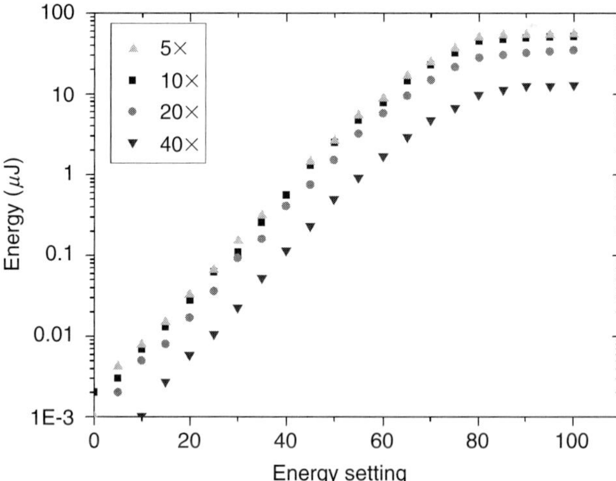

Fig. 2 Pulse energy transmitted through the microscope objectives for different settings at the laser control box of the PALM microbeam in semilogarithmic representation.

they possess large optical pupils that transmit the laser beam with little or no vignetting. The focal spot sizes were measured using a knife-edge technique, as described by Sasnett (1989) and Siegmann *et al.* (1991). The result for the 40× objective is shown in Fig. 3A. Because of the poor quality of the N_2 laser beam (Fig. 3B and C), the spot diameter (4.2 μm up to 1/e^2 irradiance values) is more than six times larger than the diffraction limited focus diameter. Nevertheless, the hot spot visible in the center of the far-field beam profile in Fig. 3C allows to produce relatively fine effects if energies very close to the ablation threshold are employed. In the present generation of microbeam systems, the N_2 laser has been replaced by a frequency-tripled Nd:YAG laser emitting at $\lambda = 355$ nm that exhibits

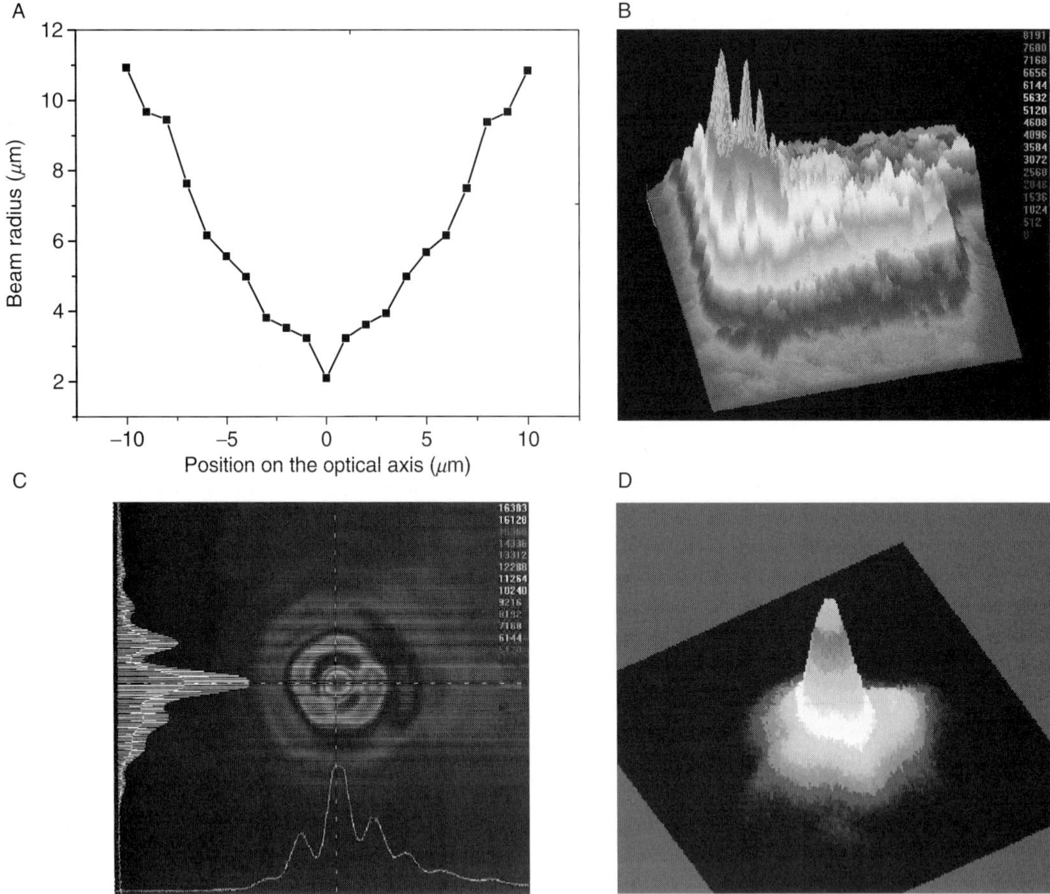

Fig. 3 (A) Spot size of the N_2 laser beam focused through the 40× objective, (B) near field and (C) far-field profiles of the N_2 laser beam, and (D) far-field beam profile for the frequency-tripled Nd: YAG laser.

a much better beam profile, as shown in Fig. 3D, and thus makes it possible to achieve a nearly diffraction limited focal spot size.

The transmission at $\lambda = 337$nm of cells, histologic material, and the polymer foils mounted below the histologic sections were measured in the microbeam setup using an Ophir PD10 detector with 1-nJ sensitivity and in a spectral photometer. All measurements were performed at low irradiance where nonlinear absorption is negligible. The measurements on the optical properties of the cultured cells [Chinese hamster ovary (CHO) cells] were performed on a single, confluent cell layer. The determination of the optical properties of the PEN foil required measurements with an integrating sphere, since PEN scatters strongly at $\lambda = 337$nm. For PEN, we measured total transmission and reflection as described by Vogel *et al.* (1991), and calculated the absorption coefficient μ_a and scattering coefficient μ_s using the Kubelka Munk theory (Cheong *et al.*, 1990; Star *et al.*, 1988). The effective attenuation coefficient μ_{eff} and optical penetration depth δ were then obtained by means of the relation (Jacques, 1993):

$$\delta = \frac{1}{\mu_{eff}} = \frac{1}{\{3\mu_a[\mu_a + \mu_s(1 - g)]\}^{1/2}} \tag{1}$$

assuming a value of $g = 0.5$ for the scattering anisotropy coefficient that is representative for moderate forward scattering in single scattering events (Cheong *et al.*, 1990).

The heat capacity of the polymer foil and the histologic specimens were determined by differential scanning calorimetry (DSC) in comparison with a sapphire standard, and the dissociation temperatures were obtained through thermogravimetric analysis (TGA), that is, by measuring the weight loss as a function of temperature. As an example, Fig. 4 shows the DSC and TGA data for PEN foil, the polymer material mounted on microscope glass slide on which the histologic specimens are usually placed. Table I presents a summary of the measurement results for the optical and thermal material properties of materials relevant for LMD and LPC of cells and histologic materials.

III. Dissection and Catapulting of Histologic Specimens

A. Mechanism of Dissection

It has been shown recently that cell surgery using pulsed visible and near-IR irradiation is plasma mediated, that is, based on nonlinear absorption via multiphoton and avalanche ionization. The irradiances required for cell surgery with ns laser pulses are $\sim 10^9$ W/mm^2, just as the optical breakdown threshold in water (Venugopalan *et al.*, 2002). For fs laser pulses, both the required irradiance for dissection and the threshold irradiance for optical breakdown in water are about two orders of magnitude larger while the breakdown energy is about three orders of magnitude smaller (Vogel *et al.*, 2005a).

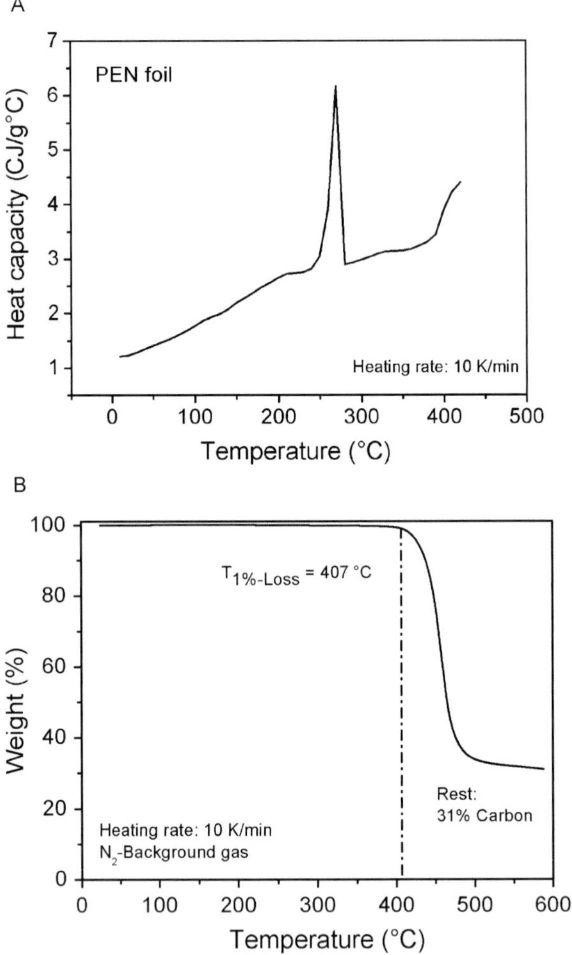

Fig. 4 Thermal properties of PEN foil. (A) Temperature-dependent heat capacity C_p. The peak at 269 °C indicates an endothermal melting transition, and the rise above 400 °C is due to dissociation. (B) Temperature-dependent weight loss for a slow heating rate, with dissociation starting at 407 °C and ending at ≈500 °C. The dissociation temperature defined by half of the total weight loss is ≈450 °C. For very fast heating rates such as in pulsed laser catapulting, it will probably be higher because dissociation is a rate process that depends on both temperature and time.

The dominant role of plasma formation for cell surgery using visible or near-IR wavelengths is not surprising because the linear absorption of water and biomolecules in this region of the optical spectrum is very small (Litjens *et al.*, 1999; Vogel and Venugopalan, 2003). Our present study revealed that plasma formation plays a key role also for the dissection of cells at UV-A wavelengths. Although the absorption increases with decreasing wavelength, it is still fairly small at $\lambda = 337$ nm

Table I

Optical Properties at 337 nm and Thermal Properties of Cells, Histologic Material, Polyethylene Naphthalate (PEN) Polymer Foil, Glass, and Water

Material	Sample thickness × (μm)	Transmission (%)	Extinction coefficient μ_{eff} (cm^{-1})	Optical penetration depth δ (μm)	Average heat capacity (kJ/K/kg)	Dissociation temperature (°C)	Heat conductivity (W/m/K)	Density (W/m*K)
Glass slide	1000	94.7	0.55	18,200	0.666	–	1.07	2500
PEN-foil	1.35	$T = 20.5$ $R = 22.4$	$\mu_a = 3{,}520$ $\mu_s = 17{,}370$ $\mu_{eff} = 11{,}360$	0.88	2.7	460	≈0.4	1.39
Teflon foil	≈25	95.8	17.2	580	1.0	–	≈0.2	2200
H&E-stained histol. specimen	≈5	7–35(15.7)	2100–5300 (3700)	1.9–4.8(2.7)	3.2	340	≈0.5	≈1000
CHO cells	≈5	93.8	127	79	4.0	150–300		≈1000
Water			0.0172	5.8×10^5	4.187	300	0.598	998

All transmission data are corrected for specular reflection, that is, they represent purely the transmission of the sample. The optical parameters for stained histologic specimen cover a certain range given by variations in staining. The values in brackets were used for the temperature calculation is Section III.C.1. Sources for data not measured in this study are water absorption, Litjens *et al.* (1999); heat conductivity, heat capacity, and density of water, glass, PEN, and Teflon, Kuchling (1991); density of PEN, www.m-petfilm.com. The dissociation temperature of water corresponds, in the present context, to the superheat limit in bubble-free liquid water (Vogel and Venugopalan, 2003), and the dissociation temperature of cells corresponds to their heterogeneous nucleation threshold (Neumann and Brinkmann, 2005; Simanowski *et al.*, 2005).

(the optical penetration depth is about 80 μm, see Table I) and even smaller at 355 nm. Therefore, the ablation threshold in unstained cells was found to be only slightly smaller than the threshold for the formation of luminescent plasma, as shown in Fig. 5A.

Stained histologic specimens and the PEN polymer foil have a much larger optical absorption coefficient than unstained cells (see Table I). For these materials, the ablation threshold is thus considerably lower than the plasma formation threshold (Fig. 5A). It has previously been speculated that ablation of biomaterial at the N_2 laser wavelength of 337 nm is based on photochemical dissociation (Schütze and Lahr, 1998). However, this is unlikely because indications for a relevant contribution of photochemical effects to ablation have been found only at much shorter wavelengths, for example with ArF laser pulses at $\lambda = 193$ nm while they were completely absent with XeCl excimer laser pulses of 308-nm wavelength (Vogel and Venugopalan, 2003). Therefore, we assume that ablation based on linear absorption at $\lambda = 337$ nm is a purely thermal process.

Figure 5A indicates that dissection of histologic specimens by N_2 laser pulses could, in principle, be performed by ablation without any contribution of plasma

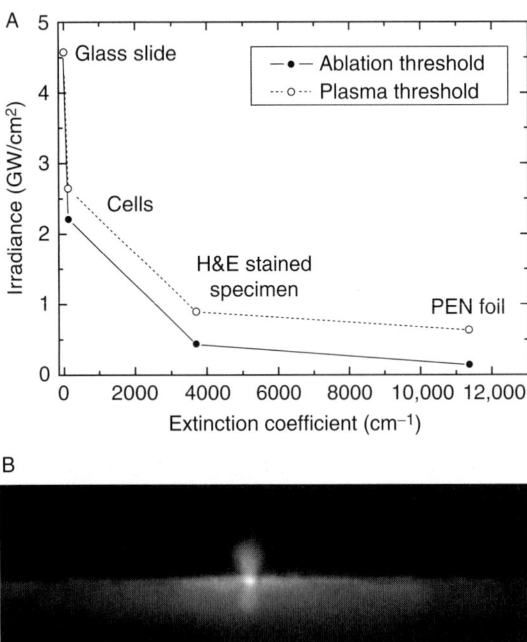

Fig. 5 (A) Thresholds for ablation and plasma formation at 337nm for various materials. (B) Image of plasma luminescence during microdissection of histologic specimens (6 pulses of 4.6 μJ). The plasma luminescence is blue at the target surface but turns red at larger distances where the temperatures are smaller and the corresponding wavelength of blackbody radiation is longer. The blue light emission adjacent to the laser focus is caused by fluorescence of the PEN polymer foil.

formation because the ablation threshold (0.15 μJ with NA = 0.6) is slightly lower than the threshold for plasma formation (0.3 μJ). However, because of the small ablation efficiency close to the threshold for material removal this dissection mode would require very many pulses which are not viable at the relatively small laser repetition rate of 30 Hz. Therefore, pulse energies of about 0.5 μJ are commonly used and focused through a 40×, NA = 0.6 microscope objective. That energy is larger than the threshold for plasma formation, and dissection is thus accompanied by blue plasma luminescence as visible in Fig. 5B. The slightly disruptive action of the plasma expansion helps to achieve clean cuts without any bridges remaining between the parts to be separated. We conclude that dissection of histologic specimens using UV laser pulses is based on plasma formation (i.e., nonlinear absorption) initiated by linear absorption.

B. Catapulting with Focused and Defocused Laser Pulses

1. Driving Forces for Catapulting with Focused Pulses

The mechanisms of catapulting of histologic specimens and live cells were analyzed by time-resolved photography. Similar techniques have been employed previously for the analysis of other laser-induced transport processes (Koulikov and Dlott, 2001; Lee *et al.*, 1992; Nakata and Okada, 1999; Papazoglou *et al.*, 2002; Rau *et al.*, 2004; Young *et al.*, 2001), and LPC by laser pulses focused through an upright microscope has been documented by high-speed cinematography with 1-ms interframing time (Elvers *et al.*, 2005). We achieved a temporal resolution better than 100 ns by using single frame photography with increasing time delay between catapulting laser pulse and the instant at which the photograph was exposed (Vogel *et al.*, 2007). The catapulted specimens were imaged in transillumination using the light of a plasma discharge lamp with 20-ns duration (Nanolite). The total magnification of the imaging system was 44×, and the images were detected using a 6-megapixel digital camera. The laser-produced pressure waves were visualized by means of a sensitive dark-field Schlieren technique (Vogel *et al.*, 2006). A frequency-doubled Nd:YAG laser beam coupled into a 300-m-long multimode fiber to destroy temporal coherence served as speckle-free light source for photography with 16-ns exposure time. To reduce the background noise and enhance the visibility of the laser-produced shock wave, a reference image was taken at time $t < 0$ and subtracted from each image recorded after the release of the catapulting laser pulse. The initial phase of the catapulting process was documented with time increments of 2–60 ns. The high time resolution allowed to deduce the laser-produced pressure from the speed of the resulting pressure waves (Lorenz, 2004; Vogel *et al.*, 1996b, 2007).

The specimen is usually located on a PEN foil backed by a transparent substrate (glass slide). Laser pulse energies commonly used for catapulting are larger than the energies employed for LMD that was already shown to rely on plasma formation. It is thus obvious that catapulting with focused laser pulse is driven by plasma formation in the confined space between specimen and substrate. The initial catapulting dynamics under these conditions is shown in Fig. 6. The luminescent plasma driving the catapulting is visible in all frames. Immediately after the laser pulse, small debris particles are ejected with high velocity, and after 270 ns, the specimen is already clearly detached from the substrate surface. At the location of plasma

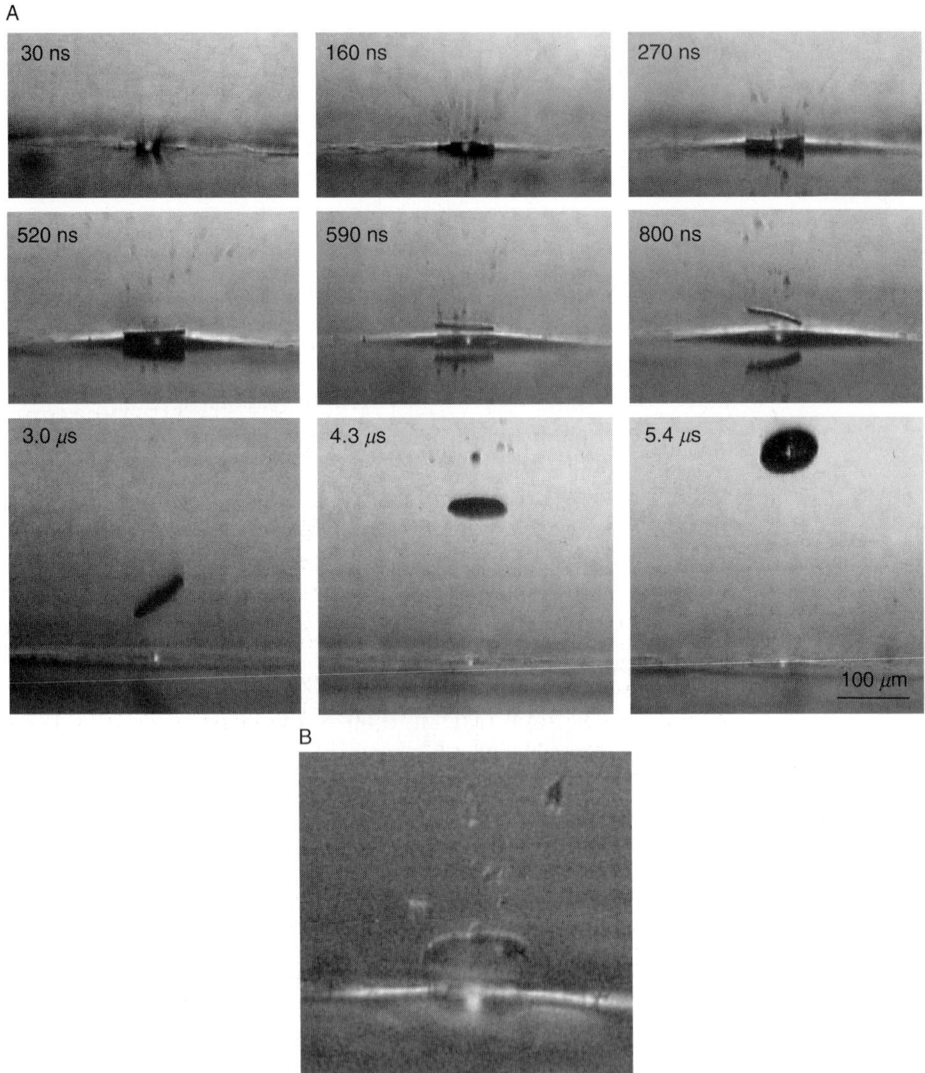

Fig. 6 (A) Initial phase of the catapulting dynamics of a specimen with 80-μm diameter from a paraffin section. 40× objective, NA = 0.6, and E = 10 μJ. (B) Enlarged view of the specimen after 1.7 μs showing the luminescent plasma driving the specimen, and the blue fluorescence of the PEN polymer foil. (See Plate 3 in the color insert section.)

formation, a hole is produced in the specimen, which is clearly visible in the image taken after 5.4 μs.

The pressure wave produced by the expansion of the laser plasma is shown in Fig. 7. An evaluation of the photographic series provided quantitative information on the propagation distance of the laser-produced pressure wave, the ejected debris, and the catapulted specimen as a function of time that is presented in Fig. 8. The initial velocity of the pressure wave amounts to 26,000 m/s, which is 76 times larger than the normal sound velocity in air and thus indicative for a strong shock wave. The plasma pressure driving the shock wave can be derived from the initial shock wave velocity v_s using the relation (Landau and Lifschitz, 1987):

$$p_{plasma} = \left[\left(\frac{7}{6}\right)\left(\frac{v_s}{c_0}\right)^2 - \left(\frac{1}{6}\right)\right]p_0 \qquad (2)$$

Here $c_0 = 345$ m/s denotes the normal sound velocity in air and $p_0 = 0.1$ MPa is the atmospheric pressure. The initial laser-produced pressure was found to be 670 MPa, a typical value for plasmas generated in a confined geometry

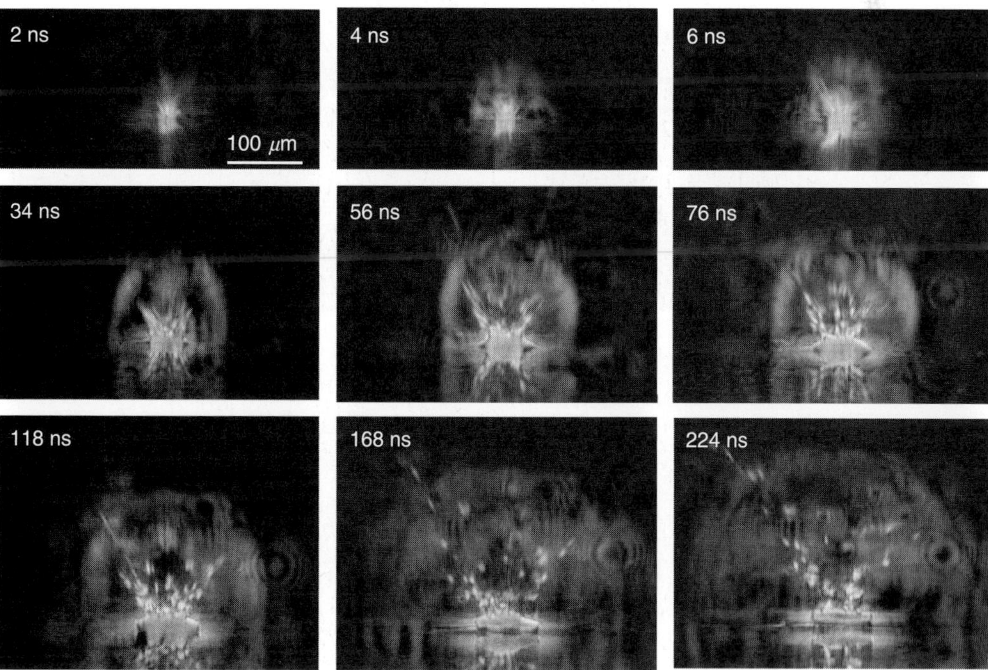

Fig. 7 Dark-field photographic series showing the first 225 ns of the catapulting dynamics. A specimen with 80-μm diameter from a paraffin section was catapulted using a 40× objective, NA = 0.6, and E = 10 μJ. The movement of the ejected particles and the propagation of the laser-induced shock wave can be recognized, followed by the detachment of the specimen after 100–200 ns.

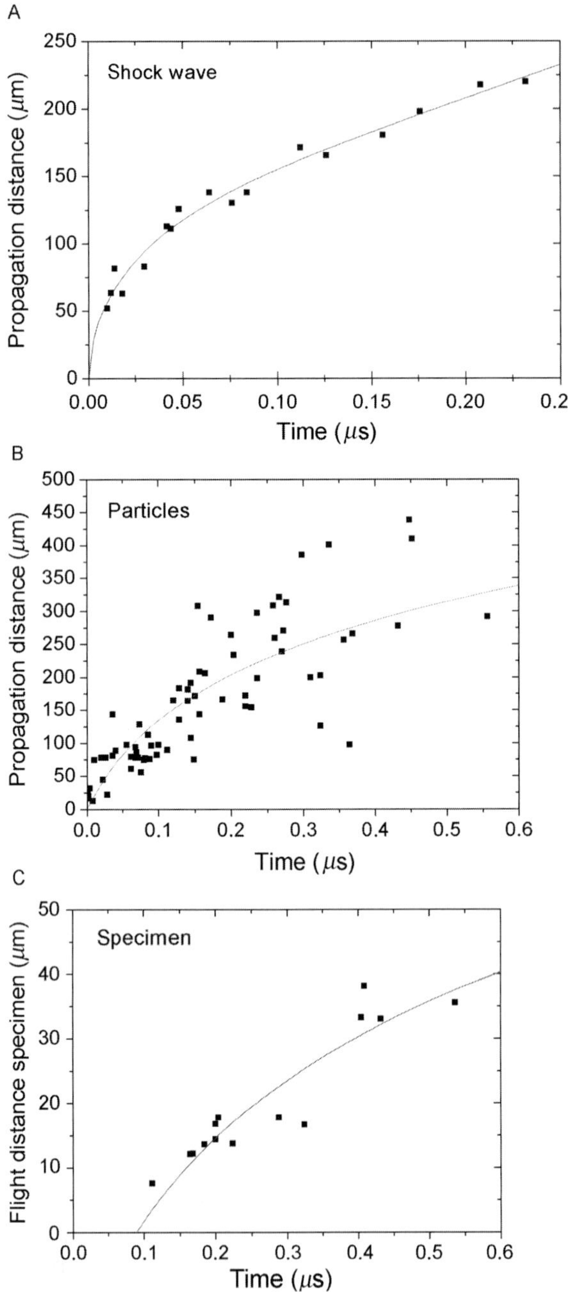

Fig. 8 Propagation distance of the laser-produced shock wave (A), the ejected particulate debris (B), and the catapulted specimen (C) as a function of time. From the slope of the d(t) curves, initial velocities of (A) 26,000 m/s, (B) 2200 m/s, and (C) 178 m/s can be deduced, which corresponds to (A) 76×, (B) 6.4×, and (C) 0.52× the sound velocity in air. 40× objective, NA = 0.6, specimen diameter 80 μm, and E = 10 μJ.

(Anderholm, 1970; Fabbro *et al.*, 1990). Similar pressure values were also obtained by analyzing the acceleration of the debris flying off from the laser focus that is visible in the dark-field images of Fig. 7. Its peak velocity after a 10-μJ laser pulse is already reached at the end of the laser pulse (i.e., after 3 ns) and amounts to 2200 m/s (Lorenz, 2004).

The initial specimen velocity in Fig. 8C is $\approx$180 m/s. In the early days of LPC, it has been speculated that the specimens are driven by the light pressure (Schütze and Lahr, 1998). We can check the validity of this hypothesis by comparing the measured specimen velocity with the speed that can be imparted by the light pressure

$$p_{\text{light}} = \frac{I}{c} \qquad (3)$$

Here I is the irradiance in the illuminated laser spot and c is the vacuum velocity of light. The accelerating force exerted by the light pressure acts during the laser pulse duration and the final velocity reached can be calculated considering the mass of the specimen. The calculation yields $v = 0.9$ mm/s for a specimen of 80-μm diameter, 7-μm thickness, and a mass density of 1 g/cm^3 that is catapulted by a 10-μJ pulse (parameters as in Fig. 8). This value is five orders of magnitude smaller than the actual velocity. The pressure exerted by the photons in the laser pulse, which is the working mechanism in laser tweezers (Ashkin, 1986; Greulich, 1999; Schütze and Clement-Sengewald, 1994), thus plays only a negligible role in LPC.

We showed that catapulting with focused pulses relies on the pressure produced by confined plasma formation. However, in spite of the large pressures involved, the catapulted specimens in Fig. 6 show little bending or other deformations. This is due to the fact that the large pressure at the focus is in vertical direction rapidly released by the hole formation in the specimen. In horizontal direction, the shock wave spreads across the specimen diameter within about 20 ns (Fig. 7). This way, it produces an approximately homogeneous elevated pressure below the specimen that lifts off from the substrate after about 100–200 ns (Figs. 6 and 7). In Fig. 6, one can see that the horizontal pressure wave also propagates under adjacent parts of the histologic section and transiently detaches them from the substrate for a few microseconds. The average pressure under the entire catapulted specimen during the acceleration phase of about 100 ns can be derived from the velocity reached after this phase, the acceleration time, and the estimated mass of the specimen (Lorenz, 2004). It amounts to about 18 MPa for the parameters in Fig. 8, which is considerably less than the peak pressure within the laser plasma of $\approx$670 MPa.

In some cases, the histologic specimens to be catapulted are mounted on a foil without backing by a glass substrate. Without confinement by the glass substrate, the plasma or ablation plume can freely expand. In this case, catapulting relies merely on the pressure produced by ablative recoil. Therefore, it is smaller than for plasma generation or ablation in a confined geometry, as shown in Fig. 9. The impulse coupling efficiency in laser ablation was theoretically analyzed by Phipps *et al.* (1988) and Dingus (1992, 1993). It was predicted to be smaller without

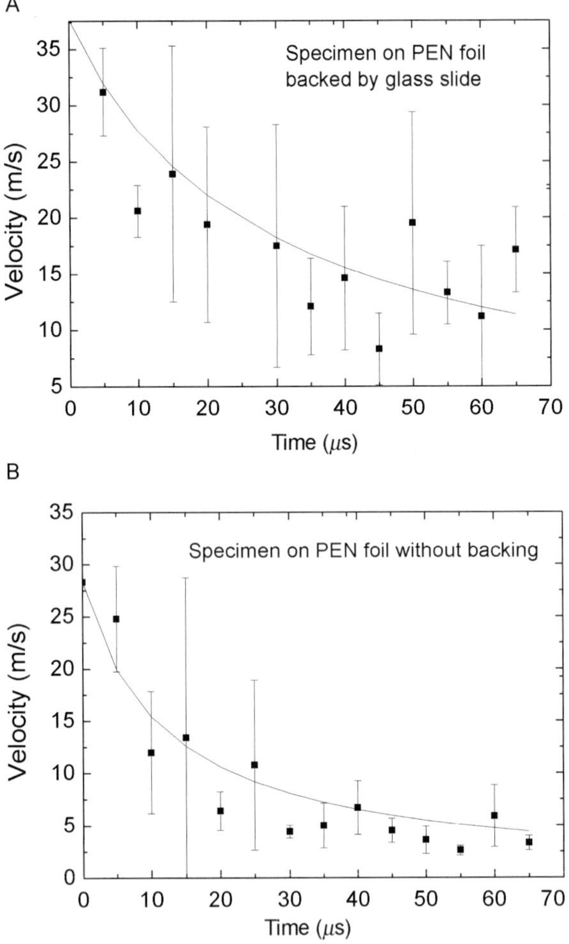

Fig. 9 Comparison of the velocities of specimens on a foil (A) with and (B) without backing by a glass substrate. Each data point refers to the average of five measurements. Specimen diameter 80 μm, 40× objective, NA = 0.6, and E = 5 μJ.

confinement of the ablation products, in agreement with our present observations and with results for aluminum foils obtained previously by Sheffield *et al.* (1966).

The dependence of the velocity on laser pulse energy for specimens that are backed by a glass slide is shown in Fig. 10. As expected, the specimen velocity increases monotonously with laser energy. In the energy range investigated, which is still close to the catapulting threshold, their relation is approximately linear. However, further above threshold we would rather expect a square root dependence corresponding to a constant conversion efficiency from laser energy into the kinetic energy of the moving specimen.

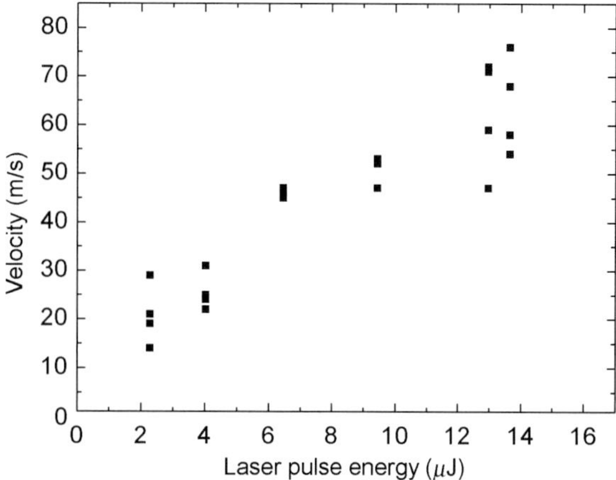

Fig. 10 Dependence of specimen velocity on laser pulse energy (average velocity values during the first 15-μs flight time). Specimens of 80-μm diameter were catapulted using a 40× objective and NA = 0.6.

For a constant laser energy of 10 μJ, the velocity of the dissectats was found to be 2.3 times larger when the laser pulses were focused through a 5× objective instead of a 40× objective. This striking difference is due to the fact that the focal spot size increases with decreasing numerical aperture (NA) of the objective. The measured focus radius of the N_2 laser beam in the PALM system was 21.4 μm for the 5×, NA = 0.25 objective compared to 2.1 μm for the 40×, NA = 0.6 objective. Because of the larger spot size, no holes were formed in the specimens during catapulting with the 5× objective, and the acceleration was faster due to the better confinement of the ablation products under the specimen.

2. Flight Trajectories of the Specimens

The catapulted specimen has to be accelerated against inertia and hydrodynamic drag. Acceleration ceases when the force exerted by the pressure below the specimen becomes smaller than the resistance originating from atmospheric pressure and hydrodynamic drag. Afterward, the specimens are slowed down by the air friction. The trajectory of the dissectats during their flight toward the microfuge cap is visible in the picture series in Fig. 11A and evaluated quantitatively in Fig. 11B.

A theoretical description of the flight trajectory $z(t)$ for the phase dominated by air friction is generally quite complicated (Elvers *et al.*, 2005; Jiménez *et al.*, 2005). However, for large particle velocities and sizes, Newtonian friction (related to the inertia of the displaced material) dominates Stokes friction (related to the medium viscosity), and it can be assumed that the hydrodynamic drag is approximately proportional to the square of the velocity. One then obtains the trajectory (Hibst, 1996; Jiménez *et al.*, 2005).

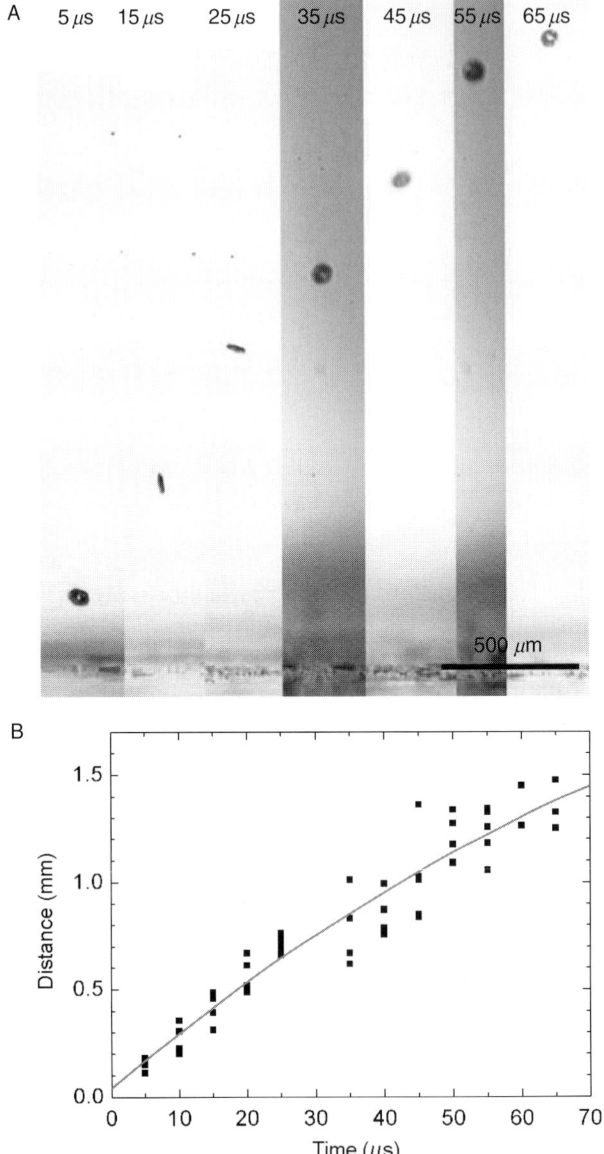

Fig. 11 Slowing of the specimens due to the friction in air. (A) shows a pictorial representation of the flight trajectory during the first 65 μs that is compiled of individual images taken at different delay times and (B) shows all measurement data on the flight distance versus delay time. The specimens had 80-μm diameter and were catapulted using a 40×, NA = 0.6 objective, and E = 6.5 μJ. The fitted curve corresponds to Eq. (4).

$$z(t) = K^{-1}\ln[kv_0(t - t_0) + 1] \quad (4)$$

where v_0 is the initial velocity at time t_0 denoting the duration of the acceleration phase, and K is a constant incorporating all material parameters including its density, cross-sectional area, and shape-related properties. Equation (4) was fitted to the measurement points in Figs. 8B, C, and 11B to obtain information on v_0 and t_0. Generally, good agreement between the trend exhibited by the experimental data and the fitted $z(t)$ curves was found. At later times >1 ms, when the specimen velocities have slowed down to $\leq$1 m/s, the hydrodynamic drag must be described by a combination of Newtonian friction ($\propto v^2$) and Stokes friction ($\propto v$), which leads to a more complex equation of motion (Elvers *et al.*, 2005).

In the experiments described above, the catapulting laser pulse was aimed at the center of the specimen to create optimum reproducibility. By contrast, in commercial microbeam systems, the region of interest is usually dissected in such a way that a small bridge is left at the location where the end of the trajectory of the cut would meet the starting point. A single pulse of larger energy is then aimed at the bridge in order to complete dissection and, at the same time, catapult the specimen. This strategy saves processing time but due to the reduced confinement of ablation products, the impulse coupling to the specimen is not as good as with laser pulses aimed at the center of the specimen. To optimize impulse coupling to the specimen, the width of the bridge should be at least equal to the diameter of the laser spot, and the spot size should be as large as possible without compromising the dissection of the bridge at the given pulse energy (Elvers *et al.*, 2005).

Another problem arising when the catapulting pulses are aimed at the rim of the specimen is that the flight trajectories are generally not in line with the optical axis of the microscope but oblique, as shown in Fig. 12. As a consequence, not all catapulted specimens arrive in the cap of the microfuge tube. In our investigations, about 93% of all dissectats would have arrived within the cap, and 65% would have reached its top part where the chance of adhesion is largest. The oblique direction of the flight trajectories is related to the fact that plasma formation at the rim of the specimen imparts an impulse not only in upward but also in lateral direction. Additionally, it causes a fast rotational movement of the specimen as visible in Fig. 13. The initial rotation frequency deduced from Fig. 13 is 500,000 rps, but it will probably vary from shot to shot. The rotation of a moving object induces a lateral force (Magnus effect) that will add to the lateral impulse imparted by the initial plasma expansion.

3. Catapulting with Defocused Pulses

To identify the gentlest way of catapulting, we investigated how the catapulting behavior depends on the size of the irradiated spot at the bottom of the specimen. The dependence of the initial velocities of the specimen and of the ejected particulate debris on spot size is presented in Fig. 14A and B. These graphs contain only a few data points because it is very time consuming to determine the initial velocity

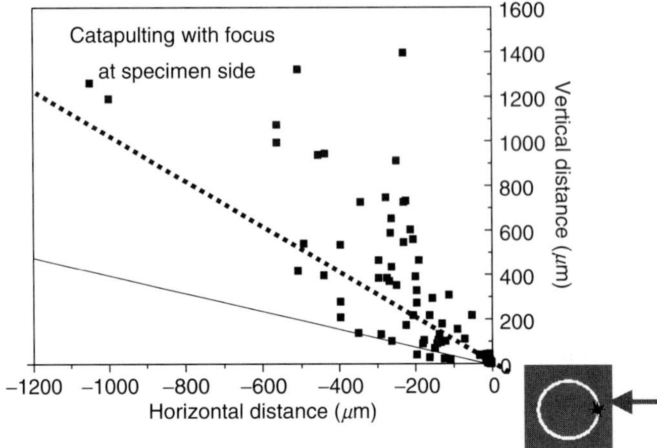

Fig. 12 Directional distribution of flight directions when the laser pulses are aimed at the rim of the specimen. Specimen diameter 80 μm, 40$\times$ objective, NA = 0.6, and E = 5 μJ. The specimens were photographed at different times after the catapulting laser pulse, and the data points show their location at these times. The flight trajectories are given by the connection between the laser focus (at 0,0) and the respective data points. The straight line delineates which specimen will reach the cap of the microfuge tube if it is located at a distance of 1 mm from the supporting slide and if the center of the cap is in line with the optical axis of the microscope objective. The dashed line indicates which specimens would reach the top of the cap, which is commonly wetted by a drop of mineral oil to improve adhesion.

values by taking series of images at different time delays as shown in Figs. 6 and 7. To cover a larger range of spot sizes, we took images at one delay time (50 μs after the laser pulse) and determined the average velocity during this flight time as shown in Fig. 14C. It is obvious that the catapulting velocities first increase and then decrease when the irradiated spot is enlarged. To facilitate the analysis of the underlying mechanistic changes, we inspected some catapulted specimens by scanning electron microscopy (SEM). The results are shown in Fig. 15.

Catapulting with tightly focused laser pulses is always associated with plasma formation and the generation of a hole through the entire specimen (Fig. 15A). With moderate defocusing, plasma formation is replaced by explosive vaporization, resulting in hole formation in the polymer foil but not in the specimen, with a zone around the hole where the foil melts and resolidifies (Fig. 15B). In this regime, the catapulting velocities are maximal (Fig. 14) because the ablation products remain confined below the specimen and cannot escape through a hole in the center of the specimen. More strongly defocused laser pulses lead to local ablation and melting of the foil, but do not perforate it (Fig. 15C). Pointlike ablated spots, but without obvious signs of surrounding melting, are also observed when the irradiated spot becomes comparable to or larger than the specimen size (Fig. 15D). In this regime, catapulting occurs still in a reproducible fashion but the velocities are much smaller than with focused and moderately defocused pulses (Fig. 14).

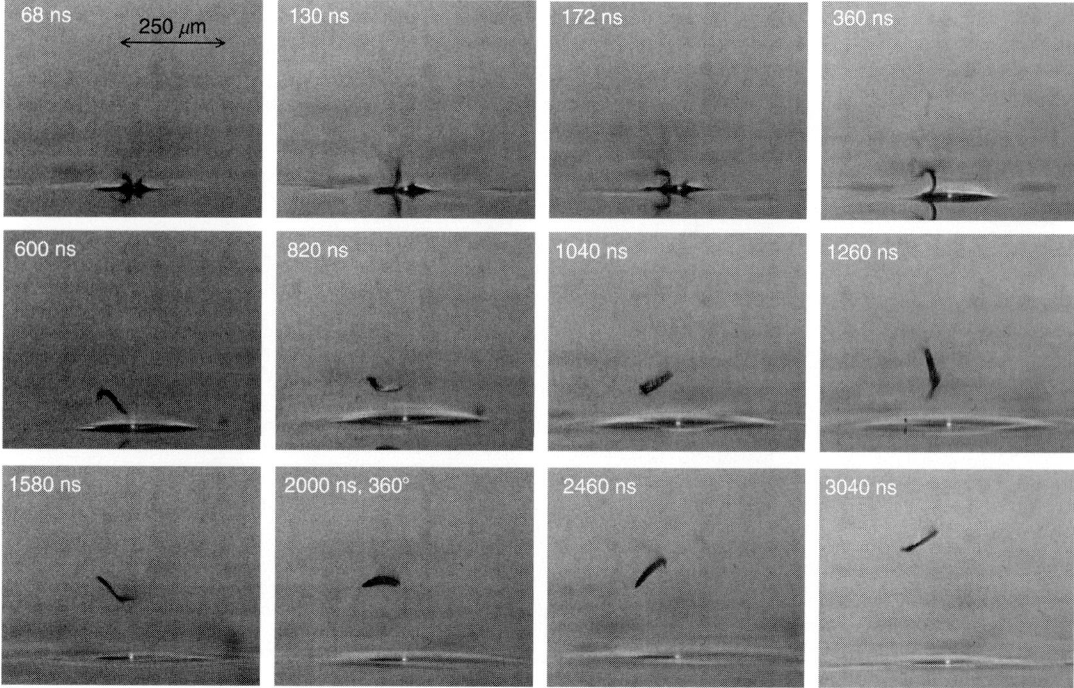

Fig. 13 Dynamics of a PEN foil specimen with 100-μm diameter that was first completely dissected and then catapulted by a 20-μJ pulse focused through a 20× objective and NA = 0.5 onto the rim of the specimen. The specimen performs one revolution within 2 μs, corresponding to a frequency of 500,000 rps.

For further analysis of the mechanisms of defocused catapulting, we calculated the average temperature rise within the irradiated spot in a layer with the thickness of the optical penetration depth, assuming a homogeneous light distribution in the irradiated spot. The calculation was based on the measured optical and thermal properties (heat capacity) listed in Table I. Because of the pronounced scattering of the PEN foil, 22.4% of the incident light are backscattered. This has been considered in calculating the temperature rise in the PEN foil, as well as the slight transmission loss pf 5.3% in the microscope glass slide. The results for the average laser-induced temperature in a PEN layer with the thickness of the optical penetration depth (0.88 μm) are presented in Fig. 16. The threshold radiant exposures for catapulting determined with different microscope objectives and laser pulse energies are listed in Table II, together with the corresponding temperatures of the light-absorbing layer. The thresholds were determined with strongly defocused laser beams, that is, in a catapulting mode for which hole formation in the specimen can be excluded.

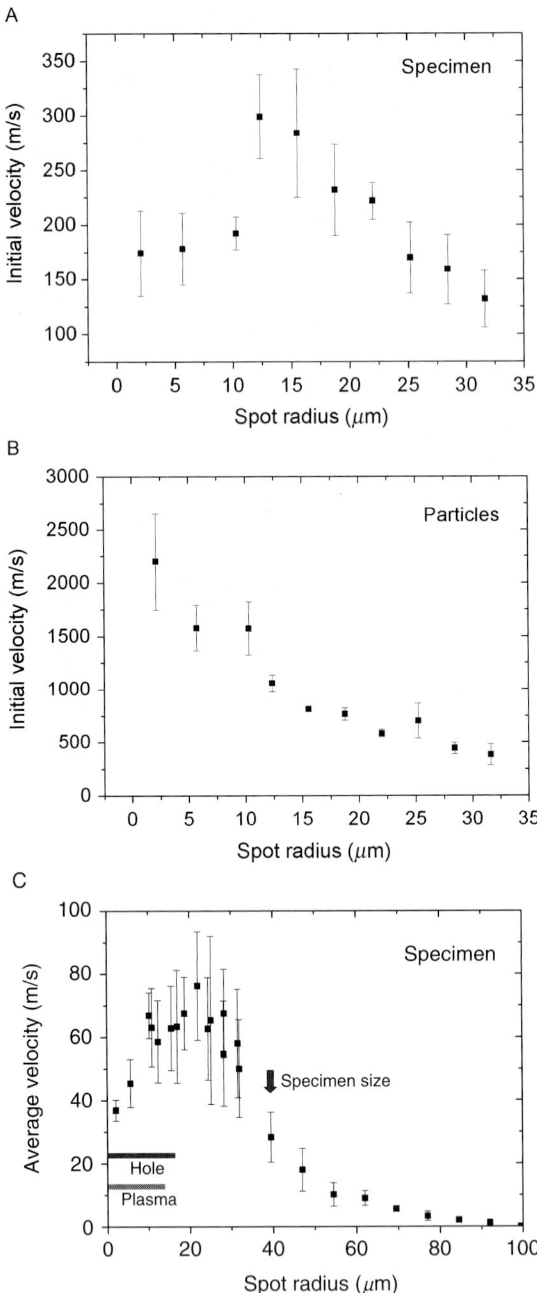

Fig. 14 Catapulting velocity as a function of laser spot size. (A) Initial velocity of the specimens, (B) initial velocity of the particulate debris, and (C) average velocity of the specimens during the first 50-μs flight time, with regimes for plasma and hole formation. 40× objective, NA = 0.6, and E = 10 μJ. The center of the irradiated spot coincided with the center of the specimen.

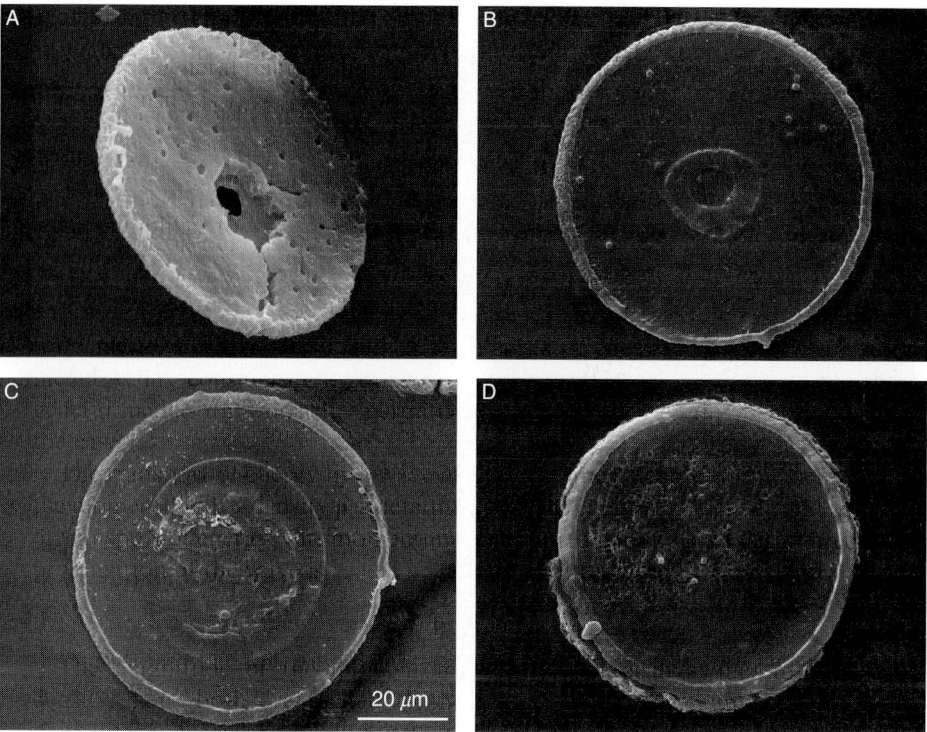

Fig. 15 Scanning electron micrographs of histologic specimens that were catapulted with irradiation of different spot sizes. The spot diameter is (A) 4, (B) 15, (C) 80, and (D) 136 μm. These values denote the diameter of the geometric cone angle of the laser beam at the location of the specimen. The actual spot size may be smaller if the irradiance distribution has an intensity peak around the optical axis.

The calculated peak temperatures for tightly focused irradiation ($\approx 10^5$ K) are unrealistically large because the calculation does not consider that the plasma expansion and adiabatic cooling already start during the laser pulse. Realistic plasma temperatures are in the order of 5000–10,000 K (Stolarski *et al.*, 1995). With defocused irradiation, for which catapulting is driven by ablation without plasma formation, adiabatic cooling by the expansion of the ablation plume reaches a significant level only toward the end of the laser pulse when most of the laser energy has been deposited (Apitz and Vogel, 2005). Therefore, the calculated peak temperatures at the bottom of the specimen are realistic. The calculated temperature rise for the spot radius of 22 μm associated with the maximum catapulting velocity is ~1400 °C. This is considerably larger than the melting temperature of PEN (269 °C for slow heating, see Fig. 4), and also larger than the dissociation temperature (462 °C for slow heating, see Table I). At the same time, it is lower than typical plasma temperatures, which is consistent with the fact that no plasma luminescence was observed. A temperature jump of 1400 °C within 3 ns will be associated with explosive

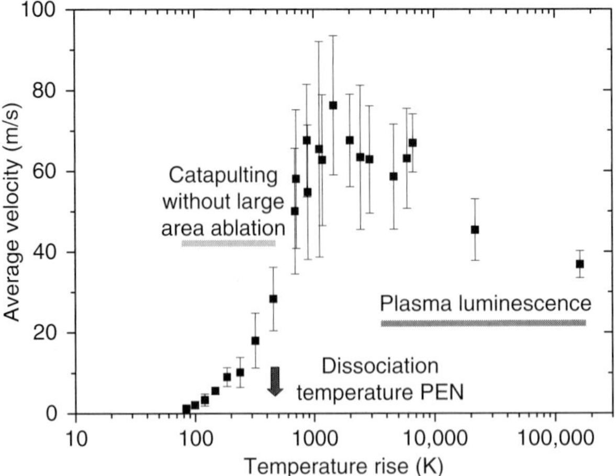

Fig. 16 Velocity data of Fig. 14C plotted as a function of the temperature rise at the bottom of the specimen. The average temperature rise in the optically absorbing layer is calculated based on the laser pulse energy (10 μJ), the measured optical penetration depth (0.88 μm), and an average value of the heat capacity of the PEN foil (2.7 J/g K, see Fig. 3). Losses by specular reflection at the glass slide, absorption in the glass (together 5.3%), and by backscattering from the PEN foil (22.4%) were taken into account.

Table II
Catapulting Thresholds for 3-ns, 337-nm Irradiation

	40× objective			5× objective		
Energy (μJ)	ΔT (K)	ϕ_{th} (J/cm^2)	I_{th} (MW/cm^2)	ΔT (K)	ϕ_{th} (J/cm^2)	I_{th} (MW/cm^2)
10	50	0.03	10	120	0.07	23.3
5	100	0.09	30	200	0.15	50

dissociation of the heated PEN material followed by a very rapid volume expansion. This is a very efficient catapulting mechanism (Fabbro *et al.*, 1990).

Interestingly, the specimens can still be catapulted when the average temperature rise in the absorbing layer is less than one-quarter of the dissociation temperature of the PEN foil. The observation that average temperatures well below the dissociation limit are sufficient for catapulting may indicate that the specimen is driven by sudden thermal expansion and deformation [similar to the working mechanism of dry laser cleaning (Luk'yanchuk, 2002; Tam *et al.*, 1992)]. Even though the actual surface movement caused by thermal expansion is very small, this expansion is achieved in the laser pulse duration of a few nanoseconds. Therefore, the acceleration by 1D surface expansion is 10^5–10^7 times larger than gravitational acceleration, leading to velocities in the order of 1 m/s at the end of the laser pulse (Arnold, 2002; Tam *et al.*, 1992). Moreover, because of the small size of the specimen, thermal expansion can also occur in lateral direction. It is stronger in the lower part of the specimen where higher temperatures are reached than the upper part. This difference results in an upward bending of the peripheral parts of the dissectat that probably contributes to its upward acceleration. However, it is unlikely that these effects alone can account for specimen velocities of up to 20 m/s that were observed with irradiation parameters for which the average temperature in the absorbing layer is below the dissection threshold. We need to consider that the beam profile of the N_2 laser is highly irregular (Fig. 3B and C), that the temperature at the irradiated specimen surface is larger than the average temperature within the optical penetration depth shown in Fig. 16, that the SEM picture in Fig. 15D) is indicative for scattered pointlike ablation, and that the volume expansion of ablated material in confined geometry is an extremely efficient catapulting mechanism (Fabbro *et al.*, 1990). Thus even when the specimen's acceleration by thermal deformation becomes relevant, it is probably still accompanied by the propelling action of ablation in hot spots of the laser beam.

While for focused beams the catapulting velocities are larger when a 5× objective is used instead of a 40× objective (see Section III.B.1), we found no significant difference between the velocities achieved at a given energy when the irradiated spot is sufficiently large to exclude hole formation in the specimen for both objectives. The threshold radiant exposure for catapulting is even slightly lower for the 40× objective (see Table II). It is important to note that the degree of defocusing required to exclude hole formation increases when lasers with better beam quality (i.e., smaller focal spot size) than the N_2 laser are used.

We conclude that catapulting with defocused pulses relies on the pressure produced by confined ablation and, for large spots, possibly also on thermal expansion and deformation of the dissectat. The relative contributions of ablation and thermoelastic forces need to be further investigated.

C. Possible Side Effects and Their Minimization

1. Thermal Effects

The high temperatures produced during plasma formation or pulsed laser ablation seem to be an obvious source for potential side effects of LMD and LPC. However, one needs to consider that any changes within the material that is disintegrated or vaporized during the cutting process do not affect the accuracy of the subsequent genomic or proteomic analysis. Of interest are only changes by heat conduction or convection which alter the remaining dissected specimen that is catapulted into the cap of the microfuge tube.

Let us first consider LMD and LPC with *tightly focused laser pulses*: Here the fraction of the specimen exposed to high temperatures is very small, and most of it is disintegrated and does not take part in the subsequent analysis. Because of the fast adiabatic cooling during the rapid expansion of the laser-induced plasma, the time

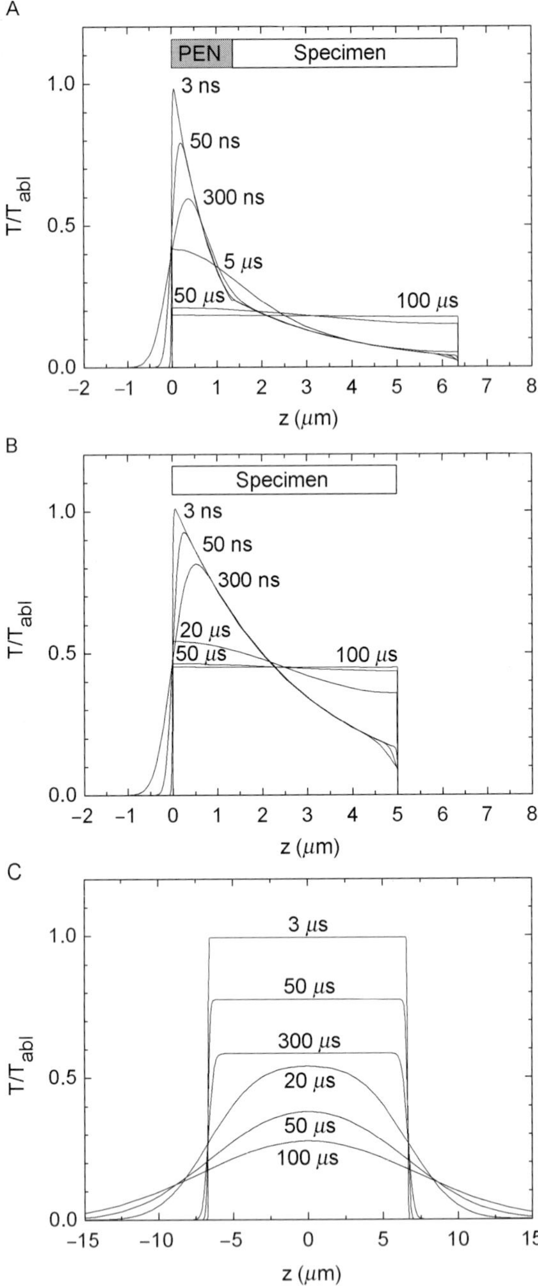

Fig. 17 Evolution of the temperature distribution for (A) H&E-stained specimen on PEN foil backed by a glass slide, homogeneously irradiated by a defocused laser pulse; (B) H&E-stained specimen on a glass slide, homogeneously irradiated by a defocused laser pulse; and (C) H&E-stained specimen on a

available for heat conduction into adjacent parts of the specimen is extremely short (<1 μs), and the heat-affected zone next to a cut is therefore very small. The width of the altered region at the rim of the dissectat in the SEM pictures of Fig. 15 amounts to 2–3 μm for the bottom side covered with PEN foil (Fig. 15B–D) but is smaller (<1 μm) for the upper side, that is for the histologic specimen itself. The fact that thermal alterations are most pronounced at the bottom side of the dissectat can be attributed to heat diffusion from the plasma through the supporting glass slide, which is a good heat conductor. The small amount of thermal damage in the histologic specimen is consistent with previous transmission electron microscopical studies of plasma-mediated dissections in ocular tissues performed with IR laser pulses of 6-ns and 40-ps duration in which the heat affected zone was found to be far below 1 μm (Niemz *et al.*, 1991; Vogel *et al.*, 1990).

The situation is less obvious when a larger fraction of the specimen is exposed to the laser radiation, that is, for moderate or strong defocusing, and when the sample is catapulted directly from a glass slide, without protection by the light-absorbing PEN foil. To assess potential side effects by thermal damage for these cases, we calculated the time evolution of the temperature distribution under different focusing conditions of the laser pulse, as shown in Fig. 17.

We used an analytical solution of the differential equations for heat diffusion in a multilayer geometry (Freund *et al.*, 1996), with layers representing glass, PEN foil, histologic specimen, and air. Glass and air were supposed to be transparent, and the absorption properties of PEN foil and histologic section were assumed to be in accordance with Lambert-Beer's law and taken from Table I. Calculations were performed for a laser pulse with rectangular temporal shape and 3-ns duration. As analytical solutions for layered geometries with different optical properties are restricted to homogeneous thermal properties of the medium, we had to use the same values for heat capacity, heat conductivity, and mass density for all layers. We employed the measured data for haematoxylin and eosin (H&E)-stained specimens listed in Table I.

We assumed that heat can diffuse from the light-absorbing PEN foil and histologic section into the adjacent glass slide until the specimen detaches from the slide during the expansion of the ablation plume, which occurs about 300 ns after the laser pulse (Fig. 6). Afterward, the calculations were continued for a thermally isolated specimen because heat conduction into the surrounding air is negligible. The adiabatic conditions were simulated by the introduction of appropriate mirror heat sources (Carslaw and Jaeger, 1959).

For catapulting without plasma formation, the temperature of the sample surface is determined by the ablation temperature T_{abl} of the sample material. For polymers, the ablation temperature corresponds to the dissociation temperatures

glass slide, irradiated by a pulse with top-hat distribution of 15-μm diameter. The plots in (A) and (B) show the temperature distribution in z-direction; the distribution in lateral direction is homogeneous. The plots in (C) present the lateral temperature distribution at the surface of the heated localized area; the temperature decay in z-direction resembles that in (B) and is not shown.

that, for slow heating, are given in Table I. At very short heat exposure times, dissociation temperatures are higher because dissociation is a chemical rate process (Pearce and Thomsen, 1995). In a similar way, the temperature required for protein denaturation increases with decreasing heat exposure time (Huttman and Birngruber, 1999; Pearce and Thomsen, 1995; Simanowski *et al.*, 2005). This rise of dissociation and denaturation temperatures could not be considered in our calculations because the rate constants for very fast heating are not yet known. To avoid this difficulty, all temperatures in Fig. 17 are not given in absolute values but normalized to the maximum temperature at the sample surface reached at the end of the laser pulse.

Figure 17 compares the temperature evolution for cases where an H&E-stained specimen is either mounted on PEN foil backed by a glass slide (Fig. 17A) or directly placed on the glass slide (Fig. 17B). In both cases the samples were homogeneously irradiated by a *strongly defocused laser beam*, and the temperature distribution in lateral direction is thus also homogenous. When the histologic specimen is mounted on PEN foil, the ablation temperature is reached at the glass–PEN interface, and the specimen itself is well protected by the strong laser light attenuation in the PEN material. The equilibrium temperature reached 50–100 μs after the end of the laser pulse amounts to only 20% of the ablation temperature, and even at the PEN-specimen interface the temperature never exceeds 35% of the PEN ablation temperature (Fig. 17A). By contrast, when the specimen is placed directly on glass, the specimen's ablation temperature (which is lower than for PEN, see Table I) must be reached at the surface of the specimen itself. Moreover, since the laser light penetrates deeper into the specimen material than into PEN, the initial temperature distribution in Fig. 17B is broader than in Fig. 17A, and the equilibrium temperature reached 50–100 μs after catapulting is higher; it amounts to 45% of the ablation temperature. Catapulting directly from glass with spatially homogeneous irradiation thus should be avoided if possible. The thermal load from tightly focused laser irradiation is much smaller, as discussed in the beginning of this section, but the amount of material transported per pulse will also be smaller than with homogeneous irradiation if no PEN foil is used.

The modeling results of Fig. 17A predicting that the histologic specimen on top of the 1.35-μm-thick PEN foil is hardly affected by heat are in good agreement with the SEM results of Fig. 15D for strongly defocused catapulting which demonstrate that only the bottom of the polymer foil is ablated. However, even when PEN foil is used, the specimen is not in all cases protected. The SEM images in Fig. 15B show that with *moderate defocusing* corresponding to 15-μm spot diameter the PEN foil is removed from the irradiated area. A part of the PEN material has probably been ablated and another part was molten and pushed aside by the pressure in the ablation cloud. Therefore, the lower surface of the histologic specimen is exposed to the PEN ablation temperature, and the temperature profiles in z-direction will thus resemble those of Fig. 17B. However, because of the small size of the irradiated spot, lateral heat diffusion now contributes to cooling as shown in Fig. 17C, and the cooling continues even after the specimen is thermally isolated form the glass slide. As a consequence, the temperature at the specimen surface drops to 28% of the ablation temperature within 100 μs, compared to 45% in Fig. 17B. Nevertheless, with moderate defocusing it cannot be excluded that the nonablated specimen volume above the perforated area and next to the molten foil is subject to thermal changes. However, for a spot diameter of 15 μm and a specimen diameter of 80 μm, this volume corresponds to only 3% of the entire specimen volume. For 25-μm spot size, the fraction increases to 10%.

2. Photodamage by UV Light

Chemical changes by UV light are, like thermal damage, only relevant for the material remaining after dissection. During dissection and catapulting with focused laser pulses, the dissectat may be irradiated by UV laser light scattered at the laser plasma and by the UV plasma luminescence. The light scattering by ns plasmas was found to involve less than 2% of the incident laser irradiation and to occur mainly in forward and backward direction (Nahen and Vogel, 1996). Therefore, the scattered laser light may, at worst, affect a very thin specimen layer at the edge of the cut that is of little relevance for LMD and LPC. The energy of the plasma radiation stays below 0.1% of the incident laser energy (Vogel *et al.*, 1999) and is thus completely irrelevant.

During defocused catapulting with an extended spot size, a large fraction of the specimen or the entire specimen is exposed to the laser light. However, the histologic section is protected by the PEN polymer foil that transmits only 20.5% of the incident light at 337 nm (Table I). Furthermore, the N_2 laser wavelength of 337 nm is far away from the peak of the action spectrum for DNA damage shown in Fig. 18 (Coohill, 2002), and the wavelength of a frequency-tripled Nd:YAG laser (355 nm) is even further away. This is consistent with investigations on the wavelength dependence of laser-induced DNA damage in lymphocytes using the comet assay (de With and Greulich, 1995). For $\lambda = 340$ nm, the detection threshold for DNA damage found by de With and Greulich (1995) corresponded to a radiant exposure of 1.5 J/cm^2. Because of the limited sensitivity of the comet assay, $\sim$300 strand breaks per cell are necessary to detect DNA damage. Hence, one single DNA strand break per cell is expected to occur after a radiant exposure of 5 mJ/cm^2. Other pathways of cell damage in the UV-A/B region of the optical spectrum involve the generation of reactive oxygen species such as H_2O_2 and OH* radical (Bertling *et al.*, 1996; Hockberger *et al.*, 1999; Tyrrell and Keyse, 1990). For broadband radiation (305–350 nm) peaking at 325 nm, significant cell killing was observed with light doses $\geq$1 J/cm^2 (Bertling *et al.*, 1996).

The total dose arriving at the PEN foil when a 10-μJ pulse irradiates a specimen with 80-μm diameter is 200 mJ/cm^2. The dose transmitted through the foil and arriving at the histologic material is 40 mJ/cm^2. This is 25 times below the threshold for cell killing reported by Bertling *et al.* (1996), but slightly above the dose causing, on average, one DNA single strand break per cell (de With and Greulich, 1995).

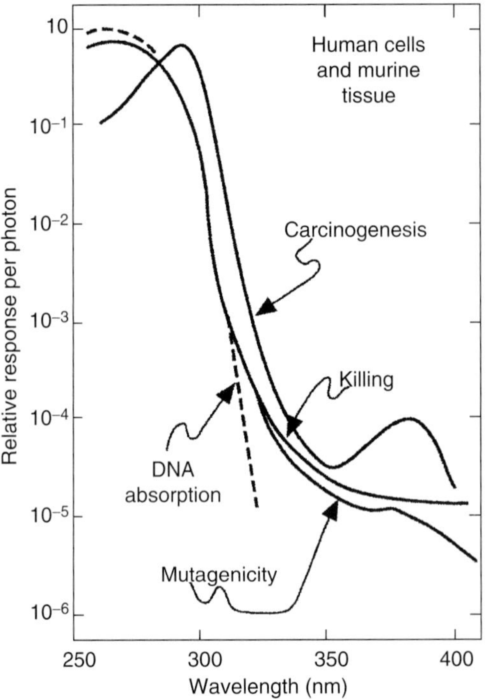

Fig. 18 UV action spectra for human cell killing, mutagenicity, DNA strand breaks, and DNA-protein cross-links. Reprinted with permission from Coohill (2002). Copyright CRC Press.

With moderate defocusing, for example for 15 μm spot diameter, the dose within the irradiated area increases by a factor of 30 to 6 mJ/cm^2, and the protection by the PEN foil is largely absent because it is ablated by the catapulting laser pulse (Fig. 15B). In this case, the risk for UV damage directly above the irradiated spot is higher because the radiant exposure exceeds the value of 1 J/cm^2 leading to significant cell damage, and largely exceeds the threshold value of 5 mJ/cm^2 for sporadic DNA strand breaks in individual cells. It is important to note, however, that for a specimen with 80-μm diameter, only 3% of the total volume will be affected by the UV irradiation of a spot with 15-μm diameter. The affected fraction is larger when histologic material is directly catapulted from glass substrates without supporting polymer foils because in this case smaller amounts of material are catapulted per pulse.

3. Mechanical Deformation and Rupture

In general, purely mechanical rupture and disintegration of a histologic sample during LPC impose no problem for subsequent genomic or proteomic analysis because only a very small fraction of the biomolecules will be affected even if the dissectat is fragmented into many large pieces. However, mechanical forces may affect the viability of live cells (see Section IV).

The specimen is accelerated to a speed of 180–350 m/s within about 200 ns (Fig. 14A). The acceleration in the initial phase of catapulting is thus enormous, being $\sim 10^8$ times larger than the gravitational acceleration. However, this does not affect the specimen's integrity as long as the accelerating forces are homogeneous because only tensile stress and shear forces lead to deformation and tearing.

The main sources of shear forces are pressure gradients, while tensile stresses may also originate from radial expansion movements (Lokhandwalla and Sturtevant, 2001; Lokhandwalla *et al.*, 2001). During focused LPC, the huge pressure gradients produced in the vicinity of the plasma lead to local rupture of the specimen, and the large absolute pressure values in this region result in the immediate ejection of debris at very high velocity. However, once the plasma pressure is partly released through the hole in the center of the specimen, the distribution of the driving forces becomes more homogeneous as the shock wave spreads laterally below the entire specimen. At this stage, the pressure gradients have become too small to cause further rupture or strong deformations, and the dissectat flies off in a disklike shape (Fig. 6).

More significant deformations may arise from moderately defocused laser pulses that do not produce a hole in the specimen. Because the irradiance is largest in the center of the irradiated spot (Fig. 3), the initial pressure distribution is inhomogeneous and the center of the specimen will bulge upward before it flies off. Such deformation can be avoided by creating a nearly homogeneous irradiance distribution across the specimen. This is possible by defocusing of the laser beam to a degree that the irradiated spot is much larger than the specimen, but that would be associated with considerable energy loss. Alternatively, a phase mask can be used to create a homogeneously illuminated spot with appropriate size.

4. Quantification of the Influence of Side Effects on the Accuracy of Proteomic and Genomic Analysis

LMD was originally developed to facilitate the histochemical analysis of cryosections (Meier-Ruge *et al.*, 1976). Correspondingly, the authors performed enzyme-histotopochemical investigations on N_2 laser dissected fragments of tissue sections to assess the quality of the procurement method. They observed a decrease of enzyme activity by 10–20% in a several microns thick region bordering the cut when dissection was performed at an energy level four times larger than that sufficient for cutting. The changes were attributed to heat diffusion from the glass support and characterized as not relevant for subsequent histochemical analysis.

At present, LMD is mostly combined with molecular biological analysis, including polymerase chain reaction (PCR) and microarrays for genomic and mass-spectrometry for proteomic studies. In qualitative and quantitative analysis of gene expression, reverse transcriptase PCR (RT-PCR) plays a prominent role. In principle, the method allows very small numbers of molecules to be detected at extreme sensitivity and high specificity (Bustin, 2000), and is therefore the preferred method for the analysis of tiny tissue fragments procured in LMD and LPC. Initially, Schütze and Lahr (1998) used formalin-fixed and paraffin-embedded tissue sections supported by a PEN foil, isolated tissue fragments by LMD and LPC, and successfully detected specific mRNA sequences in RT-PCR. Since then, it is known that numerous factors, including fixation, tissue processing, staining and labeling, and probably also laser dissection may reduce the content of mRNA in the tissue samples by several orders of magnitude and thus may critically affect quantitative analyses (von Smolinski *et al.*, 2005, 2006). A major effect on the mRNA loss can be attributed to chemical fixation in aldehydes and embedding in paraffin, as typically done in histopathology (Krafft *et al.*, 1997; Lewis *et al.*, 2001; Stanta and Schneider, 1991). As compared to unfixed cryosections, this classical tissue processing reduces the mRNA by factors of 85–99% (Abrahamsen *et al.*, 2003; von Smolinski *et al.*, 2005). In addition, conventional staining protocols (e.g., H&E) or immunohistochemical labeling (e.g., for fluorescence microscopy) also attribute to mRNA loss by factors of up to 99.8% (von Smolinski *et al.*, 2006). It must be assumed that thermal effects and those evoked by UV light during LMD and LPC can produce some additional damage to the mRNA contained in the isolated samples, but quantitative data on the absolute mRNA loss and on the relative loss compared to the factors mentioned above are still lacking. In this context, it should be noted that mechanical damage occurring during LPC of histologic specimens would not affect the accuracy of protein or gene analysis (Schütze and Lahr, 1998) because crushing of the samples is a prerequisite for many analytical techniques in biochemistry and molecular biology. In contrast, the extent of thermal and UV light damage will depend on the size of individual dissected section fragments, which vary in area between some hundred square microns for single cells and several 10,000 μm^2 for extended tissue compartments isolated in tumor biology or immunological research. As discussed in Sections II.C.1 and III.C.2, only little collateral damage is produced when large specimens are catapulted with tightly focused or strongly defocused laser pulses. The influence of heat and UV radiation is largest with moderately defocused pulses.

Besides PCR analysis in genomics, the analysis of proteins is a further interesting and promising application for laser tissue microdissection. Tiny amounts of protein are analyzed using mass spectrometry (Ai *et al.*, 2006; Fink *et al.*, 2006) or microarray techniques (Gulmann *et al.*, 2005; Niyaz *et al.*, 2005). Using an antibody-ultramicroarray, proteins can be detected with high specificity and sensitivity amounts equaling less than 10 cells (Nettikadan *et al.*, 2006). A novel method is the so-called proteohistography which combines array- and mass-spectrometric techniques and produces an image of the spatial distribution of proteins in a given tissue area (Ernst *et al.*, 2006).

The maximum likelihood for laser-induced unwanted side effects in both proteomics and genomics arises when histologic material is directly catapulted from glass substrates without supporting polymer foils. In this case, the laser beam has to be scanned across the area to be catapulted. The large number of applied pulses and the lack of a protecting polymer foil involve an increased exposure to both UV radiation and heat. Preliminary investigations revealed that mRNA recovery from cryosections decreases considerably if the material is catapulted from glass without the use of a PEN foil. To compare both methods, eight identical samples were processed both with PEN foil and without foil in the auto-LPC mode using factory settings of the instrument. The amount of intact mRNA copies of the housekeeping gene EF1α remaining in the sample after catapulting was determined using real-time PCR. After catapulting directly from glass, 35.5–40.00 amplification cycles were necessary to achieve a certain signal strength as compared to 26.1–28.2 cycles when a PEN foil was used (Dorthe von Smolinski, Institute of Anatomy, University of Lübeck, unpublished results). Since the amount of genetic material is doubled in each cycle, an increase by one cycle (1- Ct unit) corresponds to a loss of 50% specific mRNA in the sample, and an increase of 7- to 9-Ct units indicates a loss of 99.2–99.8% of specific mRNA copies. This value resembles the loss observed otherwise as a consequence of fixation, embedding, and staining. Our preliminary results thus indicate that there is still a large potential for improving laser-assisted mRNA recovery from cryosections. A systematic analysis of potential side effects of the different LMD and LPC procedures and of ways to optimize the accuracy of the genomic or proteomic analysis still needs to be performed.

IV. Dissection and Catapulting of Live Cells

For the retrieval of live cells by LMPC we used a protocol based on the use of duplex membrane dishes that was adopted from Mayer *et al.* (2002) and Stich *et al.* (2003). A 25-μm-thick Teflon membrane provides mechanical support, and a 1.35-μm-thick PEN foil conditioned with polylysine is mounted into the dish above the Teflon membrane as delineated in Fig. 19. CHO cells were cultivated on this foil until a confluent monolayer had developed. Before LMPC, the culture medium was almost completely removed such that only a thin layer of liquid (up to 40-μm thick, as determined by optical coherence tomography, OCT) remained above the cells. Then the region of interest was dissected, and the dissectat (cells and PEN foil) catapulted by a single laser pulse into the cap of a microfuge tube that had been wetted with culture medium. To test cell viability after LMPC, the cells were transferred into 12-well plates as described by Stich *et al.* (2003), and recultivated.

The original protocol involves the use of focused laser pulses for catapulting but we also performed series of experiments with defocused pulses because we hoped that this would minimize bending of the specimen and shear stresses on the cells. Some experiments were done without any fluid between Teflon membrane and PEN foil, but usually some liquid enters the space between the two membranes. Since we observed that the presence of this liquid has an influence on the catapulting

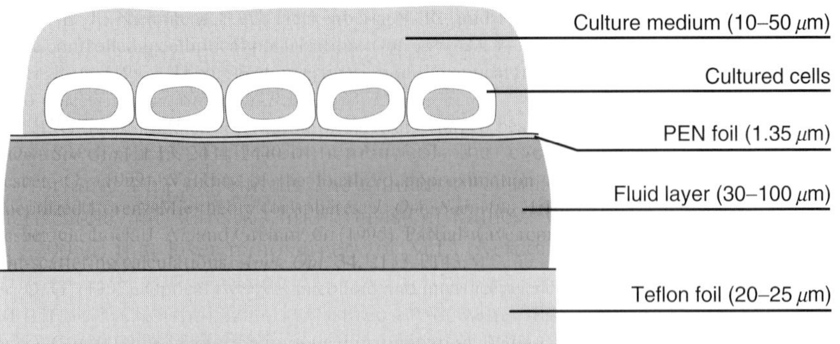

Fig. 19 Schematic presentation of the setup used for live-cell catapulting from a duplex membrane dish. Shortly before LMPC, the culture medium is almost completely removed such that only a thin layer of liquid remains above the cells.

dynamics, we performed additional experiments in which a well-defined amount of medium was injected between the membranes before LMPC. The resulting liquid layer had a thickness of 30–100 μm.

To document the dissection process before catapulting by time-resolved photography, we replaced the halogen lamp and collimation optics of the microscope by a Nanolite flash lamp combined with a Nikon 50 mm/1.2 objective used as collimator. Photographs of the dissection dynamics were taken through the microscope optics with single exposures at different time delays after the release of the dissecting laser pulse. The catapulting dynamics was photographed in side view using the setup described in Section III.B.1. To be able to photograph the initial catapulting phase without vignetting by the rim of the Petri dish, we developed a special dish with a removable rim. Teflon membrane and PEN foil were clamped around a stainless steel ring using a silicone O-ring. A flat silicone ring was placed on top of the steel ring. It was pinned down by the weight of a second stainless steel ring and thus provided a tight seal for the culture medium. After removal of most of the culture medium shortly before LMPC, the upper steel ring and the silicone seal could also be removed which offered an unobstructed view onto the cells on top of the PEN foil.

A. Dissection in a Liquid Environment

Even after removal of most of the culture medium, dissection takes place in a liquid environment. Thus the surrounding liquid confines the laser-produced plasma and the ablation products cannot freely escape. As a result, a transient cavitation bubble is formed around the laser focus (Venugopalan *et al.*, 2002; Vogel and Venugopalan, 2003, Chapter IX), which expands and collapses within a few microseconds. Because of the cavitation bubble dynamics, dissection in a liquid environment is less precise than in air. For sufficiently large laser pulse energies, the shear stress exerted by the oscillating bubble causes lysis of cells adjacent to the laser focus (Rau *et al.*, 2004, 2006) or sweeps them off the PEN foil, if their adhesion is weak.

We observed that the energy required for dissection increases in the presence of a liquid layer between Teflon membrane and PEN foil, especially when this layer is thick (≈100 μm). Adhesion between cells and foil was observed to decrease when cells continued to grow after forming a confluent layer. Adhesion could be improved by conditioning the foil with polylysine. In Fig. 20A and B, cells are swept off the PEN foil by the expanding bubble up to a distance of 20–30 μm from the laser cut even though a pulse energy of not more than 1 μJ was used for dissection. The reason is that no polylysine had been applied, and the cell density on the foil was already fairly high indicating reduced adhesion. The bubble dynamics is presented in Fig. 20C and D. Here, a larger pulse energy of ≈6 μJ was required for dissection because the liquid layer between Teflon membrane and

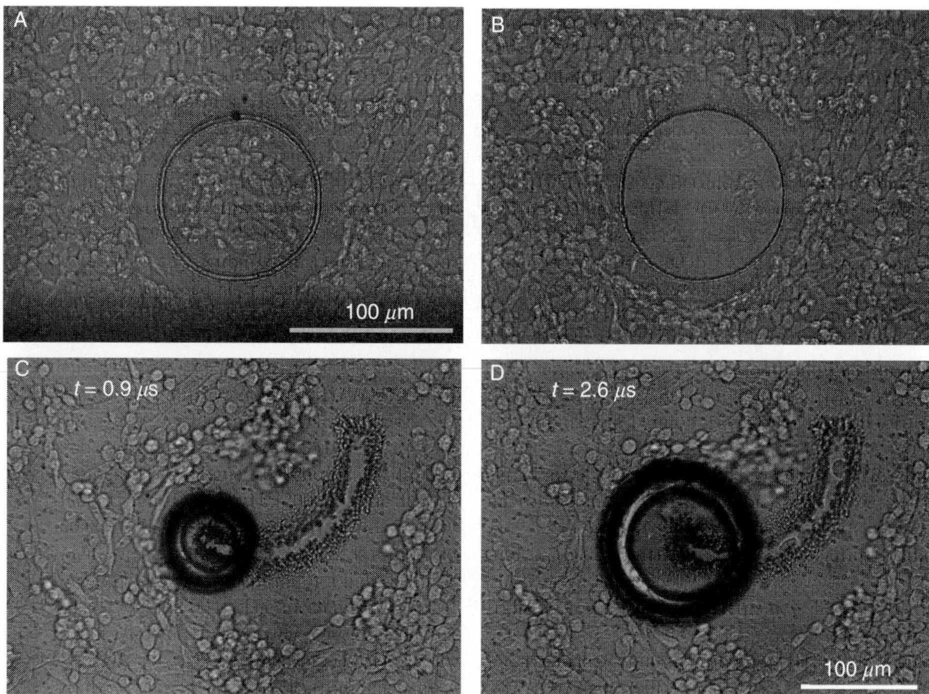

Fig. 20 Laser dissection and catapulting of CHO cells with relatively low adhesion to the PEN foil, (A) after dissection using a 20× objective, NA = 0.5, and E = 1.0 μJ, (B) after catapulting. The denudation near the cut is caused by the shear forces arising from the oscillating cavitation bubbles produced on optical breakdown in the culture medium. In (C) and (D) the cavitation bubble dynamics during dissection of CHO cells is shown at different times with (D) being the stage of maximum expansion. The dissection energy was here relatively large (E = 6 μJ) because the PEN foil was separated from the Teflon membrane below by an ≈100-μm-thick liquid layer.

PEN foil was fairly thick. The maximum bubble diameter amounts to $\approx$140 μm but the width of the denuded zone on the foil is smaller: it corresponds quite closely to the area in which the expanding bubble touches the cell layer. Under optimum conditions, the damage zone next to the cut amounts to about 5–8 μm, as shown in Fig. 21, which is considerably smaller than in Fig. 20.

We employed a 20× objective, NA = 0.5 both for dissection and catapulting, as described by Mayer *et al.* (2002) and Stich *et al.* (2003). Collateral damage during dissection can be reduced by using objectives with larger NA because they allow for plasma-mediated cutting with smaller pulse energies.

B. Catapulting of Live-Cell Populations

LMPC and recultivation of a colony of CHO cells is demonstrated in Fig. 21. To minimize the stress on the cells, we used small pulse energies for dissection and catapulting, and focused the catapulting pulse at the periphery of the specimen, as visible in Fig. 21C and D. The denuded zone at the sides of the cut is smaller than

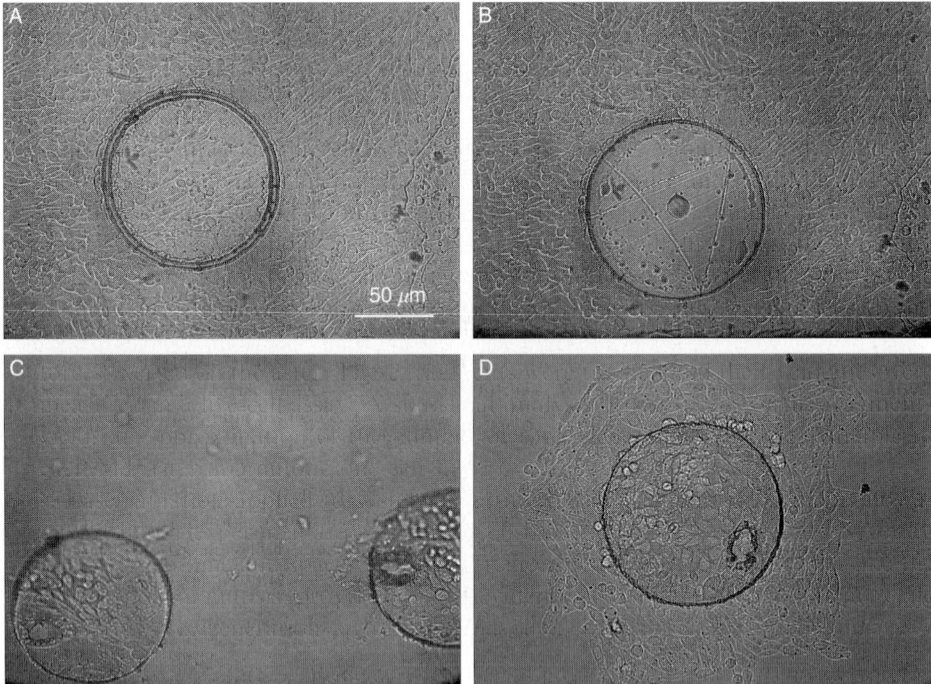

Fig. 21 Laser dissection, catapulting, and recultivation of a colony of CHO cells according to the protocol of Stich *et al.* (2003). (A) after dissection, (B) after catapulting with a focused pulse, (C) catapulted specimens in the cap of a microfuge tube, and (D) after 48 h of recultivation. We used a 20×, NA = 0.5 objective, with pulse energies of 1.0 μJ for dissection and 5.0 μJ for catapulting.

in Fig. 20 because the adhesion of the cells was better, and the liquid layer between Teflon membrane and PEN foil was thinner, which facilitates dissection.

When the laser pulse was focused into the periphery of the specimen, 16% of the catapulted specimens ($n = 60$) missed the cap for reasons already discussed in Section III.B.2. However, out of the specimens that could be transferred into a 12-well plate, 98% (all besides one) could be recultivated. Because we had observed that the use of a strongly defocused laser beam for catapulting of histologic specimens was associated with a stable flight trajectory, absence of hole formation, and moderate catapulting velocities (see Fig. 14), we tested this strategy also on live cells. We used the maximum defocusing setting on the microbeam station, corresponding to a spot diameter of 50 μm. As expected, all specimens (60 out of 60) arrived in the cap, but to our surprise the majority of the cells had been sheared off the foil, as shown in Fig. 22B and C. In only four cases (7%), recultivation was possible.

This puzzling difference is not easily understood, considering the complex geometry of membranes and liquid layers in live-cell catapulting (see Fig. 19). Therefore, we chose a stepwise approach to understand the role played by the individual layers. In the first series of experiments, a 30- to 50-μm-thick liquid layer was injected between Teflon membrane and PEN foil, but no cells or liquid was present on top of the foil. PEN dissectats of 100-μm diameter were catapulted with defocused laser pulses (50-μm spot diameter) and with pulses focused at the periphery of the specimen, and the resulting dynamics is shown in Fig. 23.

In defocused catapulting (Fig. 23A), the expanding ablation products drive the liquid below the specimen radially to the specimen's rim where it collides with the surrounding resting liquid. As a result, the liquid is pushed through the

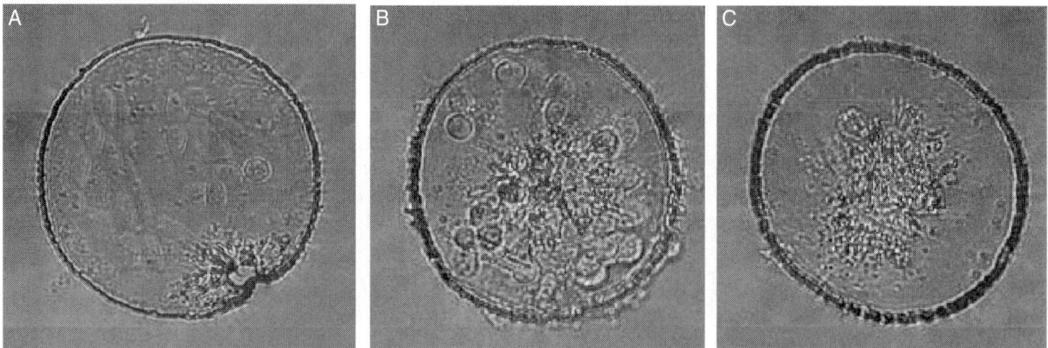

Fig. 22 Specimens after catapulting with focused (A) and moderately defocused laser pulses (B) and (C). In (A) all cells near the specimen center remained on the PEN foil but regions in the vicinity of the laser shot and at the opposite side of the specimen are denuded. In (B) some cells remained on the PEN foil, while the foil in (C) is completely denuded. The ablation pattern visible around the center of the specimen in (C) demonstrates that the intensity distribution in the catapulting laser beam is inhomogeneous.

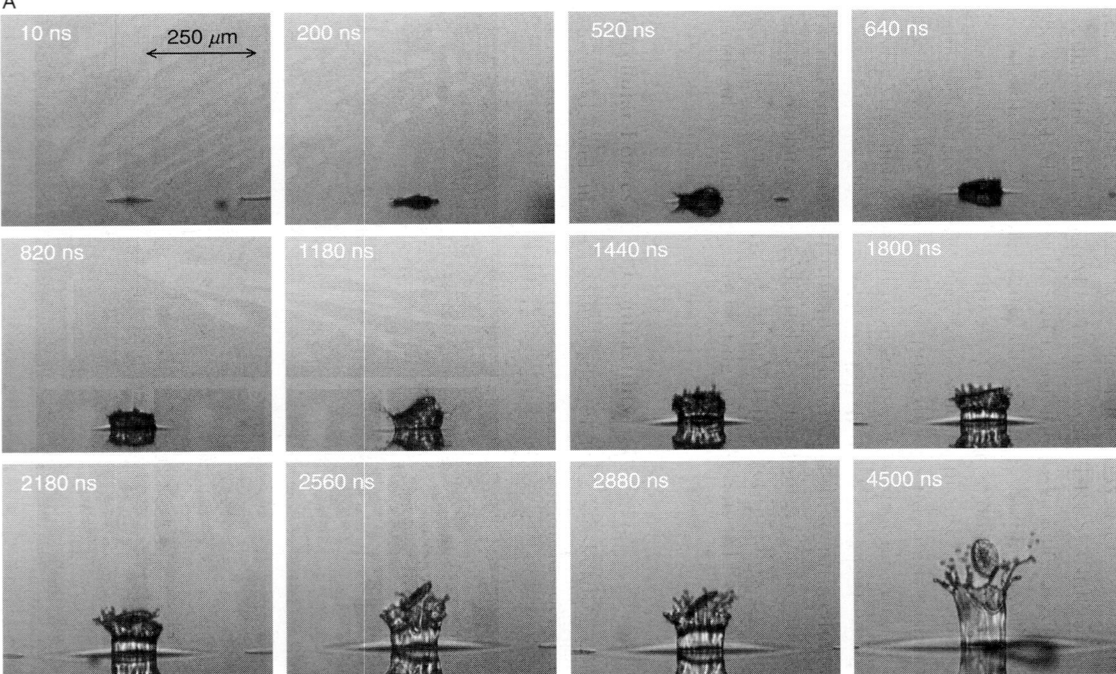
A
10 ns
250 μm
200 ns
520 ns
640 ns
820 ns
1180 ns
1440 ns
1800 ns
2180 ns
2560 ns
2880 ns
4500 ns

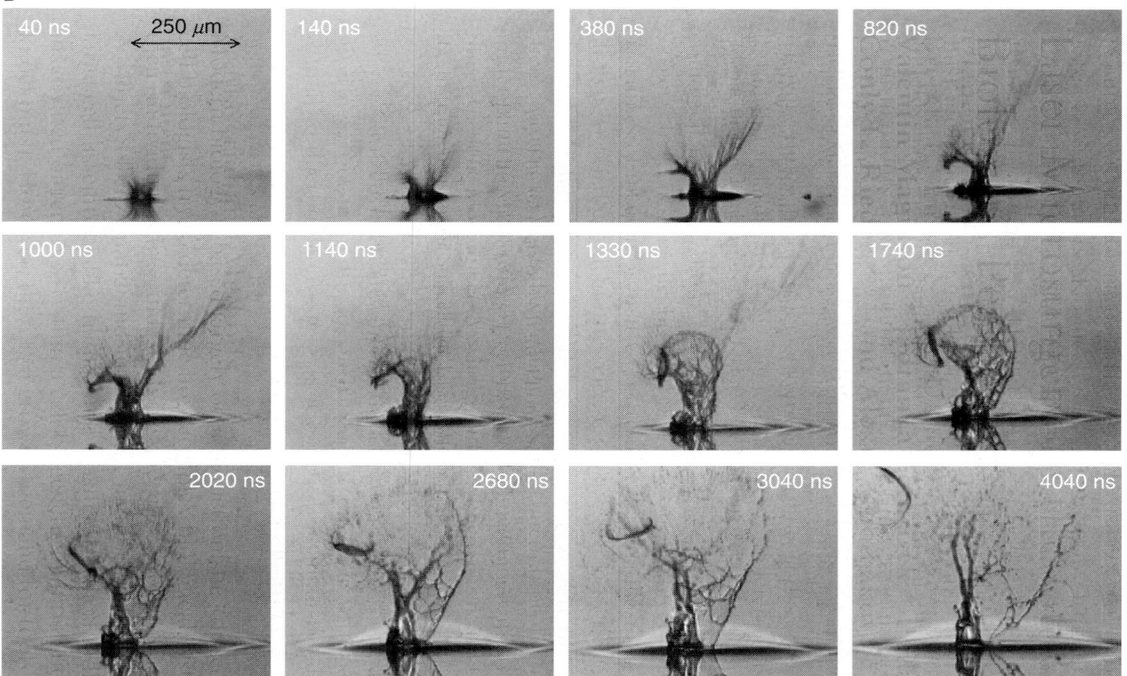

Fig. 23 Catapulting of PEN foil specimens located above a Teflon membrane, with a 30- to 50-μm-thick liquid layer between the two membranes. (A) Catapulting with defocused laser pulses aimed at the center of the specimen (50-μm spot diameter), (B) catapulting with pulses focused at the periphery of the specimen (the plasma luminescence indicates the location of the laser focus). $20\times$ objective, NA = 0.5, E = 20 μJ, and specimen diameter 100 μm. The average specimen velocity during the first 4 μs is about 50 m/s in (A) and 100 m/s in (B). The rotational movement of the specimen in (B) corresponds to a frequency of 600,000 rps during the first microsecond and 330,000 rps when averaged over the first 4 μs.

circular cut and a cylindrical splash evolves. The catapulted specimen maintains a flat, disklike shape, without bulging in the center where the laser irradiance and the ablative pressure are highest. The rim of the specimen is accelerated not only by the ablative pressure (which is lower in this region) but also by the cylindrical splash, which is driven by the ablative pressure and associated with a local concentration of kinetic energy. The combined action of ablative pressure and secondary fluid flow results in an approximately homogeneous acceleration of the sample.

By contrast, in focused catapulting (Fig. 23B), the side of the specimen at which the laser focus is located is most strongly accelerated, and the specimen assumes a fast rotational movement. Within 2 μs, it has turned by 360°, corresponding to a rotation frequency of 500,000 rps, similar to the case of catapulting from a dry substrate shown in Fig. 13. The catapulting velocity is larger than with defocused pulses ($v \approx 60$ m/s, averaged over the first 2 μs compared to $v \approx 45$ m/s) because plasma is formed at the laser focus while defocused catapulting is driven by ablation below the threshold for plasma formation.

The dynamics in Fig. 23 shows no obvious advantage of focused catapulting for live-cell retrieval and recultivation. This advantage arises only when PEN foils and cells are covered with a liquid layer such as in Fig. 24. For defocused catapulting, the upper liquid layer now largely suppresses the splash of liquid from the lower layer and the specimen is hardly accelerated at its periphery. The high pressure produced by ablation in the central region of the specimen results in an upward bulging of this region while the rim is tied down by the inertia of the liquid covering the specimen (Fig. 24A). After about 1 μs, the movement of the specimen center is slowed down by these inertial forces while the radial liquid flow in the upper liquid layer that was created during the first microsecond continues and is focused above the specimen center into an upward directed jet. Because this jet flow is faster than the movement of the specimen, it exerts a shear force on the cells. When the specimen rises out of the liquid, after a few microseconds, some fluid stays behind at the specimen's rim. This results in a flow and shear forces opposite to that of the jet flow. The combined action of these successive shear forces is probably responsible for the removal of most of the cells from the specimen that led to the low success rate of recultivation.

The catapulting dynamics induced by laser pulses focused at the periphery of the dissectats is presented in Fig. 24B. Its principal characteristics resemble the sequence of events observed without a liquid layer covering the cells (Fig. 23B). The pressure within the laser plasma is strong enough to immediately remove the upper liquid layer in the vicinity of the laser focus. This "Moses effect" gives leeway to the acceleration of the side of the specimen proximal to the laser focus. Within about 1 μs, the specimen has risen out of the culture medium and rotated by 90°, and after about 4 μs, it has propagated a distance of 220 μm (average velocity 55 m/s) and rotated by 180° (rotation frequency 125,000 rps). Because of the rotation, the fluid flow along the cells during the specimen's take off remains weak, and the shear forces acting on the cells are weaker than in the case of defocused catapulting. On the other hand, centrifugal forces come into play that increase proportional to the distance *r* from the axis of rotation. After 5–10 μs (not shown), the specimen flies free of any liquid and the rotational movement has been considerably slowed down by air friction.

Both for focused and defocused catapulting, the exact sequence of events varies with the thickness of the liquid layer above the cells (10–50 μm) but still resembles the behavior portrayed in Fig. 24. After removal of the culture medium, the liquid layer thins because of evaporation. The actual thickness on each individual photograph thus depends on the time after removal of the medium when catapulting was performed. The main consequence of an increasing layer thickness is a slowdown of the dynamics both with regard to translational and rotational specimen velocity.

C. Possible Side Effects and Their Minimization

Criteria for successful live-cell catapulting are (1) the fraction of specimens that can be recovered/collected, (2) the percentage of vital cells per specimen, and (3) the recultivation rate. Adverse factors are (1) large variations in the flight trajectories of the specimens; (2) removal of cells from the substrate by mechanical forces; and (3) damage to cells remaining on the substrate by heat, UV irradiation, or mechanical stress. While the flight trajectories were more stable with defocused catapulting, cell loss and damage were less severe when pulses focused at the rim of the specimen were used. These differences are probably due to distinctions in the mechanical effects rather than to dissimilar responses to heat and UV irradiation.

The possible sources for cellular damage by heat and UV radiation are very similar to those for histologic material that were discussed above, with the threshold values presumably being lower for living cells. Note, however, that the threshold doses for UV light-induced DNA damage and cell killing quoted in Section III.C.2 from the work of de With and Greulich (1995) and Bertling *et al.* (1996) already refer to populations of living cells. The thresholds for thermal cell damage are high for the short heat exposure times involved in catapulting, which last only a few microseconds (see Section III.C.1). Simanowski *et al.* (2005) reported that cells survived temperatures as high as 180 °C for heat exposure time of 300 μs. For heat pulses shorter than 300 μs, the threshold for cellular death was determined by the threshold for explosive vaporization that occurred at temperatures slightly above 200 °C. Thus, the most likely sources of cell damage in catapulting are mechanical effects. They may result in removal of cells from the membrane, immediate cell lysis, or in more subtle damage to cell membranes and/or organelles.

Looking at the vigorous dynamics portrayed in Figs. 23 and 24, it is quite remarkable that a large number of catapulted cells continue to proliferate in an apparently unimpeded fashion. However, it needs to be kept in mind that rapid motions per se do not necessarily cause damage. Any fluid motion can be decomposed in uniform translation, rigid rotation, and an extensional flow, and only the

A
700 ns
250 μm
780 ns
960 ns
1300 ns
1940 ns
2280 ns
2520 ns
3120 ns
4160 ns

Fig. 24 Catapulting of cell preparations (CHO cells on PEN foil) out of a duplex dish. The liquid layer between Teflon membranes and PEN foil was 30- to 50-μm thick, and a 10- to 50-μm-thick layer of culture medium covered the cells. (A) Catapulting with defocused laser pulses and (B) catapulting with pulses focused at the periphery of the specimen (the plasma luminescence indicates the location of the laser focus). 20× objective, NA = 0.5, E = 20 μJ, and specimen diameter 100 μm. The average specimen velocity during the first 4 μs is about 50 m/s in (A) and 55 m/s in (B). The rotational movement of the specimen in (B) corresponds to a frequency of 220,000 rps during the first microsecond and 125,000 rps when averaged over the first 4 μs.

latter bears damage potential. Extensional flow patterns associated with tensile stress arise from shear through pressure gradients, inertial or viscous drag, from radial expansion movements (Lokhandwalla and Sturtevant, 2001; Lokhandwalla *et al.*, 2001), or from thermoelastic effects (Paltauf and Dyer, 2003; Vogel and Venugopalan, 2003). With moderately defocused laser pulses, the irradiance is largest in the specimen center. Therefore, the initial pressure distribution is inhomogeneous, and the center of the specimen will bulge upward before it flies off due to the expansion of the bubble below the specimen. The resulting tensile stress and strain may lead to cell detachment or membrane rupture. For later phases of the catapulting dynamics the image series in Figs. 23 and 24 suggest that shear forces are also more pronounced in catapulting with a defocused rather than with a focused laser beam.

However, catapulting with pulses focused at the specimen's rim is associated with a fast rotation of the specimens that gives rise to considerable centrifugal forces that are not observed in defocused catapulting. The initial rotational velocity at the rim of a specimen with 100-μm diameter revolving by 180° in 4 μs such as in Fig. 24B is $\approx$39 m/s, and the centripetal acceleration $a = v^2/r$ at the specimen rim ($r = 50$ μm) amounts to 3×10^7 m/s^2 for $\nu_R = 125{,}000$ rps. Amazingly, the majority of the cells remain on the specimen in spite of the strong centrifugal force, probably due to its very short duration. Air friction rapidly decelerates the rotational movement, and correspondingly the centrifugal force $F = mv^2/r$ drops very fast such that strong centrifugal forces act only during a few microseconds. This time interval is apparently short enough to allow for elastic deformation avoiding rupture or detachment of cells that are located sufficiently close to the rotation axis as shown in Fig. 22A. However, Fig. 22A also indicates that cells far away from the rotation axis are sheared off—not only in the vicinity of the laser shot but also at the opposite side of the specimen rim.

In general, the damage potential of hydrodynamic effects is not only determined by the magnitude of the tensile or shear forces but also by their duration because the material must be strained before it can rupture. Rupture (or at least poration) of the cell membrane requires an areal strain larger than 2–3% (Boal, 2002; Evans *et al.*, 1976; Needham and Nunn, 1990). The deforming force must last sufficiently long to achieve this deformation. Moreover, the ultimate tensile strength (UTS) of the cell membrane or elements of the cytoskeleton may depend on the strain rate. It has been observed for tissue that, while the strain at fracture does not change significantly with strain rate, the UTS increases. The increase of the UTS is due to the fact that, under conditions of rapid deformation, there is significant viscous dissipation between matrix elements, for example collagen fibrils, and ground substance (Vogel and Venugopalan, 2003; Section II.B). It is conceivable that similar laws also apply on the cellular level. The response of cells to very large strain rates acting for very short time is still largely unexplored and requires further investigation.

Improvements of defocused catapulting can probably be achieved with stronger defocusing and a more homogeneous irradiance distribution at the specimen than in our present experiments. However, even then the advantage of the rotational movement associated with focused catapulting may still prevail if the liquid layer above the cells is relatively thick. With decreasing thickness of this layer, the catapulting dynamics will change from the scenario shown in Fig. 24A toward that of Fig. 23A which is most likely correlated with less collateral damage.

We observed that a thin liquid layer between Teflon membrane and PEN foil facilitates catapulting. However, it increases the energy requirements for cutting and thus the amount of side effects associated with dissection when it becomes too thick. The optimum thickness of this layer, which may vary with specimen size, still needs to be identified.

V. Conclusions and Outlook

A. Potential Improvements for Dissection

Three different measures can lead to finer dissections than possible with the N_2 laser employed in our experiments: (1) improvement of the beam profile, (2) reduction of the laser pulse duration, and (3) increase of the laser repetition rate. The beam profile of diode-pumped frequency-tripled Nd:YAG lasers that are incorporated in the newest generation of most commercial microbeam system is much better than that of the N_2 laser (see Fig. 3). This can lead to a considerable reduction of the focal spot size, optical breakdown energy, and cutting width, provided that the delivery optics to the focusing microscope objective maintains the good beam quality. This goal will not be achieved if the laser beam is simply coupled into the fluorescence beam path that is optimized for homogeneous illumination of the object field but not for focusing of a laser beam. Additional corrections of the spherical aberrations induced in the beam path of the fluorescence illumination are required to provide optimum focusing conditions.

An even larger reduction of the energy threshold for optical breakdown can be reached by employing shorter laser pulse durations (Vogel *et al.*, 1996a, 2005a). We observed that a reduction of the pulse duration from 6 ns to 300 fs reduces the breakdown threshold at NA = 0.9 by a factor of ~100 for UV wavelengths and even more for IR wavelengths (by A. V. and V. H., unpublished results). Use of fs lasers thus creates the potential for nanosurgery on a subcellular level (König *et al.*, 1999, 2005; Vogel *et al.*, 2005a and references therein) and for gentle optotransfection (Tirlapur and König, 2002). Already a moderate reduction of the laser pulse duration to 500 ps in combination with a good beam profile and the use of UV light ($\lambda = 355$ nm) made it possible to selectively dissect microtubules in live cells (Colombelli *et al.*, 2005).

Especially for pulse durations in the ns range, the use of UV laser pulses seems to be advantageous because the energy threshold for plasma formation decreases with decreasing wavelength. We found that the breakdown threshold for 6-ns pulses measured at NA = 0.9 is 16 times smaller for $\lambda = 355$ nm than for $\lambda = 1064$ nm, and the cavitation bubble size at threshold is even more strongly reduced (by A. V. and V. H., unpublished results). For fs pulses, the breakdown threshold was five times smaller for $\lambda = 355$ nm than for $\lambda = 1064$ nm.

When very small single-pulse energies are used for dissection, a large number of pulses are necessary to complete a cut of finite length. Therefore, the repetition rate of the laser pulses must be sufficiently large (hundreds of hertz to kilohertz) to avoid an impractical prolongation of the processing time.

B. Potential Improvements for Catapulting

The use of ultrashort laser pulses may not only improve dissection but also increase the efficiency of catapulting because (1) the linear absorption of the sample is supplemented by nonlinear mechanisms (photoionization and avalanche ionization) even for defocused laser beams; (2) the thermal expansion velocity of a heated sample increases with decreasing laser pulse duration; and (3) for sufficiently short pulses, large thermoelastic stresses are generated, and phase transitions occur at lower temperatures. All mentioned effects reduce the energy threshold for catapulting.

Nonlinear absorption will probably make it easier to catapult specimens directly from a glass slide without the use of a strongly UV-absorbing foil, and it permits to use any desired laser wavelength. By reducing the optical penetration depth of the laser light, nonlinear absorption eases the energy requirements for catapulting.

We discussed in Section III.B.3 that for strongly defocused pulses thermal expansion of the heated specimen may contribute to catapulting. This contribution increases with decreasing pulse duration (Arnold, 2002, 2003; Tam *et al.*, 1992). If the temperature rise occurs on a shorter time scale than the stress propagation time out of the heated volume, which is also the time required for thermal expansion, large thermoelastic stresses are generated in the layer absorbing the laser energy. Under such "stress confinement" conditions, part of the laser energy is transformed into elastic energy of the heated sample and the release of this energy during the subsequent expansion phase results in a considerably larger detachment velocity than mere thermal expansion.

In LPC, the light-absorbing layer is located at the bottom of the specimen that is attached to the supporting glass slide. Therefore, the detachment through thermoelastic mechanisms does not occur immediately but with a certain delay: The compressive wave generated in the absorbing layer first travels to the upper side of the specimen bordered by air. Here it is reflected as tensile stress wave because the acoustic impedance of air is much smaller than that of the specimen (Paltauf and Dyer, 2003). This tensile wave travels back into the sample, and when, after a few nanoseconds, it reaches the interface between sample and supporting substrate, it will induce or facilitate the sample's lift off from the substrate. Moreover, the tensile stress wave will reduce the temperature required for explosive vaporization in the heated layer at the bottom of the sample (Vogel and Venugopalan, 2003; Vogel *et al.*, 2005a). The latter effect increases the driving force of phase transitions involved in catapulting.

For biological materials with a sound velocity similar to that of water (1500 m/s) and for an optical penetration depth of, for example 1 μm, the stress confinement condition is fulfilled if the laser pulse duration $t_m \leq 700$ ps. In this case, the degree of stress confinement is small for ns pulses but very high for fs pulses, and it is therefore expected that the catapulting efficiency is better with fs pulses.

Indirect experimental evidence for the above considerations has been provided by Lazare *et al.* (2005) who demonstrated surface foaming of collagen and other biopolymers induced by ablation in the stress confinement regime. We were able to catapult histologic specimens 120 μm in diameter using focused IR fs pulses ($\lambda = 1040$ nm) of only 1.2 μJ (by A. V. and V. H., unpublished results). The corresponding radiant exposure averaged over the entire specimen area is only 0.01 J/cm^2, that is, one-third of the lowest value achieved with UV ns pulses (Table II). Laser printing of biomaterial with fs pulses has been demonstrated by Zergioti *et al.* (2005), and it was shown that the printing process has a better forward directionality with fs pulses than with ns pulses (Papazoglou *et al.*, 2002; Zergioti *et al.*, 2003).

It would be desirable to establish catapulting techniques for histologic samples that do not require a UV absorbing polymer foil because the foil scatters and fluoresces (see Fig. 6B) and thus impairs histochemical and fluorescence identification techniques for cells of interest. It does not help to simply omit the foil because a narrow grid of laser spots is required to catapult the specimens directly from a glass slide, which considerably increases the amount of UV-light induced and thermal damage (see Section III.C). Therefore, new types of dynamic release layers allowing for gentle catapulting will have to be explored. This research may profit from previous experience gathered in the field of laser color printing (Tolbert *et al.*, 1993), laser printing of biomaterials (Colina *et al.*, 2006), laser cleaning (Luk'yanchuk, 2002), polymers especially designed for laser ablation (Lippert *et al.*, 2003), and other fields of laser-mediated mass transfer (Barron *et al.*, 2004; Christescu *et al.*, 2004; Mathews *et al.*, 2006; Wu *et al.*, 2003).

With regard to live-cell catapulting, there is probably even more room for further improvements than with respect to LPC of histologic material. Due to the large number of layers involved in the present technique (Fig. 18), many parameters determine the catapulting dynamics. Among those, the laser spot size, laser energy, and the thickness of both liquid layers (below the PEN foil and above the cells) are especially important. On the basis of a better understanding of the complex sequence fluid dynamics induced by the laser pulse, one will be able to optimize these parameters or to develop alternative catapulting techniques. For reliable and gentle catapulting, the thickness of the liquid layer above the cells should be as small as possible without desiccation of the cells and as reproducible as possible.

Acknowledgments

This work was performed in collaboration with PALM Microlaser Technologies which provided the microbeam system. It was sponsored by the German Bundesministerium für Bildung und Forschung (BMBF) under grant number 13N8461, and, in parts, by US Air Force Office of Scientific Research under grants number FA8655-02-1-3047 and FA8655-05-1-3010. We appreciate stimulating discussions with Karin Schütze, Bernd Sägmüller, and Yilmaz Niyaz of PALM Microlaser Technologies, with Heyke Diddens (Institute of Biomedical Optics, University of Lübeck) who also provided the CHO cell line used for live-cell catapulting, and with Dorthe von Smolinski (Institute of Anatomy, University of Lübeck) who gave valuable advise on real-time RT-PCR. The laser beam profiles of Fig. 3 were provided by Carsten Lüthy (PALM), and the measurements of the thermal properties of polymer foils and histologic specimens quoted in Fig. 4 and Table I were performed by Christine Mimler (Institut für Werkstoffwissenschaften, Universität Erlangen-Nürnberg). We thank Sebastian Freidank, Ingo Apitz, Pieternel Doeswijk, Nadine Steiner, Florian Wölbeling, Helge Meyer, and Reinhard Schulz for technical assistance.

References

Abrahamsen, H. N., Steiniche, T., Nexo, E., Hamilton-Dutoit, S. J., and Sorensen, B. S. (2003). Towards quantitative mRNA analysis in paraffin-embedded tissues using real-time reverse transcriptase-polymerase chain reaction: A methodological study on lymph nodes from melanoma patients. *J. Mol. Diagn.* **5,** 34–41.

Ai, J., Tan, Y., Ying, W., Hong, Y., Liu, S., Wu, M., Qian, X., and Wang, H. (2006). Proteome analysis of hepatocellular carcinoma by laser capture microdissection. *Proteomics* **6,** 538–546.

Amy, R. L., and Storb, R. (1965). Selective mitochondrial damage by a ruby laser microbeam: An electron microscopic study. *Science* **150,** 756–758.

Anderholm, N. C. (1970). Laser-generated stress waves. *Appl. Phys. Lett.* **16,** 113–115.

Apitz, I., and Vogel, A. (2005). Material ejection in nanosecond Er:YAG laser ablation of water, liver, and skin. *Appl. Phys. A* **81,** 329–338.

Arnold, N. (2002). Dry laser cleaning of particles by nanosecond pulses: Theory. *In* "Laser Cleaning" (B. Luk'yanchuk, ed.), pp. 51–102. World Scientific Publishing, Singapore.

Arnold, N. (2003). Theoretical description of dry laser cleaning. *Appl. Surf. Sci.* **208–209,** 15–22.

Ashkin, A. (1986). Observation of a single beam gradient force trap for dielectric particles. *Opt. Lett.* **11,** 288–290.

Barron, J. A., Wu, P., Ladouceur, H. D., and Ringeisen, B. R. (2004). Biological laser printing: A novel technique for creating heterogeneous 3-dimensional cell patterns. *Biomed. Microdevices* **6,** 139–147.

Bäuerle, D. (2000). "Laser Processing and Chemistry," 3rd edn. Spinger-Verlag, Berlin, Heidelberg, New York.

Berns, M. W., Cheng, W. K., Floyd, A. D., and Ohnuki, Y. (1971). Chromosome lesions produced with an argon laser microbeam without dye sensitization. *Science* **171,** 903–905.

Berns, M. W., Olson, R. S., and Rounds, D. E. (1969). *In vitro* production of chromosomal lesions with an argon laser microbeam. *Nature* **221,** 74–75.

Bertling, C. D., Lin, F., and Girotti, A. W. (1996). Role of hydrogen peroxide in the cytotoxic effects of UVA/B radiation on mammalian cells. *J. Photochem. Photobiol.* **64,** 137–142.

Bessis, M., Gires, F., Mayer, G., and Nomarski, G. (1962). Irradiation des organites cellulaires à l'aide d'un laser à rubis. *C. R. Acad. Sci. III-Vie* **255,** 1010–1012.

Boal, D. (2002). "Mechanics of the Cell." Cambridge University Press, Cambridge, UK.

Bustin, S. A. (2000). Absolute quantification of mRNA using real-time reverse transcription polymerase chain reaction assays. *J. Mol. Endocrinol.* **25,** 169–193.

Carslaw, H. S., and Jaeger, J. C. (1959). "Conduction of Heat in Solids," 2nd edn. Oxford University Press, Oxford.

Cheong, W. F., Prahl, S. A., and Welch, A. J. (1990). A review of the optical properties of biological tissue. *IEEE. J. Quantum Electron.* **26,** 2166–2185.

Chrisey, D. B., Piqué, A., McGill, R. A., Horwitz, J. S., Ringeisen, B. R., Bubb, D. M., and Wu, P. K. (2003). Laser deposition of polymer and biomaterial films. *Chem. Rev.* **103,** 553–576.

Christescu, R., Mihaiescu, D., Socol, G., Stamatin, I., Mihailescu, I. N., and Chrisey, D. B. (2004). Deposition of biopolymer thin films by matrix-assisted pulsed laser evaporation. *Appl. Phys. A* **79,** 1023–1026.

Colina, M., Duocastella, M., Fernández-Pradas, J. M., Serra, P., and Morenza, J. L. (2006). Laser-induced forward transfer of liquids: Study of the droplet ejection process. *J. Appl. Phys.* **99,** 084909.

Colombelli, J., Reynaud, E. G., Rietdorf, J., Pepperkork, R., and Stelzer, E. H. K. (2005). Pulsed UV laser nanosurgery: Retrieving the cytoskeleton dynamics *in vivo*. *Traffic* **6,** 1093–1102.

Coohill, T. P. (2002). Uses and effects of ultraviolet radiation on cells and tissues. *In* "Lasers in Medicine" (R. W. Waynant, ed.), pp. 85–107. CRC Press, Boca Raton, London, New York, Washington, DC.

de With, A., and Greulich, K. O. (1995). Wavelength dependence of laser-induced DANN damage in lymphocytes observed by single-cell gel electrophoressis. *J. Photochem. Photobiol. B Biol.* **30,** 71–76.

Dingus, R. S. (1992). Laser-induced contained vaporization in tissue. *Proc. SPIE* **1646,** 266–274.

Dingus, R. S. (1993). Momentum induced by laser-tissue interaction. *Proc. SPIE* **1882,** 399–411.

Eltoum, I. A., Siegal, G. P., and Frost, A. R. (2002). Microdissection of histologoc sections: Past, present, and future. *Adv. Anat. Pathol.* **9,** 316–322.

Elvers, D., Remer, L., Arnold, N., and Bäuerle, D. (2005). Laser microdissection of biological tissues: Process optimization. *Appl. Phys. A* **80,** 55–59.

Emmert-Buck, M. R., Bonner, R. F., Smith, P. D., Chuaqui, R. F., Zhuang, Z., Goldstein, S. R., Weiss, R. A., and Liotta, L. A. (1996). Laser capture microdissection. *Science* **274,** 998–1001.

Ernst, G., Melle, C., Schimmel, B., Bleul, A., and von Eggeling, F. (2006). Proteohistography—direct analysis of tissue with high sensitivity and high spatial resolution using ProteinChip technology. *J. Histochem. Cytochem.* **54,** 13–17.

Evans, E. A., Waugh, R., and Melnik, L. (1976). Elastic are compressibility modulus of red cell membrane. *Biophys. J.* **16,** 585–595.

Fabbro, R., Fournier, J., Ballard, P., Devaux, D., and Virmont, J. (1990). Physical study of laser-produced plasma in confined geometry. *J. Appl. Phys.* **68,** 775–784.

Fairand, B. P., and Clauer, A. H. (1979). Laser-generation of high-amplitude stress waves in materials. *J. Appl. Phys.* **50,** 1497–1502.

Fink, L., Kwapiszewska, G., Wilhelm, J., and Bohle, R. M. (2006). Laser-microdissection for cell type- and compartment-specific analyses on genomic and proteomic level. *Exp. Toxicol. Pathol.* **57**(Suppl. 2), 25–29.

Freund, D. E., McCally, R. L., Farrell, R. A., and Sliney, D. H. (1996). A theoretical comparison of retinal temperature changes resulting from exposure to rectangular and Gaussian beams. *Lasers Life Sci.* **7,** 71–89.

Greulich, K. O. (1999). "Micromanipulation by Light in Biology and Medicine." Birkhäuser, Basel, Boston, Berlin.

Greulich, K. O., and Pilarczyk, G. (1998). Laser tweezers and optical microsurgery in cellular and molecular biology. Working principles and selected applications. *Cell. Mol. Biol.* **44,** 701–710.

Gulmann, C., Espina, V., Petricoin, E., III, Longo, D. L., Santi, M., Knutsen, T., Raffeld, M., Jaffe, E. S., Liotta, L. A., and Feldman, A. L. (2005). Proteomic analysis of apoptotic pathways reveals prognostic factors in follicular lymphoma. *Clin. Cancer Res.* **11,** 5847–5855.

Hibst, R. (1996). "Technik, Wirkungsweise und medizinische Anwendungen von Holmium- und Erbium-Lasern." Ecomed, Landsberg (in German).

Hockberger, P. E., Skimina, T. A., Centonze, V. E., Lavin, C., Chu, S., Dadras, S., Reddy, J. K., and White, J. G. (1999). Activation of flavin-containing oxidases underlies light-induced production of H_2O_2 in mammalian cells. *Proc. Natl. Acad. Sci. USA* **96,** 6255–6260.

Hopp, B., Smausz, T., Kresz, N., Barna, N., Bor, Z., Kolozsvari, L., Chrisey, D. B., Szabo, A., and Nogradi, A. (2005). Survival and proliferative ability of various living cell types after laser-induced forward transfer. *Tissue Eng.* **11,** 1817–1823.

Hosokawa, Y., Takabayashi, H., Miura, S., Shukunami, C., Hiraki, Y., and Masuhara, H. (2004). Nondestructive isolation of single cultured animal cells by femtosecond laser-induced shockwave. *Appl. Phys. A* **79,** 795–798.

Huttmann, G., and Birngruber, R. (1999). On the possibility of high-precision photothermal microeffects and the measurement of fast thermal denaturation of proteins. *IEEE J. Sel. Topics Quantum Electron.* **5,** 954–962.

Isenberg, G., Bielser, W., Meier-Ruge, W., and Remy, E. (1976). Cell surgery by laser micro-dissection: A preparative method. *J. Microsc.* **107,** 19–24.

Jacques, S. L. (1993). Role of tissue optics and pulse duration on tissue effects during high-power laser irradiation. *Appl. Opt.* **32,** 2447–2454.

Jiménez, J. L., del Valle, G., and Campos, I. (2005). A cononical treatment of some systems with friction. *Europ. J. Phys.* **26,** 711–725.

Karaiskou, A., Zergioti, I., Fotakis, C., Kapsetaki, M., and Kafetzopoulos, D. (2003). Microfabrication of biomaterials by the sub-ps laser-induced forward transfer process. *Appl. Surf. Sci.* **208–209,** 245–249.

Kehr, J. (2003). Single cell technology. *Curr. Opin. Plant Biol.* **6,** 617–621.

König, K., Riemann, I., Fischer, P., and Halbhuber, K. (1999). Intracellular nanosurgery with near infrared femtosecond laser pulses. *Cell. Mol. Biol.* **45,** 195–201.

König, K., Riemann, I., Stracke, F., and Le Harzic, R. (2005). Nanoprocessing with nanojoule near-infrared femtosecond pulses. *Med. Laser Appl.* **20,** 169–184.

Koulikov, S. G., and Dlott, D. D. (2001). Ultrafast microscopy of laser ablation of refractory materials: Ultra low threshold stress-induced ablation. *J. Photochem. Photobiol. A* **145,** 183–194.

Krafft, A. E., Duncan, B. W., Bijwaard, K. E., Taubenberger, J. K., and Lichy, J. H. (1997). Optimization of the isolation and amplification of RNA from formalin-fixed, paraffinembedded tissue: The armed forces institute of pathology experience and literature review. *Mol. Diagn.* **2,** 217–230.

Kuchling, H. (1991). "Taschenbuchder Physik," Fachbuch-Verlag GmbH, Leipzig.

Lahr, G. (2000). RT-PCR from archival single cells is a suitable method to analyze specific gene expression. *Lab. Invest.* **80,** 1477–1479.

Landau, L. D., and Lifschitz, E. M. (1987). "Fluid Mechanics," 2nd edn. Pergamon Press, Oxford.

Lazare, S., Tokarev, V., Sionkowska, A., and Wisniewski, M. (2005). Surface foaming of collagen, chitosan and other biopolymer films by KrF excimer laser ablation in the photomechanical regime. *Appl. Phys. A* **81,** 465–470.

Lee, I.-Y. S., Tolbert, W. A., Dlott, D. D., Doxtader, M. M., Foley, D. M., Arnold, D. R., and Ellis, E. W. (1992). Dynamics of laser ablation transfer imaging investigated by ultrafast microscopy. *J. Imaging Sci. Technol.* **36,** 180–187.

Lewis, F., Maughan, N. J., Smith, V., Hillan, K., and Quirke, P. (2001). Unlocking the archive—gene expression in paraffin-embedded tissue. *J. Pathol.* **195,** 66–71.

Lippert, T., Hauer, M., Phipps, C. R., and Wokaun, A. (2003). Fundamentals and applications of polymers designed for laser ablation. *Appl. Phys. A* **77,** 259–264.

Litjens, R. A. J., Quickenden, T. I., and Freeman, C. G. (1999). Visible and near-ultraviolet absorption spectrum of liquid water. *Appl. Opt.* **38,** 1216–1223.

Lokhandwalla, M., McAteer, J. A., Williams, J. C., Jr., and Sturtevant, B. (2001). Mechanical haemolysis in shock wave lithotripsy (SWL): II. *In vitro* cell lysis due to shear. *Phys. Med. Biol.* **46,** 1245–1264.

Lokhandwalla, M., and Sturtevant, B. (2001). Mechanical haemolysis in shock wave lithotripsy (SWL): I. Analysis of cell deformation due to SWL flow-fields. *Phys. Med. Biol.* **46,** 413–437.

Lorenz, K. (2004). Mechanismen des Katapultierens biologischer Strukturen mit UV-Laserpulsen. Diploma thesis, University of Applied Science Lübeck (in German)Accessible through. Accessible through vogel@bmo.uni-luebeck.de.

Luk'yanchuk, B. (ed.) (2002). "Laser Cleaning." World Scientific Publishing, Singapore.

Mathews, S. A., Augeung, C. Y., and Piqué, A. (2006). Use of Laser Direct Write in microelectronics assembly. *In* "Online Proceedings of LAMP2006. The 4th Int. Congr. Laser Advanced Materials Processing," May 16–19, 2006. Kyoto, Japan, paper 06–19. JLPS–Japan Laser Processing Society.

Mayer, A., Stich, M., Brocksch, D., Schütze, K., and Lahr, G. (2002). Going *in vivo* with laser microdissection. *Methods Enzymol.* **356,** 25–33.

Meier-Ruge, W., Bielser, W., Remy, E., Hillenkamp, F., Nitsche, R., and Unsöld, R. (1976). The laser in the Lowry technique for microdissection of freeze-dried tissue slices. *Histochem J.* **8,** 387–401.

Meredith, G. D., Sims, C. E., Soughayer, J. S., and Allbritton, N. L. (2000). Measurement of kinase activation in single mammalian cells. *Nat. Biotechnol.* **18,** 309–312.

Nahen, K., and Vogel, A. (1996). Plasma formation in water by picosecond and nanosecond Nd:YAG laser pulses—Part II: Transmission, scattering, and reflection. *IEEE J. Sel. Topics Quantum Electron.* **2,** 861–871.

Nakata, Y., and Okada, T. (1999). Time-resolved microscopic imaging of the laser-induced forward transfer process. *Appl. Phys. A* **69**(Suppl.), S275–S278.

Needham, D., and Nunn, R. S. (1990). Elastic deformation and failure of liquid bilayer membranes containing cholesterol. *Biophys. J.* **58,** 997–1009.

Nettikadan, S., Radke, K., Johnson, J., Xu, J., Lynch, M., Mosher, C., and Henderson, E. (2006). Detection and quantification of protein biomarkers from fewer than 10 cells. *Mol. Cell. Proteomics* **5,** 895–901.

Neumann, J., and Brinkmann, R. (2005). Boiling nucleation on melanosomes and microbeads transiently heated by nanosecond and microsecond laser pulses. *J. Biomed. Opt.* **10,** 024001.

Niemz, M. H., Klancnik, E. G., and Bille, J. F. (1991). Plasma-mediated ablation of corneal tissue at 1053 nm using a Nd:YLF oscillator/regenerative amplifier laser. *Lasers Surg. Med.* **11,** 426–431.

Niyaz, Y., Stich, M., Sagmüller, B., Burgemeister, R., Friedemann, G., Sauer, U., Gangnus, R., and Schütze, K. (2005). Noncontact laser microdissection and pressure catapulting: Sample preparation for genomic, transcriptomic, and proteomic analysis. *Methods. Mol. Med.* **114,** 1–24.

Paltauf, G., and Dyer, P. (2003). Photomechanical processes and effects in ablation. *Chem. Rev.* **103,** 487–518.

Papazoglou, D. G., Karaiskou, A., Zergioti, I., and Fotakis, C. (2002). Shadowgraphic imaging of the sub-ps laser-induced forward transfer process. *Appl. Phys. Lett.* **81,** 1594–1596.

Pearce, J., and Thomsen, S. (1995). Rate process analysis of thermal damage. *In* "Optical-Thermal Response of Laser-Irradiated Tissue" (A. J. Welch, and M. Van Germert, eds.), pp. 561–606. Plenum Press, New York.

Phipps, C. R., Reilly, J. P., and Campbell, J. W. C. (2000). Optimum parameters for laser launching objects into low Earth orbit. *Laser Part. Beams* **18,** 661–695.

Phipps, C. R., Turner, T. P., Harrison, R. F., York, G. W., Osborne, W. Z., Anderson, G. K., Corlis, X. F., Haynes, L. C., Steele, H. S., Spicochi, K. C., and King, T. R. (1988). Impulse coupling to targets in vacuum by KrF, HF and CO_2 lasers. *J. Appl. Phys.* **64,** 1083–1096.

Rau, K. R., Guerra, A., Vogel, A., and Venugopalan, V. (2004). Investigation of laser-induced cell lysis using time-resolved imaging. *Appl. Phys. Lett.* **84,** 2940–2942.

Rau, K. R., Quinto-Su, P. A., Hellman, A. N., and Venugopalan, V. (2006). Pulsed laser microbeam-induced cell lysis: Time-resolved imaging and analysis of hydrodynamic effects. *Biophys. J.* **91,** 317–329.

Ready, J. F. (1971). "Effects of High Power Laser Radiation." Academic Press, Orlando.

Ringeisen, B. R., Kim, H., Barron, J. A., Krizman, D. B., Chrisey, D. B., Jackman, S., Auyeung, R. Y. C., and Spargo, B. J. (2004). Laser printing of pluripotent embryonal carcinoma cells. *Tissue Eng.* **10,** 483–491.

Romain, J. P., and Darquey, P. (1990). Shock waves and acceleration of thin foils by laser pulses in confined plasma interaction. *J. Appl. Phys.* **68,** 1926–1928.

Sasnett, M. W. (1989). Propagation or multimode laser beams—the M^2 factor. *In* "The Physics and Technology of Laser Resonators" (D. R. Hall, and P. E. Jackson, eds.), pp. 132–142. Adam Hilger, Bristol, New York.

Schütze, K., and Clement-Sengewald, A. (1994). Catch and move—cut or fuse. *Nature* **368,** 667–668.

Schütze, K., and Lahr, G. (1998). Identification of expressed genes by laser-mediated manipulation of single cells. *Nat. Biotechnol.* **16,** 737–742.

Schütze, K., Pösl, H., and Lahr, G. (1998). Laser micromanipulation systems as universal tools in molecular biology and medicine. *Cell. Mol. Biol.* **44,** 735–746.

Serra, P., Colina, M., Fernández-Pradas, J. M., Sevilla, L., and Morenza, J. L. (2004). Preparation of functional DNA microarrays through laser-induced forward transfer. *Appl. Phys. Lett.* **85,** 1639–1641.

Sheffield, S. A., Rogers, J. W., and Castaneda, J. N. (1986). Velocity measurements of laser-driven flyers backed by high impedance windows. *In* "Shock Waves in Condensed Matter" (Y. M. Gupta, ed.), pp. 541–546. Plenum Press, New York, London.

Siegmann, A. E., Sasnett, M. W., and Johnston, T. F., Jr. (1991). Choice of clip levels for beam width measurements using knife-edge techniques. *IEEE J. Quantum Electron.* **27,** 1098–1104.

Simanowski, D., Sarkar, M., Irani, A., O'Connel-Rodwell, C., Contag, C., Schwettman, A., and Palanker, D. (2005). Cellular tolerance to pulsed heating. *Proc. SPIE* **5695,** 254–259.

Sims, C. E., Meredith, G. D., Krasieva, T. B., Berns, M. W., Tromberg, B. J., and Allbritton, N. L. (1998). Laser-micropipet combination for single-cell analysis. *Anal. Chem.* **700,** 4570–4577.

Stanta, G., and Schneider, C. (1991). RNA extracted from paraffin-embedded human tissues is amenable to analysis by PCR amplification. *Biotechniques* **11,**304, 306, 308.

Star, W. M., Marijnissen, J. P. A., and van Gemert, M. J. C. (1988). Light dosimetry in optical phantoms and in tissues. *Phys. Med. Biol.* **33,** 437–454.

Stich, M., Thalhammer, S., Burgemeister, R., Friedemann, G., Ehnle, S., Lüthy, C., and Schütze, K. (2003). Live cell catapulting and recultivation. *Pathol. Res. Pract.* **199,** 405–409.

Stolarski, J., Hardman, J., Bramlette, C. G., Noojin, G. D., Thomas, R. J., Rockwell, B. A., and Roach, W. P. (1995). Integrated light spectroscopy of laser-induced breakdown in aqueous media. *Proc. SPIE* **2391,** 100–109.

Tam, A. C., Leung, W. P., Zapka, W., and Ziemlich, W. (1992). Laser-cleaning techniques for removal of surface particulates. *J. Appl. Phys.* **71,** 3515–3523.

Thalhammer, S., Lahr, G., Clement-Sengewald, A., Heckl, W. M., Burgemeister, R., and Schütze, K. (2003). Laser microtools in cell biology and molecular medicine. *Laser Phys.* **13,** 681–691.

Tirlapur, U. K., and König, K. (2002). Targeted transfection by femtosecond laser. *Nature* **418,** 290–291.

Tolbert, W. A., Lee, I. Y. S., Doxtader, M. M., Ellis, E. W., and Dlott, D. D. (1993). High-speed color imaging by laser ablation transfer with a dynamic release layer: Fundamental mechanisms. *J. Imaging Sci. Technol.* **37,** 411–421.

Tyrrell, R. M., and Keyse, S. M. (1990). The interaction of UVA radiation with cultured cells. *J. Photochem. Photobiol. B Biol.* **4,** 349–361.

Venugopalan, V., Guerra, A., Nahen, K., and Vogel, A. (2002). The role of laser-induced plasma formation in pulsed cellular microsurgery and micromanipulation. *Phys. Rev. Lett.* **88,** 078103.

Vogel, A., Apitz, I., Freidank, S., and Dijkink, R. (2006). Sensitive high-resolution white-light Schlieren technique with large dynamic range for the investigation of ablation dynamics. *Opt. Lett.* **31,** 1812–1814.

Vogel, A., Busch, S., and Parlitz, U. (1996b). Shock wave emission and cavitation bubble generation by picosecond and nanosecond optical breakdown in water. *J. Acoust. Soc. Am.* **100,** 148–165.

Vogel, A., Dlugos, C., Nuffer, R., and Birngruber, R. (1991). Optical properties of human sclera, and their consequences for transscleral laser applications. *Lasers Surg. Med.* **11,** 331–340.

Vogel, A., Lorenz, K., Apitz, I., Freidank, S., and Horneffer, V. (2007). Laser catapulting of histologic specimens with focused and defocused laser pulses. *Biophys. J.* (submitted for publication).

Vogel, A., Lorenz, K., Sägmüller, B., and Schütze, K. (2005b). Catapulting of microdissected histologic specimens with focused and defocused laser pulses. *Proc. SPIE* **5863,** 8–10.

Vogel, A., Nahen, K., and Theisen, D. (1996a). Plasma formation in water by picosecond and nanosecond Nd:YAG laser pulses—Part I: Optical breakdown at threshold and superthreshold irradiance. *IEEE J. Sel. Topics Quantum Electron.* **2,** 847–860.

Vogel, A., Noack, J., Hüttmann, G., and Paltauf, G. (2005a). Mechanisms of femtosecond laser nanosurgery of cells and tissues. *Appl. Phys. B* **81,** 1015–1047.

Vogel, A., Noack, J., Nahen, K., Theisen, D., Busch, S., Parlitz, U., Hammer, D. X., Nojin, G. D., Rockwell, B. A., and Birngruber, R. (1999). Energy balance of optical breakdown in water at nanosecond to femtosecond time scales. *Appl. Phys. B* **68,** 271–280.

Vogel, A., Schweiger, P., Frieser, A., Asiyo, M., and Birngruber, R. (1990). Intraocular Nd:YAG laser surgery: Light-tissue interaction, damage range, and reduction of collateral effects. *IEEE J. Quantum Electron.* **26,** 2240–2260.

Vogel, A., and Venugopalan, V. (2003). Mechanisms of pulsed laser ablation of biological tissues. *Chem. Rev.* **103,** 577–644.

von Smolinski, D., Blessenohl, M., Neubauer, C., Kalies, K., and Gebert, A. (2006). Validation of a novel ultra-short immuno-labelling method for high quality mRNA preservation in laser microdissection and real-time RT-PCR. *J. Mol. Diagn.* **8,** 246–253.

von Smolinski, D., Leverkoehne, I., von Samson-Himmelstjerna, G., and Gruber, A. D. (2005). Impact of formalin-fixation and paraffin-embedding on the ratio between mRNA copy numbers of differently expressed genes. *Histochem. Cell Biol.* **124,** 177–188.

Wu, P. K., Ringeisen, B. R., Krizman, D. B., Frondoza, C. G., Brooks, M., Bubb, D. M., Auyeung, R. C. Y., Piqué, A., Spargo, B., McGill, R. A., and Chrisey, D. B. (2003). Laser transfer of biomaterials: Matrix-assisted pulsed laser evaporation (MAPLE) and MAPLE Direct Write. *Rev. Sci. Instrum.* **74,** 2546–2557.

Young, D., Auyeung, R. C. Y., Piqué, A., and Chrisey, D. B. (2001). Time-resolved optical microscopy of a laser-based forward transfer process. *Appl. Phys. Lett.* **78,** 3169–3171.

Zergioti, I., Karaiskou, A., Papazoglou, D. G., Fotakis, C., Kapsetaki, M., and Kafetzopoulos, D. (2005). Femtosecond laser microprinting of biomaterials. *Appl. Phys. Lett.* **86,** 163902.

Zergioti, I., Papzoglou, D. G., Karaiskou, A., Fotakis, C., Gamaly, E., and Rode, A. (2003). A comparative Schlieren imaging study between ns and sup-ps laser forward transfer of Cr. *Appl. Surf. Sci.* **208–209,** 177–180.

CHAPTER 6

Physics of Optical Tweezers

Timo A. Nieminen, Gregor Knöner, Norman R. Heckenberg, and Halina Rubinsztein-Dunlop

Centre for Biophotonics and Laser Science
School of Physical Sciences, The University of Queensland
Brisbane QLD 4072, Australia

We outline the basic principles of optical tweezers as well as the fundamental theory underlying optical tweezers. The optical forces responsible for trapping result from the transfer of momentum from the trapping beam to the particle and are explained in terms of the momenta of incoming and reflected or refracted rays. We also consider the angular momentum flux of the beam in order to understand and explain optical torques. In order to provide a qualitative picture of the trapping, we treat the particle as a weak positive lens and the forces on the lens are shown. However, this representation does not provide quantitative results for the force. We, therefore, present results of applying exact electromagnetic theory to optical trapping. First, we consider a tightly focused laser beam. We give results for trapping of spherical particles and examine the limits of trappability in terms of type and size of the particles. We also study the effect of a particle on the beam.

METHODS IN CELL BIOLOGY, VOL. 82

0091-679X/07 $35.00
DOI: 10.1016/S0091-679X(06)82006-6

This exact solution reproduces the same qualitative effect as when treating the particle as a lens where changes in the convergence or divergence and in the direction of the trapping beam result in restoring forces acting on the particle. Finally, we review the fundamental theory of optical tweezers.

I. Introduction to Optical Tweezers

Optical tweezers, or the single-beam gradient trap, grew rapidly from a novelty to a widely used tool in many fields. Apart from allowing the noncontact trapping and manipulation of microscopic particles including living cells and even organelles within cells, the ability to make noncontact measurements of forces on the order of piconewtons has enabled the measurement of mechanical properties of cells, the kinetics and properties of important biological molecules, and measurements of forces acting on particles in colloidal suspensions. In this section, we briefly review the history of optical tweezers, including coverage of new developments in trapping such as using complex optical fields or microfabricated particles which are starting to see application as new-generation optical tweezers based tools.

In the following section, we discuss the basic principles of optical tweezers. Next, we outline the fundamental theory underlying optical tweezers. Finally, we outline the computational modeling of optical tweezers.[1]

II. History

One of the first recorded speculations that light might be able to exert mechanical forces on particles was in Kepler's observations of comet tails (Keplero, 1619). Interest in this idea continued; for example John Michell, one of the pioneers of the torsion balance, attempted quantitative measurement of optical radiation pressure (Hardin, 1966) in the 1700s, and succeeded in showing that sunlight can be concentrated sufficiently to destroy one's experimental apparatus. Such ideas entered modern science with Maxwell's predictions of radiation pressure of light and electromagnetic waves (Maxwell, 1865, 1873).

However, even after the first experimental observation of radiation pressure by Ledebew (1902) and Nichols and Hull (1901, 1902, 1903a,b), the topic of electromagnetic radiation pressure remained an interesting novelty of physics, rather than an effect of practical technological value.

The primary difficulty was that the forces resulting from optical or electromagnetic radiation pressure are small compared to the power required, and are insufficient to overcome gravitational or frictional forces in most circumstances. Thus, while it was suggested that radiation pressure might be useful in environments

[1] At the end of each section, we direct the interested reader to further reading on the topic.

where such resisting forces did not need to be overcome—for example, Friedrich Zander, a pioneer in the fields of rocketry and spaceflight, suggested the use of solar sails for space propulsion in the 1920s—such applications would remain beyond practicality for many decades.

This changed in 1969 when Arthur Ashkin realized that, although the forces resulting from electromagnetic radiation were small, the forces required to move small particles were also small (Ashkin, 1997). This prompted experiments demonstrating the acceleration and trapping of small particles by radiation pressure (Ashkin, 1970). The earliest traps were not purely optical, as they involved a delicate balance between optical levitation forces and gravity. This placed a strong restriction on the forces that could be applied—radiation pressure forces greater than the weight of the levitated particle would result in its ejection from the trap. Meanwhile, gradient forces were observed in these experiments.

At this point, the history of optical trapping and manipulation of small particles splits into two branches: the trapping and cooling of atoms, which led to the award of the 1997 Nobel prize in physics (Chu, 1998; Cohen-Tannoudji, 1998; Phillips, 1998), and the pathway to optical tweezers. Single-beam trapping of microparticles using the gradient force was finally achieved (Ashkin *et al.*, 1986), and optical tweezers had arrived on the scene.

The combined ability to noninvasively trap and manipulate microscopic particles and to measure forces on the order of piconewtons rapidly led to the use of optical tweezers in a variety of biological and other fields (Block, 1992; Grier, 1997; Wright *et al.*, 1990). The new possibility of measuring forces in microscopic biological systems was a major advance, and allowed access to a wide variety of measurements such as, for example, measurements of the dynamics of single DNA molecules (Quake *et al.*, 1997). Kuo (2001) reviews the use of optical tweezers for measuring biological forces. Lang and Block (2003) list over a hundred papers reporting force measurements on biological molecules, cells, or colloidal particles.

Meanwhile, parallel developments continued, for example, the demonstration of rotational manipulation in optical tweezers by Sato *et al.* (1991) using the rotation of a nonaxisymmetric trapping beam. Continuous spinning of particles was achieved using Laguerre–Gauss mode beams which carry orbital angular momentum about the beam axis to transfer momentum to absorbing particles (He *et al.*, 1995a,b). Further work led to both the controlled two-dimensional trapping of strongly absorbing particles (Friese *et al.*, 1998b) and the alignment and spinning of birefringent particles without absorption (Friese *et al.*, 1998a). The second of these has already seen practical application (Bishop *et al.*, 2004; Knöner *et al.*, 2005).

Further outgrowth of such work on rotational manipulation included the rotation of cells and organelles (Bayoudh *et al.*, 2003; Mohanty *et al.*, 2004), and progress toward optically driven micromachines (Nieminen *et al.*, 2006).

Other interesting developments include sophisticated holographic optical tweezers (Dufresne *et al.*, 2001; Grier and Roichman, 2006; Leach *et al.*, 2004) and optical tweezers operated remotely over the internet (Botvinick and Berns, 2005).

A. Further Reading

Ashkin, the pioneer of optical trapping and optical tweezers, has described the development of optical tweezers in a number of papers (Ashkin, 1997, 2000). More developments have been reviewed by Grier (2003), Molloy *et al.* (2003), and Nieminen *et al.* (2006).

The Resource Letter by Lang and Block (2003) is an excellent starting point for further investigation of existing work in optical tweezers.

III. Basic Principles

The optical forces that are responsible for trapping in optical tweezers result from the transfer of momentum from the trapping beam to the particle. Recalling that force is the time rate of change of momentum, optical forces must be accompanied by a change in the momentum flux of the beam. Optical forces are commonly explained in terms of the momenta of incident and reflected or refracted rays. It is also possible to consider the angular momentum flux of the beam as a whole in order to understand and explain optical torques.

Consider the momentum flux of a ray of light: $p = nP/c$ in the direction of the ray, where p is the magnitude of the momentum flux, and P is the power, or energy flux; n is the refractive index of the medium the ray is moving in, and c is the speed of light in free space. The ratio of momentum to power for a ray is the same as for a section of a plane wave; if we consider a plane wave to be made up of parallel rays, we see that this must be the case. A focused beam, on the other hand, can be considered as a bundle of converging rays.

The force exerted on a particle in optical tweezers can be understood in terms of changes in the momentum flux of the trapping beam. For axial forces, the important principle is that the more convergent or divergent the beam is, the lower the momentum flux. Changes in the convergence or divergence of the beam change the axial momentum of the beam, with a resultant axial reaction force on the trapped particle. For radial forces, it is the direction of the beam as a whole that needs to be considered. If the beam is deflected in one direction, it gains transverse momentum in that direction, and the particle experiences a reaction force in the other direction.

We can think of the trapped particles as a weak positive lens. The action of such a lens on a beam is shown in Fig. 1. If the lens is at the focus of the beam, the rays pass through the center of the lens and are undeviated—the optical force is zero. If the lens is before the focus, it increases the convergence of the beam, and therefore decreases the momentum flux. The lens gains the momentum the beam loses, and there is a force in the direction of propagation. If the lens is after the focus, it decreases the divergence of the beam, and hence increases the momentum flux, resulting in a restoring force toward the focus. If the lens is displaced sideways, the beam is deflected toward the centerline of the lens, gaining lateral momentum. The lateral reaction force on the lens acts toward the beam axis.

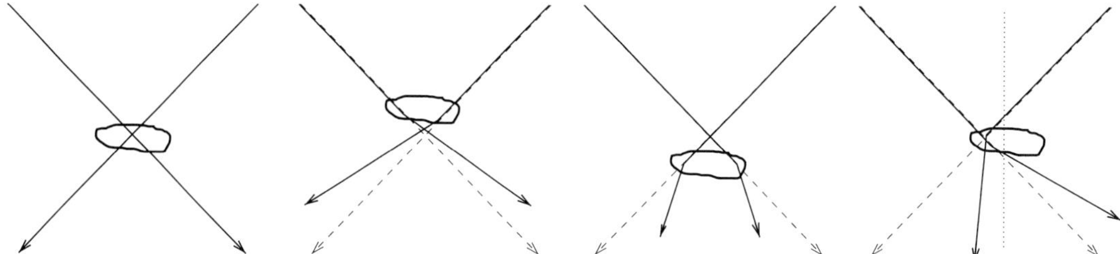

Fig. 1 The force exerted on a particle in optical tweezers can be understood in terms of changes in the momentum flux of the trapping beam. For axial forces, the important principle is that the more convergent or divergent the beam is, the lower the axial momentum flux. For radial forces, the direction of the beam is the key principle. The trapped particle can be thought of as a weak positive (converging) lens. If the lens is at the focus, the rays pass through the center of the lens and are undeviated—the optical force is zero. If the lens is before the focus, it increases the convergence of the beam, and therefore decreases the momentum flux. The lens gains the momentum the beam loses, and there is a force in the direction of propagation. If the lens is after the focus, it decreases the divergence of the beam, and hence increases the momentum flux, resulting in a restoring force toward the focus. If the lens is displaced sideways, the beam is deflected toward the centerline of the lens, gaining lateral momentum. The lateral reaction force on the lens acts toward the beam axis.

In addition to these forces, some rays will also be reflected, producing a force in the direction of propagation which will displace the particle past the focus.

This simple lens model is ideal for providing a qualitative picture, but does not provide a quantitative result for the force. First, the trapped particle interacts with the beam in a more complex manner than a simple lens. Second, most particles trapped in optics tweezers are too small for geometric optics to be a reliable approximation—the usual rule of thumb is that the smallest dimensions of the particle should be at least 10 times the wavelength of the light, preferably 20λ or more. Third, trapping occurs in the vicinity of the focus of the beam, where geometric optics fails. Despite this, many authors (Ashkin, 1992) have used the geometric optics approximation to calculate optical forces. Wright *et al.* (1993, 1994) showed that ray optics calculations are useful for large spheres.

An alternative approximate method is to consider particles that are small compared to the wavelength—the Rayleigh limit. Despite most particles trapped in optical tweezers being outside the size range where this approximation is valid, this is still a useful exercise since it provides quantitative formulas which are useful for small particles, and perhaps more importantly, have a simple physical interpretation that illuminates the forces that act on larger particles. One of the effects of an electromagnetic field in matter to is to induce a dielectric polarization, determined by the field and the permittivity of the medium. For optical problems, the refractive index is determined by the permittivity [$c/n = (\varepsilon\mu)^{1/2}$ or $c/n = (\varepsilon\mu_0)^{1/2}$ for nonmagnetic materials], so the polarization is a function of the optical properties of the medium. For the case of a sphere in a uniform field—a classic canonical problem encountered in electromagnetics courses—a simple analytical result exists: a dipole moment of

$$\mathbf{d} = 4\pi n_{\text{med}}^2 \varepsilon_0 r^3 \left(\frac{m^2 - 1}{m^2 + 2} \right) \mathbf{E} \tag{1}$$

is induced in the sphere by an electric field $\mathbf{E}$, where n_{med} is the refractive index of the surrounding medium, m is the relative refractive index of the particle, with $m = n_{\text{part}}/n_{\text{med}}$, and we use $\mathbf{d}$ for the dipole moment instead of the more usual $\mathbf{p}$ to avoid confusion with the symbols we use for momentum and power. From the energy of dipole in an applied field, we can show that the force acting on the sphere is (Harada and Asakura, 1996)

$$\begin{aligned} \mathbf{F}_{\text{grad}} &= \pi n_{\text{med}}^2 \varepsilon_0 r^3 \left(\frac{m^2 - 1}{m^2 + 2} \right) \nabla |\mathbf{E}|^2 \\ &= \frac{2\pi n_{\text{med}} r^3}{c} \left(\frac{m^2 - 1}{m^2 + 2} \right) \nabla I \end{aligned} \tag{2}$$

Since this is proportional to the gradient of the irradiance I, it is called the *gradient force*. Note that it is also proportional to the volume of the sphere (i.e., proportional to r^3). If a trapped particle is much smaller than the wavelength of the trapping beam, it will be in an approximately uniform field, and Eq. (2) is a useful approximation for the force.

If the field were static, Eq. (2) would give the total force. However, the field is time-varying, with the direction of the induced dipole moment changing twice every optical period. This time-varying dipole moment is equivalent to a current, and since a magnetic field—in this case, the magnetic field of the trapping beam—exerts a force on a current, one expects a force to act on the sphere. However, for a lossless (i.e., nonabsorbing) material, one might reasonably expect the dipole moment to be exactly in phase with the incident electric field, and hence the current to be a quarter wave out of phase with the incident wave. Inconveniently for this simple picture, this would result in a *time-averaged* force of zero. However, since we know that light reflected from the particle will result in a force, this cannot be correct. In circuit theory terms, a current quarter wave out of phase is a pure reactance, with no energy losses involved. An oscillating dipole moment, on the other hand, is the classic ideal short dipole antenna, which radiates energy. Since energy is removed from the incident beam and reradiated, momentum must also be removed from the incident beam. For a Rayleigh scatterer, the reradiation is symmetric and has a total momentum flux of zero. Hence the momentum transfer to the particle is equal to the momentum associated with the incident energy lost, and the force has a magnitude of (Harada and Asakura, 1996)

$$F_{\text{scat}} = \frac{8\pi n_{\text{med}} k^4 r^6}{c} \left(\frac{m^2 - 1}{m^2 + 2} \right) I \tag{3}$$

where k is the wavenumber of the trapping beam. This force is usually called the *scattering force*. The scattering force can also be included in the expression for the gradient force by a correction term to the polarizability, which is the ratio of the dipole moment to the applied field such that $\mathbf{d} = \alpha\mathbf{E}$, gives an effective polarizability (Draine, 1988),

$$\alpha_{\text{eff}} = \frac{\alpha_0}{1 - (2ik^3\alpha_0)/3} \tag{4}$$

in terms of the original uncorrected polarizability α_0, making the polarizability of the sphere complex. This has a similar effect to the sphere possessing a complex refractive index (i.e., the sphere being absorbing), in that it results in a force in the direction of the incident energy flux. There is one important difference, however, between a reradiating sphere and an absorbing sphere: the reradiating sphere experiences a scattering force proportional to the square of the volume (i.e., proportional to r^6), while the absorption force on a small absorbing sphere is proportional to the volume (so proportional to r^3). If a sphere is small enough, the scattering force will always be smaller than the gradient force, allowing, for example, highly reflective metallic spheres to be trapped. Absorption forces, on the other hand, scale in the same way as the gradient force, so small sizes are insufficient to allow the trapping of absorbing particles—the only mechanism by which sufficiently absorbing particles can be trapped in single-beam traps is to two-dimensionally trap them against a surface by using this absorption radiation pressure force (Rubinsztein-Dunlop *et al.*, 1998).

While the scaling of gradient forces and scattering forces with volume suggests that, given sufficiently small sizes, any nonabsorbing particles can be three-dimensionally trapped, this is not the case in practice. The trapped particle undergoes random motion due to thermal fluctuations—Brownian motion. The thermal kinetic energy of a particle in the trap is $k_B T$, where k_B is Boltzmann's constant and T is the absolute temperature. Noting that a conservative force is the gradient of a scalar function and that the gradient force is proportional to the irradiance—a scalar function—we see that the gradient force is conservative, and is the gradient of the *trapping potential*

$$U = \frac{2\pi n_{\text{med}} r^3}{c}\left(\frac{m^2 - 1}{m^2 + 2}\right) I \tag{5}$$

If this trapping potential exceeds the thermal kinetic energy significantly, the particle is unlikely to escape due to Brownian motion. We can also compare the trapping force against the average magnitude of the force due to Brownian motion, which is $12\pi r\eta k_B T$ where η is the viscosity of the fluid. Viscous drag in the medium will assist the trap in preventing escape; consideration of this is essentially equivalent to comparing the trapping potential with the thermal kinetic energy. In principle, the Brownian motion force, which is independent of the trapping power, can be overcome by the use of sufficient power, but the power required may well destroy the particle (or even one's optical system) in practice. The ultimate limit in the trapping of small particles, atom trapping, is achieved by the trapping of atoms in vacuum, or of a sufficiently cold group of atoms—the Brownian motion forces decrease with decreasing temperature.

Both the geometric optics and Rayleigh approximation provide easy to understand qualitative pictures of trapping, and useful quantitative results in their respective size regimes. However, many particles in optical trapping lie between these two regimes, and require the use of exact electromagnetic theory to calculate forces and torques. While the mathematical formulation of the problem presents a somewhat daunting barrier to entry, the practical implementation, especially for the case of spherical particles, is quite feasible, and the size range for which such methods are practical overlaps the size ranges for which the geometric optics and Rayleigh approximations are valid. We discuss such methods later in this chapter. For now, we mention that exact electromagnetic methods can yield agreement with experiment to better than within 1% (Knöner *et al.*, 2006). Overall, we recommend that some caution be used when applying results obtained using approximate methods, especially when a high level of accuracy is required.

We now proceed to present some results of applying exact electromagnetic theory to optical trapping. Since the primary component of an optical tweezers trap is a tightly focused laser beam, we first consider such beams in the context of exact electromagnetic theory. We show the equivalents of some of the above considerations using the exact theory. We next give some results for a representative of trapping of a sphere, and examine the limits of trappability—what types of particles of what sizes can be trapped in a typical optical tweezers apparatus.

Superficially, a tightly focused laser beam appears very much like a paraxial Gaussian beam with a small beam waist radius. However, if we attempt to more tightly focus the beam, the beam waist reaches a limiting size, while the theoretical paraxial Gaussian waist radius can be made as small as possible. This has been illustrated by Nieminen *et al.* (2003b). Furthermore, for a linearly polarized beam, the rotational symmetry of the intensity is broken as the waist radius approaches the limit. Components of the electric field in the direction of the beam propagation—which are assumed to be zero in the paraxial limit—approach one-third the magnitude of the transverse components. A typical cross section of the focal plane intensity of a tightly focused Gaussian beam is shown in Fig. 2; we also show the significant axial component of the electric field. For a plane polarized beam, the axial electric field in each of the two regions is π out of phase relative to the other. Interestingly, for a circularly polarized beam, the axial component forms an optical vortex, carrying orbital angular momentum (Nieminen *et al.*, 2004a).

A snapshot view of the instantaneous electric field, as opposed to the electric field amplitude, is shown in Fig. 3. The relationship between the focal volume and the wavelength of the beam is clearly visible.

The effect of a particle on the beam is shown in Fig. 4. The exact electromagnetic calculations reproduce the same qualitative effects discussed earlier, where changes in the convergence or divergence and direction of the trapping beam result in restoring forces acting on the particle. The force is shown in Fig. 5, as a function of position. Note that the radial force, as a function of radial transverse displacement,

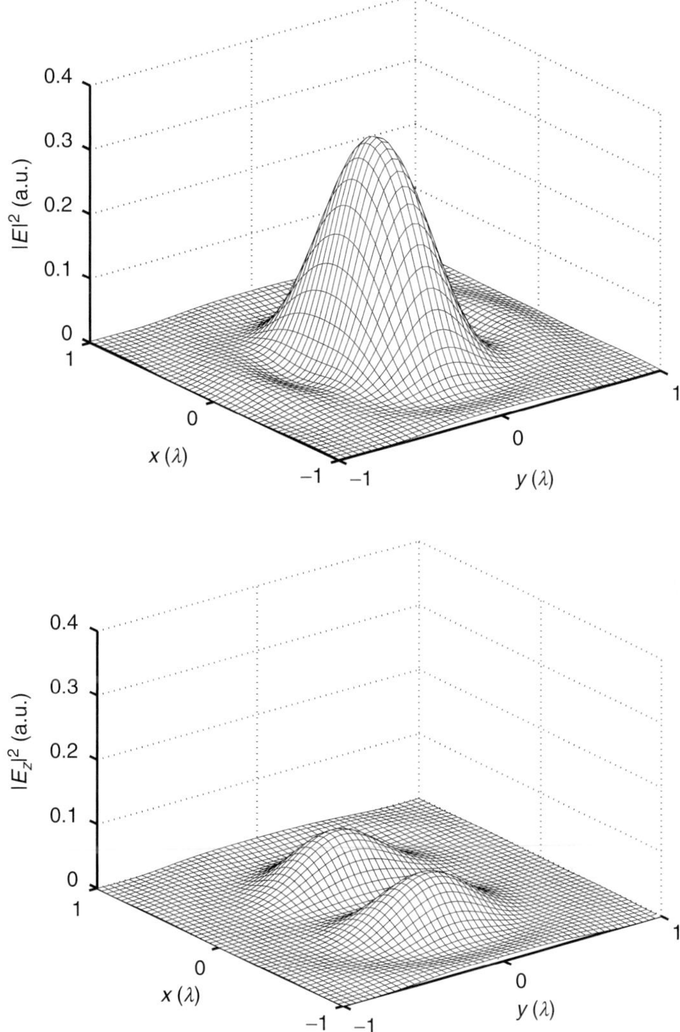

Fig. 2 Irradiance distribution in the focal plane of a tightly focused Gaussian beam is shown in the left hand plot. This beam is polarized in the x-direction, with a small but noticeable elongation of the focal spot in this direction. This results from an axial component of the electric field caused by the focusing of the beam. This component of the electric field parallel to the beam axis is shown in the right hand plot. Unsurprisingly, trap stiffnesses vary between the directions parallel to and perpendicular to the plane of polarization. While there are also small components of the electric field perpendicular to the plane of polarization, these can be safely ignored.

is symmetric, while the axial force is not symmetric as a function of axial displacement, due to the scattering force always acting in the direction of propagation of the beam. This results in the equilibrium position of any spherical particle in the trap being with its center beyond the focus of the beam. In most trapping situations, the

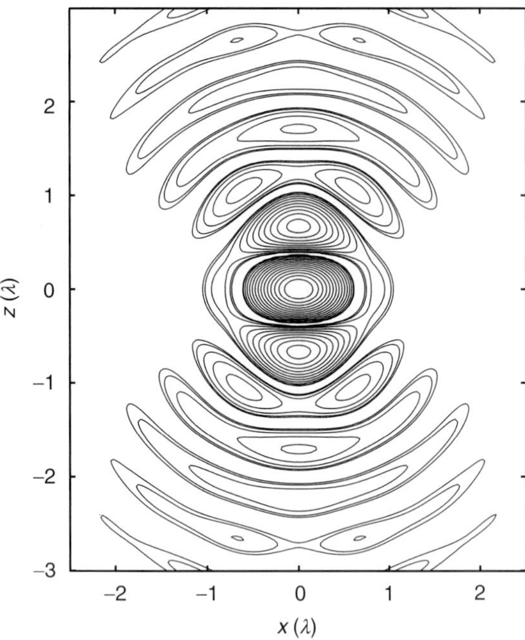

Fig. 3 A side view of the trapping beam: the magnitude of the instantaneous electric field is shown by the contours in the figure. The size of the focal volume compared to the wavelength of the light is clearly evident.

weight of the trapped particle is largely compensated by buoyancy forces, and can be neglected.

For the situation depicted in Fig. 5, the axial force is negative for a range of distances beyond the focus. This is the force responsible for axial trapping. An axial equilibrium position only exists if this force curve crosses the $F_z = 0$ axis, with a stable equilibrium in the position indicated in Fig. 5. Furthermore, the maximum force in this reverse direction is much smaller than either the maximum force in the forward direction or the maximum radial force. This is the usual case for optical tweezers. Therefore, calculation of the axial force as a function of axial displacement allows us to determine whether trapping occurs, and how strongly the particle is trapped. For a given optical tweezers setup, it is possible to calculate the trappability of and maximum reverse axial force acting on particles as a function of particle refractive index and size. In Fig. 6, we show a contour map of the maximum reverse axial restoring force as a function of size and refractive index. Where contours are absent, there is no reverse restoring force—the axial force is always in the direction of propagation, and the particle cannot be trapped. For higher refractive indices, reflection of the trapping beam becomes more important as the axial scattering force begins to push the particle out of the trap. The reflectivity of a thin layer varies periodically with thickness depending on whether

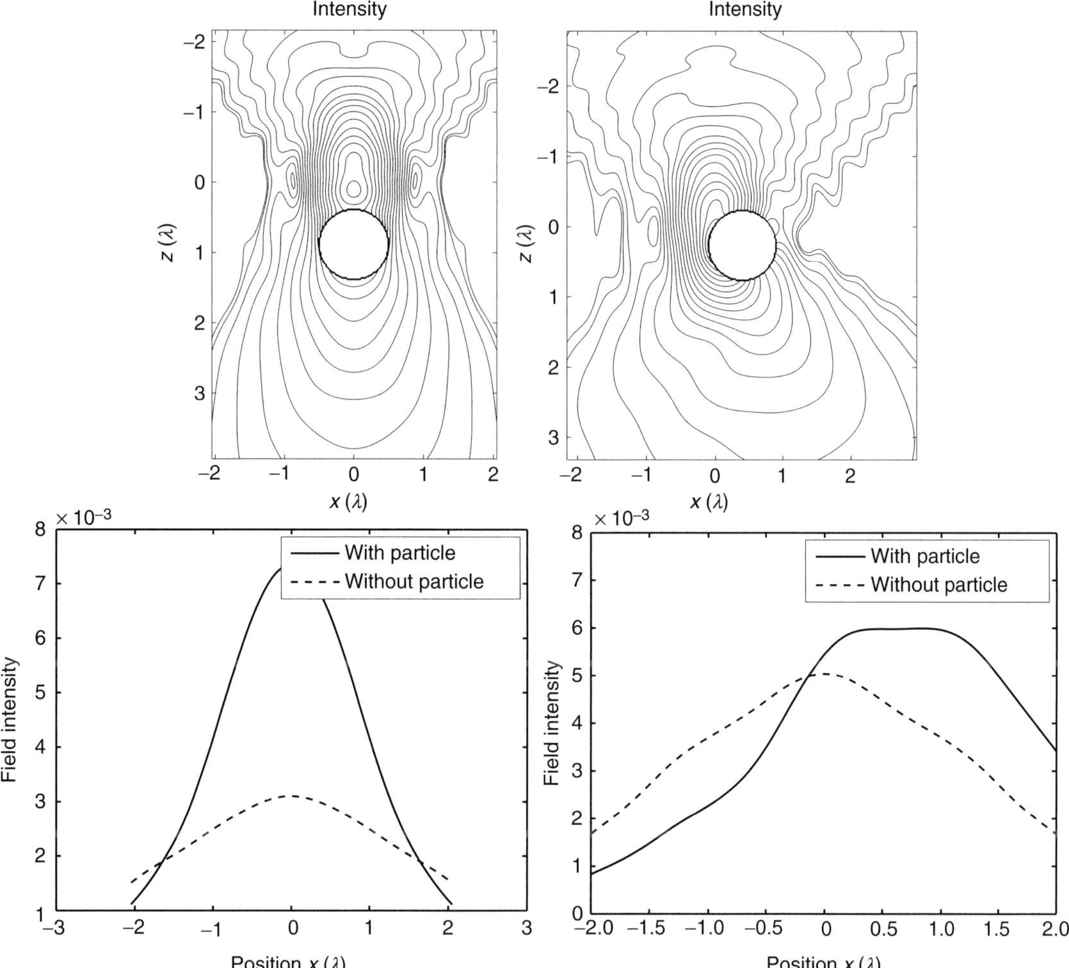

Fig. 4 Change in convergence and direction of the trapping beam due to a particle in the trap displaced away from the equilibrium position. On the left, the effect of a particle displaced along the axis of a downward-propagating beam below the equilibrium position. The contours in the upper plot show the intensity, while the lower graph shows the fields 2 μm below the focus. It can be seen that the particle acts to reduce the divergence of the beam, with an increase in the momentum flux of the beam, resulting in an upward restoring force. On the right, the effect of a particle displaced to the side of the equilibrium position is shown. Note that the beam is deflected to the right, with a resulting reaction force on the particle acting to the left. (See Plate 4 in the color insert section.)

light reflected from the front and back interfaces interferes constructively or destructively, and we can see that microspheres show similar behavior. This results in the variation in trappability with size seen in the upper region in the figure. While high-index particles such as diamond (indicated by the dashed line in the figure) are

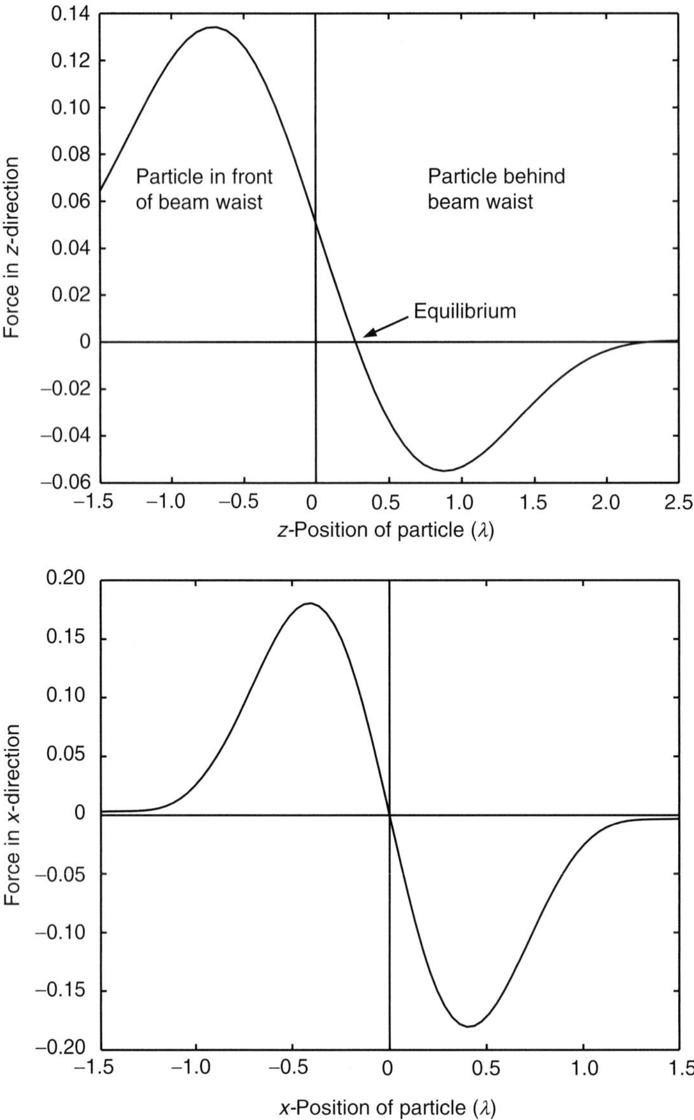

Fig. 5 Axial and radial forces. Note that the axial force is asymmetric. This is due to the contribution of the scattering force.

often considered untrappable, weak trapping should be possible for specific sizes. At the far left, we enter the Rayleigh regime, where the gradient force is proportional to r^3, while the scattering force is proportional to r^6. As a result, the trapping forces become small, but even high-index particles can be trapped as long as Brownian forces can be overcome.

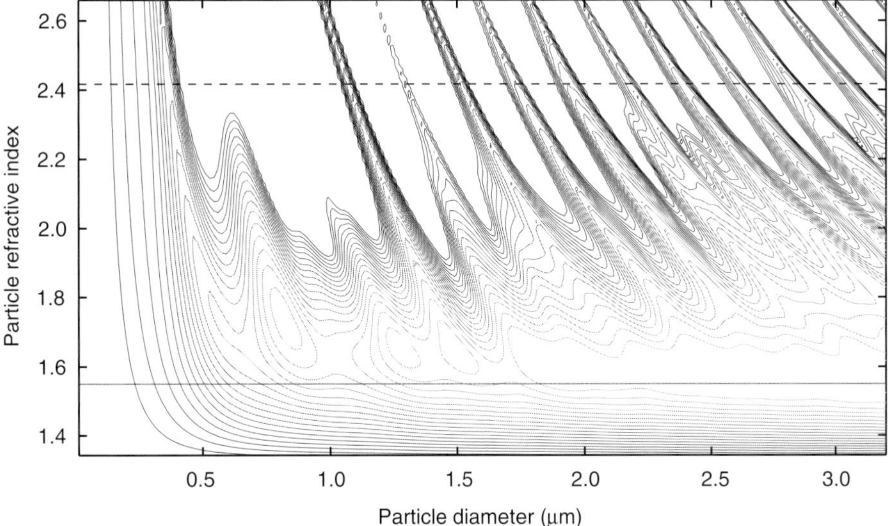

Fig. 6 Map of maximum upward axial restoring force as a function of size and refractive index. These results are for trapping in water at a free-space wavelength of 1064 nm focused by an NA = 1:3 microscope objective. Polystyrene microspheres (n = 1:55, shown by the dotted line) can be readily trapped. From this map, 1-μm polystyrene spheres are more weakly trapped than 2-μm spheres, as observed experimentally. For higher refractive indices, reflections become more important as the axial scattering force begins to push the particle out of the trap. The reflectivity of a thin layer varies with thickness depending on whether light reflected from the front and back interfaces interfere constructively or destructively. Similar behavior is seen for microspheres, resulting in the variation in trappability with size in the upper region in the figure. While high-index particles such as diamond (dashed line) are often considered untrappable, weak trapping should be possible for specific sizes. At the far left, we enter the Rayleigh regime, where the gradient force is proportional to r^3, while the scattering force is proportional to r^6. As a result, the trapping forces become small, but even high-index particles can be trapped.

A. Further Reading

The picture of a trapped particle as a lens changing the convergence or divergence and direction of the trapping beam was suggested by Ashkin *et al.* (1986). Ashkin (1992) also provided the classic quantitative treatment of optical force in terms of rays. Geometric optics has also been used for the calculation of forces and torques on nonspherical particles (Gauthier, 1997; Ukita and Nagatomi, 1997).

The use of the Rayleigh approximation to provide simple expressions for the gradient force and the scattering force appears in much of the literature (Ashkin *et al.*, 1986). The optical force acting on Rayleigh particles is well treated by Harada and Asakura (1996) and Chaumet and Nieto-Vesperinas (2000), the former using SI units (i.e., MKS), the latter Gaussian units (i.e., cgs). Harada and Asakura (1996), in particular, investigate the bounds of validity of the Rayleigh approximation for determining optical forces and show that it is accurate for particles of radius $<0.1\lambda$. An earlier study of the trapping of Rayleigh particles was undertaken by Visscher and Brakenhoff (1992).

IV. Fundamental Theory

Here we review some of the fundamental theory underlying optical tweezers. This includes the transport of energy, momentum, and angular momentum by monochromatic electromagnetic waves, and the transfer of these quantities to a particle. We do not attempt to give thorough and detailed derivations—these are readily available in the literature. Instead, we give the most important results, and explain the underlying physics. We also highlight some points that are often neglected in the literature or are points of contention.

The theory behind optical tweezers is essentially classical electromagnetic theory, so we present a brief review of the relevant basic elements of electromagnetic theory, largely as a reminder about basic concepts and terminology. After a general introduction, we will restrict ourselves to monochromatic electromagnetic waves in source-free lossless (i.e., nonabsorbing) homogeneous linear media, and we will neglect dispersion. The results we present should not uncritically be taken as being general outside these conditions. We hope that the physical explanations we offer will give an indication that a result might be more general.

The interaction between electric and magnetic fields (the electric intensity $\mathbf{E}$, electric displacement $\mathbf{D}$, magnetic intensity $\mathbf{H}$, and magnetic induction $\mathbf{B}$) and their sources—the charge density ρ and the current density $\mathbf{J}$—is described by the *Maxwell equations*:

$$\nabla \times \mathbf{E} + \frac{\partial \mathbf{B}}{\partial t} = 0 \tag{6}$$

$$\nabla \times \mathbf{H} - \frac{\partial \mathbf{D}}{\partial t} = \mathbf{J} \tag{7}$$

$$\nabla \cdot \mathbf{D} = \rho \tag{8}$$

$$\nabla \cdot \mathbf{B} = 0 \tag{9}$$

The Lorentz force law,

$$\mathbf{f} = \rho \mathbf{E} + \mathbf{J} \times \mathbf{B} \tag{10}$$

giving the force per unit volume, can be taken to be an operational definition of $\mathbf{E}$ and $\mathbf{B}$.

The properties of a material medium are contained in the constitutive relations:

$$\mathbf{D} = \varepsilon \mathbf{E} \tag{11}$$

$$\mathbf{B} = \mu \mathbf{H} \tag{12}$$

$$\mathbf{J} = \sigma \mathbf{E} \tag{13}$$

The last of these is Ohm's law. In free space (i.e., vacuum), ε and μ become ε_0 and μ_0. Since **D** and **H** are defined in terms of the sources of the fields (the charge density ρ and the current density J), ε_0 and μ_0 can be thought of as unit conversion factors. A system of units can be chosen so that they are both dimensionless and numerically equal to one; but we will use SI units throughout, where this is not the case.

The real (i.e., not complex) fields in the Maxwell equations are functions of both time and position. If the time-variation is sinusoidal, we can eliminate the time derivatives by adopting the use of complex field amplitudes, reducing the system of equations to containing functions of position only. We replace the field variables in the equations above by complex equivalents, using relations of the form

$$\mathbf{E}_{\text{actual}} = \text{Re}\{\mathbf{E}_{\text{complex}} \exp(-\text{i}\omega t)\} \tag{14}$$

Since the Maxwell equations must hold for all times, they must also hold for the imaginary part of the complex field, so there is no need to explicitly take the real part in the equations. The complex amplitudes $\mathbf{E}_{\text{complex}}$, and so on, are independent of time, so the time derivatives can be written analytically. The exponential variation with time is the same for every single term in the equation, and can therefore be eliminated from the equations. On occasion, it is necessary to recall that the complex amplitudes are not the instantaneous fields, and to use Eq. (14) to obtain the real instantaneous fields.

Dropping the complex subscript for brevity, the time-harmonic Maxwell equations in source-free media become:

$$\nabla \times \mathbf{E} - \text{i}\omega \mathbf{B} = 0 \tag{15}$$

$$\nabla \times \mathbf{H} + \text{i}\omega \mathbf{D} = 0 \tag{16}$$

$$\nabla \cdot \mathbf{D} = 0 \tag{17}$$

$$\nabla \cdot \mathbf{B} = 0 \tag{18}$$

Combining the first two of these equations, and using the constitutive relations, we obtain the vector Helmholtz equation for both the electric and magnetic fields:

$$\nabla^2 \mathbf{E} + k^2 \mathbf{E} = 0 \tag{19}$$

$$\nabla^2 \mathbf{H} + k^2 \mathbf{H} = 0 \tag{20}$$

while obtaining the further restrictions on the field that $\nabla \cdot \mathbf{E} = 0$ and $\nabla \cdot \mathbf{H} = 0$ from the last two equations. These conditions, that the fields are divergence-free, are sometimes specified by stating that the fields are *transverse*. The wavenumber k is equal to $(\varepsilon\mu)^{1/2}\omega$ so we can identify $v = (\varepsilon\mu)^{-1/2} = c/n$ as the phase speed of the wave.

We will be able to restrict ourselves to considering plane waves. For a plane electromagnetic wave, **E** and **H** are perpendicular to each other, and also to the direction of propagation. **E** and **H** are also in phase with each other, and have magnitudes related to $E\sqrt{\varepsilon} = H\sqrt{\mu}$.

The Maxwell equations as presented above are differential equations, and cannot be applied where the fields or complex amplitudes of the fields are discontinuous. This occurs at interfaces between media with different electromagnetic or optical properties. This requires us to use boundary conditions at the interface to relate solutions to the Maxwell or Helmholtz equations in the two media separated by the interface. The boundary conditions which we will need to consider are simply that the vector components of **E** and **H** parallel to the interface are the same on both sides of the interface.

V. Transport of Energy, Momentum, and Angular Momentum by Light

It is of course evident to any who use lasers, have stood in sunlight, or otherwise experienced related effects, that electromagnetic waves carry energy. Those who use optical tweezers are also familiar with the transport of momentum by electromagnetic waves.

If we consider the work done on the currents within a volume by the electric field **E** (if charges do not move, no work is done on them, and the magnetic field does no work, so this is sufficient), we can show that the instantaneous energy flux density (i.e., the power per unit area) of an electromagnetic wave is equal to the Poynting vector

$$\mathbf{S} = \mathbf{E}_{\text{actual}} \times \mathbf{H}_{\text{actual}} \tag{21}$$

For a monochromatic wave, especially at optical frequencies, we are usually interested in the time-averaged Poynting vector in terms of the complex amplitudes rather than the real fields, which is equal to

$$\mathbf{S} = \frac{1}{2}\mathbf{E} \times \mathbf{H}^* \tag{22}$$

where * denotes the complex conjugate. For a plane electromagnetic wave, the Poynting vector is always in the direction of propagation of the wave, and has a magnitude of

$$S = \frac{c\varepsilon}{2n}|E|^2 \tag{23}$$

There is a close relationship between the energy flux density (i.e., the Poynting vector), the energy density u of the wave, and the velocity **v** of energy transport:

$$\mathbf{S} = u\mathbf{v} \tag{24}$$

This is a general relationship which holds for all waves. Since the time-averaged energy density is

$$u = \frac{1}{4}(\mathbf{E} \cdot \mathbf{D}^* + \mathbf{B} \cdot \mathbf{H}^*) \tag{25}$$

which simplifies to

$$u = \frac{\varepsilon}{2}|E|^2 \tag{26}$$

for a plane wave, we can verify that the speed of energy transport in the simple media we are considering is equal to the phase speed of the wave, c/n. (For more general media, for which the permittivity can be complex, this is not necessarily the case.)

The transport of energy by a wave necessitates the transport of momentum, and the momentum flux density $\mathbf{p}$ is determined by the energy flux density and the speed of transport of energy. The momentum transported per unit area in the direction of propagation of the wave is

$$\mathbf{p} = n\mathbf{S}/c \tag{27}$$

The magnitude of the momentum flux is numerically equal to the energy density, since both are related to the energy flux by the same speed. At this point, it is necessary to mention that there is a long-standing controversy over the momentum of an electromagnetic wave in matter—whether the momentum flux density is $n\mathbf{S}/c$, $\mathbf{S}/(nc)$, or some other value—often called the Abraham–Minkowski controversy. The relationships between energy density, energy flux, and the speed at which energy is transported are general relationships valid for all waves, and therefore the momentum flux in Eq. (27) must be the total momentum flux associated with the electromagnetic wave, and is independent of the Abraham–Minkowski controversy. A review of the controversy (Pfeifer *et al.*, 2006) shows that the controversy is essentially a matter of semantics and not of physics—there is no disagreement about the total momentum associated with an electromagnetic wave in matter, only about the division of this total momentum into components called electromagnetic momentum and material momentum. Since it is the total momentum that produces physical effects, there is no way to physically distinguish between the different expressions for the electromagnetic momentum.

Finally, any transport of momentum implies a transport of angular momentum as well. In the case of electromagnetic fields, this presents some difficulties. While it is tempting to simply say that the angular momentum flux is the moment of the linear momentum flux, that is,

$$\mathbf{j} = \mathbf{r} \times \mathbf{p} = n\mathbf{r} \times \frac{\mathbf{S}}{c} \tag{28}$$

this is not correct, as it neglects the fact that electromagnetic waves can carry spin angular momentum. In general, the angular momentum flux can be divided into an orbital angular momentum flux **l** and spin **s**—named in analogy with classical mechanics where a body can both undergo orbital motion and spin about its own axis—such that

$$\mathbf{j} = \mathbf{l} + \mathbf{s} \tag{29}$$

Spin is defined as *intrinsic angular momentum*—angular momentum that is independent of the origin about which we choose to take our moments. Therefore, spin is the portion of the angular momentum which is independent of **r**, and can easily be found from an expression for the total angular momentum, even if complicated, by choosing **r** to be zero. The remainder of the angular momentum is the orbital angular momentum. The use of Eq. (28) is equivalent to stating that electromagnetic waves cannot carry spin.

The spin flux of a plane electromagnetic wave depends on its degree of circular polarization σ and is equal to

$$\mathbf{s} = \frac{\sigma \mathbf{S}}{\omega} \tag{30}$$

where ω is the optical angular frequency, and the degree of circular polarization is defined as

$$\sigma = (I_{\mathrm{L}} - I_{\mathrm{R}}) = I \tag{31}$$

where I_{L} and I_{R} are the irradiances of the left- and right-circularly polarized components (Nieminen *et al.*, 2001a). This spin flux is equivalent to the quantum mechanical result that photons carry $\pm\hbar$ spin.

However, it has been known for a long time that for fields finite in extent, the total angular momentum is given by the integral over the entire field of the moment of the linear momentum density (Humblet, 1943), and a similar result also applies to the angular momentum flux (Crichton and Marston, 2000), so that

$$\mathbf{J} = \int \mathbf{r} \times \mathbf{p} \mathrm{d}A = \int n\mathbf{r} \times \frac{\mathbf{S}}{c} \mathrm{d}A \tag{32}$$

It is natural (although incorrect in this case) to identify the integrand as the flux density of the total flux, and this has given rise to a controversy about the angular momentum content of a circularly polarized plane wave. However, since we will be dealing with physically realizable waves in practice, rather than mathematically convenient but unphysical entities such as infinite plane waves, and there is no doubt about the total angular momentum flux of such waves, we can safely ignore the controversy.

Now that we have identified the energy and momentum flux densities, and a method of obtaining the total angular momentum flux, we can consider the application of these to the question of optical trapping. To do so we need to consider non-plane waves—the incident beam essentially consists of spherical waves converging to the focus and then diverging away as spherical waves, whether refracted or reflected or not by a particle in the trap. Our method will be to consider the total energy, momentum, and angular momentum fluxes through a spherical surface centered on the focus. These directly give the power absorbed by the particle, and the optical force and torque—if, for example, the total momentum flux into the volume enclosed by the spherical surface is nonzero, this means that the total momentum inside the spherical surface is changing. Since force is the rate of change of momentum, this is the optical force acting on the particle in the trap. Similarly, the total energy and angular momentum fluxes into the enclosed sphere give the absorbed power and the optical torque. As long as the radius of the spherical surface is much greater than the wavelength, on any small portion of the spherical surface the waves will very closely approximate plane waves, and we can use the plane wave formulas for the energy, momentum, and angular momentum fluxes. We can also center the spherical surface on a particle within the trap and obtain the same results for power, force, and torque.

Since we are using a large (compared to the wavelength) radius for the spherical surface, the Poynting vector will be very nearly purely radial at the surface.

To calculate the absorbed power, it is simplest to proceed in spherical coordinates, and we can simply write the total energy flux into the sphere in terms of the radial component of the Poynting vector as

$$P_{\mathrm{abs}} = -\int r^2 S_r \, \mathrm{d}\Omega \tag{33}$$

where the integration is taken over all angles. The negative sign is a result of the desired quantity depending on the total flux *into* the sphere, while S_r is outward when positive. The radius r is the radius of the spherical surface, and appears as the factor of r^2 because the surface area of the spherical surface is proportional to r^2. Note that since the irradiance of spherical waves falls off according to an inverse square law, the Poynting vector is proportional to $1/r^2$, so the r^2 term serves to keep the total flux independent of the choice of r. If we have an analytical expression for the Poynting vector, or at least its radial component, it is usually convenient to take the limit of $r^2 S_r$ as $r \to \infty$. If the Poynting vector is calculated numerically, then the integration over the surface can be performed numerically.

For calculation of the optical force, it is most convenient to proceed with the Poynting vector expressed in Cartesian coordinates, since we usually wish to know the Cartesian components of the force. The force is given by

$$\mathbf{F} = -\int r^2 \mathbf{S} \, \mathrm{d}\Omega \tag{34}$$

Again, it can be useful to take large r limits before integrating.

The calculation of optical torque is similar in many respects, but with one crucial difference. We can write the torque as

$$\tau = -\int r^2 \mathbf{r} \times \mathbf{S}\, d\Omega \tag{35}$$

but we must make sure that, if taking large r limits, that the limit of $r^2\mathbf{r} \times \mathbf{S}$ is taken, rather than taking the moment of the limit of $r^2\mathbf{S}$, which would always yield a torque of zero.

While these formulas can be used to obtain the quantities of most interest in optical trapping, they are of little use unless we know the fields **E** and **H** which we require to find the Poynting vector on the spherical surface. Determining these fields is the subject of the next section. Before passing on to this, however, we will consider some further details of the interaction between the trapping beam and a particle within the trap.

First, if the particle in the trap is nonabsorbing, the absorbed power will be zero, and there will be no need to calculate the energy flux, except possibly to check the accuracy of numerical calculation of the fields. On the other hand, if the particle is absorbing, the absorbed power is of great interest, as, for example, it allows us to estimate the likelihood of opticution of live specimens.

Second, if the particle is completely spherically symmetric, which requires the particle to be spherical, homogeneous, and optically isotropic, the only possible mechanism allowing an optical torque to be exerted on the particle is absorption. Therefore, if one is trapping typical nonabsorbing microspheres, there will be no need to calculate torques.

Finally, if the trapped particle is stationary, then, despite any optical force exerted on it, the electromagnetic field does no work on the particle. On the other hand, if there is an optical force, and the particle is moving, or if there is an optical torque, and the particle is rotating, the trapping beam will do work on the particle. We might then ask from where, if there is no absorption, does this energy come from? The answer is that in these cases, there must be a Doppler shift in the frequency. In fact, the Doppler shift due to the reflection of a plane wave from a moving mirror can be used to obtain the energy–momentum relationship we made use of earlier in this section. In the case of rotational motion, there is a rotational frequency shift.

A. Further Reading

We have considerably simplified matters here by dealing with plane waves. In this case, the direction of transport of momentum and the direction of the transported momentum coincide. The same is true for the spin flux and the spin. This allows us to write the momentum and angular momentum fluxes simply as vectors. If these directions did not coincide, we would need to write these fluxes as 3×3 tensors—the Maxwell stress tensor and the moment of the Maxwell stress tensor.

General treatments of electromagnetic theory are many, but sources specifically dealing with the transport of energy, momentum, and angular momentum in general are relatively few. Nonetheless, the topic of transport of energy and momentum is covered in many textbooks (Griffiths, 1999; Jackson, 1999; Stratton, 1941).

The general principles of transport of energy in media by waves were first formulated by N. A. Umov in 1874; the Poynting vector is usually called the Umov vector or the Umov–Poynting vector in the Russian literature. Unfortunately, Umov's papers do not seem to be readily available in English. The transport of energy specifically by electromagnetic fields was treated 10 years later by Poynting (1884) and Heaviside. Poynting (1909) appears to have been the first to explicitly consider the transport of spin angular momentum by electromagnetic waves.

Humblet (1943) is generally credited with being the first to resolve the "angular momentum paradox," but this failed to end the controversy. Debate over this issue continues, with the fires kept well stoked by the fuel of optical rotation in laser trapping. Recent works such as Benford and Konz (2004), Nieminen (2004), Stewart (2005), and Zambrini and Barnett (2005) convey the nature of the modern debate.

Further mention should be made of orbital angular momentum. Optical orbital angular momentum is often described as being somehow special. However, any laser beam or other electromagnetic beam possesses orbital angular momentum if a suitable origin about which to take moments is chosen. For example, most macroscopic measurements of radiation pressure, such as those of Lebedew and Nichols and Hull, have actually been measurements of orbital angular momentum—such experiments typically make use of a torsion pendulum and measure a torque, and hence the orbital angular momentum of the beam about the axis of the pendulum. Since the orbital angular momentum transfer in such a case is equal to two times the product of the lever arm of the mirror and the linear momentum of the beam, the linear momentum can be measured.

This does not alter the fact that the orbital angular momentum of certain types of beams, those possessing optical vortices, is special—these beams carry a total angular momentum about the beam axis that is nonzero. Equation (28) should give the orbital angular momentum flux, and generalizing this to the beam as a whole, we expect that if we take moments about a point through which the beam axis passes, the orbital angular momentum would be zero.

However, for special types of beams, we find that the orbital angular momentum density about the beam axis is nonzero, and the total orbital angular momentum density about the beam axis is also nonzero. This is a truly remarkable result, and such beams have been discussed in a number of papers (Allen *et al.*, 1999, 2003). One especially interesting feature of such beam is that while the orbital angular momentum *density* depends on the choice of origin about which moments are taken (by definition), the *total* angular momentum about the beam axis—which would be zero for "conventional" beams such as a typical Gaussian beam—is independent on this choice (O'Neil *et al.*, 2002).

We also note that while there is a monosyllabic technical term for "intrinsic angular momentum" (i.e., "spin"), there is currently no equivalent term for "orbital angular momentum." This results in the use of, for example, "OAM," which can equally mean "optical angular momentum" or "orbital angular momentum." We believe that the optical angular momentum is of sufficient importance in optical trapping and other applications to merit its own specific terminology, and recommend, and look forward to, the adoption of the term "whorl."

We can also note that there are two distinct pictures of the transfer of optical energy, momentum, and angular momentum to a particle. Above, we consider these in terms of fluxes through a surface. We could also have considered them in terms of the Lorentz force law, finding the work done on currents by the electromagnetic field to find the power, and the force and torque exerted on charge, current, and dielectric polarization densities to find the optical force and torque. Both approaches must yield the same results. Our flux-based approach allows the problem to be reduced to two-dimensional integrals over a surface, while a Lorentz force-based approach requires three-dimensional volume integrals. The theoretical equivalence of the two approaches has been shown to hold numerically (Kemp *et al.*, 2005) and the Lorentz force approach has also been exploited to determine the internal stresses within a particle (Hoekstra *et al.*, 2001).

The transfer of angular momentum to a particle through scattering is strongly dependent on the symmetry of the particle and the electromagnetic field. We review this topic in Nieminen *et al.* (2004c).

Finally, Atkinson (1935) discusses the early work on rotational frequency shifts. Some of the modern papers include Courtial *et al.* (1998) and Padgett (2004).

VI. Computational Modeling of Optical Tweezers

As noted earlier, and shown in Fig. 7, optical tweezers usually fall into the gap between the regimes of applicability of large particle approximations such as geometric optics and small particle approximations such as Rayleigh scattering. This size range, where dimensions of the particle are comparable to the wavelength of the light, is often called the *resonance region*. As a result, computational modeling of optical tweezers usually requires resort to direct solution of the Maxwell equations (primarily time-domain methods) or the vector Helmholtz equation (frequency-domain methods). As noted earlier, the vector Helmholtz equation follows from the Maxwell equations for the case of monochromatic illumination in linear isotropic media, with the further restriction that the solutions must be divergence-free (i.e., $\nabla \cdot \mathbf{E} = 0$ and $\nabla \cdot \mathbf{H} = 0$) also following from the Maxwell equations. While special cases such as particles with nonlinear optical response, anisotropic particles, and trapping by short-pulse lasers might require direct time- or frequency-domain solution of the Maxwell equations, solution of the Helmholtz equation is almost always sufficient.

For modeling the trapping of spherical particles (by which we also mean that the particles are homogeneous and optically linear and isotropic), the existence of the analytical Lorenz–Mie solution for scattering by a sphere means that optical forces and torques can be calculated with relatively little effort. However, Lorenz and Mie dealt with the case of plane wave illumination, and optical tweezers require

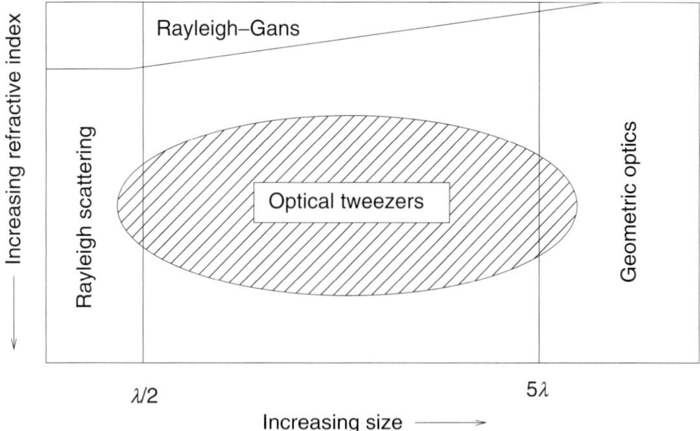

Fig. 7 Map of regimes of validity of some computational methods for calculating the scattering of light by particles. Note that typical particles trapped in optical tweezers fall outside the regimes of applicability of the approximate methods, and require direct solution of the Maxwell equations or the vector Helmholtz equation.

a tightly focused beam. The extension of the original Lorenz–Mie theory to arbitrary illumination is usually called *generalized Lorenz–Mie theory* (GLMT).

Lorenz–Mie theory is based on the representation of the incident and scattered electromagnetic waves as sums of vector spherical wavefunctions (VSWFs). The incident field can be written as

$$\mathbf{E}_{\text{inc}} = \sum_{n=1}^{\infty} \sum_{m=-n}^{n} a_{nm}\mathbf{M}_{nm}^{(3)} + b_{nm}\mathbf{N}_{nm}^{(3)} \tag{36}$$

and the scattered field as

$$\mathbf{E}_{\text{scat}} = \sum_{n=1}^{\infty} \sum_{m=-n}^{n} p_{nm}\mathbf{M}_{nm}^{(1)} + q_{nm}\mathbf{N}_{nm}^{(1)} \tag{37}$$

where n is the radial mode index, m is the azimuthal mode index, a_{nm}, b_{nm}, p_{nm}, and q_{nm} are the mode amplitudes of the incident and scattered fields (often called expansion coefficients or multipole coefficients), and $\mathbf{M}_{nm}^{(3)}$, $\mathbf{N}_{nm}^{(3)}$, $\mathbf{M}_{nm}^{(1)}$, and $\mathbf{N}_{nm}^{(1)}$ are the vector spherical wavefunctions of the third and first type, respectively. While the above expressions for the fields involve infinite sums, the sum for the scattered field produced by a sphere of radius r can be truncated at a finite $n_{\max}$ approximately equal to the size parameter (kr, the product of the wavenumber and the radius). For spheres on the order of the wavelength in size, only a modest number of terms are required. Where the illumination is also finite in width, the sum for the incident field also converges similarly.

The orthogonality properties of the vector spherical wavefunctions—the behavior of products of VSWFs integrated over a spherical surface—lead to two important results. First, there is no coupling between different modes when the scatterer is a homogeneous isotropic sphere. The scattered mode amplitudes can be simply found from the incident wave mode amplitudes, with

$$p_{nm} = a_n a_{nm} \tag{38}$$

$$q_{nm} = b_n b_{nm} \tag{39}$$

where a_n and b_n are the Mie scattering coefficients. This is the key result of Lorenz–Mie theory.

The second result following from the orthogonality properties of the VSWFs is a simplification of the calculation of the optical force and torque. Recalling that the absorbed power, force, and torque can be found by integration of the Poynting vector over a spherical surface surrounding the particle [Eqs. (34) and (35)], we might at first expect that it is necessary to perform these integrations numerically by calculating the electric and magnetic fields over a grid of points on a sphere, which would be a computationally intensive task. However, if one substitutes expressions (36) and (37) for the fields in this calculation, the resulting formulas involve integrals of products of VSWFs, the values of which are known from the orthogonality properties. Therefore, formulas for the absorbed power, force, and torque involving sums over the mode amplitudes and their products result, which can be readily calculated. A thorough treatment of this is given by Farsund and Felderhof (1996). Unfortunately, there is no universal normalization or sign and parity conventions for the VSWFs, and care is required to check that the formulas used for power, force, and torque calculations match the VSWFs used.

In the past, the practical application of Lorenz–Mie theory was highly labor intensive—calculation of the Mie coefficients requires the calculation of Ricatti–Bessel functions or spherical Bessel and Hankel functions and their derivatives, while calculation of the field (which requires evaluating the VSWFs) adds the calculation of spherical harmonics and their angular derivatives. The advent of the modern computer, and just as importantly, routines or packages that can calculate the special functions required, has greatly eased what was once an onerous computational burden.

The one remaining requirement is the calculation of the mode expansion coefficients of the trapping beam. For modeling the trapping of spherical particles, this is the most theoretically difficult and computationally intensive component of the overall task. While this is a far from trivial task, there are a number of ways in which it can be done. It is possible to perform an integral transform (Maia Neto and Nussenzweig, 2000; Mazolli *et al.*, 2003). This is well suited to trapping by a Gaussian beam. A very fast approximate method is the localized approximation (Gouesbet, 1996, 1999; Gouesbet *et al.*, 1995; Lock, 2004a,b). Our method of choice uses an overdetermined point-matching scheme (Nieminen *et al.*, 2003b) which is suitable for arbitrary beams and provides stable and robust numerical performance and convergence.

Lorenz–Mie theory is only applicable to homogeneous isotropic spheres, and while such spheres are often the particle of choice in optical trapping, other types of particles are also trapped. For example, living cells often fail to accommodate our desire for sphericity. A number of approaches are possible. Most simply, we can assume that we can use an equivalent sphere to approximately model the trapped object. A slightly more sophisticated approach is to model suitable objects as layered spheres, for example a cell as a spherical nucleus surround by a spherical layer of cytoplasm—the basic Lorenz–Mie solution is readily extensible to concentric layered spheres. However, the rotational symmetry of a trapped object strongly affects the optical torque exerted on it. If we are interested in rotational manipulation, spherical approximations completely fail us. In addition, equivalent spheres will not give the correct optical force (Nieminen *et al.*, 2001b) so are not useful beyond obtaining estimates of the force. Thus, it is sometimes necessary to resort to more general methods of calculating scattering.

In principle, any method of solution that is valid in the resonance region will suffice. However, in practice, the requirements of modeling optical tweezers restrict the range of methods that are feasible. In particular, one typically wants to be able to calculate the force acting on the trapped particle at different positions within the trap, and even when simply trying to find the equilibrium position of the particle in the trap along the beam axis, dozens of calculations are required. In calculating three-dimensional maps of the force acting on the particle, the number of calculations can easily exceed a thousand. This requirement for efficiency is probably the reason that so little modeling of optical tweezers has used general methods such as the finite-difference time-domain method (FDTD).

Fortunately, a method is available that meets these requirements for efficiency. Recall that the orthogonality of the VSWFs means than there is no coupling between modes for a spherical scatterer, resulting in each scattered mode being given directly by the incident mode and the relevant Mie coefficient. This means that the Mie coefficients completely describe the scattering properties of the sphere at that wavelength, independently of the details of the illumination. As noted above, once the Mie coefficients are found, they can be used for repeated calculations. For the more general case of a scatterer with linear optical properties, but otherwise arbitrary, expressions (36) and (37) are still valid. Taking coupling between modes into account, the equivalent of Eqs. (38) and (39) can be written for a general scatterer:

$$p_{n'm'} = \sum_{n=1}^{n_{\max}} \sum_{m=-n}^{n} A_{n'm'nm} a_{nm} + B_{n'm'nm} b_{nm} \tag{40}$$

$$q_{n'm'} = \sum_{n=1}^{n_{\max}} \sum_{m=-n}^{n} C_{n'm'nm} a_{nm} + D_{n'm'nm} b_{nm} \tag{41}$$

where the infinite sums have been truncated at a finite $n_{\max}$. If we write the incident and scattered mode amplitudes as column vectors **a** and **p**, the above expressions for the scattered field can be written compactly as a matrix–vector product:

$$\mathbf{p} = \mathbf{Ta} \tag{42}$$

The matrix **T** is called the transition matrix, or simply the *T*-matrix, and just as the Mie coefficients are independent of the illumination, so is the *T*-matrix. Therefore, if one can calculate the *T*-matrix for a trapped object, it can be used repeatedly for calculating the optical force and torque for different positions or orientations within the trap.

Calculation of the *T*-matrix is significantly more involved than calculation of the Mie coefficients, but fast methods are available for homogeneous rotationally symmetric particles, which includes a large number of shapes of interest. For example, we have used the *T*-matrix method to model the optical rotation of glass cylinders in optical tweezers (Bishop *et al.*, 2003) and the alignment and rotational manipulation of chloroplasts (Bayoudh *et al.*, 2003).

While Fig. 7 suggests that, especially for biological particles which often have a relative refractive index ($m = n_{\text{particle}}/n_{\text{medium}}$) close to 1, methods such as anomalous diffraction or other methods valid for low-contrast scatterers could be used, these methods do not appear to have been used for modeling optical tweezers. This may result from having to deal with the highly nonparaxial illumination required for optical tweezers, while such methods are almost always implemented for the calculation of scattering of plane waves. This is also, understandably, the case for almost all computational modeling of scattering, both in the resonance region or otherwise. While many light and electromagnetic scattering computer codes are available, one should not expect to be able to use them for the modeling of optical tweezers without a significant investment in modification of the original codes to accommodate the focused beam.

A. Further Reading

The classic treatment of Lorenz–Mie theory is by van de Hulst (1957). The topic is briefly covered in many books on electromagnetic theory, for example, by Stratton (1941) and Jackson (1999). The subject of scattering by spheres has long been of interest, and the early history of the field is reviewed by Logan (1965). The original papers were by Lorenz (1890) and Mie (1908).

The application of Lorenz–Mie theory to modeling the optical trapping of spheres in Gaussian beams has been comprehensively covered by Mazolli *et al.* (2003), following an earlier, shorter, paper by Maia Neto and Nussenzweig (2000). Earlier modeling of trapping using Mie theory includes the work of Ren *et al.* (1996).

The *T*-matrix method was first introduced by Waterman (1965, 1971). A widely used *T*-matrix code was developed by Michael Mishchenko, and is publicly available at http://www.giss.nasa.gov/~crmim/. A variety of light scattering computer codes, and other resources for electromagnetic and light scattering, are available at http://www.t-matrix.de/.

Our own computational methods for modeling optical tweezers are described in Nieminen *et al.* (2003a,b, 2004b).

References

Allen, L., Barnett, S. M., and Padgett, M. J. (eds.) (2003). "Optical Angular Momentum." IOP, Bristol.

Allen, L., Padgett, M. J., and Babiker, M. (1999). The orbital angular momentum of light. *Prog. Opt.* **39,** 291–372.

Ashkin, A. (1970). Acceleration and trapping of particles by radiation pressure. *Phys. Rev. Lett.* **24,** 156–159.

Ashkin, A. (1992). Forces of a single-beam gradient laser trap on a dielectric sphere in the ray optics regime. *Biophys. J.* **61,** 569–582.

Ashkin, A. (1997). Optical trapping and manipulation of neutral particles using lasers. *Proc. Natl. Acad. Sci. USA* **94,** 4853–4860.

Ashkin, A. (2000). History of optical trapping and manipulation of small-neutral particle, atoms, and molecules. *IEEE J. Select. Top. Quantum Electron.* **6,** 841–856.

Ashkin, A., Dziedzic, J. M., Bjorkholm, J. E., and Chu, S. (1986). Observation of a single-beam gradient force optical trap for dielectric particles. *Opt. Lett.* **11,** 288–290.

Atkinson, R.d'E. (1935). Energy and angular momentum in certain optical problems. *Phys. Rev.* **47,** 623–627.

Bayoudh, S., Nieminen, T. A., Heckenberg, N. R., and Rubinsztein-Dunlop, H. (2003). Orientation of biological cells using plane polarised Gaussian beam optical tweezers. *J. Mod. Opt.* **50,** 1581–1590.

Benford, G., and Konz, C. (2004). Reply to comment on "Geometric absorption of electromagnetic angular momentum." *Opt. Commun.* **235,** 231–232.

Bishop, A. I., Nieminen, T. A., Heckenberg, N. R., and Rubinsztein-Dunlop, H. (2003). Optical application and measurement of torque on microparticles of isotropic nonabsorbing material. *Phys. Rev. A* **68,** 033802.

Bishop, A. I., Nieminen, T. A., Heckenberg, N. R., and Rubinsztein-Dunlop, H. (2004). Optical microrheology using rotating laser-trapped particles. *Phys. Rev. Lett.* **92,** 198104.

Block, S. M. (1992). Making light work with optical tweezers. *Nature* **360,** 493–495.

Botvinick, E. L., and Berns, M. W. (2005). Internet-based robotic laser scissors and tweezers microscopy. *Microsc. Res. Tech.* **68,** 65–74.

Chaumet, P. C., and Nieto-Vesperinas, M. (2000). Time-averaged total force on a dipolar sphere in an electromagnetic field. *Opt. Lett.* **25,** 1065–1067.

Chu, S. (1998). The manipulation of neutral particles. *Rev. Mod. Phys.* **70,** 685–706.

Cohen-Tannoudji, C. N. (1998). Manipulating atoms with photons. *Rev. Mod. Phys.* **70,** 707–719.

Courtial, J., Robertson, D. A., Dholakia, K., Allen, L., and Padgett, M. J. (1998). Rotational frequency shift of a light beam. *Phys. Rev. Lett.* **81,** 4828–4830.

Crichton, J. H., and Marston, P. L. (2000). The measurable distinction between the spin and orbital angular momenta of electromagnetic radiation. *Electron. J. Differ. Eq., Conf.* **4,** 37–50.

Draine, B. T. (1988). The discrete-dipole approximation and its application to interstellar graphite grains. *Astrophys. J.* **333,** 848–872.

Dufresne, E. R., Spalding, G. C., Dearing, M. T., Sheets, S. A., and Grier, D. G. (2001). Computer-generated holographic optical tweezer arrays. *Rev. Sci. Instrum.* **72,** 1810–1816.

Farsund, Ø., and Felderhof, B. U. (1996). Force, torque, and absorbed energy for a body of arbitrary shape and constitution in an electromagnetic radiation field. *Phys. A* **227,** 108–130.

Friese, M. E. J., Nieminen, T. A., Heckenberg, N. R., and Rubinsztein-Dunlop, H. (1998a). Optical alignment and spinning of laser trapped microscopic particles. *Nature* **394,** 348–350; [Erratum in *Nature* **395,** 621.

Friese, M. E. J., Nieminen, T. A., Heckenberg, N. R., and Rubinsztein-Dunlop, H. (1998b). Optical torque controlled by elliptical polarization. *Opt. Lett.* **23,** 1–3.

Gauthier, R. C. (1997). Theoretical investigation of the optical trapping force and torque on cylindrical micro-objects. *J. Opt. Soc. Am. B* **14,** 3323–3333.

Gouesbet, G. (1996). Exact description of arbitrary-shaped beams for use in light-scattering theories. *J. Opt. Soc. Am. A* **13,** 2434–2440.

Gouesbet, G. (1999). Validity of the localized approximation for arbitrary shaped beams in the generalized Lorenz-Mie theory for spheres. *J. Opt. Soc. Am. A* **16,** 1641–1650.

Gouesbet, G., Lock, J. A., and Gréhan, G. (1995). Partial-wave representations of laser beams for use in light-scattering calculations. *Appl. Opt.* **34,** 2133–2143.

Grier, D. G. (1997). Optical tweezers in colloid and interface science. *Curr. Opin. Coll. Interface Sci.* **2,** 264–271.

Grier, D. G. (2003). A revolution in optical manipulation. *Nature* **424,** 810–816.

Grier, D. G., and Roichman, Y. (2006). Holographic optical trapping. *Appl. Opt.* **45,** 880–887.

Griffiths, D. J. (1999). "Introduction to Electrodynamics," 3rd edn. Prentice Hall, Upper Saddle River, NJ.

Harada, Y., and Asakura, T. (1996). Radiation forces on a dielectric sphere in the Rayleigh scattering regime. *Opt. Commun.* **124,** 529–541.

Hardin, C. L. (1966). The scientific work of the Reverend John Michell. *Ann. Sci.* **22,** 27–47.

He, H., Friese, M. E. J., Heckenberg, N. R., and Rubinsztein-Dunlop, H. (1995a). Direct observation of transfer of angular momentum to absorptive particles from a laser beam with a phase singularity. *Phys. Rev. Lett.* **75,** 826–829.

He, H., Heckenberg, N. R., and Rubinsztein-Dunlop, H. (1995b). Optical particle trapping with higher-order doughnut beams produced using high efficiency computer generated holograms. *J. Mod. Opt.* **42,** 217–223.

Hoekstra, A. G., Frijlink, M., Waters, L. B. F. M., and Sloot, P. M. A. (2001). Radiation forces in the discrete-dipole approximation. *J. Opt. Soc. Am. A* **18,** 1944–1953.

Humblet, J. (1943). Sur le moment d'impulsion d'une onde électromagnétique. *Physica* **10,** 585–603.

Jackson, J. D. (1999). "Classical Electrodynamics," 3rd edn. John Wiley, New York.

Kemp, B. A., Grzegorczyk, T. M., and Kong, J. A. (2005). Ab initio study of the radiation pressure on dielectric and magnetic media. *Opt. Express* **13,** 9280–9291.

Keplero, J. (1619). "De cometis libelli tres," Augustae Vindelicorum, Augsburg (in Latin).

Knöner, G., Parkin, S., Heckenberg, N. R., and Rubinsztein-Dunlop, H. (2005). Microrheology using dual beam optical tweezers and ultrasensitive force measurements. *Proc. SPIE* **5736,** 73–80.

Knöner, G., Parkin, S., Nieminen, T. A., Heckenberg, N. R., and Rubinsztein-Dunlop, H. (2006). Measurement of refractive index of single microparticles. *Phys. Rev. Lett.* **97,** 157402.

Kuo, S. C. (2001). Using optics to measure biological forces and mechanics. *Traffic* **2,** 757–763.

Lang, M. J., and Block, S. M. (2003). Resource Letter: LBOT-1: Laser-based optical tweezers. *Am. J. Phys.* **71,** 201–215.

Leach, J., Sinclair, G., Jordan, P., Courtial, J., Padgett, M. J., Cooper, J., and Laczik, Z. J. (2004). 3D manipulation of particles into crystal structures using holographic optical tweezers. *Opt. Express* **12,** 220–226.

Ledebew, P. (1902). An experimental investigation of the pressure of light. *Astrophys. J.* **15,** 60–62.

Lock, J. A. (2004a). Calculation of the radiation trapping force for laser tweezers by use of generalized Lorenz–Mie theory. I. Localized model description of an on-axis tightly focused laser beam with spherical aberration. *Appl. Opt.* **43,** 2532–2544.

Lock, J. A. (2004b). Calculation of the radiation trapping force for laser tweezers by use of generalized Lorenz–Mie theory. II. On-axis trapping force. *Appl. Opt.* **43,** 2545–2554.

Logan, N. A. (1965). Survey of some early studies of the scattering of plane waves by a sphere. *Proc. IEEE* **53,** 773–785.

Lorenz, L. (1890). Lysbevægelsen i og uden for en af plane Lysbølger belyst Kugle. *Videnskabernes Selskabs Skrifter* **6,** 2–62.

Maia Neto, P. A., and Nussenzveig, H. M. (2000). Theory of optical tweezers. *Europhys. Lett.* **50,** 702–708.

Maxwell, J. C. (1865). A dynamical theory of the electromagnetic field. *Philos. Trans. R. Soc. Lond.* **155,** 459–512.

Maxwell, J. C. (1873). "A Treatise on Electricity and Magnetism." (in two volumes) Clarendon Press, Oxford.

Mazolli, A., Maia Neto, P. A., and Nussenzveig, H. M. (2003). Theory of trapping forces in optical tweezers. *Proc. R. Soc. Lond. A* **459,** 3021–3041.

Mie, G. (1908). Beiträage zur Optik trüber Medien, speziell kolloidaler Metallösungen. *Annalen der Physik* **25,** 377–445.

Mohanty, S. K., Uppal, A., and Gupta, P. K. (2004). Self-rotation of red blood cells in optical tweezers: Prospects for high throughput malaria diagnosis. *Biotechnol. Lett.* **26,** 971–974.

Molloy, J. E., Dholakia, K., and Padgett, M. J. (2003). Optical tweezers in a new light. *J. Mod. Opt.* **50,** 1501–1507. (preface to special issue).

Nichols, E. F., and Hull, G. F. (1901). A preliminary communication on the pressure of heat and light radiation. *Phys. Rev.* **13,** 307–320.

Nichols, E. F., and Hull, G. F. (1902). Pressure due to light and heat radiation. *Astrophys. J.* **15,** 62–65.

Nichols, E. F., and Hull, G. F. (1903a). The pressure due to radiation. *Phys. Rev.* **17,** 26–50, 91–104.

Nichols, E. F., and Hull, G. F. (1903b). The pressure due to radiation. *Astrophys. J.* **17,** 315–351.

Nieminen, T. A. (2004). Comment on "Geometric absorption of electromagnetic angular momentum," C. Konz, G. Benford. *Opt. Commun.* **235,** 227–229.

Nieminen, T. A., Heckenberg, N. R., and Rubinsztein-Dunlop, H. (2001a). Optical measurement of microscopic torques. *J. Mod. Opt.* **48,** 405–413.

Nieminen, T. A., Rubinsztein-Dunlop, H., Heckenberg, N. R., and Bishop, A. I. (2001b). Numerical modelling of optical trapping. *Comput. Phys. Commun.* **142,** 468–471.

Nieminen, T. A., Rubinsztein-Dunlop, H., and Heckenberg, N. R. (2003a). Calculation of the *T*-matrix: General considerations and application of the point-matching method. *J. Quant. Spectrosc. Radiat. Transfer* **79–80,** 1019–1029.

Nieminen, T. A., Rubinsztein-Dunlop, H., and Heckenberg, N. R. (2003b). Multipole expansion of strongly focussed laser beams. *J. Quant. Spectrosc. Radiat. Transfer* **79–80,** 1005–1017.

Nieminen, T. A., Heckenberg, N. R., and Rubinsztein-Dunlop, H. R. (2004a). Angular momentum of a strongly focussed Gaussian beam. arXiv:physics/0408080.

Nieminen, T. A., Heckenberg, N. R., and Rubinsztein-Dunlop, H. R. (2004b). Computational modelling of optical tweezers. *Proc. SPIE* **5514,** 514–523.

Nieminen, T. A., Parkin, S. J., Heckenberg, N. R., and Rubinsztein-Dunlop, H. (2004c). Optical torque and symmetry. *Proc. SPIE* **5514,** 254–263.

Nieminen, T. A., Higuet, J., Knöner, G., Loke, V. L. Y., Parkin, S., Singer, W., Heckenberg, N. R., and Rubinsztein-Dunlop, H. (2006). Optically driven micromachines: Progress and prospects. *Proc. SPIE* **6038,** 237–245.

O'Neil, A. T., MacVicar, I., Allen, L., and Padgett, M. J. (2002). Intrinsic and extrinsic nature of the orbital angular momentum of a light beam. *Phys. Rev. Lett.* **88,** 053601.

Padgett, M. J. (2004). The mechanism for energy transfer in the rotational frequency shift of a light beam. *J. Opt. A: Pure Appl. Opt.* **6,** S263–S265.

Pfeifer, R. N. C., Nieminen, T. A., Heckenberg, N. R., and Rubinsztein-Dunlop, H. (2006). Two controversies in classical electromagnetism. *Proc. SPIE* **6326,** 93260H.

Phillips, W. D. (1998). Laser cooling and trapping of neutral atoms. *Rev. Mod. Phys.* **70,** 721–741.

Poynting, J. H. (1884). On the transfer of energy in the electromagnetic field. *Philos. Trans. R. Soc. Lond.* **175,** 343–361.

Poynting, J. H. (1909). The wave motion of a revolving shaft, and a suggestion as to the angular momentum in a beam of circularly polarised light. *Proc. R. Soc. Lond. A* **82,** 560–567.

Quake, S. R., Babcock, H., and Chu, S. (1997). The dynamics of partially extended single molecules of DNA. *Nature* **388,** 151–154.

Ren, K. F., Gréhan, G., and Gouesbet, G. (1996). Prediction of reverse radiation pressure by generalized Lorenz-Mie theory. *Appl. Opt.* **35,** 2702–2710.

Rubinsztein-Dunlop, H., Nieminen, T. A., Friese, M. E. J., and Heckenberg, N. R. (1998). Optical trapping of absorbing particles. *Adv. Quantum Chem.* **30,** 469–492.

Sato, S., Ishigure, M., and Inaba, H. (1991). Optical trapping and rotational manipulation of microscopic particles and biological cells using higher-order mode Nd:YAG laser beams. *Electron. Lett.* **27,** 1831–1832.

Stewart, A. M. (2005). Angular momentum of the electromagnetic field: The plane wave paradox resolved. *Eur. J. Phys.* **26,** 635–641.

Stratton, J. A. (1941). "Electromagnetic Theory." McGraw-Hill, New York.

Ukita, H., and Nagatomi, K. (1997). Theoretical demonstration of a newly designed micro-rotator driven by optical pressure on a light incident surface. *Opt. Rev.* **4,** 447–449.

van de Hulst, H. C. (1957). "Light Scattering by Small Particles." Wiley, New York.

Visscher, K., and Brakenhoff, G. J. (1992). Theoretical study of optically induced forces on spherical particles in a single beam trap. I. Rayleigh scatterers. *Optik* **89,** 174–180.

Waterman, P. C. (1965). Matrix formulation of electromagnetic scattering. *Proc. IEEE* **53,** 805–812.

Waterman, P. C. (1971). Symmetry, unitarity, and geometry in electromagnetic scattering. *Phys. Rev. D* **3,** 825–839.

Wright, W. H., Sonek, G. J., Tadir, Y., and Berns, M. W. (1990). Laser trapping in cell biology. *IEEE J. Quantum Electron.* **26,** 2148–2157.

Wright, W. H., Sonek, G. J., and Berns, M. W. (1993). Radiation trapping forces on microspheres with optical tweezers. *Appl. Phys. Lett.* **63,** 715–717.

Wright, W. H., Sonek, G. J., and Berns, M. W. (1994). Parametric study of the forces on microspheres held by optical tweezers. *Appl. Opt.* **33,** 1735–1748.

Zambrini, R., and Barnett, S. M. (2005). Local transfer of optical angular momentum to matter. *J. Mod. Opt.* **52,** 1045–1052.

PART III

Laser Scissors/Ablation

CHAPTER 7

Laser Microsurgery in the GFP Era: A Cell Biologist's Perspective

Valentin Magidson,* Jadranka Lončarek,* Polla Hergert,* Conly L. Rieder,*[,†,‡] and Alexey Khodjakov*[,†,‡]

*Division of Molecular Medicine, Wadsworth Center, Albany, New York 12201

[†]Department of Biomedical Sciences, SUNY, Albany, New York 12222

[‡]Marine Biological Laboratory, Woods Hole, Massachusetts 02543

Modern biology is based largely on a reductionistic "dissection" approach—most cell biologists try to determine how complex biological systems work by removing their individual parts and studying the effects of this removal on the system. A variety of enzymatic and mechanical methods have been developed to dissect large cell assemblies like tissues and organs. Further, individual proteins can be inactivated or removed within a cell by genetic manipulations (e.g., RNAi or gene knockouts). However, there is a growing demand for tools that allow intracellular manipulations at the level of individual organelles. Laser microsurgery is ideally suited for this purpose and the popularity of this approach is on the rise among cell biologists. In this chapter, we review some of the applications for laser microsurgery at the subcellular level and describe practical requirements for laser microsurgery instrumentation demanded in the field. We also outline a

0091-679X/07 $35.00
DOI: 10.1016/S0091-679X(06)82007-8

relatively inexpensive but versatile laser microsurgery workstation that is being used in our laboratory. Our major thesis is that the limitations of the technology are no longer at the level of the laser, microscope, or software, but instead only in defining creative questions and in visualizing the target to be destroyed.

> At last in an incredible manner he [Archimedes] burned up the whole Roman fleet. For by tilting a kind of mirror toward the sun he concentrated the sun's beam upon it; and owing to the thickness and smoothness of the mirror he ignited the air from this beam and kindled a great flame, the whole of which he directed upon the ships that lay at anchor in the path of the fire, until he consumed them all.[1]

I. History of the Field

A. The Genesis of "Micro-Photo-Surgery"

The origins of Cell Biology as a branch of "natural philosophy" can be traced to the English polymath Robert Hooke who, using a hand-crafted, leather and gold-tooled compound light microscope (LM), published a book containing elaborate drawings of magnified objects which he called *Micrographia*. In this book, which became an immediate best-seller (and has been reprinted countless times), Hooke used the term "cell" to describe the repeating units seen in magnified slices of cork that resembled the monk cells of a monastery. Ironically, these repeating units were not actual cells but rather just the cellulose walls that surround cells in plant tissues. It took another 175 years of optical development and exploration, before Schleiden and Schwann (1839) convincingly asserted that cells are the fundamental building blocks of all life. In 1855, the Prussian physician and politician Virchow postulated that cells arise only from preexisting cells by reproduction and cannot be formed *de novo* from amorphous "living matter" or "protoplasma." This principle, which Virchow eloquently formulated in Latin: "*Omnis cellula e cellula* (every cell [stems] from a cell) became a key principle of modern biology. Virchow also viewed the body as a "cell state in which each cell is a citizen," and he considered disease to be simply "a conflict between the citizens of the state, caused by outer forces" (http://www.whonamedit.com/doctor.cfm/912.html). This breakthrough concept, coupled with the widespread availability of the compound LM, resulted in a new science of "cytology" in the late nineteenth century that was initially dedicated to categorizing the various types of cells and their sometimes-visible contents. The subsequent rapid accumulation of morphological data generated by this endeavor spawned an immense appreciation as to the true complexity and diversity of cells, which was brilliantly summarized by E. B. Wilson in his fundamental opus *The Cell in Development and Heredity* (Wilson, 1925).

[1] *Dio's Roman History*, translated by Earnest Cary Loeb Classical Library, Harvard University Press, Cambridge, 1914, Vol. II, p. 171.

By the early twentieth century, methods were available for removing many types of cells from an organism and culturing them as individuals in dishes (http://caat.jhsph.edu/pubs/animal_alts/appendix_c.htm). Near this time, cytologists also began seeking ways to dissect cells, so that they could determine the relationships between, and functions of their constituents. Toward this end, it quickly became apparent that approaches based on mechanical dissection were seldom successful, since they usually killed the specimen by rupturing the surrounding membrane (or cell wall). The search was on for a noninvasive approach that would allow scientists to selectively remove or destroy individual intracellular components without killing the cell outright.

Although the opening description of Archimedes' use of focused sunlight in 212 BC to destroy the Roman navy may be fanciful, the notion that light can be used to destroy objects is not. Like the concave mirrors allegedly used by Archimedes, microscopes also can focus light of a powerful illuminator on a tiny spot whose size is limited only by diffraction. Thus, it was natural for those working with microscopes to ultimately apply Archimedes' principle to manipulate cells. The first to intentionally use a focused light beam to destroy chosen cellular components appears to be Sergey Tschachotin (1883–1973), who in 1912 developed a method which came to be called "micro-photo-surgery" (Tschachotin, 1912). In his initial approach (Fig. 1), Tschachotin routed an appropriate wavelength of UV light through a quartz prism into the condenser lens of an LM. He controlled the area illuminated within the specimen by placing an aperture of the appropriate diameter in the object-conjugated plane (Fig. 1B). The only high-intensity light available to Tschachotin at the time was that of magnesium sparks (Fig. 1C), which were then commonly used in photography. Fortunately, magnesium sparks contain a heavy 280-nm (UV) line that proved to be very useful for micro-photo-surgery. However, Tschachotin had to first solve the nontrivial problem of how to determine the position of the invisible UV microbeam in the field of view. He did this by using drops of fluorescein on a standard glass slide to see the beam via its induced fluorescence in the visible spectrum, and marked the position of the focused beam on the microscope eyepiece. Once this calibration step was completed, Tschachotin substituted the fluorescein slide with slides of real cells and the cell of choice was moved into position under the beam. In later years, Tschachotin used mercury arc lamps as the light source and uranium glass to visualize the UV. Remarkably, we still use many of the tricks Tschachotin developed almost a century ago for routine alignment of the laser microsurgery workstation housed in our laboratory.

Tschachotin's list of achievements, which span over 60 years, is remarkable, and many of his experiments have become standard repeats for each new generation of microbeam researchers. Not only did he conduct the first study on the reaction of cells to different wavelengths of irradiation, but he also reported that micro-irradiation of sea urchin eggs induces parthenogenesis. He discovered photosensitization by noting that irradiating *Paramecia* with 310-nm light did not induce any detectable reaction unless they were first incubated in eosin. In a remarkable

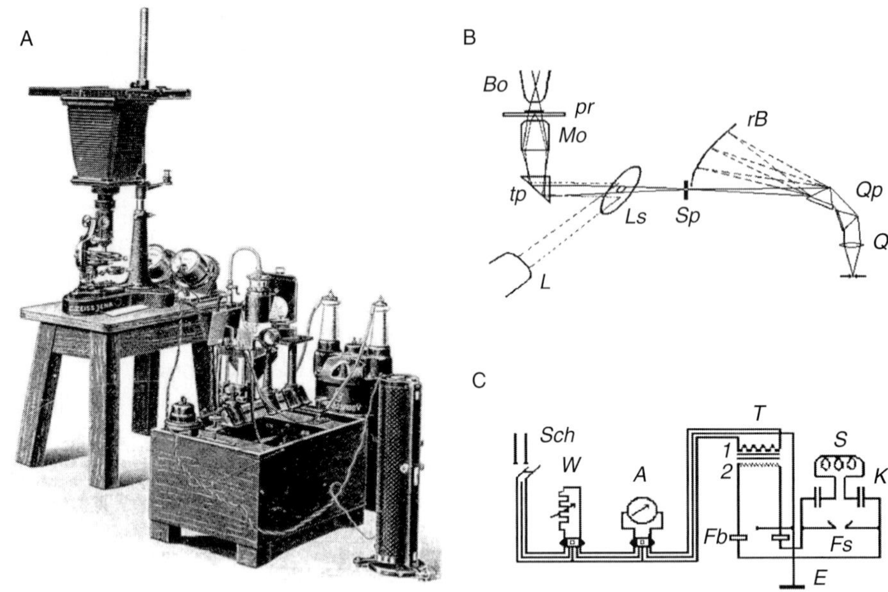

Fig. 1 One of the early micro-photo-surgery systems of Sergey Tschachotin. (A) Overview of the instrument. (B) General optical layout of the system: light generated by magnesium sparks is monochromatized by quartz prisms and selected wavelengths directed toward the microscope condenser. An aperture controls the size of the irradiated area. (C) Spark generator. Magnesium powder was ignited by the electrodes (Fb). The electric layout included a switch (Sch) for the step-up transformer (T) with a rheostat (W) and Amperemeter (labeled A), solenoid inductor (S) with two capacitors (labeled K) and, for safety reasons, an alternate spark site (Fs). Adapted from Tschachotin, 1938.

live-cell study, he proved that the pigment spot seen in *Euglena gracilis* is responsible for *Euglena*'s reaction to light (i.e., it is the "eye" of the cell). Most of Tschachotin's work was done in Italy and France, although he also worked in Denmark, Germany, Croatia, and the United States. When Hitler's forces occupied France (1939), he was thrown into a concentration camp for writing *Le Viol des Foules par la Propagande Politique* (The Rape of the Crowds by Political Propaganda). He was released 8 months later after a direct petition from prominent German cytologists. In 1958, Tschachotin returned to Russia where he died in 1973. Because Tschachotin remained active in science until his departure, he was able to witness in his latter years the transformation of his micro-photo-surgery approach into laser microsurgery (Posudin, 1995).

B. The Middle Years: Laser-Based Microirradiation

As ingenious as Tschachotin was, the success of his work was limited by the fact that the brightfield optics used in microscopes during his time generated very limited contrast in live nonstained cells. As a result, the number of components (organelles) that could be clearly delineated within the cell was usually restricted to two: the nucleus and the cytoplasm. Tschachotin's experiments on the eye (pigment spot) of *E. gracilis* were, in a way, forced on him because the eye is the only naturally opaque organelle in this protist. Indeed, the fact that one cannot selectively destroy a target if it is not visible became the major obstacle to a more widespread use of micro-photo-surgery.

The "can't see, can't destroy" stalemate changed abruptly in the early 1950s with Frits Zernike's invention of phase-contrast LM, for which he won the Nobel Prize in 1953 (Zernike, 1955). Shortly thereafter other methods for generating contrast in living cells were also developed to a useful state, including polarization and differential interference contrast microscopy. These new optics allowed many organelles and subcellular structures to be visualized within the nucleus and cytoplasm of living cells including the nucleoli, mitochondria, stress fibers, and cilia/flagella of interphase cells, as well as the chromosomes and spindle fibers of dividing cells. When coupled in a creative manner to a UV microbeam and a cinematographic system, these new imaging modes allowed researchers to cut or destroy selected components, and then to follow the subsequent behavior of the cell. Zirkle and colleagues (Uretz *et al.*, 1954; Zirkle and Bloom, 1953), for example, combined a UV microbeam with a phase-contrast microscope to prove (the already well-known fact) that destroying the kinetochore region of a chromosome inhibits motion of the chromosome. In a more biologically successful project, Forer (1965) combined a UV microbeam and polarization light microscopy to show that once severed, spindle (kinetochore) fibers in the cranefly spermatocyte regrew from the chromosome to the spindle pole. These original experiments have since been repeated on more than one occasion, although on different cell types, with increasingly sophisticated UV and then visible laser microbeams (LaFountain, Jr., *et al.*, 2001; Maiato *et al.*, 2004; Spurck *et al.*, 1990).

The ability to visualize and thus target for destruction of many intracellular organelles resurrected biologist's interest in micro-photo-surgery. However, it immediately became apparent that there was a major problem of using lamp-generated UV light for this approach: since chromatin (DNA) efficiently absorbs 280-nm light, all cellular systems are extremely sensitive to UV, which leads to unavoidable nonspecific (and not always easy to define) side effects (reviewed in Khodjakov *et al.*, 1997b). This problem has been largely overcome with modern UV laser-based microirradiation system that allows the beam to be tightly focused (Colombelli *et al.*, 2005). However, in microirradiation systems based on conventional light sources the irradiated area was defined by the size of the aperture used in the conjugated plane, and this often exceeded that required for the specific task. Thus, in the middle years "cell surgeons" could see their targets with relative clarity, but they still had to operate with "dull" scalpels. This dullness prompted investigators to experiment with alternative irradiation sources, often choosing those that simply already existed in a laboratory. These included, for example, the proton and α-particle beams of Zirkle and Bloom (1953) and Winson (1965), respectively. The latter study is a perfect example of what biologists are willing to cope with on the quest to understand how cells work. Because α-particles cannot be focused by light-microscope optics, Winson and Kuzin (1965) positioned slices of mica with small (several micrometer in diameter) holes in front of living cells so that certain parts of a cell were protected from the beam while others were exposed. This elaborate approach revealed that the deleterious effects of α-particles were due to their interaction with the nucleus (i.e., DNA damage). However, although potentially useful, proton and α-particle irradiation systems were so cumbersome that they had little chance of becoming a standard technique in cell biology.

A significant breakthrough in micro-photo-surgery came with the invention of lasers in 1959, and their commercialization in the 1960s, which provided a ready source of powerful and highly focusable light beams. The first laser "microirradiation" study on living cells can be traced to 1962 when Bessis *et al.* (1962), working in Paris (France), conducted a series of investigations on the effects of irradiating cell structures with a low-power Ruby Red laser. The conclusions of this work were that, in essence, nonnaturally pigmented cells did not respond to irradiation unless they were first sensitized by adding an exogenous chromophore (e.g., acridine orange, acridine red, alcian blue, psoralens, coumarins, Janus B green). This conclusion was then confirmed at the EM level, which also revealed that under the appropriate vital staining and laser power conditions, restricted and selective damage was created at the irradiated site (Storb *et al.*, 1966).

In 1969, Michael Berns and his colleagues at the University of California (Irvine) showed, using an argon laser coupled to a phase-contrast microscope, that very small lesions could be easily placed at predetermined sites on selected chromosomes (Berns *et al.*, 1969). Encouraged by their initial successes, this team began a series of studies, based on UV and later visible spectrum laser beams, on how cells react to the selective removal of various structures from the nucleolar organizer/primary constriction (Berns and Cheng, 1971; Berns *et al.*, 1970b, 1972; Ohnuki *et al.*, 1972) to the centrosome region (Berns and Richardson, 1977; Berns *et al.*, 1977; Peterson and Berns, 1978). These studies were conducted on cells sensitized with chemical fluorophores such as acridine orange. However, Strahs and Berns (1979) discovered that stress fibers, mitochondria, and other organelles could be selectively cut or destroyed by 100- to 150-ns pulses of 532- or 537-nm (green) laser light, obtained from a Q-switched neodymium-doped yttrium aluminum garnet (Nd:YAG) laser, without any prior sensitization treatment. Under these circumstances the lesions created were identical to those produced by UV irradiation. The salient conclusion of this landmark study was that the interaction of light with biological systems is nonspecific and not restricted to a particular class of molecules. Thus, short pulses of highly focusable visible-spectrum laser light can be used to ablate a wide range of cell components.

The exact mechanism by which short pulses of visible laser light destroy biological components in the absence of photosensitization remains controversial (see other chapters in this book and a brief discussion later in this chapter). Regardless of the mechanism, lasers can be used to selectively destroy any structure clearly visible by light microscopy in a living cell with minimum collateral damage and without need for prior sensitization. Once this became clear, researchers began to use laser microsurgery to ablate most of the more conspicuous organelles within tissue culture cells, often just to prove that they could. A prime example here is the mitochondrion: when viewed by phase-contrast or differential contrast microscopy (DIC) LM, these thin wormlike structures suddenly disappear (i.e., the contrast between the structure and the surroundings is lost) when irradiated with laser pulses of sufficient energy (e.g., a single 7-ns 532-nm pulse from the workstation used in our laboratory). This occurs due to localized rupture of the mitochondrial membranes so that constituents of the mitochondrion are expelled into the cytoplasm (Fig. 2; also see Khodjakov *et al.*, 2004b). Since first reported by Berns *et al.* (1970a), this result has been subsequently confirmed by many others (Adkisson *et al.*, 1973; Moreno *et al.*, 1973; Rattner *et al.*, 1976; Salet, 1972). In fact, it appears that punching holes in mitochondria with new and improved lasers, without providing any additional insight into how these organelles work, has become a

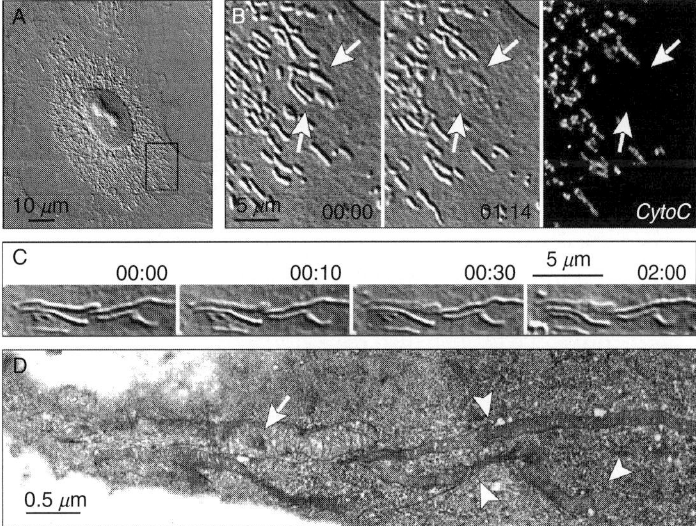

Fig. 2 Rupturing the membrane of a single mitochondrion by laser microsurgery. (A) An interphase CV-1 cell (green monkey kidney) in culture. (B) Enlarged view of the area boxed in (A). Individual mitochondria appear as wormlike structures several micrometers long and diffraction limited in width. Irradiation of mitochondria with 7-ns pulses of 532-nm laser light results in the disappearance of the refractive-index gradient between the irradiated mitochondrion and surrounding cytoplasm [arrows indicate a group of five mitochondria individually irradiated (1 pulse/mitochondrion) between 00:00 and 01:14]. Immunostaining reveals that microirradiation results in the disappearance of intramitochondrial proteins, like cytochrome *C*, from the irradiated mitochondria. (C) A single mitochondrion was irradiated in a PtK1 cell (rat kangaroo kidney) (arrow in 00:00). (D) Same-cell serial-section electron microscopy analysis reveals that the irradiated mitochondrion is swollen (arrow) and its matrix is much less dense than in the surrounding, nonirradiated mitochondria (arrowheads). This is consistent with the loss of intramitochondrial proteins from the irradiated mitochondria. Time in minutes:seconds. See Khodjakov *et al.* (2004b) for more details.

litmus test for proving the utility of a new system. Just last year there were at least three independent reports of this exact experiment (Colombelli *et al.*, 2005; Shen *et al.*, 2005; Shimada *et al.*, 2005).

Initially, the complexity of laser microirradiation systems restricted their distribution to just a few institutions specializing primarily in laser physics. However, in early 1980s Michael Berns started the *la*ser *m*icrobeam *p*rogram (LAMP) at the University of California (Irvine) which was (and remains) sponsored by the NIH Center for Research Resources as a National Biotechnology Resource. This facility provided Cell Biologists throughout the country with an opportunity to explore the applicability of laser microsurgery to their particular research programs. Giving biologists with little background in Physics unrestricted access to costly and sophisticated laser microsurgery workstations rapidly led to important new biological findings. In a very influential paper, for example, McNeill and Berns (1981) showed, by selectively destroying just one of the two sister kinetochores on a prometaphase chromosome, that the velocity with which a kinetochore moves is independent of the mass associated with it. This study also implied that the mechanism that moves chromosomes during spindle assembly is the same that moves them poleward during anaphase. In another notable study conducted at the LAMP facility, Rieder *et al.* (1986) reported that severed chromosome arms are ejected from the forming mitotic spindle in animal cells, meaning that they are under a constant away-from-the-pole pushing force (i.e., so called polar winds or polar ejection force; Fig. 3). Since their discovery, the polar ejection forces have become an important part of modeling how chromosome position is governed during mitosis (reviewed in Kapoor and Compton, 2002). Subsequent laser microsurgery studies proved that the forces acting on chromosome arms differ dramatically between animal cells, where spindle assembly is driven by the centrosome, and plant cells that lack this organelle (Fig. 4; also see Khodjakov *et al.*, 1996).

In the early 1990s, a DIC-based Nd:YAG laser microsurgery workstation was constructed in the Rieder laboratory at the Wadsworth Center (Albany, NY). This system (Cole *et al.*, 1995) was patterned after the phase-contrast systems developed by Berns and colleagues, and it quickly proved that ready access to laser microsurgery could be an enormous benefit to a group of cell biologists. The Rieder laboratory had a long-standing interest in studying mitosis, cell cycle regulation, and the microtubule cytoskeleton. The extensive biological experience of this group allowed its members to formulate a number of questions that could only be answered by laser microsurgery. During the next several years this workstation was used to demonstrate, for example, that the spindle assembly checkpoint monitors kinetochore attachment (Rieder *et al.*, 1995), that chromosomes containing a single kinetochore can congress to the equator of the forming spindle (Khodjakov *et al.*, 1997a), and that entry into mitosis in vertebrate cells is guarded by a DNA damage checkpoint that reverses the cell cycle when triggered during early prophase (Rieder and Cole, 1998). Further, this same instrument was also used by other biologists to address a number of questions in systems as diverse as fungi to cranefly (Inoue *et al.*, 1998; LaFountain, Jr., *et al.*, 2001, 2002; Orokos *et al.*, 2000).

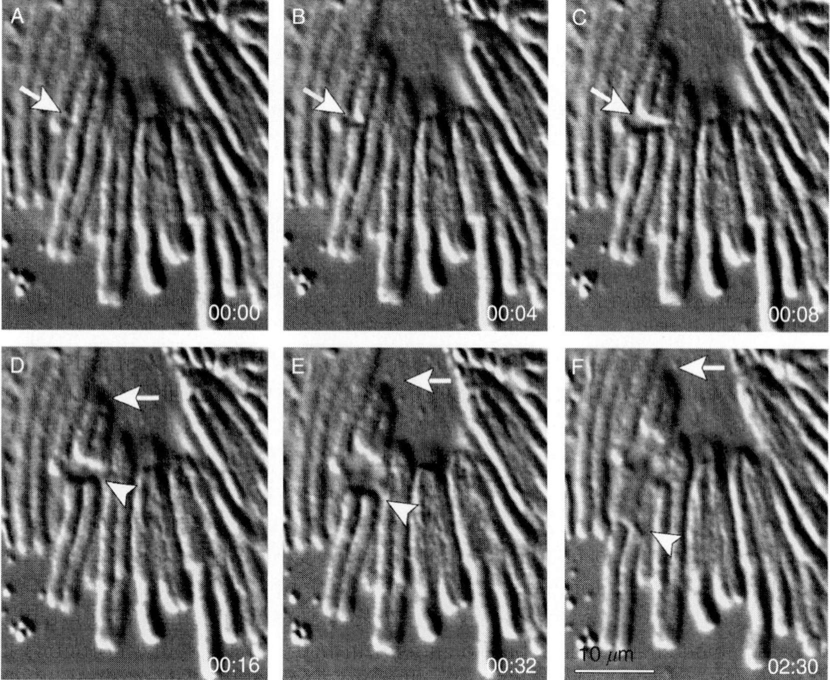

Fig. 3 Severing chromosome arms during mitosis in an animal (newt lung) cell. Selected frames from a time-lapse recording of an early prometaphase cell containing a monopolar spindle. 7-ns, 532-nm, 10-Hz laser pulses were used to separate the arms of one chromosome (arrows in A–C) from the centromere. The operation took ~5 s (~50 laser pulses). Once severed, the chromosome arms were ejected away from the spindle pole (arrowheads in D–F), while the central fragment containing the kinetochore moved closer to the spindle pole (arrows in D–F). Experiments like this proved the existence of a "spindle ejection force" or "polar winds" that act on chromosome arms (Rieder *et al.*, 1986). Time in minutes:seconds.

As emphasized early on by Berns *et al.* (1981), Gaussian laser beams can actually generate a central "hot spot" inside the Airy disk. Thus, it is possible to select a beam energy at which damage to the specimen is restricted only to the "hot spot" at the peak intensity in the center of the Airy disk. This allows the resolution of laser microsurgery to surpass the Raleigh criterion which restricts the resolution of light microscopy. In practical turns, this means that lasers have been developed to the point where the "sharpness of the scalpel" ceases to be a limiting factor. With a properly conditioned laser beam and minimal practice, it is relatively easy to destroy every target that can be clearly delineated within a cell. Thus, by the late 1990s, the major remaining impediment in laser microsurgery became the fact that, even with modern contrast-enhanced DIC or phase-contrast microscopes, researchers are limited in their ability to see organelles in live cells. For this reason, most of the biologically meaningful experiments conducted during the "middle years" of micro-photo-surgery were aimed at solving problems related to large and/or high-contrast

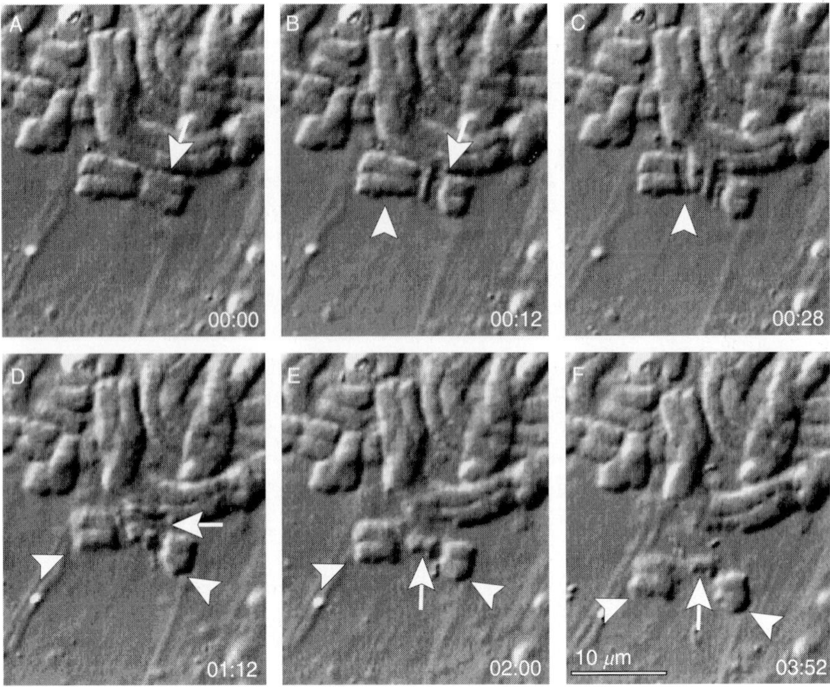

Fig. 4 Generating chromosome fragments with and without kinetochores during mitosis in a plant (endosperm of lily) cell. In this example, the laser microbeam was first used to sever the chromosome arms on the right (arrows in A and B) and left (arrowheads in B and C) sides of the centromere region, and then to slice the centromere in between the sister kinetochores (arrow in D). The entire operation (three cuts) took about 1 min and required ~200 laser pulses (7-ns, 532-nm). In sharp contrast to animal cells (Fig. 3), in plants chromosome fragments containing kinetochores (arrows in D–F) as well as the chromosome arms (arrowheads in D–F) move toward the spindle pole with similar velocities. This reflects a dramatic difference in the distribution of forces during mitosis in plants and animals (Khodjakov *et al.*, 1996). Time in minutes:seconds.

structures like chromosomes, nuclei, nucleoli, and mitochondria. Although, there were multiple attempts throughout the 1980s and 1990s to operate on other less conspicuous organelles like the centrosome (Berns and Richardson, 1977; Hyman, 1989; Koonce *et al.*, 1984), the interpretation of these studies was always clouded by the fact that centrosomes are not visible, and thus their boundaries cannot be defined, in living vertebrate somatic cells. This, in turn, meant that the success or failure of the experiment could only be evaluated after the fact by fixing the cell for a subsequent serial-section electron microscopy analyses. Not surprisingly, the intense labor behind this same cell correlative LM/EM approach (Rieder and Cassels, 1999) severely limited the range of useful questions that could be cleanly answered by laser microsurgery.

In summary, by the end of the twentieth century it was evident that to extend the utility of laser microsurgery to the cell biologists a more direct method was needed to visualize components in living cells that were otherwise not visible because of their small size and/or physical properties, as well as to instantaneously assay the success or failure of an operation.

C. The Modern Era: A Synergy of Laser Microsurgery and GFP Imaging

In 1992, Doug Prasher and colleagues, working at the Woods Hole Oceanographic Institute (Massachusetts) successfully cloned green fluorescent protein (GFP), the small (238 amino acid) molecule responsible for the bioluminescence of the jellyfish *Aequorea victoria* (Prasher *et al.*, 1992). Shortly thereafter Martin Chalfie, Doug Prasher, and others reported that GFP can be used for monitoring gene expression in prokaryotic (*Escherichia coli*) and eukaryotic (*Caenorhabditis elegans*) cells (Chalfie *et al.*, 1994). They also predicted that it could be fused with other proteins to report their presence and location. The next year this prediction became a reality when several groups demonstrated that GFP-chimeras could be successfully used to illuminate mitochondria (Rizzuto *et al.*, 1995). This started the "GFP revolution" in Cell Biology. The extent and speed with which this revolution changed how biologists study cells is readily apparent from a simple search of databases like PubMed. As of August of 2006, queries for GFP yielded ~20,000 experimental papers and about 600 reviews!

Since 1995, a large number of fluorescent proteins have been constructed and their utility for studying cells demonstrated (see Giepmans *et al.*, 2006 for review). This fluorescence-tagging technology can be used to visualize practically any macromolecular assembly in living cells ranging from yeast to vertebrates (Fig. 5). For our purposes it makes otherwise invisible small organelles suitable targets for laser microsurgery.

In 1997, the first proof-of-concept paper describing the laser ablation of GFP-labeled organelles was published (Khodjakov *et al.*, 1997b). From this study it was clear that combining GFP-labeling with laser microsurgery produces a synergistic approach that allows one to achieve the precision of laser ablation that was never dreamt possible. Clear examples of this capability are illustrated by a series of studies that we conducted on the centrosomes in mammalian cells.

The centrosome is a minute organelle (Fig. 6) that, although absent in higher plants, is present as a single copy in all animal somatic cells (Ou and Rattner, 2004). When present, it acts as the principal microtubule-organizing center, but this function is not essential since a normal microtubule array can be assembled via centrosome-independent mechanisms. Yet, since mutations in core centrosomal components are lethal to the organism, this organelle clearly plays one or more essential vital functions (Basto *et al.*, 2006; Bettencourt-Dias *et al.*, 2005) which many generations of cell biologists have sought to identify. Another mystery of the centrosome is that it is built around two "centrioles"—complex macromolecular assemblies that replicate in a typical semiconservative fashion. As a result of this replication pattern, each centrosome contains one older (mother) centriole that was formed at least two cell cycles previously and one younger (daughter) centriole that

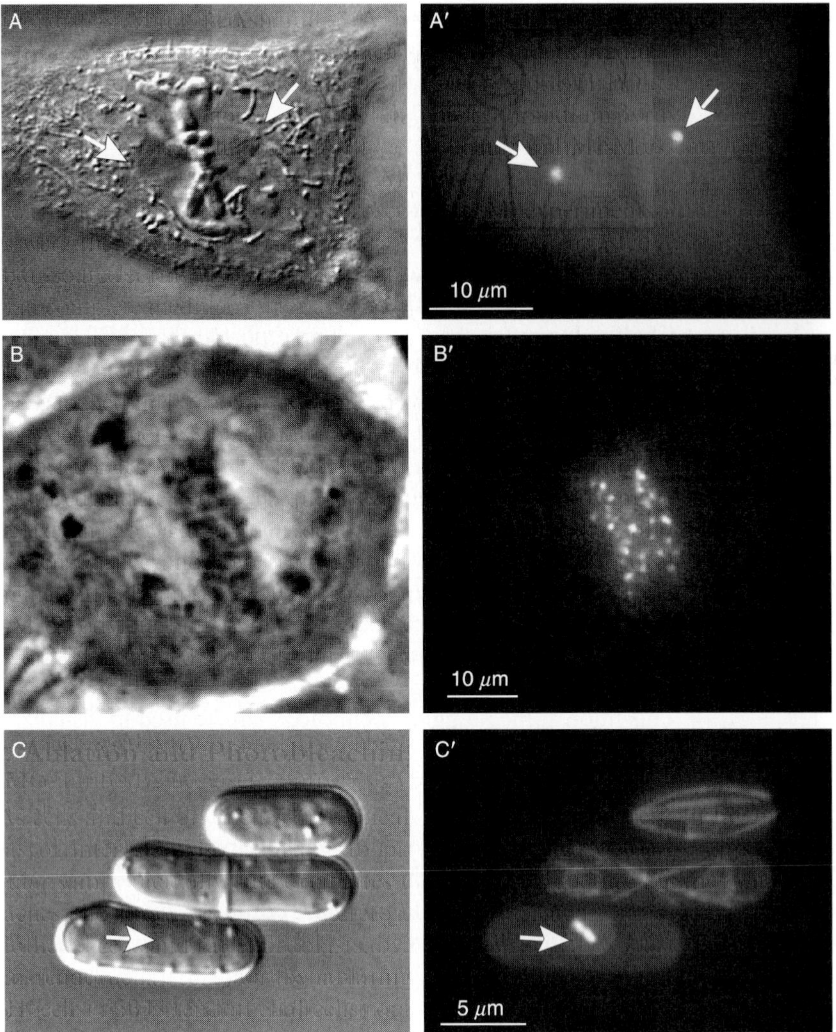

Fig. 5 Examples of normally invisible intracellular structures that can be readily seen after GFP-labeling. (A–A′) A PtK1 (rat kangaroo kidney) cell in mitosis as visualized by DIC (A) or fluorescence (A′) microscopy. Although the approximate position of the two centrosomes in mitotic cells can be inferred from the DIC image (arrows), expression of a γ-tubulin/GFP fusion precisely delineates their boundaries (A′). (B–B′) A U2OS (human osteosarcoma) cell in metaphase of mitosis, as viewed by phase-contrast (B) and fluorescence (B′) microscopy. Normally kinetochores are not visible by phase-contrast or DIC LM (B). However, after labeling with a CENP-B/GFP fusion protein they appear as paired bright dots associated with the chromosomes (B′). (C–C′) Neither microtubules nor nuclei are reproducibly seen in yeast (S. pombe) cells by DIC (C). However, simultaneously expressing Tub1(α-tubulin)/GFP and Uch2p (ubiquitin C-terminal hydrolase)/GFP fusion proteins in these cells clearly reveals these structures (C′). Arrows indicate the position of intranuclear mitotic spindle which is undetectable in DIC but clearly delineated in fluorescence. (A, B, B′, and C) = individual focal planes. (A′ and C′) = maximal-intensity projections through the entire cell volume collected at 0.2-μm Z-steps.

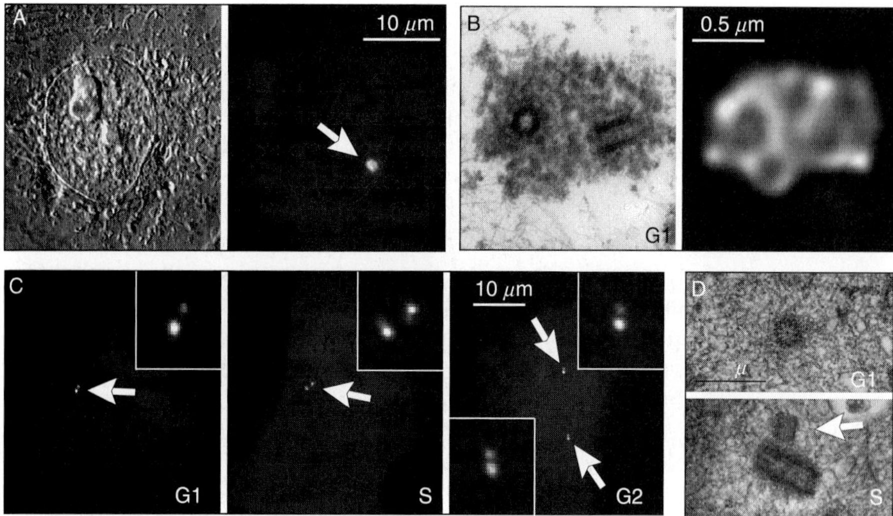

Fig. 6 Structural organization of the centrosome in vertebrate somatic cells. (A) During interphase the centrosome is not detectable in live cells by DIC (left) but can be visualized via expression of GFP fusion proteins, for example, γ-tubulin/GFP (right). PtK1 cell. (B) During G1 the centrosome consists of two cylindrical structures, termed centrioles (left) surrounded by a cloud of amorphous pericentriolar material (PCM). γ-Tubulin (along with other components responsible for microtubule nucleation) resides in the PCM (right). Left image is a 500-nm thick section of a PtK1 cell treated with 10-μM Nocodazole (to depolymerize microtubules) and permeabilized with Triton X-100 prior to fixation. Right image is a maximal intensity projection of the same centrosome stained with an anti-γ-tubulin antibody. The raw dataset collected with 100× 1.4 NA PlanApo lens was deconvolved using super-resolution algorithms in the "AutoDeblur" software (AutoQuant, Watervliet, NY). (C) Centrioles visualized via expression of centrin-1/GFP fusion in CHO-K1 cells. G1 cells contain two individual centrioles. As cells enter S period the centrioles replicate and centrin dots become doubled (inset in S). During G2 daughter centrioles elongate which is manifested by increasing separation between the doubled centrin dots (insets in S and G2). (D) An individual centriole during G1 and replicating centrioles during mid-S in 100-nm thin EM section. Arrow indicates a short daughter centriole attached to the wall of the mother.

was formed in the last cell cycle. The list of mysteries associated with the centrosome is extensive (it has been called *The Central Enigma of Cell Biology*—Wheatley, 1982) and, as outlined above (also see Uzbekov *et al.*, 1995), there have been numerous unsuccessful attempts to remove it from living cells via UV and later laser irradiations. The limited success of past attempts can be ascribed partly to the subresolutional size of the centrosome, but also to the fact that it lacks a sharp natural boundary (like a surrounding membrane) to separate it from its surroundings (Fig. 6). However, it is now quite easy to delineate the entire centrosome, or just its component centrioles, by simply expressing fusions between the appropriate centrosomal/centriolar proteins and GFP (Fig. 6). Once a centrosome is so labeled, it becomes simple to destroy, without the ambiguity of previous studies, with just several laser pulses (Khodjakov *et al.*, 1997b).

Using this GFP/laser microsurgery approach, we subsequently proved that animal cells form a functional bipolar spindle when both centrosomes are ablated before mitosis (Khodjakov *et al.*, 2000). This finding overturned the 125-year-old dogma that centrosomes are required for spindle formation in animal somatic cells. Additional microsurgery studies further revealed that the assembly of new centrioles is not limited to the semiconservative replication pathway, disproving another common belief in centrosomal biology. Instead in vertebrate somatic cells, centrioles can also form via a *de novo* assembly pathway (Khodjakov *et al.*, 2002; La Terra *et al.*, 2005). Under some conditions, this pathway results in the simultaneous assembly of too many centrioles, suggesting that its activation contributes to the increase of centrosome numbers and to the chromosomal instability seen in many cancer cells (reviewed in Fukasawa, 2005; Nigg, 2002; Salisbury *et al.*, 2004).

In addition to their biological significance, our centrosome ablation studies also revealed that laser microsurgery has now advanced to a precision that is remarkable: we can now reproducibly ablate just one of the two centrioles within a centrosome, with no detectable damage to the other centriole situated only 500 nm away (Fig. 7),

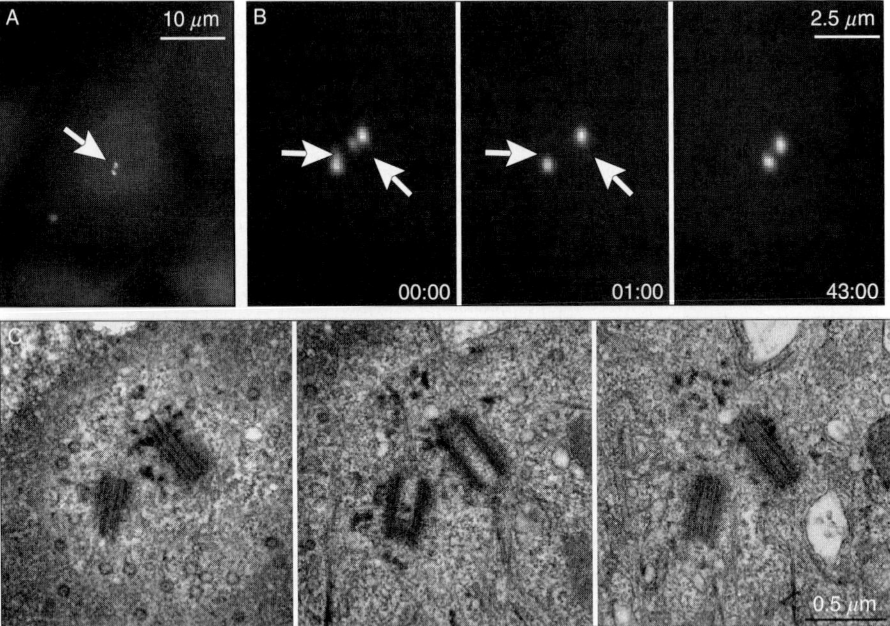

Fig. 7 Ablation of individual centrioles within diplosomes. (A) A HeLa (human epithelial) cell during S period (similar to the stage shown in Fig. 6C–S and D–S). The centrosome (arrow) is labeled via centrin/GFP expression. (B) A higher-magnification view of the centrosome reveals that both mother centrioles have already developed short daughters (arrows). Both daughter centrioles were irradiated (arrows in 00:00 and 00:01) with short series of laser pulses (~10 per centriole), and 43 min later the cell was fixed for EM analysis. (C) Serial-section EM revealed that both daughter centrioles were completely ablated while mother centrioles remained structurally intact. Time in minutes:seconds.

using a laser workstation assembled in-house (see Section II). In practical terms, this means that at this time neither the "sharpness" of the microbeam scalpel nor the ability to see the targets are problematic for most live-cell microsurgery studies. Laser microsurgery has matured to the point where the current demands are to improve the automation and user-friendliness of the modern instruments.

II. Instrumentation for Subcellular Laser Microsurgery

A. Laser Microsurgery Workstations for Biologists: Practical Considerations

The ability to delineate otherwise invisible structures in the living cell allows microsurgery studies to be conducted that were previously impossible. As a result, laser microsurgery is becoming a much more popular tool in cell biology as evidenced by the fact that within last 2 years it has been used to generate exciting results in all of the major model systems, for example, *Schizosaccharomyces pombe* (Khodjakov *et al.*, 2004a; Tolic-Norrelykke *et al.*, 2004), *C. elegans* (Bringmann and Hyman, 2005; Yanik *et al.*, 2004), *Dictyostelium* (Brito *et al.*, 2005), *Drosophila* (Maiato *et al.*, 2004, 2005), mammals (Botvinick *et al.*, 2004; Colombelli *et al.*, 2005; La Terra *et al.*, 2005), and plants (Reinhardt *et al.*, 2005). Once small, the club of "laser surgeons" is growing rapidly.

There are already relatively simple laser microsurgery systems available on the market (MicroPoint System—Photonics Science, Arlington Heights, IL). Further, it has been demonstrated that commercial multiphoton microscopes equipped with Ti:Sapphire lasers (e.g., Zeiss LSM510) can be used for laser microsurgery (Galbraith and Terasaki, 2003). However, most contemporary laser ablation studies are still conducted on systems assembled in-house that are based on different types of lasers and differ dramatically in their design. Often, this results in disagreements between groups regarding the type of instrumentation needed for a particular task. In this regard, there are numerous claims put in the literature that certain types of lasers (usually the most expensive ones) provide superior "precision" (size of ablated area) while other types should never be used for live-cell work. At one time, we were guilty of this pretentious selectivity by stating that UV lasers are inferior to our favorite 532-nm green light because they "unavoidably induce DNA damage" in live cells (Khodjakov *et al.*, 1997b). However, this claim is no longer valid as shown by the Steltzer group who found that very sensitive PtK1 cells continue to progress through normal cell cycles after they were operated on with a picosecond UV laser beam (Colombelli *et al.*, 2005). It has been claimed that femtosecond-range lasers provide much superior resolution when compared with the nanosecond- and picosecond-range lasers (Konig *et al.*, 2001; Shen *et al.*, 2005), and it has even been suggested that microsurgery with femtosecond lasers should be termed "nanosurgery" (Konig *et al.*, 1999).

One goal of this chapter is to demystify the technique of laser microsurgery by emphasizing that it can be successfully conducted using a wide range of pulsed lasers. Although the physical mechanisms by which laser pulses destroy structures in live cells can differ between nanosecond and femtosecond lasers (Calmettes and Berns, 1983; Rau *et al.*, 2006; Venugopalan *et al.*, 2002; Vogel *et al.*, 2005), the important point is that the biological consequences of organelle ablation appear to be the same (see discussion in Botvinick *et al.*, 2004).

Under conditions of extremely short (femtoseconds) pulse durations, it is generally accepted that ablation occurs through multiphoton absorption (Schaffer *et al.*, 2002; Vogel *et al.*, 2005). On the other hand, for relatively long (nanosecond) pulses multiphoton absorption is highly improbable in such materials as water or glass. As a result, it is mostly assumed that in this case ablation is based on pressure wave propagation and/or cavitation bubble dynamics (Rau *et al.*, 2006). Superfluously, this implies that the damage inflicted by nanosecond lasers is less localized than that generated by ultrashort pulses. However, in practice we find that the size of the damage inflicted by 532-nm nanosecond pulses in live cells can be as small as 250–300 nm. As mentioned above, our laser microsurgery workstation is capable of ablating individual centrioles inside a centrosome (Fig. 7). We can also cut cytoskeletal elements immediately adjacent to plasma membrane (Fig. 8) which would be impossible if the mechanism was based on the propagation of a pressure wave. Thus, the precision of nanosecond ablation in living cells equals that achieved with femtosecond lasers (Kumar *et al.*, 2006; Shen *et al.*, 2005).

Further, the total energy delivered to the cell during an operation is similar between femtosecond and nanosecond systems. Although most of our operations are done with ~10–20 pulses (@20 Hz), multiple pulses are only needed to ensure that the often-moving target is solidly hit. In fact, a single ~1-μJ pulse of 532-nm laser (8 ns) is sufficient to rupture an individual mitochondria (Fig. 2) or cut microtubules (Fig. 10). With femtosecond pulses rupturing a mitochondrion require several hundreds of 2-nJ pulses (Shen *et al.*, 2005) which amasses to

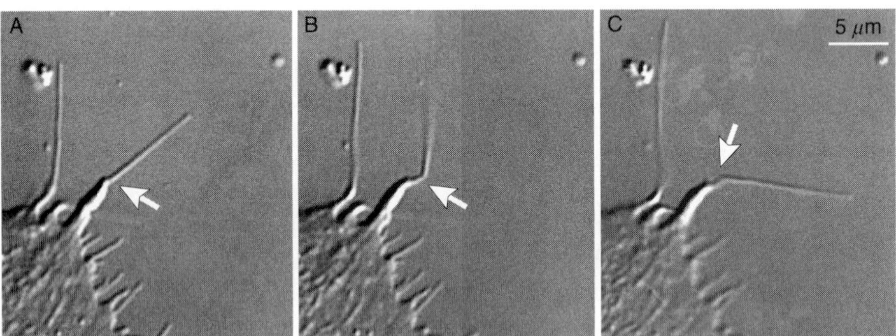

Fig. 8 Cutting cytoskeletal elements beneath the plasma membrane with nanosecond laser pulses. In this example actin filaments inside a filopodium were sliced with ~5–10 pulses of 532-nm laser light. The typical diameter of a filopodium is just 0.2–0.4 μm, and thus the actin bundle is immediately adjacent to the plasma membrane. Nevertheless, the beam aimed at the center of this organelle (arrows) does not rupture the membrane revealing that damage inflicted by nanosecond lasers is highly localized.

roughly the same total energy (~1 μJ). Cutting microtubules with femtosecond pulses also required 1.5 μJ (1000 of 1.5-μJ pulses; Heisterkamp *et al.*, 2006).

The truth of the matter is that surgeons rarely think of how the scalpel cuts. Whether the object is annihilated by plasma or destroyed by a shock wave—the salient point is that it disappears. Thus far, there are no indications that in live cells the precision of near infrared femtosecond-laser ablations is any different from that achieved with green nanosecond or UV picosecond pulses. Where a direct comparison can be made, there appears to be little difference. A good example is illustrated by two recent microsurgery studies on the dynamics of spindle microtubules in fission yeast (Khodjakov *et al.*, 2004a; Tolic-Norrelykke *et al.*, 2004). Although, one study employed a femtosecond two-photon confocal system (Sacconi *et al.*, 2005) and the other a 532-nm nanosecond laser (described in this chapter), the outcomes were quite similar. This being the case, for all practical purposes the real-life resolution and precision of the beams used in laser microsurgery are identical for nano- through femtosecond lasers.

For practitioners in the field, or for those who want to get involved, it often makes more sense to use the least expensive and most user-friendly system possible that will do the required job(s). In this regard, the high cost of purchasing and maintaining a femtosecond system currently prohibits its use as a "personal" instrument within an average-size biology laboratory. Obviously, there are specific applications where the use of femtosecond lasers is necessary. For example, near infrared femtosecond lasers have much better penetration depth and thus are indispensable for in-tissue ablations (Chung *et al.*, 2006; Yanik *et al.*, 2004). However, as outlined above, for laser microsurgery applications in relatively thin preparations, ranging from monolayers of cultured animal cells to yeast, relatively inexpensive green-light nanosecond-pulse lasers provide a more economical alternative. Importantly, small nanosecond-range lasers can be easily retrofitted to a research-grade inverted microscope, and such an upgrade can be performed with modest funds and a reasonable effort in the typical cell-biology laboratory environment.

B. General Layout of a Versatile Low-Cost Laser Microsurgery Workstation

Below we describe the layout of the system currently used in our laboratory (Fig. 9). Our design is based on the Nikon TE2000E2 microscope; however, in principle the same layout can be used to couple a laser to any research-grade inverted microscope. Although the total cost of our system is ~$250K, most of the costs are for the microscope, spinning-disk confocal attachment, CCD cameras, and peripheral devices not related to the laser. The cost of the laser, optical elements, and mechanic components necessary to upgrade a high-end imaging workstation to a laser microsurgery system is currently ~$30–35K.

Our system is based on open-space in which both the laser and the microscope are situated on a vibration-isolation table, and the output of the laser is steered toward the microscope by a series of front-surface mirrors. In addition to being

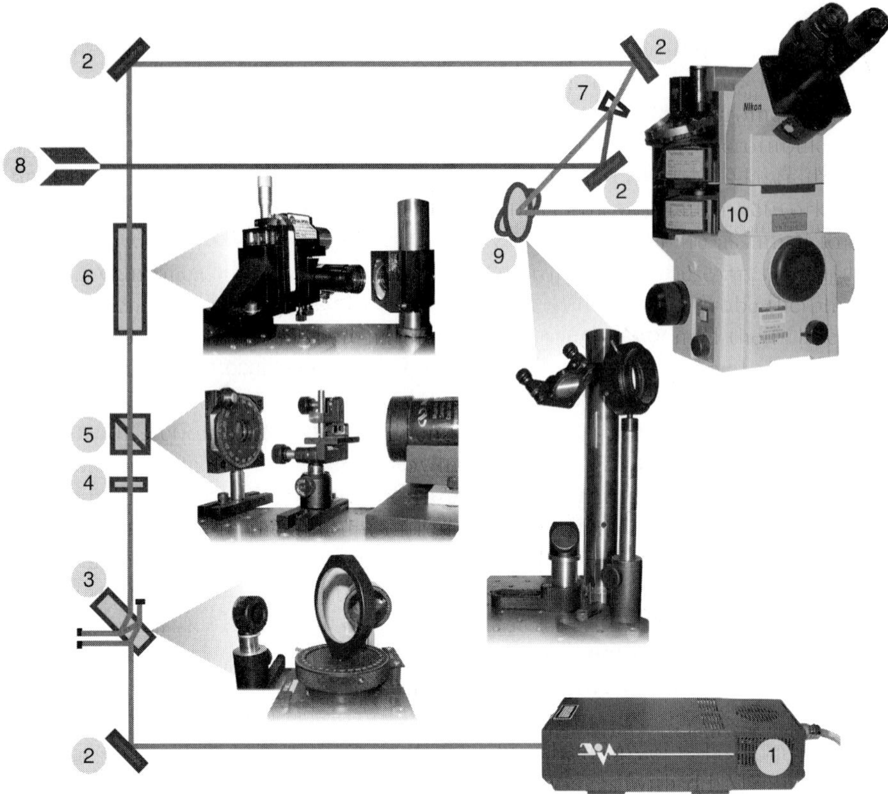

Fig. 9 Schematic layout of a basic laser microsurgery system. (1) Nd:YAG laser (Diva II, Thales Lasers, Paris, France) produces 532-nm, 8-ns pulses with about 5 mJ of light energy in each pulse. The laser is run at 20 Hz. (2) The beam is steered toward the microscope with front-surface dielectric mirrors. (3) Initial attenuation of the beam is achieved by the double reflection on a tilted uncoated glass parallel window. The level of attenuation can be adjusted (before the rest of the system is aligned) by varying the angle of incidence. (4) A 532-nm zero-order half-wave quartz waveplate in a rotatable mount to control beam polarization. (5) Glan-laser calcite polarizer permits only vertically polarized light to pass through. The combination of (4) and (5) allows us to precisely tune the energy of laser pulses reaching the microscope. (6) Focusable zoom beam expander (Special Optics, Wharton, NJ) mounted with two translational and two angular adjustment controls. (7) Beam combiner for simultaneous delivery of the ablation and 488-nm CW photobleaching beams. Conditioning of the photobleaching beam (8) is achieved in the way similar to that of the ablation beam. By adding additional beam combiners at this point additional laser beams can be delivered to the microscope (for example, 405-nm CW beam for photoactivation of PA-GFP). (9) Two-mirror periscope allows for elevating the beam to match the level of microscope's epiport and to rotate the polarization plane of the beam. (10) The beam enters the microscope through the lower epiport and is steered toward the objective lens with a custom-made dichroic mirror (525dcsp, Chroma, Brattleboro, VT). Not shown: mechanical shutter (Uniblitz, Vincent Associates, Rochester, NY) positioned between (3) and (4).

the least expensive option, all of the optical elements necessary for beam conditioning, as well as diagnostic equipment like a beam profiler and photo-detectors, can be easily placed at any point in the optical path. Further, an open-space layout allows one to easily deliver several laser beams to the same microscope. Indeed, our current system is also capable of diffraction-limited photobleaching with a continuous-wave 488-nm laser (Fig. 10). However, it is important to emphasize that open-beam laser systems require thoughtful considerations for laser safety. For starters, the system must be housed in a dedicated room that is accessible to only trained personnel. Fortunately, as a rule high-end imaging workstations are already housed in a dedicated space so that compliance with laser safety is not burdensome.

A variety of commercially available lasers are perfectly suited for laser microsurgery applications. We currently use a diode-pumped air-cooled Q-switched Nd: YAG laser (Diva II; Thales Lasers, Paris, France) which was chosen for its highly focusable ($M^2 < 1.2$) true Gaussian-profile beam, pulse-to-pulse stability, small size (14.5 × 6 × 3.9 in.), and a very reasonable cost (currently under \$25K). This laser operates in TEM_{00} mode outputting 8-ns 532-nm pulses at up to 20-Hz repetition rate.

One common misconception among cell biologists is that microsurgery can only be conducted with a very powerful laser. In fact, the single most important parameter that needs to be considered when choosing a laser is the quality of the beam. Unfortunately, most of the high beam-quality lasers available on the market produce at least three orders of magnitude more power than needed for diffraction-limited laser ablations in live cells. As a result, the laser beam needs to be significantly attenuated before it can be delivered to the specimen which, in principle, can be achieved in the laser head. However, in our laboratory we choose not to change any major laser operation settings because it could affect the pulse-to-pulse stability and pulse width. Instead we attenuate the beam in two stages between the laser and the microscope. For the first stage (~500-fold) we use an uncoated 12-mm thick parallel-surfaces window (Newport Corporation, fused silica with $\lambda/20$ flatness) tilted ~45° with respect to the laser beam. When passing through this window, a small portion of the beam reflects on both the front and then the rear surface of the window. Because of the large thickness of the window, the part of the original beam that passes straight through and the part that undergoes two internal reflections become spatially separated. This separation allows us to block the straight-through high-power beam with a beam-stop while directing the reflected beam toward the microscope (Fig. 9). By adjusting the tilt of the window, we can adjust the level of attenuation at this stage so that only 10 μJ/pulse is steered toward the microscope. This double-reflection approach also improves the polarization purity of the beam: the orthogonal polarization component (noise) that is inevitably present in the original beam is largely transmitted without internal reflection because the angle between the beam and the window is close to the Brewster angle.

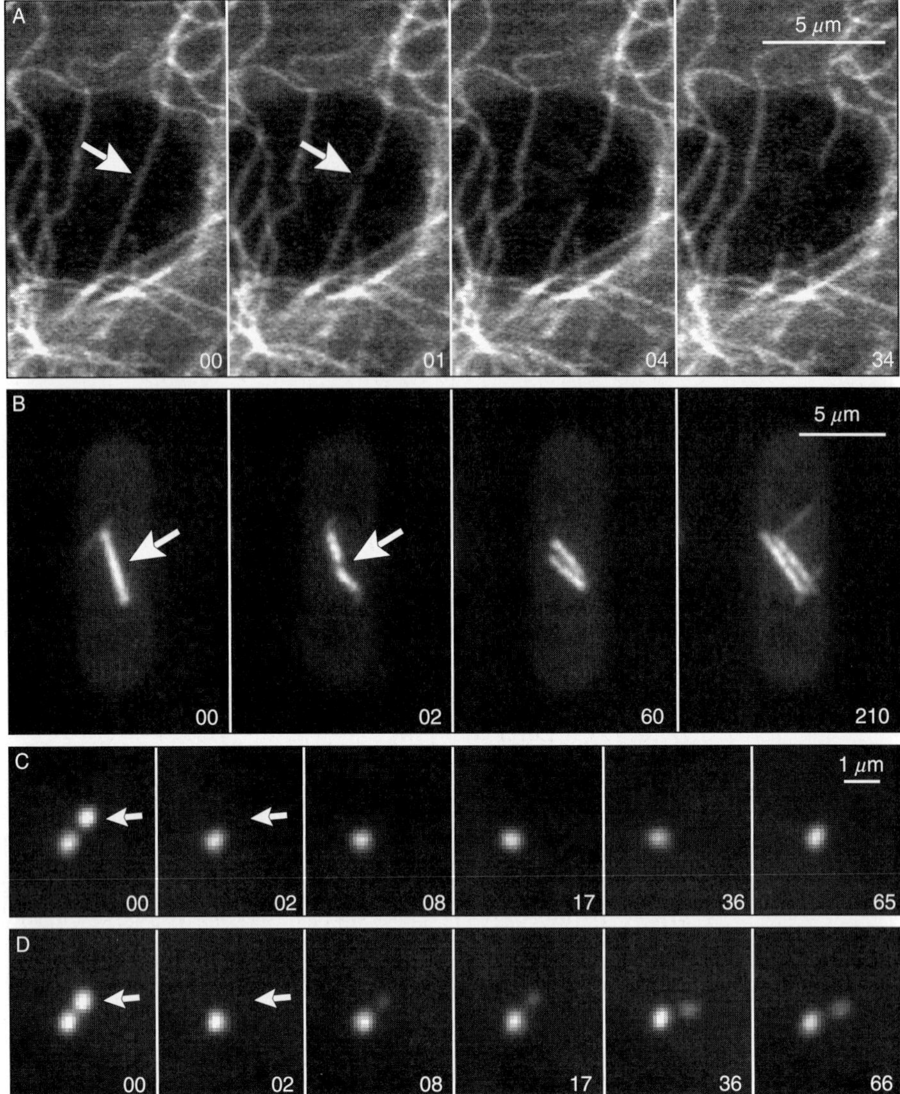

Fig. 10 Examples of typical microsurgery (A–C) and photobleaching (D) experiments that can be used for evaluating the capabilities of a laser microsurgery workstation. (A) Severing an individual microtubule. A single 1.5-μJ (measured before the beam expander, Fig. 9) 8-ns pulse cuts an α-tubulin/GFP-labeled microtubule in a PtK2 cell (arrow). Notice that after the cut one of the exposed ends depolymerizes rapidly (the "plus" end), while the other remains stable (the "minus" end). (B) Severing a mitotic spindle in fission yeast (*S. pombe*) expressing Tub1/GFP. Similar to (A) except in this case a series of pulses (<10) was used to guarantee that the target is solidly hit. The halves of the spindle remain stable and eventually elongate (not shown, see Khodjakov *et al.*, 2004a). (C–D) Comparison between laser ablation (C) and photobleaching (D) of an individual centriole in a human mammary epithelial (HMEC) cell. Irradiation with a series of 2.5-μJ 532-nm pulses (~10) completely destroys the centriole

The second attenuation step is achieved using an adjustable polarization rotator (half-wave plate) followed by a fixed Glan laser polarizer (Thorlabs). This approach allows us to attenuate the beam power approximately fivefold without significantly degrading the polarization purity of the beam. Further, the rotatable half-wave plate allows us to precisely tune the power of the beam, which is monitored immediately after the Glan polarizer with a laser power meter (818J-09B detector, Newport Corporation). Because all optical elements below this point remain constant, adjusting the power to a fixed value (currently ~2.5 μJ on our system) allows us to compensate for any day-to-day fluctuations in the laser output (surprisingly common even in $25K lasers!). This is critical for ensuring that the energy delivered to the specimen remains constant. Although monitoring the energy of the beam before it passes through the objective lens does not reveal the absolute value of the energy delivered to the specimen, our method is convenient and quite reproducible. Here, it is noteworthy that achieving precise measurements of the light energy focused in the central spot of high-NA oil immersion lens are not a trivial task. We therefore empirically adjust the energy by monitoring the biological effects of laser irradiation (Fig. 10).

Because modern research-grade microscopes utilize infinity-corrected optics, creating a diffraction-limited spot in the focal plane of the objective lens is actually quite simple. All that needs to be done is to deliver a collimated beam to the back aperture of the objective. We achieve this by first expanding the attenuated beam with a focusable zoom beam-expander (Special Optics, 2–8× zoom, λ/8 beam distortion) mounted on a four-axis adjustable platform. The zoom expander allows us to precisely match the diameter of the beam to the size of the back aperture of the objective lens and to adjust beam collimation so that it focuses in the imaging plane. The expanded beam is steered toward the microscope with a series of front-surface mirrors mounted on standard adjustable mounts (Thorlabs) for precise beam alignment. The exact number of mirrors needed depends on the relative positions of the laser and the microscope on the optical table. The best way to minimize the footprint of the system is to mount the laser behind the microscope, facing away from the position of the microscope operator. Thus, the beam needs to be wrapped around the table (with several mirrors, Fig. 9), and also elevated to the height of the epiport on the microscope. The latter task is achieved by a two-mirror periscope to change the beam direction in the horizontal plane by 45°, which results in the rotation of the polarization plane by the same angle. This rotation is necessary for laser microsurgery on microscopes equipped for DIC. Most modern DIC microscopes utilize Wollaston prisms that are mounted just below the objective lens and oriented 45° with respect to the "left–right" axis of the microscope.

(arrows in C). In sharp contrast, irradiation ~200-ms irradiation with a 488-nm CW laser beam bleaches centrin/GFP molecules within the irradiated centrioles (arrows in D). Due to exchange of molecules between the irradiated centriole and the cytoplasmic pool GFP fluorescence gradually recovers. Time is in seconds in (A and B), and in minutes in (C and D).

Because the direction of polarization on all commercial lasers is either vertical (90°) or horizontal (0°), rotating the beam by 45° allows us to match the polarization of the beam to the sheer direction of the Wollaston prism so that the beam passes through the prism without major distortions.

The alignment of the laser beam in the layout described above is quite easy and can be achieved interactively by monitoring the position and shape of the beam in real-time via the same CCD camera used for imaging the specimen. For convenience, and to prevent potential damage to the CCD during alignment procedures, we use an additional neutral-density filter (OD3) that can be temporarily inserted into the beam at any point between the beam expander and the microscope. Alignment is then achieved by tilting the mirrors so that: (1) the focused beam becomes positioned in the center of the field of view of the objective lens and (2) its shape is symmetric indicating that the beam is centered and coaxial with the optical axis of the microscope. This type of alignment does not require an in-depth understanding of optics or laser physics, and graduate students or postdocs with minimal training in microscopy can easily perform it.

One of the salient features of our layout is that the focused laser beam remains stationary in the center of the field of view. Thus, for aiming the target needs to be moved into the beam. We achieve this by using a precise electronically controlled microscope stage (Ludl Electronic Products, Hawthorne, NY). This is less than an ideal approach because it makes ablation of large areas inside the cell virtually impossible. However, it is perfectly suited for ablating individual small objects such as centrosomes, kinetochores, or mitochondria. Further, the stage translation approach works fairly well for irradiating linear paths as needed when cutting across a chromosome or cytoskeletal assembly (e.g., actin filaments or microtubules within the spindle). In this type of operation, the operator opens the shutter and drives the stage (via joystick controls) along either the x- or y-axis. More sophisticated laser microsurgery systems usually employ special hardware and software that allow the beam to scan the field of view (see Botvinick and Berns, 2005; Colombelli *et al.*, 2005, and Chapter 1 by Berns, this volume). This beam-scanning approach is obviously more versatile; however, it significantly increases the cost of the system and cannot be easily installed on an in-house assembled system.

One important consideration in designing a laser microsurgery workstation is that most laser ablations are now conducted in cells labeled with fluorescent proteins that are imaged either in wide-field epifluorescence or confocal mode. In epifluorescence, illumination of the object is achieved through the objective lens instead of a dedicated condenser and therefore, delivery of the laser beam to the back aperture of the lens must not interfere with the epifluorescence excitation. This presents an interesting problem because the 532-nm wavelength of the ablation laser is longer than both the 488-nm excitation and 510-nm emission peaks of the most common GFP isoform. Thus, the standard dichroic mirror used for imaging GFP fluorescence is transparent to the 532-nm wavelength and cannot be used to steer the laser beam towards the lens. Further, most multicolor dichroic mirrors do not perform well, particularly when the peak intensity of the laser pulses exceeds the intensity of the fluorescence excitation and emission light by several orders of magnitude. Until recently, the most common way to deal with this problem was to use two different dichroic mirrors mounted individually in two different filter cubes: one for observations and one for ablations. There are however two severe limitations associated with this approach. First, most filter-cube turrets on off-the-shelf microscopes are not sufficiently reproducible to cycle filter cubes between the exact same positions. As a result, after a full cycle the orientation of mirrors with respect to the laser beam has changed slightly which in turn shifts the position of the laser beam in the imaging plane. This irreproducibility is particularly prominent in faster (and usually less-precise) turrets. More precise changers tend to be much slower, and this creates the second problem inherent in the switch-mirror approach. Because two dramatically different filter cubes are used for imaging and laser ablation, the target (specimen) cannot be observed during irradiation. This feature makes it much more difficult to precisely aim the laser beam at the target as the latter often changes position during the 1–2 s required for switching the filter cubes. Further, the immediate response of the irradiated structure to the beam cannot be observed. Fortunately, some of the modern research-grade inverted microscopes can now be equipped with two independent filter-cube assemblies that provide for independent deliveries of the epifluorescence excitation and the ablation-laser beam. This feature, which we utilize in our system, makes it possible to steer the 532-nm beam of the cutting laser toward the lens by a stationary dichroic mirror positioned in the lower filter-cube turret, while the top (motorized) turret hosts the standard filter cube for imaging GFP-fluorescence (Fig. 9). This layout allows us to avoid problems associated with "wobbling" of the laser beam and at the same time to continuously observe the target during the operation. Finally, our system can operate both as a wide-field fluorescence and a spinning-disk confocal microscope. In the latter case, the top filter cube is rotated out of the optical path, and the excitation light of the confocal head illuminates the specimen through the laser dichroic (Fig. 9).

As noted above, an added bonus of using the open-space layout is that it supports the delivery of several laser beams to the same microscope simultaneously. This can be achieved by situating additional lasers (e.g., 488-nm CW laser for GFP photobleaching or 405-nm CW laser for photoactivation of PA-GFP) on the same table and by steering their beams toward the same epiport of the microscope in the manner described above. Combining the beams can be easily achieved by using either a conventional 50/50 beam-splitter cube or a round wedge-prism (Thorlabs part number PS814), which is less costly and provides a higher quality beam because of the smaller number of reflective surfaces. In this design, the ablation beam undergoes distortion on only two surfaces with little attenuation. The bleaching beam hits the surface of this window at sharp angle, so that the front-surface reflection follows the same path toward the microscope as the transmitted ablation beam (Fig. 9). The intensity of such a reflected beam (~4%) is more than sufficient for photobleaching a diffraction-limited spot which requires about a hundred microwatts with a PlanApo 100× 1.4 NA objective. The ability to photobleach and ablate intracellular components in the same cell has provided valuable information on the dynamics of microtubules in yeast and animal cells (Khodjakov *et al.*, 2004a; Kumar *et al.*, 2006; Magidson *et al.*, 2006; Maiato e*t al.*, 2004, 2005).

The laser microsurgery/photobleaching system described in this chapter requires minimal maintenance costs (although we highly recommend purchasing a comprehensive service contract to cover potential malfunctions of the laser). However, it is surprisingly versatile as evident from the number of illustrations presented in this chapter.

Acknowledgments

We acknowledge use of Wadsworth Center's EM core facility. Our work is sponsored by grants from the NIH (GM59363 to A.K. and GM40198 to C.L.R.) and HFSP (RGP0064 to A.K.). Construction of the laser microsurgery workstation was supported in par by Summer Research Fellowship from Nikon/ Marine Biological Laboratory (2003 to A.K.). We thank Dr. Zhenye Yang for the images used in Fig. 5B. Requests for technical details of our system should be addressed to Dr. Valentin Magidson (valentin@wadsworth.org). The authors declare that they have no competing financial interests.

References

Adkisson, K. P., Baic, D., Burgott, S. L., Cheng, W. K., and Berns, M. W. (1973). Argon laser micro-irradiation of mitochondria in rat myocardial cells in tissue culture. IV. Ultrastructural and cytochemical analysis of minimal lesions. *J. Mol. Cell. Cardiol.* **5,** 559–564.

Basto, R., Lau, J., Vinogradova, T., Gardiol, A., Woods, C. G., Khodjakov, A., and Raff, J. W. (2006). Flies without centrioles. *Cell* **125,** 1375–1386.

Berns, M. W., Aist, J., Edwards, J., Strahs, K., Girton, J., McNeill, P., Rattner, J. B., Kitzes, M., Hammer-Wilson, M., Liaw, L. H., Siemens, A., Koonce, M., *et al.* (1981). Laser microsurgery in cell and developmental biology. *Science* **213,** 505–513.

Berns, M. W., and Cheng, W. K. (1971). Are chromosome secondary constrictions nucleolar organizers? A re-examination using a laser microbeam. *Exp. Cell Res.* **69,** 185–192.

Berns, M. W., Floyd, A. D., Adkisson, K., Cheng, W. K., Moore, L., Hoover, G., Ustick, K., Burgott, S., and Osial, T. (1972). Laser microirradiation of the nucleolar organizer in cells of the rat kangaroo (Potorous tridactylis). Reduction of nucleolar number and production of micronucleoli. *Exp. Cell Res.* **75,** 424–432.

Berns, M. W., Gamaleja, N., Olson, R., Duffy, C., and Rounds, D. E. (1970a). Argon laser micro-irradiation of mitochondria in rat myocardial cells in tissue culture. *J. Cell. Physiol.* **76,** 207–213.

Berns, M. W., Ohnuki, Y., Rounds, D. E., and Olson, R. S. (1970b). Modification of nucleolar expression following laser micro-irradiation of chromosomes. *Exp. Cell Res.* **60,** 133–138.

Berns, M. W., Rattner, J. B., Brenner, S., and Meredith, S. (1977). The role of the centriolar region in animal cell mitosis. A laser microbeam study. *J. Cell Biol.* **72,** 351–367.

Berns, M. W., and Richardson, S. M. (1977). Continuation of mitosis after selective laser microbeam destruction of the centriolar region. *J. Cell Biol.* **75,** 977–982.

Berns, M. W., Rounds, D. E., and Olson, R. S. (1969). Effects of laser micro-irradiation on chromosomes. *Exp. Cell Res.* **56,** 292–298.

Bessis, M., Gires, F., Mayer, G., and Nomarski, G. (1962). Irradiation des organites cellulaires a l'aide d'un Laser a rubis. *Comptes Rendus de l Academie des Sciences* **255,** 1010–1012.

Bettencourt-Dias, M., Rodrigues-Martins, A., Carpenter, L., Riparbelli, M., Lehmann, L., Gatt, M. K., Carmo, N., Ballox, F., Callaini, G., and Glover, D. M. (2005). SAK/PLK4 is required for centriole duplication and flagella development. *Curr. Biol.* **15,** 2199–2207.

Botvinick, E. L., and Berns, M. W. (2005). Internet-based robotic laser scissors and tweezers microscopy. *Microsc. Res. Tech.* **68,** 65–74.

Botvinick, E. L., Venugopalan, V., Shah, J. V., Liaw, L. H., and Berns, M. W. (2004). Controlled ablation of microtubules using a picosecond laser. *Biophys. J.* **87,** 4203–4212.

Bringmann, H., and Hyman, A. A. (2005). A cytokinesis furrow is positioned by two consecutive signals. *Nature* **436,** 731–734.

Brito, D. A., Strauss, J., Magidson, V., Tikhonenko, I., Khodjakov, A., and Koonce, M. P. (2005). Pushing forces drive the comet-like motility of microtubule arrays in *Dictyostelium*. *Mol. Biol. Cell* **16,** 3334–3340.

Calmettes, P. P., and Berns, M. W. (1983). Laser-induced multiphoton processes in living cells. *Proc. Natl. Acad. Sci. USA* **80,** 7197–7199.

Chalfie, M., Tu, Y., Euskirchen, G., Ward, W. W., and Prasher, D. C. (1994). Green fluorescent protein as a marker for gene expression. *Science* **263,** 802–805.

Chung, S. H., Clark, D. A., Gabel, C. V., Mazur, E., and Samuel, A. D. (2006). The role of the AFD neuron in C. elegans thermotaxis analyzed using femtosecond laser ablation. *BMS Neurosci.* **7,** 30.

Cole, R. W., Khodjakov, A., Wright, W. H., and Rieder, C. L. (1995). A differential interference contrast-based light microscopic system for laser microsurgery and optical trapping of selected chromosomes during mitosis *in vivo*. *J. Microsc. Soc. Am.* **1,** 203–215.

Colombelli, J., Reynaud, E. G., Rietdorf, J., Pepperkork, R., and Stelzer, E. H. (2005). *In vivo* selective cytoskeleton dynamics quantification in interphase cells induced by pulsed ultraviolet laser nanosurgery. *Traffic* **6,** 1093–1102.

Forer, A. (1965). Local reduction of spindle fiber birefringence in living Nephrotoma suturalis (Loew) spermatocytes induced by ultraviolet microbeam irradiation. *J. Cell Biol.* **25,** 95–117.

Fukasawa, K. (2005). Centrosome amplification, chromosome instability and cancer development. *Cancer Lett.* **230,** 6–19.

Galbraith, J. A., and Terasaki, M. (2003). Controlled damage in thick specimens by multiphoton excitation. *Mol. Biol. Cell* **14,** 1808–1817.

Giepmans, B. N., Adams, S. R., Ellisman, M. H., and Tsien, R. Y. (2006). The fluorescent toolbox for assessing protein location and function. *Science* **312,** 217–224.

Heisterkamp, A., Maxwell, I. Z., Mazur, E., Underwood, J. M., Nickerson, J. A., Kumar, J., and Ingber, D. E. (2006). Pulse energy dependence of subcellular dissection by femtosecond laser pulses. *Optics Express* **13,** 3690–3696.

Hyman, A. A. (1989). Centrosome movement in the early divisions of *Caenorhabditis elegans:* A cortical site determining centrosome position. *J. Cell Biol.* **109,** 1185–1193.

Inoue, S., Yoder, O. C., Turgeon, B. G., and Aist, J. R. (1998). A cytoplasmic dynein required for mitotic aster formation *in vivo*. *J. Cell Sci.* **111,** 2607–2614.

Kapoor, T. M., and Compton, D. A. (2002). Searching for the middle ground: Mechanisms of chromosome alignment during mitosis. *J. Cell Biol.* **157,** 551–556.

Khodjakov, A., Cole, R. W., Bajer, A. S., and Rieder, C. L. (1996). The force for poleward chromosome motion in Haemanthus cells acts along the length of the chromosome during metaphase but only at the kinetochore during anaphase. *J. Cell Biol.* **132,** 1093–1104.

Khodjakov, A., Cole, R. W., McEwen, B. F., Buttle, K. F., and Rieder, C. L. (1997a). Chromosome fragments possessing only one kinetochore can congress to the spindle equator. *J. Cell Biol.* **136,** 229–240.

Khodjakov, A., Cole, R. W., Oakley, B. R., and Rieder, C. L. (2000). Centrosome-independent mitotic spindle formation in vertebrates. *Curr. Biol.* **10,** 59–67.

Khodjakov, A., Cole, R. W., and Rieder, C. L. (1997b). A synergy of technologies: Combining laser microsurgery with green fluorescent protein tagging. *Cell Motil. Cytoskeleton* **38,** 311–317.

Khodjakov, A., La Terra, S., and Chang, F. (2004a). Laser microsurgery in fission yeast: Role of the mitotic spindle midzone in anaphase B. *Curr. Biol.* **14,** 1330–1340.

Khodjakov, A., Rieder, C., Mannella, C. A., and Kinnally, K. W. (2004b). Laser micro-irradiation of mitochondria: Is there an amplified mitochondrial death signal in neural cells? *Mitochondrion* **3,** 217–227.

Khodjakov, A., Rieder, C. L., Sluder, G., Cassels, G., Sibon, O. C., and Wang, C. L. (2002). *De novo* formation of centrosomes in vertebrate cells arrested during S phase. *J. Cell Biol.* **158,** 1171–1181.

Konig, K., Riemann, I., Fischer, P., and Halbhuber, K. J. (1999). Intracellular nanosurgery with near infrared femtosecond laser pulses. *Cell. Mol. Biol.* **45,** 195–201.

Konig, K., Riemann, I., and Fritzsche, W. (2001). Nanodissection of human chromosomes with near-infrared femtosecond laser pulses. *Optics Lett.* **26,** 819–821.

Koonce, M. P., Cloney, R. A., and Berns, M. W. (1984). Laser irradiation of centrosomes in newt eosinophils: Evidence of centriole role in motility. *J. Cell Biol.* **98,** 1999–2010.

Kumar, J., Maxwell, I. Z., Heisterkamp, A., Polte, T. R., Lele, T. P., Salanga, M., Mazur, E., and Ingber, D. E. (2006). Viscoelastic retraction of single living stress fibers and its impact on cell shape, cytoskeletal organization, and extracellular matrix mechanics. *Biophys. J.* **90,** 3762–3773.

La Terra, S., English, C. N., Hergert, P., McEwen, B. F., Sluder, G., and Khodjakov, A. (2005). The *de novo* centriole assembly pathway in HeLa cells: Cell cycle progression and centriole assembly/maturation. *J. Cell Biol.* **168,** 713–720.

LaFountain, J. R., Jr., Cole, R. W., and Rieder, C. L. (2002). Partner telomeres during anaphase in crane-fly spermatocytes are connected by an elastic tether that exerts a backward force and resists poleward motion. *J. Cell Sci.* **115,** 1541–1549.

LaFountain, J. R., Jr., Oldenbourg, R., Cole, R. W., and Rieder, C. L. (2001). Microtubule flux mediates poleward motion of acentric chromosome fragments during meiosis in insect spermatocytes. *Mol. Biol. Cell* **12,** 4054–4065.

Magidson, V., Chang, F., and Khodjakov, A. (2006). Regulation of cytokinesis by spindle-pole bodies. *Nat. Cell Biol.* **8,** 891–893.

Maiato, H., Khodjakov, A., and Rieder, C. L. (2005). *Drosophila* CLASP is required for the incorporation of microtubule subunits into fluxing kinetochore fibres. *Nat. Cell Biol.* **7,** 42–47.

Maiato, H., Rieder, C. L., and Khodjakov, A. (2004). Kinetochore-driven formation of kinetochore fibers contributes to spindle assembly during animal mitosis. *J. Cell Biol.* **167,** 831–840.

McNeill, P. A., and Berns, M. W. (1981). Chromosome behavior after laser microirradiation of a single kinetochore in mitotic PtK2 cells. *J. Cell Biol.* **88,** 543–553.

Moreno, G., Salet, C., and Vinzens, F. (1973). Etude en microscpie electronique des mitochondries de cellules en culture de tissus apres micro-irradiation par laser. *Journal de Microscopie et de Biologie Cellulaire* **16,** 269–278.

Nigg, E. A. (2002). Centrosome aberrations: Cause or consequence of cancer progression? *Nat. Rev. Cancer* **2,** 815–825.

Ohnuki, Y., Rounds, D. E., Olson, R. S., and Berns, M. W. (1972). Laser microbeam irradiation of the juxtanucleolar region of prophase nucleolar chromosomes. *Exp. Cell Res.* **71,** 132–144.

Orokos, D. D., Cole, R. W., and Travis, J. L. (2000). Organelles are transported on sliding microtubules in Reticulomyxa. *Cell Motil. Cytoskeleton* **47,** 296–306.

Ou, Y., and Rattner, J. B. (2004). The centrosome in higher organisms: Structure, composition, and duplication. *Int. Rev. Cytol.* **238,** 119–182.

Peterson, S. P., and Berns, M. W. (1978). Evidence for centriolar region RNA functioning in spindle formation in dividing PTK2 cells. *J. Cell Sci.* **34,** 289–301.

Posudin, Yu. I. (1995). Biophysisist Sergei Tschachotin Natl. Agricult. Univ. Publ Kiev, pp. 1–92.

Prasher, D. C., Eckenrode, V. K., Ward, W. W., Prendergast, F. G., and Cormier, M. J. (1992). Primary structure of the Aequorea victoria green-fluorescent protein. *Gene* **111,** 229–233.

Rattner, J. B., Lifsics, M., Meredith, S., and Berns, M. W. (1976). Argon laser microirradiation of mitochondria in rat myocardial cells. VI. Correlation of contractility and ultrastructure. *J. Mol. Cell. Cardiol.* **8,** 239–248.

Rau, K. R., Quinto-Su, P. A., Hellman, A. N., and Venugopalan, V. (2006). Pulsed laser microbeam-induced cell lysis: Time-resolved imaging and analysis of hydrodynamic effects. *Biophys. J.* **91,** 317–329.

Reinhardt, D., Frenz, M., Mandel, T., and Kuhlemeier, C. (2005). Microsurgical and laser ablation analysis of leaf positioning and dorsoventral patterning in tomato. *Development* **132,** 15–26.

Rieder, C. L., and Cassels, G. (1999). Correlative light and electron microscopy of mitotic cells in monolayer cultures. *Methods Cell Biol.* **61,** 297–315.

Rieder, C. L., and Cole, R. W. (1998). Entry into mitosis in vertebrate somatic cells is guarded by a chromosome damage checkpoint that reverses the cell cycle when triggered during early but not late prophase. *J. Cell Biol.* **142,** 1013–1022.

Rieder, C. L., Cole, R. W., Khodjakov, A., and Sluder, G. (1995). The checkpoint delaying anaphase in response to chromosome monoorientation is mediated by an inhibitory signal produced by unattached kinetochores. *J. Cell Biol.* **130,** 941–948.

Rieder, C. L., Davison, E. A., Jensen, L. C., Cassimeris, L., and Salmon, E. D. (1986). Oscillatory movements of monooriented chromosomes and their position relative to the spindle pole result from the ejection properties of the aster and half-spindle. *J. Cell Biol.* **103,** 581–591.

Rizzuto, R., Brini, M., Pizzo, P., Murgia, M., and Pozzan, T. (1995). Chimeric green fluorescent protein as a tool for visualizing subcellular organelles in living cells. *Curr. Biol.* **5,** 635–642.

Sacconi, L., Tolic-Norrelykke, I. M., Antolini, R., and Pavone, F. S. (2005). Combined intracellular three-dimensional imaging and selective nanosurgery by a nonlinear microscope. *J. Biomed. Optics* **10,** 14002.

Salet, C. (1972). A study of beating frequency of a single myocardial cell. 1. Q-switched laser micro-irradiation of mitochondria. *Exp. Cell Res.* **73,** 360–366.

Salisbury, J. L., D'Assoro, A. B., and Lingle, W. L. (2004). Centrosome amplification and the origin of chromosomal instability in breast cancer. *J. Mammary Gland Biol. Neoplasia* **9,** 275–283.

Schaffer, C. B., Nishimura, N., Glezer, E. N., Kim, A. M. T., and Mazur, E. (2002). Dynamics of femtosecond laser-induced breakdown in water from femtoseconds to microseconds. *Optics Express* **10,** 196–203.

Shen, N., Datta, D., Schaffer, P., LeDuc, P., Ingber, D. E., and Mazur, E. (2005). Ablation of cytoskeletal filaments and mitochondria in live cells using a femtosecond laser nanoscissor. *Mech. Chem. Biosys.* **2,** 17–25.

Shimada, T., Watanabe, W., Matsunaga, S., Higashi, T., Ishii, H., Fukui, K., Isobe, K., and Itoh, K. (2005). Intracellular disruption of mitochondria in a living HeLa cell with a 76-MHz femtosecond laser oscillator. *Optics Express* **13,** 9869–9880.

Spurck, T. P., Stonington, O. G., Snyder, J. A., Pickett-Heaps, J. D., Bajer, A., and Mole-Bajer, J. (1990). UV microbeam irradiations of the mitotic spindle. II. Spindle fiber dynamics and force production. *J. Cell Biol.* **111,** 1505–1518.

Storb, R., Amy, R. T., Wertz, B., Fauconnier, B., and Bessis, M. (1966). An electron microscope study of vitally stained single cells irradiated with a ruby laser microbeam. *J. Cell Biol.* **31,** 11–29.

Strahs, K. R., and Berns, M. W. (1979). Laser microirradiation of stress fibers and intermediate filaments in non-muscle cells from cultured rat heart. *Exp. Cell Res.* **119,** 31–45.

Tolic-Norrelykke, I. M., Sacconi, L., Thon, G., and Pavone, F. S. (2004). Positioning and elongation of the fission yeast spindle by microtubule-based pushing. *Curr. Biol.* **14,** 1181–1186.

Tschachotin, S. (1912). Die Mikroskopische Strahlenstichmethode, Eine Zelloperationsmethode. *Biol. Zentalbl.* **32,** 623–630.

Tschachotin, S. (1938). Die Mikrostrahlstichmethode und andere Methoden des zytologischen Mikroexperimentes. *In* "Methoden der allgemeinen vergleichenden Physiologie" (E. Abderhalden, ed.), (Viena: Urban & Schwarzenberg), pp. 877–958.

Uretz, R. B., Bloom, W., and Zirkle, R. E. (1954). Irradiation of parts of individual cells. II. Effects of an ultraviolet microbeam focused on parts of chromosomes. *Science* **120,** 197–199.

Uzbekov, R. E., Votchal, M. S., and Vorobjev, I. A. (1995). Role of the centrosome in mitosis: UV micro-irradiation study. *J. Photochem. Photobiol.: B Biol.* **29,** 163–170.

Venugopalan, V., Guerra, A., III, Nahen, K., and Vogel, A. (2002). Role of laser-induced plasma formation in pulsed cellular microsurgery and micromanipulation. *Phys. Rev. Lett.* **88,** 078103.

Vogel, A., Noack, J., Huttmann, G., and Paltauf, G. (2005). Mechanisms of femtosecond laser microsurgery of cells and tissues. *Appl. Phys. B* **81,** 1015–1047.

Wheatley, D. N. (1982). "The Centriole: A Central Enigma of Cell Biology," pp. 1–232. Elsevier, Amsterdam.

Wilson, E. B. (1925). "The Cell in Development and Heredity," pp. 1–1232. The Macmillan Co., New York.

Winson, A. A. (1965). Apparatus for local irradiation of animal cells with microbeams of alpha particles. *Radiobiologiia (Russian)* **5,** 752–756.

Winson, A. A., and Kuzin, A. M. (1965). Synthesis of DNA during irradiation of the cytoplasm and nucleus of HeLa cells with a micro-pencil of alpha-particles. *Doklady Akademii Nauk SSSR* **165,** 933–936.

Yanik, M. F., Cinar, H., Cinar, H. N., Chisholm, A. D., Jin, Y., and Ben-Yakar, A. (2004). Neurosurgery: Functional regeneration after laser axotomy. *Nature* **432,** 822.

Zernike, F. (1955). How I discovered phase contrast. *Science* **121,** 345–349.

Zirkle, R. E., and Bloom, W. (1953). Irradiation of parts of individual cells. *Science* **117,** 487–493.

CHAPTER 8

Investigating Relaxation Processes in Cells and Developing Organisms: From Cell Ablation to Cytoskeleton Nanosurgery

Julien Colombelli, Emmanuel G. Reynaud, and Ernst H. K. Stelzer

Light Microscopy Group
Cell Biology and Biophysics Unit
European Molecular Biology Laboratory (EMBL)
D-69117 Heidelberg, Germany

METHODS IN CELL BIOLOGY, VOL. 82

0091-679X/07 $35.00
DOI: 10.1016/S0091-679X(06)82008-X

Dynamic microscopy of living cells and organisms alone does not reveal the high level of complexity of cellular and subcellular organization. All observable processes rely on the activity of biochemical and biophysical processes and many occur at a physiological equilibrium. Experimentally, it is not trivial to apply a perturbation that targets a specific process without perturbing the overall equilibrium of a cell. Drugs and more recently RNAi certainly have general and undesired effects on cell physiology and metabolism. In particular, they affect the entire cell. Pulsed lasers allow to severe biological tissues with a precision in the range of hundreds of nanometers and to achieve ablation on the level of a single cell or a subcellular compartment.

In this chapter, we present an efficient implementation of a picosecond UV-A pulsed laser-based nanosurgery system and review the different mechanisms of ablation that can be achieved at different levels of cellular organization. We discuss the performance of the ablation process in terms of the energy deposited onto the sample and compare our implementation to others recently employed for cellular and subcellular surgery. Above the energy threshold of ionization, we demonstrate how to achieve single-cell ablation through the induction of mechanical perturbation and cavitation in living organisms. Below this threshold, we induce cytoskeleton severing inside live cells. By combining nanosurgery with fast live-imaging fluorescence microscopy, we show how the apparent equilibrium of the cytoskeleton can be perturbed regionally inside a cell.

I. Introduction

Dynamic microscopy of living cells and organisms alone does not reveal the high level of complexity of cellular and subcellular organization. All observable processes rely on the activity of many biochemical and biophysical processes and many occur at a physiological equilibrium. Scientists have always used perturbation methods that interfere with a physiological equilibrium. The goals were and still are to understand and measure the physical, biochemical, or biological parameters sustaining this balance and to identify their role and influence. For instance, chemical drugs that inhibit the performance of specific subcellular components, for example nocodazole that destabilizes microtubules (MTs), have contributed immensely to a better understanding of the relation between subcellular compartments and their roles in signaling pathways. However, it is not trivial to apply a perturbation that targets a specific process without perturbing the overall equilibrium of a cell. Drugs and more recently RNAi certainly have general and undesired effects on cell physiology and metabolism. In particular, they affect the entire cell.

Pulsed lasers allow to severe biological tissues with a precision in the range of hundreds of nanometers. They provide an ideal method for highly localized perturbation procedures. With a pulsed UV laser, nonlinear effects ionize the tissue in the focal volume of the laser beam. The localized damage provokes an ablation on the level of a single cell or a subcellular compartment.

In this chapter, we present an efficient implementation of a picosecond UV-A laser nanosurgery system and review the different mechanisms of ablation that can be achieved at different levels of cellular organization. We discuss the performance of the ablation process in terms of the energy delivered to the sample and compare our implementation to others recently employed for cellular and subcellular surgery. Above the energy threshold of ionization, we demonstrate how to achieve single-cell ablation through the induction of mechanical perturbation and cavitation in living organisms. Below this threshold, we induce cytoskeleton severing inside live cells. By combining nanosurgery with fast live-imaging fluorescence microscopy, we show how the apparent equilibrium of the cytoskeleton can be perturbed regionally inside a cell. In particular, subcellular tension can be released by dissecting stress fibers and provide evidence for the not so obvious forces exerted by the actomyosin complex. Similarly, we show that nanosurgery of single MTs leads to their destabilization and that their recovery, or rescue, follows the principles of dynamic instability *in vivo*.

II. Subnanosecond UV-A Pulsed Laser Applied in a Conventional Wide-Field Microscope and Combined with FRAP

A. Diffraction-Limited Nanosurgery by Plasma Formation

Laser surgery of biological tissue has been studied and applied for over 30 years, essentially since pulsed lasers were invented. The quality and the availability of pulsed lasers have dramatically improved during the past 10 years. Very high optical intensities are delivered during ultrashort time periods, which range from nanoseconds (10^{-9} s) down to a few tens of femtoseconds (10^{-15} s). A review (Vogel and Venugopalan, 2003) summarizes the mechanisms of interaction between pulsed lasers and biological tissues. By tightly focusing laser pulses with a high numerical aperture (NA) lens, optical intensities can reach different orders of magnitude of MW/cm^2 up to TW/cm^2 and induce ionization of the irradiated medium. Pulsed laser irradiation induces the release of free electrons via an interplay of multiphoton and avalanche ionization, which leads to the formation of a plasma. The phenomenon is known as *plasma-induced ablation* or *optical breakdown*. Plasma formation occurs above a well-defined intensity threshold. It is a material property and is usually accompanied by destructive mechanical and thermal side effects, which depend on the pulse length and the excess of the applied intensity above its threshold.

1. Diffraction-Limited Coupling for a Minimal Spatial Beam Extent

The quality and the extent of the focused laser beam are crucial for achieving submicron precision in laser surgery. The efficiency of an instrument relies heavily on the quality of its lasers optical coupling into the microscope. High-quality beams from solid-state lasers allow one to conserve their Gaussian profile when used with high-NA objective lenses. Their lateral beam extent can reach the physical minimum, that is, the diffraction limit. Since plasma formation arises at a precise threshold in intensity, diffraction-limited coupling is crucial to reduce the beam's extent and, therefore, the total energy delivered to the targeted sample. If the 2-sigma diameter of a laser beam is equivalent to the entrance aperture of a lens, the focal diameter in the sample is most likely twice as wide ($\sim$NA) and four times as deep ($\sim NA^2$). In our system, the diffraction limit is achieved by expanding the collimated beam diameter to almost twice the diameter of the back aperture of the objective lens.

Beam aberrations usually cause severe changes to the electromagnetic field distribution in the focus and can show several strong maxima. These secondary maxima, or side lobes, can cause damage not only close to the vicinity of the focus but also several hundreds of nanometers away. The main aberration to consider, and to reduce, originates from the flatness of the beam wavefront. As shown previously (Colombelli *et al.*, 2004), the flatness of optical surfaces plays a major role in conserving the beam wavefront quality. In our implementation, all surfaces reach a requested flatness of $\lambda/10$ (a tenth of the design wavelength). Optical aberrations from lens systems can also prevent a perfect diffraction-limited beam focusing. Therefore, one needs to keep in mind that coupling lasers through the regular optics of a conventional microscope can result in undesired aberrations, which in turn increase the beam extent. A compromise we adopted was to optimize the coupling through the fluorescence port of the microscope in order to get a good diffraction limit for one objective lens, that is, with one specific beam expansion prior to the objective.

Another important feature for an optimal beam extent is the choice of a suitable wavelength. The spatial extent of a focused laser beam is proportional to the wavelength, short wavelength laser sources are therefore preferable to achieve the smallest possible focal volume. The use of UV-A laser light is a good compromise that allows focusing to the energy into a small volume as well as easy coupling through the regular optics of a conventional microscope. We use the third harmonic of an Nd:YAG pulsed laser (355-nm wavelength) and achieve a theoretical lateral beam extent of 319 nm with a Zeiss C-Apo 63×/1.2-W water immersion lens. Besides, commercial objective lenses are generally not corrected for chromatic aberrations down to 355 nm and the axial focus position is often not conjugated with the image plane. Correcting for these aberrations by modifying the beam expansion prior to the objective would automatically result in an enlargement of the lateral focus extent and the loss of the diffraction limit. It is therefore necessary to use a piezoelectric objective positioner (Colombelli *et al.*,

2004) to quickly displace the focus prior to laser ablation. The necessary axial displacement in the UV (355 nm) ranges from a couple of microns for water immersion lenses (less than 2 μm for a Zeiss C-Apo 63×/1.2 W) to up to a few tens of microns for water dipping or air immersion lenses.

2. Plasma Formation with Subnanosecond Pulses

From theoretical simulations of plasma formation in aqueous media (Noack and Vogel, 1999; Vogel and Noack, 2001; for a review see Vogel and Venugopalan, 2003; Vogel *et al.*, 2005), it is known that the photochemical reaction leading to the ionization of an aqueous medium is heavily nonlinear and should be divided into two phases. First, the simultaneous absorption of several photons from the incident laser pulse by the medium produces quasi-free electrons (*multiphoton ionization*). Subsequently, these electrons gain kinetic energy, or accelerate, by further absorption of photons from the same laser pulse and generate new free electrons by impact with other molecules (*impact ionization*). This second reaction is believed to be mainly responsible for the desired surgical effect and explains why the duration of laser pulses is a central issue when trying to understand the kinetics of plasma expansion. It was shown experimentally that the intensity threshold for plasma formation decreases by three orders of magnitude when decreasing the pulse duration from 10 ns down to 100 fs (i.e., six orders of magnitude). Since the energy delivered to the medium is the product of the pulse duration and the optical power, femtosecond laser pulses require 100 times less energy to achieve plasma formation than nanosecond pulses. From an experimental point of view, this could mean that laser surgery is performable at energy levels that are orders of magnitude lower when using ultrashort laser pulses rather than pulses in the nanosecond range. The latter are used in most commercially available systems (Willingham, 2002). A recent interest in laser surgery has hence been directed toward the shortest possible laser pulses in order to reduce down to the minimum usable levels of energy. Publications (Botvinick *et al.*, 2004; Heisterkamp *et al.*, 2005; Sacconi *et al.*, 2005; Tirlapur and König, 2002; Watanabe *et al.*, 2004) reporting intracellular surgery follow this principle and the trend is the use of femtosecond lasers, traditionally used in multiphoton microscopy.

In our implementation of a laser nanosurgery system, we chose to use subnanosecond pulses (close to 470 ps) of a frequency-tripled Nd:YAG UV solid-state laser, which is simple to operate and considerably less expensive than femtosecond systems. An advantage of using pulsed UV irradiation over IR was shown theoretically and experimentally in a previous study (Vogel and Noack, 2001). Simulations of the temporal increase of the free electron density at the plasma formation intensity threshold showed a much smoother rate of increase with UV as compared to IR. An explanation for this phenomenon could be the difference in photon energy required to perform multiphoton ionization of water. With UV at 355 nm, already two photons can recombine to efficiently generate ions whereas six are required when using IR lasers. In consequence, the transition to optical breakdown was shown to be more abrupt with IR sources, characterized by a "step" increase in free electron density at the threshold, whereas with UV, the threshold is gradually reached. As we will discuss later in Section III.B, the choice of short subnanosecond UV pulses allows one to work in an intermediate regime where, although the energy required for plasma formation is still high as compared to femtosecond systems, the free electron density, and therefore the plasma energy density, rises more gradually and performs smooth ablations even at low plasma densities.

A simple method for measuring the plasma extent in the object plane is shown in Fig. 1. The pulsed 470-ps UV beam was focused inside a conventional glass coverslip and the visible effects of a single pulse were observed in bright-field transmission mode for different optical powers. The minimum optically detectable effect occurred at a pulse energy of 0.4 ± 0.1 μJ deposited in the medium (Fig. 1A). At this intensity threshold, the formation of plasma in transparent glass induces irreversible changes of the medium properties at the focus position. The glass refractive index changes inside the plasma volume and its dimensions can therefore be visualized and measured. We reported a lateral extension of 0.45 ± 0.05 μm in the focal plane and an axial plasma length of 2.50 ± 0.25 μm. This efficient ablation volume in glass is 5.2× greater than the focal volume (Colombelli *et al.*, 2004), which is a consequence of plasma propagation and represents an improvement due to the low UV wavelength over other reported values at 532 and 1064 nm (Venugopalan *et al.*, 2002). Well above this threshold, the excess energy deposited into the glass is partly absorbed by the plasma in front of the beam waist and gives rise to an increase in free electron and energy densities. This induces a phenomenon called *plasma shielding* and results in very large damage due to glass melting and propagation of mechanical shock waves, as shown in Fig. 1C. Using a similar method of glass nanopatterning, other authors could show single laser pulse precision well below the diffraction limit with lateral extents of around 30 nm (Joglekar *et al.*, 2004) with femtosecond lasers.

3. Three-Dimensional Laser Surgery

A fundamental aspect of pulsed laser surgery for biological applications, which is often not well demonstrated, is the ability to induce plasmas in three dimensions in order to penetrate biological tissues at different levels and with high axial precision. Figure 1D shows how a three-dimensionally structured pattern can be formed in a glass volume by focusing the laser pulses at different axial locations. The same approach can be applied in thick live tissues or organism. No photomechanical damage occurs in the out-of-focus planes. The only critical parameters for achieving the deep penetration of tissues are therefore the laser wavelength and the intrinsic scattering and absorption of the targeted tissue. Near IR light or visible light is the best compromise because the absorption by water is minimal. With a UV pulsed laser, we could achieve efficient cellular ablation through at least 150 μm of live embryos (e.g., in *Platynereis dumerilii*, data not shown). On the basis of the same principle, an additional feature that is routinely used during laser

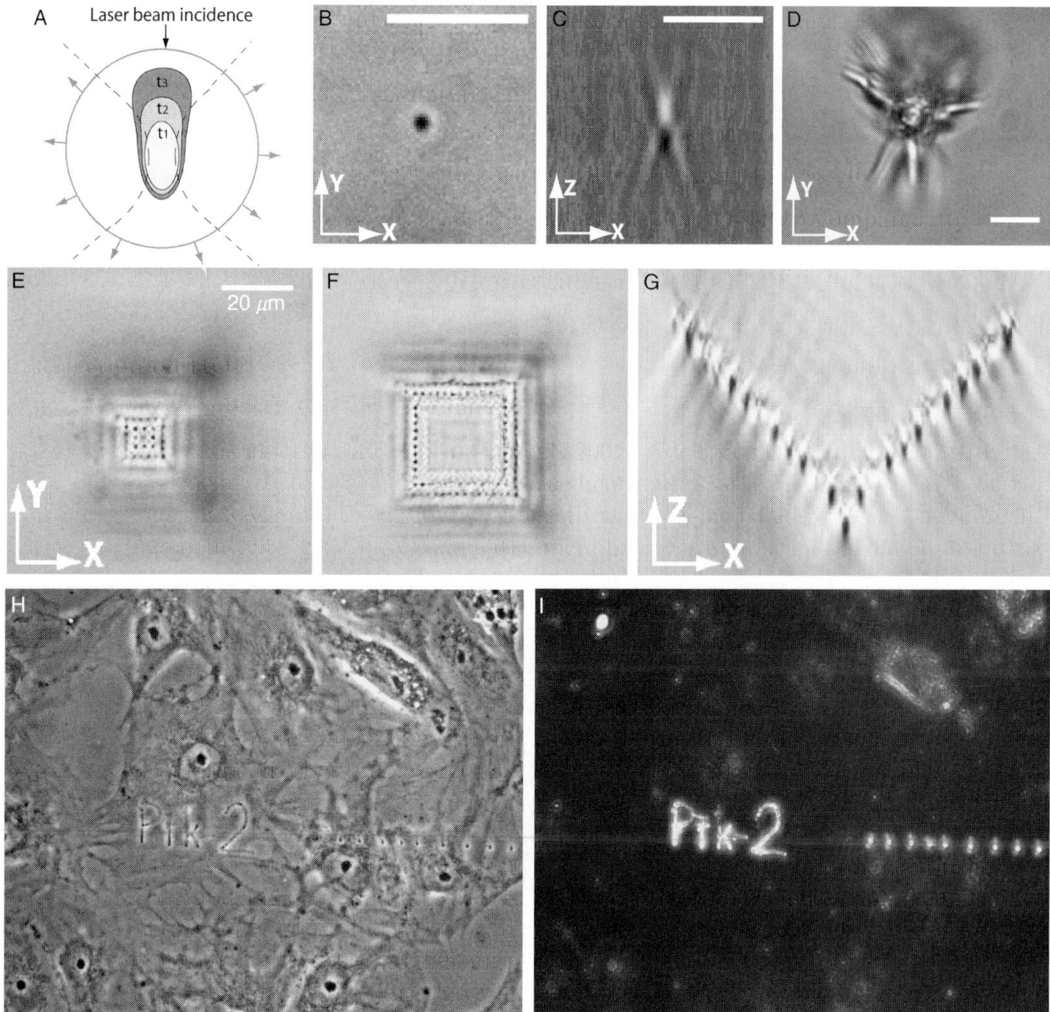

Fig. 1 Plasma formation in glass. By focusing one laser pulse into the volume of a simple glass coverslip, the optical appearance of the latter can be permanently modified. The refractive index of the glass, inside the focal volume after ionization, is modified and the plasma volume becomes visible in transmission microscopy, acting thus as some sort of a micro lens. (A) A schematic representation of the plasma propagation. At three consecutive time points during the laser pulse duration, the plasma ignites (t1) and then expands toward the laser incidence (t2) to finally reach its maximal size (t3). The visible nanopattern induced by one laser pulse of 470 ps is shown laterally in (B) and axially in (C). Well above the energy threshold of plasma formation, secondary effects like mechanical shock wave propagation are induced as shown in (D). Images (E–G) demonstrate the three-dimensional property of plasma formation with a pyramid-like pattern throughout the glass volume shown laterally in two different planes (E and F) and axially (median plane of the pattern projected in G). The capability to inscribe information into the glass well below the sample of interest is used on a regular basis during laser ablation experiments. Image (H) shows Ptk2 cells in a culture glass-bottom dish, imaged in phase contrast. The characters "Ptk2" were inscribed 5 μm below the cell plane without affecting the cell viability, and a line of laser pulses can be used toward the edge of the dish in order to find the area of interest more easily. The same pattern is shown in dark field in (I). (A) Reused with permissions from Colombelli *et al.* (2006). (B) to (G) reused with permissions from Colombelli *et al.* (2004), Copyright 2004, American Institute of Physics.

surgery experiments consists in hardcoding information in the glass volume underneath the sample itself, as shown in Fig. 1E and F. Such an experimental protocol allows for tracking single cells among numerous populations after laser surgery for further analysis under different experimental conditions (e.g., a different microscope).

B. Laser Dissection Versus FRAP

The application of FRAP (fluorescence recovery after photobleaching) only affects the visibility of the cell's fluorophores distribution. The cell's metabolic processes are unaffected and hence the cell continues to behave as it behaved prior to the bleaching process. If the cell was in a state of equilibrium prior to the bleaching process, it will continue to be in this state after the bleaching has occurred. So ideally, FRAP allows one to study the reformation of the equilibrium of the fluorophore distribution. FRAP data analyses usually ignore the toxic side effects of flurophore degradation, which have been known to exist for many years.

The application of laser surgery introduces a dramatic change to the current metabolic condition of a cell. It potentially damages biopolymers, lipid membranes, and entire organelles in such a manner that they lose their functionality. Hence, as long as the cell has not been too severely damaged, one tends to observe a relaxation process toward a new equilibrium. One of the main differences between the applications of FRAP and of laser cutters is, therefore, that in the first case we assume that the equilibrium remains unaffected while in the latter we induce a new equilibrium.

We implemented a FRAP setup in the same microscope system in order to combine and compare FRAP and laser surgery experiments. We use a continuous multiline argon ion laser to perform photobleaching. The coupling of the beam is identical to that of the pulsed UV beam path.

III. From Microsurgery to Nanosurgery: From Cell Ablation to Intracellular Manipulation

In cell and developmental biology, we consider two ways of performing laser surgery. The first takes advantage of the mechanical and thermal secondary effects induced by plasma expansion in order to operate single-cell ablation at the level of several microns. Usually called microsurgery, this procedure is already possible with nanosecond laser pulses. Nanosurgery, on the other hand, addresses more particularly the dissection of subcellular compartments. The inactivation or severing of subcellular structures or organelles is possible without affecting the plasma membrane or the viability of the cell if the ablation process is thoroughly controlled. We focus here on providing a qualitative description of the major differences between cell ablation and intracellular surgery and describe the most recent applications in cell and developmental biology.

A. Cell Ablation by Photomechanical Perturbation

Developmental biology concentrates on the study of the differentiation stages (morphogenesis or organogenesis) that lead to the formation of an individual from a single cell. In this respect, remarkable progress was provided by experimental approaches of cellular ablation (Berger, 1998; Hutson *et al.*, 2003; Kiehart *et al.*, 2000; Montell *et al.*, 1991; Supatto *et al.*, 2005). Unlike the classical methods using micromanipulation needles with a tenth of millimeter precision, laser surgery has allowed for the targeted destruction of single cells located inside developing embryos as well as cell extensions such as axons.

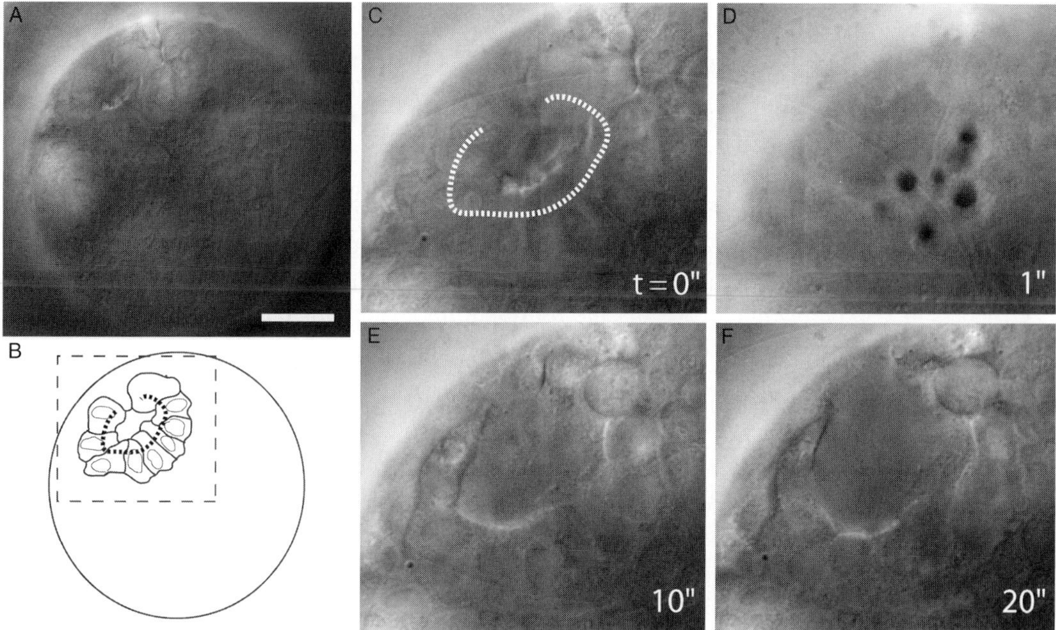

Fig. 2 Cell ablation by plasma and cavitation formation in a living *P. dumerilii* embryo. The structured group of cells called *stomodeum* (precursor to the development of the adult's mouth) forms after 25 h of development (rosette shape in A), schematically designed in (B). The laser focus is scanned along the outline of each cells (target dashed line in B). (C) shows a zoomed view of the target group of cells. Lethal cell permeabilization is achieved by plasma formation at a reasonably high energy level (ca. twice the threshold energy of plasma formation) in order to induce cavitation (bubbles in D). The targeted cells leak out and the cell outlines rapidly vanish (E–F), respectively, showing the ablated organ 10 and 20 s after ablation. Scale bar = 30 μm. Reused with permissions from Colombelli *et al.* (2006).

The ablation process is illustrated during the development of the marine worm *P. dumerilii* (Fig. 2; Colombelli *et al.*, 2006). By inducing a series of microplasmas with a pulse energy twice higher than the energy threshold of bubble formation, biological tissues can be vaporized, and the resulting formation of transient and destructive cavitation bubbles, shown in Fig. 2D, induces cell lysis by membrane perforation. Thus, the complete ablation of a group of cells can be performed without affecting the embryonic membrane and the surrounding cells, which are very well preserved. Further analysis on development of the severed embryo provides important information about the expression patterns of the genes that are involved in the development of the organisms (by immunofluorescence or *in situ* hybridization).

Such an approach can also be used to study the developing *Drosophila melanogaster* (i.e., the fruit fly drosophila embryo). Supatto *et al.* (2005) have demonstrated that the targeted destruction of a few dorsal cells with a femtosecond laser during gastrulation induces the inhibition of lateral tissue migration. They could correlate the resulting differences in the laser-ablated phenotype to the expression of a mechanosensitive protein. Other studies (Hutson *et al.*, 2003; Kiehart *et al.*, 2000) have aimed at identifying the different forces involved in the dorsal closure of *D. melanogaster* and could demonstrate the synchronized and relative contribution of different cell layers during this complex morphogenetic process. Finally, microsurgery can be used to remove not only single cells but also part of a cell without affecting cell viability, as demonstrated by Yanik *et al.* (2004) in *Caenorhabditis elegans*. They performed *in vivo* axotomy in adult worms by femtosecond laser surgery. This work opened the way to studies of axon regeneration and neuronal behavior of developing organism lacking defined neurons.

Cell ablation and cellular amputation by laser surgery have become classical experimental protocols and represent important advances in the study of developmental and cellular processes.

B. Subcellular Nanosurgery: High Reactivity of Certain Subcellular Targets Below the Plasma Formation Threshold

The understanding of the physical principles underlying laser surgery greatly improved in the past years. The description of plasma-mediated ablation (for review see Vogel and Venugopalan, 2003) has set a theoretical framework to which experimentalists can now refer. However, experimental data reporting energy thresholds for intracellular surgery still show wide discrepancies. In particular, values of irradiance and energy (Botvinick *et al.*, 2004) used for intracellular surgery are well below the simulated threshold values in aqueous medium (Vogel and Venugopalan, 2003 and Vogel *et al.*, 2005). These allow for the existence of a subthreshold regime for intracellular damage. In Table I, we list the different laser parameters used in recent studies to achieve intracellular surgery of subcellular structure like the mitotic spindle in yeast, a mitochondrion or a single MT in mammalian cells. While the efficiency of intracellular surgery can be described in terms of the lowest possible optical pulse

Table I
Comparison of Eight State-of-the-Art Pulsed Laser Systems for *In Vivo* Intracellular Surgery[a]

Intracellular target (reference)	Energy per pulse (nJ)	Total number of pulses to achieve surgery	Total energy deposited on the sample (μJ)	Pulse duration	Peak wavelength (nm)	Repetition rate (kHz)
Plant chloroplast (Tirlapur and König, 2002)	0.375	10^6	390	170 fs	800	80,000
Yeast spindle (Khodjakov *et al.*, 2004)	500	40	20	7 ns	532	0.01
Yeast spindle (Sacconi *et al.*, 2005)	0.05	12×10^6	600	200 fs	840	80,000
Mitochondrion (Watanabe *et al.*, 2004)	3	250	0.75	150 fs	800	1
Mitochondrion (Shimada *et al.*, 2005)	0.39	2.4×10^6	936	145 fs	800	76,000
Microtubule (Botvinick *et al.*, 2004)	0.01	228,000	2.28	80 ps	532	76,000
Microtubule (Heisterkamp *et al.*, 2005)	1.5	1000	1.5	200 fs	790	1
Microtubule (Colombelli *et al.*, 2005a)	50	2	0.1	470 ps	355	0.1

[a]All values were directly reported from the considered publications and/or calculated. Adapted from Colombelli *et al.* (2005b). Copyright (2005), with permission from Elsevier.

energy deposited at the sample, Table I shows quite clearly that the total deposited energy (number of pulses multiplied by the pulse energy) varies greatly between the studies. Vogel *et al.* (2005) investigated in great details the mechanisms of plasma formation in the femtosecond time regime and showed that depending on the repetition rate, or the laser pulse frequency, the efficiency of ablation and the underlying physical effect differ. Historically, plasma formation was defined by luminescent emission from the plasma. The most recent studies cited above have set the threshold of plasma formation as the free electron density necessary to induce thermoelastic breakdown of the irradiated volume, which is characterized by the formation of transient cavitation bubbles. Experimental data is still missing to define the precise ablation threshold of biological intracellular targets surgery. However, depending on the subcellular target, a very low free electron density could be sufficient to perturb the molecular stability of structures such as MTs, which are very unstable dimer assemblies (see Section IV.A), without reaching the thermoelastic breakdown. The existence of such a regime at low free electron densities could explain the extremely low values applied experimentally.

Taken together, the most recent theoretical and experimental results let us conclude that subcellular laser surgery can be performed along several strategies.

First, decreasing the pulse duration from nanoseconds to hundreds of femtoseconds allows for the use of lower energy levels (Section II). Using high repetition rates (typically 80 MHz with femtosecond lasers), surgical effects can be achieved well below plasma formation at low free electron density (Vogel *et al.*, 2005). Second, with longer pulse durations and even shorter wavelengths (532 nm and 80 ps, Botvinick *et al.*, 2004; 355 nm and 470 ps, Colombelli *et al.*, 2005a), experimental values for intracellular surgery show that a subthreshold regime exists. Shorter wavelengths allow for a more gradual increase in free electron density (Vogel and Noack, 2001). The combination with longer pulses could therefore sustain a pure photochemical interaction regime in which low free electron density can be applied over longer time periods, providing the plasma with a higher energy but without reaching the thermoelastic breakdown threshold. This regime should be sufficient to severe certain subcellular targets like MTs.

To illustrate the phenomenon of intracellular surgery without apparent mechanical damage, we chose to perform a combined nanosurgery and FRAP experiment. Ptk2 epithelial cells were simultaneously transfected for the expression of a red F-actin marker (monomeric RFP) and a green membrane marker (pEYFP-Mem). As shown in Fig. 2, we generated FRAP stripes in the EYFP channel by using the 488-nm line of a continuous argon laser and measured the fluorescence recovery. We then repeated the same experiment with our pulsed UV laser for nanosurgery and measured the fluorescence recovery after bleaching caused by the UV to the EYFP. It is clear from Fig. 2 that the same level of bleaching is reached for the chosen parameters and that the cytoskeleton target (stress fibers) are only bleached with the continuous laser and severed using the pulsed UV laser. However, and surprisingly, the membrane fluorescence recovery, characterized by a recovery time constant at half maximum τ (White and Stelzer, 1999) and a mobile fraction, is very similar in both cases. A slight increase in the mobile fraction was observed (data not shown) with the pulsed UV laser, together with a slight and surprising decrease in recovery time ($\tau_{1/2} = 4.94 \pm 1.80$ s, $n = 12$ cells for regular FRAP and $\tau_{1/2} = 4.19 \pm 1.14$ s, $n = 10$ cells for pulsed UV FRAP). The pEYFP-Mem encodes a fusion protein consisting of a neuromodulin fragment that contains a signal for posttranslational palmitoylation and targets EYFP to the plasma membrane. The unexpected increase in mobility, and decrease in recovery time, of the EYFP marker after nanosurgery can be understood in the light of theories regarding the confinement of membrane diffusion proteins by cortactin (for comment see Abbott, 2005). There are reasonable evidences that the actin network is severed after laser nanosurgery and, therefore, cortactin or the actin meshwork at the membrane cannot act as an obstacle to membrane diffusion. This explains the slight decrease in recovery time (Fig. 3).

We conclude from this simple experiment that applying a low-energy UV pulse series for nanosurgery has less impact on the membrane integrity than on the cytoskeleton stability. This experiment illustrates that an absolute energy threshold for efficient intracellular surgery is not obvious. The applicable ablation procedure depends also on the subcellular target. The cytoskeleton, due to its intrinsic poly-

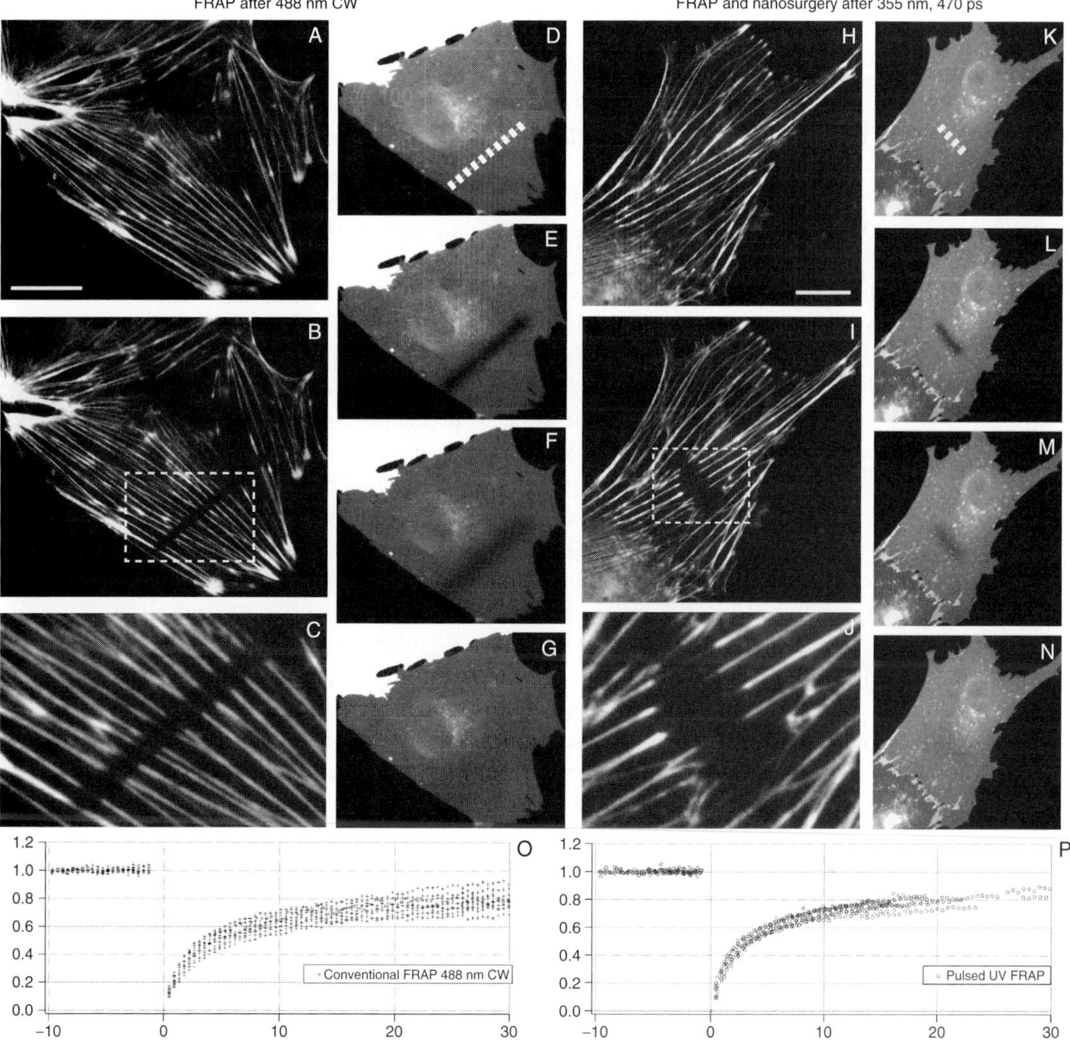

Fig. 3 Intracellular surgery: high reactivity of cytoskeleton versus membrane. We illustrate the phenomenon of nanosurgery with a comparison of conventional and simultaneous FRAP and nanosurgery experiments. The figure shows a comparison of FRAP affecting the membrane of Ptk2 cells with two different lasers. Regular FRAP with a continuous argon laser line at 488 nm is achieved in (A–G). FRAP with a nanosurgery pulsed laser unit (470 ps pulses, 355 nm) is achieved in (H–N). Cells were doubly transfected with mRFP actin and pEYFP-Mem, which is a membrane marker. Bleaching was recorded for 30 s in both experiments. *Conventional FRAP*: the bleach region is a 2 μm-wide and up to 40 μm-long stripe, performed with a laser power at the sample of about 50 mW. (A) shows a preview of the cell in the RFP channel before bleaching. (B) shows the cell 30 s after bleaching. The dashed square indicates the zoom region shown in (C). Sequence (D–G) shows the bleaching sequence with a preview in (D) and the dashed line targets for the laser path. *FRAP + nanosurgery*: The same sequences are shown in (H–N) for the FRAP experiment with a UV pulsed laser on a different cell. Laser pulses with an

merization properties, is an ideal target that can be easily severed, or destabilized, as compared to lipidlike structures that are intrinsically more stable.

IV. Cytoskeleton Nanosurgery

Very early experiments in the 1960s reported intracellular severing of subcellular compartments such as mitochondria or chromosomes. Researchers have tried to analyze the intracellular behavior on localized laser irradiation for many years. In particular, a lot of interest was assigned to the cytoskeleton behavior in interphase or mitosis and pioneering research was performed by M. W. Berns who studied the forces exerted by astral MTs (Aist *et al.*, 1993) or the role of the centriole during mitosis (Berns and Richardson, 1977). Interestingly, the advent of fluorescent probes or fused fluorescent proteins (e.g., GFP), as well as that of short and ultrashort pulsed lasers that deliver very high optical intensities and of highly sensitive detection technologies (e.g., CCD cameras), allows researchers now to investigate much more detailed cellular mechanisms.

The cytoskeleton is a highly dynamic network based on different families of proteins that assemble in polymer-like chains: actin filaments, MTs, and intermediate filaments. They polymerize with a very high turnover and allow the cell to exert forces involved in cell mechanics. Cytokinesis, cell migration, and mitosis are the fundamental cellular processes governed by the cytoskeleton. Live imaging cannot always reveal the high molecular and mechanical complexity involved in those processes as many counteracting dynamics can result in an apparent static image or at equilibrium. Applied to the cytoskeleton, intracellular laser surgery is presented here as a powerful perturbation approach to unravel the complex and collaborative mechanisms that sustain the cytoskeleton overall equilibrium.

A. Toward Single-Filament Dynamics Resolution: MT Dynamic Instability After Laser Severing or the Relaxation of the MT Lattice

The study of MT networks is defined by the concept of dynamic instability (Desai and Mitchison *et al.*, 1997), originally *in vitro*, that describes MTs in a stochastic

---

energy of 50 nJ fill a stripe of 2-μm thickness and are spaced about 150 nm laterally. Image (J) is a zoomed area of (I) and clearly shows that the actin stress fibers are dissected as they retract. Graph (O) plots the normalized recovery curves from the CW conventional FRAP experiment on 12 cells. Plot (P) shows the recovery curves after simultaneous bleaching and nanosurgery to 10 cells. Results show that fluorescence recovery after nanosurgery is slightly faster than for the conventional FRAP (15% faster, see text for a discussion). Curves were double normalized and fitted according to the empirical diffusion formula for stripe bleaching (Ellenberg *et al.*, 1997; White and Stelzer, 1999), where the intensity is expressed by $I(t) = I_{\text{final}}(1 - [w^2/(w^2 + 4\pi Dt)]^{1/2})$, with w the stripe width, D the diffusion coefficient, I_{final} the mobile and for which the half maximum of recovery time is expressed by $\tau_{1/2} = 0.75\ w^2/\pi D$. Fitting was realized on a region of interest of $2 \times 3\ \mu\text{m}^2$ in the center of the bleach region. Scale bars = 20 μm. (See Plate 5 in the color insert section.)

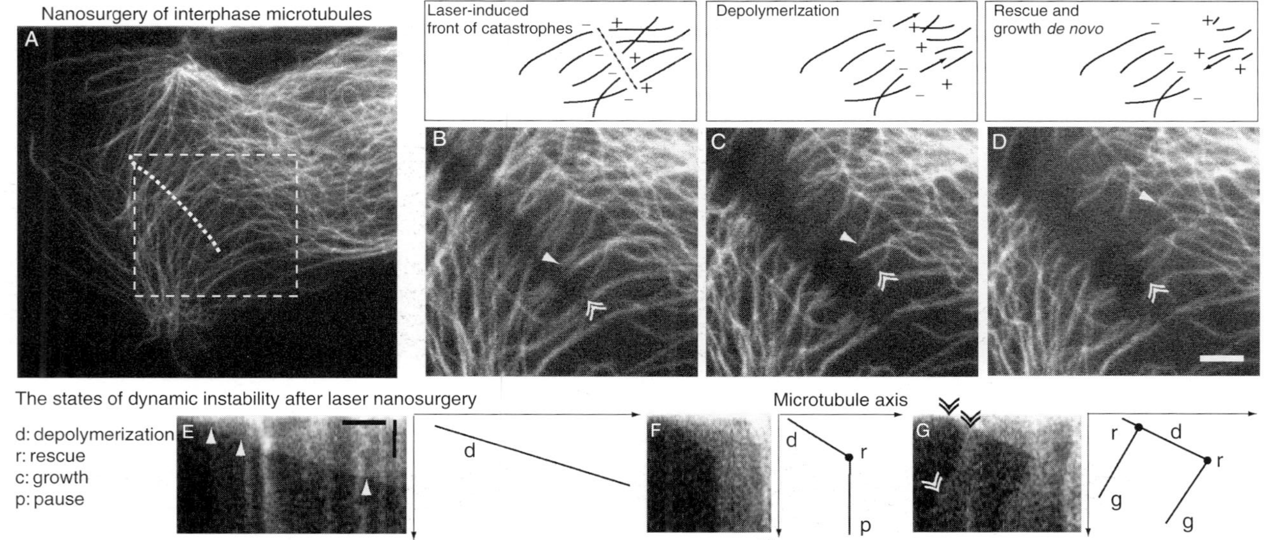

Fig. 4 Nanosurgery of interphase MTs induces dynamic instability. (A) An overview of a Ptk2 cell stably expressing YFP α-tubulin and subject to pulsed laser nanosurgery. The dashed square indicates the zoomed area shown in (B–D), the thicker dashed line shows the laser target path. Sequence (B–D) shows the behavior of interphase MTs after nanosurgery and a schematic view represents the orientation of MTs [minus (–) and plus (+) ends]. The laser severing of MTs induces catastrophe in the middle of targeted MTs. New plus and minus ends are generated and depolymerization occurs from the plus ends. Subsequently, MTs pass through all possible phases of dynamic instability and the trajectories of a few MTs are plotted on kymographs in (E–F): depolymerization in (E), pause in (F), and rescue and growth *de novo* in (G). Arrows show the MT tip positions in the images and correlate with the time–space positions shown in the kymographs. It was shown that MTs dynamics after nanosurgery follow physiological values of dynamic instability and that, therefore, dynamic instability can be induced regionally in a cell (Colombelli *et al.*, 2005a). Adapted with permissions from Colombelli *et al.* (2006). Scale bar for (D) 5 μm and (E) 5 μm horizontal and 30 s vertical.

succession of growth and shrinkage phases with transition states called rescue, from shrinkage to growth, and catastrophe, from growth to shrinkage. However, *in vivo* studies by time-resolved imaging have encountered severe practical limits. A major problem was the ability to resolve single fluorescent MTs only at the thin cell edge. In consequence, the spatial confinement of the events reduces the probability of performing a measurement. The number of evaluated events was not sufficient to derive relevant statistics and consistent values. We have shown (Colombelli *et al.*, 2005a) that laser nanosurgery of MTs overcomes these practical limits. As illustrated in Fig. 4, severing MTs with a pulsed laser induces catastrophes in a very reproducible manner. The stages following this induced catastrophe were shown to follow the principles of dynamic instability, that is severed MTs show new plus and minus ends and depolymerize from the plus end, then undergo rescue, growth *de novo*, and further catastrophe.

These results correlate well with modern theories of dynamic instability. Artificial catastrophe away from the MT tip shows that rescue is a "free-cap" process, that is, it occurs without the influence of cap proteins. A theoretical study (Janosi *et al.*, 2002) proposed a purely structural model for MT capping and interpreted rescue events as a result of random thermal fluctuations relaxing the GDP MT lattice. Our measurements of rescue events performed in the middle of the GDP MT lattice could be an experimental evidence of the random nature of MT rescue. Detailed image analysis via kymographs also provides the evidence for pause states, a controversial third state that is difficult to observe *in vivo*, after shrinkage. Further quantitative analysis after laser-induced catastrophe also revealed the influence of fluorescent markers on MT dynamics as the high repeatability of the surgery process allows for higher statistics and, therefore, to discriminate between finer dynamic differences.

B. Quantitative Analysis of MT Temperature Behavior

The normal temperature gradient of the body goes from about 37°C deep in the body's core down to about 29°C at the skin's surface and transiently reaches up to 39°C during fever periods. Some cells are exposed to lower temperature in the dermis and the epidermis while others such as lymphocytes are constantly circulating between these thermally changing compartments and are thus required to function at a variety of different physiological temperatures. The migration velocity of neutrophil fibroblasts and keratocytes vary with temperature (Hartmann-Petersen *et al.*, 2000; Nahas *et al.*, 1971; Ream *et al.*, 2003; Thurston and Palcic, 1987) as well as the direction of migration toward the higher temperature in the case of trophoblastic cells (Higazi *et al.*, 1996). This thermotaxis behavior is also observed in *C. elegans* (Yamada and Ohshima, 2003).

We further illustrate the high repeatability of the laser-induced MT depolymerization processes in Table II where we show a temperature study of MT polymerization dynamics as studied with laser nanosurgery in epithelial cells. Temperatures of 34, 37, and 39°C were applied in a culture microscopy chamber.

Table II
Temperature Dependence of MT Polymerization Dynamics Induced by Laser Nanosurgery and Compared to TIRF Measurements[a]

T (°C)	Nanosurgery-induced shrinkage rate (μm/min)	TIRF shrinkage rate (μm/min)	Rescue frequency after nanosurgery (s^{-1})	Control MT growth rate (μm/min)
39	32.9 ± 7.9 (193) +17%	39.7 ± 10.8 (121) +17.8%	–	13.6 ± 4.4 (162) +18.3%
37	28.1 ± 5.6 (38)	33.7 ± 14.4 (406)	0.151 ± 0.145 (296)	11.5 ± 3.9 (98)
34	19.5 ± 3.9 (106) −31%	22.5 ± 10.24 (273) −33%	0.125 ± 0.083 (357) −17.2%	8.7 ± 3.7 (72) −24%

[a]Numbers in brackets represent the number of MTs studied for each measurement. Percentage values refer to the difference relative to the 37°C measurement.

Results show that a difference of three degrees below the physiological temperature lets the laser-induced depolymerization drop by 31%. Shrinkage measurements were performed with total internal reflection microscopy (TIRF) and showed a very consistent value dropping by 33% at 34°C, as compared to 37°C. Interestingly, the rescue frequency, measured as the inverse value of the time between a laser-induced catastrophe and the next rescue (visible by the appearance of a new tip, Colombelli *et al.*, 2005a), also dropped from 37 to 34°C, by 17%. At 39°C, shrinkage rates increase were observed, of 17% and 17.8% as measured with laser nanosurgery and TIRF, respectively.

Migration of eukaryotic cells is the result of complex dynamic processes such as cytoskeletal organization, transduction of extracellular signals, and interactions with the substrate (Vicente-Manzanares *et al.*, 2005). Cells in motion can remodel their MT array rapidly in response to an extracellular signal. Selective stabilization in the direction of the movement of MTs is an early event in the generation of cellular asymmetry (Gundersen and Bulinski, 1988; Kaverina *et al.*, 1998). Moreover, the maintenance of the polarized distribution of actin-dependent protrusion at the leading edge required an intact MT cytoskeleton (Salmon *et al.*, 2002). MT dynamics can be driven and controlled along gradient of tenth of microns long (Bastiaens *et al.*, 2006). Therefore, it is easy to conceive that in the case of trophoblastic cells seeking the maternal blood circulation to establish the placenta (Higazi *et al.*, 1996), local variation of MT polymerization speed may affect their interactions with adhesion sites, destabilizing them on the coldest side of the cell promoting migration toward the opposite direction. The effect of temperature on MT dynamics is of importance when considering migration of cells in a body where temperature varies.

C. Relaxation of Intracellular Forces: Probing Spindle Mechanics in Mitosis

Very early experiments of intracellular laser surgery on fungi (Aist and Berns, 1981) provided the first evidences that spindle elongation during the second phase of mitosis, or anaphase, could involve external pulling forces applied by astral MTs. This experimental approach became a classical way to discriminate between the different forces applied to mitotic asters during spindle elongation. In mammalian cells, laser severing of the spindle zone (Aist *et al.*, 1991; Spurck *et al.*, 1990) leads to an increase in the poles' separation velocity, and damage to one of the poles does not prevent spindle elongation. It is widely accepted now that astral MTs pull on mitotic asters and contribute to its elongation and alignment. More recent laser spindle severing experiments showed a similar mechanism in yeast *Saccharomyces pombe* (Khodjakov *et al.*, 2004; Tolić-Nørrelykke *et al.*, 2004) and the nematode *C. elegans* (Grill *et al.*, 2001) providing also candidates for the interaction between MTs and the cell cortex, for example dynein motors.

We studied the positioning of individual MT asters (Grill *et al.*, 2001, 2003). This process is particularly important for asymmetric spindle positioning, where both MT asters are eccentrically positioned within the cell, directing the cleavage furrow to an off center position and allowing for the creation of one larger and one smaller daughter cells. This type of asymmetric cell division is inherent to any developing organism as they participate in the generation of cell fate diversity. An MT aster is a polar structure with astral MTs emanating from the centrosome, which is at the center of the aster. The minus ends of MTs are located at the centrosome, and their plus ends reach out to the cortex. A minus-end-directed MT motor (such as dynein) anchored to the cortex can capture an astral MT and start to walk toward the MT's minus end. It will thereby exert a pulling force on the aster. As astral MTs point out from the centrosome in all directions, these pulling forces determine the positioning behavior of an MT aster. Now if the centrosome is disintegrated and broken up into smaller fragments, tension is released and these fragments start to move toward the cortex, thus allowing visualization of the forces exerted on the aster. These fragments were tracked and their velocities were measured in a live-cell assay. A statistical analysis of the speeds of the fragments yields an approximation of the number of force generators that are available to each MT aster in a mitotic one-cell stage *C. elegans* embryo.

Laser ablation applied to systems exerting forces of different origins is nowadays a major experimental protocol in cell biology and cell mechanics. Severing selectively in time and space the specific regions where forces are exerted, like the spindle pole in mitosis, provides unique evidences to discriminate and identify the actors of force generation.

D. Breaking the Interactions Between the Cytoskeleton and Organelles

Eukaryotic organisms rely on intracellular transport to well-placed organelles and other components within their cells. Organelles often move bidirectionally, employing both plus-end- and minus-end-directed motors (Welte, 2004). Organelle transport is vital, especially for cell types such as neurons where the distances between sites of organelle biogenesis, function, and recycling or degradation can be vast (Hollenbeck and Saxton, 2005). The MT network is one of the main transport systems, but also a maintenance system as demonstrated in the case of the endoplasmic reticulum (ER) that uses MT motor proteins to adopt and maintain its extended, reticular organization (Lane and Allan, 1999). Along the same line, the organization of the secretory pathway and the transfer of components between them uses MT motors for both directions of movement. They are simultaneously present on vesicles and intermediate compartments, but their activity is coordinated so that when plus-end motors are active, minus-end motors are not and vice versa (Appenzeller-Herzog and Hauri, 2006). The ER exit sites (ERES) are the sites of

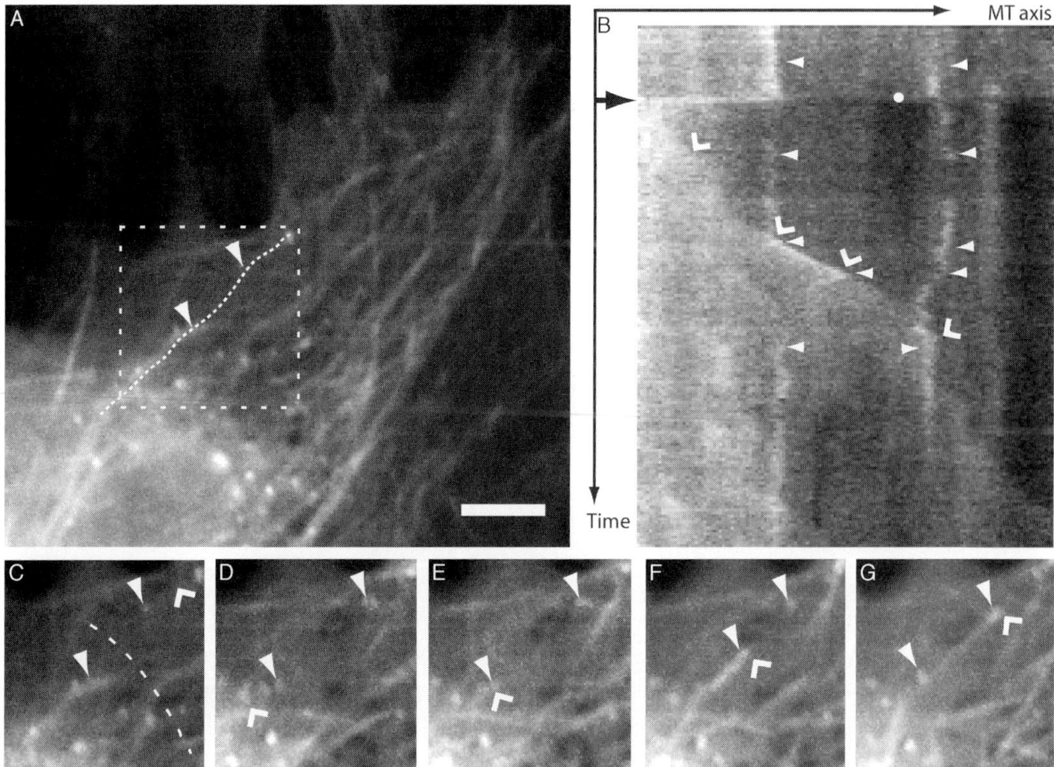

Fig. 5 Interaction of MTs with organelles: ERES. (A) An overview of a Vero cell doubly transfected with EB3 GFP and YFP-Sec23A. ERES form dotlike structures. Sequence (B–F) shows the regrowth of an MT toward two ERES after surgery. ERES are free of MTs (after they depolymerized) after surgery. Thin arrows show the position of the two ERES and the right angle arrow follows the MT plus tip. A kymograph in (G) shows the time–space evolution of the MT along its axis. It clearly shows the MT regrowing *de novo* toward the two ERES and even displacing the first one along a relatively long distance (about 5 μm). Scale bars = 10 μm.

vesicle budding at the ER. For efficient transport toward the Golgi apparatus, they must be coupled to MTs and recruit the necessary motors and accessory factors to initiate vesicle motility. Recent evidences pointed out a role for the COPII coat to recruit dynactin. Dynactin-labeled MT plus ends translocate through the cytoplasm in the direct proximity of the ERES (Watson *et al.*, 2005). However, the molecular clue that target MT orient toward ERES is still to be identified.

The study of MT dynamics in relation to organelle targeting is based on drugs with wide effects over the entire cell (e.g., nocodazole, taxol) generating an important disturbance of the complete cell system. As shown in Fig. 5, it is perfectly possible to use laser nanosurgery to study a precise targeting system at the single ERES level, this allows easier procedures and higher numbers for statistical analyses. After inducing a front of catastrophe anywhere in the cell, *de novo* growth of MTs can be followed and correlated to ERES location, thereby showing the direct and targeted interaction of MTs toward these intracellular organelles. Moreover, the coupling of FRAP and laser nanosurgery may allow the study of coat proteins turnover before and after MT interaction allowing us to precisely dissect the molecular machinery attracting or grabbing MT to enhance ER to Golgi transport.

E. Relaxation of Intracellular Forces: Cell Contractility

A second, and major, component of the cytoskeleton consists of the actin network and is organized in different filament scaffolds depending on the cellular

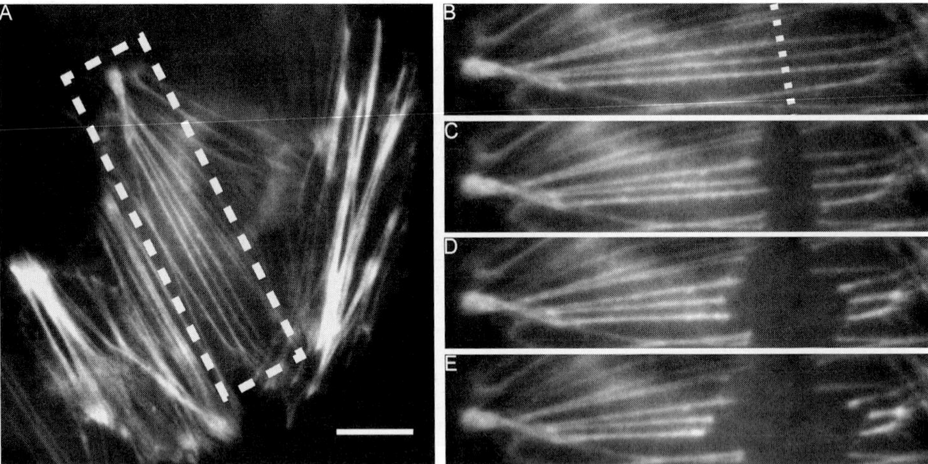

Fig. 6 Stress fiber release after laser nanosurgery. Stress fibers exert contractile forces between adhesion sites in cultured cells. The image sequence shows a Ptk2 mammalian cell expressing GFP F-actin undergoing nanosurgery. (A) Overview of the cell, the dashed square refers to the zoomed area showed in (B–E) and the dashed line to the laser target. It is clear from the figure that stress fibers retract by contracting from the new end of the released stress fiber. Adapted with permissions from Colombelli *et al.* (2006). Scale bar = 10 μm.

region and activity. Actin polymerization is actively responsible for cell motility, cytokinesis, cell polarity, tension, and adhesion processes. In particular, when seeded under culture conditions, actin filaments can align and form contractile bundles, called stress fibers, by interacting with other actin-related proteins, for example α-actinin or myosin II. As early as 1982, first experiments of laser surgery on stress fibers (Koonce *et al.*, 1982) showed interesting behavior of these bundles on laser treatment. The same experiment was performed by several authors (Colombelli *et al.*, 2005a; Heisterkamp *et al.*, 2005; Kumar *et al.*, 2006; Lele *et al.*, 2006) to demonstrate stress fiber contractility with fluorescent imaging. Figure 6 illustrates the behavior of stress fibers following laser severing. These actin bundles have a very complex structure made of the assembly of polymerized actin filaments, myosin light chains, α-actinin, tropomyosin, and other proteins. However, fluorescent imaging alone results in very slow motion of stress fibers that are indeed very stable in epithelial cells and FRAP experiments could show recovery times in the order of tens of minutes (Edlund *et al.*, 2001). Stress fibers retract after laser severing, suggesting the complete release of the exerted forces along the bundle. As already shown, the retraction and compaction of stress fibers could tell us more about their internal structure. The strength of applying laser surgery here lies in the possibility to perturb irreversibly the equilibrium. The dynamics of mechanosensitive proteins at focal adhesions, for example zyxin, was shown to be affected by the release of stress fibers (Lele *et al.*, 2006) dependent forces and mathematical models start to emerge (Kumar *et al.*, 2006) to simulate their retraction. We believe that this experimental protocol will become crucial in understanding the role of a battery of mechanosensitive proteins (Vogel and Sheetz, 2006; Yoshigi *et al.*, 2005) and help to connect biochemical pathways to mechanical behavior and cell mechanotransduction. Indeed, most of the cellular processes like cell contractility, cell–substrate, or cell–cell adhesions are based on very organized filaments interactions regulated by biochemical pathways. Laser severing allows now for a controlled perturbation of those pathways, regionally with a submicron precision, by releasing intracellular forces in defined places.

V. Material and Methods

A. Nanosurgery Instrument

The instrument is based on a Zeiss inverted microscope Axiovert 200M. A pulsed UV laser Nd:YAG at $\lambda = 355$ nm (third harmonic of $\lambda_D = 1064$ nm) is expanded by means of telecentric optics to meet the diffraction limit with a Zeiss C-Apo 63×/1.2 W lens. The pulse is estimated to have a duration of 470 ps and its theoretical beam diameter in the focal plane is 319 nm. A galvanometer pair scan unit guides the laser beam across a field of $100 \times 100\ \mu m^2$. Coupling into the microscope is achieved via the fluorescence illumination path. The field aperture slider has been modified and contains a dichroic mirror reflecting only the laser wavelength of 355 nm. The laser power is controlled by an acousto-optical tunable filter. To correct for chromatic aberrations at the UV-A wavelengths and to perform laser surgery in three dimensions, the objective is mounted on a piezoelectric positioner. To control the irradiation sequence, the user adjusts the following parameters: pulse energy (up to 8.8 μJ), number of pulses, repetition rate (up to 1 kHz), and focus z-position correction. A custom target shape is defined on a live window across the full irradiation field, and the scan controller board converts the graphical coordinates to angular coordinates that position the laser beam pulses inside the sample, that is in three dimensions. The setup offers all commonly used microscopy contrast modes [fluorescence, differential interference contrast (DIC), phase contrast, as well as confocal microscopy] and also permits the use of different objective lenses.

Wild-type Ptk2 cells were grown in DMEM supplemented with 10% fetal bovine serum under standard tissue culture conditions. Ptk2 cells stably expressing tagged tubulin-YFP were maintained in the presence of 0.45 mg/ml G-418 sulfate. All cell-culture media and sera were obtained from Invitrogen (Carlsbad, CA). Cells were plated onto glass bottom dishes (MatTek, Ashland, MA). Actin-mRFP, actin-EGFP, pEYFP-Mem, EYFP-Sec23A, and EB3-EGFP expression constructs (0.6 μg each) were transfected using 2.6 μl of FuGene 6 Transfection Reagent (Roche, Indianapolis, IN) per dish. In all our experiments, only low-expressing cells were considered for dynamical study.

VI. Concluding Remarks

Laser nanosurgery is increasingly used in new applications that require qualitative as well as quantitative studies of a wide range of intracellular processes. In particular, with the advent of modern fluorescent imaging techniques, highly dynamic processes such as cytoskeleton polymerization can be accurately studied in time and space and correlated with the behavior of subcellular organelles or well-defined subcellular compartments. The advantage of laser nanosurgery is its well controlled and noninvasive capability of perturbing specific and dynamic processes through the targeted severing of subcellular structures that are normally at equilibrium. Particularly important is that the effects can be applied very precisely in time and three-dimensional space. This makes nanosurgery a unique way to push cellular processes out of a physiological equilibrium and allows the reproducible and quantitative study of cellular pathways that are under no condition observable at a physiological equilibrium. If it is carefully implemented and applied and assuming the usage of cost-effective optical parts and laser sources to make it available to more laboratories, laser nanosurgery will find an ever increasing number of applications in cell and developmental biology. The high accuracy in space and in time of pulsed laser nanosurgery will allow the design of new approaches to the regional study of cell mechanics, achieving understanding of cellular compartmentalization. It is unfortunately not yet considered a classical approach to cell mechanics rheology (for review see Heidemann and Wirtz, 2004), unlike optical or magnetic tweezers or atomic force microscopy (AFM), but is definitely adapted to high-accuracy subcellular studies, in particular to processes involving forces.

Acknowledgments

The authors wish to thank Professor Alfred Vogel for invaluable discussions and critical comments on the chapter, Dr. Detlev Arendt and Sebastian Klaus for *P. dumerilii* experiments and for kindly sharing data. We are grateful to Alfons Riedinger and Brigitte Joggerst for technical support and Carl Zeiss (Germany) for providing the microscope apparatus. This research was partially supported by the VDI-TZ and the German Ministry for Research and Development (BMBF) by grant FKZ 13N8287.

References

Abbott, A. (2005). Hopping fences. *Nature* **433,** 680–682.

Aist, J. R., Bayles, C. J., Tao, W., and Berns, M. W. (1991). Direct experimental evidence for the existence, structural basis and function of astral forces during anaphase B *in vivo*. *J. Cell Sci.* **100,** 279–288.

Aist, J. R., and Berns, M. W. (1981). Mechanics of chromosome separation during mitosis in Fusarium (fungi imperfecti): New evidence from ultrastructural and laser microbeam experiments. *J. Cell Biol.* **91,** 446–458.

Aist, J. R., Liang, H., and Berns, M. W. (1993). Astral and spindle forces in PtK2 cells during anaphase B: A laser microbeam study. *J. Cell Sci.* **104,** 1207–1216.

Appenzeller-Herzog, C., and Hauri, H. P. (2006). The ER-Golgi intermediate compartment (ERGIC): In search of its identity and function. *J. Cell Sci.* **119,** 2173–2183.

Bastiaens, P., Caudron, M., Niethammer, P., and Karsenti, E. (2006). Gradients in the self-organization of the mitotic spindle. *Trends Cell Biol.* **16**(3), 125–134.

Berger, F. (1998). Cell ablation studies in plant development. *Cell. Mol. Biol. (Noisy le-Grand)* **44,** 711–719.

Berns, M. W., and Richardson, S. M. (1977). The role of the centriolar region in animal cell mitosis. *J. Cell Biol.* **72,** 351–367.

Botvinick, E. L., Venugopalan, V., Shah, J. V., Liaw, L. H., and Berns, M. W. (2004). Controlled ablation of microtubules using a picosecond laser. *Biophys. J.* **87,** 4203–4212.

Colombelli, J., Grill, S. W., and Stelzer, E. H. K. (2004). Ultraviolet diffraction limited nanosurgery of live biological tissues. *Rev. Sci. Instr.* **75,** 472–478.

Colombelli, J., Reynaud, E. G., Rietdorf, J., Pepperkok, R., and Stelzer, E. H. K. (2005a). *In vivo* cytoskeleton dynamics quantification in interphase cells induced by UV pulsed laser nanosurgery. *Traffic* **6**(12), 1093–1102.

Colombelli, J., Reynaud, E. G., and Stelzer, E. H. K. (2005b). Subcellular nanosurgery with a sub-nanosecond pulsed UV laser. *Med. Laser Appl.* **20,** 217–222.

Colombelli, J., Pepperkok, R., Stelzer, E. H. K., and Reynaud, E. G. (2006). Laser nanosurgery in cell biology. *Med. Sci.(Paris)* **22**(6–7), 651–658.

Desai, A., and Mitchison, T. J. (1997). Microtubule polymerization dynamics. *Annu. Rev. Cell. Dev. Biol.* **13,** 83–117.

Edlund, M., Lotano, M. A., and Otey, C. A. (2001). Dynamics of α-actinin in focal adhesions and stress fibers visualized with α-actinin green fluorescent protein. *Cell Motil. Cytoskeleton* **48**(3), 190–200.

Ellenberg, J., Siggia, E. D., Moreira, J. E., Smith, C. L., Presley, J. F., Worman, H. J., and Lippincott-Schwartz, J. (1997). Nuclear membrane dynamics and reassembly in living cells: Targeting of an inner nuclear membrane protein in interphase and mitosis. *J. Cell Biol.* **138,** 1193–1206.

Grill, S. W., Gönczy, P., Stelzer, E. H. K., and Hyman, A. A. (2001). Polarity controls forces governing asymmetric spindle positioning in the *Caenorhabditis elegans* embryo. *Nature* **409,** 630–633.

Grill, S. W., Howard, J., Schäffer, E., Stelzer, E. H. K., and Hyman, A. A. (2003). The distribution of active force generators controls mitotic spindle position. *Science* **301,** 518–521.

Gundersen, G. G., and Bulinski, J. C. (1988). Selective stabilization of microtubules oriented toward the direction of cell migration. *Proc. Natl. Acad. Sci. USA* **85**(16), 5946–5950.

Hartmann-Petersen, R., Walmod, P. S., Berezin, A., Berezin, V., and Bock, E. (2000). Individual cell motility studied by time-lapse video recording: Influence of experimental conditions. *Cytometry* **40**(4), 260–270.

Heidemann, S. R., and Wirtz, D. (2004). Towards a regional approach to cell mechanics. *Trends Cell Biol.* **14**(4), 160–166.

Heisterkamp, A., Maxwell, I. Z., Mazur, E., Underwood, J. M., Nickerson, J. A., Kumar, S., and Ingber, D. E. (2005). Pulse energy dependence of subcellular dissection by femtosecond laser pulses. *Opt. Express* **13,** 1390–1396.

Higazi, A. A., Kniss, D., Manuppello, J., Barnathan, E. S., and Cines, D. B. (1996). Thermotaxis of human trophoblastic cells. *Placenta* **17**(8), 683–687.

Hollenbeck, P. J., and Saxton, W. M. (2005). The axonal transport of mitochondria. *J. Cell Sci.* **118,** 5411–5419.

Hutson, M. S., Tokutake, Y., and Chang, M. S. (2003). Forces for morphogenesis investigated with laser microsurgery and quantitative modelling. *Science* **300,** 145–149.

Janosi, I. M., Chretien, D., and Flyvbjerg, H. (2002). Structural microtubule cap: Stability, catastrophe, rescue, and third state. *Biophys. J.* **83**(3), 1317–1330.

Joglekar, A. P., Liu, H. H., Meyhöfer, E., Mourou, G., and Hunt, A. J. (2004). Optics at critical intensity: Applications to nanomorphing. *Proc. Natl. Acad. Sci. USA* **101,** 5856–5861.

Kaverina, I., Rottner, K., and Small, J. V. (1998). Targeting, capture, and stabilization of microtubules at early focal adhesions. *J. Cell Biol.* **142**(1), 181–190.

Khodjakov, A., La Terra, S., and Chang, F. (2004). Laser microsurgery in fission yeast: Role of the mitotic spindle midzone in anaphase B. *Curr. Biol.* **14,** 1330–1340.

Kiehart, D. P., Galbraith, C. G., Edwards, K. A., Rickoll, W. L., and Montague, R. A. (2000). Multiple forces contribute to cell sheet morphogenesis for dorsal closure in *Drosophila*. *J. Cell Biol.* **149,** 471–490.

Koonce, M. P., Strahs, K. R., and Berns, M. W. (1982). Repair of laser-severed stress fibers in myocardial non-muscle cells. *Exp. Cell Res.* **141**(2), 375–384.

Kumar, S., Maxwell, I. Z., Heisterkamp, A., Polte, T. R., Lele, T. P., Salanga, M., Mazur, E., and Ingber, D. E. (2006). Viscoelastic retraction of single living stress fibers and its impact on cell shaoe, cytoskeletal organization, and extracellular matrix mechanics. *Biophys. J.* **90**(10), 3762–3773.

Lane, J. D., and Allan, V. J. (1999). Microtubule-based endoplasmic reticulum motility in *Xenopus laevis*: Activation of membrane-associated kinesin during development. *Mol. Biol. Cell* **10**(6), 1909–1922.

Lele, T. P., Pendse, J., Kumar, S., Salanga, M., Karavitis, J., and Ingber, D. E. (2006). Mechanical forces alter zyxin unbinding kinateics within focal adhesions of linving cells. *J. Cell. Physiol.* **207**(1), 187–194.

Montell, D. J., Keshishian, H., and Spradling, A. C. (1991). Laser ablation studies of the role of the *Drosophila oocyte* nucleus in pattern formation. *Science* **254,** 290–293.

Nahas, G. G., Tannieres, M. L., and Lennon, J. F. (1971). Direct measurement of leukocyte motility: Effects of pH and temperature. *Proc. Soc. Exp. Biol. Med.* **138,** 350–352.

Noack, J., and Vogel, A. (1999). Laser-induced plasma formation in water at nanosecond to femtosecond time scales: Calculation of thresholds, absorption coefficients, and energy density. *IEEE J. Quantum Electron.* **35,** 1156–1167.

Ream, R. A., Theriot, J. A., and Somero, G. N. (2003). Influences of thermal acclimation and acute temperature change on the motility of epithelial wound-healing cells (keratocytes) of tropical, temperate and Antarctic fish. *J. Exp. Biol.* **206,** 4539–4551.

Sacconi, L., Tolić-Nørrelykke, I., Antolini, R., and Pavone, F. S. (2005). Combined intracellular threedimensional imaging and selective nanosurgery by a nonlinear microscope. *J. Biomed. Opt.* **10,** 014002.

Salmon, W. C., Adams, M. C., and Waterman-Storer, C. M. (2002). Dual-wavelength fluorescent speckle microscopy reveals coupling of microtubule and actin movements in migrating cells. *J. Cell Biol.* **158,** 31–37.

Shimada, T., Watanabe, W., Matsunaga, S., Higashi, T., Ishii, H., Fukui, K., Isobe, K., and Itoh, K. (2005). Intracellular disruption of mitochondria in a living HeLa cell with a 76-MHz femtosecond laser oscillator. *Opt. Express* **13**(24), 9869–9880.

Spurck, T. P., Stonington, O. G., Snyder, J. A., Pickettheaps, J. D., Bajer, A., and Molebajer, J. (1990). UV Microbeam irradiations of the mitotic spindle. 2. Spindle Fiber dynamics and force production. *J. Cell Biol.* **111**(4), 1505–1518.

Supatto, W., Debarre, D., Moulia, B., Brouzes, E., Martin, J. L., Farge, E., and Beaurepaire, E. (2005). *In vivo* modulation of morphogenetic movements in *Drosophila* embryos with femtosecond laser pulses. *Proc. Natl. Acad. Sci. USA* **102,** 1047–1052.

Tirlapur, U. K., and König, K. (2002). Femtosecond near-IR laser pulses as a versatile non-invasive tool for intra-tissue nanoprocessing in plants without compromising viability. *Plant J.* **31**(3), 365–374.

Tolić-Nørrelykke, I. M., Sacconi, L., Thon, G., and Pavone, F. S. (2004). Positioning and elongation of the fission yeast spindle by microtubule-based pushing. *Curr. Biol.* **14,** 1181–1186.

Thurston, G., and Palcic, B. (1987). 3T3 cell motility in the temperature range 33 degrees C to 39 degrees C. *Cell Motil. Cytoskeleton* **7**(4), 361–367.

Venugopalan, V., Guerra, A., Nahen, K., and Vogel, A. (2002). Role of laser-induced plasma formation in pulsed cellular microsurgery and micromanipulation. *Phys. Rev. Lett.* **88**(7), 078103.

Vicente-Manzanares, M., Webb, D. J., and Horwitz, A. R. (2005). Cell migration at a glance. *J. Cell Sci.* **118,** 4917–4919.

Vogel, A., and Noack, J. (2001). Numerical simulation of optical breakdown for cellular surgery at nanosecond to femtosecond time scales. *Proc. SPIE* **4260,** 83–93.

Vogel, A., Noack, J., Huttman, G., and Paltauf, G. (2005). Mechanisms of femtosecond laser nanosurgery of cells and tissues. *Appl. Phys. B* **81**(8), 1015–1047.

Vogel, A., and Venugopalan, V. (2003). Mechanisms of pulsed laser ablation of biological tissues. *Chem. Rev.* **103,** 577–644.

Vogel, V., and Sheetz, M. (2006). Local force and geometry sensing regulate cell functions. *Nat. Rev. Mol. Cell Biol.* **7**(4), 265–275.

Watanabe, W., Arakawa, N., Matsunaga, S., Higashi, T., Fukui, K., Isobe, K., and Itoh, K. (2004). Femtosecond laser disruption of subcellular organelles in a living cell. *Opt. Express* **12,** 4203–4213.

Watson, P., Forster, R., Palmer, K. J., Pepperkok, R., and Stephens, D. J. (2005). Coupling of ER exit to microtubules through direct interaction of COPII with dynactin. *Nat. Cell Biol.* **7**(1), 48–55.

Welte, M. A. (2004). Bidirectional transport along microtubules. *Curr. Biol.* **14**(13), R525–R537.

White, J., and Stelzer, E. H. K. (1999). Photobleaching GFP reveals protein dynamics inside live cells. *Trends Cell Biol.* **9,** 61–65.

Willingham, E. (2002). Laser microdissection systems. *The Scientist* **16,** 42.

Yamada, Y., and Ohshima, Y. (2003). Distribution and movement of *Caenorhabditis elegans* on a thermal gradient. *J. Exp. Biol.* **206,** 2581–2593.

Yanik, M. F., Cinar, H., Cinar, H. N., Chisholm, A. D., Jin, Y., and Ben Yakar, A. (2004). Functional regeneration after laser axotomy. *Nature* **432,** 822.

Yoshigi, M., Hoffman, L. M., Jensen, C. C., Yost, H. J., and Beckerle, M. C. (2005). Mechanical force mobilizes zyxin from focal adhesions to actin filaments and regulates cytoskeletal reinforcement. *J. Cell Biol.* **171**(2), 209–215.

CHAPTER 9

Fs-Laser Scissors for Photobleaching, Ablation in Fixed Samples and Living Cells, and Studies of Cell Mechanics

Alexander Heisterkamp,* Judith Baumgart,* Iva Z. Maxwell,† Anaclet Ngezahayo,‡ Eric Mazur,† and Holger Lubatschowski*

*Laser Zentrum Hannover, Hollerithallee 8, D-30419 Hannover, Germany

†Department of Engineering and Applied Science, Harvard University, Cambridge, Massachusetts 02138

‡Institute of Biophysics, Leibniz University Hannover, D-30419 Hannover, Germany

The use of ultrashort laser pulses for microscopy has steadily increased over the past years. In this so-called multiphoton microscopy, laser pulses with pulse duration around 100 femtoseconds (fs) are used to excite fluorescence within the samples. Due to the high peak powers of fs lasers, the absorption mechanism of the laser light is based on nonlinear absorption. Therefore, the fluorescence signal is highly localized within the bulk of biological materials, similar to a confocal microscope. However, this nonlinear absorption mechanism can not only be used for imaging but for selective alteration of the material at the laser

0091-679X/07 $35.00
DOI: 10.1016/S0091-679X(06)82009-1

focus: The absorption can on one hand lead to the excitation of fluorescent molecules of fluorescently tagged cells by the simultaneous absorption of two or three photons or on the other hand, in case of higher order processes, to the creation of free-electron plasmas and, consequently, plasma-mediated ablation. Typical imaging powers are in the range of tens of milliwatts using 100-fs pulses at a repetition rate of 80–90 MHz, while pulse energies needed for ablation powers are as low as a few nanojoules when using high numerical aperture microscope objectives for focusing the laser radiation into the sample.

Since the first demonstration of this technique, numerous applications of fs lasers have emerged within the field of cellular biology and microscopy. As the typical wavelengths of ultrashort laser systems lie in the near infrared between 800 and 1000 nm, high penetration depth can be achieved and can provide the possibility of imaging and manipulating the biological samples with one single laser system.

I. Introduction

Ultrashort laser pulses can lead to very high peak power up to the TW/cm^2, when the laser radiation is focused by high numerical aperture (NA) objectives. These high peak intensities can be used to induce nonlinear absorption effects at the focus region, even in materials that are under one-photon conditions transparent, confining the interaction to very small volumes of only a few femtoliters. As a consequence, two main applications of ultrashort laser pulse have evolved over the past years, imaging, for example multiphoton microscopy, as first realized in living cells by Denk *et al.* (1990), and micromachining within the bulk of different materials by plasma-mediated ablation, as used by Stern *et al.* (1989) and Juhasz *et al.* (1999), to ablate corneal tissue or even bulk materials like fused silica (Chichkov *et al.*, 1996; Schaffer *et al.*, 2001). In multiphoton microscopy, the laser beam is scanned within the biological sample, inducing the excitation of fluorescent molecules by the simultaneous absorption of two or three photons. The image is reconstructed by collecting the induced fluorescence spot by spot, analogue to confocal microscopy.

For these imaging purposes the average laser powers are typically in the range of several milliwatts up to tens of milliwatts, when focusing with objectives of 0.8 or higher NA. When higher laser power is applied, an interplay of multiphoton absorption and cascade ionization, induced by absorption of inverse bremsstrahlung, leads to the creation of high-density plasmas, which results in a very precise and localized plasma-mediated ablation of the material at the laser focus (Vogel and Venugopalan, 2003).

The first group to use short pulses for micro ablation was Michael Berns in the late 1980s or early 1990s (Berns *et al.*, 1981; Tao *et al.*, 1988). In the following years, various applications of laser microbeams in cell biology evolved, as for example the precise dissection of chromosomes (Djabali *et al.*, 1991; Liang *et al.*, 1993; Monajembashi *et al.*, 1986).

Fs lasers have been introduced to cellular dissection by Karsten Koenig by demonstrating as well the dissection of chromosomes in a living cell (Koenig *et al.*, 1999).

While studying the side effects of ultrashort pulse illumination of biological samples, he found that after exceeding a certain threshold, very precise ablation within the bulk of the sample is possible, coining the term nanosurgery. Nowadays, fs lasers are increasingly used to micromanipulate and ablate nanoscale structures in living cells, for example dissection and precise ablation of chromosomes (Koenig, 2000) or ablation of single organelles like mitochondria (Shen *et al.*, 2005; Watanabe *et al.*, 2004). In another application, fs-laser pulses can be used to selectively transfect cells with foreign DNA as shown by Tirlapur and Koenig (2002).

In contrast to UV-laser scissors, the application of fs-laser systems, which can achieve a comparable cutting precision within biological samples (Aist *et al.*, 1993; Stern *et al.*, 1989), offers two main advantages: First, the ablation mechanism is based on nonlinear absorption, allowing the precise ablation of subcellular components deep inside the bulk of tissue or cell cultures. Although ns or ps pulses can be employed for multiphoton absorption as well, the energy delivered to the sample can be minimized by using ultrashort laser pulses (Vogel and Venugopalan, 2003), reducing mechanical and thermal side effects. By fine-tuning of the parameters, ablation dimensions below 100 nm are achievable (Koenig, 2000), leading to the term nanoscissors. As the wavelength of an ultrashort laser is typically in the near infrared, for example centered at 800 nm, high penetration depths of up to several hundreds of micrometers into tissues are possible (So *et al.*, 1998). The amount of laser radiation which is absorbed in regions outside the focus is negligible. Thus, applications even in whole organisms, like *Caenorhabditis elegans* (Chung *et al.*, 2005; Yanik *et al.*, 2004), and drosophila embryos (Supatto *et al.*, 2005), or even working in the brains of living rats (Nishimura *et al.*, 2006) become possible. The second advantage is the possible combination of imaging and ablation, as one fs-laser system can be used simultaneously for both imaging and cutting of cellular samples. Moreover, the nonlinear nature of the optical absorption allows the treatment of any transparent sample, regardless of its linear absorption coefficient as the multiphoton absorption is limited to the focus region. With pulse duration of about 100 fs, only a few nanojoules of energy are necessary to achieve ablation. As possibly harmful effects like cavitation and heat deposition scale with the amount of applied laser energies (Vogel *et al.*, 2005), ultrashort laser applications offer a way to manipulate live cells with very low side effects, the low energy limits collateral damage to the very vicinity of the laser focus and reduces the possibility that the cells will be injured or killed.

II. Methods

A. Optical Setup

Since the laser pulse energies needed to achieve a breakdown are low compared to the typical output powers of standard fs-laser systems, so-called oscillator laser systems with repetition rates of 80–90 MHz and pulse energies of several nanojoules are usually used. Thus, the laser power of such systems is typically several hundreds of milliwatts. Our laser system (Coherent Chameleon) emits laser pulses of 140-fs pulse duration and allows the tuning of the wavelength of the laser between 715 and 955 nm. The laser radiation is coupled into a microscope objective, for example a Zeiss Achroplan C water immersion objective with an NA of 0.8, focusing the laser beam to a theoretical spot size of roughly 600 nm at a wavelength of 800 nm. As the laser threshold for inducing multiphoton processes is intensity dependent, the pulse duration should be kept as low as possible, leading to a minimum amount of energy delivered to the sample, minimizing the collateral damage to the sample. Moreover, one has to adjust the laser parameters correctly, as laser parameters like repetition rate and pulse energy are very sensitive with respect to the collateral damage within the sample. At high repetition rates, pulse-to-pulse interaction can lead to heat accumulation within the sample, as observed in fused silica (Schaffer *et al.*, 2001). Therefore, the application of MHz pulses for cutting has to be controlled by a fast shutter, preventing the heating of the sample by consecutive pulses, or by an acousto-optic modulator to lower the repetition rates to the kHz regime. Additionally, nonlinear propagation effects inside the objective and the sample can stretch the pulse duration, making pulse width well below 100 fs not well usable. The dispersion of our objective leads to an effective pulse duration inside the sample of roughly 220 fs (Koenig, 2000).

In our setup, the laser radiation is guided over a computer-controlled scanning system (Cambridge Technologies, MA, 6210) for acquiring multiphoton images with the possibility of synchronous ablation by a second laser beam. The setup is shown in Fig. 1. An acousto-optic modulator (by APE, Berlin, Pulse Select) is used to divide the repetition rate in the cutting beam down to kHz, whereas the zero-order diffracted beam is used for multiphoton imaging. This imaging beam is directed over the scanning mirrors to scan the laser beam over the back aperture of the objective. The image acquisition and scanner system is computer controlled by a LabView program described by Tsai *et al.* (2002). The setup for the multiphoton imaging consists of different dichroic mirrors and a photomultiplier tube (Hamamatsu, R6357, Japan) for the detection of the fluorescence and a CCD camera for white light images. Additionally, a UV lamp offers the possibility of acquiring conventional fluorescence images. The second laser beam, the cutting beam, could be whether chosen to be fixed, taking continuously multiphoton images, or it could be guided over the scanning mirrors, allowing more complicated cutting pattern or to target multiple points within the sample. A three-dimensional (3D) computer-controlled positioning of the sample is enabled by a piezo-controlled positioning stage (Thorlabs Inc., NJ, Nanomax 311) with a resolution of 20 nm.

B. Cell Preparation

Bovine capillary endothelial (BCE) cells (passage 10–15) were maintained at 37 °C in 5% CO_2 on tissue culture dishes in a complete medium composed of low-glucose Dulbecco's modified Eagle's medium (DMEM; Gibco-BRL, Grand Island, NY) supplemented with 10% fetal calf serum (FCS) (Hyclone, Logan, Utah),

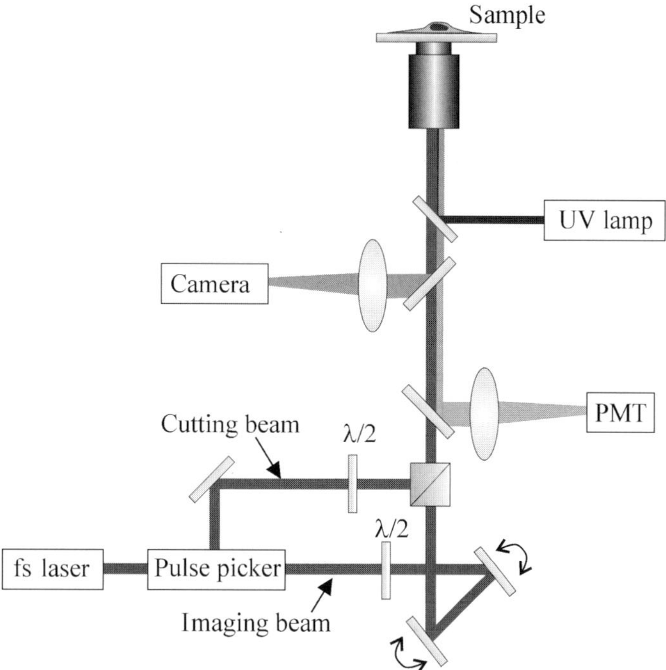

Fig. 1 The laser radiation is guided in two paths toward the objective, an imaging path at 90-MHz repetition rate and at low power, the other path at higher powers and low repetition rates for disruption of organelles. The half wave plates are used in combination with beam splitter cubes to attenuate the laser beams.

10-mM HEPES (JRH-Biosciences, Lenexa, KS), and glutamine (0.292 mg/ml)/penicillin (100 U/ml)/streptomycin (100 g/ml) as previously described by Kumar *et al.* (2006). The granulosa cells were cultivated the same way and were labeled, for the live-cell experiments targeting the mitochondria, with 1-mmol/liter MitoTracker Orange buffered solution (Invitrogen GmbH, Karlsruhe, Germany), compounded with dimethylsulfoxide (DMSO). The dye was added to the nutrient solution at the rate of 1:500 for 45 min. The linear absorption maximum of MitoTracker Orange is at 554 nm and the emission maximum is at 576 nm.

The rat fibroblasts were resistant for 5-bromo-2′-deoxyuridine. The cells were incubated for 10 min with 1.7 μg/ml DAPI (4′,6-diamidino-2-fenipentol) to label the cell nucleus. The linear absorption maximum of DAPI is at 358 nm and the emission maximum is at 461 nm.

For the live-cell experiments targeting the cytoskeleton, cells were transfected for 48 h with an adenoviral vector system encoding enhanced green fluorescent protein (EGFP)-tagged G-tubulin (for microtubules) or with yellow fluorescent protein (YFP)-tagged G-actin (for actin fibers), trypsinized (Trypsin-EDTA, Gibco by Invitrogen GmbH, Karlsruhe, Germany), harvested, and seeded onto glass-bottomed

35-mm dishes (MatTek, Ashland, MA) in complete medium. Prior to imaging, cells were transferred into a nonfluorescent, CO_2-independent medium (pH 7.3) containing (in mM): $CaCl_2$ (1.26), $MgSO_4$ (0.81), KCl (5.36), KH_2PO_4 (0.44), NaCl (137), Na_2HPO_4 (0.34), D-glucose (5.55), L-glutamine (2.0), sodium pyruvate (1.0), HEPES (20.0), 1% bovine serum albumin, 10% calf serum, and MEM essential and nonessential amino acids (Sigma, St. Louis, MO).

For the transmission electron microscopy (TEM) experiments with fixed cells, the cells were then trypsinized (Trypsin–EDTA, Gibco), harvested, and seeded onto -carbon-coated formvar on Embra TEM finder grids (Electron Microscope Sciences, PA) in complete medium. After allowing the cells to attach and spread for 12–24 h, the cells were fixed in 4% formaldehyde (electron microscopy grade) in phosphate buffered saline (PBS) for 40 min, permeabilized in 0.1% Triton X-100 in PBS for 5 min, blocked in 1% bovine serum albumin in PBS for 1 h, and stained for either actin (Alexa Fluor488 phalloidin, Molecular Probes, Eugene, OR) or nuclear DNA (Hoechst 33348, Molecular Probes), all at room temperature. Afterward the cells were stored and treated in an aqueous solution (PBS). Following laser treatment, cells were fixed in 2.5% glutaraldehyde in 0.1-M cacodylate buffer, pH 7.4, at 4°C for 1 h and then washed and stored in this buffer at 4°C until processing. Prior to TEM imaging with a Philips CM-10 microscope, cells were fixed in 1% osmium tetroxide in 0.1-M cacodylate buffer, pH 7.4, at 4°C for 30 min, washed in the same buffer, dehydrated in graded ethanol solutions, critical point dried, and carbon coated.

III. Ablation and Photobleaching in Fixed Samples

As cells and cell organelles are typically visualized by fluorescent tagging of the areas of interest, it is important to prove that the laser energy applied to the cellular sample actually cuts or ablates the targeted structure and not only photobleaches the fluorescent dye. Therefore, we performed studies on fixed bovine endothelial cells and rat1 fibroblasts, by cutting within the cytoskeleton stained by actin (endothelial cells) or by ablating within the nucleus of the cell, stained by Hoechst 33348 (endothelial cells) or by DAPI (rat1 fibroblasts).

Figure 2 shows the fluorescence from the actin network within an endothelial cell after it has been irradiated along five parallel lines with various pulse energies (Heisterkamp *et al.*, 2005). The sample was translated at a speed of $\sim$0.7 μm/s corresponding to roughly 15,000 pulses per line. The image shows that the fluorescence intensity following irradiation depends strongly on pulse energy. At 1.8 nJ, the effect of irradiation is barely visible in the fluorescence image, see fluorescence intensity plot in small inset of Fig. 2. Increasing the pulse energy to 2.2 nJ produces a clear decrease in fluorescence with a width of 240 nm. At higher energy, the width of this decrease in fluorescence scales with pulse energy from 360 nm at 2.8 nJ to 500 nm at 3.5 nJ and 600 nm at 4.4 nJ.

Similar results and thresholds for laser ablation are found in rat1 fibroblast, when targeting the cells' nucleus. Figure 3A shows the nucleus of a fixed

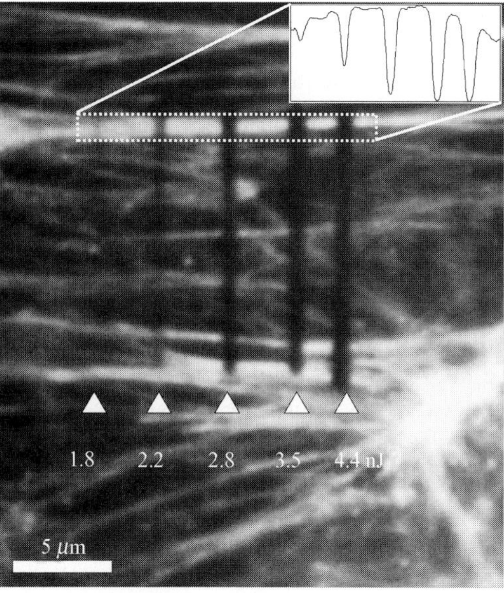

Fig. 2 Fluorescence microscope image of laser cuts through YFP-labeled actin fibers in a fixed endothelial cell obtained by irradiation (Heisterkamp *et al.*, 2005).

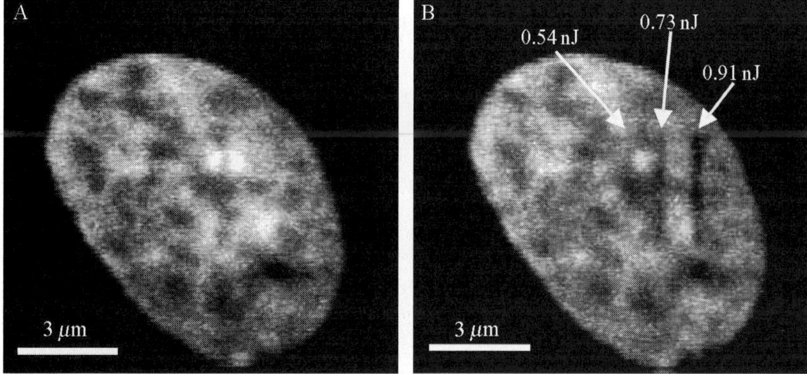

Fig. 3 Fixed rat1 fibroblast, the nucleus is DAPI labeled. (A) The nucleus before manipulation and (B) the same nucleus after manipulation by different pulse energies: 0.54, 0.73, and 0.91 nJ. The images were taken by two-photon microscopy.

fibroblast, stained by DAPI for nuclear DNA. After applying the laser pulses at three different energies of 0.54, 0.73, and 0.91 nJ, ablated channels within the nucleus can be seen, Fig. 3B, showing a clear line at 0.91 nJ, a faint cut at 0.73 nJ, and no change in fluorescence at 0.54 nJ.

To determine the onset of plasma-mediated ablation from the threshold for laser-induced bleaching of the dyes, we compared fluorescent micrographs of laser-irradiated cells with images taken by TEM. Figure 4A shows the fluorescence image of the nucleus of an endothelial cell, stained by Hoechst 33348, with laser cuts at 1.45, 1.8, and 2.3 nJ. Again, the thickness and visibility of the cut scales with laser intensity. In Fig. 4B, the same nucleus is shown in TEM. Apparently, the faint cut at 1.45 nJ is not visible in the TEM, thus, the observed line in Fig. 4A must be due to photobleaching.

At higher energies, clear cuts can be seen in both images, see cuts at 1.8 and 2.3 nJ. The data allow us to define three regimes of irradiation: no interaction (no damage visible in either image), photobleaching without material loss (only the fluorescence image shows a change), and ablation of material (both images show cuts). For pulse energies above 10–15 nJ, a much larger part of the cell is ablated (not shown) most likely due to cavitation. Part of the energy delivered to the sample cannot be dissipated through thermal diffusion, producing rapid, local increase in material temperature, leading to an explosive expansion of the material and, thus, damage far from the laser focus (Vogel *et al.*, 2005).

Although the thresholds vary from sample to sample and especially from dye to dye, see for comparison thresholds in Figs. 2 and 3, we found the energy threshold of ablation to be at most 20% higher than the photobleaching threshold. In other words, if energies exceed 1.2 times the threshold at which fluorescence disappears, one can be assured of ablation. The TEM and fluorescence microscopy measurements reveal that the ablation width depends strongly on pulse energy, with pulse energies between 1.2 and 1.7 nJ producing material loss as small as 200 nm. Above

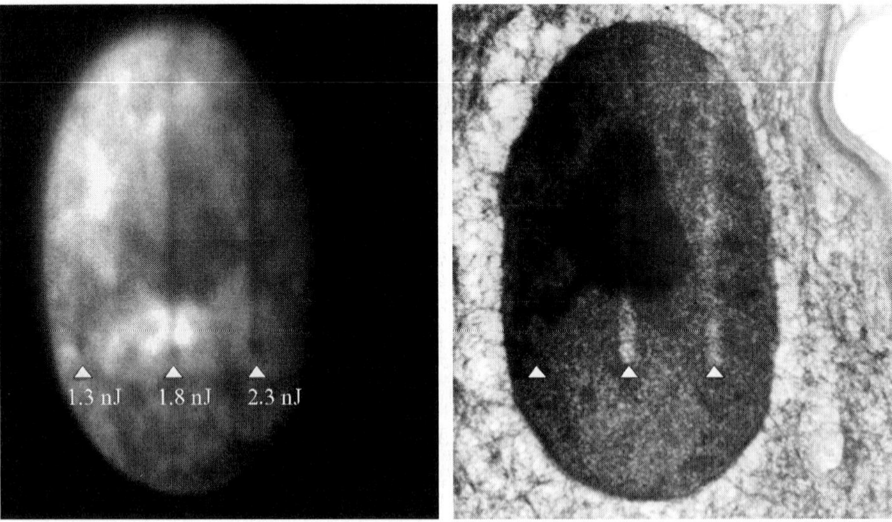

Fig. 4 Fluorescent (A) and transmission electron microscopy (B) images of the same nucleus of a bovine endothelial cell, processed by fs laser (Heisterkamp *et al.*, 2005).

1.7 nJ, the ablation width increases with energy; around 3 nJ the ablation width is ~1 mm. If such energies are applied to live cells, micromanipulation at resolution below a micrometer within the cells is possible.

IV. Applications in Living Cells

A. Cell Organelle Ablation

Once the thresholds for ablation in fixed cells are established, one can move to living cells and target submicron regions inside a single cell. In a first set of experiments, we targeted mitochondria in endothelial cells and granulosa cells stained by MitoTracker Orange to study the induction of cell death, so-called apoptosis. Imaging the cell's reaction to the disruption by multiphoton microscopy imparts information about the sequences of intracellular processes. In order to observe the viability of the cell in dependence to mitochondria disruption, the cell had to be observed over a longer period up to 1 h with both fluorescence imaging and bright field microscopy imaging. The mitochondria disruption was realized at pulse energies between 0.7 and 1.0 nJ and with a speed of ~14 μm/s at a repetition rate of the cutting beam of 90 MHz. The reaction of the cell was observed by comparing the images before and after manipulation.

Typically, a cell changes its volume when it initiates apoptosis and forms vesicles within its volume, which can be seen in the bright field microscopy image and in the fluorescence image by an important change of the mitochondria arrangement. Figure 5A shows living endothelia cells whose mitochondria are MitoTracker

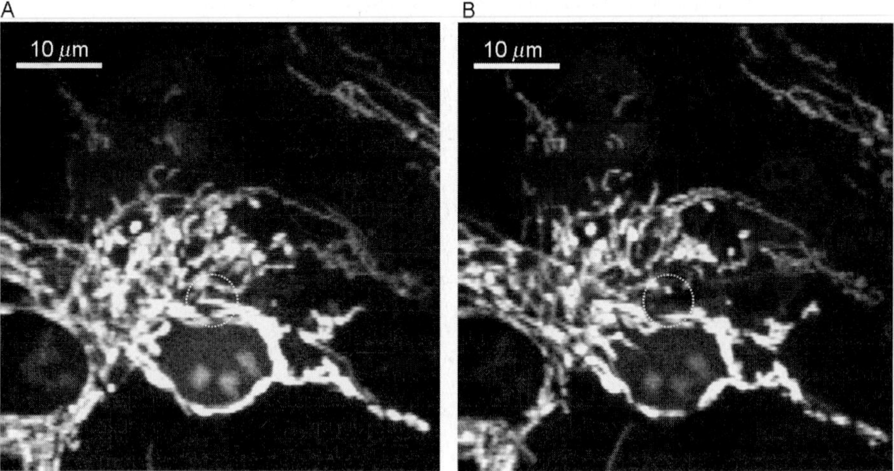

Fig. 5 MitoTracker Orange labeled endothelia cells before (A) and after (B) manipulation. The mitochondrion pointed highlighted by the circle was disrupted at a pulse energy of 1 nJ. Images were taken by two-photon microscopy, in parallel to the laser cutting.

Orange labeled. In Fig. 5B, the ablation of a single mitochondrion (see circle) was realized at pulse energies of 1 nJ.

In none of our studies, the ablation of a single mitochondrion led to the induction of cell death. Apparently, to induce cell death, one has to severe multiple mitochondria. Thus, we targeted several mitochondria within different granulosa cells, as displayed in Fig. 6, showing two cells before ablation (A), the targeted mitochondria (B) within the cells and after ablation (C). Figure 6D provides a white light image of the two cells. The difference in the fluorescence image between Fig. 6A and C is shown in Fig. 6E by digital subtraction of the images. Clearly, the areas of targeted ablation coincide with the areas of change in fluorescence, although some changes outside the laser manipulation are visible, likely due to movement of the cell and reorganization of the mitochondria network. To further investigate the process of apoptosis, a LabView program to target the mitochondria in the whole cell in 3D is currently developed and further studies comparing cancerous with healthy cells are underway.

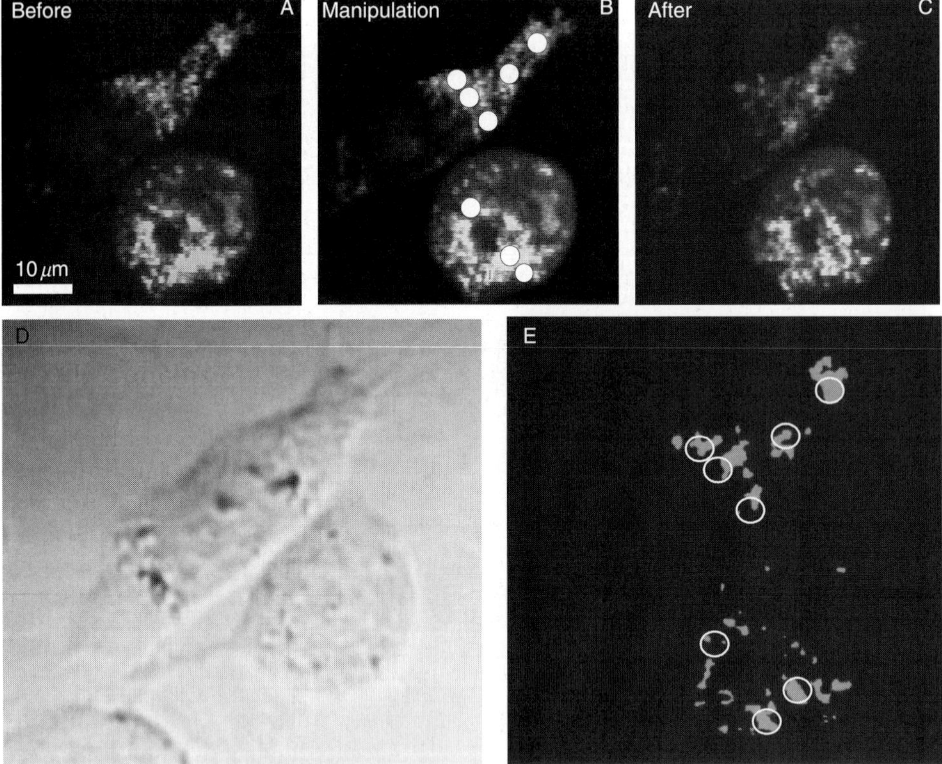

Fig. 6 Multiphoton microscope images of two granulosa cells by MitoTracker Orange. (A) Before laser irradiation, (B) laser-targeted mitochondria, (C) after laser irradiation. (D) White light image of the cells, (E) digital subtraction of (A) and (C).

B. Cell Mechanics

Another application of fs-laser scissors in cell biology is cell mechanics, meaning the manipulation at parts of the cytoskeleton of a cell in order to gather more information about the structural integrity and mechanical stress and tension within a cell. Thus, we tried to cut single microtubules in a live bovine endothelial cell. Microtubules consist of protein subunits of tubulin, which form a stiff hollow tube with typical diameters of about 25–30 nm and length of up to several 10 μm and are part of the cytoskeleton. If such a structure is cut by a laser, it depolymerizes rapidly due to its dynamic instability. This can be seen by the fluorescence microscopy images in Fig. 7 (Heisterkamp *et al.*, 2005). The laser was aimed at a curved

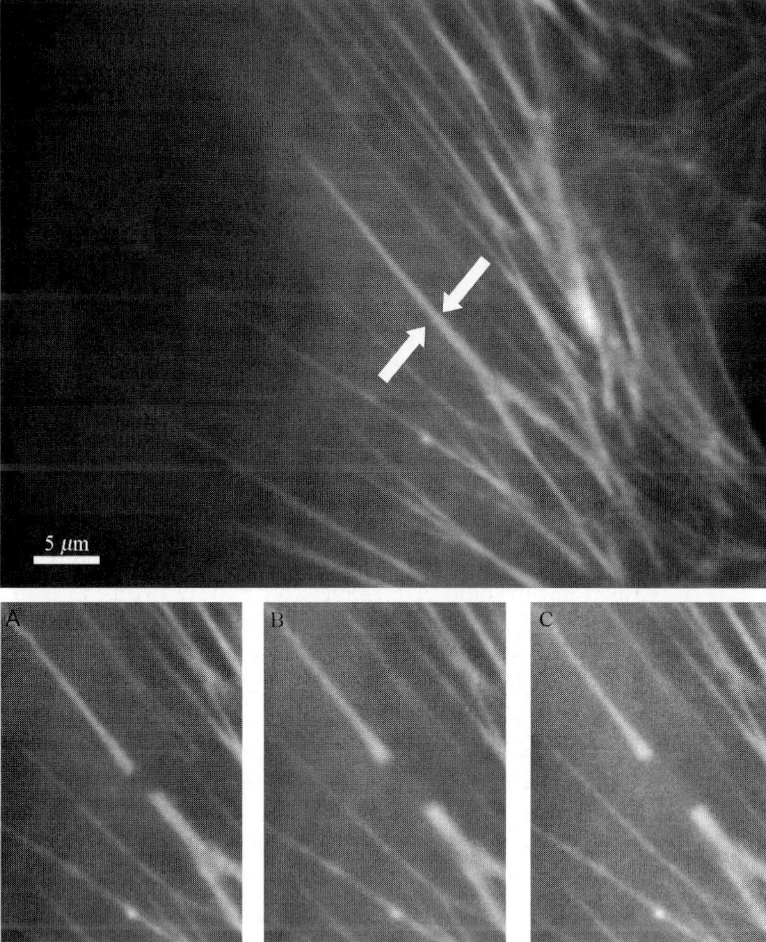

Fig. 7 Cuts in the cytoskeleton of a live endothelial cell stained by YFP for actin taken fluorescent microscopy. Cell before and after laser cutting (A) = 0.5 s, (B) = 5 s, and (C) = 7.5 s, showing the retraction of the fiber ends.

microtubule in the middle of the image (see arrow), lying above the cells' nucleus (dark area in the image). After cutting, both ends of the microtubule snapped back and started to depolymerize, as shown in the right of Fig. 7, taken ~16 s after the laser cut. Both ends of the microtubule are already several micrometers apart, proving that the microtubule is clearly cut. Nevertheless, other microtubules close to the ablation site stay intact, showing the very low side effects of fs-laser ablation.

In a different set of experiments, we tried to target the actin fiber network of the BCE cells, which form another part of the cytoskeleton and are assumed to bear tension. Many measurements of the mechanical properties of these stress fibers have already been done *in vitro*; however, there has been no tool to probe the physical properties of actin fibers in their natural environment. We aimed the fs laser at a single fiber and measured its retraction after cutting. Figure 8 displays a time lapse of the stress fiber before and at 0.5 s (A), 5 s (B), and 7.5 s (C) after laser application. After the dissection of the fiber, both ends snap back in a rapid movement due to the tension resting on the fiber and slowing down until reaching their end position several micrometers apart from the half-micrometer-wide laser spot. By fitting mechanical models to the amount of retraction and imaging the change in shape of the whole cell, information about the tension within a single actin fiber and within the actin network of a cell can be gathered (Kumar *et al.*, 2006).

V. Summary

In this chapter, we have summarized several applications of fs-laser nanoscissors in the field of biology. Comparing fluorescent images with TEM images, it is important to note that the threshold for ablation is roughly 20% higher than the threshold for photobleaching. Thus, laser energies clearly above the notice of a visible effect in fluorescent microscopy should be chosen to ablate structures within living cells. However, the side effects and cutting dimensions scale with the amount of laser energy applied to the target, providing highest precision close to the ablation threshold. The achievable cutting precision is in the range of 200 nm at pulse energy of ~1 nJ. One major advantage of fs-laser systems is their possible combination or integration into multiphoton imaging system, allowing the simultaneous manipulation and imaging by a single laser. However, as a disadvantage, the current cost of an fs-laser system remains quite high when compared to ns- or ps-laser systems. Thus, the use of fs-laser systems is often limited to groups, which are already equipped with a multiphoton microscope and which are using the fs-laser cell surgery as an additional feature of their microscope. Moreover, the minimized side effects of fs-laser systems are not necessary in various applications of cell biology and lower survival rates are quite acceptable, as for example shown by Paterson *et al.* (2006), by successfully transfecting Chinese hamster ovarian cells using a simple violet diode laser. Nevertheless, the application of multiphoton microscope systems is rapidly growing and the development of low-cost fs-laser systems remains an ongoing process, which is why the use of fs-laser systems in cell biology will probably steadily increase in the near future.

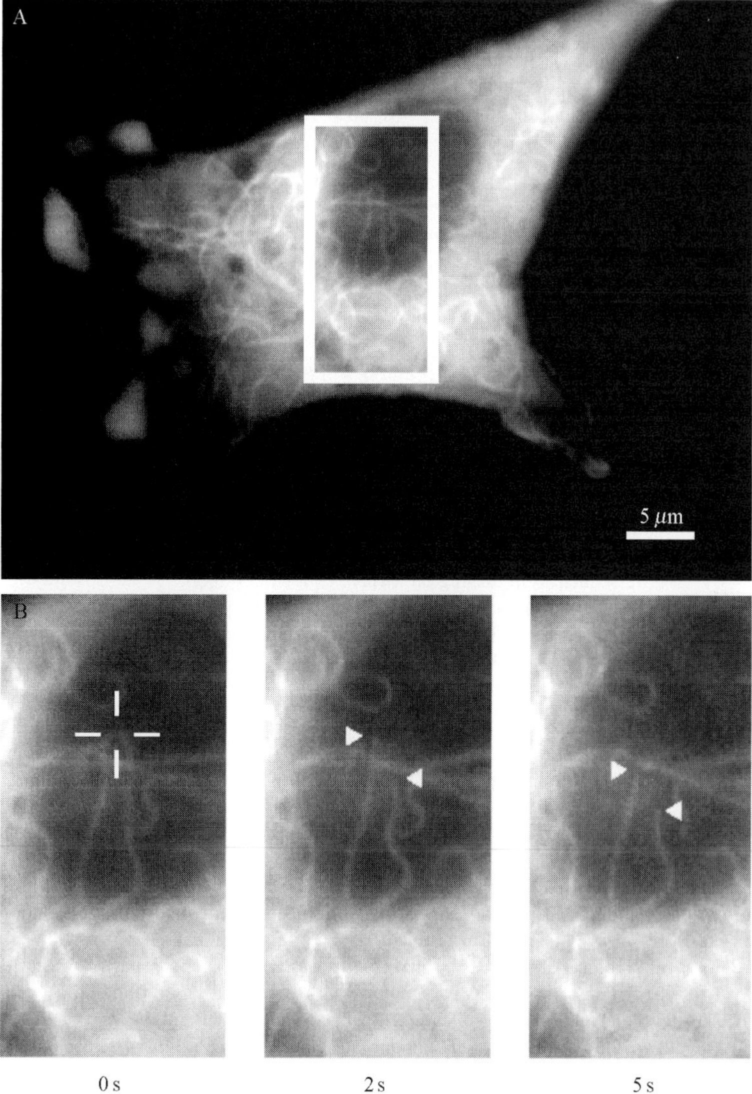

Fig. 8 (A) Fluorescence microscope image of GFP-labeled microtubule network in an endothelial cell. (B) Time-lapse sequence showing rapid retraction of microtubule due to depolymerization. The cross hair shows the position targeted by the laser; the triangles show the retracting ends of the microtubule.

However, open question remains, concerning the optimum repetition rate of such fs-laser scissors, as the basic mechanism of plasma-mediated ablation and possible pulse-to-pulse interaction are topic of actual discussions.

Acknowledgments

The author would like to thank the editors of this book for the helpful comments on the manuscript. Parts of this work were funded by the Deutsche Forschungsgemeinschaft (DFG). The author has no financial or proprietary interest in any of the procedures or devices used in this chapter.

References

Aist, J. R., Liang, H., and Berns, M. W. (1993). Astral and spindle forces in PTK2 cells during anaphase B: A laser microbeam study. *J. Cell Sci.* **104,** 1207–1216.

Berns, M. W., Aist, J., Edwards, J., Strahs, K., Girton, J., McNeil, P., Rattner, J. B., Kitzes, M., Hammerwilson, M., Liaw, L. H., Slemens, A., Koonce, M., *et al.* (1981). Laser microsurgery in cell and developmental biology. *Science* **213,** 505–513.

Chung, S. H., Clark, D. A., Gabel, C. V., Mazur, E., and Samuel, A. (2005). The role of the AFD neuron in *C. elegans* thermotaxis analyzed using femtosecond laser ablation. *BMC Neurosci.* **7,** 30.

Chichkov, B., Momma, C., Nolte, S., von Alvensleben, F., and Tuennermann, A. (1996). Femtosecond, picosecond and nanosecond laser ablation. *Appl. Phys. A* **63,** 109–115.

Denk, W., Strickler, J. H., and Webb, W. W. (1990). Two-photon laser scanning fluorescence microscopy. *Science* **248,** 73–76.

Djabali, M., Nguyen, C., Biunno, I., Oostra, B. A., Mattei, M. G., Ikeda, J. E., and Jordan, B. (1991). Laser microdissection of the fragile X region: Identification of cosmid X clones and of conserved sequences in this region. *Genomics* **10,** 1053–1060.

Heisterkamp, A., Maxwell, I. Z., Mazur, E., Underwood, J. M., Nickerson, J. A., Kumar, S., and Ingber, D. E. (2005). Pulse energy dependence of subcellular dissection by femtosecond laser pulses. *Opt. Express* **13,** 3690–3696.

Juhasz, T., Loesel, F. H., Kurtz, R. M., Horvath, C., Bille, J. F., and Mourou, G. (1999). Corneal refractive surgery with femtosecond lasers. *IEEE J. Sel. Top. Quantum Electron.* **5**(4), 902–910.

Koenig, K. (2000). Multiphoton microscopy in life sciences. *J. Microsc.* **200,** 83–104.

Koenig, K., Riemann, I., Fischer, P., and Halbhuber, K. (1999). Intracellular nanosurgery with near infrared femtosecond laser pulses. *Cell. Mol. Biol.* **45,** 192–201.

Kumar, S., Maxwell, I. Z., Heisterkamp, A., Polte, T. R., Lele, T. P., Salanga, M., Mazur, E., and Ingber, D. E. (2006). Viscoelastic retraction of single living stress fibers and its impact on cell shape, cytoskeletal organization, and extracellular matrix mechanics. *Biophys. J.* **90,** 3762–3773.

Liang, H., Wright, W. H., Cheng, S., He, W., and Berns, M. W. (1993). Micromanipulation of chromosomes in PTK2 cells using laser microsurgery (optical scalpel) in combination with laser induced optical force (optical tweezers). *Exp. Cell Res.* **204,** 110–120.

Monajembashi, S., Cremer, S., Cremer, T., Wolfrum, J., and Greulich, K. O. (1986). Microdissection of human chromosomes by a laser microbeam. *Exp. Cell. Res.* **167**(1), 262–265.

Nishimura, N., Schaffer, C. B., Friedman, B., Tsai, P. S., Lyden, P. D., and Kleinfeld, D. (2006). Targeted insult to subsurface cortical blood vessels using ultrashort laser pulses: Three models of stroke. *Nat. Meth.* **3,** 99–108.

Paterson, L., Agate, B., Comrie, M., Ferguson, R., Lake, T. K., Morris, J. E., Carruthers, A. E., Brown, C. T. A., Sibbett, W., Bryant, P. E., Gunn-Moore Riches, F. C., and Dholakia, K. (2006). Photoporation and cell transfection using a violet diode laser. *Opt. Express* **13**(2), 595–600.

Schaffer, C. B., Brodeur, A., Garcia, J. F., and Mazur, E. (2001). Micromachining bulk glass by use of femtosecond laser pulses with nanojoule energy. *Opt. Lett.* **26,** 93–95.

Shen, N., Datta, D., Schaffer, C. B., Le Duc, P., Ingber, D. E., and Mazur, E. (2005). Ablation of cytoskeletal filaments and mitochondria in live cells using a femtosecond laser microscissors. *Mech. Chem. Biosyst.* **2,** 17–26.

So, P. T. C., Kim, H., and Kochevar, I. E. (1998). Two-photon deep tissue *ex vivo* imaging of mouse dermal and subcutaneous structures. *Opt. Express* **3**(9), 339–350.

Stern, D., Schoenlein, R., Puliafito, C., Dobi, E., Birngruber, R., and Fujimoto, J. (1989). Corneal ablation by nanosecond, picosecond, and femtosecond lasers at 532 nm and 625 nm. *Arch Ophthalmol.* **107**(4), 587–592.

Supatto, W., Debarre, D., Moulia, B., Brouzes, E., Martin, J., Farge, E., and Beaurepaire, E. (2005). *In vivo* modulation of morphogenetic movements in *Drosophila* embryos with femtosecond laser pulses. *Proc. Natl. Acad. Sci. USA* **102,** 1047–1052.

Tao, W., Walter, R. J., and Berns, M. W. (1988). Laser-transected microtubules exhibit individuality of regrowth, however most free new ends of the microtubules are stable. *J. Cell Biol.* **107**(3), 1025–1035.

Tirlapur, U. K., and Koenig, K. (2002). Targeted transfection by femtosecond laser. *Nature* **448,** 290–291.

Tsai, P. S., Nishimura, N., Yoder, E. J., White, A., Dolnick, E., and Kleinfeld, D. (2002). Principles, design and construction of a two photon scanning microscope for *in vitro* and *in vivo* studies. *In* "Methods for *In Vivo* Optical Imaging" (R. Frostig, ed.), pp. 113–171. CRC Press, Boca Raton, Florida.

Vogel, A., and Venugopalan, V. (2003). Mechanisms of pulsed laser ablation of biological tissues. *Chem. Rev.* **103,** 577–644.

Vogel, A., Noack, J., Huettmann, G., and Paltauf, G. (2005). Mechanisms of femtosecond laser nanosurgery of cells and tissues. *Appl. Phys. B* **81**(8), 1015–1047.

Watanabe, W., Arakawa, N., Matsunaga, S., Higashi, T., Fukui, K., Isobe, K., and Itoh, K. (2004). Femtosecond laser disruption of subcellular organelles in a living cell. *Opt. Express* **12,** 4203–4213.

Yanik, M. F., Cinar, H., Cinar, H. N., Chisholm, A. D., Jin, Y., and Ben-Yakar, A. (2004). Neurosurgery: Functional regeneration after laser axotomy. *Nature* **432,** 822.

CHAPTER 10

Cellular Laserfection

Kate Rhodes, Imran Clark, Michelle Zatcoff, Trisha Eustaquio, Kwame L. Hoyte, and Manfred R. Koller

Cyntellect, Inc., San Diego, California 92121

Many studies in modern biology often rely on the introduction of a foreign molecule (i.e., transfection), be it DNA plasmids, siRNA molecules, protein biosensors, labeled tracers, and so on, into cells in order to answer the important questions of today's science. Many different methods have been developed over time to facilitate cellular transfection, but most of these methods were developed to work with a specific type of molecule (usually DNA plasmids) and none work well enough with difficult, sensitive, or primary cells to meet the needs of current life science researchers. A novel procedure that uses laser light to gently permeabilize large number of cells in a very short time has been developed and is described in detail in this chapter. This method allows difficult cells to be efficiently transfected in a high-throughput manner, with a wide variety of molecules, with extremely low toxicity.

METHODS IN CELL BIOLOGY, VOL. 82

0091-679X/07 $35.00
DOI: 10.1016/S0091-679X(06)82010-8

I. Introduction

Understanding the functional processes of living cells is one of the main endeavors of modern biology. Steps toward achieving this understanding often rely on the introduction of foreign material into a cell to study the resulting effect on intracellular processes. The ability to introduce DNA into cells enables the study of gene expression (Ashkenazi and Dixit, 2006; Glatt *et al.*, 2005; Jones *et al.*, 2006; Lefkowitz and Shenoy, 2006; Pastinen and Hudson, 2006), mutational analyses (Bienko *et al.*, 2005), gene regulatory elements (Donaldson *et al.*, 2005) and has allowed the production of specific proteins that can then be purified and studied (Gartsbein *et al.*, 2006; Matoba *et al.*, 2006) or used for therapeutic purposes (Butler, 2005; Wurm, 2004). Introduction of small interfering RNA (siRNA) into cells has revolutionized the study of single-gene function (Hong *et al.*, 2005; McManus *et al.*, 2002; Sen and Blau, 2006; Xia *et al.*, 2002) and gene interaction cascades (Bouwmeester *et al.*, 2004; Matheny *et al.*, 2004). Introduction of proteins, biosensors, ions, and other small molecules into live cells has great potential to increase our knowledge of intracellular activity and regulation (Du and Macara, 2004; Giuliano and Taylor, 1998; Jones *et al.*, 2004; Lalonde *et al.*, 2005). Although numerous chemical (Bonetta, 2005; Boussif *et al.*, 2001; Nagy and Watzele, 2006) and mechanical (Fischer *et al.*, 2006; Mehier-Humbert and Guy, 2005; Prentice *et al.*, 2005; Sambrook and Russell, 2001) methods have been devised to get molecules through cellular membranes, no single method can introduce all types of macromolecules of interest into cells types with low impact on innate cellular processes. By using the relatively gentle, contact-free energy of laser light, cellular membranes can be transiently permeabilized, thereby allowing the passage into the cell of virtually any molecule present in the surrounding medium.

A handful of studies has been performed over the past two decades in which laser light was tightly focused, using optics with high numerical aperture (NA) and high magnification, to puncture a small hole in the cell membrane (Krasieva *et al.*, 1998; Mohanty *et al.*, 2003; Paterson *et al.*, 2005; Schneckenburger *et al.*, 2002; Shirahata *et al.*, 2001; Stevenson *et al.*, 2006; Terakawa *et al.*, 2006; Tirlapur and Konig, 2002). Other published studies have used laser irradiation aimed at the cell substrate to create a shock wave that optoporates molecules into nearby cells (Knoll *et al.*, 2004a,b; Krasieva *et al.*, 1998; Mohanty *et al.*, 2003; Sagi *et al.*, 2003; Soughayer *et al.*, 2000; Terakawa *et al.*, 2006). This chapter describes an alternative method that uses optics with low NA creating a defocused laser spot that can be used to transfect many cells very quickly (Clark *et al.*, 2006). This method, referred to here as cellular laserfection, enables the transfection of siRNA, plasmids, protein, and other small molecules in a high-throughput manner with high efficiency and high cell viabilities.

II. Rationale

A. Current Methods for Cell Transfection

Compounds and macromolecules that are desirable to transfect into cells include various nucleic acids (e.g., DNA plasmids, aptamers, siRNA, and so on), polypeptide biosensors, proteins, and quantum dots. Unfortunately, these reagents do not pass easily through intact cell membranes. The importance of transfection, and lack of an ideal procedure, has resulted in the development of numerous techniques.

The most commonly used transfection methods include cationic reagents (Sambrook *et al.*, 1989), liposomes (Eldstrom *et al.*, 2000; Hirko *et al.*, 2003), electroporation (Beebe *et al.*, 2003), viruses (Bridge *et al.*, 2003; Sambrook *et al.*, 1989), biolistics (O'Brien and Lummis, 2002), and microinjection (Jensen *et al.*, 2003). While this list is not meant to be exhaustive, many unlisted methods are closely related to one of those listed. An ideal transfection method would have the attributes listed in Table I and an overview of the current methods with respect to these attributes is shown in Table II.

B. Laser-Mediated Delivery of Reagents into Cells

The use of lasers to deliver reagents into cells was first reported by Tsukakoshi *et al.* (1984), and later confirmed in other laboratories (Guo *et al.*, 1995; Shirahata *et al.*, 2001; Tao *et al.*, 1987; Tirlapur and Konig, 2002). The mechanism underlying laser-mediated transfection is unknown. It is likely that different mechanisms are involved depending on the optical arrangement of the instrumentation used (Fig. 1). The more commonly used method, termed optoinjection, opens a relatively small physical hole (1–5 μm) in the cellular membrane by a tightly focused laser (<1-μm spot size) (Kurata *et al.*, 1986). A related technique, termed optoporation, focuses the laser on the culture substrate, and the resulting shock wave causes a temporary permeabilization in the membranes of nearby cells (an analogy of the two methods would be a bullet vs a hand grenade). The advantage of optoporation

Table I
Desirable Attributes for a Cell-Transfection Method

- Simple (i.e., minimal reagent preparation or cell manipulations)
- Efficient (i.e., delivers reagent to large percentage of cells)
- Robust (i.e., not significantly influenced by cell type, molecule of interest, or experimental setting)
- Nontoxic (i.e., not result in significant cell death or change in physiology)
- Fast (i.e., not requiring long incubation times, to enable high-throughput experimentation)
- Amenable to automation, miniaturization, and multiplexing
- Low cost (i.e., per test cost, amortized for instruments)

Table II
Overview of Commonly Used Cell-Transfection Methods

Method	Simple	Efficient	Robust	Nontoxic	Fast	Automation	Low cost
Cationic reagents	+++	−	−	−	+	+++	+++
Liposomes	+++	+	+	+	+	+++	++
Electroporation	++	+	++	−	++	+	++
Viruses	−	++	−	+	+	+++	−
Biolistics	+	+	++	−	++	+	+
Microinjection	−	+++	+++	++	−	−	−

+++, method exhibits this desired attribute to a high extent; ++, method exhibits this desired attribute to a moderate extent; +, method exhibits this desired attribute to some extent; −, method does not exhibit this desired attribute.

is that numerous cells can be loaded with a single laser pulse, thereby alleviating somewhat the current throughput limitation of optoinjection. However, the considerable disadvantages are that significant cell death occurs, and cells at varying distances from the shock wave are loaded to different extents (Krasieva *et al.*, 1998; Soughayer *et al.*, 2000). Optoinjection apparently only affects cells that are targeted by the laser and it occurs with laser wavelengths that are not lethal to the cell (Krasieva *et al.*, 1998). Therefore, it would appear that high cell viability could be maintained following optoinjection (Stevenson *et al.*, 2006). A disadvantage of published optoinjection approaches is the relatively low throughput that can be achieved with existing instrumentation. Further, existing studies have only utilized high-magnification/high-NA optics which are costly and impose significant limitations. For example, those experimental systems were limited to glass substrates with very short working distances (e.g., coverslips). Use of plastic substrates is not possible due to the damage caused by the laser pulses through the high-NA optics. Further, significant cell motion induced by the laser pulses using published experimental conditions makes systematic cell processing impossible.

The prior limitations of optoinjection have been overcome with the discovery of novel optoinjection conditions which are different from the previously published literature (Clark *et al.*, 2006). To distinguish this method from reported laser transfection methods it will be referred to here as cellular laserfection. In this method, less expensive, low-NA optics and lower magnifications (3×–5×) were used, and the laser beam was relatively defocused compared to previous optoinjection methods. The 1/e spot size of the Gaussian beam is larger than a single cell (>10 μm), thereby irradiating the entire cell or multiple cells (depending on cell density) in a single pulse. In addition, lower irradiance is used. This method is employed in both the LEAP and HOP instruments (developed and built at Cyntellect, Inc., San Diego, CA), and has made laserfection a commercially viable technique for the first time. Laserfection implemented on LEAP and HOP achieves many of the desired attributes shown in Table I.

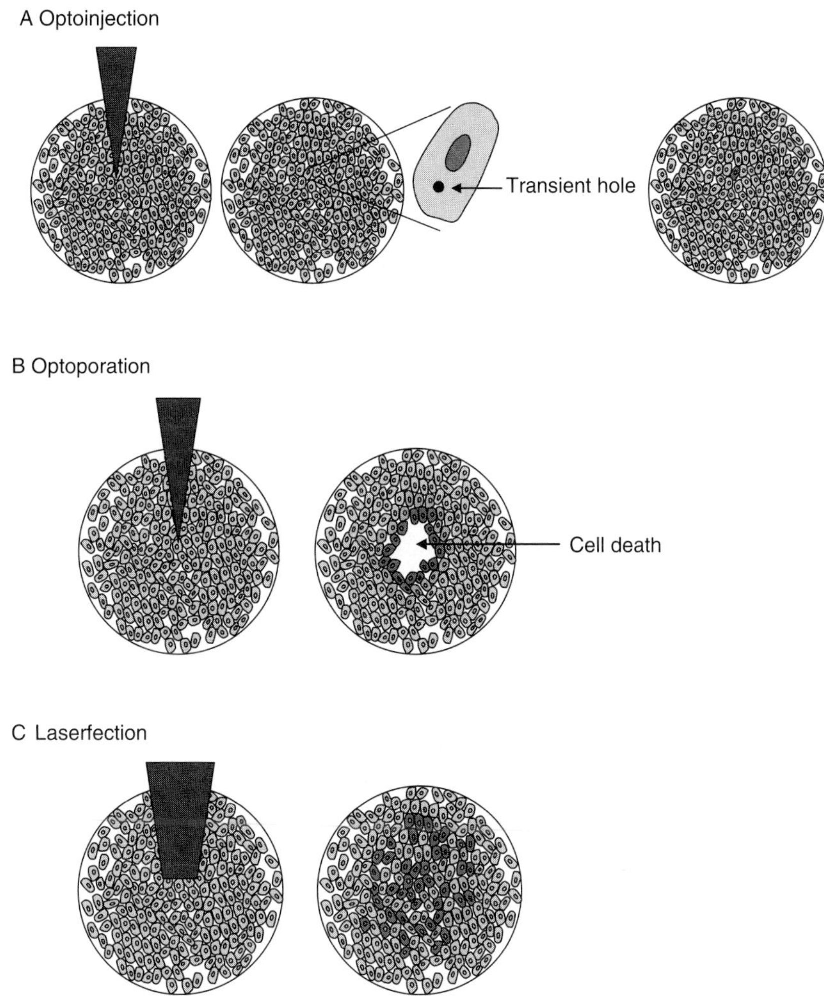

Fig. 1 Three different modes of transfection using laser irradiation. (A) Optoinjection occurs when a tightly focused laser beam punctures a hole in the membrane of a single cell. This hole can be visualized as a 1- to 5-μm spot (Kurata, *et al.*, 1986; Palumbo, 1996; Stevenson *et al.*, 2006). Each cell must be individually processed. (B) With optoporation, a high-energy laser beam focused on the culture substrate causes a shock wave that kills all the cells in the proximity of the laser spot. At some distance from the laser spot, the cells are transfected. This method causes a great deal of cell death and different degrees of cell loading based on the distance from the laser spot. (C) Laserfection occurs when a defocused laser spot of relatively low power (between $5e^{10}$ and $1e^{11}$ W/cm^2) pulses multiple times over a population of cells. Little cell death is observed and many cells are transfected quickly by this method. (See Plate 6 in the color insert section.)

III. Materials and Methods

A. The LEAP System

Cyntellect has developed a novel laser-based cell-processing system, called LEAP™ (laser-enabled analysis and processing), for high-speed cell imaging and laser-based manipulation. LEAP images cells at upto 10^5 per second (Fig. 2) and laser irradiates specific cells at upto 10^3 per second allowing various cell manipulations such as cell death (i.e., cell purification) or laserfection (i.e., cell transfection). Several novel capabilities have been demonstrated on LEAP, including: (1) ultrahigh-throughput cell imaging, (2) *in situ* purification of cells with high yield and purity (Hanania *et al.*, 2005; Koller *et al.*, 2004), (3) *in situ* laserfection of various macromolecules into cells with high efficiency and low toxicity (Clark *et al.*, 2005; Szaniszlo *et al.*, 2006), and (4) chromophore-assisted laser inactivation (i.e., protein knockout, CALI) (Brandes *et al.*, 2004). High-throughput processing is obtained by using a large field-of-regard (FOR) F-theta lens with high-speed galvanometers to both scan a large surface area to obtain images, and to steer the laser beam to hit target cells, all without moving the cell culture substrate. Importantly, the F-theta lens has a flat-field correction that maintains a constant laser spot size (and corresponding irradiance) as it is scanned across the FOR. The engineering design relies on proven optics, microelectronics, and software technologies. The heart of the design, the optical subassembly, is shown in Fig. 2B. Above the optical subassembly is a stage subassembly (not shown) that provides precision x–y–z movement of plates with respect to the scanning lens so that an entire plate may be processed. The relatively large FOR of the F-theta lens significantly reduces the number of stage movements and refocusing events required as compared with microscope-based imaging, resulting in a 10- to 100-fold speed improvement. Adding the laser to this optical platform enables rapid cell manipulation across the large FOR. This approach has the added advantage of exploiting the rapid and continual improvements in speed, cost, and reliability that are being made in the semiconductor materials manufacturing markets, where the same technologies are used for applications such as disk drive texturing and resistor trimming.

LEAP has the following important features:

- LED-based bright-field illumination
- Stabilized hybrid halide lamp excitation (350–680 nm, 8 position wheel)
- Megapixel CCD camera for high-throughput image capture (8 position wheel)
- Adjustable magnification (3×, 5×, 10×, and 20×)
- Rapid autofocus mechanism
- Multiwavelength cell-targeting lasers (to achieve cell purification, laserfection, CALI, and so on)
- Automated laser beam steering (upto 1000 cells/s) and
- Closed-system noncontact *in situ* cell imaging and processing

A

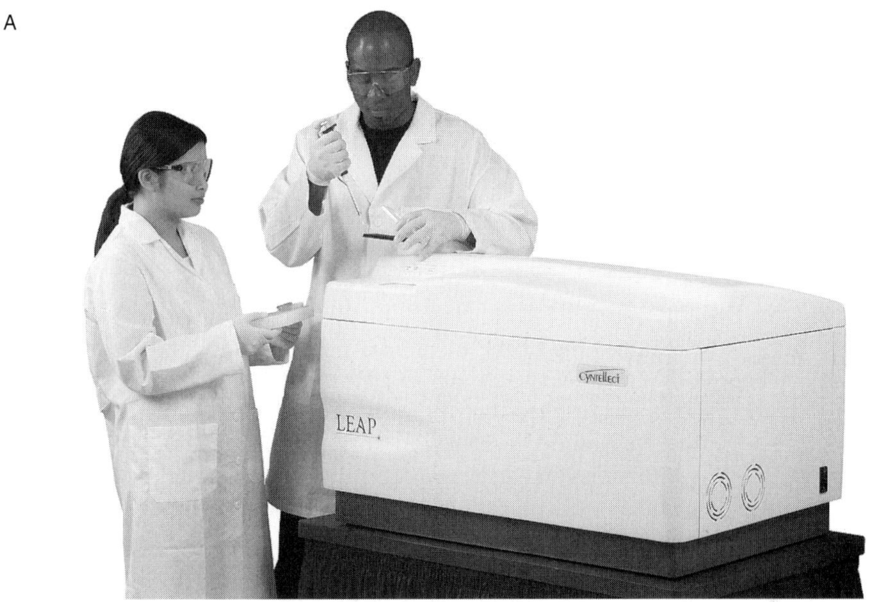

B

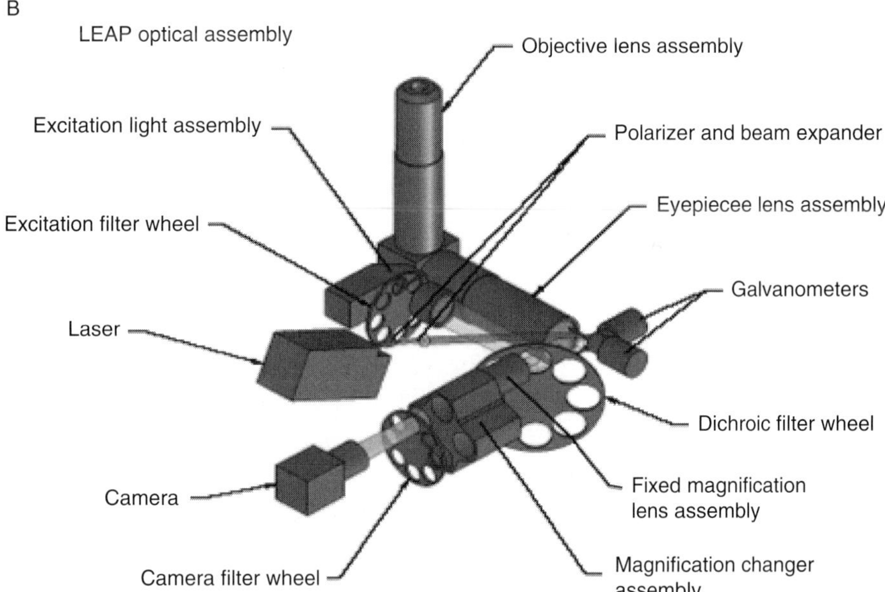

Fig. 2 (A) The commercial LEAP system and (B) diagram of the optical subassembly in the LEAP system.

More detailed description of this technology can be found in several published patents (Koller *et al.*, 2003; Palsson *et al.*, 2003a,b).

Although laserfection on LEAP has clear advantages over existing methods, LEAP is a relatively complex high-cost instrument as compared to other transfection devices. To address this issue, Cyntellect has developed a simple low-cost device specifically for laserfection (called HOP™ for *h*igh-throughput *op*toinjector). HOP will make laserfection available to a larger customer base, enabling researchers in diverse fields to utilize this important technology to overcome limitations of current transfection methods.

B. Laserfection on LEAP

1. Targeted Versus Nontargeted Laserfection

Studies on LEAP have used both targeted and nontargeted laserfection methods. In the targeted mode, LEAP identifies the *x–y* coordinates of each cell in a well and applies the desired number of pulses directly at each cell. In the nontargeted mode, laserfection is performed without individual cell imaging, image processing, and laser targeting, thereby eliminating the need for many of the complex optical elements of LEAP and improving throughput. This nontargeted mode is employed on HOP, also known as "grid processing." During grid processing, single or multiple laser pulses are fired in a two-dimensional grid pattern covering the entire area of a well without regard to specific cell locations. To determine the optimal density of the grid, several densities were tested based on the theoretical calculations diagrammed in Fig. 3.

To test grid spacing, a 14 × 14 μm^2 grid was used to ensure that the ~14-μm-diameter laser beam achieved a near-direct hit on every cell in the well (Fig. 4A). Surprisingly, the results seen with this approach were somewhat better than using the same laser conditions directly targeted at each cell (Clark *et al.*, 2005). Noting that the distribution of energy within the laser spot was Gaussian, and further,

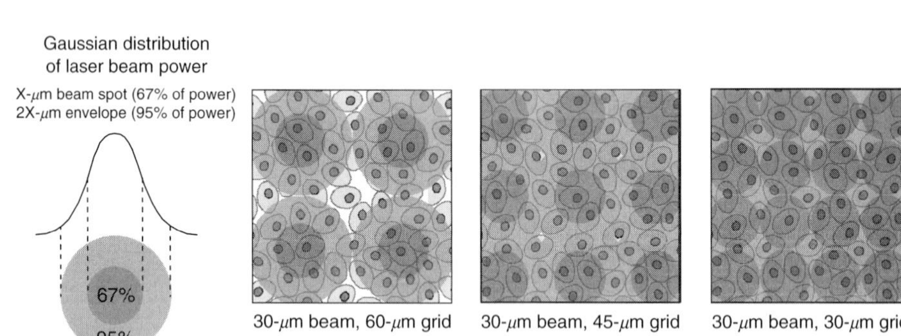

Fig. 3 Interaction between laser spot size, grid spacing, and number of shots. The laser beam has a Gaussian distribution, such that 67% of its energy is in the nominal spot diameter (30 μm here), and an additional 28% is in the adjacent annular ring (out to 60 μm here). Preliminary data indicate that direct cell hits with the peak energy center of the beam are more lethal, such that grid spacing larger than the nominal beam diameter will result in irradiation of more cells with less lethal effects. Throughput is also affected, as the number of shots required decreases with larger beams and grids.

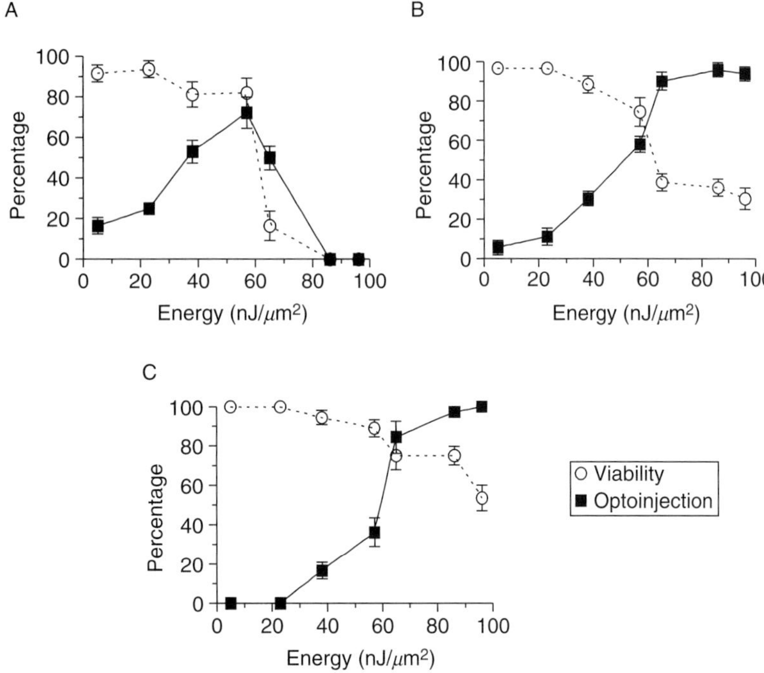

Fig. 4 Laserfection of the cell-impermeable dye Sytox Green. HeLa cells grown in 384-well plates at ~500 cells per well were laserfected on LEAP utilizing a 532-nm pulsed laser. The energy of each laser pulse was controlled by a variable neutral density filter wheel. Viability was based on the percentage of cells that resisted subsequent Sytox Blue staining. Laserfection efficiency was based on the percentage of total viable cells remaining that demonstrated obvious nuclear staining with Sytox Green. (A–C) Single laser pulses were fired in a two-dimensional grid pattern covering the entire area of a well without regard to specific cell locations. Three different grid spacings were tested with each radiant exposure, including: (A) 14 μm, (B) 28 μm, and (C) 56 μm. Mean values (±SEM) from five independent experiments (each performed in triplicate) are shown. See Fig 4B for representative image of cells laserfected with Sytox Green. Reprinted from Clark *et al.* (2006).

that 37% of the laser's energy falls outside the nominal 14-μm-diameter spot, it was hypothesized that direct shots at cells with the center of the beam (i.e., the peak irradiance spot) were more lethal than cell irradiance that occurred away from the center of the beam (Fig. 3). This hypothesis led to testing larger grid spacings of 28 and 56 μm, thereby reducing the number of direct cell hits with the center of the beam while still allowing cells to be irradiated with substantial peripheral energy. The 28-μm grid spacing improved laserfection efficiency to near 100% with cell viability improved to 40% (Fig. 4B). The 56-μm grid spacing using higher radiant exposure per point produced the best results, achieving 85–100% laserfection efficiency with up to 80% cell viability. Using this grid-based approach, more than one cell was affected with each laser pulse, improving throughput of the process to ~2000–5000 cells/s (depending on plated cell density). Importantly, a variety of

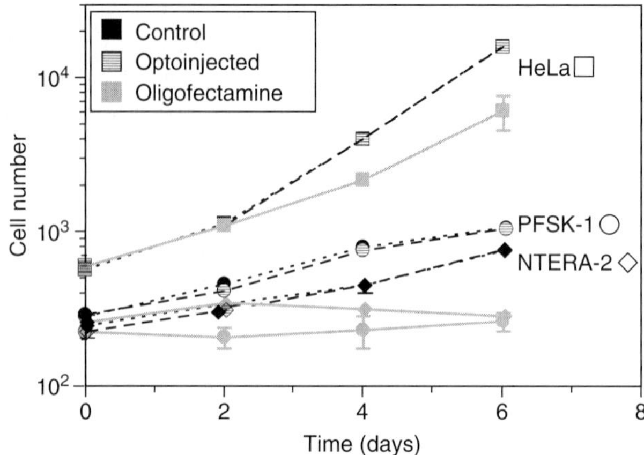

Fig. 5 Outgrowth of cells following laserfection or Oligofectamine transfection. Exponential growth was not affected by laserfection, whereas Oligofectamine inhibited cell outgrowth long after it was removed from the culture. Reprinted from Clark *et al.* (2006).

laserfected cell types (HeLa, NTERA-2, PFSK-1) exhibited growth rates equivalent to nonlaserfected control cells, whereas the same cells exposed to a commonly used lipid delivery reagent (Oligofectamine™) exhibited significantly retarded growth (Fig. 5; Clark *et al.*, 2005).

2. Multiple Laser Pulses

The LEAP laser is relatively high power and can achieve laserfection with single pulses (see above). The rapid kinetics of laserfection, combined with the knowledge that lethal laser effects are not additive at this wavelength (Koller *et al.*, 2004), led to the hypothesis that increased laserfection efficiency might be achieved with repeated sublethal laser pulses without affecting cell viability. Cells were therefore laserfected using multiple pulses of the laser at relatively low radiant exposures. Experiments were performed in which a varied number of laser pulses were delivered to cells in 1-ms intervals. Groupings of laser pulses were also repeated a variable number of times with 500-ms intervals between pulse groupings. Results from a radiant exposure setting of 23 nJ/μm^2 per laser pulse are shown (Fig. 6), which previously gave only modest laserfection efficiency, but high cell viability, when delivered as a single pulse. Laserfection efficiency increased significantly as the number of laser pulses was increased. At 30 pulses, laserfection efficiency was increased twofold with only a 15% decline in cell viability as compared with one pulse. Beyond 30 pulses in one grouping, increased laserfection efficiency was offset by an equivalent decrease in cell viability. However, smaller groupings of laser pulses repeated several times achieved greater than 90% laserfection efficiency while maintaining very high cell viability (Fig. 6).

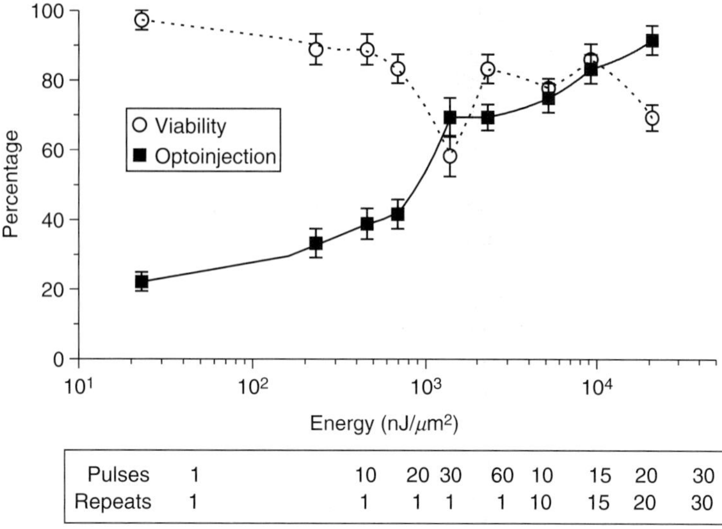

Pulses	1	10	20	30	60	10	15	20	30
Repeats	1	1	1	1	1	10	15	20	30

Fig. 6 Laserfection of Sytox Green using multiple sequential laser pulses. HeLa cells were laserfected on LEAP using multiple pulses of the laser at a radiant exposure setting of 23 nJ/μm^2. Pulses were spaced 1 ms apart. Repeated groupings of pulses were spaced 500 ms apart. Mean values ($\pm$SEM) from five independent experiments (each performed in triplicate) are shown. Reprinted from Clark *et al.* (2006).

Extremely high levels of cumulative radiant exposure, over 20 μJ/μm^2, could be delivered to cells by this approach. These data indicate that very low light energy delivered in multiple small groupings of repeated pulses produced more desirable laserfection results as compared with higher light energy delivered in a single pulse or in a single large grouping of low-energy pulses.

3. Broad Compound and Macromolecule Capability

Figure 7 demonstrates successful laserfection of six different reagents including ions, various nucleic acids, proteins, and quantum dots. Different ions were laserfected into NIH-3T3 (Zn^{2+} shown in Fig. 7A), 293T, HeLa, and primary rat cardiomyocyte cells. Studying the effects of changes in intracellular ion concentrations on cell physiology has broad applications (Rudolf *et al.*, 2004; Wehrens and Marks, 2004), and laserfection represents the only reversible (i.e., transient) and targeted method for rapid delivery of ions into specific cells. Also shown is sequential targeted laserfection of two different molecules (Sytox Green and Sytox Blue) into different subsets of cells within the same well (Fig. 7B) and laserfection of tetramethylrhodamine-dextran (3 kDa) into HeLa cells (Fig. 7C). An enhanced green fluorescent protein (EGFP) expressing plasmid (Fig. 7D) and a biosensor protein of 55 kDa (Fig. 7E) were successfully laserfected into viable cells as well. To demonstrate the capability of laserfection with even the most unusual macromolecules, fluorescent semiconductor nanocrystals (Chan *et al.*, 2002) were laserfected

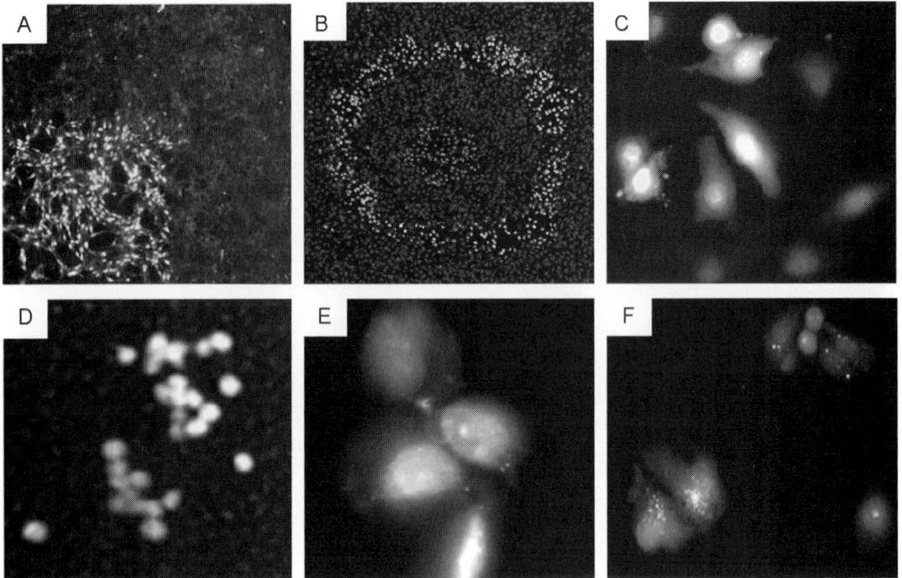

Fig. 7 Laserfection of a variety of compounds into diverse cell types on LEAP. (A) NIH/3T3 cells were preloaded with the Zn^{2+}-sensitive dye RhodZin-1, 50-μM $ZnCl_2$ buffer was added, and cells were laserfected using grid-based laser irradiation. The grid was placed over a small defined region within the bottom left of the field-of-view to demonstrate specificity of laserfection. (B) HeLa cells were counter stained with Cell Tracker Orange. A subset of cells was laserfected in a ring pattern in the presence of Sytox Green. Cells were rinsed and then another subset of cells was laserfected in a small circle in the presence of Sytox Blue. (C) HeLa cells were laserfected with dextran (3000 MW) conjugated to tetramethylrhodamine. Cells were counter stained with Cell Tracker Green and nonviable cells were detected with Sytox Blue. (D) 293T cells were laserfected with a 4.3-kb GFP-encoding plasmid and assessed the next day for GFP expression. (E) 293T cells were laserfected with a 55-kDa protein consisting of the Cdc42-binding domain of WASP conjugated to an I-SO dye (red) and were counter stained for viable cells using calcein green. Fluorescence of the I-SO dye was increased on binding to activated Cdc42 (Nalbant *et al.*, 2004). (F) HepG2 (human hepatocyte) cells were laserfected with semiconductor nanocrystals that fluoresce at 525 nm. Reprinted from Clark *et al.* (2006). (See Plate 7 in the color insert section.).

into HepG2 (human hepatocyte; Fig. 7F) and PC-3 (human prostate epithelial, not shown) cells.

C. The HOP Prototype Instrument Platform

The HOP prototype has the following important features:

- Small benchtop device
- LED-based bright-field illumination
- Megapixel CCD camera for high-throughput image capture
- Rapid autofocus mechanism

- Single-wavelength cell-targeting laser (optimized for laserfection)
- Rapid grid-based processing
- Automated cell-counting capability
- Automated laser beam steering (10 kHz) and
- Closed-system noncontact *in situ* cell imaging and processing

Table III shows the features of laserfection on HOP compared to LEAP. Because the HOP system performs only laserfection, the HOP instrument size is smaller and the optical design is greatly simplified (Fig. 8).

D. Laserfection on HOP

1. Superior siRNA Knockdown

RNA interference (RNAi; Tuschl, 2001) is a specific posttranscriptional gene silencing pathway mediated by intracellular dsRNAs of 21–23 bp in length known as siRNA. Since the discovery of effective siRNA-mediated gene silencing in mammalian cells (Elbashir *et al.*, 2001), there has been significant validation and enormous interest in the approach from both academic and corporate researchers, for both discovery and therapeutic applications (Shuey *et al.*, 2002; Thompson, 2002). There have now been many reports in a number of useful model systems, such as inhibition of HIV infection/replication in T cells (Novina *et al.*, 2002), silencing of MAP2 and YB-1 expression in neurons (Krichevsky and Kosik, 2002), and even whole-genome knockdown approaches in lower organisms (Kamath *et al.*, 2003) and humans (Berns *et al.*, 2004; Kolfschoten *et al.*, 2005).

RNAi experiments are designed to determine specific gene function through targeted knockdown. For these experiments to be useful, relevant cell types must be used and off-target effects and nonspecific toxicity due to the transfection method must be minimized. The practice of RNAi would substantially benefit from an efficient and nontoxic delivery method, particularly if it was effective in primary cells and other difficult to transfect cell types (e.g., Jurkat, HL-60, and so on). In addition, studies have linked endosomal processing of siRNAs with the toll-like receptor (TLR)-mediated induction of cytokines and interferon (Hornung *et al.*, 2005; Marques and Williams, 2005; Sioud, 2005) (Fig. 9A). Therefore, a nonchemical-based transfection method that bypasses the endosome would be valuable for this area of research. Transfection using laser irradiation very likely

Table III
Overview of Optoinjection Attributes on LEAP and HOP

Method	Simple	Efficient	Robust	Nontoxic	Fast	Automation	Low cost
LEAP	+	++	+++	+++	+	+++	–
HOP	+++	+++	+++	+++	+++	+++	++

See footnote of Table II.

A

B

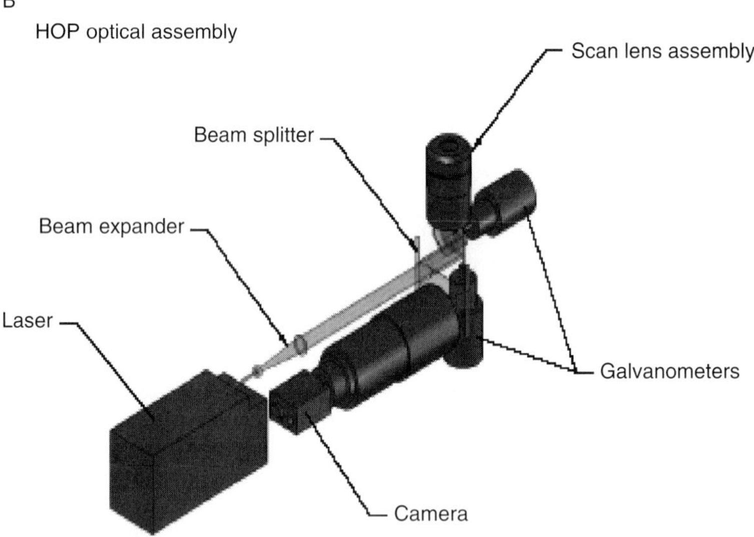

Fig. 8 (A) The prototype HOP system. (B) Diagram of the optical subassembly in the HOP prototype instrument. HOP, like LEAP, uses an F-theta scanning lens and galvanometer mirrors instead of a biological microscope, but has greatly simplified optics as compared to LEAP.

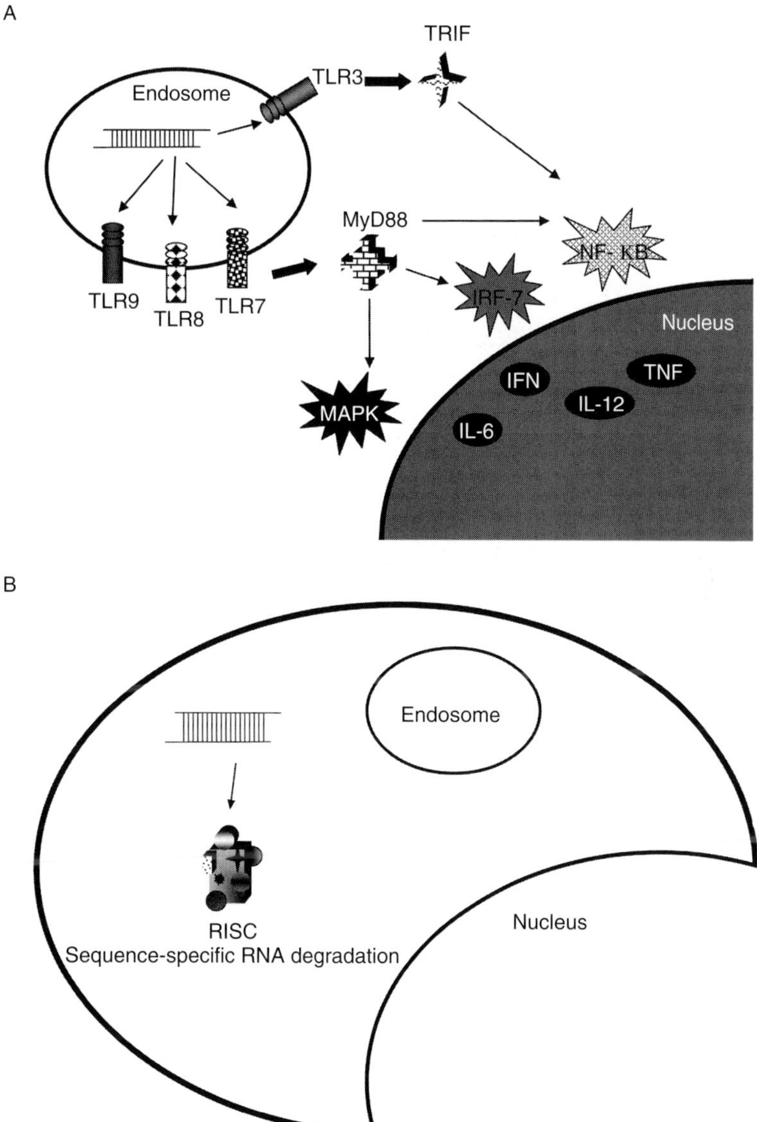

Fig. 9 (A) Chemical transfection methods result in endosomal sequestration of the siRNA product, which has been shown to activate TLRs and induce a potent cytokine response. (B) Mechanical transfection methods, like laserfection, likely bypass the endosome and therefore do not activate an immune response signal cascade.

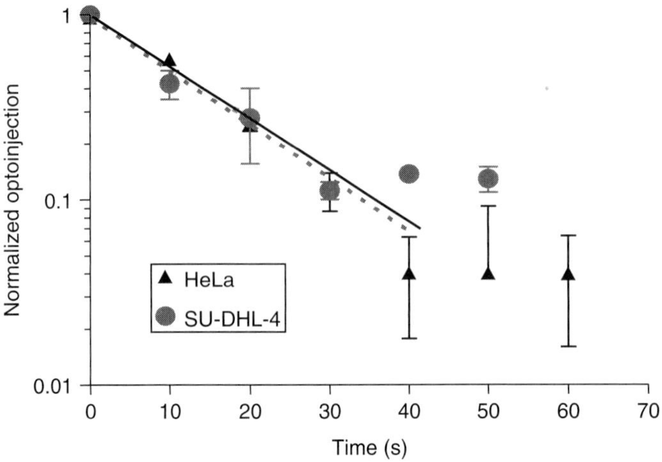

Fig. 10 Kinetics of laserfection. HeLa cells and SU-DHL-4 (human B-cells) were irradiated using a $56 \times 56\ \mu m^2$ grid spacing with a single laser pulse per point at an energy setting of 65 nJ/μm^2. Sytox Green was then added to the well at various times post-irradiation. Laserfection efficiency was based on the percentage of total viable cells remaining that demonstrated obvious nuclear staining with Sytox Green. All data were normalized to the zero time point. The best-fit line was an exponential decay with a time constant of ~0.065 s^{-1}. Mean values (±SEM) from three independent experiments (each performed in triplicate) are shown.

causes siRNAs to enter the cytoplasm without the formation of an endosome (Fig. 9B). The membrane opens for a short period of time, not long enough for an endosome to form (Fig. 10).

To test the effectiveness of laserfection for RNAi, several cell types were transfected with siRNA pools for GAPDH (Ambion, Austin, TX) (Fig. 11). Knockdown was determined relative to a commercially validated negative-scrambled siRNA pool (Dharmacon, Lafayette, CO). Protein knockdown was measured using a fluorescent-based commercially available GAPDH activity assay (Ambion) normalized for cell number. In HeLa cells, both Lipofectamine 2000 and laserfection caused significant protein knockdown. Untreated samples were also assessed in which siRNAs were added to the cells but they were not processed on HOP. These samples showed no statistically relevant changes in GAPDH activity. In Jurkat and HL-60 cells, siRNA transfection with Lipofectamine 2000 was unable to induce protein knockdown, whereas laserfection induced knockdown of up to 95%. Notably, the viability of these cell types was unaffected by laser processing, whereas Lipofectamine 2000 reduced cell viability after only 24 h (Fig. 12).

2. Transfection *In Situ*

Another notable benefit of laserfection is that cells can be transfected without trypsinization or exposure to chemical reagents that are toxic. This attribute reduces stress on sensitive primary cell types that generally do not fare well in protocols using chemical- or electrical-based transfection methods (Koller and Papoutsakis, 1994). Figure 13 shows the typical morphology of primary rat brain hippocampus neurons after laserfection versus Lipofectamine-mediated transfection.

IV. Discussion

Laserfection is a novel method that enables delivery of diverse compounds and macromolecules into a variety of different cell types with high efficiency and low toxicity, thereby overcoming many limitations of existing delivery methods. In ongoing studies at Cyntellect, the laserfection process is being studied and optimized in considerable detail using a large number of cells on both the LEAP and HOP instrumentation platforms. Key parameters for optimization of optoinjection are being identified, resulting in a robust and versatile method for delivering a variety of substances into numerous cell types.

The results thus far indicate that the parameters of laser energy delivery are key to the success of laserfection: relatively low energies applied in a series of multiple pulses lead to efficient optoinjection and high cell viability. In addition, grid-based targeting has been shown to be a fast and effective laserfection method, apparently by simultaneously targeting several regions around the periphery of the cell, and thereby delivering partial doses of multiple laser pulses to each cell. This result leads directly to greatly increased optoinjection throughput on the HOP platform, as the laser beam diameter can be made large enough to encompass many cells with each pulse of laser light.

The mechanism underlying laserfection is not yet known. It has been hypothesized that optoinjection creates a physical hole in the membrane when the laser is tightly focused on a portion of the cell (Kurata *et al.*, 1986). The data do not support this mechanism for laserfection, as the laser beam diameter is effective even when it is larger than the target cell. Optoporation is a method that is related to optoinjection in which the laser is focused on the culture substrate, resulting in a shock wave that causes a temporary permeabilization in the membranes of nearby cells. Optoporation causes transfection of numerous cells with a single laser pulse. However, significant cell death occurs in the area around the laser spot, and cells at varying distances from the center of the shock wave are loaded to different extents (Krasieva *et al.*, 1998; Kurata *et al.*, 1986). Laserfection does not produce data consistent with the shock wave model since the effective laser energies are below the shock wave threshold, little if any cell death occurs and gradients of cell loading are not observed. The method shown here, in which repeated sublethal energy pulses delivered to entire cells (or simultaneously to multiple cells) induce transient changes in overall cell membrane permeability, appears to represent a mechanism of action that is different from the hole-punching or shock wave models. Kinetic measurements done at Cyntellect have shown the same transient response from a variety of cell lines and suggest that optoinjection is a physicochemical process. The unique combination of the laser wavelength and the energies used for laserfection are outside of the range

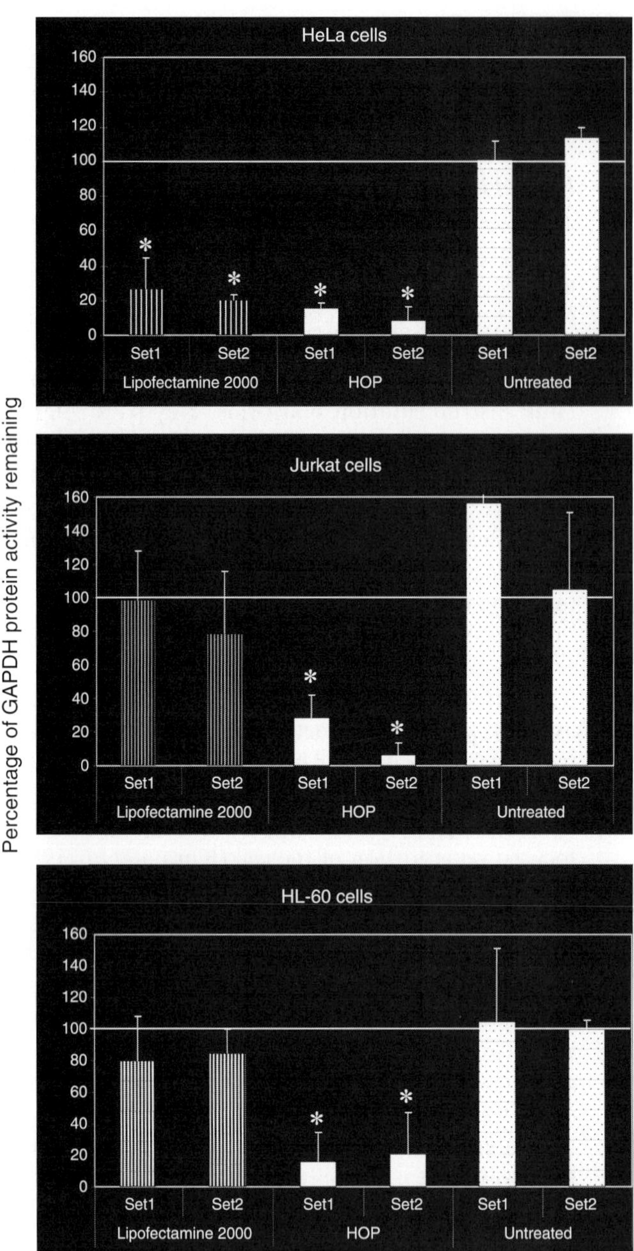

Fig. 11 siRNA knockdown of GAPDH protein activity using laserfection compared to Lipofectamine 2000 in (A) HeLa, (B) Jurkat, and (C) HL-60 cells. Cells were seeded into wells of a C-lect 384-well plate and laserfected on HOP with siRNA targeted to GAPDH (Dharmacon). After 48 h, cells in each well were counted on LEAP, lysed, and the protein was harvested. GAPDH activity was measured using the KDAlert fluorescent assay kit (Ambion). Knockdown was calculated relative to GAPDH activity in

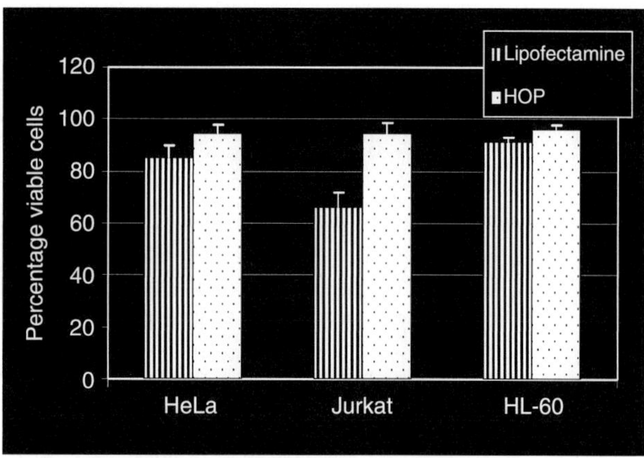

Fig. 12 Viability of HeLa, Jurkat, and HL-60 cells 24 h after siRNA transfection using Lipofectamine 2000 or HOP. Dead cells were stained with propidium iodide and live cells were labeled using calcein-AM. Cells were counted using LEAP.

known to cause biological damage to cells and tissues (other than retinal) (Niemz, 1996), consistent with the low degree of cell death observed.

Practical implementation of laserfection requires the ability to perform it in a high-throughput setting. A limitation of previously published optoinjection or optoporation approaches (Guo *et al.*, 1995; Palumbo *et al.*, 1996; Shirahata *et al.*, 2001; Tao *et al.*, 1987; Tirlapur and Konig, 2002; Tsukakoshi *et al.*, 1984) is the relatively low throughput that can be achieved with microscope-based instrumentation. Additionally, we have observed that significant cell motion and/or damage are induced by laser pulses using the previously published experimental conditions (not shown), thereby making orderly and efficient cell processing challenging, if not impossible. These limitations have been overcome through the F-theta scanning lens and automated high-throughput laser beam steering methodologies implemented on LEAP and HOP. Targeted *in situ* laserfection linked with laser-mediated cell purification (Koller *et al.*, 2004) on LEAP represents a powerful combination of cell manipulation capabilities that has not been available before. The ability to selectively deliver different macromolecules into specific cells in a well and eliminate unwanted cells (e.g., untransfected, or lacking a certain phenotype or desired response), all performed *in situ* in an iterative fashion, opens up a new realm of experimental possibilities. Furthermore, the success of the grid-based approach

---

cells transfected with a negative scrambled siRNA pool (Dharmacon) and was normalized for total cell number per well. Lipofectamine transfections were performed following the manufacturer's recommended protocol (Invitrogen). Each bar represents the average knockdown from a set of six replicate wells. Two independent sets were run for each cell type. Asterisks (*) indicate a p value of <0.05 relative to the negative control. The untreated bars show the knockdown when cells are bathed in siRNA at the same concentration and for the same amount of time as those processed on HOP.

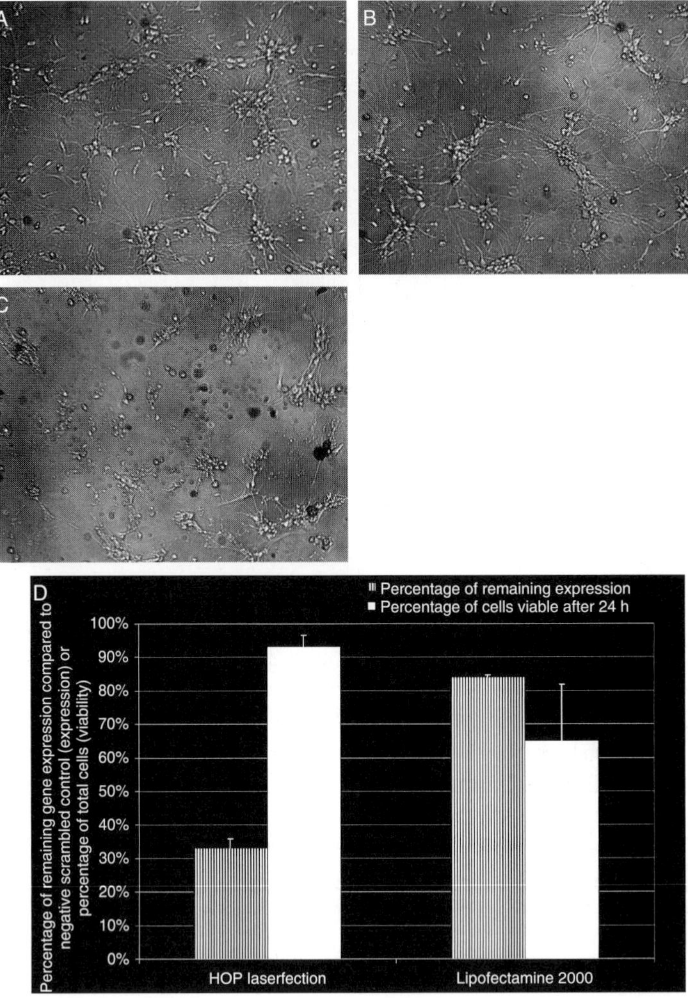

Fig. 13 Morphology of primary rat brain hippocampus neurons (Cambrex) (A) before transfection, (B) after laserfection with HOP, and (C) after transfection with Lipofectamine. (D) Knockdown, morphology, and viability were significantly superior after laserfection as compared with lipid-mediated transfection.

used on HOP establishes the prospect of low-cost, high-throughput transfection which can be used for a myriad of screening applications.

The applicability of optoinjection to a wide range of substances and cell types has broad implications for different fields of study. For example, the increasing interest in siRNA has resulted in new demand for oligonucleotide delivery methods, particularly for cell types refractory to existing methods. In these RNAi-based studies, the goal is to specifically determine the function of a single gene or to understand how a gene interacts in a protein activity cascade. Any background gene modulation due to the delivery method, as has been shown to occur (Scacheri *et al.*, 2004; Scherer and Rossi, 2003), most acutely with chemical-based methods that act via endosomes, will interfere with the goal of the study. Further, many important genes of interest are only active in cell types that are difficult to transfect (e.g., immune cells, neurons, and so on). Another example of an emerging need for macromolecule delivery is the use of semiconductor nanocrystals, many applications of which require intracellular delivery (Gao *et al.*, 2004). Optoinjection has the potential to address needs in these rapidly growing fields of research.

V. Summary

In summary, there is broad and growing interest in delivering cell-impermeable compounds and macromolecules into living cells for use as biomarkers, to elucidate basic cell mechanisms, and to investigate potential druglike effects. With the advent of sequencing the human genome, the ensuing cataloguing of the human proteome and other related technological developments, an immense quantity of information must now be placed in context to further our understanding of human cell physiology and the treatment of pathological conditions. The challenge that the life science community now faces is the functional assessment of gene products and candidate therapeutic molecules, not only in isolation but also in the context of a living cell. The simple, robust, and versatile loading of compounds and macromolecules into living cells via optoinjection provides a unique tool for addressing these complex biological issues.

Acknowledgments

This work was supported by NIH SBIR grant number R41 GM074436 and STTR grant number DMI-0321740.

References

Ashkenazi, A., and Dixit, V. M. (2006). Death receptors: Signaling and modulation. *Science* **281,** 1305–1308.

Beebe, S. J., Fox, P. M., Rec, L. J., Willis, E. L., and Schoenbach, K. H. (2003). Nanosecond, high-intensity pulsed electric fields induce apoptosis in human cells. *FASEB J.* **17,** 1493–1495.

Berns, K., Hijmans, E. M., Mullenders, J., Brummelkamp, T. R., Velds, A., Heimerikx, M., Kerkhoven, R. M., Madiredjo, M., Nijkamp, W., Weigelt, B., Agami, R., Ge, W., *et al.* (2004). A large-scale RNAi screen in human cells identifies new components of the p53 pathway. *Nature* **428,** 431–437.

Bienko, M., Green, C. M., Crosetto, N., Rudolf, F., Coull, B. Z. B., Kannouche, P., Wider, G., Peter, M., Lehmann, A. R., Hofmann, K., and Dikic, I. (2005). Ubiquitin-binding domains in y-family polymerase regulate translation synthesis. *Science* **310,** 1821–1824.

Bonetta, L. (2005). The inside scoop—evaluating gene delivery methods. *Nat. Methods* **2,** 875–882.

Boussif, O., Lezoualch, F., Zanta, M. A., Mergny, M. D., Scherman, D., Demeneix, B., and Behr, J.-P. (2001). A versatile vector for gene and oligonucleotide transfer into cells in culture and *in vivo*: Polyethylenimine. *Proc. Natl. Acad. Sci. USA* **92,** 7297–7301.

Bouwmeester, T., Bauch, A., Ruffner, H., Angrand, P. O., Bergamini, G., Croughton, K., Cruciat, C., Eberhard, D., Gagneur, J., Ghidelli, S., Hopf, C., Huhse, B., *et al.* (2004). A physical and functional map of the human TNF-α/NF-κB signal transduction pathway. *Nat. Cell Biol.* **6,** 97–105.

Brandes, R., Gray, W., Fieck, A., and Koller, M. R. (2004). Fluorophore-assisted laser inactivation (FALI) of GPCR function. *Biophys. J.* **86,** 147a.

Bridge, A. J., Pebernard, S., Ducraux, A., Nicoulaz, A. L., and Iggo, R. (2003). Induction of an interferon response by RNAi vectors in mammalian cells. *Nat. Genet.* **34,** 263–264.

Butler, M. (2005). Animal cell cultures: Recent achievements and perspectives in the production of biopharmaceuticals. *Appl. Microbiol. Biotechnol.* **68,** 283–291.

Chan, W. C., Maxwell, D. J., Gao, X., Bailey, R. E., Han, M., and Nie, S. (2002). Luminescent quantum dots for multiplexed biological detection imaging. *Curr. Opin. Biotechnol.* **13,** 40–46.

Clark, I. B., Hanania, E. G., Stevens, J., Gallina, M., Fieck, A., Brandes, R., Palsson, B. O., and Koller, M. R. (2005). Optoinjection for efficient targeted delivery of a broad range of compounds and macromolecules into diverse cell types. *J. Biomed. Opt.* (in press).

Clark, I. B., Hanania, E. G., Stevens, J., Gallina, M., Fieck, A., Brandes, R., Palsson, B. O., and Koller, M. R. (2006). Optoinjection for efficient targeted delivery of a broad range of compounds and macromolecules into diverse cell types. *J. Biomed. Opt.* **11,** 014034-1 to 014034-8.

Donaldson, I. J., Chapman, M., Kinston, S., Landry, J. R., Knezevic, K., Piltz, S., Buckley, N., Green, A. R., and Gottgens, B. (2005). Genome-wide identification of cis-regulatory sequences controlling blood and endothelial development. *Hum. Mol. Genet.* **14,** 595–601.

Du, Q., and Macara, I. G. (2004). Mammalian PINS is a conformational switch that links NUMA to heterotrimeric G proteins. *Cell* **119,** 503–516.

Elbashir, S. M., Harborth, J., Lendeckel, W., Yalcin, A., Weber, K., and Tuschl, T. (2001). Duplexes of 21-nucleotide RNAs mediate RNA interference in cultured mammalian cells. *Nature* **411,** 494–498.

Eldstrom, J. R., La, K., and Mathers, D. A. (2000). Polycationic lipids translocate lipopolysaccharide into HeLa cells. *Biotechniques* **28,** 510–516.

Fischer, A., Stanke, J., Omar, G., Askwith, C., and Burry, R. (2006). Ultrasound-mediated gene transfer into neuronal cells. *J. Biotechnol.* **122,** 393–411.

Gao, X., Cui, Y., Levenson, R. M., Chung, L. W. K., and Nie, S. (2004). *In vivo* cancer targeting and imaging with semiconductor quantum dots. *Nat. Biotechnol.* **22,** 969–976.

Gartsbein, M., Alt, A., Hashimoto, K., Nakajima, K., Kuroki, T., and Tennenbaum, T. (2006). The role of protein kinase c delta activation and STAT3 Ser727 phosphorylation in insulin-induced keratinocyte proliferation. *J. Cell Sci.* **119,** 470–481.

Giuliano, K. A., and Taylor, D. L. (1998). Fluorescent-protein biosensors: New tools for drug discovery. *Trends Biotechnol.* **16,** 135–140.

Glatt, S. J., Everall, I. P., Kremen, W. S., Corbeil, J., Sasik, R., Khanlou, N., Han, M., Liew, C. C., and Tsuang, M. T. (2005). Comparative gene expression analysis of blood and brain provides concurrent validation of selenbp1 up-regulation in schizophrenia. *Proc. Natl. Acad. Sci. USA* **102,** 15533–15538.

Guo, Y., Liang, H., and Berns, M. W. (1995). Laser-mediated gene transfer in rice. *Physiol. Plant.* **93,** 19–24.

Hanania, E. G., Fieck, A., Stevens, J., Bodzin, L. J., Palsson, B. O., and Koller, M. R. (2005). Automated *in situ* measurement of cell-specific antibody secretion and laser-mediated purification for rapid cloning of highly-secreting producers. *Biotechnol. Bioeng.* **91,** 872–876.

Hirko, A., Tang, F., and Hughes, J. A. (2003). Cationic lipid vectors for plasmid DNA delivery. *Curr. Med. Chem.* **10,** 1185–1193.

Hong, J.-H., Hwang, E. S., McManus, M. T., Amsterdam, A., Tian, Y., Kalmukova, R., Mueller, E., Benjamin, T., Spiegelman, B. M., Sharp, P. A., Hopkins, N., and Yaffe, M. B. (2005). TAZ, a transcriptional modulator of mesenchymal stem cell differentiation. *Science* **309,** 1074–1078.

Hornung, V., Guenthner-Biller, M., Bourquin, C., Ablasser, A., Schlee, M., Uematsu, S., Noronha, A., Manoharan, M., Akira, S., de Fougerolles, A., Endres, S., and Hartmann, G. (2005). Sequence-specific potent induction of IFN-α by short interfering RNA in plasmacytoid dendritic cells through TLR7. *Nat. Med.* **11,** 263–270.

Jensen, K. D., Nori, A., Tijerina, M., Kopeckova, P., and Kopecek, J. (2003). Cytoplasmic delivery and nuclear targeting of synthetic macromolecules. *J. Control. Release* **87,** 89–105.

Jones, D. C., Wein, M. N., Oukka, M., Hofstaetter, J., Glimcher, M., and Glimcher, L. (2006). Regulation of adult bone mass by the zinc finger adapter protein schnurri-3. *Science* **312,** 1223–1227.

Jones, J. T., Myers, J. W., Ferrell, J. E., and Meyer, T. (2004). Probing the precision of the mitotic clock with a live-cell fluorescent biosensor. *Nat. Biotechnol.* **22,** 306–312.

Kamath, R. S., Fraser, A. G., Dong, Y., Poulin, G., Durbin, R., Gotta, M., Kanapin, A., Le Bot, N., Moreno, S., Sohrman, M., Welchman, D. P., Zipperlen, P., *et al.* (2003). Systematic functional analysis of the *Caenorhabditis elegans* genome using RNAi. *Nature* **421,** 231–237.

Knoll, T., Sagi, S., Trojan, L., Schaaf, A., Alken, P., and Michel, M. S. (2004a). *In vitro* and *ex vivo* gene delivery into proximal tubular cell by means of laser energy—a potential approach for curing cystinuria? *Urol. Res.* **32,** 129–132.

Knoll, T., Trojan, L., Langbein, S., Sagi, S., Alken, P., and Michel, M. S. (2004b). Impact of holmium: YAG and neodymium: YAG lasers on the efficacy of DNA delivery in transitional cell carcinoma. *Lasers Med. Sci.* **19,** 33–36.

Kolfschoten, I. G. M., van Leeuwen, B., Berns, K., Mullenders, J., Beijersbergen, R. L., Bernards, R., Voorhoeve, P. M., and Agami, R. (2005). A genetic screen identifies PITX1 as a suppressor of RAS activity and tumorigenicity. *Cell* **121,** 849–858.

Koller, M. R., Hanania, E. G., Stevens, J., Eisfeld, T. M., Sasaki, G. C., Fieck, A., and Palsson, B. O. (2004). High-throughput laser-mediated *in situ* cell purification with high purity and yield. *Cytometry A* **61,** 153–161.

Koller, M. R., Palsson, B. O., and Eisfeld, T. M. (2003). Method for inducing a response in one or more targeted cells. US Patent **6,** 642,018.

Koller, M., and Papoutsakis, E. (1994). "Cell Adhesion in Animal Cell Culture." Marcel Dekker, New York, NY.

Krasieva, T. B., Chapman, C. F., LaMorte, V. J., Venugopalan, V., Berns, M. W., and Tromberg, B. J. (1998). Mechanisms of cell permeabilization by laser microirradiation. *Proc. SPIE* **3260,** 38–44.

Krichevsky, A. M., and Kosik, K. S. (2002). RNAi functions in cultured mammalian neurons. *Proc. Natl. Acad. Sci. USA* **99,** 11926–11929.

Kurata, S., Tsukakoshi, M., Kasuya, T., and Ikawa, Y. (1986). The laser method for efficient introduction of foreign DNA into cultured cells. *Exp. Cell Res.* **162,** 372–378.

Lalonde, S., Ehrhardt, D. W., and Frommer, W. B. (2005). Shining light on signaling and metabolic networks by genetically encoded biosensors. *Curr. Opin. Plant Biol.* **8,** 574–581.

Lefkowitz, R. J., and Shenoy, S. K. (2006). Transduction of receptor signals by beta-arrestins. *Science* **308,** 513–517.

Marques, J. T., and Williams, B. R. G. (2005). Activation of the mammalian immune system by siRNAs. *Nat. Biotechnol.* **23,** 1399–1405.

Matheny, S. A., Chen, C., Kortum, R. L., Razidlo, G. L., Lewis, R. E., and White, M. A. (2004). Ras regulates assembly of mitogenic signalling complexes through the effector protein IMP. *Nature* **427,** 256–260.

Matoba, S., Kang, J.-G., Patino, W. D., Wragg, A., Boehm, M., Gavrilova, O., Hurley, P. J., Bunz, F., and Hwang, P. M. (2006). P53 regulates mitochondrial respiration. *Science* **312,** 1650–1653.

McManus, M. T., Haines, B. B., Dillon, C. P., Whitehurst, C. E., van Parijs, L., Chen, J., and Sharp, P. A. (2002). Small interfering RNA-mediated gene silencing in T lymphocytes. *J. Immunol.* **169,** 5754–5760.

Mehier-Humbert, S., and Guy, R. (2005). Physical methods for gene transfer: Improving the kinetics of gene delivery into cells. *Adv. Drug Deliv. Rev.* **57,** 733–753.

Mohanty, S. K., Sharma, M., and Guptal, P. K. (2003). Laser-assisted microinjection into targeted animal cells. *Biotechnol. Lett.* **25,** 2123–2136.

Nagy, V., and Watzele, M. (2006). FuGENE 6 transfection reagent: Minimizing reagent-dependent side effects as analyzed by gene-expression profiling and cytotoxicity assays. *Nat. Methods* **3,** iii–v.

Nalbant, P., Hodgson, L., Kraynov, V., Toutchkine, A., and Hahn, K. M. (2004). Activation of endogenous Cdc42 visualized in living cells. *Science* **305,** 1615.

Niemz, M. H. (1996). "Laser-Tissue Interactions: Fundamentals and Applications." Springer-Verlag, Berlin.

Novina, C. D., Murray, M. F., Dykxhoorn, D. M., Beresford, P. J., Riess, J., Lee, S.-K., Collman, R. G., Lieberman, J., Shankar, P., and Sharp, P. A. (2002). SiRNA-directed inhibition of hiv-1 infection. *Nat. Med.* **8,** 681–686.

O'Brien, J., and Lummis, S. C. (2002). An improved method of preparing microcarriers for biolistic transfection. *Brain Res. Protoc.* **10,** 12–15.

Palsson, B. O., Koller, M. R., and Eisfeld, T. M. (2003a). Method and apparatus for selectively targeting specific cells within a cell population. US Patent **6,** 514,722.

Palsson, B. O., Koller, M. R., and Eisfeld, T. M. (2003b). Method and apparatus for selectively targeting specific cells within a mixed cell population. US Patent **6,** 534,308.

Palumbo, G., Caruso, M., Crescenzi, E., Tecce, M. F., Roberti, G., and Colasanti, A. (1996). Targeted gene transfer in eukaryotic cells by dye-assisted laser optoporation. *J. Photochem. Photobiol. B* **36,** 41–46.

Pastinen, T., and Hudson, T. J. (2006). Cis-acting regulatory variation in the human genome. *Science* **306,** 647–650.

Paterson, L., Agate, B., Comrie, M., Ferguson, R., Lake, T. K., Morris, J. E., Carruthers, A. E., Brown, C. T. A., Sibbett, W., Bryant, P. E., Gunn-Moore, F., Riches, A., *et al.* (2005). Photoporation and cell transfection using a violet diode laser. *Opt. Express* **13,** 595–600.

Prentice, P., Cuschieri, A., Dholakia, K., Prausnitz, M., and Campbell, P. (2005). Membrane disruption by optically controlled microbubble cavitation. *Nat. Phys.* **1,** 107–110.

Rudolf, R., Mongillo, M., Rizzuto, R., and Pozzan, T. (2004). Looking forward to seeing calcium. *Nat. Rev. Mol. Cell Biol.* **4,** 579–586.

Sagi, S., Knoll, T., Trojan, L., Schaaf, A., Alken, P., and Michel, M. S. (2003). Gene delivery into prostate cancer cells by holmium laser application. *Prostate Cancer Prostatic Dis.* **6,** 127–130.

Sambrook, J., Fritsch, E. F., and Maniatis, T. (1989). "Molecular Cloning: A Laboratory Manual," 2nd edn. Cold Spring Harbor Press, Cold Spring Harbor, NY.

Sambrook, J., and Russell, D. W. (2001). "DNA Transfection by Electroporation." Cold Spring Harbor Laboratory Press, Cold Spring Harbor, NY.

Scacheri, P. C., Rozenblatt-Rosen, O., Caplen, N. J., Wolfsberg, T. G., Umayam, L., Lee, J. C., Hughes, C. M., Shanmugam, K. S., Bhattacharjee, A., Meyerson, M., and Collins, F. S. (2004). Short interfering RNAs can induce unexpected and divergent changes in levels of untargeted proteins in mammalian cells. *Proc. Natl. Acad. Sci. USA* **101,** 1892–1897.

Scherer, L. J., and Rossi, J. J. (2003). Approaches for the sequence-specific knockdown of mRNA. *Nat. Biotechnol.* **21,** 1457–1465.

Schneckenburger, H., Hendinger, A., Sailer, R., Strauss, W. S. L., and Schmitt, M. (2002). Laser-assisted optoporation of single cells. *J. Biomed Opt.* **7,** 410–416.

Sen, G. L., and Blau, H. M. (2006). A brief history of RNAi: The silence of the genes. *FASEB J.* **20,** 1293–1299.

Shirahata, Y., Ohkohchi, N., Itagak, H., and Satomi, S. (2001). New technique for gene transfection using laser irradiation. *J. Investig. Med.* **49,** 184–190.

Shuey, D. J., McCallus, D. E., and Giordano, T. (2002). RNAi: Gene silencing in therapeutic intervention. *Drug Discov. Today* **7,** 1040–1046.

Sioud, M. (2005). Induction of inflammatory cytokines and interferon responses by double-stranded and single-stranded siRNAs is sequence-dependent and requires endosomal localization. *J. Mol. Biol.* **348,** 1079–1090.

Soughayer, J. S., Krasieva, T., Jacobson, S. C., Ramsey, J. M., Tromberg, B. J., and Allbritton, N. L. (2000). Characterization of cellular optoporation with distance. *Anal. Chem.* **72,** 1342–1347.

Stevenson, D., Agate, B., Tsampoula, X., Rischer, P., Brown, C. T. A., Sibbett, W., Riches, A., Gunn-Moore, F., and Dholakia, K. (2006). Femtosecond optical transfection of cells: Viability and efficiency. *Opt. Express* **14,** 7125–7133.

Szaniszlo, P., Rose, W. A., Wang, N., Reece, L. M., Tsulaia, T. V., Hanania, E. G., Elferink, C. J., and Leary, J. F. (2006). Scanning cytometry with a leap: Laser-enabled analysis and processing of live cells *in situ*. *Cytometry A* **69,** 641–651.

Tao, W., Wilkinson, J., Stanbridge, E. J., and Berns, M. W. (1987). Direct gene transfer into human cultured cells facilitated by laser micropuncture of the cell membrane. *Proc. Natl. Acad. Sci. USA* **84,** 4180–4184.

Terakawa, M., Sato, S., Ashida, H., Aizawa, K., Uenoyama, M., Masaki, Y., and Obara, M. (2006). *In vitro* gene transfer to mammalian cells by the use of laser-induced stress waves: Effects of stress wave parameters, ambient temperature, and cell types. *J. Biomed. Opt.* **11,** 0140261–0140267.

Thompson, J. D. (2002). Applications of antisense and siRNAs during preclinical drug development. *Drug Discov. Today* **7,** 912–917.

Tirlapur, U. K., and Konig, K. (2002). Targeted transfection by femtosecond laser. *Nature* **418,** 290–291.

Tsukakoshi, M., Kurata, S., Nominya, Y., Ikawa, Y., and Kasuya, T. (1984). A novel method of DNA transfection by laser microbeam cell surgery. *Appl. Phys.* **35,** 135–140.

Tuschl, T. (2001). RNA interference and small interfering RNAs. *Chembiochem* **2,** 239–245.

Wehrens, X. H. T., and Marks, A. R. (2004). Novel therapeutic approaches for heart failure by normalizing calcium cycling. *Nat. Rev. Drug Discov.* **3,** 565–573.

Wurm, F. M. (2004). Production of recombinant protein therapeutics in cultivated mammalian cells. *Nat. Biotechnol.* **22,** 1393–1398.

Xia, H., Mao, Q., Paulson, H. L., and Davidson, B. L. (2002). SiRNA-mediated gene silencing *in vitro* and *in vivo*. *Nat. Biotechnol.* **20,** 1006–1010.

CHAPTER 11

Chromophore-Assisted Laser Inactivation

Diane Hoffman-Kim,* Thomas J. Diefenbach,† Brenda K. Eustace,‡ and Daniel G. Jay†

*Department of Molecular Pharmacology, Physiology, and Biotechnology
Center for Biomedical Engineering, Brown University
Providence, Rhode Island 02912

†Department of Physiology, Tufts University School of Medicine
Boston, Massachusetts 02111

‡Vertex Pharmaceuticals Inc., Cambridge
Massachusetts 02140

The major challenge of the post-genome world is ascribing *in situ* function to the myriad of proteins expressed in the proteome. This challenge is met by an arsenal of inactivation strategies that include RNAi and genetic knockout. These are powerful approaches but are indirect with respect to protein function and are subject to time

0091-679X/07 $35.00
DOI: 10.1016/S0091-679X(06)82011-X

delays before onset and possible genetic compensation. This chapter describes two protein-based inactivation approaches called chromophore-assisted laser inactivation (CALI) and fluorophore-assisted light inactivation (FALI). For CALI and FALI, light inactivation is targeted via photosensitizers that are localized to proteins of interest through antibody binding or expressed domains that are fluorescent or bind fluorescent probes. Inactivation occurs when and where the cells or tissues are irradiated and thus CALI and FALI provide an unprecedented level of spatial and temporal resolution of protein inactivation. Here we provide methods for the labeling of antibodies and setup of light sources and discuss controls, advantages of the technology, and potential pitfalls. We conclude with a discussion on a number of new technologies derived from CALI that combine molecular genetic approaches with light-induced inactivation that provide new tools to address *in situ* protein function.

I. Introduction

This volume has highlighted the physical interaction of light with molecules and cells. However, the ability of light of specific wavelengths to generate photochemical changes in cells is also of high utility in cell biology. This chapter describes chromophore-assisted laser inactivation (CALI), a technology developed in the late 1980s as a means of photodestruction of specific proteins in living cells to help ascribe their *in situ* function (Jay, 1988). CALI uses antibodies to target photosensitizers to specific proteins such that excitation of the photosensitizer results in light-induced free radical damage to the antibody-bound protein but not other nearby proteins. Since its inception, parallel technologies have been developed including fluorophore-assisted light inactivation (FALI) to facilitate high-throughput applications also developed in this laboratory (Beck *et al.*, 2002). Here we will describe the rationale of CALI and FALI and how they work, how they have been applied, methods for their application, and required controls. We will conclude with a discussion on advantages and potential pitfalls of these technologies and recent progress to combine CALI with molecular genetics that have great promise in addressing cellular function.

II. Rationale for CALI

Advances in the genomics field have afforded scientists the opportunity to consider and explore the roles of the estimated 30,000 genes in the human genome and the ~1 million protein products into which these genes are translated. With this opportunity has come the challenge of how to rigorously and efficiently integrate large quantities of genomic and proteomic information. This data is invaluable; however, a critical goal remains to move beyond the admittedly large task of cataloging the proteins expressed within a particular cell at a particular point in time, toward the more exciting albeit much more difficult aim of determining the function of each protein. Obtaining and utilizing this knowledge to elucidate protein function in the context of the dynamic cellular environment will further our comprehension of cellular biology.

Current approaches toward determining protein function include analyzing changes in the protein profile following a treatment of interest, via mass spectrometry (Li *et al.*, 2003), or via antibody or ligand microarrays (Templin *et al.*, 2003). Protein–protein interactions can also be identified by utilizing yeast two-hybrid screens (Li *et al.*, 2004) or yeast GFP-protein libraries (Huh *et al.*, 2003). This information often provides suggestions but not definitive demonstrations of function within the cell. Protein function can be tested by a number of approaches. Molecular genetics strategies include gene knockout, gene overexpression, and most recently, protein-knockdown approaches that are RNA-based (Dorsett and Tuschl, 2004). Selective gene silencing via RNA interference (RNAi) has been employed for functional analysis in *Caenorhabditis elegans*, *Drosophila*, and mammalian systems in studies focused on development and aging (Lee *et al.*, 2003), cell growth and morphology (Boutros *et al.*, 2004; Kiger *et al.*, 2003), cell signaling (Berns *et al.*, 2004), apoptosis (Aza-Blanc *et al.*, 2003), cytokinesis, and proteolysis (Silva *et al.*, 2004). However, limitations in applying RNAi to dissecting protein function include its inability to distinguish between posttranslationally modified proteins, its susceptibility to functional compensation, and its slow onset. Also of concern is the finding that off-target effects of RNAi can severely perturb neuronal structure and function (Alvarez *et al.*, 2006). Small molecule inhibitors block protein function quickly, but large libraries often need to be screened in efforts to find selective, effective inhibitors, which is not always possible (Bredel and Jacoby, 2004). Function-blocking antibodies can also be effective and may be useful for therapeutics, but less than 1% of the antibodies raised block function specifically and efficiently. CALI provides a unique alternative to these approaches by achieving direct, acute inactivation of protein function *in situ*.

III. Principle of CALI

Molecular function can be inactivated by light when the molecule of interest binds to a photosensitizer, in some cases via a binder molecule, and the photosensitizer absorbs the light. Through this absorption energy is transferred to inactivate the molecule of interest. In traditional CALI (Jay, 1988), the protein of interest is bound to an antibody covalently coupled to a photosensitive dye, where the dye is malachite green (MG) isothiocyanate. To inactivate a membrane-bound molecule, dye-coupled antibodies are bound to the surface of the cell, and to inactive an intracellular molecule, dye-coupled antibodies are loaded into the cell by trituration, scraping, microinjection, or electroporation. When the cell is irradiated with laser light of 620 nm wavelength, the MG absorbs the light, energy is transferred, and short-lived hydroxyl radicals are formed with a half-maximal radius of activity of 15 Å and a lifetime of 1 ps. The 620 nm wavelength is not significantly absorbed by cellular components. Achieving inactivation via CALI requires sufficient energy of irradiation, sufficient concentrations of both the chromophore and the protein of interest, the presence of oxygen, and the use of a laser.

CALI is an attractive approach in principle for high-throughput applications. To realize its potential, an alternate light inactivation approach was developed to overcome its laser-based limitations, that is its 2 mm laser spot size. FALI uses diffuse light (300 W) rather than laser light. The same types of antibodies can be employed as are used in CALI, but fluorescein isothiocyanate (FITC) acts as the chromophore, conventional light sources provide the light, and singlet oxygen species are generated with a longer lifetime of 200 ns and thus a larger associated half-maximal radius of inactivation of 40 Å (Beck *et al.*, 2002). Since the average intramolecular distance between proteins within a cell is 80 Å (Linden *et al.*, 1992), FALI can be utilized in cells, with requisite attention paid to possible effects on nearest neighbor proteins. Since it is possible to carry out FALI using diffuse light emitted by a slide projector, FALI can inactivate multiple samples concurrently in parallel, in multiwell plates. If these inactivation approaches are coupled to libraries of binders (potentially including antibodies, chemical compounds, or complementary RNA aptamers), high-throughput inactivation can be achieved.

IV. Applications of CALI

CALI was developed to acutely inactivate and thereby investigate protein function *in situ* with high spatial and temporal control. CALI experiments performed on the subunits of the T cell receptor complex demonstrated the high level of spatial specificity of the technique, wherein the function of the targeted subunit was disrupted while nearest neighbor subunits remained unaffected (Liao *et al.*, 1995). The CALI approach has been effective in probing the functions of proteins that lack effective pharmacological inhibitors or genetically engineered models. CALI has been performed on molecules of diverse types and across multiple organisms. Enzymes, cell surface proteins, cytoskeletal proteins, signal transduction molecules, and transcription factors have all been inactivated successfully, and among these include molecules with roles in nerve growth and regeneration, cancer, and prion proteins (for review see Eustace *et al.*, 2002; Table I). For many molecules, CALI experiments clarified function where other approaches were only able to suggest phenotypes. CALI has helped to sort out the distinct functions of members of the myosin family (Diefenbach *et al.*, 2002; Wang *et al.*, 1996, 2003). It has elucidated the role of pp60c-src in neurite outgrowth (Hoffman-Kim *et al.*, 2002), a role that a knockout mouse model was unable to resolve, likely because of functional compensation. Traditional CALI and double-CALI served to determine the functions of two neural cell adhesion molecules, L1 and NCAM-180, whose functions had been difficult to study in knockout models due to complications with adhesion (Takei *et al.*, 1999). CALI has found particular utility in experiments with neurons, which

Table I
Proteins Inactivated by CALI-Based Technologies

Enzymes	
	Acetylcholinesterase (Jay, 1988)
	Alkaline phosphatase (Jay, 1988)
	β-galactosidase (Jay, 1988)
	β-galactosidase (Bulina *et al.*, 2006)
	Caspase 3 (Rubenwolf *et al.*, 2002)
Signal transduction molecules	
	Calcineurin (Chang *et al.*, 1995)
	Calmodulin (Yan *et al.*, 2006)
	Cyclophilin A (H. Y. Chang, D. G. Jay, unpublished data)
	Focal adhesion kinase (Rajfur *et al.*, 2002)
	IP3 receptor (Takei *et al.*, 1998; Yogo *et al.*, 2004)
	MAP kinase (Rubenwolf *et al.*, 2002)
	p59fyn (Hoffman-Kim *et al.*, 2002)
	Phospholipase C δ 1 pleckstrin homology domain (Bulina *et al.*, 2006)
	pp60c-src (Hoffman-Kim *et al.*, 2002)
	Synaptotagmin I (Marek and Davis, 2002)
Surface proteins	
	α chain of the T cell receptor (Liao *et al.*, 1995)
	β chain of the T cell receptor (Liao *et al.*, 1995)
	β1 integrin (Beck *et al.*, 2002)
	Calcium channel (L-type) (Tour *et al.*, 2003)
	Calcium channel (N-type) (Guo *et al.*, 2006)
	CD44 (Hauptschein *et al.*, 2005)
	CD155 (Sloan *et al.*, 2005)
	Connexin43 (Tour *et al.*, 2003)
	Drosophila fasciclin (H. Keshishain, D. G. Jay, unpublished data)
	Drosophila patched protein (Schmucker *et al.*, 1994)
	ε Chain of the T cell receptor (Liao *et al.*, 1995)
	Ephrin-A5 (Sakurai *et al.*, 2002)
	Fas receptor (Rubenwolf *et al.*, 2002)
	FMRF amide receptor (Feigenbaum *et al.*, 1996)
	Grasshopper fasciclin I (Diamond *et al.*, 1993; Jay and Keshishian, 1990)
	Grasshopper fasciclin II (Diamond *et al.*, 1993)
	Hsp90α (Eustace *et al.*, 2004)
	L1 (Takei *et al.*, 1999)
	MAG (Wong *et al.*, 2003)
	NCAM (Takei *et al.*, 1999)
	PI_3 receptor type III (Takei *et al.*, 1998)
	Prion protein (Graner *et al.*, 2000)
	RGM (Muller *et al.*, 1996)
Transcription factors	
	Drosophila engrailed (R. Schroeder, D. G. Jay, unpublished data)
	Drosophila even skipped (Schroeder *et al.*, 1996)
	Tribolium even skipped (Schroeder *et al.*, 1999)
Cytoskeletal proteins	
	α-actinin (Rajfur *et al.*, 2002)
	Actin (A. M. Sydor, D. G. Jay, unpublished data)
	Ezrin (Lamb *et al.*, 1997)
	GAP-43 (K. A. Vancura, D. G. Jay, unpublished data)
	Hamartin (TSC I; Lamb *et al.*, 2000)
	Kinesin (Surrey *et al.*, 1998)
	MAP 1b (Mack *et al.*, 2000)
	Myosin 1β (Diefenbach *et al.*, 2002; Wang *et al.*, 2003)
	Myosin V (Wang *et al.*, 1996)
	Radixin (Castelo and Jay, 1999)
	Talin (Sydor *et al.*, 1996)
	Tau (Liu *et al.*, 1999)
	Vinculin (Sydor *et al.*, 1996)
	Zyxin (Jay, 2000)

are less amenable than other cells to transfection. The roles of many key proteins in neurite outgrowth and growth cone motility have been elucidated by combining microscope-focused CALI (micro-CALI; laser spot size of 10 μm) with time-lapse microscopy of single-moving neurons (for review see Buchstaller and Jay, 2000).

CALI performed *in vivo* has demonstrated its ability to target proteins at specific points in development or during key dynamic processes. CALI of fasciclin I resulted in decreased bundling of pioneer axons in grasshopper larvae (Diamond *et al.*, 1993), and inactivation of patched protein by CALI in *Drosophila* led to alterations in cell fate (Schmucker *et al.*, 1994). In the chick visual system, CALI of ephrin-A5, a repulsive axon guidance cue in the optic tectum, inhibited its restriction of axons projecting from the retina to the tectum (Sakurai *et al.*, 2002), and CALI of inhibitory myelin-associated glycoprotein promoted retinal ganglion cell regeneration following an optic nerve crush in organ culture (Wong *et al.*, 2003). In many cases, CALI experiments *in vitro* and *in vivo* have confirmed the corresponding genetic loss of function experiments.

V. Methods

A. Antibody Preparation and Labeling with MGITC and FITC

CALI and FALI are best performed using nonfunction-blocking antibodies with high affinity to the target of interest. Traditional monoclonal or polyclonal antibodies can be used for inactivation, as well as recombinant single-chain variable fragments (scFvs). Also, ligands for receptors, streptavidin (for biotinylated enzymes), and small molecules have been used for inactivation with high specificity.

Malachite green isothiocyanate (MGITC) and FITC are used for labeling and react with amino groups (e.g., lysine residues) on antibodies to form a stable thioester. The antibody solution should not contain other components with free amino groups, such as Tris- or glycine-based buffers, to avoid nonproductive side reactions with isothiocyanate.

1. MGITC Labeling

An unavoidable competing side reaction is MGITC hydrolysis and this product can precipitate due to π-stacking interactions. This side reaction is reduced somewhat by keeping the labeling reaction at a high pH (antibody in 0.5 M $NaHCO_3$, pH 9.5), resuspending the MGITC dye in dry dimethylsultoxide (DMSO), keeping the MGITC at high concentrations in the labeling mix (by adding MGITC in aliquots), and adding bovine serum albumin (BSA) to prevent precipitation. MG-labeled BSA has not been found to cause damage to surrounding protein *in vitro* or *in vivo*, and thus can be used effectively to reduce hydrophobic MG interactions.

2. Solutions

1. *Phosphate buffer solution (PBS) buffer*: 15 liter of 0.1 M PBS buffer.
2. *Hank's balanced salt solution*: Ca^{2+}/Mg^{2+}-free, 15 ml (HBSS, Gibco, Cat. No. 24020117, Invitrogen Corporation).
3. *10 mg/ml MGITC*: Weigh 1 mg MGITC (Cat. No. M-689, Molecular Probes, Inc.) and resuspend in 0.1 ml DMSO in a 1.5 ml centrifuge tube immediately before use.
4. *BSA (Cat. No. B2518, Sigma-Aldrich Corporation)*: 20 mg/ml stock (weigh 20 mg and resuspend in 1 ml PBS) to combine with the antibody solution.
5. *Antibody solution*: Typically, 100 μg to 1 mg antibody in less than 1-ml of buffer that does not contain amino groups (e.g., PBS).
6. *1.0 M $NaHCO_3$ stock*: To make 100 ml, take 8.2 g of $NaHCO_3$ and dissolve in less than 100 ml of distilled, deionized water. Add 1 M NaOH dropwise to bring pH to 9.5 and bring volume up to 100 ml with water.

3. Steps

1. Antibodies are dialyzed against PBS buffer (pH 7.4) overnight at 4°C.
2. Antibodies can be concentrated by centrifugation (see step 10 below). Typically, between 100 μg and 1 mg of antibody is labeled at a time. The volume of antibody should be minimized (<500 μL), as too dilute a reaction mixture reduces labeling reaction efficiency.
3. An equal (1:1) amount of BSA can be added to the antibody solution as a stabilizing reagent.
4. One hundred microliters of 1 M $NaHCO_3$ (pH 9.5) is added to the antibody solution. The high pH favors the labeling reaction over hydrolysis.
5. A 1:5 weight-to-weight ratio of MGITC to protein is added to the antibody solution to yield an optimal labeling ratio of ~5 MGITC:1 antibody molecule (or 1 MGITC:1 Fab or scFv). MGITC is added to the antibody solution in three equal aliquots dispensed every 5 min, which favors the labeling reaction over the hydrolysis reaction. A rocker (Cat. No. 14-512-28, Lab-Line Maxi Rotator, Fisher Scientific Company) is used to gently mix the reaction mixture during incubation. After the last aliquot, the mixture is incubated for an additional 15 min. The free MGITC label in solution is green. As the labeling reaction proceeds MGITC reacts with amino groups to form the thioester and the solution changes color to blue. The hydrolysis product is purple and forms an insoluble precipitate.
6. After incubation, the mixture is centrifuged at high speed (6400 rpm) on a mini centrifuge (Cat. No. EF4241A, Costar WX4241A, Daigger Laboratory Supplies) for 30 s to pellet any precipitated hydrolyzed dye.

7. Unbound MGITC and buffer salts are separated from the reaction mixture using a prepacked, 5-cc Sephadex G25 gel filtration column (PD-10, Cat. No. 17-0851-01, Amersham Biosciences Corporation) using a medium suitable for cell loading as an eluant, such as with or without calcium and magnesium.
8. Usually ~2 ml of eluant is collected. Depending on the initial concentration, the antibody solution can be concentrated in Centricon tubes (30,000 MW cutoff; preblocked with 1-mg/ml BSA) using a centrifuge (Sorval RC-5, or equivalent) at 4200 rpm at room temperature. The eluant should be kept above 15 °C to avoid precipitation of residual, free MGITC, which is toxic to cells.
9. Once concentrated, labeling ratio is calculated by taking the optical density of the labeled solution, to determine dye concentration, at 620 nm (molar absorptivity = 150,000 M^{-1} cm^{-1}) and dividing that value by the antibody concentration. The molecular weight of IgG is ~150,000, and BSA is ~66,000. We aim to reach a labeling ratio of four to eight dye molecules per IgG or approximately one to two dyes per scFv (a much smaller protein). While this estimate is inaccurate, it is used as a relative standard when determining the labeling versus efficacy for CALI and FALI.
10. The concentrated labeled antibody solution is aliquoted into use-appropriate volumes (~100–200 μL), quick-frozen and stored at −80 °C for less than 6 months.

4. FITC Labeling

Hydrolyzed FITC does not readily aggregate, and thus labeling is easier and performed in a slightly different manner. The steps are the same as MGITC labeling except for the following differences. BSA is not added before labeling since hydrophobic interactions between FITC molecules are less common. Instead, BSA is added after labeling (to 1 mg/ml) to stabilize the FITC-labeled antibody during storage. A freshly prepared 10 mg/ml solution of FITC (Cat. No. F-143, Molecular Probes, Inc., Eugene, OR) is made in dry DMSO and added to the antibody solution (in 0.5 M $NaHCO_3$, pH 9.5) in a 1:5 weight-to-weight ratio. As with MGITC labeling, we generally label between 100 μg and 1 mg of protein in up to 1 ml of total reaction solution. The FITC solution is added all at once to the antibody solution, and is incubated with constant rocking for 1 h at room temperature. Free hydrolyzed dye is separated from FITC-labeled antibody using PD10 gel filtration with the same protocol as described above for MGITC. For FALI, the elution buffer must be phenol red-free. Phenol red is a pH indicator added to many culture media. It is an efficient quencher of singlet oxygen species and must be avoided for efficient FALI-generated inactivation. A hand-held UV lamp is used to detect fluorescence in the eluant to collect the fractions containing FITC-labeled protein (~2 ml). As for MG labeling, the goal is to have four to six FITC molecules per IgG molecule and this can be calculated by dividing the optical density at 494 nm (molar absorptivity of 68,000 M^{-1} cm^{-1}) by the antibody concentration.

5. Multiplex FITC Labeling

Given the high-throughput application of FALI, we developed FITC labeling of many antibodies in parallel. We routinely label 48 scFv molecules at one time in a 96-well plate, and full-chain antibodies could be done in a similar way. A solution containing between 50 and 100 μg of antibody in 200 μl is added to a standard 96-well plate along with 2 μl of 10 mg/mL FITC in DMSO and 25 μl of 1 M $NaHCO_3$, pH 9.5. This reaction is incubated for 2 h at room temperature with rocking. The free dye is removed using G-25 spin columns and the associated multiplex-24 plate apparatus (Pharmacia, Piscataway, NJ). The labeling ratio is determined by taking the optical density of the FITC-labeled antibody solution at 494 nm in a 96-well Spectrapor Plus fluorescence plate reader (Tecan Corporation, Durham, NC).

B. CALI: Laser Setup and Irradiation Conditions (Fig. 1)

1. Equipment

Tunable Surelite optical parametric oscillator (OPO) Nd:YAG laser (Continuum Scientific Service)

Right-angle prism, holder, and mounting rods (Newport Corporation, Fountain Valley, CA)

Convex lens and lens holder

Laser parameters: wavelength = 620 nm, peak power = 56 Mw/cm^2, spot size = 2 mm, pulse width = 3 ns, frequency = 10 Hz, energy/pulse = 15 mJ.

2. Methods

The low efficiency of MG as a photosensitizer requires very high power at a wavelength not absorbed by cellular components. To achieve this power, we employ pulsed tunable dye lasers. We currently use a tunable Surelite OPO Nd:YAG laser (Continuum Scientific Service) for CALI, but other laser sources with comparable power are equally effective. For example, we also use an Nd:YAG-driven dye laser (GCR-11 with HG-2 doubling crystal, PDL-2; Spectra Physics Corporation, Mountain View, CA). For the GCR11 dye laser, we use 630 nm instead of 620 nm (the λ_{max} for MG-labeled protein) because much higher power can be generated at 630 nm using DCM as the circulating dye. To measure laser power, a beam splitter directs a fraction (1/7) of the light to a laser light meter (Model JD500, Molectron Detector, Inc.). The remaining light is directed downward by total internal reflection through a right-angle prism and focused using a 25 mm diameter planoconvex lens (Cat. No. SPX016AR.14, Newport Corporation) to a 2 mm diameter spot (typically the spot is a 2 mm $\times$ 4 mm oval). Spot size and shape is recorded on preexposed black photographic paper (Polaroid Corporation), and adjusted using

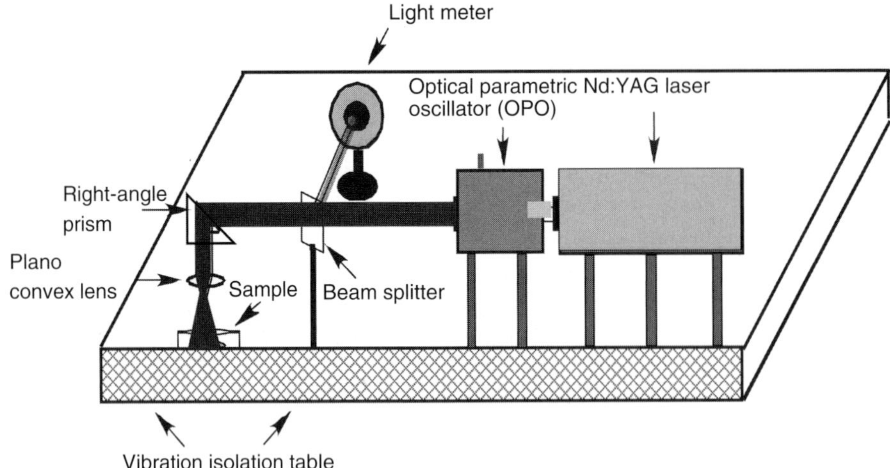

Fig. 1 Schematic for CALI apparatus. An Nd:YAG pulsed laser drives an OPO laser to generate 620 nm laser pulses directed downward onto the sample below with a right angle prism. A small fraction (1/7) of the light is directed perpendicularly via a beam splitter to a laser light meter.

a diaphragm (Cat. No. MH-2P, Newport Corporation) placed between the output of the dye laser and beam splitter. To confirm laser efficacy at the specimen plane, control *in vitro* experiments (e.g., CALI of β-galactosidase) are used to establish suitable irradiation times. Generally, irradiation times for experiments are typically 2–5 min but one should perform a dose–response curve for efficacy.

Protein or suspended cell samples to be used for CALI *in vitro* are placed in the wells of a Nunc-transferrable solid phase plate (Nunc International, Roskilde, Denmark). The wells of the plate are ~2 mm in diameter, and the laser beam should be centered as accurately as possible in the well. The samples are irradiated for 2–5 min, depending on the assay to be performed. *In vitro* experiments are usually pulsed for 5 min, while assays involving cells are usually limited to 2 min. Samples are incubated on ice during irradiation to reduce sample heating.

The short pulse width of the Nd:YAG laser produces a very high peak power (megawatts) and extreme caution is employed, with special concern for eye protection. Protective goggles are always worn, and beam blockers are placed to prevent stray reflections. Contact with skin should also be avoided. Investigators are advised to carefully follow laser safety protocols provided with the laser system.

C. FALI: Slide Projector Setup and Irradiation Conditions (Fig. 2)

1. Equipment

Ektographic III slide projector (Kodak Corporation, Rochester, NY)
Brilliant blue filter #69 (Roscolux Corporation, Stamford, CT)

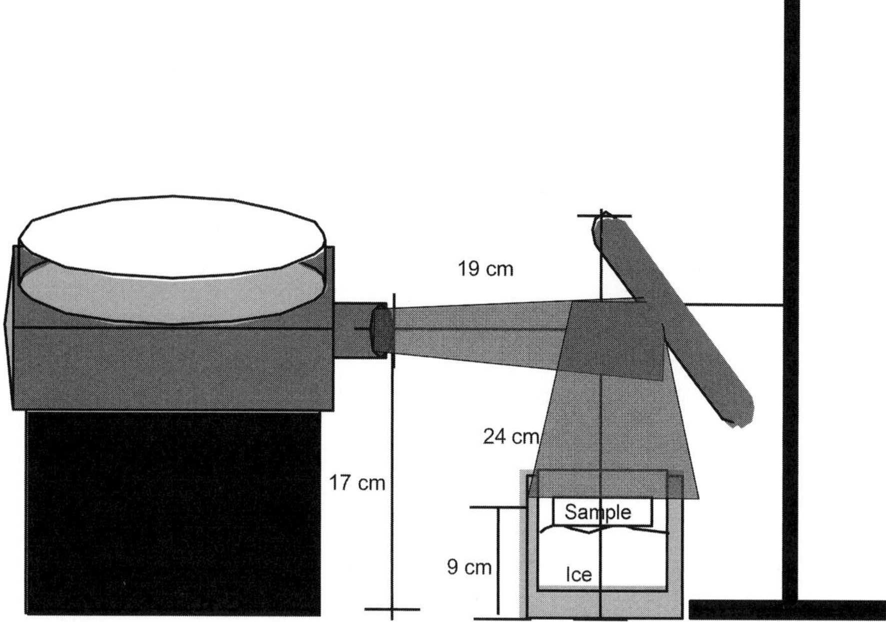

Fig. 2 Schematic for FALI apparatus. A blue light beam generated by a 300 W slide projector light beam is directed using a 45° angle mirror samples below that are incubated on ice.

Rectangular mirror, holder, and mounting rods (Newport Corporation, Fountain Valley, CA)

Light parameters: power = 300 W.

2. Methods

Since FITC absorbs visible light (494 nm) and is ~50-fold more efficient compared with MGITC for inactivation, FALI can be performed with a variety of continuous wave light sources for FALI. We have used continuous wave laser light (Argon ion), diffuse light from an ordinary 60 W light bulb, or from a 300 W slide projector for FALI. Routinely, we use a 300 W slide projector containing a blue filter (Brilliant blue filter #69, Roscolux Corporation, Stamford, CT) in the slide slot that limits the transmission of light so that greater than 50% of the transmitted light is between 420 and 500 nm. The system has to be configured to ensure uniform illumination over the entire sample, and adequate precautions need to be taken to prevent thermal damage, such as cooling of the sample during irradiation or increasing the distance of the sample from the light source. The slide projector is set up such that the light is directed onto a mirror that is oriented at a 45° angle ~19 cm from the projector and 10 cm from the sample. The light is thus projected downward onto the sample. Many samples can be illuminated at once if a multiwell plate is used, since the projected light is not focused into a beam. However, the irradiation time is much longer when using a diffuse light source than when using focused laser light. One should perform a dose–response curve for FALI. For example, the $t_{1/2}$ for inactivation is ~10 min for FALI of β-galactosidase. In cellular assays, we illuminate samples for 1 h incubated on ice to ensure maximal inactivation.

VI. Controls for CALI

Numerous controls are required to confirm that the cellular effects derive from the specific CALI treatment and not from the light, the chromophore, or the antibody. Experimental analysis must also take into account the possibility of residual protein activity, which could mask the effects of CALI. In addition, functional proteins can enter into the local region of CALI either by diffusion from unirradiated regions or by *de novo* synthesis.

To confirm the specific loss of function by CALI on a protein target, the following control experiments can be performed:

1. Cells without antibody are exposed to laser light (no antibody, laser control) to exclude an effect of laser light alone.
2. Cells with preimmune MG-labeled IgG are exposed to laser light (IgG control) as an antibody specificity control and MG-labeled antibody control.
3. Cells with MG-labeled specific antibody without exposure to laser light (MG-antibody, no laser control) serve as a control to show that the antibody employed is not function-blocking.
4. Cells with an MG-labeled Fab fragment derived from an anti-Fc antibody (MG-anti-Fc) coupled to nonlabeled primary antibody against the protein target and then exposed to laser light serve as a control for nonspecific free radical damage to neighboring proteins. In this control, the MG is effectively positioned ~100 Å away from the antibody-binding site (Sakurai *et al.*, 2002). Since the half-maximal radius of inactivation of MG is 15 Å (Liao *et al.*, 1995), laser irradiation should cause no significant effect due to additional distance between the chromophore and the antigenic site.
5. CALI of the purified protein *in vitro* followed by an assay for biochemical activity can be used to demonstrate that CALI results in loss of protein function.

VII. Advantages of CALI

CALI is both spatially restricted, with a small radius of inactivation that limits inactivation specifically to the bound protein of interest, as well as temporally restricted, as disruption occurs only following irradiation. This is useful in showing when a particular protein is required for a cellular process or developmental event

(Sakurai *et al.*, 2002). As an acute light inactivation method, it has the additional advantage of eliminating the complications of functional compensation. In addition, with the use of specific antibodies, CALI can distinguish between isoforms that arise from posttranslational modification. CALI also provides numerous versatile options for testing function in response to distinct cellular dynamics. The spatial resolution of CALI is unique among current loss of function strategies and it is possible using microscope-focused lasers (micro-CALI) to inactivate function in subcellular regions such as the nerve growth cone (Chang *et al.*, 1995; Wang *et al.*, 1996).

VIII. Potential Pitfalls of CALI

In order to interpret CALI results and ascribe a protein's function, a number of caveats to the approach need to be considered. Antibody selection is critical; suitable antibodies exhibit high selectivity and affinity for their protein target, and do not inherently inhibit protein function.

The utility of CALI is determined in part by the retention of damaged protein in the cell region irradiated. Diffusion of unirradiated protein, degradation of damaged protein, or synthesis of new protein will determine the half-life of the effect of CALI in cells and tissues. Retention of labeled antibody will also determine the utility of CALI and will be cell-type specific. For proteins subjected to CALI thus far, efficacy of CALI has been observed typically up to 10–15 h after loading (Diamond *et al.*, 1993; Sakurai *et al.*, 2002), while over 15 h after loading, recovery can occur by *de novo* synthesis (Sakurai *et al.*, 2002).

Another important consideration in the efficiency of CALI is the specificity of the antibody. The antibody may be highly specific for its intended antigen, sufficiently abundant inside or outside the cell to saturate that antigen, and optimally labeled with MGITC or FITC, but may still not have an effect on protein function if it binds to a site on the protein that is not essential in the biological function being tested. Polyclonal antibodies that recognize a greater number of epitopes on a single protein could partially circumvent this limitation, as could the use of more than one monoclonal antibody. Indeed, when an array of scFvs that recognize different epitopes of β-galactosidase were used for FALI, their inactivation effects were clearly additive (Beck *et al.*, 2002). In addition, addressing protein function at the level of more than one epitope can be helpful in elucidating the function of distinct protein domains.

Abundance of the native protein should be considered when determining the effectiveness of CALI. Since CALI is a knockdown strategy employing a limited number of antibodies per cell, typically less than 100% of the protein targeted is inactivated in an entire cell. Thus, it is technically difficult to inactivate a highly abundant intracellular protein. In spite of this limitation, most proteins will occupy a small percentage of the total protein in a given cell type, and antibodies targeted against extracellular domains of integral membrane proteins would not be so limited.

IX. CALI Versus FALI

Thus far, the major application of FALI has been for high-throughput screening of functional requirement (Eustace *et al.*, 2004; Hauptschein *et al.*, 2005; Sloan *et al.*, 2004). The other advantage is that FALI does not require a laser and may be performed at low cost without highly specialized instrumentation. There are a number of considerations when selecting CALI versus FALI. First, FALI requires 488 nm excitation wavelength of light, a wavelength of higher energy that is absorbed more readily by cells and tissues compared to 620 nm light used for CALI. This necessitates brief illumination periods for FALI, and cell viability should be verified when longer periods of illumination are used. MGITC requires a longer excitation wavelength (620 nm), which not only has lower energy, but is also a wavelength not significantly absorbed by biological material. The ~2.5-fold greater half-maximal radius of damage of FALI compared with CALI leads to a greater likelihood of collateral damage to neighboring proteins and this has already been reported (Guo *et al.*, 2006; Hauptschein *et al.*, 2005). To our knowledge, there is not yet an example of nearest neighbor damage for CALI. FITC has been reported to be 50 times more efficient than MGITC in *in vitro* assays (Surrey *et al.*, 1998). However, intracellularly, the efficacy of CALI or FALI will be determined in part by the ability of a cell to scavenge hydroxyl radical or singlet oxygen, respectively, in addition to the greater half-maximal radius of inactivation of FITC compared to MGITC. Damage resulting from hydroxyl radical is likely to occur through hydroxylation of aromatic amino acids forming a hydroxycyclohexadienyl radical on the aromatic ring (Halliwell and Gutteridge, 1990), though given the high reactivity and low specificity of the hydroxyl radical, other reactions are possible. FALI causes methionine oxidation and protein cross-linking as demonstrated by mass spectrometry analysis after FALI of calmodulin (Yan *et al.*, 2006).

X. New Developments in CALI Technology

Progress in CALI technology has expanded the range and utility of the approach. The use of binder molecules has developed beyond nonfunction-blocking IgG, IgM, Fab fragments, and single-chain antibodies for traditional CALI to include, for example, small molecule binders to the FMRF-amide receptor (Feigenbaum *et al.*, 1996) and to the IP3 receptor (Inoue *et al.*, 2001). In addition, MG-binding aptamers have been employed for the inactivation of RNA via CALI (Grate and Wilson, 1999). CALI has also been performed with scFv phage display antibodies that contain the phage coat protein and the

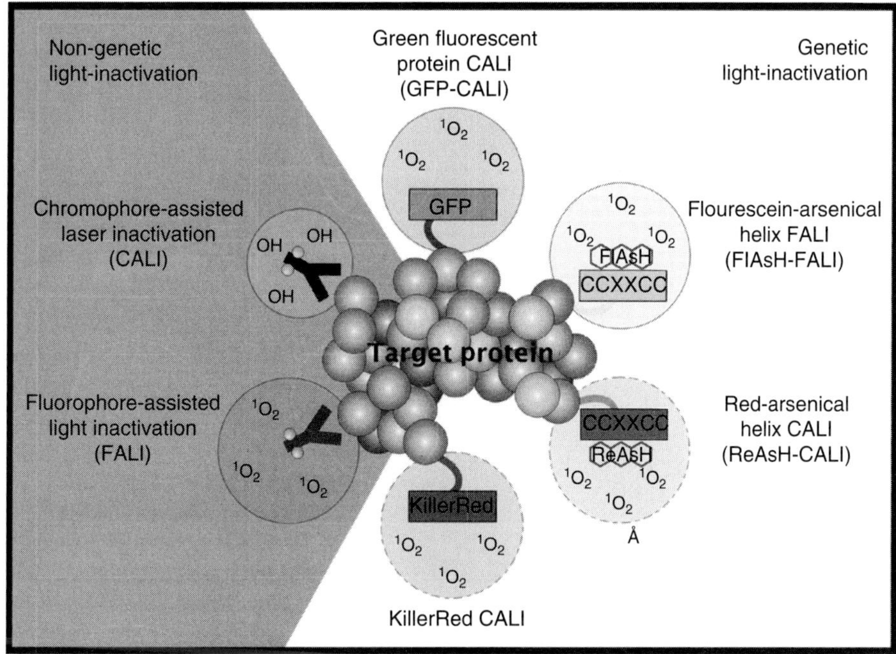

Fig. 3 A summary of CALI-derived technologies. Direct light inactivation technologies. The radius of inactivation is 15 Å for CALI, 40 Å for FALI, GFP-CALI, and FlAsH-FALI, and not yet determined for ReAsH-CALI or for KillerRed CALI. (See Plate 8 in the color insert section.)

antigen-binding site of immunoglobulin G molecules (Beck *et al.*, 2002). Significantly, use of combinatorial chemistry and phage display libraries augments the application of CALI to include the realm of high-throughput exploration of the proteome. A study used a Boyden chamber assay to identify molecules that influenced tumor invasion (Eustace *et al.*, 2004); analogous high-throughput approaches could be employed to find molecules involved in such functions as cell proliferation, motility, apoptosis, and others.

Recent advances in CALI technology have extended its applicability to systems amenable to molecular genetics approaches, with the development of new photosensitizers and new ways for those photosensitizers to bind molecules of interest (Fig. 3). Genetically targeted photoactivators include GFP (Rajfur *et al.*, 2002), FlAsH-EDT_2 (a fluorescein derivative with two As(III) substituents on the 4′ and 5′ positions; Adams *et al.*, 2002; Griffin *et al.*, 1998), ReAsH-EDT_2 (a red biarsenical fluorophore, also with two As(III) substituents on the 4′ and 5′ positions; Tour *et al.*, 2003), and KillerRed (a dimeric red homologue of GFP; Bulina *et al.*, 2006). While GFP from *Aequorea victoria* is genetically encoded and CALI against GFP-fusion proteins has been employed to investigate the roles of α-actinin and focal adhesion kinase (Rajfur *et al.*, 2002), its efficiency as a photoactivator is relatively low, presumably because its protein shell shields the fluorophore and increases the time required to generate singlet oxygen (Surrey *et al.*, 1998). In more efficient processes, genetically targeted FlAsH and ReAsH specifically target expressed proteins by genetically tagging them with a tetracysteine motif, which is then subsequently recognized by a biarsenical dye. It is important to note that endogenous proteins are not targeted by these approaches. The FlAsH and ReAsH dyes only fluoresce when bound to the cysteine on the protein of interest and are shorter than GFP, so are less likely to interfere with protein processing (Marek and Davis, 2002). One caveat is that some degree of nonspecific binding to endogenous tetracysteine is possible.

FlAsH-FALI using incandescent light disrupted synaptotagmin I in the neuromuscular junction of syt I null *Drosophila* larvae *in situ*, showing that FlAsH is effective as both a protein marker and a photosensitizer (Marek and Davis, 2002). FlAsH-FALI requires the use of a wavelength of light that is absorbed by cells and has associated concerns regarding nonspecific effects; as a result, ReAsH-CALI was developed to be excitable by longer wavelengths (593 nm; Tour *et al.*, 2003). ReAsH-CALI has been shown to require only 10–30 s of irradiation with a 150 W xenon arc lamp to generate maximal inactivation of connexin-43, a component of gap junctions (Tour *et al.*, 2003). The recently developed KillerRed-CALI combines the advantages of GFP-CALI and ReAsH-CALI. Light-induced killing of *Escherichia coli* and eukaryotic cells and inactivation of fusion proteins to β-galactosidase and phospholipase C δ 1 pleckstrin homology domain have been realized with this approach. KillerRed is fully genetically encoded and more efficient than GFP, inactivation can be achieved with relatively long wavelengths (585 nm), and it has very low associated levels (~2%) of nonspecific inactivation (Bulina *et al.*, 2006). Taken together, the developments in this area advance the utility of CALI and generate valuable connections to molecular genetic approaches.

XI. Conclusions

CALI-based technology is a highly specific means of eliminating protein function in real time, under tight spatial and temporal control. Recent advances in the development of binder molecules have moved the technology into the arena of molecular genetic approaches. CALI and its progeny form an extremely versatile tool set that can be applied to *in vitro* systems, to individual live cells, to multiple live cells in a high-throughput manner, and to *in vivo* systems and organisms. Future innovations to further enhance the approach will likely include more efficient reactive dye molecules to increase the potency of inactivation, and more comprehensive libraries of binders to address wider protein populations, thereby propelling the level of inquiry toward a highly complex assessment of dynamic, interactive protein function.

Acknowledgments

The authors thank Vivian Fong for assistance with figure preparation. DGJ was supported by grants from NIH.

References

Adams, S. R., Campbell, R. E., Gross, L. A., Martin, B. R., Walkup, G. K., Yao, Y., Llopis, J., and Tsien, R. Y. (2002). New biarsenical ligands and tetracysteine motifs for protein labeling *in vitro* and *in vivo*: Synthesis and biological applications. *J. Am. Chem. Soc.* **124,** 6063–6076.

Alvarez, V. A., Ridenour, D. A., and Sabatini, B. L. (2006). Retraction of synapses and dendritic spines induced by off-target effects of RNA interference. *J. Neurosci.* **26,** 7820–7825.

Aza-Blanc, P., Cooper, C. L., Wagner, K., Batalov, S., Deveraux, Q. L., and Cooke, M. P. (2003). Identification of modulators of TRAIL-induced apoplosis via RNAi-based phenotypic screening. *Mol. Cell* **12,** 627–637.

Beck, S., Sakurai, T., Eustace, B. K., Beste, G., Schier, R., Rudert, F., and Jay, D. G. (2002). Fluorophore-assisted light inactivation: A high-throughput tool for direct target validation of proteins. *Proteomics* **2,** 247–255.

Berns, K., Hijmans, E. M., Mullenders, J., Brummelkemp, T. R., Velds, A., Heimerikx, M., Kerkhoven, R. M., Madiredjo, M., Nijkamp, W., Weigelt, B., Agami, R., Ge, W., *et al.* (2004). A large-scale RNAi screen in human cells identifies new components of the p53 pathway. *Nature* **428,** 431–437.

Boutros, M., Kiger, A. A., Armknecht, S., Kerr, K., Hild, M., Koch, B., Haas, S. A., Consortium, H. F., Paro, R., and Perrimon, N. (2004). Genome-wide RNAi analysis of growth and viability in *Drosophila* cells. *Science* **303,** 832–835.

Bredel, M., and Jacoby, E. (2004). Chemogenomics: An emerging strategy for rapid target and drug discovery. *Nat. Rev. Genet.* **5,** 262–275.

Buchstaller, A., and Jay, D. G. (2000). Micro-scale chromophore-assisted laser inactivation of nerve growth cone proteins. *Microsc. Res. Tech.* **48,** 97–106.

Bulina, M. E., Chudakov, D. M., Britanova, O. V., Yanushevich, Y. G., Staroverov, D. B., Chepurnykh, T. V., Merzlyak, E. M., Shkrob, M. A., Lukyanov, S., and Lukyanov, K. A. (2006). A genetically encoded photosensitizer. *Nat. Biotechnol.* **24,** 95–99.

Castelo, L., and Jay, D. G. (1999). Radixin is involved in lamellipodial stability during nerve growth cone motility. *Mol. Biol. Cell* **10,** 1511–1520.

Chang, H. Y., Takei, K., Sydor, A. M., Born, T., Rusnak, F., and Jay, D. G. (1995). Asymmetric retraction of growth cone filopodia following focal inactivation of calcineurin. *Nature* **376,** 686–690.

Diamond, P., Mallavarapu, A., Schnipper, J., Booth, J., Park, L., O'Connor, T. P., and Jay, D. G. (1993). Fasciclin I and II have distinct roles in the development of grasshopper pioneer neurons. *Neuron* **11,** 409–421.

Diefenbach, T. J., Latham, V. M., Yimlamai, D., Liu, C. A., Herman, I. M., and Jay, D. G. (2002). Myosin 1c and myosin IIB serve opposing roles in lamellipodial dynamics of the neuronal growth cone. *J. Cell Biol.* **158,** 1207–1217.

Dorsett, Y., and Tuschl, T. (2004). siRNAs: Applications in functional genomics and potential as therapeutics. *Nat. Rev. Drug Discov.* **3,** 318–329.

Eustace, B. K., Buchstaller, A., and Jay, D. G. (2002). Adapting chromophore-assisted laser inactivation for high throughput functional proteomics. *Brief. Funct. Genomic. Proteomic.* **1,** 257–265.

Eustace, B. K., Sakurai, T., Stewart, J. K., Yimlamai, D., Unger, C., Zehetmeier, C., Lain, B., Torella, C., Henning, S. W., Beste, G., Scroggins, B. T., Neckers, L., *et al.* (2004). Functional proteomic screens reveal an essential extracellular role for hsp90 alpha in cancer cell invasiveness. *Nat. Cell Biol.* **6,** 507–514.

Feigenbaum, J. J., Choubal, M. D., Crumrine, D. S., and Kanofsky, J. R. (1996). Receptor inactivation by dye-neuropeptide conjugates. 2. Characterization of the quantum yield of singlet oxygen generated by irradiation of dye-neuropeptide conjugates. *Peptides* **17,** 1213–1217.

Graner, E., Mercadante, A. F., Zanata, S. M., Martins, V. R., Jay, D. G., and Brentani, R. R. (2000). Laminin-induced PC-12 cell differentiation is inhibited following laser inactivation of cellular prion protein. *FEBS Lett.* **482,** 257–260.

Grate, D., and Wilson, C. (1999). Laser-mediated, site-specific inactivation of RNA transcripts. *Proc. Natl. Acad. Sci. USA* **96,** 6131–6136.

Griffin, B. A., Adams, S. R., and Tsien, R. Y. (1998). Specific covalent labeling of recombinant protein molecules inside live cells. *Science* **281,** 269–272.

Guo, J., Chen, H., Puhl, H. L., III, and Ikeda, S. R. (2006). Fluorophore-assisted light inactivation produces both targeted and collateral effects on N-type calcium channel modulation in rat sympathetic neurons. *J. Physiol.* **576,** 477–492.

Halliwell, B., and Gutteridge, J. M. (1990). Role of free radicals and catalytic metal ions in human disease: An overview. *Methods Enzymol.* **186,** 1–85.

Hauptschein, R. S., Sloan, K. E., Torella, C., Moezzifard, R., Giel-Moloney, M., Zehetmeier, C., Unger, C., Ilag, L. L., and Jay, D. G. (2005). Functional proteomic screen identifies a modulating role for CD44 in death receptor-mediated apoptosis. *Cancer Res.* **65,** 1887–1896.

Hoffman-Kim, D., Kerner, J. A., Chen, A., Xu, A., Wang, T. F., and Jay, D. G. (2002). pp60(c-src) is a negative regulator of laminin-1-mediated neurite outgrowth in chick sensory neurons. *Mol. Cell. Neurosci.* **21,** 81–93.

Huh, W. K., Falvo, J. V., Gerke, L. C., Caroll, A. S., Howson, R. W., Weissman, J. S., and O'Shea, E. K. (2003). Global analysis of protein focalization in budding yeast. *Nature* **425,** 686–691.

Inoue, T., Kikuchi, K., Hirose, K., Iino, M., and Nagano, T. (2001). Small molecule-based laser inactivation of inositol 1,4,5-trisphosphate receptor. *Chem. Biol.* **8,** 9–15.

Jay, D. G. (1988). Selective destruction of protein function by chromophore-assisted laser inactivation. *Proc. Natl. Acad. Sci. USA* **85,** 5454–5458.

Jay, D. G. (2000). The clutch hypothesis revisited: Ascribing the roles of actin-associated proteins in filopodial protrusion in the nerve growth cone. *J. Neurobiol.* **44,** 114–125.

Jay, D. G., and Keshishian, H. (1990). Laser inactivation of fasciclin I disrupts axon adhesion of grasshopper pioneer neurons. *Nature* **348,** 548–550.

Kiger, A., Baum, B., Jones, S., Jones, M., Coulson, A., Echeverri, C., and Perrimon, N. (2003). A functional genomic analysis of cell morphology using RNA interference. *J. Biol.* **2,** 27.

Lamb, R. F., Ozanne, B. W., Roy, C., McGarry, L., Stipp, C., Mangeat, P., and Jay, D. G. (1997). Essential functions of ezrin in maintenance of cell shape and lamellipodial extension in normal and transformed fibroblasts. *Curr. Biol.* **7,** 682–688.

Lamb, R. F., Roy, C., Diefenbach, T. J., Vinters, H. V., Johnson, M. W., Jay, D. G., and Hall, A. (2000). The TSC1 tumour suppressor hamartin regulates cell adhesion through ERM proteins and the GTPase Rho. *Nat. Cell Biol.* **2,** 281–287.

Lee, S. S., Lee, R. Y., Fraser, A. G., Kamath, R. S., Ahringer, J., and Ruvkun, G. (2003). A systematic RNAi screen identifies a critical role for mitochondria in *C. elegans* longevity. *Nat. Genet.* **33,** 40–48.

Li, J., Steen, H., and Gygi, S. P. (2003). Protein profiling with cleavable isotope-coded affinity tag (cICAT) reagents: The yeast salinity stress response. *Mol. Cell Proteomics* **2,** 1198–1204.

Li, S., Armstrong, C. M., Bertin, N., Ge, H., Milstein, S., Boxem, M., Vidalain, P. O., Han, J. D., Chesneau, A., Hao, T., Goldberg, D. S., Li, N., *et al.* (2004). A map of the interactome network of the metazoan *C. elegans. Science* **303,** 540–543.

Liao, J. C., Berg, L. J., and Jay, D. G. (1995). Chromophore-assisted laser inactivation of subunits of the T-cell receptor in living cells is spatially restricted. *Photochem. Photobiol.* **62,** 923–929.

Linden, K. G., Liao, J. C., and Jay, D. G. (1992). Spatial specificity of chromophore assisted laser inactivation of protein function. *Biophys. J.* **61,** 956–962.

Liu, C. W., Lee, G., and Jay, D. G. (1999). Tau is required for neurite outgrowth and growth cone motility of chick sensory neurons. *Cell Motil. Cytoskeleton* **43,** 232–242.

Mack, T. G., Koester, M. P., and Pollerberg, G. E. (2000). The microtubule-associated protein MAP1B is involved in local stabilization of turning growth cones. *Mol. Cell. Neurosci.* **15,** 51–65.

Marek, K. W., and Davis, G. W. (2002). Transgenically encoded protein photoinactivation (FlAsH-FALI): Acute inactivation of synaptotagmin I. *Neuron* **36,** 805–813.

Muller, B. K., Jay, D. G., and Bonhoeffer, F. (1996). Chromophore-assisted laser inactivation of a repulsive axonal guidance molecule. *Curr. Biol.* **6,** 1497–1502.

Rajfur, Z., Roy, P., Otey, C., Romer, L., and Jacobson, K. (2002). Dissecting the link between stress fibres and focal adhesions by CALI with EGFP fusion proteins. *Nat. Cell Biol.* **4,** 286–293.

Rubenwolf, S., Niewohner, J., Meyer, E., Petit-Frere, C., Rudert, F., Hoffmann, P. R., and Ilag, L. L. (2002). Functional proteomics using chromophore-assisted laser inactivation. *Proteomics* **2,** 241–246.

Sakurai, T., Wong, E., Drescher, U., Tanaka, H., and Jay, D. G. (2002). Ephrin-A5 restricts topographically specific arborization in the chick retinotectal projection *in vivo*. *Proc. Natl. Acad. Sci. USA* **99,** 10795–10800.

Schmucker, D., Su, A. L., Beermann, A., Jackle, H., and Jay, D. G. (1994). Chromophore-assisted laser inactivation of patched protein switches cell fate in the larval visual system of *Drosophila*. *Proc. Natl. Acad. Sci. USA* **91,** 2664–2668.

Schroeder, R., Jay, D. G., and Tautz, D. (1999). Elimination of EVE protein by CALI in the short germ band insect Tribolium suggests a conserved pair-rule function for even skipped. *Mech. Dev.* **80,** 191–195.

Schroeder, R., Tautz, D., and Jay, D. G. (1996). Chromophore-assisted laser inactivation of even skipped in *Drosophila* phenocopies genetic loss of function. *Dev. Genes Evol.* **206,** 86–88.

Silva, J. M., Mizuno, H., Brady, A., Lucito, R., and Hannon, G. J. (2004). RNA interference microarrays: High-throughput loss-of-function genetics in mammalian cells. *Proc. Natl. Acad. Sci. USA* **101,** 6548–6552.

Sloan, K. E., Eustace, B. K., Stewart, J. K., Zehetmeier, C., Torella, C., Simeone, M., Roy, J. E., Unger, C., Louis, D. N., Ilag, L. L., and Jay, D. G. (2004). CD155/OVR plays a key role in cell motility during tumor cell invasion and migration. *BMC Cancer* **4,** 73.

Sloan, K. E., Stewart, J. K., Treloar, A. F., Matthews, R. T., and Jay, D. G. (2005). CD155/PVR enhances glioma cell dispersal by regulating adhesion signaling and focal adhesion dynamics. *Cancer Res.* **65,** 10930–10937.

Surrey, T., Elowitz, M. B., Wolf, P. E., Yang, F., Nedelec, F., Shokat, K., and Leibler, S. (1998). Chromophore-assisted light inactivation and self-organization of microtubules and motors. *Proc. Natl. Acad. Sci. USA* **95,** 4293–4298.

Sydor, A. M., Su, A. L., Wang, F. S., Xu, A., and Jay, D. G. (1996). Talin and vinculin play distinct roles in filopodial motility in the neuronal growth cone. *J. Cell Biol.* **134,** 1197–1207.

Takei, K., Chan, T. A., Wang, F. S., Deng, Y., Rutishauser, U., and Jay, D. G. (1999). The neural cell adhesion molecules L1 and NCAM-180 act in different steps of neurite outgrowth. *J. Neurosci.* **19,** 9469–9479.

Takei, K., Shin, R. M., Inoue, T., Kato, K., and Mikoshiba, K. (1998). Regulation of nerve growth mediated by inositol 1,4,5-trisphosphate receptors in growth cones. *Science* **282,** 1705–1708.

Templin, M. F., Stoll, D., Schwenk, J. M., Potz, O., Kramer, S., and Joos, T. O. (2003). Protein microarrays: Promising tools for proteomic research. *Proteomics* **3,** 2155–2166.

Tour, O., Meijer, R. M., Zacharias, D. A., Adams, S. R., and Tsien, R. Y. (2003). Genetically targeted chromophore-assisted light inactivation. *Nat. Biotechnol.* **21,** 1505–1508.

Wang, F. S., Liu, C. W., Diefenbach, T. J., and Jay, D. G. (2003). Modeling the role of myosin 1c in neuronal growth cone turning. *Biophys. J.* **85,** 3319–3328.

Wang, F. S., Wolenski, J. S., Cheney, R. E., Mooseker, M. S., and Jay, D. G. (1996). Function of myosin-V in filopodial extension of neuronal growth cones. *Science* **273,** 660–663.

Wong, E. V., David, S., Jacob, M. H., and Jay, D. G. (2003). Inactivation of myelin-associated glycoprotein enhances optic nerve regeneration. *J. Neurosci.* **23,** 3112–3117 [Erratum in: *J. Neurosci.* **23,** 5391, 2003].

Yan, P., Xiong, Y., Chen, B., Negash, S., Squier, T. C., and Mayer, M. U. (2006). Fluorophore-assisted light inactivation of calmodulin involves singlet-oxygen mediated cross-linking and methionine oxidation. *Biochemistry* **45,** 4736–4748.

Yogo, T., Kikuchi, K., Inoue, T., Hirose, K., Iino, M., and Nagano, T. (2004). Modification of intracellular Ca2+ dynamics by laser inactivation of inositol 1,4,5-trisphosphate receptor using membrane-permeant probes. *Chem. Biol.* **11,** 1053–1058.

CHAPTER 12

Investigation of Laser-Microdissected Inclusion Bodies

Naomi S. Hachiya and Kiyotoshi Kaneko

Department of Neurophysiology, Tokyo Medical University
6-1-1 Shinjuku, Shinjuku-ku
Tokyo 160-8402, Japan

We established a novel combinatorial method of laser microdissection system and immunoblot analysis in combination with a novel unfolding chaperone (oligomeric Aip2p/Dld2p) that enables us to examine the molecular profile of proteins in the microscopic regions of interest. As a model system for analyzing inclusion

0091-679X/07 $35.00
DOI: 10.1016/S0091-679X(06)82012-1

bodies associated with various diseases such as Alzheimer's disease, Parkinson's disease, and prion diseases including bovine spongiform encephalopathy (BSE), we applied this novel method to examine brain samples of patients with Pick's disease, a type of progressive presenile dementia with intraneuronal lesions denoted as Pick bodies (PBs) whose major structural components are tau proteins. After boiling in Laemmli's sample buffer according to the established immunoblotting procedures, 500–2000 PBs were initially applied onto SDS-PAGE gels; however, only faint signals were obtained. Remarkably, only one Pick body was sufficient to illustrate an immunoblot signal; this indicates that pretreatment with oligomeric Aip2p/Dld2p enhances the immunoblot sensitivity by more than a 100-fold. This unprecedented property of laser microdissection combined with oligomeric Aip2p/Dld2p may have further potential applications. For example, a number of proteomic strategies for such inclusion bodies depend on liquid chromatography-tandem mass spectrometry (LC-MS/MS); however, sample preparation methods typically involve the use of detergents and chaotropic agents that often interfere with chromatographic separation and/or electrospray ionization. However, the use of oligomeric Aip2p/Dld2p would not interfere with the LC-MS/MS procedures. Therefore, it might significantly facilitate nanoscale analysis, which is often hindered by the aggregation property of the target proteins present under various analytical conditions, particularly, when the sample protein is present in minor quantities.

I. Introduction

Laser microdissection technology has become indispensable for clarifying the changes of protein components and abnormality of protein functions associated with human diseases (Lehmann *et al.*, 2000; Simone *et al.*, 2000). The development of a laser microdissection methodology allows us to manipulate microstructures at microscopic regions of interest *in situ* (Tanaka *et al.*, 2002). Although immunohistochemical analysis has been widely used for the characterization of microstructures under various conditions and disorders at a light microscopic level, immunoblot analysis has been indispensable in the analysis of proteins at a macroscopic level (Laemmli, 1970). At present, no analytical method equivalent to immunoblotting has been developed against targets for examination under the microscope. To overcome this issue, we established a novel combinatorial method of laser microdissection system (Tanaka *et al.*, 2002) and immunoblot analysis that enables us to examine the molecular profile of proteins in the microscopic regions of interest.

As a model system for analyzing inclusion bodies associated with various diseases, we applied this novel method to examine brain samples of patients with Pick's disease, a type of progressive presenile dementia, characterized by a frontotemporal cortical atrophy, widespread white matter degeneration, and intraneuronal lesions denoted as Pick bodies (PBs) whose major structural components are tau proteins

(Buee *et al.*, 2000; Hardin and Schooley, 2002). Pick's disease accounts for 5% of all dementias (Zhukareva *et al.*, 2002). Abnormally phosphorylated tau proteins have been investigated by classical biochemical methods, and a tau doublet (Tau 60 and Tau 64) was detected from the brain homogenates in these patients (Arai *et al.*, 2001; Delacourte *et al.*, 1998; Zhukareva *et al.*, 2002) but not from normal brain homogenates because the tau proteins are not abnormally phosphorylated in normal brain (Sergeant *et al.*, 2005).

On the other hand, use of laser-microdissected samples largely depends on the highly sensitive protein detection methods because of the limited sample availability and the absence of *in vitro* amplification steps for proteins (Martinet *et al.*, 2004). Furthermore, these inclusion bodies generally possess extensive aggregation properties that often have a negative impact on the immunoblot assay. Unfortunately, the effectiveness of conventional procedures including sample pretreatment with chemical-denaturing agents or detergents was negligible. In an attempt to overcome this problem, oligomeric actin-interacting protein 2 (Aip2p) (Amberg *et al.*, 1995)/ D-lactate dehydrogenase protein 2 (Dld2p) (Chelstowska *et al.*, 1999; Flick and Konieczny, 2002) was used as a nonchemical denaturant (Hachiya *et al.*, 2004a,b, c). Dld2p (Chelstowska *et al.*, 1999; Flick and Konieczny, 2002) was initially identified as Aip2p using a two-hybrid screen to search for proteins that interact with actin (Amberg *et al.*, 1995). Exclusively in an oligomeric form, Aip2p/Dld2p exhibits robust protein conformation-modifying activity (Hachiya *et al.*, 2004c); Aip2p/ Dld2p was isolated from *Saccharomyces cerevisiae* by utilizing an *in vitro* protein conformation-modifying assay that measures the factor-dependent increase in protease susceptibility of a substrate as a criterion for activity (Hachiya *et al.*, 1993, 1994, 1995).

Oligomeric Aip2p/Dld2p possesses a unique novel oligomeric grapple-like structure of 10–12 subunits with an ATP-dependent opening that is required for protein conformation-modifying activity (Hachiya *et al.*, 2004a,b). ATP regulates the opening and closing of the "gate" that forms the opening within the oligomeric Aip2p/Dld2p where it binds to the substrate while in the open form. Notably, the oligomeric Aip2p/Dld2p could target both properly folded and highly aggregated pathogenic proteins, and thus, it exhibited no obvious substrate specificity for its binding and robust protein conformation-modifying activity. Thus, oligomeric Aip2p/Dld2p was able to modify the conformation of highly aggregated pathogenic polypeptides such as the recombinant prion protein (rPrP) in the β-sheet form, α-synuclein , and amyloid β (1–42) in the presence of ATP *in vitro* (Hachiya *et al.*, 2004b).

In this procedure, oligomeric Aip2p/Dld2p and 1-mM ATP are mixed in a reaction tube containing the collected PBs, and then the sample is incubated for 60 min at 30 °C. The current approach with the laser microdissection technology in combination with the novel unfolding chaperone (oligomeric Aip2p/Dld2p) allows us to analyze the single inclusion bodies such as PBs in the order of several micrometers in radius.

II. Materials and Methods

A. Patients with Pick's Disease

After obtaining the informed consent, the samples from the frontal and temporal cortexes of four patients with sporadic Pick's disease [patient 1 (sPiD1), female, 55 year; patient 2 (sPiD2), female, 76 year; patient 3 (Y337), female, 71 year; and patient 4 (Y332), male, 72 year] were obtained from the National Center of Neurology and Psychiatry, the Brain Research Institute, Niigata University (Ohkubo *et al.*, 2006), or the Nippon Medical School (Hachiya *et al.*, 2005). The tissues were placed directly in a deep freezer at −80 °C (sPiD1, Y337, and Y332) or immediately quick frozen in cold isopentane and kept in a deep freezer at −80 °C (sPiD2) until use. The procedures followed were in accordance with the institutional ethical standards on human experimentation.

B. Laser Microdissection System

A new laser microdissector was developed to excise the microstructures under a microscope (Tanaka *et al.*, 2002). A diagram of a new laser microdissector constructed by combining a UV Laser System HCL-2100SUV (HOYA Corp., Tokyo, Japan) and an inverted research microscope IX70 (Olympus Imaging Corp., Tokyo, Japan) is shown in Fig. 1A; *1*, UV laser head; *2*, slit; *3*, dichroic mirror; *4*, halogen lamp; *5*, objective (UV laser); *6*, sample slide; *7*, objective (observation); *8*, prism (light path switch); and *9*, CCD camera. The dissector produced a rectangular space of variable size (1–56 μm) on a tissue section with a single shot of a UV laser beam at 266 nm. Tissues surrounding the target were widely burnt off by the consecutive laser shots (Fig. 1B).

The slide preparations for the current protocol were prepared by NexES® automated immunohistochemistry-staining system (Ventana Medical Systems, Inc., Tucson, AZ) by using a mouse monoclonal antibody against phosphorylation-dependent tau proteins (AT8, 1:200) (Ohkubo *et al.*, 2006). Anti-tau AT8 (phosphorylation-dependent monoclonal antibody specific to phosphorylated Ser 202/Thr 205) and AT100 (specific to phosphorylated Thr 212/Ser 214) were purchased from Innogenetics, Ghent, Belgium. The dissected samples were collected with distilled water by using the Cell Tram Oil® hydraulic manual microinjector (Eppendorf, Hamburg, Germany).

C. Sarkosyl Insoluble Fractions of Brain Samples and Their Dephosphorylation

A frozen total brain sample (200 mg) was homogenated with 500 μl of ice-cold extraction buffer [50-mM Tris–chloride (pH 7.4), 0.8-M NaCl, 1-mM EGTA, 10% sucrose, and 1/1000 (w/v) protease inhibitor cocktail (Sigma-Aldrich Japan K.K., Tokyo, Japan)] and 1/10 volume of glass beads (Ohkubo *et al.*, 2006). After 20,000 × *g* centrifugation for 10 min at 4 °C, the supernatant was incubated with 1% sodium *N*-lauroyl sarcosinate. Sarkosyl insoluble fractions were collected

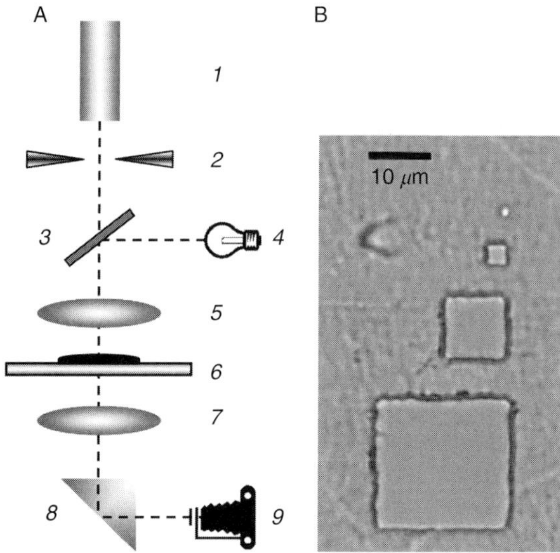

Fig. 1 (A) A diagram of a new laser microdissector that has been constructed by combining a UV Laser System HCL-2100SUV (HOYA Corp.) and an inverted research microscope IX70 (Olympus Imaging Corp.). *1*, UV laser head; *2*, slit; *3*, dichroic mirror; *4*, halogen lamp; *5*, objective (UV laser); *6*, sample slide; *7*, objective (observation); *8*, prism (light path switch); and *9*, CCD camera. (B) Square spaces burnt in a muscle frozen section by the dissector. Scale bar represents 10 μm. From the top: 1-, 3-, 10-, and 20-μm squares. Adapted from Tanaka *et al.* (2002).

by 182,000 × *g* for 30 min at 4 °C, and half of the sarkosyl insoluble material was suspended in 50-mM Tris–chloride (pH 7.4) for immunoblot analysis of total brain sample (Goedert *et al.*, 1992). The remaining sarkosyl insoluble material was used for the following dephosphorylation study. The sample was treated with 4-M guanidine hydrochloride for 1 h at room temperature, followed by overnight dialysis at 4 °C against 50-mM Tris–chloride (pH 8.8), 0.1-mM EDTA, and 0.1-mM phenylmethylsulfonyl fluoride (PMSF). The sample was then incubated for 4 h at 67 °C with 5-U/ml *Escherichia coli* alkaline phosphatase (type III-N; Sigma, Tokyo, Japan). The incubated sample was purified for 1 h at room temperature with 1.5-M ammonium sulfate solution to remove alkaline phosphatase. After 20,000 × *g* centrifugation for 10 min at 4 °C, the pellet was suspended in 50-mM Tris–chloride (pH 8.8) and boiled for 10 min at 95 °C. The sample was then loaded onto 7.5% SDS-PAGE gels and then transferred onto nitrocellulose membranes. The nitrocellulose membranes were blocked with 5% nonfat milk in phosphate-buffered saline containing 0.05% Tween 20 (PBS-T) and incubated with 1:10,000 hydroxytryptamine-7 (HT7) (Innogenetics) specific to amino acid residues 159–163 of tau in PBS-T. Immunodecorated bands were incubated with 1:10,000 horseradish peroxidase-conjugated anti-mouse IgG antibody in PBS-T, visualized by ECL plus (Amersham Biosciences, Uppsala, Sweden), and analyzed using the VersaDoc (Bio-Rad Laboratories, Hercules, CA).

D. Construction, Expression, and Purification of Six Recombinant Tau Isoforms as Marker Proteins

The classical Goedert's method (Goedert *et al.*, 1992) was modified as follows. Human full-length cDNA clones encoding six human tau protein isoforms were subcloned into the *Eco*RI and *Xho*I sites of the pBluescript II SK + plasmid (Ohkubo *et al.*, 2006). Following cleavage with *Eco*RI and *Xho*I, the resulting cDNA fragments were subcloned downstream of the T7 RNA polymerase promoter into *Eco*RI and *Xho*I cut expression plasmid pET11a, and the recombinant plasmids were transformed into *E. coli* BL21 (DE3) cells. The *E. coli* cells were grown to an optical density (OD) of 0.6–1.0 at 600 nm. The expression was induced by adding IPTG to a final concentration of 0.4 mM. After shaking for 3 h at 30 °C, the cells were collected by centrifugation. The *E. coli* pellets were suspended in 50-mM PIPES (pH 6.8), 1-mM DTT, 1-mM EDTA, and protease inhibitors cocktail (Sigma-Aldrich Japan K.K.) and sonicated for 10 min on ice using Branson Sonifier 250. The homogenates were centrifuged at 182,000 × *g* for 30 min at 4 °C, and the resultant supernatant was loaded onto a phosphocellulose-packed column equilibrated in the extraction buffer. After exhaustively washing in the same buffer, the protein was eluted batchwise in 3-ml aliquots of extraction buffer containing 0.5-M NaCl. Fractions 2 and 3, which contained the recombinant tau proteins, were pooled and precipitated with 1.5-M ammonium sulfate. The pellet was washed in 50-mM PIPES (pH 6.8), 1-mM DTT, 1-mM EDTA, and protease inhibitors cocktail (Sigma-Aldrich Japan K.K.) and dialyzed overnight against 50-mM MES and 1-mM DTT (pH 6.5). After centrifugation, the dialysate was loaded onto a Mono S HR 5/5 column (Amersham Biosciences). The column was washed with 50-mM MES, 1-mM DTT, and 50-mM NaCl (pH 6.5), and then the protein was eluted using a 100- to 300-mM NaCl gradient in 50-mM MES and 1-mM DTT (pH 6.5). The column fractions were screened by gel electrophoresis and Quick-CBB (Coomassie brilliant blue) PLUS stain (Wako Chemicals, Osaka, Japan), and the peak tau fractions were pooled. Subsequently, these fractions were loaded onto a Superdex 200 HR 10/30 column (Amersham Biosciences). The column was washed with 50-mM MES, 1-mM DTT, and 150-mM NaCl (pH 6.5). After screening the column fractions by the same methods as described above, the peak tau fractions were pooled, frozen by liquid nitrogen, and stored in deep freezer until use. After six tau isoforms were collected, purified, and their concentrations were measured with densitometry against bovine serum albumin (BSA), equal volumes of each of the six tau isoforms were mixed together. Of this mixture, 5 μl was run alongside the dephosphorylated sample as the molecular weight (MW) marker.

E. Preparation of Oligomeric Aip2p/Dld2p with a Robust Protein Conformation-Modifying Activity

1. Yeast Strains and Antibodies

Wild-type yeast strains (ATCC24657, the wild-type strain; ATCC96099 and ATCC96100 were mated to obtain diploid cells) used in this study were purchased from American Type Culture Collection, Manassas, VA. The protease-deficient strain

SH2777 was gifted by Dr. Harashima, Osaka University. Anti-actin antibody was purchased from Chemicon, Temecula, CA. Anti-Aip2p/Dld2p antibody was raised against the synthetic peptide corresponding to the C-terminal 15-amino acid residues of Aip2p/Dld2p (VHYDPNGILNPYKYI), and these amino acid residues were coupled through a C-terminal cysteine residue to BSA (Hachiya *et al.*, 2004a,b,c).

2. Purification of Hexahistidine-Tagged Aip2p/Dld2p

In an attempt to obtain sufficient quantities of Aip2p/Dld2p, the protein was prepared from the expression strain of yeast under control of the ADH promoter. The C-terminal hexahistidine-tagged *YDL178w* gene was amplified by PCR and inserted into the aureobasidin A (Ab A) selective expression vector pAUR123 (TaKaRa Biomedicals, Shiga, Japan). The protease-deficient strain SH2777 was transformed by this plasmid, and the transformants were grown on YPD plates containing 0.5 μg/ml of Ab A. The inoculated medium (8 liter) was incubated overnight at 30 °C to an OD of 1–2 at 600 nm. The cells were collected, resuspended in 4 volumes of buffer B [50-mM NaPi (pH 8.0), 150-mM NaCl, and 10-mM imidazole], crushed using glass beads, and centrifuged at 10,000 rpm for 10 min at 4 °C. The supernatants were collected and ultracentrifuged at 100,000 $\times$ *g* at 4 °C for 1 h. The precipitate was resuspended, passed through an Ni-NTA agarose column (QIAGEN, K.K., Hilden, Germany) equilibrated in buffer B, and subsequently eluted with buffer B containing 0.5-M imidazole. The eluted fractions were dialyzed against buffer C [10-mM HEPES–KOH (pH 7.4), 50-mM NaCl, and 1-mM DTT], applied to a Mono Q ion-exchange column (AKTA system, Amersham Pharmacia Biotech, Piscataway, NJ) equilibrated with buffer C, and eluted with a linear NaCl gradient (100–500 mM). Immunoreactive fractions were dialyzed against buffer D [50-mM NaPi (pH 7.5), 10-mM NaCl, and 1-mM Mg $(OAc)_2$] and finally passed through a Superdex 200 gel filtration column equilibrated with buffer D.

3. Preparation of Substrate Proteins

Amyloid β (1–28 and 1–42), cytochrome *c*, DNase I, malate dehydrogenase (MDH), and mitochondrial superoxide dismutase (SOD) were purchased from Sigma chemical, Temecula, CA. α-Synuclein was purchased from Chemicon. Heavy meromyosin (HMM), luciferase, and the mature form of invertase were purchased from Wako Chemicals. The rabbit muscle actin was purchased from Molecular Probes, Inc., Carlsbad, CA. The rabbit muscle G-actin (2 μM) was polymerized in the high-salt buffer [10-mM Tris–chloride (pH 8.0), 100-mM KCl, and 2-mM $MgCl_2$] at 37 °C for 2 h and used as F-actin. The gene fragments of hexahistidine-tagged prepro-alpha-factor (ppαF) and pro-alpha-factor (proαF) were amplified by PCR, inserted into pET11a plasmid, expressed in *E. coli* BL21 (DE3) using the pET system, and purified according to the manufacturer's protocol (QIAGEN, K.K.). Purified ppαF was dialyzed against buffer A [10-mM HEPES–KOH (pH 7.4), 1-mM DTT, and 1-mM Mg $(OAc)_2$], and subsequently used in the trypsin-susceptibility protein conformation-modifying assay. The rPrP was purchased from Prionics AG, Schlieren-Zurich, Switzerland. The PrP solubilized in PBS was maintained at 4 °C until circular dichroism detected over 50% of the β-contents in the rPrP, and then it was used as "PrP in the β-sheet form."

4. Trypsin-Susceptibility Assay

Assays (in a total volume of 200 μl) were initiated by adding 200 ng of polymerized rabbit muscle actin to buffer E containing 1-mM ATP and 500 ng of hexahistidine-tagged Aip2p/Dld2p, and incubated at 30 °C for 15 min. After incubation, the samples were treated with trypsin (0.2 μg/ml) at 16 °C for 15 min. The reaction was terminated by incubation with soybean trypsin inhibitor (0.4 μg/ml) on ice for 5 min; the proteins were TCA precipitated using tRNA carrier, and then subjected to SDS-PAGE and immunoblotting. In order to detect actin, affinity-purified polyclonal rabbit anti-actin antibody was used as the primary antibody and horseradish peroxidase-linked IgG (ICN Pharmaceuticals, Inc., Costa Mesa, CA) as the secondary antibody. The immunoreactive bands were visualized by ECL-plus and analyzed using a Fluoro-S MAX MultiImager (Bio-Rad Laboratories).

F. Immunoblot Analysis and Protein Quantification

Immunoblot analyses were performed as follows (Hachiya *et al.*, 2005; Laemmli, 1970). Total brain homogenates (TBH) (10–40 μg) or laser-microdissected PBs (500 pieces) were solubilized in 500 μl of ice-cold extraction buffer [Tris–chloride (pH 7.4), 0.8-M NaCl, 1-mM EGTA, 10% sucrose, and 1/1000 (w/v) protease inhibitor cocktail (Sigma-Aldrich Japan K.K.) with 1% sodium *N*-lauroyl sarcosinate (sarkosyl)]. The sarkosyl insoluble fractions were collected by centrifugation at 182,000 $\times$ g for 30 min at 4 °C, and then suspended in 50-mM Tris–chloride (pH 7.4). Samples were pretreated with 8-M urea (Wako Chemicals), 6-M guanidine hydrochloride (Nacalai Tesque, Kyoto, Japan), or 2% SDS (Wako Chemicals) and followed by TCA precipitation to denature or untangle the proteins. Pretreatment with oligomeric Aip2p/Dld2p was performed as previously described (Hachiya *et al.*, 2004a,b,c). In brief, 1–500 ng of oligomeric Aip2p/Dld2p was mixed with the sarkosyl insoluble fraction of 1–500 PBs at a ratio of 1 ng per 1 PB in the presence of 1-mM ATP for 60 min at 30 °C in a total volume of 20 μl. The samples were then loaded onto 12% SDS-PAGE gels and transferred onto 0.22-μm nitrocellulose membranes in 25-mM Tris solution, 190-mM glycine, 0.01% SDS, and 20% methanol at 400 mA for 40 min at 4 °C. Membranes were blocked using 4% BSA in PBS-T, incubated with 1:1000 (unless otherwise indicated) AT8 and AT100 in PBS-T overnight at 4 °C, washed with PBS-T several times at room temperature, and then incubated with 1:10,000 horseradish peroxidase-conjugated anti-mouse IgG antibody (Amersham Biosciences) in PBS-T for 1 h at room temperature. After washing the membranes, the immunodecorated bands were visualized using ECL-plus (Amersham Biosciences) and then analyzed using a Fluor-S MAX MultiImager or VersaDoc.

The protein concentration of the PBs that were pretreated with oligomeric Aip2p/Dld2p was measured by using a spectrophotometer (Tecan Japan, Kanagawa, Japan) at 595 nm in combination with a protein assay system (Bio-Rad Laboratories) according to the manufacturer's instructions. Oligomeric Aip2p/Dld2p was applied at a ratio of 1 ng per 1 PB, and the value was subtracted later.

III. Results

A. Laser Microdissection of the PB

Immunostained PBs of $\sim$10 μm^3 (Fig. 2A) were dissected by our laser microdissection system (Olympus Imaging Corp.) coupled to HOYA laser cutter, HCL2100 (30 mJ/pulse, 266 nm; Hoya Corp.). The dissected samples were collected with distilled water by using the Cell Tram Oil® hydraulic manual microinjector (Eppendorf) (Fig. 2B; Ohkubo *et al.*, 2006). The laser microdissection system combined with the sample collector facilitated the dissection of targets, and thus, up to 500 PBs were collected at each collection procedure over a period of *one* day (Hachiya *et al.*, 2005).

First, the brain homogenates of the patients with Pick's disease were tested according to the established common immunoblotting analysis. In the TBH of sPiD1, for example, tau migrated as two major bands of 60 and 64 kDa (Tau 60 and Tau 64) by common immunoblot (Fig. 2C, lane 2) as previously described (Arai *et al.*, 2001; Delacourte *et al.*, 1998). After dephosphorylation by bacterial alkaline phosphatase (type III-N; Sigma), tau in sarkosyl insoluble fractions appeared as two major bands that align with either 3 repeat tau (3R-tau) 0N, that is 3R-tau with no N-terminal amino acid inserts (dephosphorylated Tau 60) or 3R-tau 1N, that is 3R-tau with 29-amino acid inserts encoded by exon 2 (dephosphorylated Tau 64) (Fig. 2D, lane 2).

B. Robust Protein Conformation-Modifying Activity of Oligomeric Aip2p/Dld2p

The electron microscopic observation of the purified oligomeric Aip2p/Dld2p by using rotary-shadowing method (Tsukita, 1985) revealed that it possesses an unusual grapple-like oligomeric structure of $\sim$10 nm in diameter within the opening (Fig. 3, upper panels) (Hachiya *et al.*, 2004a). Furthermore, the negative-staining observation (Nonomura *et al.*, 1975) revealed that oligomeric Aip2p/Dld2p adopted at least two different states that corresponded to either an "open state" in the presence of ATP or a "closed state" in the absence of ATP (Fig. 3, lower panels). Oligomeric Aip2p/Dld2p contains two Walker-type B ATP-binding motifs (ZZZZD, where Z is a hydrophobic residue) (Walker *et al.*, 1982) located at amino acid residues from 142 to 146 and from 189 to 193, which are conserved among several Mg^{2+} nucleotide-binding proteins.

Subsequently, gel filtration chromatography was performed to determine the number of Aip2p/Dld2p monomers present in the oligomeric complex. Oligomeric

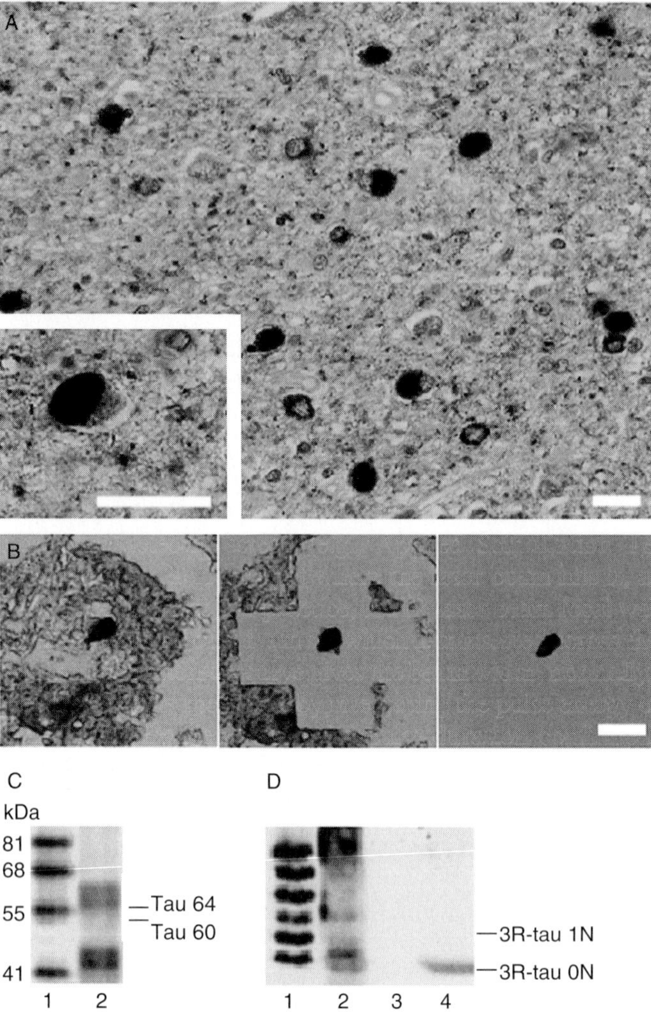

Fig. 2 Temporal cortex of patient 1 with sporadic Pick's disease (sPiD1). (A) Cryosection of 10-μm thickness. PBs of sPiD1 stained with AT8 (purple) and hematoxylin (blue). Scale bar represents 20 μm. (B) Cryosection of sPiD1 of 2-μm thickness. PBs isolated from the section by the laser microdissection system. Scale bar represents 20 μm. (C) Immunoblot analysis of sPiD1 with AT8 and AT100 antibodies. MW marker (Dr. Western, Oriental Yeast, Tokyo, Japan; lanes 1 and 3) and temporal brain homogenates (40 μg, lane 2). (D) Immunoblot analysis of sarkosyl insoluble fraction of temporal brain homogenates of sPiD1 after dephosphorylation by bacterial alkaline phosphatase (8 μg, lane 2), blank lane (lane 3), and bacterial alkaline phosphatase alone (lane 4) with HT7 antibody. MW marker (six recombinant tau isoforms in lane 1; 67, 62, 59, 54, 52, and 48 kDa). Adapted from Ohkubo *et al.* (2006). (See Plate 9 in the color insert section.)

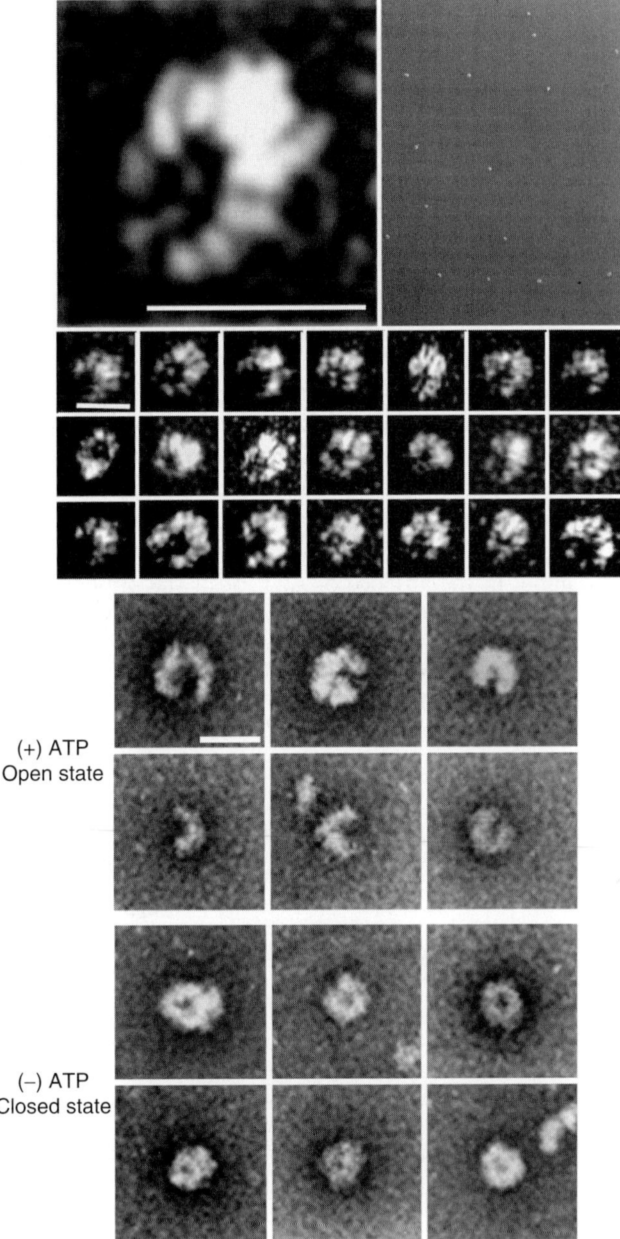

Fig. 3 A grapple-like oligomeric structure of Aip2p/Dld2p. (A) Upper panels: purified hexahistidine-tagged oligomeric Aip2p/Dld2p (100 μg/ml) was used as a specimen for low-angle-shadowing electron microscopy. Scale bars represent 10 nm. Lower panels: negative staining of oligomeric Aip2p/Dld2p in the "open state" and "closed state." In the presence of ATP, oligomeric Aip2p/Dld2p opens its gate [(+) ATP] and without ATP the gate is closed [(−) ATP]. Scale bar represents 10 nm. Adapted from Hachiya *et al.* (2004a).

Aip2p/Dld2p eluted as a single peak with an apparent MW of ~700 kDa. Since the MW of Aip2p/Dld2p in the monomeric form was ~60 kDa (Hachiya *et al.*, 2004c), it is apparent that Aip2p/Dld2p forms an oligomeric structure consisting of 10–12 subunits. The secondary structure prediction (Lupas *et al.*, 1991) indicated that Aip2p/Dld2p contains a coiled-coil domain in its C-terminal region. Consistent with this observation, the monomeric Aip2p/Dld2p in which a C-terminal coiled-coil region had been truncated failed to exhibit the protein conformation-modifying activity.

Further investigations revealed that the oligomeric Aip2p/Dld2p modified the conformation of all proteins examined thus far, including actin, DNase I, the mature form of invertase, proαF, and mitochondrial SOD in the presence of ATP (Hachiya *et al.*, 2004b). Thus, no specific substrates that were obvious have been identified as being responsible for the protein conformation-modifying activity of oligomeric Aip2p/Dld2p *in vitro*. However, in the absence of ATP, the substrate was protected from trypsin digestion in the same manner as in the control, suggesting that the protein conformation-modifying activity is ATP dependent. Taken together, substrate proteins that entered through the grapple-shaped opening were caught in it, where its conformation was modified in the presence of ATP (Fig. 4). On the other hand, in the absence of ATP or in the presence of ADP, the opening was closed, thereby preventing the entry of the protein substrates into the opening.

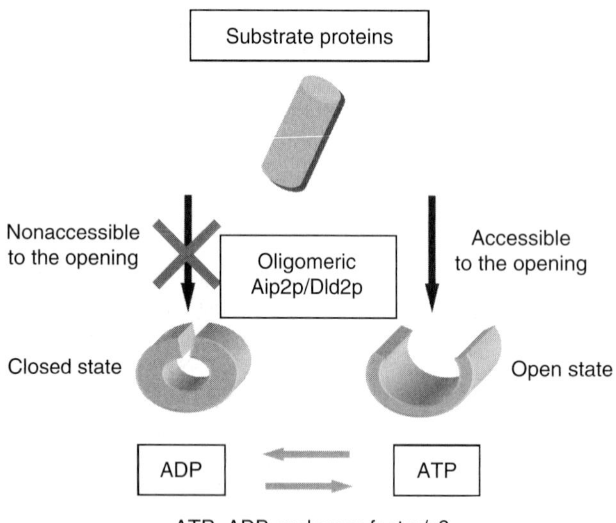

Fig. 4 The ATP-dependent open and closed state model of oligomeric Aip2p/Dld2p. ATP-bound oligomeric Aip2p/Dld2p exhibits a protein conformation-modifying activity (open state), whereas F-actin is unable to penetrate ATP-unbound oligomeric Aip2p/Dld2p (closed state). Adapted from Hachiya *et al.* (2004a). (See Plate 10 in the color insert section.)

In addition to the aforementioned indirect biochemical analyses suggesting that oligomeric Aip2p/Dld2p possesses protein conformation-modifying activity, more direct evidence of this activity was sought through the use of low-angle rotary shadowing (Tsukita, 1985). The rabbit skeletal muscle HMM has a characteristic structure consisting of two globular heads and one tail (Fig. 5A, HMM). Following incubation with oligomeric Aip2p/Dld2p in the presence of 1-mM ATP, HMM heads unfolded and adopted an extended globular structure, while the helical tail became longer and thinner (Fig. 5A, Aip2p/Dld2p-HMM). Trypsin susceptibility of HMM also increased after the incubation with oligomeric Aip2p/Dld2p (Fig. 5A, right panel).

Interestingly, even when the pathogenic polypeptides, such as the rPrP in the β-sheet form, α-synuclein, or amyloid β (1–42) peptide, were tested as substrates in the trypsin-susceptibility assay, it was found that trypsin susceptibility increased in the presence of oligomeric Aip2p/Dld2p. Although these highly aggregated pathogenic polypeptides (Fig. 5B, left panels) were resistant to 2.5-μg/ml trypsin, digestion was significant in the presence of only 200-ng/ml trypsin following incubation with oligomeric Aip2p/Dld2p (Fig. 5B, right panels).

C. Extensive Aggregation Property of PB

After boiling in Laemmli's sample buffer according to established immunoblotting procedures, 500 PBs were initially applied onto SDS-PAGE gels (Hachiya *et al.*, 2005; Laemmli, 1970). However, only faint and blurred signals were obtained with the anti-tau antibodies AT8 and AT100 (Fig. 6, lane 4) in comparison with 10–40 μg of the TBH (Fig. 6, lanes 2 and 3). Immunostaining of the entire gel including the loading wells and the stacking gel revealed no additional immunoblot signals that may have developed due to the extensive aggregation property of the PBs. Further increases in the number of PBs applied (up to 2000) could not improve the signal intensity.

The effect of chemical denaturants or detergents including 6-M guanidine hydrochloride, 8-M urea, and 2% SDS was then determined (Hachiya *et al.*, 2005). The use of the aforementioned chaotropic agents, however, did not improve the immunoblot signals (Fig. 6, lanes 5–7). In fact, the signal intensities diminished to some extent, possibly due to the presence of phosphorylated tau bound to the walls of the tube after removing the chaotropic agents prior to loading onto the SDS-PAGE gels (Kaneko *et al.*, 1995).

D. Marked Increase in Immunoblot Signals of Laser-Microdissected PB

Although negative results were obtained following the use of the aforementioned chemical denaturants and detergents, we demonstrated that oligomeric Aip2p/Dld2p could modify the conformation of highly aggregated pathogenic polypeptides (Hachiya *et al.*, 2004b, 2005). Hence, the PBs were pretreated with oligomeric Aip2p/Dld2p prior to loading onto SDS-PAGE gels. Notably, immunoblot analyses

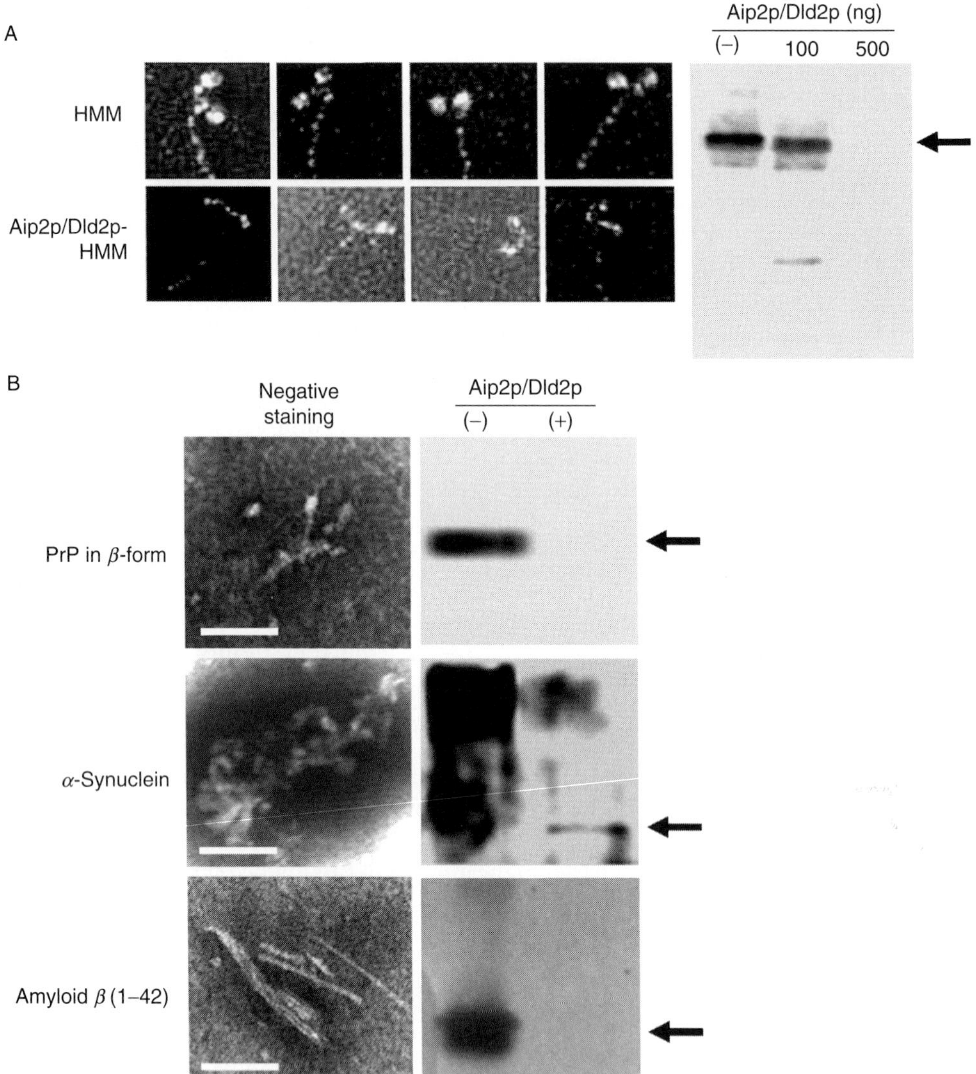

Fig. 5 Oligomeric Aip2p/Dld2p displays broad substrate specificity *in vitro*. (A) HMM is unfolded by oligomeric Aip2p/Dld2p. HMM (100 μg/ml) was incubated with or without oligomeric Aip2p/Dld2p (100 μg/ml) in buffer A containing 1-mM ATP as described in Section 2. Following incubation, each mixture was subjected to low-angle-shadowing electron microscopy. Scale bar represents 50 nm. Right panel represents the increased trypsin susceptibility of oligomeric Aip2p/Dld2p-treated HMM (100 and 500 ng). HMM was immunostained with anti-HMM polyclonal antibody. (B) Trypsin susceptibility of oligomeric Aip2p/Dld2p-treated highly aggregated pathogenic proteins is dramatically increased. rPrP in the β-sheet form (20 μg), α-synuclein (20 μg), and amyloid β (1–42) peptide (60 μg) were used as specimens for negative staining (left panels). PrP in β-sheet form (300 ng), α-synuclein (200 ng), and amyloid β (1–42) peptide (400 ng) were used for the trypsin-susceptibility assay (right panels). PrP and α-synuclein were immunostained with anti-PrP polyclonal antibody K1 (1:200) and anti-α-synuclein antibody, respectively. Amyloid β (1–42) peptide was silver stained according to the manufacturer's instruction (Wako Chemicals). Scale bars represent 100 nm. Adapted from Hachiya *et al.* (2004b).

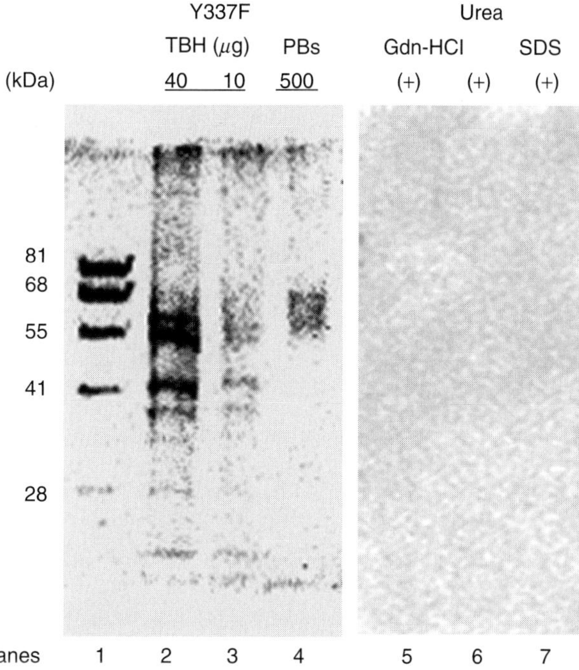

Fig. 6 Immunoblot analyses of PBs pretreated with chemical denaturants or detergents. About 500 PBs were used for each trial. Lane 1: MW marker (Dr. Western, Oriental Yeast); lanes 2 and 3: TBH of Y337F (40 and 10 μg, respectively); and lanes 4–7: 500 laser-microdissected PBs of Y337F with no pretreatment (lane 4), 6-M Guanidine hydrochloride (Gdn-HCl, lane 5), 8-M urea (lane 6), and 2% SDS pretreatment (lane 7). Samples were stained with anti-tau AT8 (1:1000) and AT100 (1:1000). Adapted from Hachiya *et al.* (2005).

of 500 PBs from Y337F, Y332F, and Y332T demonstrated discrete bands stained with anti-tau AT8 and AT100 antibodies following the pretreatment with oligomeric Aip2p/Dld2p (Fig. 7). In a serial dilution assay, 1/500 PBs (equivalent to one PB) was detected (Fig. 7, upper panel, lanes 4–8; lower panel, lanes 2–10). Oligomeric Aip2p/Dld2p was also detected in the same reaction mixtures using anti-Aip2p/Dld2p antibody (Fig. 7, upper panel, lanes 12–16), but did not cross-react with anti-tau AT8 and AT100 antibodies (Fig. 7, upper panel, lane 10) (Hachiya *et al.*, 2005).

These immunoreactive bands migrated slightly faster than those associated with the 500 PBs processed without oligomeric Aip2p/Dld2p pretreatment (Fig. 7, upper panel, lane 2) (Hachiya *et al.*, 2005). One possible explanation is that pretreatment with oligomeric Aip2p/Dld2p might permit the detection of the phoshorylated form of 60-kDa tau (Tau 60) (Arai *et al.*, 2001; Delacourte *et al.*, 1998; Zhukareva *et al.*, 2002). On the other hand, only the phosphorylated form of 69-kDa tau (Tau 69) is negligibly detected after boiling in Laemmli's sample buffer according to the classical immunoblotting procedures. Whether the different tau isoforms could account for the faster migration pattern observed remains to be determined.

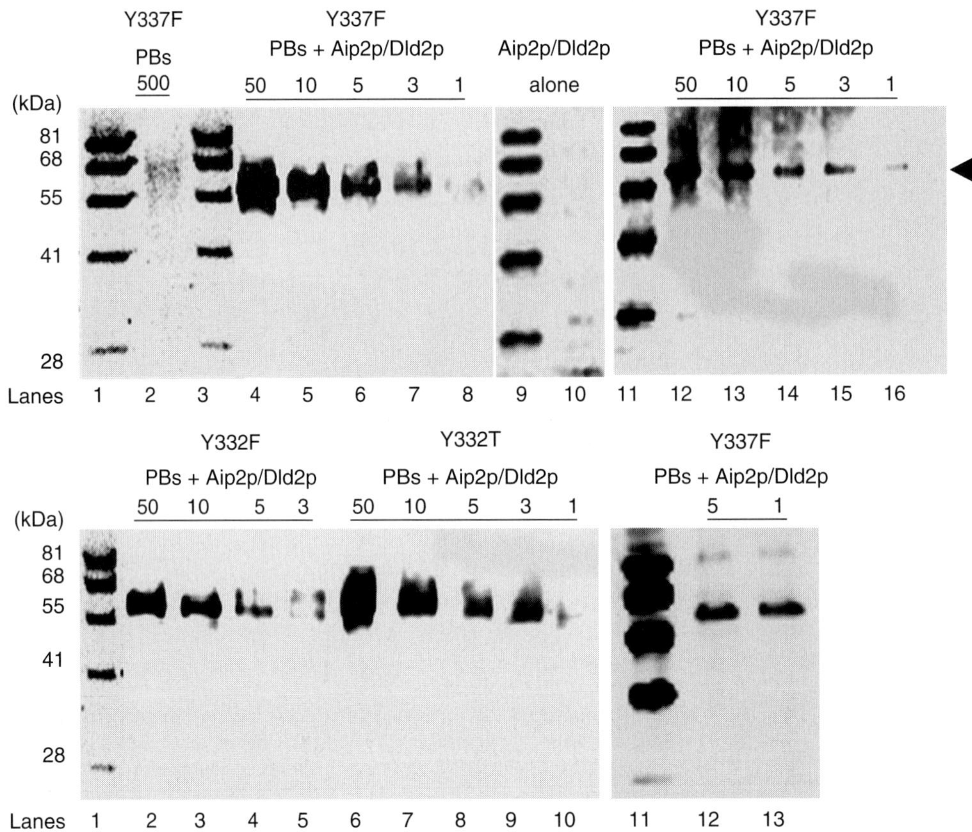

Fig. 7 Immunoblot analyses of laser-microdissected PBs from Y337F (frontal cortex), Y332F (frontal cortex), and Y332T (temporal cortex). Upper panels: MW marker (Dr. Western, Oriental Yeast, lanes 1, 3, 9, and 11), 500 laser-microdissected PBs of Y337F (lane 2), and serial dilutions equivalent to 50, 10, 5, 3, and 1 PBs of Y337F (lanes 4–8 and 12–16). Lane 2 represents sample without oligomeric Aip2p/Dld2p pretreatment, whereas lanes 4–8 and 12–16 represent samples with oligomeric Aip2p/Dld2p pretreatment. Lane 10: 50 ng of Aip2p/Dld2p alone. Lanes 2, 4–8, and 10 were stained with anti-tau AT8 (1:1000) and AT100 (1:1000), whereas lanes 12–16 were stained with anti-Aip2p/Dld2p polyclonal antibody. Arrowhead indicates the position of Aip2p/Dld2p (MW = 58 kDa). Lower panels: MW marker (Dr. Western, Oriental Yeast, lanes 1 and 11), serial dilutions of 500 PBs of Y332F equivalent to 50, 10, 5, and 3 PBs (lanes 2–5), and those of Y332T equivalent to 50, 10, 5, 3, and 1 PBs (lanes 6–10). Lanes 12 and 13: 5 and single PBs of Y337F, respectively. Samples in lower panels were pretreated with oligomeric Aip2p/Dld2p and stained with anti-tau AT8 (1:1000) and AT100 (1:1000). Adapted from Hachiya *et al.* (2005).

It should be noted that a single PB directly pretreated with oligomeric Aip2p/Dld2p was sufficient to yield an immunoblot signal (Fig. 7, lower panel, lane 13), indicating that pretreatment with oligomeric Aip2p/Dld2p enhanced the immunoblot signal by more than a 100-fold. Transmission electron microscopy using uranyl acetate negative staining of laser-microdissected PBs revealed that they were

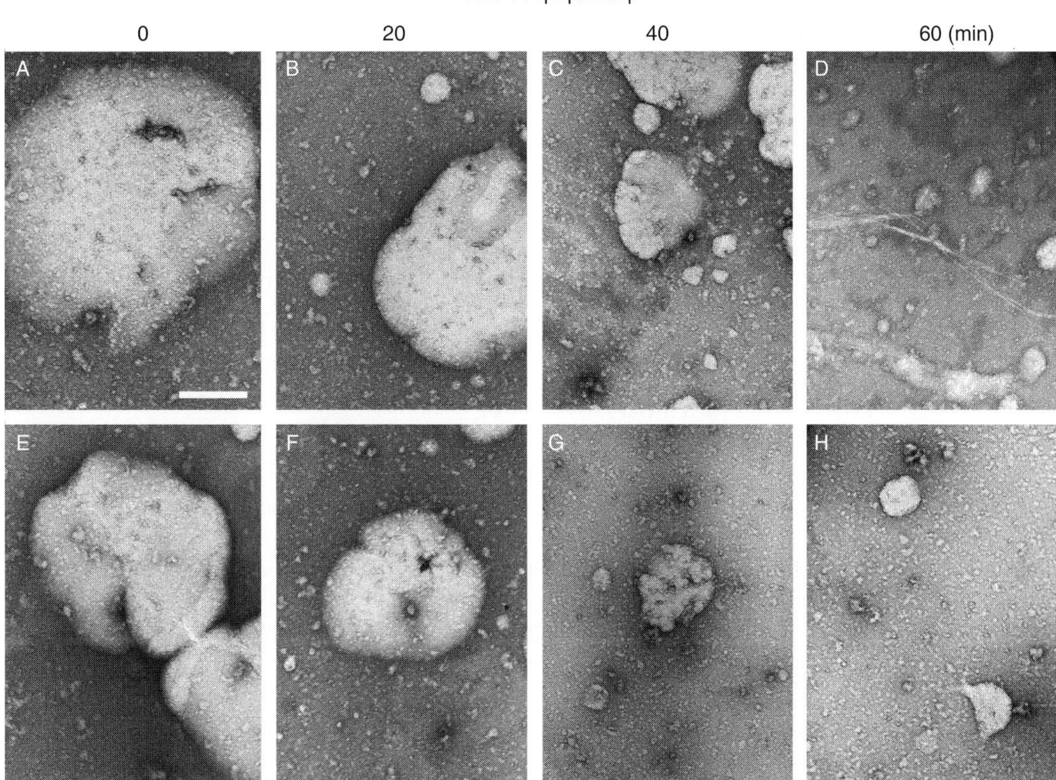

Fig. 8 Transmission electron microscopy using uranyl acetate-negative staining of laser-microdissected PBs prior to (panels A and E) and following oligomeric Aip2p/Dld2p pretreatment for 20 min (panels B and F), 40 min (panels C and G), and 60 min (panels D and H). Five hundred PBs of Y332F were used as specimens for the negative staining. Scale bar represents 4 μm. Adapted from Hachiya *et al.* (2005).

untangled following the pretreatment with oligomeric Aip2p/Dld2p, whereas the average diameter of the PBs decreased markedly from 10–15 μm to <1 μm in a time-dependent manner (Fig. 8).

E. Protein Quantification of the PBs

Although protein quantification of the highly aggregated proteins such as the PBs has been considerably problematic to date, pretreatment with oligomeric Aip2p/Dld2p has permitted the ready quantification of the protein content of the PBs (Table I) (Hachiya *et al.*, 2005). The protein concentrations of sarkosyl insoluble fractions were 0.8 ng/PB (Y332T), 1.1 ng/PB (Y332F), and 2.8 ng/PB (Y337F). Since the average diameters of the PBs were 10 μm (Y332) and 15 μm (Y337), the relative densities of the PBs were 1.6–2.2 (Y332) and 1.6 (Y337).

Table I
Quantitative Analyses of Pick Bodies (Hachiya *et al.*, 2005)

	Y332T	Y332F	Y337F
Total protein (ng/Pick body)	0.8	1.1	2.8
Average diameter (μm)	10	10	15
Relative density	1.6	2.2	1.6

The protein concentration of sarkosyl insoluble fractions was measured following pretreatment with oligomeric Aip2p/Dld2p, and the relative density of the Pick bodies was calculated. Samples of the Frontal (Y337F and Y332F) and temporal (Y332T) cortexes of two patients with sporadic Pick's disease were analyzed.

IV. Discussion and Summary

In general, immunohistochemical method as well as immunoblot techniques are indispensable and mutually complementary. Simultaneously, such combinations of morphological and biochemical techniques significantly complement the existing histopathological methods and have a great potential for investigating normal or abnormal microstructures associated with various conditions and disorders. Our novel combinatorial method targets proteins related to specific regions of interest at the micrometer order, and it exclusively permits the collection of information pertaining to the molecular profile such as MW of target proteins under the microscope *in situ*.

During our investigations, we noticed that the laser-microdissected PBs exhibited only faint and blurred immunoblot signals with anti-tau AT8 and AT100 antibodies even after the pretreatment with chemical denaturants or detergents, presumably resulting from the extensive aggregation property. In fact, this is extremely crucial when only a minimum quantity of target protein is available.

The protein conformation-modifying activity of oligomeric Aip2p/Dld2p can modify the conformation of highly aggregated pathogenic polypeptides (Hachiya *et al.*, 2004b). Therefore, the PBs were pretreated with oligomeric Aip2p/Dld2p to overcome the extensive aggregation property. Following the pretreatment, 500 ng of oligomeric Aip2p/Dld2p (MW = ~700 kDa) was mixed with 500 PBs that consisted of abnormally phosphorylated tau (MW = 58 kDa), indicating that the stoichiometry of oligomeric Aip2p/Dld2p:phosphorylated tau is ~1:10.

The inclusion bodies, which might protect against toxicity (Tanaka *et al.*, 2004), have also been associated with various protein conformation disorders such as Alzheimer's disease (Lustbader *et al.*, 2004), Parkinson's disease (Greenamyre and Hastings, 2004), and prion diseases including bovine spongiform encephalopathy (BSE) (Mayer *et al.*, 2000). The use of oligomeric Aip2p/Dld2p with our combinatorial method provides significant improvement in the investigation of normal or abnormal microstructures associated with various conditions and disorders with extremely enhanced sensitivity.

This unprecedented property of laser microdissection combined with oligomeric Aip2p/Dld2p may have further potential applications. For example, proteomic strategies depend on liquid chromatography-tandem mass spectrometry (LC-MS/MS), which should permit us to identify the protein components of these inclusion bodies that are unidentified to date. However, sample preparation methods typically involve the use of detergents and chaotropic agents that often interfere with the chromatographic separation and/or electrospray ionization (Blonder *et al.*, 2002). However, the use of oligomeric Aip2p/Dld2p would not interfere with the LC-MS/MS procedures. Therefore, it might significantly facilitate nanoscale analysis, which is often hindered by the aggregation property of target proteins present under various analytical conditions, particularly, when the sample protein is present in minor quantities.

Acknowledgments

We are indebted to T. Ohkubo and Y. Sakasegawa for their contributions in this study and to Y. Kozuka and K. Watanabe for providing technical assistance. We are also grateful to G. Schatz, S. B. Prusiner, and K. Mihara for their critical discussions and to I. Wada for providing helpful comments. This work was supported by grants from the Ministry of Health, Labor and Welfare, Japan; the Ministry of Education, Culture, Sports, Science, and Technology, Japan; the Ministry of Agriculture, Forestry, and Fisheries, Japan; Exploratory Research for Advanced Technology (ERATO); and the Core Research for Evolutional Science and Technology (CREST) of the Science and Technology Agency, Japan.

References

Amberg, D. C., Basart, E., and Botstein, D. (1995). Defining protein interactions with yeast actin *in vivo*. *Nat. Struct. Biol.* **2,** 28–35.

Arai, T., Ikeda, K., Akiyama, H., Shikamoto, Y., Tsuchiya, K., Yagishita, S., Beach, T., Rogers, J., Schwab, C., and McGeer, P. L. (2001). Distinct isoforms of tau aggregated in neurons and glial cells in brains of patients with Pick's disease, corticobasal degeneration and progressive supranuclear palsy. *Acta Neuropathol. (Berl).* **101,** 167–173.

Blonder, J., Goshe, M. B., Moore, R. J., Pasa-Tolic, L., Masselon, C. D., Lipton, M. S., and Smith, R. D. (2002). Enrichment of integral membrane proteins for proteomic analysis using liquid chromatography-tandem mass spectrometry. *J. Proteome Res.* **1,** 351–360.

Buee, L., Bussiere, T., Buee-Scherrer, V., Delacourte, A., and Hof, P. R. (2000). Tau protein isoforms, phosphorylation and role in neurodegenerative disorders. *Brain Res. Brain Res. Rev.* **33,** 95–130.

Chelstowska, A., Liu, Z., Jia, Y., Amberg, D., and Butow, R. A. (1999). Signalling between mitochondria and the nucleus regulates the expression of a new D-lactate dehydrogenase activity in yeast. *Yeast* **15,** 1377–1391.

Delacourte, A., Sergeant, N., Wattez, A., Gauvreau, D., and Robitaille, Y. (1998). Vulnerable neuronal subsets in Alzheimer's and Pick's disease are distinguished by their tau isoform distribution and phosphorylation. *Ann. Neurol.* **43,** 193–204.

Flick, M. J., and Konieczny, S. F. (2002). Identification of putative mammalian D-lactate dehydrogenase enzymes. *Biochem. Biophys. Res. Commun.* **295,** 910–916.

Goedert, M., Spillantini, M. G., Cairns, N. J., and Crowther, R. A. (1992). Tau proteins of Alzheimer paired helical filaments: Abnormal phosphorylation of all six brain isoforms. *Neuron* **8,** 159–168.

Greenamyre, J. T., and Hastings, T. G. (2004). Biomedicine. Parkinson's—divergent causes, convergent mechanisms. *Science* **304,** 1120–1122.

Hachiya, N., Alam, R., Sakasegawa, Y., Sakaguchi, M., Mihara, K., and Omura, T. (1993). A mitochondrial import factor purified from rat liver cytosol is an ATP-dependent conformational modulator for precursor proteins. *EMBO J.* **12,** 1579–1586.

Hachiya, N., Komiya, T., Alam, R., Iwahashi, J., Sakaguchi, M., Omura, T., and Mihara, K. (1994). MSF, a novel cytoplasmic chaperone which functions in precursor targeting to mitochondria. *EMBO J.* **13,** 5146–5154.

Hachiya, N., Mihara, K., Suda, K., Horst, M., Schatz, G., and Lithgow, T. (1995). Reconstitution of the initial steps of mitochondrial protein import. *Nature* **376,** 705–709.

Hachiya, N. S., Ohkubo, T., Kozuka, Y., Yamazaki, M., Mori, O., Mizusawa, H., Sakasegawa, Y., and Kaneko, K. (2005). More than a 100-fold increase in immunoblot signals of laser-microdissected inclusion bodies with an excessive aggregation property by oligomeric actin interacting protein 2/D-lactate dehydrogenase protein 2. *Anal. Biochem.* **347,** 106–111.

Hachiya, N. S., Sakasegawa, Y., Jozuka, A., Tsukita, S., and Kaneko, K. (2004c). Interaction of D-lactate dehydrogenase protein 2 (Dld2p) with F-actin: Implication for an alternative function of Dld2p. *Biochem. Biophys. Res. Commun.* **319,** 78–82.

Hachiya, N. S., Sakasegawa, Y., Sasaki, H., Jozuka, A., Tsukita, S., and Kaneko, K. (2004a). Oligomeric Aip2p/Dld2p forms a novel grapple-like structure and has an ATP-dependent F-actin conformation modifying activity *in vitro*. *Biochem. Biophys. Res. Commun.* **320,** 1271–1276.

Hachiya, N. S., Sakasegawa, Y., Sasaki, H., Jozuka, A., Tsukita, S., and Kaneko, K. (2004b). Oligomeric Aip2p/Dld2p modifies the protein conformation of both properly-folded and misfolded substrates *in vitro*. *Biochem. Biophys. Res. Commun.* **323,** 339–344.

Hardin, S., and Schooley, B. (2002). A story of Pick's disease: A rare form of dementia. *J. Neurosci. Nurs.* **34,** 117–122.

Kaneko, K., Peretz, D., Pan, K. M., Blochberger, T. C., Wille, H., Gabizon, R., Griffith, O. H., Cohen, F. E., Baldwin, M. A., and Prusiner, S. B. (1995). Prion protein (PrP) synthetic peptides induce cellular PrP to acquire properties of the scrapie isoform. *Proc. Natl. Acad. Sci. USA* **92,** 11160–11164.

Laemmli, U. K. (1970). Cleavage of structural proteins during the assembly of the head of bacteriophage T4. *Nature* **227,** 680–685.

Lehmann, U., Bock, O., Glockner, S., and Kreipe, H. (2000). Quantitative molecular analysis of laser microdissected paraffin-embedded human tissues. *Pathobiology* **68,** 202–208.

Lupas, A., Van Dyke, M., and Stock, J. (1991). Predicting coiled coils from protein sequences. *Science* **252,** 1162–1164.

Lustbader, J. W., Cirilli, M., Lin, C., Xu, H. W., Takuma, K., Wang, N., Caspersen, C., Chen, X., Pollak, S., Chaney, M., Trinchese, F., Liu, S., *et al.* (2004). ABAD directly links Abeta to mitochondrial toxicity in Alzheimer's disease. *Science* **304,** 448–452.

Martinet, W., Abbeloos, V., Van Acker, N., De Meyer, G. R., Herman, A. G., and Kockx, M. M. (2004). Western blot analysis of a limited number of cells: A valuable adjunct to proteome analysis of paraffin wax-embedded, alcohol-fixed tissue after laser capture microdissection. *J. Pathol.* **202,** 382–388.

Mayer, M. P., Schroder, H., Rudiger, S., Paal, K., Laufen, T., and Bukau, B. (2000). Multistep mechanism of substrate binding determines chaperone activity of Hsp70. *Nat. Struct. Biol.* **7,** 586–593.

Nonomura, Y., Katayama, E., and Ebashi, S. (1975). Effect of phosphates on the structure of the actin filament. *J. Biochem. (Tokyo)* **78,** 1101–1104.

Ohkubo, T., Sakasegawa, Y., Toda, H., Kishida, H., Arima, K., Yamada, M., Takahashi, H., Mizusawa, H., Hachiya, N. S., and Kaneko, K. (2006). Three-repeat Tau 69 is a major tau isoform in laser-microdissected Pick bodies. *Amyloid* **13,** 1–5.

Sergeant, N., Delacourte, A., and Buee, L. (2005). Tau protein as a differential biomarker of tauopathies. *Biochim. Biophys. Acta* **1739,** 179–197.

Simone, N. L., Paweletz, C. P., Charboneau, L., Petricoin, E. F., III, and Liotta, L. A. (2000). Laser capture microdissection: Beyond functional genomics to proteomics. *Mol. Diagn.* **5,** 301–307.

Tanaka, M., Kim, Y. M., Lee, G., Junn, E., Iwatsubo, T., and Mouradian, M. M. (2004). Aggresomes formed by alpha-synuclein and synphilin-1 are cytoprotective. *J. Biol. Chem.* **279,** 4625–4631.

Tanaka, T., Ito, T., Furuta, M., Eguchi, C., Toda, H., Wakabayashi-Takai, E., and Kaneko, K. (2002). *In situ* phage screening. A method for identification of subnanogram tissue components *in situ*. *J. Biol. Chem.* **277,** 30382–30387.

Tsukita, S. (1985). Desmocalmin: A calmodulin-binding high molecular weight protein isolated from desmosomes. *J. Cell Biol.* **101,** 2070–2080.

Walker, J. E., Saraste, M., Runswick, M. J., and Gay, N. J. (1982). Distantly related sequences in the alpha- and beta-subunits of ATP synthase, myosin, kinases and other ATP-requiring enzymes and a common nucleotide binding fold. *EMBO J.* **1,** 945–951.

Zhukareva, V., Mann, D., Pickering-Brown, S., Uryu, K., Shuck, T., Shah, K., Grossman, M., Miller, B. L., Hulette, C. M., Feinstein, S. C., Trojanowski, J. Q., and Lee, V. M. (2002). Sporadic Pick's disease: A tauopathy characterized by a spectrum of pathological tau isoforms in gray and white matter. *Ann. Neurol.* **51,** 730–739.

CHAPTER 13

In Situ Analysis of DNA Damage Response and Repair Using Laser Microirradiation

Jong-Soo Kim,*,[1] Jason T. Heale,* Weihua Zeng,* Xiangduo Kong,* Tatiana B. Krasieva,† Alexander R. Ball, Jr.,* and Kyoko Yokomori*

*Department of Biological Chemistry, School of Medicine
University of California, Irvine, California 92697

†Beckman Laser Institute, Department of Surgery
Laser Microbeam and Medical Program
University of California, Irvine, California 92697

[1] Present address: Neuroscience Research Institute, Gachon Medical School, Namdong-gu, Incheon 405.

0091-679X/07 $35.00
DOI: 10.1016/S0091-679X(06)82013-3

A proper response to DNA damage is critical for the maintenance of genome integrity. However, it is difficult to study the *in vivo* kinetics and factor requirements of the damage recognition process in mammalian cells. In order to address how the cell reacts to DNA damage, we utilized a second harmonic (532 nm) pulsed Nd:YAG laser to induce highly concentrated damage in a small area in interphase cell nuclei and cytologically analyzed both protein recruitment and modification. Our results revealed for the first time the sequential recruitment of factors involved in two major DNA double-strand break (DSB) repair pathways, non-homologous end-joining (NHEJ) and homologous recombination (HR), and the cell cycle-specific recruitment of the sister chromatid cohesion complex cohesin to the damage site. In this chapter, the strategy developed to study the DNA damage response using the 532-nm Nd:YAG laser will be summarized.

I. Introduction

A. General Background on DNA Double-Strand Break Response and Repair

Genome integrity is continually threatened by endogenous metabolic products produced during normal cellular respiration, by errors that arise during DNA replication and recombination and by exogenous exposure to DNA-damaging agents. The resulting DNA damage, if not repaired, can lead to the accumulation of mutations ranging from single nucleotide changes to chromosomal rearrangements and loss. Organisms have developed an array of DNA repair mechanisms acting in concert with damage sensing and checkpoint-signaling pathways to protect cells from the deleterious effects of permanent genetic changes that can lead to cancer, developmental abnormality, and cell death (Fig. 1). Different insults to DNA are recognized by lesion-specific repair factors, which invoke distinct repair pathways including nucleotide excision repair (NER), base excision repair (BER), mismatch repair (MMR), and double-strand break (DSB) repair. Error-prone repair, incomplete repair due to excess damage, and perturbation of DNA repair or checkpoint function all cause accumulation of mutations and genomic instability. Mutations of genes involved in DNA damage response and repair processes

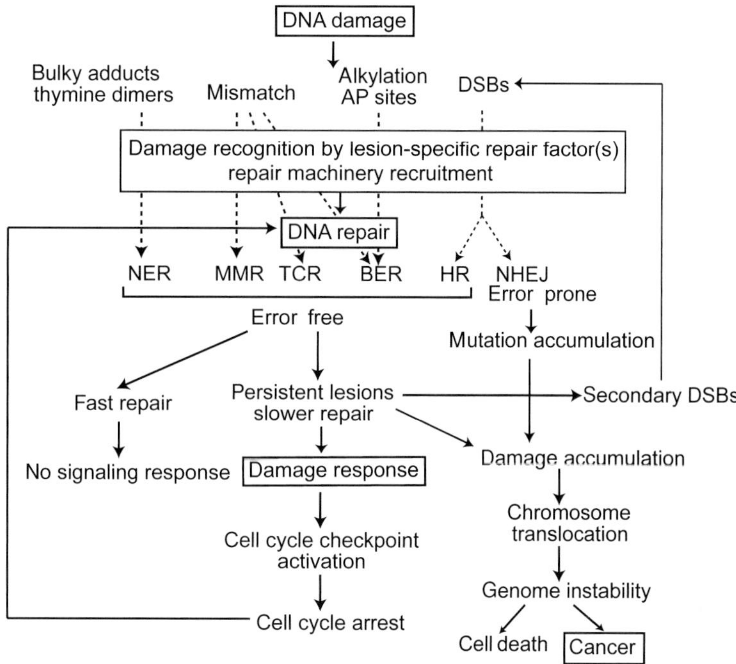

Fig. 1 An integrated model of DNA damage recognition, signaling, and repair. DNA lesions are recognized by lesion-specific repair factors and are repaired by one (or more) of several pathways including NER, BER, MMR, and DSB repair. Each of these pathways is composed of proteins encoded by 20–40 genes. Certain types of lesion under suitable conditions will be repaired promptly without any significant induction of DNA damage response signals. However, if unrepaired lesions persist, DNA damage signaling is transduced resulting in nuclear-wide responses affecting other nuclear processes such as DNA replication, cell cycle progression, and transcription. If the damage is irreversible, cells may go into permanent cell cycle arrest as is often seen in fibroblast cells, or possibly enter apoptosis. In cancer cells, however, this scheme may be abrogated due to mutations in critical factors, and cells continue to proliferate despite the unrepaired damage with resultant loss of genome integrity (e.g., secondary mutations due to damage accumulation, chromosome translocation, and/or aneuploidy).

were found to cause cancers and cancer-prone developmental abnormalities (Lehmann, 2003; Taylor, 2001).

During the past decade, there have been huge advances in identifying genes important for DNA damage response and repair. Mutagenesis studies in yeast and mammalian cells as well as identification of naturally occurring mutations in human diseases with cancer predisposition have resulted in the discoveries of a large number of DNA repair and checkpoint genes and their roles in repairing different classes of DNA damage. However, many aspects of the molecular processes underlying DNA damage recognition, checkpoint signaling, and DNA repair inside the living cell remain elusive and are still under active investigation. Furthermore, genetic studies in yeast and rodent cells revealed species-specific differences in repair factor requirements (Dasika *et al.*, 1999; Di Virgilio and Gautier, 2005; Tauchi *et al.*, 2002a), emphasizing the importance of studies in mammalian cells, particularly human cells, to understand the impact of DNA repair processes on human health.

1. DSB Damage Checkpoint Activation and Repair

DNA DSBs are particularly detrimental to the cell. Arising from exogenous factors such as ionizing radiation, or by DNA replication fork collapse, DSBs can lead to chromosome translocations, which are frequently associated with developmental disorders and cancer. Furthermore, the DSB repair machinery is normally required during meiosis (meiotic recombination) and in immune cells (immunoglobulin gene rearrangement). Thus, normal function of the DSB repair machinery is critical for human genomic homeostasis.

Two major events are evoked by the presence of DSBs: DNA checkpoint activation and the actual repair of DNA breaks (Fig. 2). DNA checkpoint activation offers a mechanism to stall the cell cycle and DNA replication, as well as modulate RNA transcription, allowing for effective DNA repair while preventing the proliferation of cells harboring damaged DNA. There are multiple checkpoints throughout the cell cycle, including the G1/S, S, G2/M, and mitotic spindle checkpoints. One unrepaired DSB is sufficient to cause cell cycle delay (Lee *et al.*, 1998).

*A*taxia-*t*elangiectasia, *m*utated (ATM), a member of the PI3-kinase-related protein kinase (PIKK) family (Abraham, 2004), and its homologue ATR are major DSB signal transducers and are critical for activating the interphase checkpoints.

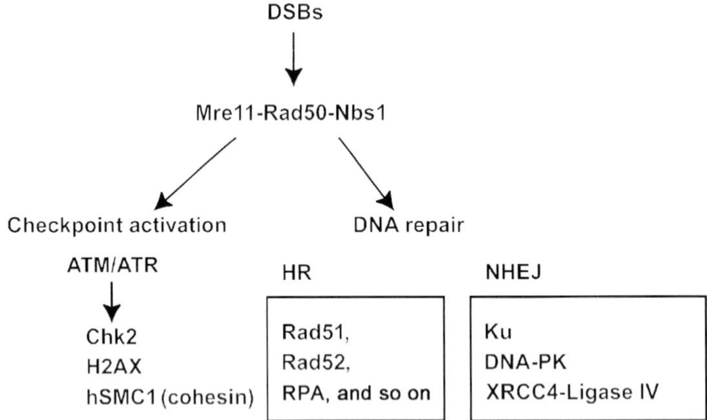

Fig. 2 DSB damage induces checkpoint activation and DNA repair. The Mre11 complex (Mre11-Rad50-Nbs1) is one of the earliest factors to recognize DSBs and plays a role in both checkpoint activation and DNA repair. ATM/ATR kinases are activated in response to damage, which phosphorylate downstream target proteins such as Chk2, H2AX, and the cohesin component hSMC1. Two major repair pathways (HR and NHEJ) with distinct sets of proteins are involved in DSB repair.

ATM is activated in response to DSBs and phosphorylates itself as well as many different downstream target proteins to transmit checkpoint signals. These target proteins include checkpoint kinases 1 and 2 (Chk1 and Chk2), the histone variant H2AX, and the structural maintenance of chromosomes 1 (SMC1) protein of the sister chromatid cohesion complex "cohesin" (Niida and Nakanishi, 2006).

For repairing DSBs, two major repair pathways have been identified, which involve distinct sets of repair proteins (Fig. 2) (Lieber *et al.*, 2003; West, 2003). DNA-PKcs and Ku are recruited to DNA ends to initiate the non-homologous end-joining (NHEJ) cascade followed by the recruitment of the XRCC4/ligase IV complex. The NHEJ pathway detects DSBs and simply re-ligates the ends with nonspecific nucleotide addition or deletion, thereby presenting the risk of introducing errors. In contrast, the homologous recombination (HR) repair pathway utilizes the homologous template (i.e., sister chromatid) for copying and restoring the damaged region accurately. Rad51 and RPA are essential factors in the HR pathway in vertebrate cells, and are recruited to single-stranded DNA (ssDNA) regions at broken DNA ends to catalyze invasion of ssDNA into the homologous DNA template (Baumann *et al.*, 1996; Sung, 1994). Since the sister chromatids can only be generated after DNA replication in S phase, HR takes place only in S/G2 phase of the cell cycle. The trimeric complex containing Mre11, Rad50, and Nbs1 (the Mre11 complex) functions at an early stage of HR (Paull and Gellert, 1998) as well as in ATM-mediated checkpoint activation in mammalian cells (Carson *et al.*, 2003; Costanzo *et al.*, 2004; Horejsi *et al.*, 2004; Lee and Paull, 2004, 2005; Uziel *et al.*, 2003). Thus, checkpoint and repair factors act together to protect the genome against DSBs.

ATM is mutated in the pleiotropic genetic disorder ataxia-telangiectasia (A-T), which predisposes patients to cancer (Shiloh, 2001). ATM is a tumor suppressor protein also mutated in many cancers. Mutant cell lines established from A-T patients exhibit sensitivity to ionizing radiation, chromosome instability, and radioresistant DNA synthesis (RDS), an S phase checkpoint defect. Mutations in Mre11 and Nbs1 were found in AT-like disorder (ATLD) and Nijmegen breakage syndrome (NBS), respectively (Carney *et al.*, 1998; Stewart *et al.*, 1999; Varon *et al.*, 1998). These two diseases share similar cellular phenotypes with A-T and predispose the affected patients to cancer, emphasizing the notion that DNA repair and checkpoint pathways are critical tumor-suppressing processes (Tauchi *et al.*, 2002b).

2. Understanding the DSB Damage Recognition Process

Proper recognition of DSBs is critical for initiating the repair and checkpoint-signaling processes. However, it was difficult to examine the DNA damage response events inside the cell. Although genetic examination of cell viability following damage identifies critical genes and genetic interactions important for overall repair and recovery at the cellular level, specific damage recognition events at the molecular level cannot be addressed by this method. Important outstanding issues that require further investigation include the kinetics and cell cycle specificity of repair and checkpoint factor recruitment to the damage site, their upstream factor requirement, the mechanism of pathway choice (i.e., how the cell decides between NHEJ and HR), and the mechanism of checkpoint signal dissemination (i.e., how checkpoint signaling originating at the damage site spreads throughout the entire nucleus). Different strategies used to address these questions are discussed below.

B. Cytological Analysis of the DNA Damage Response by Examination of Ionizing Radiation-Induced Foci and Its Limitations

It was observed early on by immunofluorescent staining using antibodies against repair and checkpoint factors that irradiation of mammalian cells induces protein redistribution in the cell nucleus and that repair and checkpoint factors often form foci or speckles, termed *i*onizing *r*adiation-*i*nduced *f*oci (IRIF) (Shiloh, 2003). IRIF typically overlap with phosphorylated H2AX, a marker of DSBs (Section I.D), indicating that IRIF are formed around DNA damage sites. On the basis of this notion, numerous studies have used IRIF as indicators of factor recruitment to the damage sites. However, some studies have raised questions about this phenomenon. First, typical IRIF for the Mre11 complex were prominent at 6–8 h after ionizing radiation, even though most of the damage repair occurs within the first 30 min, suggesting that IRIF formation does not necessarily mirror the kinetics of DNA repair (Carney *et al.*, 1998; Maser *et al.*, 1997; Mirzoeva and Petrini, 2001; Nelms *et al.*, 1998). A study using a laser system (see below) provided compelling evidence that the initial recruitment to the damage site is distinct from IRIF formation for the Mre11 complex (Celeste *et al.*, 2003). Second, IRIF formation appears to be restricted to HR factors and selected checkpoint proteins. For example, Chk2 kinase associates with DNA damage sites transiently and does not form IRIF (Lukas *et al.*, 2003). Furthermore, although there are clear *in vitro* and *in vivo* evidence that NHEJ factors are recruited to DSBs and are involved in DNA repair, none of them form discernable IRIF, which hindered the analysis of the damage response process of these factors *in vivo* (Lisby *et al.*, 2004). The problem with ionizing radiation or any other agent that induces DSBs is that one never knows for certain where exactly in the nucleus or on the chromosome the actual damage is being induced. Thus, it was essential to develop a method to induce damage at a defined region in the cell nucleus, or in the genome, so that it would be possible to follow factor recruitment and/or modification at the damage site *in vivo*.

C. Chromatin Immunoprecipitation Analysis of the Endonuclease-Induced DSB Site

One approach was to induce a DSB using a rare DNA endonuclease at a specific target sequence in the genome, and detect factor recruitment by chromatin immunoprecipitation (ChIP) using antibodies against DNA repair factors and specific

PCR primers corresponding to the DSB site and neighboring regions (for detailed method, see Sugawara and Haber, 2006). The HO and I-Sce I endonucleases were used successfully for this purpose in yeast and mammalian cells, respectively (Kondo *et al.*, 2001; Murr *et al.*, 2006). By this method, one can examine the recruitment of factors that do not form IRIF and determine how far each factor spreads from the original DSB site. In addition, one can study upstream factor requirements. By making the endonuclease expression inducible, it is also possible to examine the recruitment kinetics (Shroff *et al.*, 2004). However, the results obtained in these systems reflect average steady-state data at each time point from a mixed cell population, in which the nuclease action and cellular response may not be completely synchronous. Furthermore, antibodies suitable for ChIP must be available. Thus, an alternative, preferably single cell, analysis could greatly complement this approach.

D. Different Selective Irradiation Systems to Study the Molecular Responses to DNA Damage *In Vivo*

The first groundbreaking study that made an attempt to analyze factor recruitment to the damage site *in vivo* was by John Petrini's group, who utilized gamma irradiation through grids to irradiate cells in a strip pattern (Nelms *et al.*, 1998). Immunofluorescent analysis of these cells using antibody against the Mre11 complex revealed that the complex is indeed recruited to the irradiated area within 30 min of irradiation.

In another significant advance, several laboratories took advantage of a previous study demonstrating that UV light can induce DSBs on bromodeoxyuridine (BrdU)-sensitized DNA (Limoli and Ward, 1993). William Bonner's group used a UV laser to induce damage in the cell nucleus presensitized with BrdU and demonstrated that phosphorylation of a histone H2A variant, H2AX, at serine 139 (termed "γH2AX") occurs specifically at the damage site and surrounding areas, establishing that the antibody specific for γH2AX can be used to detect DSB sites in the cell (Rogakou *et al.*, 1998, 1999). This was the first demonstration that DSBs not only cause recruitment of repair factors but also evoke changes in the surrounding chromatin. On the basis of this work, it is now widely accepted that γH2AX is a DSB marker (Celeste *et al.*, 2002; Fernandez-Capetillo *et al.*, 2002; Limoli *et al.*, 2002; Paull *et al.*, 2000; Petersen *et al.*, 2001). Several laboratories began to use a UV-A laser in combination with halogenated nucleotides [such as BrdU, chlorodeoxyuridine (CldU), and iododeoxyuridine (IdU)] or a DNA-intercalating agent (Hoechst 33258) for *in situ* detection of factor recruitment to damage sites in mammalian cells (Table I). In addition, a system utilizing α-particle irradiation was used for similar purposes (Aten *et al.*, 2004). A comprehensive review of different laser systems is also available (Lukas *et al.*, 2005).

Table I
List of Subnuclear Irradiation Systems

Gridded gamma irradiation (Nelms *et al.*, 1998)
Damage site clustering: α-particle irradiation (Aten *et al.*, 2004)
UV, 390 nm/Hoechst dye 33258 (Bradshaw *et al.*, 2005; Rogakou *et al.*, 1999)
UV, 337 nm/BrdU, CldU, IdU (Lukas *et al.*, 2003; Tashiro *et al.*, 2000)
UV, 365 nm/1,5-Dihydroxyisoquinoline (DIQ) (Lan *et al.*, 2004)
A second harmonic pulsed Nd:YAG laser (mitosis) (Mikhailov *et al.*, 2002)
A second harmonic pulsed Nd:YAG laser (interphase) (Kim *et al.*, 2002a)

E. Damage Induction by the 532-nm Nd:YAG Laser

We utilized a second harmonic (532 nm) pulsed Q-switched neodymium:yttrium aluminum garnet (Nd:YAG) laser microbeam. The advantages of this system are as follows: (1) it does not require the preincorporation of a chromophore or halogenated nucleotide analogue, such as BrdU, into chromosomal DNA as required for the UV laser system, thus exerting no effect on chromatin packing, (2) it is possible to focus the beam to a less than half a micrometer diameter spot in the cell nucleus, and (3) this laser produces visible green light that is not absorbed by the cell, that is it can be effectively focused inside the nucleus and any collateral damage resulting from beam scattering is minimal (see Chapter 1 by Berns, this volume, for review of the Nd:YAG laser system for subcellular surgery; Berns *et al.*, 1981).

In Section II, the basic protocol for laser cutting of cells and the major results obtained are summarized.

II. Nd:YAG Laser Setup and Sample Preparation

A. System Setup

The basic laser system is assembled as shown in Fig. 3. The second harmonic beam (532 nm) of a Nd:YAG laser (Continuum, Surelite I-10) was coupled into an inverted microscope system (Olympus, IX81) via the side-port illumination path. A 532-nm dichroic mirror (Chroma Technology, Rockingham, VT) was preinstalled inside of the microscope. The system parameters used (2–3 μJ/pulse energy after objective, 4- to 6-ns pulse duration, 7.5 Hz) correspond to irradiances just above the estimated threshold values of optical breakdown in water sufficient to cause highly localized damage confined to the focal plane of the objective (area affected is typically less than 200–300 nm in diameter) (Krasieva *et al.*, 1998; Venugopalan *et al.*, 2002). Since the cell areas exposed to the unfocused beam receive energies below the threshold for optical breakdown, and there is little or no absorption by the cell at 532 nm, the only region affected by the laser will be at the point of focus. This is in contrast to UV-A that can be absorbed by the cell membrane and causes oxidative

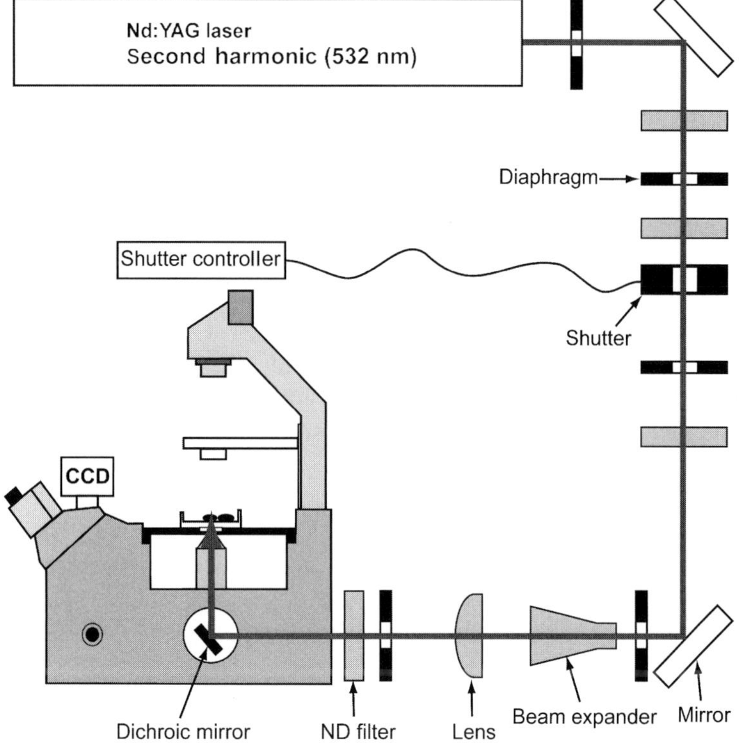

Fig. 3 The Nd:YAG laser system configuration (see text).

stress and cell damage. Electron microscopy studies confirmed that the damage is structurally confined to the focused area (Berns *et al.*, 1998; Chan *et al.*, 2001; Venugopalan *et al.*, 2002). Damage is most likely caused by ionization (optical breakdown) of the medium (water) and plasma formation (Berns *et al.*, 1998; Chan *et al.*, 2001; Venugopalan *et al.*, 2002). The expansion of cavitation bubbles caused by plasma generates transient photomechanical pressure, which results in breakage of molecular bonding and numerous DNA breaks in a confined area.

The laser was set at the same height as the side port of the microscope. A high-magnification oil-immersion objective (UPlanFl 100×, NA 1.3) was used for focusing the laser beam into the subnuclear area. Both 40× and 63× objectives have been used in the UV/BrdU system (Bradshaw *et al.*, 2005; Lukas *et al.*, 2003; Rogakou *et al.*, 1999; Tashiro *et al.*, 2000). A special sample holder was designed for our experiments. An external beam expander (Thorlabs, Newton, NJ, BE05X-A) was used to increase the beam size enough to fill the back aperture of the objective. Several types of antireflection-coating neutral-density filters were used to adjust the amount of transmitted light. Using the plano-convex antireflection-coating lenses with different focal lengths, the desired laser focus and spot size in the cells were achieved. All components of the system were secured to a custom antivibration table (Technical Manufacturing Corporation, MA).

1. The Laser Beam Alignment Procedure

1. Turn on the laser and keep the power as low as possible. Confirm that the shutter is closed.
2. Remove any microscope objective and place a piece of paper on the stage. Place a beam dump over the objective hole on the microscope stage.
3. Align the laser beam with mirrors so that the laser beam is centered on the dichroic mirror. Check the beam to ensure that it is exiting the microscope perfectly vertical.
4. Using a beam expander and a plano-convex lens, align the beam again. Ensure that the beam size overfills the back aperture of the objective (>5 mm in diameter).
5. Reinstall the objective and put a sample coverslip on the stage above the lens. Focus on the surface of the coverslip.
6. Adjust the spot size by adjusting the position of the lens (this can be a time-intensive procedure).
7. Use neutral-density filters, attenuator, and/or diaphragms as needed.

B. Sample Preparation

Sample preparation procedures for the analysis of the DNA damage response and repair factors using laser microirradiation are shown in Fig. 4.

1. Materials

- 35 mm × 10 mm cell culture dishes (BD Falcon, Cat. No. 351008, BD Biosciences, San Jose, CA).
- Grid glass coverslips (Bellco Biotechnology, Vineland, NJ, Cat. No. 1916-92525).
- Silicon glue.
- Biosafety cabinet equipped with UV lamp.
- Phenol red-free cell culture media.
- SNBP buffer (0.02% Saponin, 0.05% NaN_3, 1% BSA in 1× PBS).
- Fixative (4% paraformaldehyde in PBS, methanol, and so on).

2. Procedures

Culture dish preparation

1. Machine an opening (18- to 20-mm diameter) on the bottom of the tissue culture dish.
2. Rinse the coverslips with distilled water (DW) and keep dry.

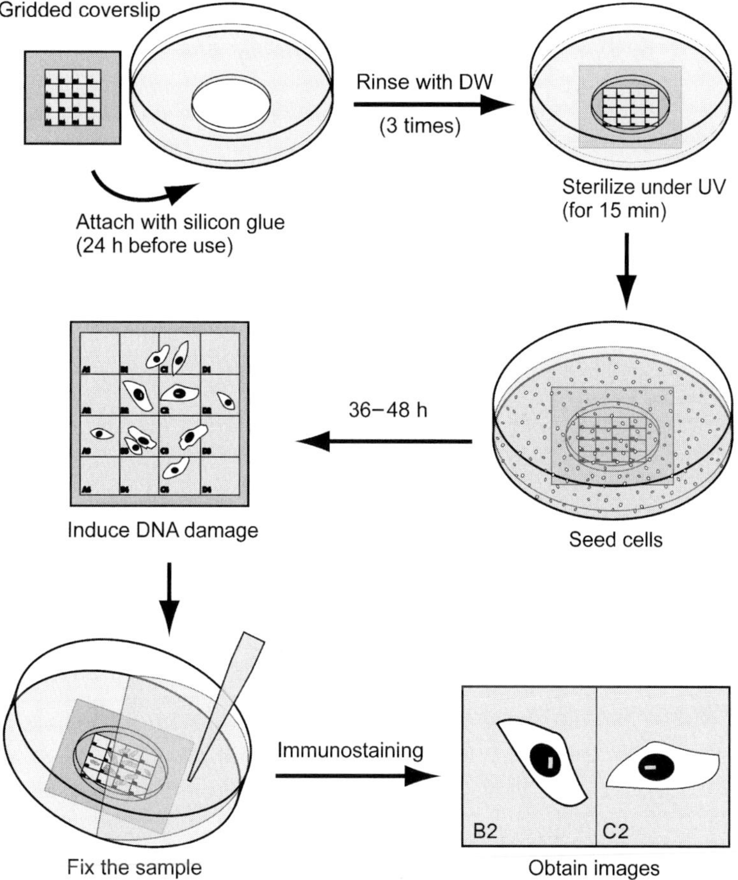

Fig. 4 Sample preparation for laser microirradiation.

3. Secure a gridded coverslip with silicon glue over the opening on the bottom of the dish.
4. Allow 24 h for silicon hardening and then wash dish thoroughly with sterile DW.
5. Sterilize dishes for 15 min in a culture hood with a UV source.

Cell preparation

Any adherent cells can be used. We have used various cells, including HeLa, IMR90, and mutant fibroblast cells (Kim *et al.*, 2002a, 2005).

1. Prepare fresh cells with a low passage number (36–48 h before DNA damage induction).

2. Decide on an appropriate dilution to allow for 36- to 48-h growth with a final cell confluence of 40–50% (complete confluency impairs reading of the gridded coverslip).
3. Seed cells in the dish and disperse well to avoid clustering at the center of the coverslip.
4. Prior to cutting, replace the media with phenol red-free media at least 0.5–1 h before DNA damage induction.
5. Proceed with laser microirradiation.
6. Remove culture media and rinse gently twice with SNBP buffer (cells can be displaced if washed too vigorously).
7. Fix sample with appropriate fixative.

C. Image Analysis

1. Immunofluorescence Analysis Protocol

1. Rinse the sample with fresh SNBP buffer.
2. Incubate in blocking solution (SNBP buffer with 4% serum and 0.1% fish gelatin) at 37 °C for 15 min. Note that appropriate serum must be selected based on the species of animal from which the secondary antibody is derived.
3. Carefully remove blocking solution.
4. Incubate with primary antibody diluted in SNBP buffer with 1% serum and 0.05% fish gelatin at 37 °C for 30 min.
5. Wash three times with SNBP buffer at room temperature; 10 min for each wash.
6. Incubate with fluorochrome-conjugated secondary antibody solution at 37 °C for 30 min.
7. Wash three times with SNBP buffer, as in step 5.
8. Store in SNBP for visualization or repeat steps 4–7 for double staining.

2. Real-Time Analysis

Instead of detecting the endogenous protein by antibody, a green fluorescent protein (GFP) fusion of the DNA repair factor of interest can be expressed by transient transfection or in a stable cell line, and damage site recruitment can be analyzed in real time in live cells. This approach was taken in mammalian cells using the UV laser system as well as in low-dose ionizing radiation experiments in yeast (Bradshaw *et al.*, 2005; Lisby *et al.*, 2004; Lukas *et al.*, 2003). Similar experiments can be performed using a 532-nm Nd:YAG laser system combined with a heating unit (to attain a 37 °C environment) and CO_2 chamber that encloses the microscope.

1. Construct the mammalian expression plasmid containing the protein of interest fused to GFP (e.g., pIRES neo3, Clontech, Mountain View, CA). An expression plasmid containing only GFP is used as a negative control.

2. Seed the appropriate adherent cells (e.g., HeLa) onto the coverslip (see above).
3. After 24 h, transfect the expression plasmid with a lipofection reagent according to the manufacturer's instructions.
4. Begin equilibrating the CO_2 chamber and heating unit (5% CO_2 at 37 °C) overnight or for at least 12 h prior to cutting.
5. At 24-h posttransfection, identify and mark the GFP-positive cells under the fluorescent microscope. Minimize the UV exposure time since this may cause undesired cell damage.
6. Expose the identified cells to the 532-nm Nd:YAG laser as described above.
7. Following damage induction, perform time-lapse fluorescent-imaging analysis at selected time intervals.
8. Perform similar experiments using the GFP-only control. Damage induction should not result in any clustering of the GFP signal at the damage site.

We have observed variability in the recruitment efficiency depending on the factor being analyzed. If visualization of the GFP clustering at the damage site is difficult in live cells, damaged cells can be fixed and stained with antibody specific for GFP (Clontech). Although this no longer offers the advantage of real-time analysis, the recruitment of the recombinant protein can still be confirmed and can in turn be used as the basis for deletion analysis to map the domain(s) required for damage site targeting *in vivo* (Bradshaw *et al.*, 2005; Lan *et al.*, 2005).

III. Analysis of the 532-nm Laser-Induced Damage

A. Detection of DNA Breaks at the Site of Laser Damage by TUNEL Assay

The Nd:YAG laser was developed to microablate cells or subcellular structures and was previously used to study the opposing forces of spindles in mitotic cells by cutting condensed mitotic chromosomes in live cells (Liang *et al.*, 1994). However, whether the 532-nm Nd:YAG laser can induce DNA damage that can be recognized by cellular checkpoint and repair machinery in interphase-decondensed chromatin was unknown. We first tested whether the laser induces DNA breaks in the interphase nucleus. A TUNEL assay, which utilizes TdT to incorporate BrdU at DNA ends, was used to detect strand breaks in conjunction with a fluorescent antibody against BrdU (Fig. 5A) (Kim *et al.*, 2002a). A similar analysis was used in the aforementioned grid irradiation experiments (Nelms *et al.*, 1998). Our results confirmed that the second harmonic 532-nm Nd:YAG laser can indeed induce DNA strand breaks. It should be noted that the damage site became visible as a dark line due to a change in refractive index under the phase-contrast microscope, thus serving as a visible marker for the damage site (Fig. 5). This type of optical change was described for laser-induced lesions (Berns *et al.*, 1998).

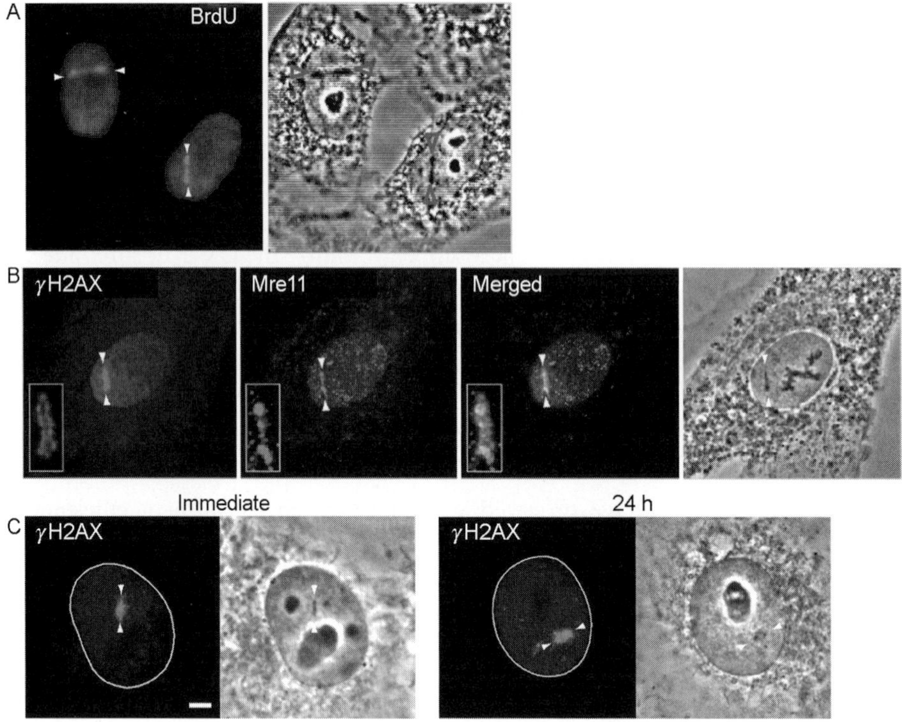

Fig. 5 Analysis of damage induced by the Nd:YAG laser. (A) Detection of DNA breaks by TdT-mediated BrdU incorporation. BrdU is stained by anti-BrdU antibody. (B) Mre11 and γH2AX signals at the damage sites after 4 min of damage induction. (C) γH2AX signal persists at 24-h p.d. Fig. 5C reproduced from *The Journal of Cell Biology, 2005, 170: 341–347.* Copyright 2005 The Rockefeller University Press. (See Plate 11 in the color insert section.)

B. Presence of DSB Indicated by the Presence of γH2AX and the Mre11 Complex

To further confirm the presence of DSBs, we used antibodies specific for γH2AX and Mre11 protein as markers for DSBs. Immediately following laser damage, γH2AX and Mre11 recruitment are obvious, indicating the presence of DSB damage (Fig. 5B) (Kim *et al.*, 2005). Similar observations were made with Rad50 and Nbs1, the remaining two components of the Mre11 complex (data not shown).

The second harmonic Nd:YAG laser damages not only DNA but also associated proteins such as histones within the focused area. This is a feature associated with the 532-nm Nd:YAG laser system, but not the UV/BrdU system in which sensitized DNA is selectively cut. Consistent with this, the actual 532-nm Nd:YAG laser-damaged site was devoid of the γH2AX signal (Fig. 5B). In contrast, the Mre11 signal accumulated inside of the damaged area surrounded by the γH2AX signal, indicating that the Mre11 complex is newly recruited to fragmented DNA present in the damaged area (Fig. 5B). The broader distribution of γH2AX is consistent with previous observations using the UV/BrdU system that the γH2AX response

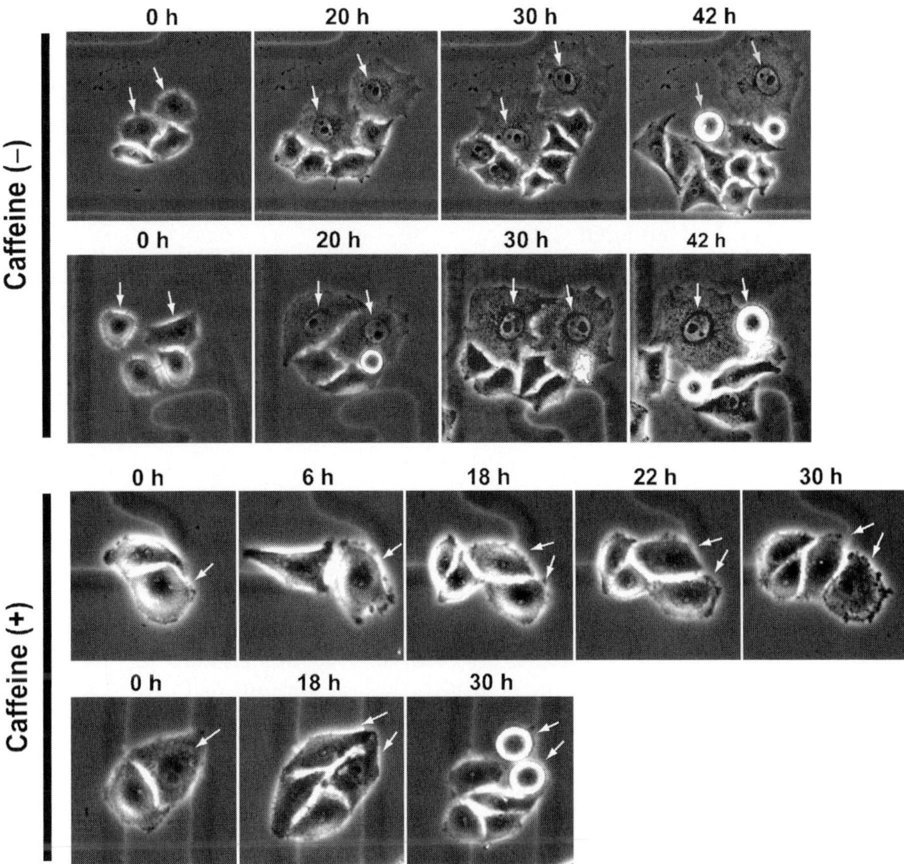

Fig. 6 Caffeine-sensitive cell cycle delay following laser damage. Cells released from double-thymidine block were preincubated with or without caffeine (2 mM, Sigma) for 1 h before damage induction (Blasina *et al.*, 1999). Laser-damaged cells (indicated by yellow arrows) on gridded coverslips were photographed at the times indicated. Reproduced from *The Journal of Cell Biology, 2005, 170: 341–347.* Copyright 2005 The Rockefeller University Press.

A. Methods

1. Antibodies

We used mouse monoclonal antibodies specific for Mre11, Rad50 and ATM (GeneTex, San Antonio, TX), Ku70 (Novus Biologicals, Littleton, CO), PARP1 (Trevigen, Gaithersburg, MD), and DNA-PKcs (Abcam Limited, Cambridge, UK); rabbit polyclonal antibodies specific for hMre11 (Oncogene Research Products, Cambridge, MA), Ligase IV and RPA (Chemicon International, Temecula, CA), γH2AX (Upstate, Charlottesville, VA), phospho-Chk2(Thr68) (Cell Signaling Technology, Beverly, MA), phospho-hSMC1(Ser966) (Bethyl Laboratories, Montgomery, TX), and polyclonal and monoclonal anti-Nbs1 antibodies (Novus occurs over megabases adjacent to the DSB site (Rogakou *et al.*, 1998, Similar distribution patterns of γH2AX and the Mre11 complex were obser ChIP analysis of the HO endonuclease cut site in yeast (Shroff *et al.*, 2004). together, these results confirm that the cellular response to 532-nm Nd:YA(damage resembles that of other damaging systems in mammalian and yeas

C. The Fate of the Damaged Cells

The fact that H2AX phosphorylation was observed indicates that the mediated checkpoint pathway was activated (Fig. 5B). γH2AX was ob immediately after damage induction and persisted even after 24-h postd (p.d.), indicating the presence of unrepaired damage in the irradiated (Fig. 5C). This persistent H2AX phosphorylation raises the concern that tl damage is so severe that cell death may occur. To address this, laser-damag were observed under the microscope and it was confirmed that they are vi at least 64-h p.d. (data not shown). Furthermore, we found that laser induces checkpoint-dependent cell cycle delay, such that damaged cells re interphase while control cells undergo mitosis (Fig. 6) (Kim *et al.*, 2005). W adjacent undamaged cells continued to divide, damaged cells remained i phase for more than 40 h, some eventually entering mitosis [Fig. 6; Caffe Importantly, these arrested cells do not end up in "mitotic catastroph eventually go through cell division without cell death. Furthermore, wh were treated with caffeine, which inhibits PIKK activity (including that o ATR), all the damaged cells successfully completed mitosis within the first 1 manner similar to the adjacent undamaged cells [Fig. 6; Caffeine (+)]. These demonstrate that although some DSBs persist, damage induced by the not lethal. Instead, a checkpoint-dependent cell cycle delay occurs. Theref 532-nm laser-induced damage, under the conditions described above, is com to gamma irradiation.

IV. Analysis of Checkpoint Activation and DSB Repair Factor Targeting

Using a laser microirradiation system, it is possible to directly visualiz recruitment and protein modifications in the cell nucleus using specific ant By inducing damage at different cell cycle stages and by fixing the dama at different time points p.d., the kinetics and cell cycle specificity of rec to, and/or activation of factors at, the damage site can be examined. Furth one can determine real-time kinetics and map the domain(s) required for dar targeting using fluorescent fusion proteins and mutants. Finally, by using cells or small interfering RNA (siRNA)-mediated protein depletion, one ca which factors are required for the recruitment of additional downstrea factors.

Biologicals, Littleton, CO, and BD Bioscience, San Diego, CA, respectively). Goat polyclonal anti-ATR antibody was from Santa Cruz Biotechnology (Santa Cruz, CA). Donkey anti-mouse IgG and anti-goat IgG conjugated with AlexaFlour 488 and goat anti-rabbit IgG conjugated with AlexaFlour 546 were from Molecular Probes (Eugene, OR). Cy3-conjugated goat anti-rabbit IgG was from Jackson ImmunoResearch Laboratories (West Grove, PA).

2. Cell Synchronization

HeLa cells were synchronized by double-thymidine block in combination with nocodazole as described previously with slight modification (Gregson *et al.*, 2001). Briefly, cells were incubated in 2-mM thymidine for 17 h, then rinsed and incubated in DMEM for 9 h. The 2-mM thymidine was added again for 15 h and cells were rinsed and grown in DMEM for an appropriate length of time for each cell cycle stage. The efficiency of synchronization at S phase by double-thymidine block was assessed by FACS analysis (Gregson *et al.*, 2001). More than 80% of the cells entered M phase synchronously 13 h after the release from thymidine, indicating proper synchronization (Heale *et al.*, 2006). For G1 phase, cells in mitosis following a single-thymidine block were continuously monitored to confirm cell division. They were subjected to laser microirradiation 5 h after M phase.

Several different methods can be employed to confirm the cell cycle stage of individual cells. We typically identify mitotic cells under the light microscope as cells that have a rounded morphology and are less adherent to the bottom of the plate compared to the flat interphase cells, and follow these cells until they divide and re-enter interphase. Thus, it is highly likely that these cells are in G1 phase. We also perform immunostaining for PCNA, noting that S phase-synchronized cells are clearly positive for PCNA while the G1 phase cells are negative (Fig. 7). S/G2 phase cells can also be identified by cyclin staining (e.g., Cyclin A or B1) (Bekker-Jensen *et al.*, 2006).

B. Damage Checkpoint Response

In addition to the caffeine-sensitive cell cycle delay, checkpoint activation in response to 532-nm Nd:YAG laser-induced damage was further validated by detection of the ATM/ATR checkpoint kinases and their downstream target phosphorylation using antibodies specific for phosphorylated forms of the ATM/ATR target proteins (Kim *et al.*, 2005). With ATM/ATR protein recruitment and H2AX phosphorylation (γH2AX) (Fig. 5B and C), phosphorylation of Chk2 was also examined using phosphor-specific antibodies (Kim *et al.*, 2005). Furthermore, phosphorylation of the cohesin component hSMC1 was also detected (data not shown). Thus, the 532-nm Nd:YAG laser-induced damage evokes checkpoint responses similar to ionizing radiation.

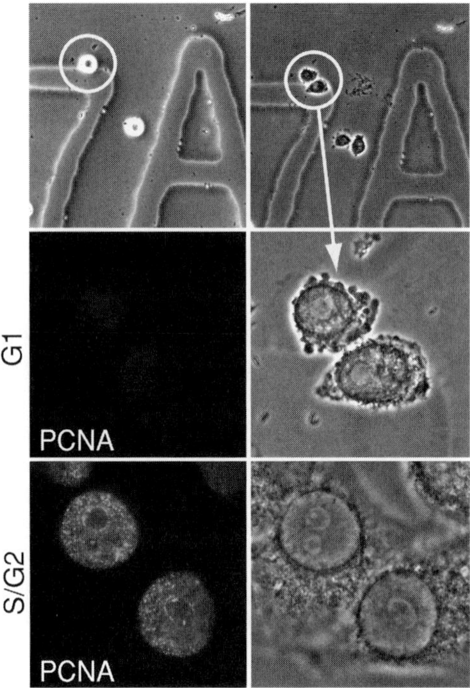

Fig. 7 Confirmation of G1 cell cycle stage. A metaphase cell was identified under the microscope on a gridded coverslip. After 5 h, the cell divided (G1). Staining of these cells with anti-PCNA antibody showed no significant signal. In contrast, cells at 3 h after release from the double-thymidine block (S/G2) exhibited a strong PCNA signal. Reproduced from *The Journal of Cell Biology, 2005, 170: 341–347.* Copyright 2005 The Rockefeller University Press.

C. Recruitment Kinetics

1. Temporal Relationship Between NHEJ and HR Factor Accumulation at the Damage Site

We found that various factors involved in different pathways exhibit distinct kinetics of damage site recruitment and retention (Fig. 10) (Kim *et al.*, 2005). The Mre11 complex was recruited to the damage sites within the first 20 min of damage induction in both G1 and S/G2 phases (Fig. 8A). Since HR prefers the use of sister chromatids as templates in vertebrate cells, HR repair is considered to be the main pathway for postreplicative repair during late S/G2 phase (Rothkamm *et al.*, 2003; Sonoda *et al.*, 1999; Takata *et al.*, 1998). However, the factors involved in NHEJ (Ku, DNA-PK, and Ligase IV) immediately clustered to the damage site in both G1 and S/G2 phases with similar efficiency, suggesting that the NHEJ pathway is activated equally well in both cell cycle phases (Fig. 8A). As mentioned above, NHEJ factor recruitment to the damage site has not been observed previously due to lack of discernable focus formation by these factors following conventional DNA-damaging agents such as ionizing radiation (Section I.B). It was thought that the number of molecules needed at the damage site is too low

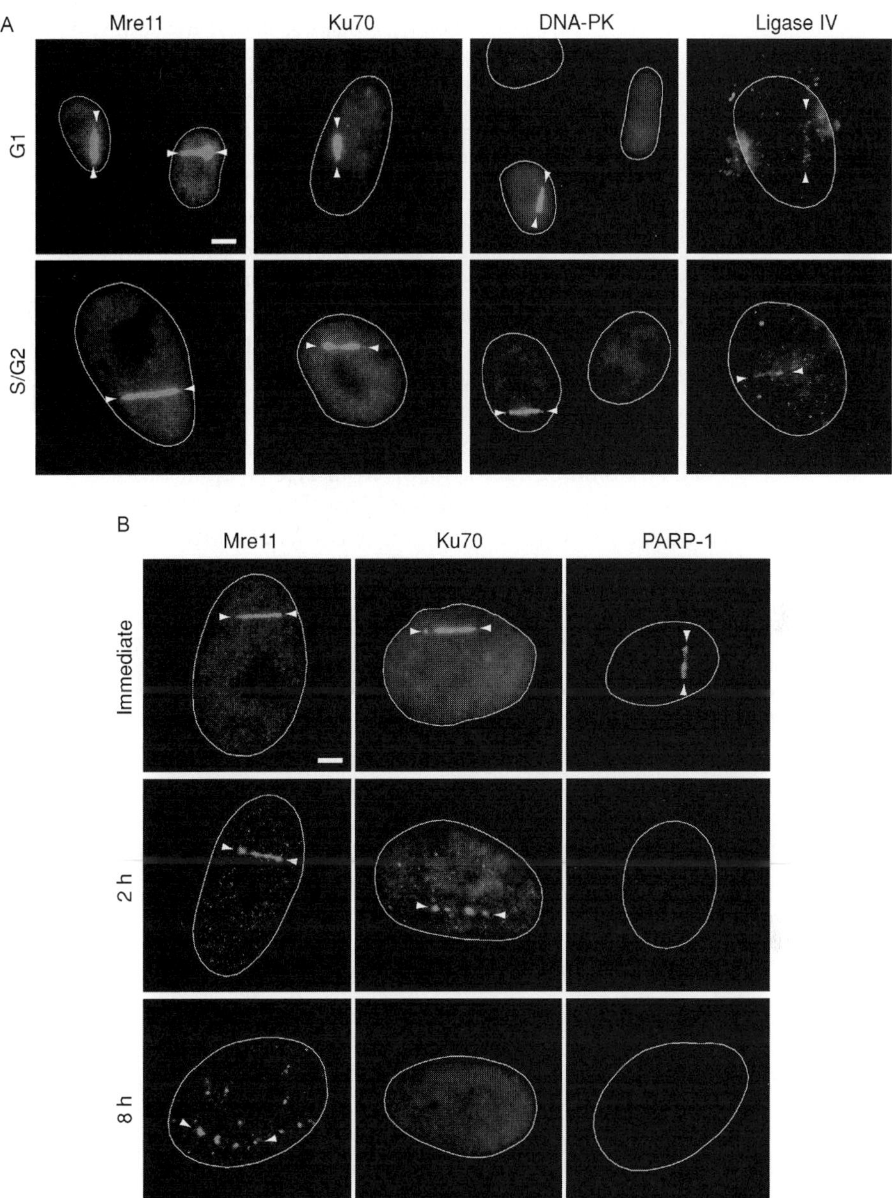

Fig. 8 Damage site association of the Mre11 complex and NHEJ factors. (A) Recruitment of the Mre11 complex and NHEJ factors in G1 and S/G2 phases. G1 and S/G2 phase cells were damaged and fixed within the first 20-min p.d. for staining. (B) A time-course analysis of localization of Mre11, Ku, and PARP-1 at damage sites. Cells were fixed at indicated time points p.d. and stained. Reproduced from *The Journal of Cell Biology, 2005, 170: 341–347.* Copyright 2005 The Rockefeller University Press.

to detect (Lisby *et al.*, 2004). Thus, the use of the Nd:YAG laser offered a unique opportunity to study the recruitment of NHEJ factors *in vivo*.

Interestingly, Ku left the damage sites earlier than the Mre11 complex, irrespective of the cell cycle stage (Fig. 8B; data not shown). By 8-h p.d., the Ku signal was absent from the damage sites. In contrast, the Mre11 complex was found to persist at the apparently unrepaired lesions for at least up to 24 h, behaving as a marker of DNA damage (Fig. 8B). PARP1 is known to be one of the earliest factors to recognize DNA damage, but its immediate autoribosylation results in its dissociation from the damage site (Ziegler and Oei, 2001). Consistent with this, PARP1 accumulated at the damage site immediately following damage, and was absent from the damage site at 2-h p.d. (Fig. 8B). In S/G2 phase, the HR factor Rad51 was also recruited to the damage sites (Fig. 9). However, its appearance at the damage site was delayed compared to NHEJ factor recruitment and persisted as long as the Mre11 complex. A similar timing of recruitment was observed with RPA, an ssDNA-binding factor involved in HR (Kim *et al.*, 2005). On the basis of these results, we proposed that NHEJ factor assembly precedes that of HR factors, even in S/G2 phase (Fig. 10). This is consistent with previous reports suggesting that NHEJ precedes HR in DSB repair (Delacote *et al.*, 2002; Frank-Vaillant and Marcand, 2002). We hypothesize that this delay is due to the end processing of broken DNA ends necessary for Rad51/RPA loading. Thus, the results suggest that NHEJ and HR are not parallel pathways that compete against each other. Rather, NHEJ functions as an immediate early repair pathway while the HR factors make a more prolonged attempt to repair persistent DNA lesions. These partially overlapping, but complementary, roles of the two pathways could explain the compensatory (originally interpreted as competitive) and cooperative functions of NHEJ and HR in DSB repair (Couedelle *et al.*, 2004; Delacote *et al.*, 2002; Fukushima *et al.*, 2001; Mills *et al.*, 2004; Pierce *et al.*, 2001; Takata *et al.*, 1998).

2. Recruitment of HR Factors in G1 Phase

Although IRIF formation of Rad51 is restricted to S/G2 phase (Lisby *et al.*, 2004), our study revealed that Rad51 accumulates at the laser-induced damage sites in G1 phase with kinetics similar to that in S/G2 phase (Fig. 11A). This was surprising since the expression level of Rad51 protein is low in G1 phase and HR occurs in S/G2 phase using sister chromatid templates. A similar observation was made with RPA (Kim *et al.*, 2005). The results indicate that Rad51 and RPA are capable of detecting processed ssDNA ends at DSB sites throughout interphase in human cells. Interestingly, a similar G1 phase recruitment of RPA, but not Rad51, was observed in yeast following ionizing radiation (Lisby *et al.*, 2004). The presence of dense DSBs induced by the 532-nm laser may explain the observed Rad51 signal in G1 phase, which was not detected as IRIF previously (Jakob *et al.*, 2002; Lisby *et al.*, 2004; Tashiro *et al.*, 2000). Curiously, several studies using lasers suggested that IRIF formation involves more than the initial damage site factor recruitment. For example, although Nbs1 is required for IRIF formation by Mre11-Rad50 (Carney *et al.*, 1998), it is not required for recruitment of Mre11-Rad50 to laser-induced

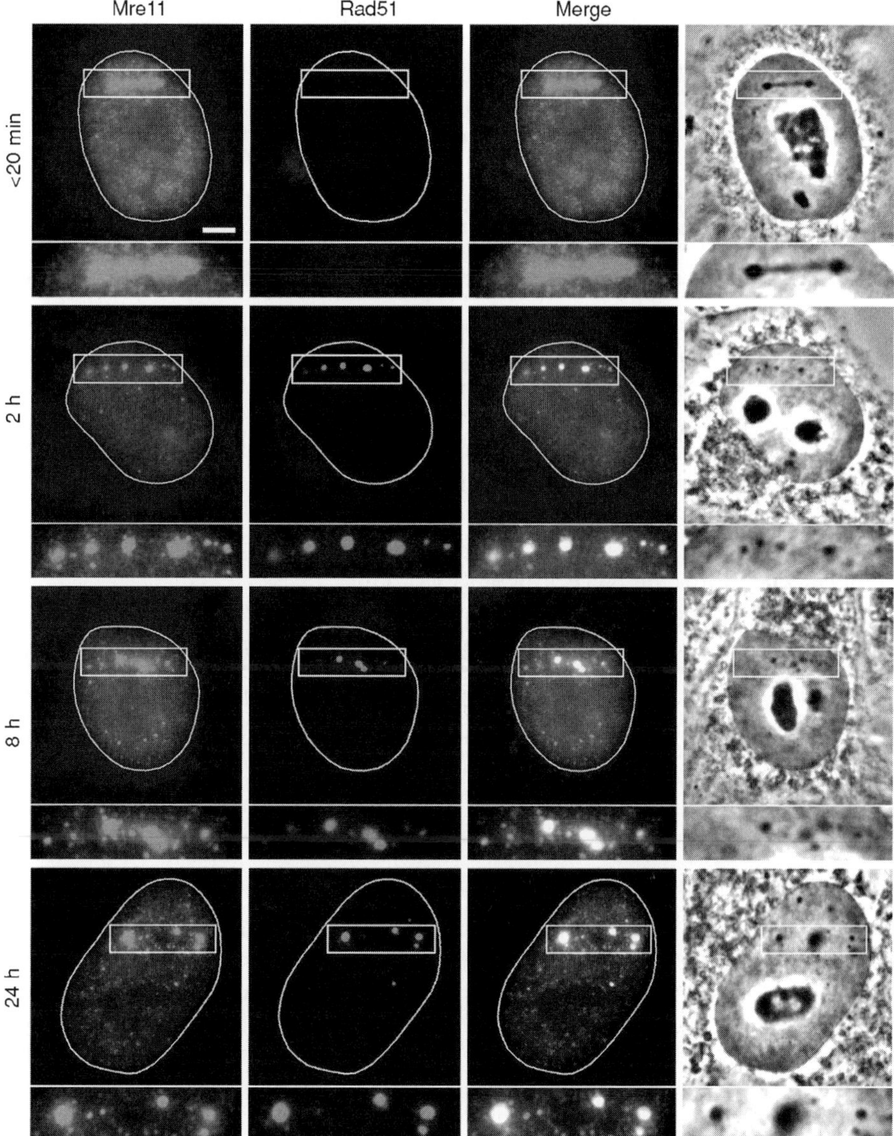

Fig. 9 A time-course analysis of Mre11 and Rad51 clustering at damage sites. S/G2 cells were fixed at the indicated time points p.d. Light microscope images of damaged cells are also shown. Damage sites typically appear as a dark line under the light microscope due to a change in the refractive index (Berns *et al.*, 1998). Damage sites under high magnification are shown below each panel. Reproduced from *The Journal of Cell Biology, 2005, 170: 341–347.* Copyright 2005 The Rockefeller University Press. (See Plate 12 in the color insert section.)

damage sites (Kim *et al.*, 2002a). H2AX is required for Nbs1 focus formation (Celeste *et al.*, 2002), but is dispensable for the recruitment of Nbs1 to UV/BrdU-induced damage sites (Celeste *et al.*, 2003). Thus, while S/G2-specific Rad51 IRIF may reflect

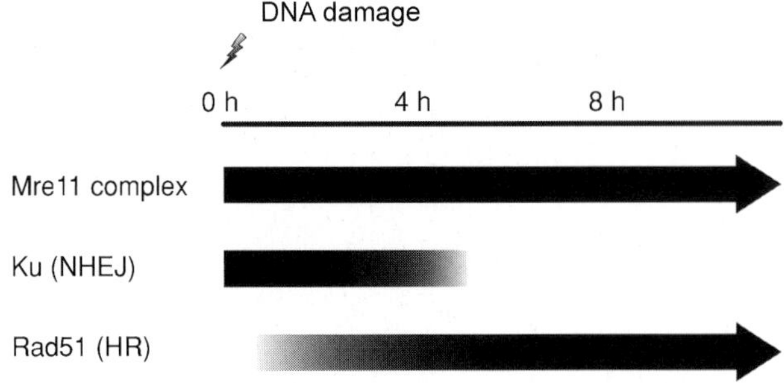

Fig. 10 A summary of damage site recruitment of the Mre11 complex, NHEJ, and HR factors. Reproduced from *The Journal of Cell Biology, 2005, 170: 341–347.* Copyright 2005 The Rockefeller University Press.

ongoing HR, Rad51 and RPA may nonetheless associate with ssDNA at damage sites without initiating the actual strand invasion process until the homologous templates are available.

3. S/G2 Phase-Specific Recruitment of Cohesin

On laser damage induction in synchronized HeLa cells, recruitment of the Mre11 complex was observed in both G1 and S/G2 phases (Fig. 8A). However, cohesin recruitment to damage is limited to S and G2 phases, and not G1 phase (Fig. 11B). Thus, cohesin requires a S/G2-specific factor(s) and/or modification(s) for damage site recognition. This is consistent with the notion that cohesin establishes local sister chromatid template pairing to facilitate postreplicative HR repair (Sjögren and Nasmyth, 2001).

D. Analysis of Factor Requirements Using Mutant Cells

Using cytological analysis of damage site targeting as an assay, it is possible to determine what factor(s) is required for damage site recognition of a given protein of interest. This has been done effectively using UV/BrdU systems (Bekker-Jensen *et al.*, 2005; Bradshaw *et al.*, 2005; Lan *et al.*, 2005). Utilizing full-length and deleted versions of a tagged protein of interest for both cytological and biochemical interaction analyses, it is possible to understand the molecular mechanism of damage site recognition by determining exactly what factor interacts with which part of the protein to mediate damage site targeting.

1. Methods and Cell Lines

Two approaches can be used to determine factor requirement(s): use of preexisting mutant cells or induced protein depletion by siRNA. There are well-characterized primary cells or established cell lines derived from patients, in which certain DNA repair genes are mutated. We have used several of these cell lines for our own work. Wild-type human fibroblast IMR-90 cells and isogenic

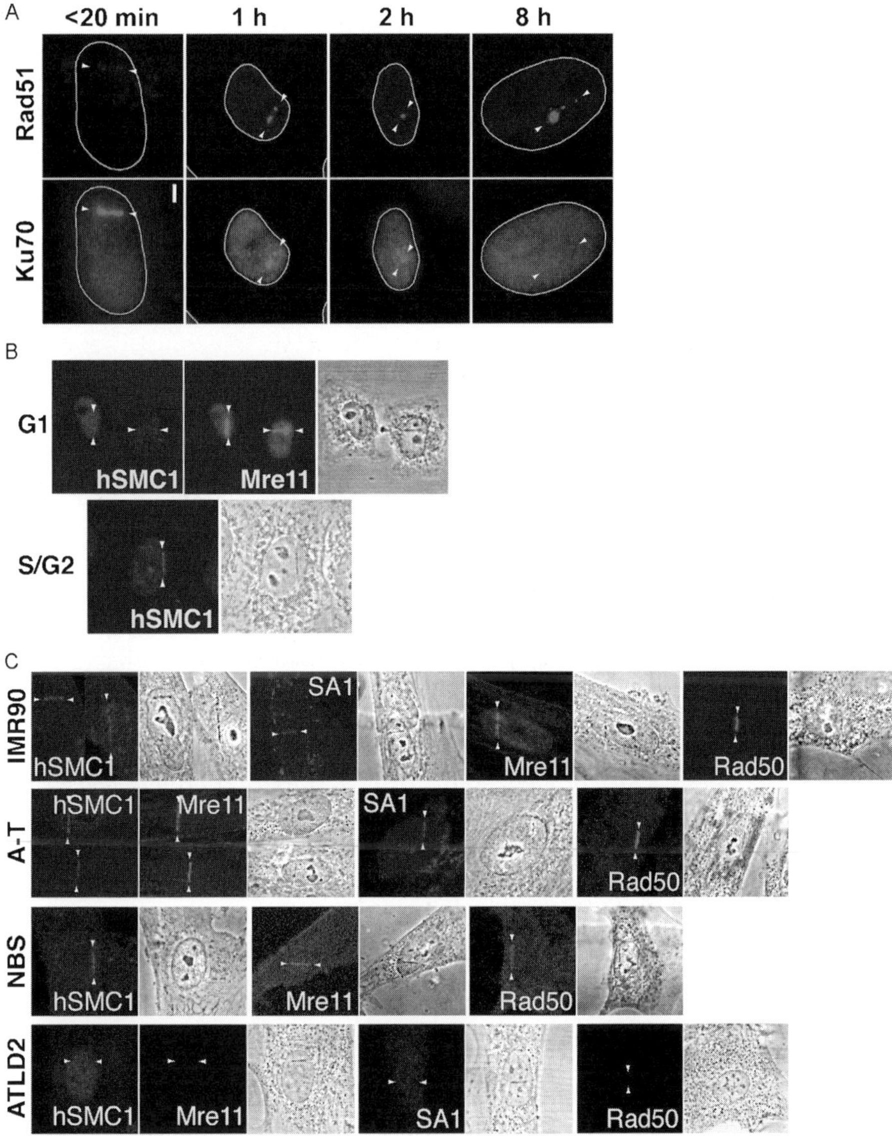

Fig. 11 Cell cycle specificity of factor recruitment to the damage site. (A) Damage site clustering of Ku and Rad51 in G1 phase. G1 HeLa cells were damaged and fixed at indicated time points p.d. (B) Cell cycle-specific recruitment of cohesin to the laser-induced damage sites. G1, S, and G2 cells were laser-damaged and localization of cohesin was determined by staining with anti-hSMC1 antibody. An hSMC1 signal was detectable in S and G2 phase cells, but not in G1. (C) Recruitment of cohesin, Mre11, and Rad50 in fibroblasts. The antibodies used for staining are indicated. Cell lines used are IMR90, A-T, NBS, and ATLD2 as indicated. Fig. 11A reproduced from *The Journal of Cell Biology, 2005, 170: 341–347.* Copyright 2005 The Rockefeller University Press. (See Plate 13 in the color insert section.)

human glioma cell lines either lacking (M059J) or expressing (M059K) DNA-PKcs were obtained from American Type Culture Collection (Manassas, VA). Mre11 mutant fibroblast ATLD cells were derived from ATLD patients (Stewart *et al.*, 1999). The human A-T fibroblast cell line (GM02052D) was obtained from the Coriell Cell Repositories (Camden, NJ). In addition, various knockout and mutant mouse embryonic fibroblasts (MEFs) can be used. As an alternative to mutant cells, proteins can be depleted by transfecting siRNA or a short hairpin RNA (shRNA)-encoding plasmid. To identify transfected cells for subsequent laser irradiation, the transfected siRNA can be conjugated with fluorescent dye or the shRNA-encoding vector can coexpress GFP.

2. Mre11/Rad50-Dependent Recruitment of Cohesin to Laser-Induced DNA Damage

The requirement for cohesin targeting to the damage site was tested in mutant fibroblast cells derived from A-T, NBS, and ATLD patients (Fig. 11C) (Kim *et al.*, 2002a). Although hSMC1 phosphorylation by ATM/Nbs1 was shown to be important for the S phase checkpoint (Kim *et al.*, 2002b; Yazdi *et al.*, 2002), our results showed that SMC1 phosphorylation is not required for the initial recruitment of cohesin to the site of damage. Conversely, a study using a UV/BrdU system showed that hSMC1 phosphorylation occurs in the residual cohesin at the damaged area even in G1 phase (Bekker-Jensen *et al.*, 2006). Thus, *de novo* recruitment of cohesin is not required for hSMC1 phosphorylation by ATM, indicating that cohesin recruitment to the damage site and hSMC1 phosphorylation are two independent events. As described above, since the 532-nm laser damages both DNA and proteins in the focused area, any protein detected at the damage site must be the result of *de novo* recruitment. In contrast, cohesin failed to cluster at the damage site in Mre11-mutated ATLD2 cells (Fig. 11C). In these cells, the level of Rad50 is also downregulated, likely due to a secondary effect of the Mre11 mutation (Stewart *et al.*, 1999). Cohesin proteins are present in these cells at wild-type levels. These results indicate that either Mre11 or Rad50 (or both) plays a crucial role in recruitment of cohesin to the sites of DNA damage. Interestingly, specific recruitment of cohesin to the damage site and its dependence on the Mre11 complex was recently confirmed in *Saccharomyces cerevisiae* using ChIP analysis of cohesin binding at a HO endonuclease-induced DSB site (Ström *et al.*, 2004; Ünal *et al.*, 2004). These data provide further strong support that the 532-nm Nd:YAG laser system produces physiologically relevant results.

V. Special Considerations

Although it has not been formally determined how many DSBs are created within the irradiated area by the 532-nm Nd:YAG laser system, the results described above indicate that the DNA must be highly fragmented, creating numerous DSBs within the small nuclear region where the laser beam is focused. The supporting data include (1) strong BrdU incorporation at the damaged region by TdT, which labels DNA ends, (2) clear recruitment of NHEJ factors in both G1 and S/G2 phases and cohesin in S/G2 phase, which do not form large protein clusters (i.e., IRIF) in response to ionizing radiation, and (3) recruitment of HR factors Rad51 and RPA to the damage site in G1 phase, in which they do not usually form IRIF. Furthermore, detected protein signals at the damage site must reflect newly recruited proteins since 532-nm laser also damages proteins in the focused area. These results indicate that it is possible to observe the *de novo* recruitment of non-IRIF-forming factors to the damage site using this system.

Interestingly, the damage site recruitment of non-IRIF-forming factors such as those described above were not well reproduced in the UV/BrdU system compared to IRIF-forming factors, suggesting that the UV/BrdU system under the conditions used cannot induce damage in as dense a manner as the 532-nm Nd:YAG laser (Bekker-Jensen *et al.*, 2006). This is not surprising since the efficiency of incorporation of the sensitizing agent (e.g., BrdU) limits the number of breaks within the irradiated area.

In order to observe the recruitment of the non-IRIF-forming factors, a higher dose of UV without sensitizing agent was required, which apparently also caused loss of DAPI staining (DNA) and impairment of 53BP1 recruitment, suggesting that the damage response is no longer physiological (Bekker-Jensen *et al.*, 2006). Thus, some concern was raised about whether damage induced by the 532-nm laser is physiologically relevant, and whether the observed clustering of the non-IRIF-forming factors is due to some nuclear structural damage and abnormal cellular responses. Since no sensitizing agent was required for this type of UV damage, the mechanism of DNA damage is unclear and is difficult to compare with damage induced by a 532-nm laser. However, high-dose UV may be toxic to the cell since UV-A is known to be absorbed by cellular membranes causing oxidative stress and cell damage. This is in contrast to the 532-nm laser, which is not absorbed by the cell, and thus the damage is restricted to the focused area in the nucleus. Consistent with this, the abnormalities observed with the high-dose UV are not seen with the 532-nm Nd:YAG laser: there was no visible change in DAPI staining after 532-nm laser irradiation (Fig. 12A), and clear 53BP1 clustering was observed following damage induction (Fig. 12B). Importantly, the damage site recruitment of factors not exhibiting IRIF formation was confirmed in yeast using different methods, indicating that it is not unique to the 532-nm laser-induced damage. Ku and cohesin recruitment was demonstrated by ChIP analysis at a cleaved HO endonuclease site (Shim *et al.*, 2005; Ström *et al.*, 2004; Ünal *et al.*, 2004). RPA recruitment to the damage site in G1 phase was also demonstrated in response to ionizing radiation using a GFP fusion in yeast (Lisby *et al.*, 2004).

Estimated numbers of DSBs created by the UV laser with BrdU (or other sensitizing agents) per nucleus seem to differ between laboratories, which may be due to different energy levels and/or interpretational variability. In some studies, damage was estimated to be equivalent to that induced by 80-Gy gamma irradiation, or ~2800 DSBs per nucleus (Bradshaw *et al.*, 2005; Paull *et al.*, 2000). Another group

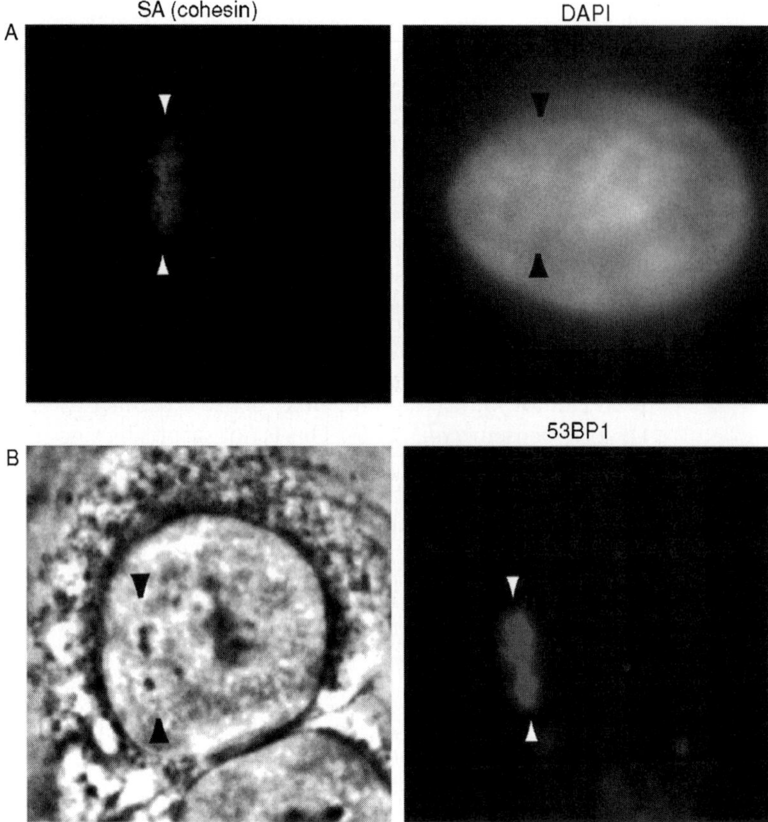

Fig. 12 DAPI and 53BP1 staining of a cell damaged by the Nd:YAG laser. In contrast to the UV/BrdU system (Bekker-Jensen *et al.*, 2006), the Nd:YAG laser, which allows detection of NHEJ factor and cohesin recruitment, does not cause any visible change in DAPI staining (A) and induces 53BP1 clustering (B) similar to conventional DSB damage.

concluded that the induced damage was equivalent to 3- to 10-Gy (Bekker-Jensen *et al.*, 2006). Since these estimates came from several different laboratories using different laser systems, sensitizing agents, UV wavelengths, and energy levels, it would be informative to assess whether all of these systems induce damage and a damage response in a similar manner. Nonetheless, IRIF formation appears to be prerequisite for effective detection of factors at the UV/BrdU-induced damage site, according to the study by Jiri Lukas's laboratory (Bekker-Jensen *et al.*, 2006), while the 532-nm Nd:YAG laser system can be used to observe the recruitment of factors regardless of their IRIF-forming capability (Kim *et al.*, 2002a, 2005).

Taken together, the 532-nm laser appears to serve as a unique system to study the recruitment of DSB factors regardless of their IRIF-forming capability by artificially creating high-density, but nonlethal, DNA breaks. However, it is important to further characterize the nature of the 532-nm laser-induced damage and better evaluate the number of DSBs. Another critical goal is to confirm the results obtained with the 532-nm laser using alternative methodologies.

VI. Conclusion

We confirmed the presence of DNA breaks at the 532-nm Nd:YAG laser-induced damage sites, and observed specific recruitment of multiple factors known to be involved in the DNA DSB response with specific kinetics. In addition, we demonstrated that damaged cells undergo transient checkpoint-dependent cell cycle arrest but eventually go through cell division, indicating that this type of damage causes neither irreversible destruction of nuclear structure nor cell death. Furthermore, the 532-nm laser appears to be particularly suitable to study the recruitment of NHEJ factors and cohesin without having deleterious effects on the cell compared to the UV/BrdU system, which appears to induce DSBs at a lower density. Therefore, the 532-nm Nd:YAG laser microirradiation system is a useful tool to dissect the process of DNA damage recognition and repair *in vivo*.

There are now multiple approaches available to study DNA damage response in the cell, such as conventional DNA-damaging methods (e.g., ionizing radiation), endonuclease-ChIP, and different microirradiation systems, including the 532-nm Nd:YAG laser. Each system has its own advantages and disadvantages, which should be properly recognized. For the 532-nm laser, one must be cautious and carefully interpret the physiological significance of the massive, though highly confined, damage in the nucleus. Parallel studies using these different systems should facilitate significant advances in our understanding of the cellular DNA damage response.

Acknowledgments

This work was supported in part by the Laser Microbeam and Medical Program (LAMMP) (NIH RR01192) and grants from the NIH CA100710 and the DOD Breast Cancer Research Program (DAMD17-03-1-0436) to K.Y.

References

Abraham, R. T. (2004). PI 3-kinase related kinases: "Big" players in stress-induced signaling pathways. *DNA Repair* **3,** 883–887.

Aten, J. A., Stap, J., Krawczyk, P. M., van Oven, C. H., Hoebe, R. A., Essers, J., and Kanaar, R. (2004). Dynamics of DNA double-strand breaks revealed by clustering of damaged chromosome domains. *Science* **303,** 92–95.

Baumann, P., Benson, F. E., and West, S. C. (1996). Human Rad51 protein promotes ATP-dependent homologous pairing and strand transfer reactions *in vitro*. *Cell* **87,** 757–766.

Bekker-Jensen, S., Lukas, C., Kitagawa, R., Melander, F., Kastan, M. B., Bartek, J., and Lukas, J. (2006). Spatial organization of the mammalian genome surveillance machinery in response to DNA strand breaks. *J. Cell Biol.* **173,** 195–206.

Bekker-Jensen, S., Lukas, C., Melander, F., Bartek, J., and Lukas, J. (2005). Dynamic assembly and sustained retention of 53BP1 at the sites of DNA damage are controlled by Mdc1/NFBD1. *J. Cell Biol.* **170,** 201–211.

Berns, M. W., Aist, J., Edwards, J., Strahs, K., Girton, J., McNeill, P., Rattner, J. B., Kitzes, M., Hammer-Wilson, M., Liaw, L.-H., Siemens, A., Koonce, M., *et al.* (1981). Laser microsurgery in cell and developmental biology. *Science* **213,** 505–513.

Berns, M. W., Tadir, Y., Liang, H., and Tromberg, B. (1998). Laser scissors and tweezers. *Methods Cell Biol.* **55,** 71–98.

Blasina, A., Price, B. D., Turenne, G. A., and McGowan, C. H. (1999). Caffeine inhibits the checkpoint kinase ATM. *Curr. Biol.* **9,** 1135–1138.

Bradshaw, P. S., Stavropoulos, D. J., and Meyn, M. S. (2005). Human telomeric protein TRF2 associates with genomic double-strand breaks as an early response to DNA damage. *Nat. Genet.* **37,** 193–197.

Carney, J. P., Maser, R. S., Olivares, H., Davis, E. M., Le Beau, M., Yates, J. R., III, Hays, L., Morgan, W. F., and Petrini, J. H. (1998). The hMre11/hRad50 protein complex and Nijmegen breakage syndrome: Linkage of double-strand break repair to the cellular DNA damage response. *Cell* **93,** 477–486.

Carson, C. T., Schwartz, R. A., Stracker, T. H., Lilley, C. E., Lee, D. V., and Weitzman, M. D. (2003). The Mre11 complex is required for ATM activation and the G2/M checkpoint. *EMBO J.* **22,** 6610–6620.

Celeste, A., Fernandez-Capetillo, O., Kruhlak, M. J., Pilch, D. R., Staudt, D. W., Lee, A., Bonner, R. F., Bonner, W. M., and Nussenzweig, A. (2003). Histone H2AX phosphorylation is dispensable for the initial recognition of DNA breaks. *Nat. Cell Biol.* **5,** 675–679.

Celeste, A., Petersen, S., Romanienko, P. J., Fernandez-Capetillo, O., Chen, H. T., Sedelnikova, O. A., Reina-San-Martin, B., Coppola, V., Meffre, E., Dilippantonio, M. J., Redon, C., Pilch, D. R., *et al.* (2002). Genomic instability in mice lacking histone H2AX. *Science* **296,** 922–927.

Chan, K. F., Pfefer, T. J., Teichman, J. M. H., and Welch, A. J. (2001). A perspective on laser lithotripsy: The fragmentation processes. *J. Endourol.* **15,** 257–273.

Costanzo, V., Paull, T., Gottesman, M., and Gautier, J. (2004). Mre11 assembles linear DNA fragments into DNA damage signaling complexes. *PLoS Biol.* **2,** E110.

Couedelle, C., Mills, K. D., Barchi, M., Shen, L., Olshen, A., Johnson, R. D., Nussenzweig, A., Essers, J., Kanaar, R., Li, G. C., Alt, F. W., and Jasin, M. (2004). Collaboration of homologous recombination and nonhomologous end-joining factors for the survival and integrity of mice and cells. *Genes Dev.* **18,** 1293–1304.

Dasika, G. K., Lin, S.-C. J., Zhao, S., Sung, P., Tomkinson, A., and Lee, E. Y.-H. P. (1999). DNA-damage-induced cell cycle checkpoints and DNA strand break repair in development and tumorigenesis. *Oncogene* **18,** 7883–7899.

Delacote, F., Han, M., Stamato, T. D., Jasin, M., and Lopez, B. S. (2002). An XRCC4 defect or Wortmannin stimulates homologous recombination specifically induced by double-strand breaks in mammalian cells. *Nucleic Acids Res.* **30,** 3454–3463.

Di Virgilio, M., and Gautier, J. (2005). Repair of double-strand breaks by nonhomologous end joining in the absence of Mre11. *J. Cell Biol.* **171,** 765–771.

Fernandez-Capetillo, O., Chen, H.-T., Celeste, A., Ward, I., Romanienko, P. J., Morales, J. C., Naka, K., Xia, Z., Camerini-Otero, R. D., Motoyama, N., Carpenter, P. B., Bonner, W. M., *et al.* (2002). DNA damage-induced G2-M checkpoint activation by histone H2AX and 53BP1. *Nat. Cell Biol.* **4,** 993–997.

Frank-Vaillant, M., and Marcand, S. (2002). Transient stability of DNA ends allows nonhomologous end joining to precede homologous recombination. *Mol. Cell.* **10,** 1189–1199.

Fukushima, T., Takata, M., Morrison, C., Araki, R., Fujimori, A., Abe, M., Tatsumi, K., Jasin, M., Dhar, P. K., Sonoda, E., Chiba, T., and Takeda, S. (2001). Genetic analysis of the DNA-dependent protein kinase reveals an inhibitory role of Ku in late S–G2 phase DNA double-strand break repair. *J. Biol. Chem.* **276,** 44413–44418.

Gregson, H. C., Schmiesing, J. A., Kim, J.-S., Kobayashi, T., Zhou, S., and Yokomori, K. (2001). A potential role for human cohesin in mitotic spindle aster assembly. *J. Biol. Chem.* **276,** 47575–47582.

Heale, J. T., Ball, A. R., Schmiesing, J. A., Kim, J. S., Kong, X., Zhou, S., Hudson, D., Earnshaw, W. C., and Yokomori, K. (2006). Condensin I interacts with the PARP-1-XRCC1 complex and functions in DNA single-stranded break repair. *Mol. Cell* **21,** 837–848.

Horejsi, Z., Falck, J., Bakkenist, C. J., Kastan, M. B., Lukas, J., and Bartek, J. (2004). Distinct functional domains of Nbs1 modulate the timing and magnitude of ATM activation after low doses of ionizing radiation. *Oncogene* **23,** 3122–3127.

Jakob, B., Scholz, M., and Taucher-Scholz, G. (2002). Characterization of CDKN1A (p21) binding to sites of heavy-ioninduced damage: Colocalization with proteins involved in DNA repair. *Int. J. Radiat. Biol.* **78,** 75–88.

Kim, J.-S., Krasieva, T. B., Kurumizaka, H., Chen, D. J., Taylor, A. M., and Yokomori, K. (2005). Independent and sequential recruitment of NHEJ and HR factors to DNA damage sites in mammalian cells. *J. Cell Biol.* **170,** 341–347.

Kim, J.-S., Krasieva, T. B., LaMorte, V. J., Taylor, A. M. R., and Yokomori, K. (2002a). Specific recruitment of human cohesin to laser-induced DNA damage. *J. Biol. Chem.* **277,** 45149–45153.

Kim, S.-T., Xu, B., and Kastan, M. B. (2002b). Involvement of the cohesin protein, Smc1, in Atm-dependent and independent responses to DNA damage. *Genes Dev.* **16,** 560–570.

Kondo, T., Wakayama, T., Naiki, T., Matsumoto, K., and Sugimoto, K. (2001). Recruitment of Mec1 and Ddc1 checkpoint proteins to double-strand breaks through distinct mechanisms. *Science* **294,** 867–870.

Krasieva, T. B., Chapman, C. F., LaMorte, V. J., Venugopalan, V., and Tromberg, B. J. (1998). Mechanisms of cell permeabilization by laser microirradiation. *Proc. Soc. Photo-opt. Instrum. Eng.* **3260,** 38–44.

Lan, L., Nakajima, S., Komatsu, K., Nussenzweig, A., Shimamoto, A., Oshima, J., and Yasui, A. (2005). Accumulation of Werner protein at DNA double-strand breaks in human cells. *J. Cell Sci.* **118,** 4153–4162.

Lan, L., Nakajima, S., Oohata, Y., Takao, M., Okano, S., Masutani, M., Wilson, S. H., and Yasui, A. (2004). *In situ* analysis of repair processes for oxidative DNA damage in mammalian cells. *Proc. Natl. Acad. Sci. USA* **101,** 13738–13743.

Lee, S. E., Moore, J. K., Holmes, A., Umezu, K., Kolodner, R. D., and Haber, J. E. (1998). Saccharomyces Ku70, Mre11/Rad50, and RPA proteins regulate adaptation to G2/M arrest after DNA damage. *Cell* **94,** 399–409.

Lee, J. H., and Paull, T. T. (2004). Direct activation of the ATM protein kinase by the Mre11/Rad50/Nbs1 complex. *Science* **304,** 93–96.

Lee, J. H., and Paull, T. T. (2005). ATM activation by DNA double-strand breaks through the Mre11-Rad50-Nbs1 complex. *Science* **308,** 551–554.

Lehmann, A. R. (2003). DNA repair-deficient diseases, xeroderma pigmentosum, Cockayne syndrome and trichothiodystrophy. *Biochemie* **85,** 1101–1111.

Liang, H., Wright, W. H., Rieder, C. L., Salmon, E. D., Profeta, G., Andrews, J., Liu, Y., Sonek, G. J., and Berns, M. W. (1994). Directed movement of chromosome arms and fragments in mitotic newt lung cells using optical scissors and optical tweezers. *Exp. Cell Res.* **213,** 308–312.

Lieber, M. R., Ma, Y., Pannicke, U., and Schwarz, K. (2003). Mechanism and regulation of human non-homologous DNA end-joining. *Nat. Rev. Mol. Cell Biol.* **4,** 712–720.

Limoli, C. L., Giedzinski, E., Bonner, W. M., and Cleaver, J. E. (2002). UV-induced replication arrest in the xeroderma pigmentosum variant leads to DNA double-strand breaks, γ-H2AX formation, and Mre11 relocalization. *Proc. Natl. Acad. Sci. USA* **99,** 233–238.

Limoli, C. L., and Ward, J. F. (1993). A new method for introducing double-strand breaks into cellular DNA. *Radiat. Res.* **134,** 160–169.

Lisby, M., Barlow, J. H., Burgess, R. C., and Rothstein, R. (2004). Choreography of the DNA damage response: Spatiotemporal relationships among checkpoint and repair proteins. *Cell* **118,** 699–713.

Lukas, C., Bartek, J., and Lukas, J. (2005). Imaging of protein movement induced by chromosomal breakage: Tiny "local" lesions pose great "global" challenges. *Chromosoma* **114,** 146–154.

Lukas, C., Falck, J., Bartkova, J., Bartek, J., and Lukas, J. (2003). Distinct spatiotemporal dynamics of mammalian checkpoint regulators induced by DNA damage. *Nat. Cell Biol.* **5,** 255–260.

Maser, R. S., Monsen, K. J., Nelms, B. E., and Petrini, J. H. (1997). hMre11 and hRad50 nuclear foci are induced during the normal cellular response to DNA double-strand breaks. *Mol. Cell. Biol.* **17,** 6087–6096.

Mikhailov, A., Cole, R. W., and Rieder, C. L. (2002). DNA damage during mitosis in human cells delays the metaphase/anaphase transition via the spindle-assembly checkpoint. *Curr. Biol.* **12,** 1797–1806.

Mills, K. D., Ferguson, D. O., Essers, J., Eckersdorff, M., Kanaar, R., and Alt, F. W. (2004). Rad54 and DNA ligase IV cooperate to maintain mammalian chromatid stability. *Genes Dev.* **18,** 1283–1292.

Mirzoeva, O. K., and Petrini, J. H. J. (2001). DNA damage-dependent nuclear dynamics of the Mre11 complex. *Mol. Cell. Biol.* **21,** 281–288.

Murr, R., Loizou, J. I., Yang, Y. G., Cuenin, C., Li, H., Wang, Z. Q., and Herceg, Z. (2006). Histone acetylation by Trrap-Tip60 modulates loading of repair proteins and repair of DNA double-strand breaks. *Nat. Cell Biol.* **8,** 91–99.

Nelms, B. E., Maser, R. S., MacKay, J. F., Lagally, M. G., and Petrini, J. H. (1998). *In situ* visualization of DNA double-strand break repair in human fibroblasts. *Science* **280,** 590–592.

Niida, H., and Nakanishi, M. (2006). DNA damage checkpoints in mammals. *Mutagenesis* **21,** 3–9.

Paull, T. T., and Gellert, M. (1998). The 3′ to 5′ exonuclease activity of Mre11 facilitates repair of DNA double-strand breaks. *Mol. Cell* **1,** 969–979.

Paull, T. T., Rogakou, E. P., Yamazaki, V., Kirchgessner, C. U., Gellert, M., and Bonner, W. M. (2000). A critical role for histone H2AX in recruitment of repair factors to nuclear foci after DNA damage. *Curr. Biol.* **10,** 886–895.

Petersen, S., Casellas, R., Reina-San-Martin, B., Chen, H. T., Difilippantonio, M. J., Wilson, P. C., Hanitsch, L., Celeste, A., Muramatsu, M., Pilch, D. R., Redon, C., Ried, T., *et al.* (2001). AID is required to initiate Nbs1/g-H2AX focus formation and mutations at sites of class switching. *Nature* **414,** 660–665.

Pierce, A. J., Hu, P., Han, M., Ellis, N., and Jasin, M. (2001). Ku DNA end-binding protein modulates homologous repair of double-strand breaks in mammalian cells. *Genes Dev.* **15,** 3237–3242.

Rogakou, E. P., Boon, C., Redon, C., and Bonner, W. M. (1999). Megabase chromatin domains involved in DNA double-strand breaks *in vivo*. *J. Cell Biol.* **146,** 905–915.

Rogakou, E. P., Pilch, D. R., Orr, A. H., Ivanova, V. S., and Bonner, W. M. (1998). DNA double-stranded breaks induce histone H2AX phosphorylation on serine 139. *J. Biol. Chem.* **273,** 5858–5868.

Rothkamm, K., Kruger, I., Thompson, L. H., and Lobrich, M. (2003). Pathways of DNA double-strand break repair during the mammalian cell cycle. *Mol. Cell. Biol.* **23,** 5706–5715.

Shiloh, Y. (2001). ATM and ATR: Networking cellular responses to DNA damage. *Curr. Opin. Genet. Dev.* **11,** 71–77.

Shiloh, Y. (2003). ATM and related protein kinases: Safeguarding genome integrity. *Nat. Rev. Cancer* **3,** 155–168.

Shim, E. Y., Ma, J. L., Oum, J. H., Yanez, Y., and Lee, S. E. (2005). The yeast chromatin remodeler RSC complex facilitates end joining repair of DNA double-strand breaks. *Mol. Cell. Biol.* **25,** 3934–3944.

Shroff, R., Arbel-Eden, A., Pilch, D., Ira, G., Bonner, W. M., Petrini, J. H., Haber, J. E., and Lichten, M. (2004). Distribution and dynamics of chromatin modification induced by a defined DNA double-strand break. *Curr. Biol.* **14,** 1703–1711.

Sjögren, C., and Nasmyth, K. (2001). Sister chromatid cohesion is required for postreplicative double-strand break repair in *Saccharomyces cerevisiae*. *Curr. Biol.* **11,** 991–995.

Sonoda, E., Sasaki, M. S., Morrison, C., Yamaguchi-Iwai, Y., Takata, M., and Takeda, S. (1999). Sister chromatid exchanges are mediated by homologous recombination in vertebrate cells. *Mol. Cell. Biol.* **19,** 5166–5169.

Stewart, G. S., Maser, R. S., Stankovic, T., Bressan, D. A., Kaplan, M. I., Jaspers, N. G. J., Raams, A., Byrd, P. J., Petrini, J. H., and Taylor, A. M. R. (1999). The DNA double-strand break repair gene hMRE11 is mutated in individuals with an ataxia-telangiectasia-like disorder. *Cell* **99,** 577–587.

Ström, L., Lindroos, H. B., Shirahige, K., and Sjögren, C. (2004). Postreplicative recruitment of cohesin to double-strand breaks is required for DNA repair. *Mol. Cell* **16,** 1003–1015.

Sugawara, N., and Haber, J. E. (2006). Repair of DNA double-strand breaks: *In vivo* biochemistry. *Methods Enzymol.* **408,** 416–429.

Sung, P. (1994). Catalysis of ATP-dependent homologous DNA pairing and strand exchange by yeast RAD51 protein. *Science* **265,** 1241–1243.

Takata, M., Sasaki, M. S., Sonoda, E., Morrison, C., Hashimoto, M., Utsumi, H., Yamaguchi-Iwai, Y., Shinohara, A., and Takeda, S. (1998). Homologous recombination and non-homologous end-joining pathways of DNA double-strand break repair have overlapping roles in the maintenance of chromosomal integrity in vertebrate cells. *EMBO J.* **17,** 5497–5508.

Tashiro, S., Walter, J., Shinohara, A., Kamada, N., and Cremer, T. (2000). Rad51 accumulation at sites of DNA damage and in postreplicative chromatin. *J. Cell Biol.* **150,** 283–291.

Tauchi, H., Kobayashi, J., Morishima, K., van Gent, D. C., Shiraishi, T., Verkaik, N. S., vanHeems, D., Ito, E., Nakamura, A., Sonoda, E., Takata, M., Takeda, S., *et al.* (2002a). Nbs1 is essential for DNA repair by homologous recombination in higher vertebrate cells. *Nature* **420,** 93–98.

Tauchi, H., Matsuura, S., Kobayashi, J., Sakamoto, S., and Komatsu, K. (2002b). Nijmegen breakage syndrome gene, NBS1, and molecular links to factors for genome stability. *Oncogene* **21,** 8967–8980.

Taylor, A. M. (2001). Chromosome instability syndromes. *Best Pract. Res. Clin. Haematol.* **14,** 631–644.

Ünal, E., Arbel-Eden, A., Sattler, U., Shroff, R., Lichten, M., Haber, J. E., and Koshland, D. (2004). DNA damage response pathway uses histone modification to assemble a double-strand break-specific cohesin domain. *Mol. Cell* **16,** 991–1002.

Uziel, T., Lerenthal, Y., Moyal, L., Andegeko, Y., Mittelman, L., and Shiloh, Y. (2003). Requirement of the MRN complex for ATM activation by DNA damage. *EMBO J.* **22,** 5612–5621.

Varon, R., Vissinga, C., Platzer, M., Cerosaletti, K. M., Chrzanowska, K. H., Saar, K., Beckmann, G., Seemanova, E., Cooper, P. R., Nowak, N. J., Stumm, M., Weemaes, C. M. R., *et al.* (1998). Nibrin, a novel DNA double-strand break repair stem cell lethality, abnormal embryonic development, and sensitivity protein, is mutated in Nijmegen breakage syndrome. *Cell* **93,** 467–476.

Venugopalan, V., Guerra, A., III, Nahen, K., and Vogel, A. (2002). Role of laser-induced plasma formation in pulsed cellular microsurgery and micromanipulation. *Phys. Rev. Lett.* **88,** 078103.

West, S. C. (2003). Molecular views of recombination proteins and their control. *Nat. Rev. Mol. Cell Biol.* **4,** 435–445.

Yazdi, P. T., Wang, Y., Zhao, S., Patel, N., Lee, E. Y.-H. P., and Qin, J. (2002). SMC1 is a downstream effector in the ATM/NBS1 branch of the human S-phase checkpoint. *Genes Dev.* **16,** 571–582.

Ziegler, M., and Oei, S. L. (2001). A cellular survival switch: Poly(ADP-ribosyl)ation stimulates DNA repair and silences transcription. *Bioessays* **23,** 543–548.

CHAPTER 14

Laser Effects in the Manipulation of Human Eggs and Embryos for *In Vitro* Fertilization

Yona Tadir*,† and Diarmaid H. Douglas-Hamilton‡

*Beckman Laser Institute, University of California, Irvine, California 92612

†Department of Obstetrics and Gynecology
Rabin Medical Center, Tel Aviv University, Israel

‡Hamilton Thorne Biosciences
Beverly, Massachusetts 01915

0091-679X/07 $35.00
DOI: 10.1016/S0091-679X(06)82014-5

Gamete manipulations using laser micro beams were introduced in 1991 and testing its application for assisted hatching occurred shortly thereafter. This procedure has now become an accepted modality of penetrating or reducing the thickness of the zona pellucida in human *in vitro* fertilization (IVF). Lasers used in earlier work are summarized. Although the earliest lasers used pulses as long as 15 ms, the simplest and safest laser presently used in this application is the high-power 1480-nm In GaAsP diode, used in pulses with duration typically <1 ms. Since prevention of damage to the blastomeres is essential, we specifically discuss this system with particular attention to safety considerations. The laser operates by its thermal effect on the zona pellucida, and the implications for embryo safety are discussed in detail. A thermal model is derived using numerical analysis and the effect on the embryo of laser beam power and pulse duration is indicated. Typical recommended protocols and operating values for various applications in the human IVF laboratory are given.

I. Introduction

Assisted hatching (AH), in which the zona pellucida (ZP) is artificially breached or thinned, has been proposed by Cohen *et al.* (1990) as a way of improving the implantation process. Studies using animal models have demonstrated that AH can significantly increase rates of embryo hatching (Khalifa *et al.*, 1992; Malter and Cohen, 1989). AH has also been shown to significantly increase hatching rates of human embryos (Dokras *et al.*, 1994). Data from patients undergoing *in vitro* fertilization (IVF) demonstrated that hCG is detectable in maternal serum earlier when AH is performed than when it is not (Liu *et al.*, 1993), suggesting that human embryos also hatch earlier when assisted.

A variety of micromanipulation techniques, including mechanical partial zona dissection (Cohen *et al.*, 1990), zona drilling using acidic Tyrode's solution (Cohen *et al.*, 1992), piezomicromanipulation (Nakayama *et al.*, 1999), and laser drilling (Neev *et al.*, 1992; Obruca *et al.*, 1994; Tadir *et al.*, 1991, 1992), have been reported to improve ART outcomes among selected groups of patients (Ali *et al.*, 2003).

Two meta-analyses of studies evaluating potential benefits of AH reported significant heterogeneity among study results (Edi-Osagie *et al.*, 2003; Sallam *et al.*, 2003), suggesting that effects of AH may differ depending on patient characteristics. Both concluded that there is strong evidence that AH increases pregnancy rates among patients with a history of previous IVF failures. Other patient populations that have been reported to benefit from AH include patients whose embryos have thick zonae, older patients (Magli *et al.*, 1998; Stein *et al.*, 1995), and patients using cryopreserved embryos (Check *et al.*, 1996; Mandelbaum *et al.*, 1994; Tucker *et al.*, 1991).

Noncontact lasers used included for example wavelengths 248, 308, and 2.1 μm (Blanchet *et al.*, 1992; Neev *et al.*, 1993, 1995), all of which have optical depth in water $\geq$0.4 mm, allowing this mode of operation (Tadir *et al.*, 1993). In most cases, the embryo is supported on a manipulator pipette in a Petri dish on an inverted microscope. Antinori *et al.* (1994) obtained successful pregnancies with a contact ErYAG laser (with absorption coefficient in water $\sim 10^4$ cm^{-1}, the beam penetrates only $\sim$1 μm of medium) necessitating contact delivery of the radiation. UV radiation is potentially mutagenic and even near infrared (IR) can be damaging (Konig *et al.*, 1996), and the inconvenience and potential contamination of contact fiber manipulation led to search for other wavelengths.

Rink *et al.* (1996) used IR light at wavelength $\lambda \sim 1480$ nm from an InGaAsP laser diode, and this wavelength is now preferred, since its high absorption in water causes rapid and precisely localized heating of the target region but is compatible with noncontact laser application. The $\lambda = 1450$–1480 nm laser is the method of choice for AH of embryos due to its noncontaminating, noncontact application and the speed, precision, and ease with which the ZP can be drilled. In the present section, we will concentrate on laser thermal interaction with the embryo for the $\lambda = 1450$–1480 nm region. The wavelength range covers the $v_3 + 2v_2$ (first level asymmetric stretch + second level bending mode) peak absorption band of the H_2O molecule (Herzberg, 1950) and is strongly absorbed in biological media, allowing the generation of intense local heating in the embryo ZP. The optical depth in water is close to 0.5 mm, allowing the IR beam to reach the embryo edge ($\sim$60 μm above the floor of the Petri dish) relatively undiminished, but the beam does not propagate significantly beyond the embryo. The ZP rapidly liquefies in the heated region but not elsewhere, facilitating later embryo hatching as well as access to the blastomeres or blastocyst.

The present discussion will concentrate on the InGaAsP laser diode and its beam properties applied to ZP penetration.

II. Laser Beam

A. Epifluorescent Illumination

The most convenient method of conveying the laser radiation to the target is through the optical objective. The IR laser radiation travels in the opposite direction to the visible light and reaches the target in a manner analogous to fluorescent illumination except that the radiation is concentrated on the target and does not diffusely illuminate it. The laser Gaussian beam will contract down to a focal waist at the focal point on the target, and the diffraction-limited diameter will be close to $4\lambda f/\pi D$. In typical practical cases, the collimated beam diameter incident on the objective is $D \sim 2$ mm, and for a 40×, focal length $f = 5$ mm objective, the diffraction-limited diameter for the $\lambda = 1480$ nm beam would be 4.7 μm at the focal waist. In practice, the measured value of the beam diameter at the e^{-2} intensity is typically 4.5–5 μm. Since the embryo or oocyte ZP is 14- to 18-μm thick, the focused beam is much smaller than the ZP. Its effect on the ZP extends outside the beam focal diameter by heat conduction and the interaction is thermal.

B. Laser Beam Properties

One type of embryo-manipulating laser beam is the integrated laser + objective ZILOS-tk (Hamilton Thorne Biosciences, Beverly, MA), which uses a $\lambda = 1480$ nm beam set to maximum power 300 mW in single pulses with maximum duration of 3 ms. Typical pulse duration is 500 μs. A useful way to make the IR beam visible is to use a fluorescent detector screen (e.g., Applied Scintillation Technologies, Harlow, UK, which gives a green spot where the laser beam hits it, and which is easily visible for a 100-μJ pulse). In practice, the IR laser beam focal waist diameter can be measured by focusing on a 1-μm slit which is translated across the focus. The width of the transmitted intensity curve gives the beam width, normally determined at the e^{-2} level. Its power is measured by integrating the pulse energy on a thermal detector (Ophir Nova, Danvers, MA), for pulse duration confirmed by oscilloscope. Characteristic measured beam waist parameters are of focal diameter 4.5 μm and power 300 mW at the air focus during the pulse. The beam is attenuated by passing through the 60–75 μm of medium between Petri dish floor and embryo edge, reducing its power by ~0.82, so that the power reaching the ZP target is typically $P \sim 246$ mW. At this power pulse durations, 200–600 μm are regarded as safe for an IR beam incident tangentially on the ZP of human embryos: maximum incident energy is less than 0.15 mJ.

C. Laser Effects

The ZP will be treated as having the same absorptance and thermal properties as water, which is a close approximation. Laser radiation of wavelength $\lambda \sim 1480$ nm is strongly absorbed by water, with a room temperature optical depth of ~0.5 mm. During the laser pulse, the medium in and near the beam focus is rapidly heated to temperatures above 100°C (in the absence of a nucleation surface, the water does not boil before the pulse ends). Heat is rapidly lost by conduction from the heated region during and after the laser pulse, convection being too slow and radiation being negligible at the temperatures involved (Douglas-Hamilton and Conia, 2001). The conducted energy increases the temperature in the vicinity of the laser beam, which induces dissolution of the ZP as observed in embryo and oocyte manipulation. The absorption coefficient in H_2O at $\lambda = 1480$ nm is given by Hale and Querry (1973) and Driscoll and Vaughan (1978) as $\alpha_{1480} = 21$ cm^{-1}. This value may be somewhat low: measured with a nominal $\lambda = 1480$ nm InGaAsP laser (Mitsubishi ML96125, Tokyo, Japan), we obtained the transmitted energy absorption coefficient in water as $\alpha_{1480} = 27.6 \pm 0.7$ cm^{-1}. A similar value was reported by Rink *et al.* (1996) in embryo medium. The effect of higher absorptance would be to increase proportionately αP, the beam power deposition density, and the local temperature changes produced will scale with αP. The variation of the absorption coefficient with temperature has not been measured: it may be expected to decrease with temperature and be lower than its value at room temperature. In the present analysis, we will use the more conservative Hale and Querry (1973) absorptance value to calculate peak temperatures. We note that the local H_2O absorptance maximum is given at ~1450 nm, where the absorptance is given as close to 27 cm^{-1}.

The absorptance and refractive index of liquid water at the superheated temperatures produced by the laser pulse have not been determined, and in the following we assume we can use values extrapolated from Hale and Querry (1973) and IAPWS (Dooley, 1977). One can expect, however, that at superheated temperatures a given molecule will be less affected by hydrogen bonds from its H_2O neighbors, and this will increase the relative strength of its molecular bond, increasing vibrational frequency. Absorption IR bands would be shifted to lower wavelengths, and absorption on the long wavelength side of a stretching mode (v_1, v_3) absorption peak would be reduced while that on the high wavelength side would be increased, as discussed by Luck (1965) and observed at the OH stretch frequency by Kazarian and Martirosyan (2002). We can expect a shift to higher frequency for the stretch component of the $v_3 + 2v_2$ asymmetric stretch + bending mode IR absorption band at 1450 nm (Herzberg, 1950). Consequently, the $v_3 + 2v_2$ absorption coefficient at 1480 nm is likely to decrease with temperature. Note that in this case the peak temperature reached will be correspondingly reduced. Reduced absorptance with a form of temperature-induced transparency would result in lower heating at and around the focal point. The temperatures derived for constant absorptance are therefore likely to be upper limits. However, since detailed measurements of absorptance in the $v_3 + 2v_2$ band at superheated temperatures are not available, in order to estimate thermal effects we will ignore the probable decrease of absorption with temperature. We use the measured values of absorptance and obtain the upper limit on laser-induced temperature.

As temperature rises from 20 to above 100°C, the thermal conductivity of liquid water is reported as rising from 6×10^{-3} to a maximum of 6.8×10^{-3} W/cm/°C before falling below 6×10^{-3} W/cm/°C just below the critical point (CRC Handbook, 1984). We will approximate conductivity with the value $K = 6 \times 10^{-3}$. Peak temperatures would be reduced proportionately by higher K, but since the high temperature region is restricted to that very near the beam, there would not be a significant change due to this slight conductivity dependence on temperature.

Since we are interested in predicting safe regions for cells and embryos, the assumptions made here are conservative, and errors will be in the direction of overpredicting temperature.

III. Numerical Analysis

A. Thermal Transfer

The laser beam deposits energy along its path and heats the medium it passes through. In the present case, we have an embryo which is normally at or near the floor of a Petri dish in an inverted microscope, and the beam of IR radiation emerges upward from the floor as a converging cone and is focused on to the ZP, at the rim of the embryo, typically 50–75 μm above the floor. It heats up the medium in the vicinity of the beam, and the heated medium causes thermal dissolution of the ZP. We need to predict the local heating as function of position and time. In order to derive the thermal history during laser pulse turn-on and after turn-off, numerical analysis must be used.

Labeling the medium density as ρ, heat capacity C_p (for water $\rho C_p = 4.18$ J/cm^3/°C), absorption coefficient α, temperature T, and thermal conductivity K, the thermal diffusivity for water is $\kappa = K/\rho C_p \approx 1.44 \times 10^{-3}$ cm^2/s. Writing the power source function as S, with units W/cm^3, the temperature distribution at time t is described by the thermal diffusion equation:

$$\frac{\partial T}{\partial t} = \nabla \cdot (\kappa \nabla T) + \frac{S}{\rho C_p} \tag{1}$$

In the present case we assume radial symmetry, so that the temperature depends only on the radial distance r from the optical axis, and on the distance z along the axis from the focal point. Taking the diffusivity as approximately constant gives:

$$\frac{\partial T}{\partial t} = \kappa \left[\frac{1}{r} \frac{\partial}{\partial r} \left(\frac{r \partial T}{\partial r} \right) + \frac{\partial^2 T}{\partial z^2} \right] + \frac{S}{\rho C_p} \tag{2}$$

B. Source Function

The source function S represents the heating caused by the beam, and in the converging and diverging regions it includes terms representing the Gaussian beam cross section and the cumulative beam attenuation. With origin at the focal point we can divide the source function into three regions: converging cone (S_+), waist (S_0), and diverging cone (S_-). The laser converges and diverges at an angle θ, where the effective numerical aperture is NA $= n \sin \theta$ (with refractive index $n = 1.33$ for water), and the beam waist radius is a. Typically, with a collimated laser beam 1.3 mm in diameter, the effective numerical aperture is NA $= 0.26$, giving $\theta = 12.6°$. Normalizing the intensity to that at the waist, the attenuation terms are symmetric and the converging and diverging source functions may be expressed for the Gaussian beam as:

$$S_{\pm} = \frac{\alpha P}{2\pi(1-\cos\theta)}\frac{1}{r^2+z^2}\mathrm{e}^{-2[\arctan(r/z)/\theta]^2}\mathrm{e}^{\pm\alpha(r^2+z^2)^{1/2}} \qquad (3)$$

while the source function in the laser waist region is:

$$S_0 = \frac{\alpha P}{\pi a^2}\mathrm{e}^{-2(r/a)^2} \qquad (4)$$

Two-dimensional finite element analysis (FEA) has been used to solve Eqs. (2–4) and derive the temperature as function of time and position (Douglas-Hamilton and Conia, 2001). Since the microscope is focused on the edge of the embryo, the focal plane bisects the embryo and the maximum temperatures will be on the focal plane. We are interested in the predictions of temperature in the focal plane normal to the optical axis drawn through the focal point, since this will give maximum values of temperature and maximum impact on other cells in the vicinity.

Results from FEA of a 600-μs pulse at wavelength $\lambda = 1480$ nm and power 245 mW are shown in Fig. 1. The white lines outline the beam. The (maximum) NA = 0.6 beam travels from right to left with optic axis parallel to the abscissa, 4 μm per division, and is approximated as focusing at the central point, then

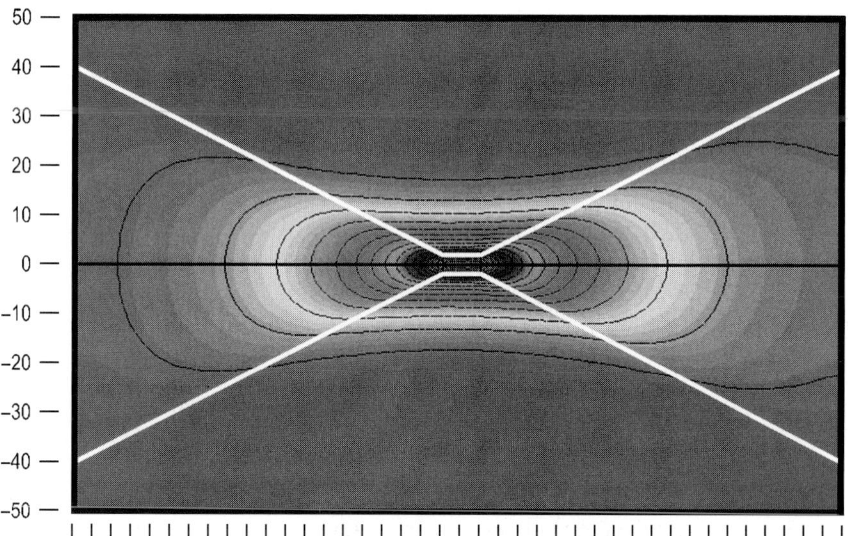

Fig. 1 Finite element analysis of $\lambda = 1480$ nm laser heating of aqueous medium. Beam power is 245 mW, pulse duration 600 μs, and initial temperature 37 °C. Focused beam with NA = 0.6 travels from right to left, outlined by the white lines. Scale on ordinate in micrometers and contours represent 20 °C intervals. In the vicinity of the waist, the heating effect is almost independent of axial position.

diverging. The ordinate is the radial distance in micrometers from the optic axis. The focal waist is approximated as 4 μm in diameter, and the source functions in Eqs. (3) and (4) are used in the conical and waist regions, respectively. Initial temperature is 37°C, and the contours represent 20°C intervals. The red color represents $T \leq 75^\circ$ and the entire ellipsoidal region within the yellow color is at superheated temperature. The laser therefore produces an ellipsoid of superheated water with diameter ~18 μm and length ~92 μm. Even for maximum NA it can be seen that the isotherms are almost parallel to the optic axis, and that beam intensity attenuation has only a slight effect on the isotherm contours as the beam proceeds to the left. The isotherms will be more parallel at lower effective NA. The FEA results for focal plane temperature for laser pulses in the range 0.2–3 ms are very close to those obtained by solving the one-dimensional case in which z dependence is ignored, and the infinite cylindrical source approximation gives results very close to those from the detailed FEA. In other words, for thermal analysis the double-cone focal geometry of the laser focal region is equivalent to a one-dimensional heating: far from the beam, the narrow cones leading to and away from the focus give the same heating as would a narrow collimated beam. We can therefore use one-dimensional cylindrical source FEA to predict temperatures near the pulsed laser beam. This is faster and will be outlined in the next section.

C. Time-Dependent One-Dimensional Solution

The thermal diffusion equation, assuming constant diffusivity and independent of z and azimuth, in cylindrical coordinates is:

$$\frac{\partial T}{\partial t} = \frac{\kappa}{r}\frac{\partial}{\partial r}\left(\frac{r\partial T}{\partial r}\right) + \frac{S_0}{\rho C_p} \tag{5}$$

In the present case of a laser beam heating an aqueous medium, the source function from Eq. (4) above is appropriate, since we may neglect the converging and diverging aspects of the beam. We may ignore the slow dependence of K, ρ, and C_p on temperature, and will assume α is also independent of temperature. Expression (5) may be rewritten as a difference equation with step-size Δr and Δt in radius and time. The space is divided into N concentric shells and the system integrated for n time steps. We assume that at each time step, a linear relation exists between the temperature of any shell and the temperature of the adjacent exterior shell. We can apply a Gauss–Seidel interpolation to determine the value of the linear coefficients α and β for temperature at each shell, from the outside in. After the coefficients have been determined, the temperature is extrapolated out from the center to the edge, using these coefficients. The procedure is then repeated for the next time step.

D. One-Dimensional Numerical Analysis Summary

We will summarize the interpolation method used for the analysis.

Define the radius as $r_i = i\,\Delta r$ and the time as $t_j = j\,\Delta t$. The temperature at the position and time represented by (i, j) is then $T_{i,j}$, and the index limits are $0 < i < N$, and $0 < j < n$. Writing the thermal diffusivity as $\kappa = K/\rho C_p$, we define:

$$a_i = -\kappa \frac{\Delta t}{\Delta r}\left[\frac{1}{\Delta r} - \frac{1}{r_i + r_{i-1}}\right] \quad \text{and} \quad c_i = -\kappa \frac{\Delta t}{\Delta r}\left[\frac{1}{\Delta r} + \frac{1}{r_i + r_{i-1}}\right] \tag{6}$$

The Gauss–Seidel inward iteration for the temperature linear coefficients is then:

$$\alpha_{i-1} = \frac{-a_i}{1 + 2\kappa\Delta t/\Delta r^2 + \alpha_i c_i} \tag{7}$$

and

$$\beta_{i-1,j} = \frac{T_{i,j} + \frac{\Delta t}{\rho C_p} S_{i,j} - \beta_{i,j} c_i}{1 + 2\kappa\Delta t/\Delta r^2 + \alpha_i c_i} \tag{8}$$

While α is time independent, β depends on the thermal distribution calculated in the preceding step. The steps proceed from the boundary condition values at the external radius, which are $\alpha_{N-1} = 0$ and $\beta_{N-1} = T_0$. At the end of the downward iteration, α and β have been recalculated over all space. Both are then used to obtain the next time step values of temperature, over all radii. We get:

$$T_{i+1,j+1} = \alpha_i T_{i,j} + \beta_{i,j} \tag{9}$$

Here the central value of the temperature is that giving zero radial gradient at the axis, which is:

$$T_{0,j+1} = \frac{\beta_0}{1 - \alpha_0}$$

By repeated applications of the algorithm for each time step, we can now rapidly obtain the temperature history over the entire domain in space and time.

E. Error Ratio

While the Gauss–Seidel iteration is nominally stable against diffusion rate errors (caused by the time step exceeding the thermal diffusion time between layers), the error obtained in the numerical calculation scheme is monitored by continuously computing the total laser absorbed energy and comparing it with the sum of the total internal energy in the domain plus the time integral of the energy leaking out at the maximum radius. The ratio of laser absorbed energy to internal + outconducted energy must be close to unity if the integration is correct. In the results reported here we use 64,000 time steps and 100 radial steps, for an integration time of 6 ms and maximum radius $R_{max} = 100$ μm. The error ratio is 0.996, corresponding to a valid self-consistent integration with error $\leq$0.4%.

F. Scaling with Beam Power and Pulse Duration

From Eq. (1), it can be seen that the temperature rise is directly proportional to the beam power, since the source function varies linearly with beam power in Eqs. (3) and (4). Therefore, once a numerical solution for temperature history has been obtained for a particular power level and pulse duration, at a particular distance from the axis, the corresponding values for all power levels are available by linear scaling. The pulse lengths (0.2–6 ms) are short enough so that the small increase of temperature due to thermal diffusion following the end of the laser pulse may be neglected. In calculating maximum temperatures by performing the numerical solution for a certain pulse length, we obtain the temperature history (and maximum temperatures) for all shorter pulse lengths.

G. Thermal History

The temperature history within and near the beam is shown in Fig. 2 for a 250-mW beam, pulse duration 600 μs. The temperature falls off rapidly away from the center during the pulse. Cooling when the pulse is complete is extremely rapid due to the small size of the heated region. In practice, the length of time at high temperature is comparable to the pulse duration. The integration error is $<$0.4%.

In practice, the peak temperature reached and the duration of high-temperature exposure are biologically most significant. The requirement must be to minimize both peak temperature and exposure time for cells neighboring the drilling point.

The axial temperature peak in this case is predicted to be over 350°C, and higher laser power or pulse duration would result in higher predicted peak temperature. If it reaches the critical temperature for water ($T_c = 374$°C), we can expect extremely rapid homogeneous nucleation of vapor bubbles, followed by phase explosion (Miotello and Kelly, 1999). In fact, as described below, phase explosion has only been observed with superheated water in contact with a surface, but not with the laser beam focused in water, implying that the absorption coefficient may be significantly lower at superheated temperatures than at 25°C. Since the H_2O density near the critical point is ~0.3, the absorptance is unlikely to be greater than 30% of its value at room temperature.

H. Linear Approximation

Safety precautions require that blastomeres in the vicinity of the target should not be significantly heated. In general the distance of any cell from the laser beam is at least several beam diameters. We have estimated the temperature at such distances around the beam during the laser pulse in the above analysis. However, in practice, a simpler approach is adequate and we can avoid the FEA. At

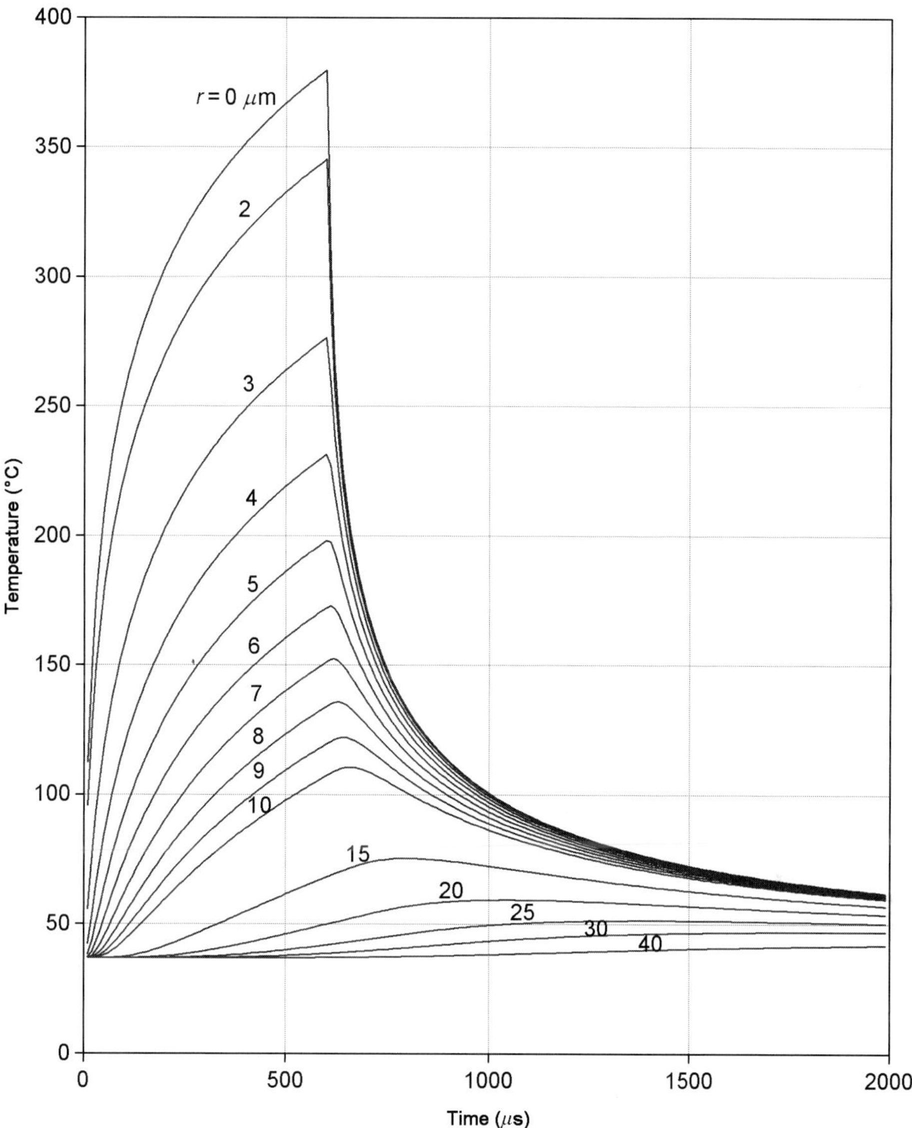

Fig. 2 One-dimensional thermal history of aqueous medium heated by a $\lambda = 1480$ nm laser pulse. The laser pulse duration is 0.6 ms and beam power is 250 mW at target. Beam diameter at focus is 4 μm. Temperatures are given for the radial distances from optic axis indicated, as function of time over 2 ms. The strong temperature gradient and extremely rapid cooling after beam turn-off are apparent.

distances large compared to the beam waist radius at the focal point, the beam diameter may be ignored compared to the radial distance and thermal effects, which are well represented by a line source approximation: the laser beam is regarded as equivalent to a very fine wire emitting heat at a known rate for a known time (Carslaw and Jaeger, 1959). The laser heating may be approximated by an analytic series, avoiding the need for FEA. Equation (2) is integrated for a short-pulse point source, then over z for a line source and over t for a line source with power input rate constant over time. This gives the temperature rise at any time t measured from the beginning of the pulse, at distance r from the optic axis, which may be expressed in terms of the exponential integral as:

$$\Delta T = \frac{\alpha P}{4\pi K} F\left(\frac{r^2}{4\kappa t}\right) \tag{10}$$

where F is the exponential integral, with P the laser beam power (watts), α the absorption coefficient of the beam in aqueous media, r the radial distance from the optical axis, t the time, K the thermal conductivity of the medium, and κ its thermal diffusivity.

With $\gamma = 0.57722$ (Euler's constant), the exponential integral is conveniently expressed as a series (Abramowitz and Stegun, 1966):

$$F(x) = -\gamma - \ln x - \sum_{n=1}^{\infty} \frac{(-x)^n}{nn!}$$

The series converges rapidly and in the present case the upper limit $n = 10$ suffices.

Equation (10) is a close approximation to the solution derived from the detailed FEA, and provides a more convenient way to estimate temperature without numerical integration. We can therefore use it to estimate the maximum temperatures reached by local blastomeres near the beam. The temperature error is $<2\%$ for the region $r > 3a$, where the beam radius is a. For regions up to and within the beam, the temperature will be higher, the approximation fails and numerical integration results should be used. But in practical cases Eq. (10) is adequate for temperature prediction.

We use the linear approximation in constructing the isotherms shown in Fig. 3. We take the measured focal beam radius of 2 μm, and use the beam absorptance for water, $\alpha_{1480} = 21\ \text{cm}^{-1}$, and mean thermal conductivity for water $K = 6 \times 10^{-3}$ W/cm/°C. The maximum temperature reached during the pulse is given as function of the pulse duration and the radial distance r in the focal plane. The experimental setup used to drill and measure the diameters shown will be discussed in the next section.

IV. Laser Drilling

A. Laser and System

A ZILOS-tk laser was mounted on a Leica DMIL microscope. The laser fits in the objective socket on the microscope turret and was set to produce 220 mW. Following reflection and absorption in the medium, this was estimated to deliver 180 mW to the embryo ZP. Pulse duration was monitored using a germanium

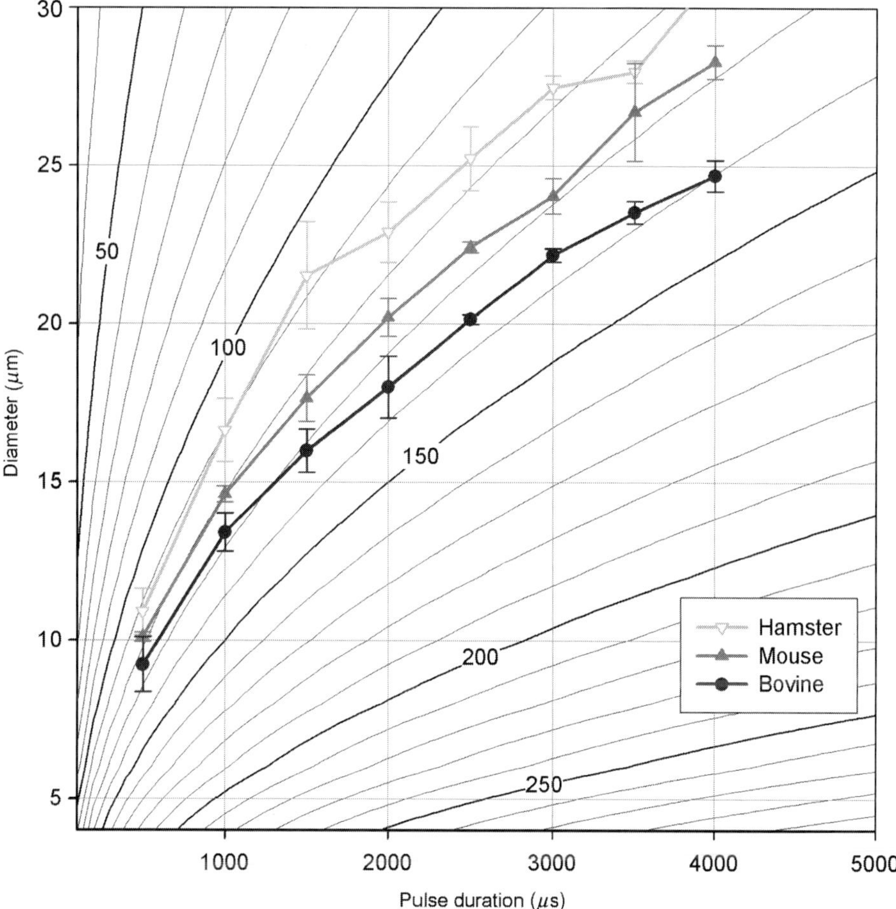

Fig. 3 Isotherms of the 220-mW incident beam (180 mW at target). The isotherms are computed from Eq. (10), and the results averaged over four hamster and mouse embryos and four bovine oocytes are shown.

detector set to observe scattered $\lambda = 1480$ nm radiation from the target, and the duration of the germanium detector signal was measured on a CRT for each hole drilled. The beam energy was measured by placing a modified pyrometer (Nova detector, Ophir Inc., Peabody, MA) over the laser objective and determining the beam energy at air focus for a 6-ms pulse.

B. Drilled Hole Diameter Measurement

The laser code of the ZILOS-tk provides vectors for measuring objects on the screen, so that the embryo diameter and drilled hole diameter can be measured. The magnification setting was calibrated and verified using a standard Ronchi slide with line width 100 μm. At least three vectors were drawn to determine each drilled hole diameter, which was recorded with the species, laser power, and laser pulse duration data.

C. Absorption Coefficient and Delivered Power

We are assuming that the medium has an absorption coefficient (α) close to that of water at wavelength $\lambda = 1480$ nm, for which we have $\alpha = 21$ cm^{-1} (Hale and Querry, 1973). So for an embryo of diameter 150 μm resting on the bottom of the Petri dish, the beam power reaching the embryo will be attenuated by passage through ~75 μm of medium, resulting in a drop to ~85% of its incident intensity. Allowing for reflection at the Petri dish surface reduces the beam at target to ~82% of its focal intensity in air. The power of the beam at target will therefore be ~82% of the power at air focus.

D. Animal Embryos and Oocytes

Mouse and hamster embryos frozen at the 2–4 cell stage were obtained from Embryotech Laboratories, Inc., Wakefield, MA, and bovine oocytes provided courtesy of Charletta Ohlrich, Trans Ova Genetics, Sioux Center, IA. After thawing they were placed on a Petri dish, the Petri dish was closed with its cover and transferred to the microscope. Experimental irradiation and drilled hole diameter measurement was typically complete within 2 h of thawing the embryos.

E. Animal Protocol

The laser was used at the power settings giving power at target shown in Table I. The smaller mouse and hamster embryos require lower distance traveled by the beam through the medium, and a beam of given incident power is attenuated less by the time it reaches the target for these embryos than for the bovine oocyte.

The actual beam power at target increases slightly for smaller embryos. The power levels in Table I were used in the present experiments.

Table I
Estimated Beam Power (mW) at Target on Embryo/Oocyte for Three Species

Embryo	Diameter (μm)	Power at target (mW)
Mouse	99	187
Hamster	115	184
Bovine	163	175

We have shown previously (Douglas-Hamilton and Conia, 2001) that in order to safely create a given diameter of drilled hole, short pulses at high power are preferable to long pulses at low power, to minimize peripheral heating of blastomeres. We suggested using pulses with duration <5 ms. Accordingly, in the experiments described here we examine the low-pulse-duration region. The pulse duration is varied from 0.5–6 ms. Four embryos from each species (four oocytes in the bovine case) were examined. For the successively increasing pulse durations, neighboring aim points clockwise around each embryo ZP were selected. Each data point is therefore the average of at least four embryos (oocytes for bovine) for each species.

F. Results

The drilling diameters obtained for incident beam power 220 mW are indicated in Fig. 3. Beam power actually reaching target is estimated in Table I. The mean drilled diameters for the four embryos of each species are indicated, with error bars at one standard deviation. The bovine oocyte drilled diameter follows the 130–140 °C isotherms, while the hamster embryo ZP is softer and falls between the 110–120 °C isotherms. Isotherms giving the peak temperature reached at each position and pulse duration are superimposed on the graphs. The peak pulse temperature isotherms are obtained from expression (10), as outlined in previous sections. The initial temperature of the embryo medium is taken as 37 °C. The bovine oocyte appears to have the toughest ZP, with the hamster embryo being drilled to a larger diameter at the same laser fluence: the hamster ZP is liquefied at lower temperature.

G. Human Protocol: Typical Laser Protocols in the Human IVF Laboratory

1. Stevens (2006)

Take the G1 oil dish to the microscope stage. Slide the oil dish to the center of the stage and center the embryo in the microscopes field of view.

1. Focus on the embryo on high power (20×). Orient the embryo so that hatching will be done at a gap between blastomeres or at an area of fragmentation (Fig. 4).

It is preferable to orient the embryo when using the laser so that the area to be hatched will be between twelve and six o'clock.

2. Once the embryo is in place, turn the objective wheel to get to the laser objective. Focus on the embryo. Follow the appropriate procedure below for the laser and program being used.

Hamilton Thorne Laser: Make sure the pulse of the laser is set to 800 μs. Move the holding pipette (HP) so that the part of the embryo to be hatched is directly under the red circle on the computer screen (this is the area that will be hatched).

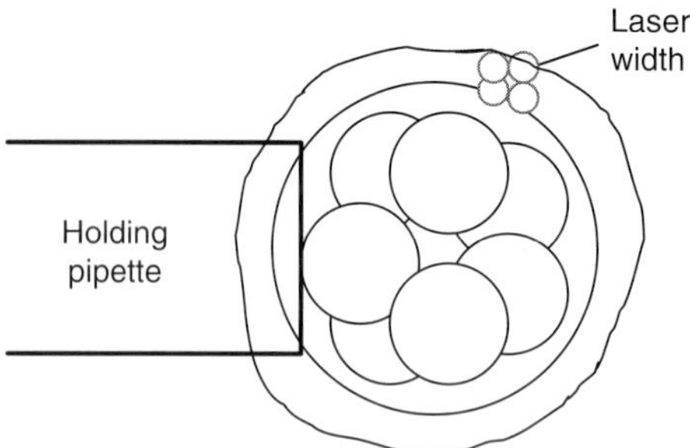

Fig. 4 Typical human hatching protocol. The glass pipette is holding an eight cell stage embryo during exposure to laser shots.

Begin on the outer edge of the zona. When ready, press "fire" with the mouse cursor on the computer screen. Move the HP so that the next pulse breaks the inner membrane. Keep moving and firing the laser until a hole twice the width of the red circle (20–30 μm) is made in the zona and the inner membrane is broken along this width. It usually takes four to six laser shots to make this hole.

2. Sapir and Fisch (2006)

a. Laser-Assisted Hatching (LAH)

AH is performed to enhance implantation in selective cases: patients of advanced age (>40 years old), following multiple failed IVF attempts and whenever embryos show abnormal ZP characteristics.

AH may be performed a few hours or immediately prior to ET.

Shortly before the procedure a culture dish is prepared with 10 μl drops of HEPES-buffered culture medium (to minimize pH changes during the procedure) supplemented with 10% serum. The dish is overlaid with prewarmed mineral oil.

Embryos selected for transfer are placed individually in separate micro drops and the dish is placed on the heated stage of the inverted microscope equipped with the Hamilton Thorne ZILOS-tk laser.

The use of small volume micro drops and the speed of the procedure eliminate the need for stabilizing the embryo with a holding micropipette.

The position of each embryo at a time is located using the small 4× objective and, once fully viewed, the objective wheel is rotated to the laser objective and the setting is switched for computer screen monitoring.

The laser pulse is set to 300 mW (100%) and 800 μs.

Routinely, a single opening, ~30-μm long (about 10% of the ZP circumference), is created by dissecting the full width of the ZP. The site is chosen as an area between two blastomeres, where the largest perivitelline space can be located.

By moving the stage, the red circle target is focused at the outer part of the ZP. Either the mouse or the foot pedal is used to click the "fire" button on the screen. Moving inward, four to six pulses are necessary to complete the cut through the ZP.

When the ZP is very thick or dark, "ZP thinning" (wider but superficial) is additionally performed at a second location by removing only the outer part of the envelope.

At the end of the procedure, the embryo is immediately replaced in the culture media and returned to the incubator.

b. Laser-Assisted Embryo Biopsy

Opening the ZP for embryo biopsy is done following the same steps of the AH procedure with the following differences:

A holding glass micropipette is used to stabilize the embryo throughout the procedure.

Rough edges are removed with the laser to ensure a smooth introduction and withdrawal of the biopsy pipette.

An opening of ~70 μm is breached. The biopsy pipette should be kept in view to determine when an appropriate size of the opening has been created. The biopsy pipette is then inserted through the opening to pull out a blastomere for genetic analysis.

V. Laser Settings Used in LAH Application

We summarize here settings typically used for zona drilling, reported from a number of sites.

Laser power	300 mW
Laser pulse width range	500–1000 μs
Laser pulse width typical	700 μs
Starting position	Outer edge of ZP
Drill shape	Flared
	Conical
	Channel
Drill diameter	Inner, 18–20 μm
	Outer, >30 μm

VI. Stroboscopic Image

A. Visualization

The laser pulse produces a region of highly superheated water near the focal point: but this very hot water does not boil, having nothing to nucleate on (Douglas-Hamilton and Conia, 2001). The water rapidly cools after the laser pulse. After the laser is turned off, the heat is rapidly conducted away and the focal region cools down on a timescale similar to the laser heating time, see Fig. 2.

The heated region can be examined during the heating process using a pulse-delay stroboscopic light source: a xenon flashlamp is the microscope light source and is triggered to fire at a set time interval from 1 to 3000 μs after the laser pulse initiation. The flash produced (duration <7 μs) gives an image of the medium at that time at the objective focus. In Fig. 5, the laser is focused on the ZP of a bovine oocyte. Laser power (in air) and pulse duration are 316 mW and 1500 μs. The ZP appears distorted, "pinched" near the focal point, and the region is less well illuminated than elsewhere. Both effects are due to the demagnifying "lens" produced by the thermal gradient in the medium. The bright central light at the focal point is an optical effect: the microscope illumination source aperture is imaged by the "lens." The same oocyte is shown 1 s later in the second image. By this time the ZP within the 140 °C peak temperature region has liquefied and disappeared.

The effect of the thermal lens is shown in Fig. 6 in which a 315-mW laser beam of wavelength $\lambda = 1450$ nm is focused on the floor of a 1-mm glass slide on which is a 40 μl drop of dH_2O. The beam travels upward through the glass slide. The water contains latex beads of diameter 4 μm. These beads are slightly denser than water and sink to the floor, where consecutive images give an indication of flow produced by the laser beam. Successive strobed images were taken at the times indicated in microseconds, measured from separate laser pulse onsets. Pulse length always exceeded the interval to the strobe so that the laser was continuously heating the sample up to and beyond the imaging. The visible black "bubble" is due to refractive index change in the liquid surrounding the beam focal point as the medium is heated, and represents the heated water region around the laser

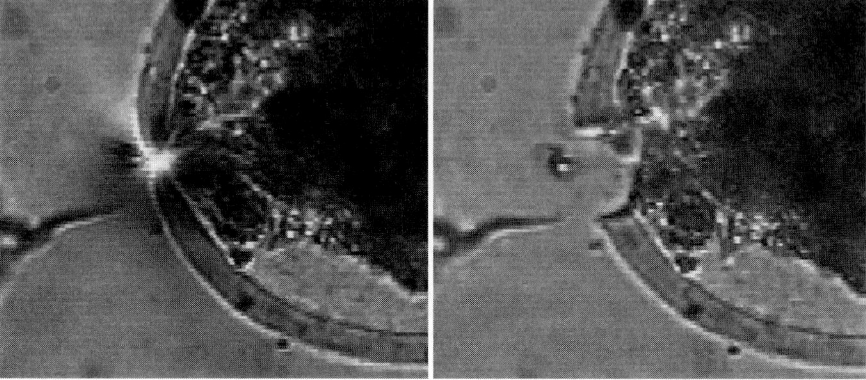

Fig. 5 Bovine oocyte with 315-mW laser pulse, duration 1500 μs. The heated water acts as a lens, and a bright image of the condenser aperture appears at the focal point in the first image. The lens "pinches" the ZP. The illumination intensity near the focal point is reduced by this effect. The second image shows the appearance 1 s later: the heated region of the ZP around the focal point has all liquefied.

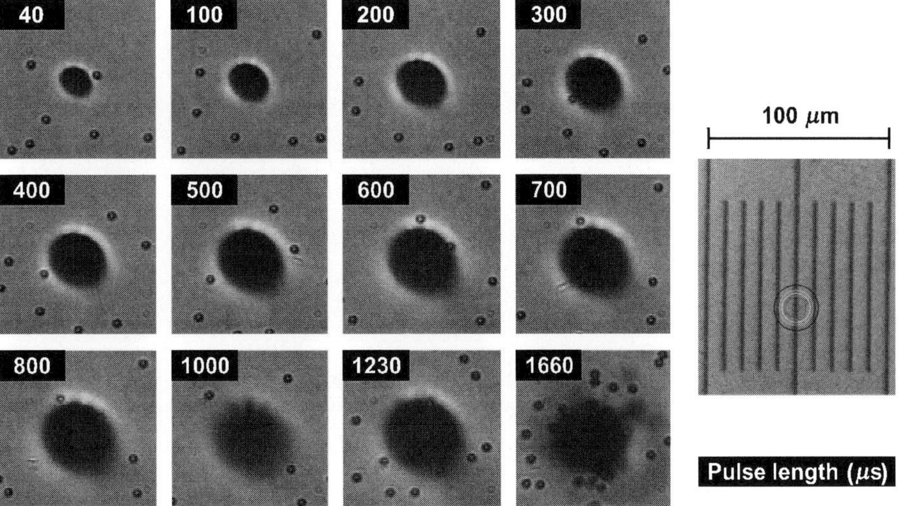

Fig. 6 Strobe images of the target region. Laser is focused on the Petri dish floor in dH_2O containing 4-μm latex beads. Time interval in microseconds after pulse initiation is given.

beam. The index gradient radial to the optic axis acts similarly to a concave lens and deflects light away from the camera. Over the range 300–1000 μs after pulse onset, the temperature estimated from Eq. (10) at the edge of the "bubble" is 56–75 °C, and higher inside. The refractive index gradient is sharp enough to exclude internal light. No phase change is apparent in this case until after 1000-μs pulse length, when the "bubble" becomes irregular and is probably a relic aftermath of explosive phase change.

B. Phase Transition

If phase change were to occur, a "phase explosion" would result with possible shock and local damage. By adjusting the timing of the stroboscopic flash delay, it is possible to capture the rapid phase explosion that occurs if the superheated water at the target volatilizes. This effect is well known in lasers with picosecond and nanosecond pulse length and is important in biological interactions such as retinal surgery where the beam is absorbed by melanosomes (Faraggi *et al.*, 2005). It is therefore important to determine whether any vapor bubble nucleation effect can take place with the lower power and much longer IR pulse lengths used for embryo manipulation. Clearly if it does, one would expect the water to change phase explosively, producing a large vapor bubble, which would then collapse as the vapor contacts cooler liquid.

A phase explosion image is shown in Fig. 7. A Petri dish was loaded with 1-ml water containing 4-μm latex beads (Accubeads, Hamilton Thorne Biosciences,

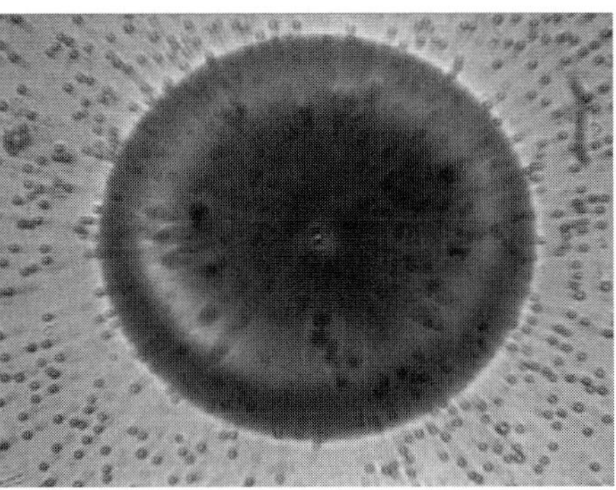

Fig. 7 Phase explosion in water in a Petri dish. A beam of 406 mW is focused on the floor and the single-shot xenon strobe illumination initiated at 700 μs into the pulse. Diameter of the vapor bubble is ~150 μm.

Beverly, MA), which were allowed to settle on the floor. A $\lambda = 1450$ nm beam of power 406 mW was focused on the floor of the Petri dish and a stroboscopic image of ~7-μs duration was made at 700 μs after laser pulse initiation. The beads move ~10 μm during the image acquisition as shown by the image streaks, and the corresponding bead velocity is 143 cm/s. The vapor bubble is not visible unless strobed illumination is used since its expansion and contraction are very rapid. The large vapor bubble (diameter 140 μm) would reach that size in ~100 μs, and we estimate recondensation time for a vapor bubble of radius r as $\tau \approx \rho L r^2 / 2K\Delta T \sim 100$ μs, where K is water thermal conductivity, L is heat of vaporization, ΔT is initial temperature difference between the bubble of water vapor (at ~100 °C) and the ambient medium, so the phase explosion is normally not visible in video imaging at 30 Hz. It exists for ~200 μs before collapsing as the vapor recondenses.

The vapor bubble can occur when a sufficiently powerful laser beam is focused on the surface of the slide or Petri dish. But it does not occur when the beam is focused on the embryo. A nucleation surface (a solid or very highly absorbing surface) is required for explosive evaporation to initiate. After collapse, small 1- to 3-μm-diameter bubbles (presumably exsolved O_2 and N_2) are present and remain visible for about 3 s before shrinking to zero as they redissolve (or move up out of the plane of focus).

C. Safety Limits

As mentioned, the phase explosion has only been observed when the laser beam is focused on the floor of the dish. It is absent if the beam is focused on the embryo ZP, some 60 μm above the floor of the chamber, and it does not occur for

$P \leq 300$ mW pulses with duration ≤ 800 μs. Therefore, in practice, phase explosion can always be avoided by maintaining limits on beam power and beam focus.

VII. Conclusions

Oocyte and embryo manipulation by IR laser provides a rapid noncontact method of drilling an opening in the ZP. Drilling is accomplished by the temperature rise caused by high absorptance of the IR beam in H_2O. A simplified analysis is adequate for predicting the resultant thermal regime around the laser beam. The analysis anticipates that very high water temperatures are produced at beam focus, with the potential for violent phase transitions (explosions). Although phase explosion has been observed when a high-power IR InGaAsP laser is focused on to an immersed surface, under normal conditions of power and duration there is no nucleation center available on embryo ZP, the heat is conducted away without phase transition, and any potential for phase explosion can be prevented by keeping beam power sufficiently low ($P \leq 300$ mW), pulse duration sufficiently short ($\tau \leq 800$ μs), and avoiding focusing the beam on solid surfaces.

Procedures where the ZP must be breached can benefit from the noncontact laser system. Applications include LAH, preimplantation genetic diagnosis, cell enucleation and cloning, and cell separation during sample extraction.

References

Abramowitz, M., and Stegun, I. A. (1966). "A Handbook of Mathematical Functions." US Department of Commerce, NBS AMS 55.

Ali, J., Rahbar, S., Burjaq, H., Sultan, A. M., Al Flamerzi, M., and Shahata, M. A. (2003). Routine laser assisted hatching results in significantly increased clinical pregnancies. *J. Assist. Reprod. Genet.* **20,** 177–181.

Antinori, S., Versaci, C., Fuhrberg, P., Panci, C., Caffa, B., and Gholami, G. H. (1994). Seventeen live births after the use of an erbium-yytrium aluminum garnet laser in the treatment of male factor infertility. *Hum. Reprod.* **9,** 1891–1896.

Blanchet, G. B., Russell, J. B., Fincher, C. R., Jr., and Portmann, M. (1992). Laser micromanipulation in the mouse embryo: A novel approach to zona drilling. *Fertil. Steril.* **57,** 1337–1341.

Carslaw, H. C., and Jaeger, J. C. (1959). "The Conduction of Heat in Solids." Clarendon Press, Oxford.

Check, J. H., Hoover, L., Nazari, A., O'Shaughnessy, A., and Summers, D. (1996). The effect of assisted hatching on pregnancy rates after frozen embryo transfer. *Fertil. Steril.* **65,** 254–257.

Cohen, J., Alikani, M., Trowbridge, J., and Rosenwaks, Z. (1992). Implantation enhancement by selective assisted hatching using zona drilling of human embryos with poor prognosis. *Hum. Reprod.* **7,** 685–691.

Cohen, J., Elsner, C., Kort, H., Malter, H., Massey, J., Mayer, M. P., and Wiemer, K. (1990). Impairment of the hatching process following IVF in the human and improvement of implantation by assisting hatching using micromanipulation. *Hum. Reprod.* **5,** 7–13.

CRC Handbook of Chemistry and Physics (1984). A ready-reference book of chemical and physical data (Robert C. Weast, ed.). CRC Press Boca Raton, FL.

Dokras, A., Ross, C., Gosden, B., Sargent, I. L., and Barlow, D. H. (1994). Micromanipulation of human embryos to assist hatching. *Fertil. Steril.* **61,** 514–520.

Dooley, R. B. (1977). "Release on the Refractive Index of Ordinary Water Substance as a Function of Wavelength, Temperature and Pressure." International Association for the Properties of Water and Steam EPRI, Palo Alto, CA.

Douglas-Hamilton, D. H., and Conia, J. (2001). Thermal effects in laser-assisted pre-embryo zona drilling. *J. Biomed. Opt.* **6,** 205–213.

Driscoll, W. G., and Vaughan, W. (1978). "Handbook of Optics." McGraw-Hill, New York.

Edi-Osagie, E., Hooper, L., and Seif, M. W. (2003). The impact of assisted hatching on live birth rates and outcomes of assisted conception: A systematic review. *Hum. Reprod.* **18,** 1828–1835.

Faraggi, E., Gerstman, B. S., and Sun, J. (2005). Biophysical effects of pulsed lasers in the retina and other tissues containing strongly absorbing particles: Shockwave and explosive bubble generation. *J. Biomed. Opt.* **10,** 064029-1–064029-10.

Hale, G. M., and Querry, M. R. (1973). Optical constants of water in the 200 nm to 200 μm wavelength region. *Appl. Opt.* **12,** 555–563.

Herzberg, G. (1950). "Molecular Spectra and Molecular Structure." Van Nostrand, New York.

Kazarian, S. G., and Martirosyan, G. G. (2002). ATR-IR spectroscopy of superheated water and *in situ* study of the hydrothermal decomposition of poly(ethylene terephthalate). *Phys. Chem. Chem. Phys.* **4,** 3759–3763.

Khalifa, E. A., Tucker, M. J., and Hunt, P. (1992). Cruciate thinning of the zona pellucida for more successful enhancement of blastocyst hatching in the mouse. *Hum. Reprod.* **7,** 532–536.

Konig, K., Tadir, Y., Patrizio, P., Berns, M. W., and Tromberg, B. J. (1996). Effects of ultraviolet exposure and near infrared laser tweezers on human spermatozoa. *Hum. Reprod.* **11,** 2162–2164.

Liu, H. C., Cohen, J., Alikani, M., Noyes, N., and Rosenwaks, Z. (1993). Assisted hatching facilitates earlier implantation. *Fertil. Steril.* **60,** 871–875.

Luck, W. A. P. (1965). Zur assoziation des Wassers: Die temperaturabhangigkeit der wasserbanden bis zum kritischen punkt. *Ber. Bunsenges. Physik. Chem.* **69,** 626.

Magli, M. C., Gianaroli, L., Ferraretti, A. P., Fortini, D., Aicardi, G., and Montanaro, N. (1998). Rescue of implantation potential in embryos with poor prognosis by assisted zona hatching. *Hum. Reprod.* **13,** 1331–1335.

Malter, H. E., and Cohen, J. (1989). Blastocyst formation and hatching *in vitro* following zona drilling of mouse and human embryos. *Gamete Res.* **24,** 67–80.

Mandelbaum, J., Plachot, M., Junca, A. M., Salat-Baroux, J., Belaisch-Allart, J., and Cohen, J. (1994). The effects of partial zona dissection on *in-vitro* development and hatching of human cryopreserved embryos. *Hum. Reprod.* **9**(Suppl. 4), 39.

Miotello, A., and Kelly, R. (1999). Laser induced phase explosion: New physical problems when a condensed phase approaches the thermodynamic critical temperature. *Appl. Phys. A: Mater. Sci. Process.* **69,** S67–S73.

Nakayama, T., Fujiwara, H., Yamada, S., Tastumi, K., Honda, T., and Fujii, S. (1999). Clinical application of a new assisted hatching method using a piezo-micromanipulator for morphologically low quality embryos in poor-prognosis infertile patients. *Fertil. Steril.* **71,** 1014–1018.

Neev, J., Gonzales, A., Licciardi, F., Alikani, M., Tadir, Y., Berns, M., and Cohen, J. (1993). Opening of the mouse zona pellucida by laser without a micromanipulator. *Hum. Reprod.* **8,** 939–944.

Neev, J., Schiewe, M. C., Sung, V. W., Kang, D., Hezeleger, N., Berns, M. W., and Tadir, Y. (1995). Assisted hatching in mouse embryos using a noncontact Ho:YSGG laser system. *J. Assist. Reprod. Genet.* **12,** 288–293.

Neev, Y., Tadir, Y., Ho, P., Asch, R., Ord, T., and Berns, M. W. (1992). Microscope-delivered UV laser zona dissection: Principles and practices. *J. Assist. Reprod. Genet.* **9,** 513–523.

Obruca, A., Strohmer, H., Sakkas, D., Menezo, Y., Kogosowski, A., Barak, Y., and Feichtinger, W. (1994). Use of lasers in assisted fertilization and hatching. *Hum. Reprod.* **9,** 1723–1726.

Rink, K., Delacrétaz, G., Salathe, R., Senn, A., Nocera, D., Germond, P., de Grandi, P., and Fakan, S. (1996). Non-contact drilling of mouse zona pellucida with an objective-delivered 1.48 μm diode laser. *Lasers Surg. Med.* **18,** 52–62.

Sallam, H. N., Sadek, S. S., and Agameya, A. F. (2003). Assisted hatching—a meta-analysis of randomized controlled trials. *J. Assist. Reprod. Genet.* **20,** 332–342.

Sapir, O., and Fisch, B. (2006). IVF Unit Rabin Medical Center, Petah Tiqva, Israel.

Stein, A., Rufas, O., Amit, S., Avrech, O., Pinkas, H., Ovadia, J., and Fisch, B. (1995). Assisted hatching by partial zona dissection of human pre-embryos in patients with recurrent implantation failure after *in vitro* fertilization. *Fertil. Steril.* **63,** 838–841.

Stevens, J. (2006). Private communication, IVF Laboratory Supervisor, Fertility Laboratories of Colorado,799 East Hampden Ave., Suite 330, Englewood, Co. 80113.

Tadir, Y., Neev, Y., and Berns, M. W. (1992). Laser micromanipulation of oocytes and sperm. *Lancet* (*Lett.*) **339,** 1424.

Tadir, Y., Neev, Y., Ho, P., and Berns, M. W. (1993). Lasers for gamete micromanipulation: Basic concepts. *J. Assist. Reprod. Genet.* **10,** 121–125.

Tadir, Y., Wright, W. H., Vafa, O., Liau, L.-H., Asch, R., and Berns, M. W. (1991). Micromanipulation of gametes using laser microbeams. *Hum. Reprod.* **6,** 1011–1016.

Tucker, M. J., Cohen, J., Massey, J. B., Mayer, M. P., Wiker, S. R., and Wright, G. (1991). Partial dissection of the zona pellucida of frozen-thawed human embryos may enhance blastocyst hatching, implantation, and pregnancy rates. *Am. J. Obstet. Gynecol.* **165,** 341–344.

CHAPTER 15

Laser Microdissection-Based Analysis of Plant Sex Chromosomes

Roman Hobza and Boris Vyskot

Laboratory of Plant Developmental Genetics
Institute of Biophysics, Academy of Sciences of the Czech Republic
CZ-612 65 Brno, Czech Republic

A recent progress in plant molecular biology has led to enormous available data of DNA sequences, including complete nuclear genomes of *Arabidopsis*, rice, and poplar. On the other hand, in plant species with more complex genomes, containing widespread repetitive sequences, it is important to establish genomic resources that help us to focus on particular part of genomes. Laser technology enables to handle with specific subcellular structures or even individual chromosomes. Here

0091-679X/07 $35.00
DOI: 10.1016/S0091-679X(06)82015-7

we present a comprehensive protocol to isolate and characterize DNA sequences derived from the sex chromosomes of white campion (*Silene latifolia*). This dioecious plant has become the most favorite model to study the structure, function, and evolution of plant sex chromosomes due to a large and distinguishable size of both the X and Y chromosomes. The protocol includes a versatile technique to prepare metaphase chromosomes from either germinating seeds or *in vitro* cultured hairy roots. Such slides can be used for laser chromosome microdissection, fluorescence *in situ*-hybridization mapping, and immunostaining. Here we also demonstrate some applications of the laser-dissected chromosome template, especially a modified FAST-FISH technique to paint individual chromosomes, and construction and screening of chromosome-specific DNA libraries.

I. Introduction

Microdissection of chromosomes and generation of chromosome-specific DNA for genomic studies is a powerful tool with wide range of applications in current cytogenetics and molecular biology. The use of chromosome-specific probes for fluorescence *in situ* hybridization (FISH) plays an important role both in clinical diagnostics and studies concerning structure, evolution, and modulation of plant and animal genomes. In comparison to animals, plants reveal a complicated model for direct application of chromosome-derived probes. An ordinary plant chromosome, contrary to the animal one, rarely contains specific DNA sequences that would enable a discrete resolution of particular chromosome by direct painting with a chromosome-specific probe. The plant nuclear genomes evolved by many multiplication events and they are formed by widespread repetitive sequences. The alternative approach to paint individual chromosomes based on chromosomal *in situ* suppression (CISS) or construction of chromosome contigs is often both ineffective and time consuming. There are few special chromosomes in plants, represented by nonrecombining sex chromosomes and B chromosomes, which have been successfully used for chromosome painting. The specific hybridization signal usually corresponds to specific repeats accumulated on a particular nonrecombining chromosome during evolution.

Laser microdissection of chromosomes has been a powerful technique both in animal and human (Monajembashi *et al.*, 1986) and plant genetics (Fukui *et al.*, 1992). Here we report an alternative approach based on laser microdissection of selected chromosomes, optimized DOP-PCR, and modified FAST-FISH that led to a specific painting of the sex chromosomes in the model dioecious plant *Silene latifolia* (white campion, formerly *Melandrium album*). This protocol enabled us not only to paint the Y chromosome that can be supposed as a specific case of accumulation of repetitive DNA motif in the nonrecombining region but also the X chromosome that undergoes recombination with its opposite X partner in meiosis of female plants. The general application of our method has to be verified in other plant models. The protocol follows several basic steps. First, germinating seeds or transformed hairy root cultures are synchronized and chromosomes are proceeded to accumulate in mitotic metaphase. Protoplasts are subsequently released by enzymatic treatment and stored in fixative. The protoplast suspension is dropped on slides (covered with polyethylene naphthalate membrane) and chromosomes are stained with Giemsa. The microdissection is then carried out by nitrogen laser using PALM MicroLaser System (P.A.L.M., GmbH, Bernried, Germany). The chromosomes are selected, captured by laser, and catapulted into a cap of Eppendorf tube. The chromosomes are amplified by modified DOP-PCR reaction and used as a probe for FISH experiments. The specific signal on chromosomes of probe origin is achieved by shortened hybridization time and low amount of probe used. Here we also report laser-based isolation of chromosomal arms of the X and Y chromosomes and discuss the application of this approach in gene mapping.

II. Design of Chromosome Analyses

There are a large number of techniques enabling to isolate plant mitotic cells. In general, it is rather difficult to work with plant material due to the presence of rigid cellulose walls. In order to study plant mitotic chromosomes, various squash techniques are most often used. However, the squash approach, as applied after a partial enzymatic hydrolysis, does not regularly enable to release free chromosomes suitable for laser microdissection. Here we describe a protocol, which is based on isolation of living protoplasts from synchronized root tips. The living round-shaped protoplasts, completely got rid of cellulose remnants, are usually fixed in an acidic fixative and later dropped on microscopic slide. This kind of slides is suitable for standard karyology, FISH, and immunofluorescence analysis, where DNA epitope is needed. If the fixed protoplasts are dropped on a nylon membrane, chromosomes can be microdissected by laser, harvested, and used as DNA template. Here we describe general protocols designed for isolation of chromosomes for different purposes.

A. Source of Plant Material

To isolate mitotic chromosomes, root tips prepared from germinating seeds or established hairy root cultures are used (Fig. 1). The seedlings represent the most versatile material, if enough large amount of seeds are available (at least 100 or more). In our experiments, dry seeds of *S. latifolia* are thoroughly rinsed in tap water, washed in 70% ethanol, and sterilized in 10% (v/v) solution of commercial bleach for 20 min. The sterilizing solution is removed and the seeds are carefully washed with sterile distilled water. At this stage, the seeds dipped in water are kept in a Petri dish in cool chamber (at about 4 °C) for at least several days to swell the seeds properly and to induce rather synchronous germination. The germination process is switched on by putting the dish into cultivation room (about 25 °C),

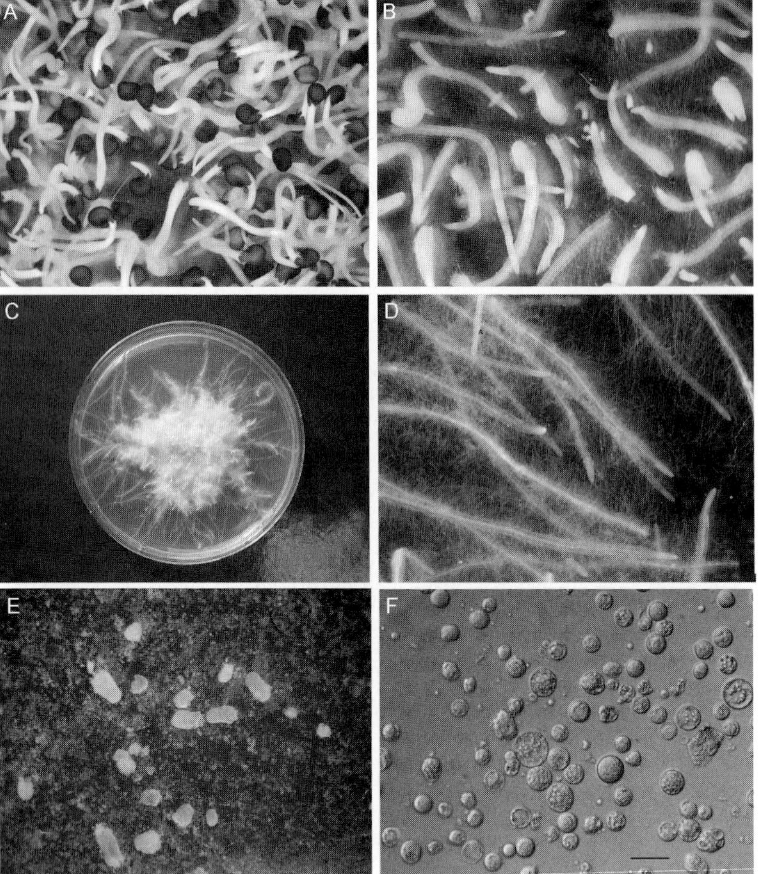

Fig. 1 Plant root material of the dioecious *S. latifolia* used to study the sex chromosomes. (A) Seedlings prepared by cultivation of sterilized seeds in distilled water for 2 days. (B) Tips of synchronized roots are cut off and incubated in the cellulase–pectinase solution to release clusters of meristematic cells. (C) Stabilized hairy root culture isolated from *S. latifolia* after infection with *Agrobacterium rhizogenes*. (D) Tips of hairy roots cultivated in liquid medium and synchronized with aphidicoline and oryzalin. (E) Root tips, either from seedlings or hairy roots, are incubated in the cellulase–pectinase solution for about 1 h, degraded by releasing terminal tips with meristematic cells. (F) The final population of root tip protoplasts, just before purification and fixation. Scale bar = 50 μm.

changing the water daily. In the case of *S. latifolia*, the seeds start to germinate in 2 days, and the roots are ready for synchronization when they are about 2- to 4-mm long. We have a similar experience also with other plant species, such as *Silene* spp. (*S. dioica, S. diclinis, S. pendula, S. vulgaris*, and *S. chalcedonica*), wheat (*Triticum aestivum*), tobacco (*Nicotiana* spp.), and cabbage (*Brassica oleracea*).

B. Synchronization of Cell Division and Protoplast Isolation

At this stage of root tip development, DNA synthesis is reversibly stopped with a drug, aphidicolin (Sigma, St. Louis, MO, 30 μM, for 12 h), to accumulate root meristematic cells before or at the S-phase. The aphidicolin is then thoroughly washed out with water, and beginning this step the procedure can be done in a nonsterile way. The seedlings are cultivated in distilled water for another 6 h to reach the stage of metaphase. Metaphases are accumulated by adding the solution of a spindle inhibitor, herbicide oryzaline (Sigma, 15 μM, for 4 h). Alternatively, a prolonged cold treatment of roots (5 °C for 24 h) can substitute both these drugs and yield a similar mitotic index. To facilitate the protoplast isolation during the following enzymatic treatment, it is envisage to cut off the remaining parts of seedlings and to discard the testa with cotyledons. Collected root tips are thoroughly washed and dipped in an enzymatic solution (Hladilova *et al.*, 1998; Veuskens *et al.*, 1995). This regularly consists of 2% (w/v) solution of commercial crude cellulase (e.g., Sigma), 1% (w/v) pectinase (Sigma), and 0.5% (w/v) sorbit to equilibrate the osmotic value. It is recommendable to add a highly active enzyme pectolyase Y-23 (MP Biomedicals, Basingstoke, UK) at a concentration about 0.2% (w/v). The acidity of this enzymatic mixture is adapted to about pH 5.0 to optimize the function of cellulolytic enzymes. The enzymatic solution is penetrated into the root tips by application of slight vacuum pump. The samples are then incubated in the enzymatic mixture at room temperature with a slow shaking.

Usually during the first hour of incubation terminal parts of root tips are gradually released and round-shaped protoplasts accumulate at the bottom of Petri dish. This process is monitored under inverted microscope with a low magnification. The plant material is filtered through a stainless mesh (wholes around 100 μm) to get rid of large nondissolved rests of roots. Alternatively, if the plant material is to be used for laser chromosome flow sorting, the root tips are mechanically homogenized to release chromosomes rather than incubated in the enzyme mixture (Fig. 2).

C. Protoplast Harvest and Fixation

The protoplasts are purified by spinning down in a washing solution in a swing-out rotor (10 min at 500 × *g*). The supernatant is discarded by Pasteur pipette connected with water vacuum pump and the sediment of protoplasts is again resuspended in a hypotonic solution W5 (Menczel *et al.*, 1982) to swell the protoplasts and spread chromosomes within. The samples are again centrifuged to be concentrated and the hypotonic solution is changed for a fixative. According to the purpose of experiments we will decide which type of fixative is to be used. The most efficient is the acidic Farmer's fixative (3 volumes of ethanol to 1 volume of ice acetic acid), which clarifies the plant samples and removes a majority of dissociable proteins. This fixative is used for standard karyology and some more recently derived techniques, such as FISH or chromosome laser microdissection.

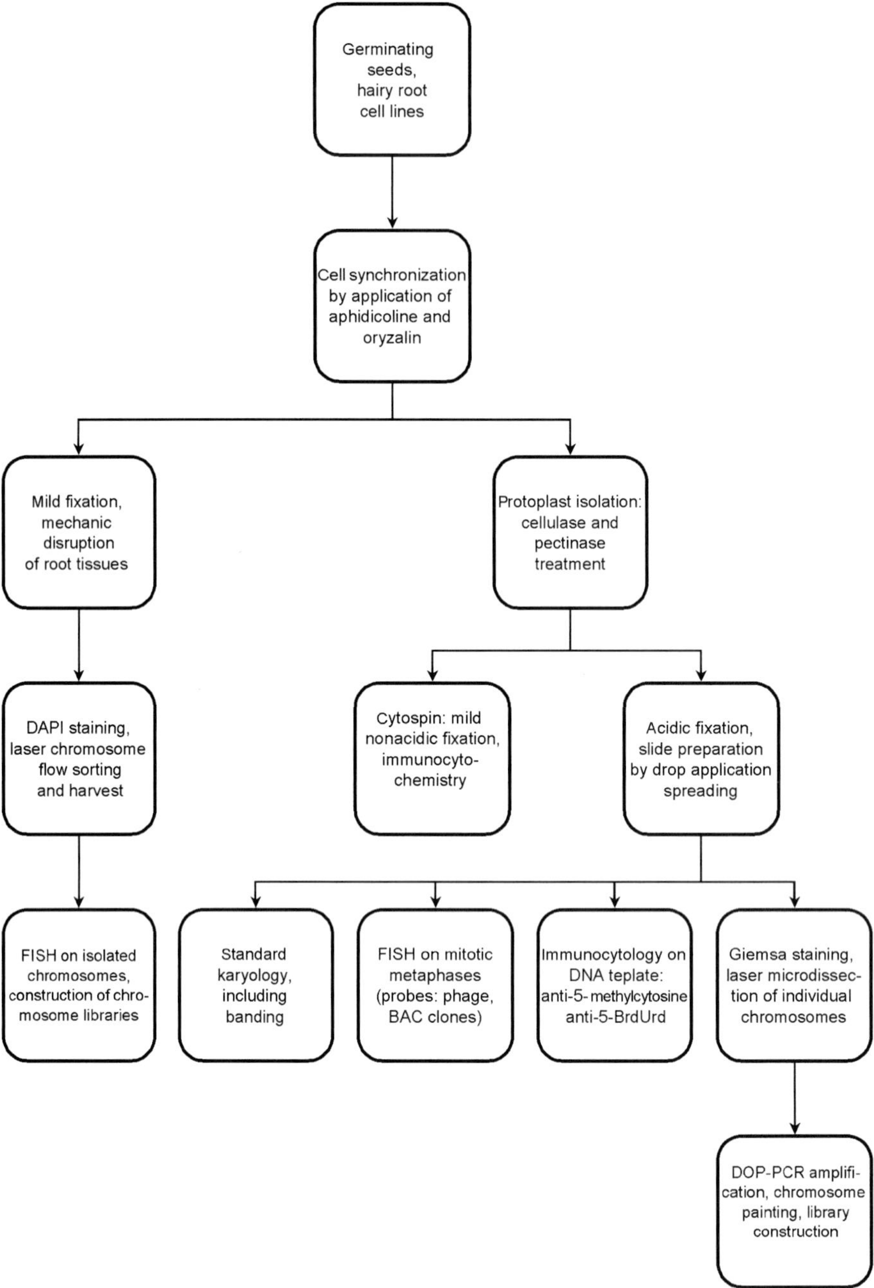

Fig. 2 Flow diagram of protoplast technologies applied to prepare mitotic chromosomes of *Silene*.

It is also applicable for immunofluorescence studies, where DNA epitope is targeted, such as detection of 5-methylcytosine or 5-bromo-2′-deoxyuridine (after feeding of cells with 5-bromo-2′-deoxyuridine, 5-BrdUrd).

D. Cytospin Technique to Detect Protein Epitopes

This is a very careful technique used to detect protein epitopes. The isolated protoplasts are resuspended in a hypotonic solution and kept in cold water bath (8 °C), until use. Usually, 100 μl of the protoplast suspension is pipetted into a column of cytocentrifuge and 500 μl of a lytic buffer is added. This lytic buffer consists of neutral formaldehyde [0.1–1% (v/v)] and Triton X-100 [2% (v/v)] in PBS (phosphate-buffered saline). The protoplasts are fixed to the slide by centrifugation at 500 $\times$ g for 10 min. It is advisable to use poly-l-lysine coated slides to increase adhesion of cell material. The slides are immediately dipped in ice-cold methanol at least for 15 min and dehydrated (Hladilova *et al.*, 1998; Vyskot *et al.*, 1999). Keep the slides wet during the whole protocol, since a drying could increase a nonspecific background. The slides are eventually stored in 50% water solution of glycerol in a fridge. After the washing of slides in 2× SSC (sodium saline citrate) solution, we can apply an antibody detecting a specific protein, for example, primary antisera or monoclonal antibody against a specifically modified histone (e.g., acetylated H4 at specific lysine positions, Siroky *et al.*, 1999; Vyskot *et al.*, 1999). This antiserum is later visualized with a labeled secondary antibody constructed against the animal in which the primary antisera was made. A series of slides are necessary as positive and negative controls to avoid misinterpretation of the immunofluorescence data received.

E. Studying DNA Replication or Cytosine Methylation Patterns

Timing of DNA replication and 5-methylcytosine distribution are important epigenetic markers indicating the chromatin status of respective domains. The early replication and low 5-methylcytosine content accompany euchromatic domains, potentially expressed, while the late-replicating regions of chromosomes show a higher density of 5-methylcytosine signals and are called heterochromatic. The kinetics of DNA replication could be advantageously studied using incorporation of a thymidine analogue, 5-BrdUrd. This drug is simply applied into the water or cultivation medium in which the seedlings or root culture are cultivated. The root tips synchronized with aphidicolin are pulsed with 5-BrdUrd, usually at the final concentration 30 μM for 30 min. This pulse is applied at different time after washing out the aphidicolin solution, and it is always finished by a careful removal of 5-BrdUrd followed by several washings with thymidine solution (150 μM) and distilled water. The protoplasts with accumulated chromosomes are fixed in the Farmer's fixative, dropped on slides, and air-dried. 5-BrdUrd is later detected using a commercial monoclonal anti-BrdUrd antibody, made in mouse, and visualized with an anti-mouse antibody, usually labeled with FITC (Vyskot *et al.*, 1999). The density of

5-methylcytosine distribution is checked using a polyclonal antibody constructed against 5-methylcytosine. This antiserum is directly applied on the fixed slides after washing with an appropriate buffer (Siroky *et al.*, 1998). Similarly, as in the case of 5-BrdUrd detection, the mouse antibody is visualized with the goat anti-mouse antibody.

III. Laser Microdissection of Plant Chromosomes

A. Flow Sorting

In complex plant genomes, containing widespread repetitive sequences, it is important to establish genomic resources that enable us to focus on a particular part of the genome. There are several methods available that are used to dissect particular chromosome or subchromosomal regions. The most powerful method in terms of quantity of DNA separated is flow sorting of chromosomes. The isolation of chromosomes by flow sorting is based on purification of individual chromosomes after quantification of fluorescence of stained chromosomes in a chromosome sorting equipment. Electrically charged droplets containing chromosome of interest are passed through an electrostatic field, separated, and subsequently collected for further experiments (Dolezel *et al.*, 2004). The main advantage of this approach is collecting a large amount of high-molecular-weight DNA suitable for direct restriction analysis and cloning. The collected chromosomes can be also used as a template for polymerase chain reaction (PCR) to physically detect the presence of gene sequences, as in the case of the X chromosome of *S. latifolia* (Kejnovsky *et al.*, 2001). In plants, the chromosome sorting was successfully used for construction of chromosome-specific BAC library from bread wheat (Safar *et al.*, 2004). The main disadvantage of the flow-sorting approach is contamination of dissected material by chromosomes of similar size and the presence of particles with the same DNA contents as sorted chromosomes (Dolezel *et al.*, 2001). In animals, the construction of chromosome-specific libraries by flow sorting can be achieved also by using somatic cell hybrid lines containing a single chromosome from genome to be sorted (Potier *et al.*, 1992).

B. Laser Microdissection

Contrary to flow sorting, both manual and laser beam-based microdissection yield a smaller amount of DNA and both are laborious and time consuming. The necessity of amplification of collected DNA can lead to bias in DNA fragments resulted from different types of PCR methods used for amplification of anonymous DNA samples. The main advantage of chromosome microdissection is nearly 100% purity of DNA. Chromosome microdissection can be used not only for direct isolation of chromosome of interest but also for generating chromosome region-specific DNA populations using preparative *in situ* hybridization (Prep-ISH) method. This method combines *in situ* hybridization of DNA population to standard mitotic chromosomes. The DNA of interest (e.g., cDNA) is ligated with adaptors containing PCR-binding sites at the termini of each molecule. Hybridization is carried out according to a standard protocol for hybridization of complex probes and is followed by microdissection of chromosome of interest. Finally, hybridized DNA from microdissected chromosome is amplified using adaptor-specific primers (Hozier *et al.*, 1994). The feasibility of this method was verified by Gracia *et al.* (1997) in experiments focusing on isolation of chromosome-specific ESTs in human. This approach can help to shed the light on gene content of specific regions of chromosomes especially in organisms with large genomes.

C. Microdissection of Plant Sex Chromosomes

In plants, chromosome microdissection was recently used to study the structure and evolution of sex chromosomes of two model species, *Rumex acetosa* and *S. latifolia*. These species possess heteromorphic sex chromosomes that can be microscopically distinguished from other chromosomes of complement (Vyskot and Hobza, 2004). Shibata *et al.* (1999) used the manual micromanipulator (ON-T1, Olympus, Tokyo, Japan) for mechanic microdissection of the *R. acetosa* Y chromosome. This experiment finally led to characterization of a Y chromosome-specific, tandemly arranged repetitive DNA sequence called *RAYSI*. The use of inverted microscope equipped with mechanic micromanipulator enabled to characterize both sex chromosome-linked, tandemly arrayed repetitive sequences (Buzek *et al.*, 1997) and isolate sex chromosome linked genes (Atanassov *et al.*, 2001; Delichere *et al.*, 1999) in *S. latifolia*.

A more sophisticated approach was used by other authors (Hobza *et al.*, 2004; Matsunaga *et al.*, 1999a; Scutt *et al.*, 1997) to dissect the sex chromosomes of *S. latifolia*. Scutt *et al.* (1997) used for microdissection mitotic metaphase chromosome preparations on polyester membranes. The examination and ablation of chromosomes was, in this experiment, carried out using ACAS 470 Cell Workstation (Meridian Instruments, Okemos, MI). The ablation of autosomes using an argon ion laser microbeam and isolation of sex chromosomes (by recovering excised fragments of polyester membrane) was controlled by movement of the microscope stage across the static laser beam. Finally, a higher power of laser beam was used to cut octagonal disk, 0.5 mm in diameter, containing chromosomes of interest. The fine forceps were used to manipulate with disks. Prior to DNA amplification of chromosomes by DOP-PCR, the chromosomes were treated with proteinase K. In this case, modified primers originally designed by Telenius *et al.* (1992) were used for amplification. To enrich DOP-PCR for DNA from sex chromosomes of *S. latifolia* rather than from extraneous contaminating DNA, further subtraction experiments were carried out. Biotinylated genomic DNA (gDNA) and reverse transcriptase PCR products were mixed with DOP-PCR, denatured, and incubated in hybridization buffer for

20 h to allow reassociation. The mixture of hybridization solution was further incubated with streptavidin-coated magnetic particles to bind biotinylated molecules. These particles were magnetically captured and finally stringently washed. The elution of enriched sex chromosome-derived DNA was followed by second step of PCR with nested specific primer derived from original DOP sequence.

The approach used by Matsunaga *et al.* (1999a) was based on semiautomatic microdissection of desired chromosomes. First, chromosomes were placed on laser-penetrable film (DuraSeal, Diversified Biotech, Boston, MA) and stained by Giemsa. In this case, the regions surrounding chromosome of interest were automatically sublimated by a UV laser microbeam of the PALM laser microscope system, using "Laser Sweeper" software (Carl Zeiss Vision, Tokyo, Japan). Finally, burning of the region except the sex chromosome(s) was performed manually. The membrane with remaining chromosomes was manually picked up from the glass and placed into a PCR tube. In this case, topoisomerase I was added for 1 h at 37°C to treat chromosomes before further amplification steps. The chromosomes were finally amplified by modified primers originally designed for DOP-PCR.

The last experiments that were carried out to microdissect the sex chromosomes of *S. latifolia* successfully employed fully automatic approach by use of the PALM MicroLaser system (Hobza *et al.*, 2004). The first step is application of protoplast suspensions containing synchronized chromosomes in metaphase on slides (slide is covered with polyethylene naphthalate membrane). Sufficient dispersion of chromosomes is achieved both by dropping the suspension and by cooling the target slide. The microdissection equipment consists of a 337-nm nitrogen laser that is coupled to the light path of an inverted microscope and focused through an oil immersion objective with high numerical aperture to yield a spot size of less than 1 μm in diameter. An energy of 1.5–11.7 μJ per pulse is used for microdissection and of 2 μJ per pulse for catapulting. All procedures in this experiment are controlled by computer (movement of the microscopic stage, micromanipulation, and catapulting of chromosomes) (Schermelleh *et al.*, 1999). All procedures are adapted from experiments performed by Kubickova *et al.* (2002) to dissect chromosomes of farm animals. The membrane around chromosome of interest is cut and chromosome is than catapulted by a single laser pulse directly into the cap of an Eppendorf tube (Fig. 3). This approach enabled minimalization of possible contamination by undesired extraneous DNA. Contrary to previously described experiments, chromosomes are directly used (without any enzymatic treatment) for amplification by DOP-PCR with regular primers designed by Telenius *et al.* (1992).

D. Methods to Amplify Dissected DNA

Analysis of particular chromosomes or chromosome-specific regions based on laser microdissection is a key approach both in basic research and in clinical diagnostics. Recent advent in PCR techniques led to development of different

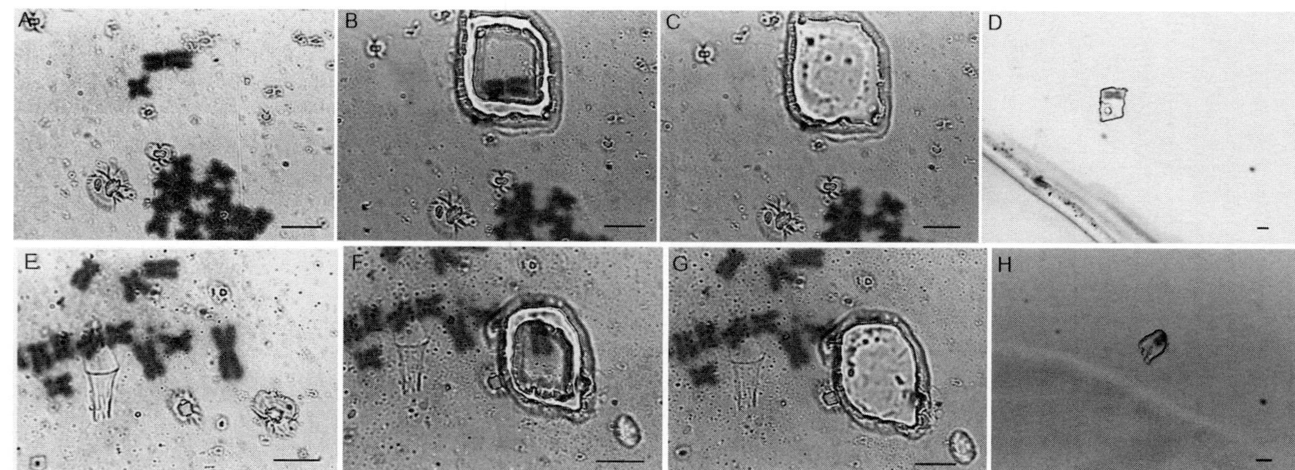

Fig. 3 Laser microdissection of *S. latifolia* metaphase X chromosome and the q arm of the X chromosome. Metaphase protoplasts were dropped on a polyethylene naphthalate membrane and stained with Giemsa. A suitable X chromosome was localized under the inverted microscope (A). The membrane was cut around the selected chromosome using a laser microbeam (B) and the chromosome was catapulted by a single laser pulse (C) into the cap of a PCR tube (D). The same approach was used for microdissection of the q arm of the X chromosome (E–H). Scale bars = 10 μm.

methods to amplify anonymous samples presented by complex chromosomal, subchromosomal, or other gDNA. Among mostly used methods are LAM-PCR (Albani *et al.*, 1993; Chen and Armstrong, 1995), DOP-PCR (Telenius *et al.*, 1992), PEP-PCR and I-PEP-PCR (Dietmaier *et al.*, 1999; Zhang *et al.*, 1992), and MDA (Dean *et al.*, 2002; Hosono *et al.*, 2003). Linker adaptor-mediated PCR (LAM-PCR or LA-PCR) is based on restriction enzyme digestions of microdissected chromosomes and subsequent ligation of suitable adaptors to digested DNA. Amplification of adaptor–DNA–adaptor constructs is carried out by adaptor-nested primers. The use of this method is limited by occurrence of particular restriction enzyme site within the analyzed region. Nevertheless, the feasibility of this method was shown, for example, by construction of plasmid from microdissected short arm of maize chromosome 6 (Stein *et al.*, 1998) and by amplification of microdissected chromosomes from rice and wheat (Zhou *et al.*, 2000). The comparison of LAM-PCR and DOP-PCR for comparative genomic hybridization of human chromosomes showed that very low amount of template DNA for PCR yielded both false positive and false negative results by DOP-PCR, but comparative genomic hybridization (CGH) could be analyzed correctly by using LAM-PCR (Pirker *et al.*, 2004).

DOP-PCR originally developed by Telenius *et al.* (1992) takes advantage of partially degenerated oligonucleotides together with initial low annealing temperature of PCR. DOP-PCR was successfully used for chromosome-specific library construction both in plant (Arumuganathan *et al.*, 1994; Jamilena *et al.*, 1995; Macas *et al.*, 1996) and animal species (Grimm *et al.*, 1997; Weikard *et al.*, 1997; Xiao *et al.*, 1996). Another and more general use of DOP-PCR is preparation of suitable chromosome-specific probes for cytogenetic studies. The DOP-PCR is a useful tool of choice to generate probes from microdissected chromosomes in the field of human clinical diagnostics to study chromosomal aberrations (Aubele *et al.*, 2000; Engelen *et al.*, 2000; Junker *et al.*, 1999). The painting DOP-PCR-derived chromosomal probes also enable to study, for example, rearrangements during evolution of animal genomes (Muller *et al.*, 1997; Scalzi and Hozier, 1998; Yang *et al.*, 1995).

Another method to amplify anonymous DNA is primer-extension-preamplification-PCR (PEP-PCR). This method was introduced by Zhang *et al.* (1992) and contrary to DOP-PCR utilizes totally degenerated PCR primers 15 nucleotides long. The original PEP-PCR protocol was later modified (improved PEP-PCR, I-PEP-PCR) by Dietmaier *et al.* (1999) by mainly adding proofreading *Pwo* polymerase to avoid amplification errors. The study comparing value of DOP-PCR, PEP-PCR, and I-PEP-PCR for amplification of flow-sorted and microdissected tumor cells shows that efficiency rate to amplify DNA from five cells is 100% for I-PEP-PCR, 33% for PEP-PCR, and 20% for DOP-PCR (Dietmaier *et al.*, 1999). PEP-PCR and I-PEP-PCR were successfully used for amplification DNA in human diagnostics (Tsai, 1999), in genotyping analysis of bovine embryos (Chrenek *et al.*, 2001), and to amplify DNA from microdissected single pollen grains in plants (Matsunaga *et al.*, 1999b).

The most recently introduced method for large-scale amplification of anonymous DNA is multiple displacement amplification (MDA). The method is based on the strand displacement amplification by use of Φ29 DNA polymerase and random exonuclease-resistant hexamer primers (Dean *et al.*, 2002). The main advantage of this method is minimal bias and generation of long DNA molecules (>10 kb) from a small amount of DNA (Lasken and Egholm, 2003). The processivity of Φ29 DNA polymerase is highest of all known polymerases (Blanco *et al.*, 1989) with error rate of about 1 in 10^6–10^7 (Esteban *et al.*, 1993).

IV. Application of Laser-Isolated Plant Chromosomes

A. FISH Techniques

The use of microdissected and amplified chromosomal or subchromosomal DNA is a method of choice to study the structure and evolution of both plant and animal genomes. Chromosome painting probes are currently commercially available for many animal species and represent a useful tool to study genome changes in tumor biology, pre- and postnatal diagnostics, and other fields of clinical diagnostics (Reid *et al.*, 1998). In plants, the use of complex subgenomic probes often lead to nonspecific FISH signal on all chromosomes due to the differences of complexity of genomes and organization of repetitive sequences compared to animals (Schmidt and Heslop-Harrison, 1998; Schubert *et al.*, 2001). Although chromosome painting was successfully carried out to specifically label wheat telosome 5L using probe from microdissected chromosome arm (Vega *et al.*, 1994), next experiments revealed that in this case the chromosome painting was rather a misinterpretation of an artifact (Fuchs *et al.*, 1996). To avoid hybridization of dispersed repeats, the CISS technique is often used. This method is based on blocking of repetitive DNA either before hybridization, when labeled probe is incubated with an excess of highly repetitive fraction of gDNA (cot-1), or by suppression of cross-hybridization with repetitive sequences by competitor DNA during hybridization. The main exceptions to the rule that plant chromosomes do not allow application of chromosome painting techniques are plant B chromosomes (Houben *et al.*, 2001) and evolutionary advanced sex chromosomes (Hobza *et al.*, 2004; Shibata *et al.*, 1999). Schubert *et al.* (2001) carried out comprehensive experiments using CISS-hybridization on *Haplopappus gracilis* and *Crepis capillaris* possessing morphologically distinguishable chromosomes. For this experiment, 40 microdissected chromosomes of each chromosome type were amplified by DOP-PCR. The hybridization was carried out using a probe derived from one amplified chromosome whereas 100-fold excess of unlabeled DOP-PCR amplified DNA from the rest of chromosomes of whole complement was used as a competitor during 2-h prehybridization. In this study, different combinations of probe/competitor ratio were used, but the only result of competitor presence was a weaker signal rather than the specific hybridization on the target chromosome.

We have designed a protocol to use microdissected X and Y chromosomes from *S. latifolia* for chromosome painting (Hobza *et al.*, 2004). Standard conditions for *in situ* hybridization gave us spread signal on all chromosomes even when we used suppressor unlabeled DNA in 100-fold excess. We have combined five hybridization parameters to get chromosome-specific signal: hybridization time, concentration of the probe, concentration of suppressor DNA, temperature of hybridi zation, and stringency of washing step. Finally, we found that by combination of short hybridization time (1 h) and low concentration of the probe (30 ng), we can paint not only the Y chromosome that can accumulate chromosome-specific repetitive DNA but also the X chromosome in *S. latifolia* (Fig. 4). This experiment revealed that sex chromosomes of *S. latifolia* that are supposed to be at the early stage of their evolution from a regular pair of autosomes, already largely differ in their DNA composition.

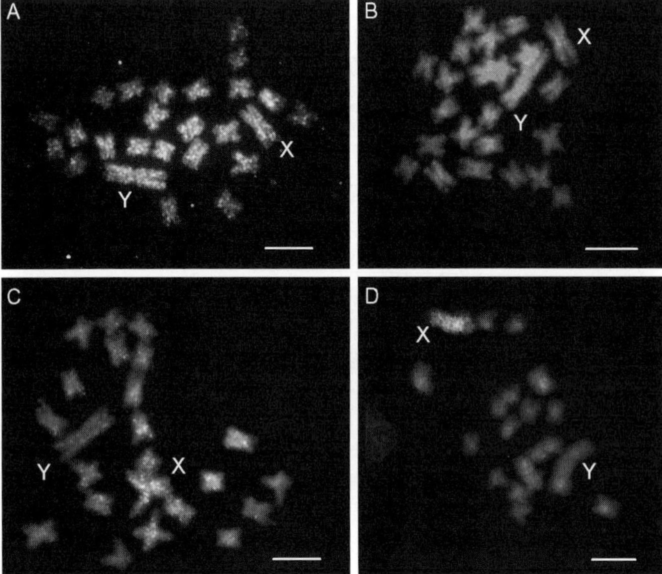

Fig. 4 FISH patterns on the male metaphase chromosomes of *S. latifolia* when using a different amount of DNA probe and hybridization time. Microdissected X chromosomes were used as a template for DOP-PCR. After amplification, DNA was labeled with Cy3 (here in yellow) and chromosomes were counterstained with DAPI (here in red). (A) Standard hybridization procedure. Strong hybridization signal was obtained on all the chromosomes when 100 ng of the probe was used (16-h hybridization). (B) High stringency of hybridization yields a very weak signal on all the chromosomes (6 ng of the probe, 1-h hybridization). (C) A slight tendency toward different chromosome labeling. The X chromosome is labeled on the entire length, while other chromosomes had only weak signals (15 ng of probe, 16-h hybridization). (D) Painting of the X chromosome when optimum amount of probe and hybridization time is used (30 ng of the probe per slide, hybridized for 1 h). The sex chromosomes indicated. Scale bars = 10 μm. (See Plate 14 in the color insert section.)

B. Chomosome-Specific DNA Libraries and Other Applications

The microdissection of chromosomes can be further used in experiments focused on physical mapping of DNA sequences and construction and screening of chromosome-specific libraries. Before the advent of amplification methods, for library construction from chromosome-specific DNA, direct cloning methods were employed. In animals, a technique to isolate a chromosome-specific DNA from microdissected chromosomes was applied to clone polytene chromosomes of *Drosophila* (Scalenghe *et al.*, 1981). Later, a similar approach of chromosome microdissection was successfully used for cloning of human chromosomes (Ludecke *et al.*, 1989).

In plants, Sandery *et al.* (1991) focused on cloning of rye chromosome B by incubation of dissected chromosomes with restriction enzyme and ligated DNA into λgt10 arms. The phage library was further successfully used to screen for sequences common to both A and B chromosomes. The construction of chromosome-specific library from microdissected chromosomes already based on amplification step was carried out by Jung *et al.* (1992). In this experiment, the authors used monosomic addition line of sugar beat (*Beta vulgaris*) with a chromosome originated from *Beta patellaris*. The microdissection of chromosomes was followed by restriction digestion and ligation of DNA pieces into a suitable plasmid vector. The ligated fragments were amplified using vector-specific primers, PCR products incubated with a second restriction enzyme and cloned into a new vector. The characterization of the library revealed mean size of inserts about 130 bp.

The construction of chromosome-specific libraries using LA-PCR amplification step was originally developed by Albani *et al.* (1993) to construct DNA library of wheat chromosome arms. A potential robustness of the method was showed in experiments of Chen and Armstrong (1995) that constructed a DNA library comprising 500,000 recombinant clones from a single microdissected oat chromosome. The library contained 59% low copy sequences with the mean size of inserts about 650 bp.

The feasibility of DOP-PCR for construction and screening of chromosome-specific libraries was comprehensively shown by Macas *et al.* (1996). Microdissection followed by DOP-PCR was, for example, used to construct centromere-proximal region-specific plasmid library of Japanese red pine, *Pinus densiflora* (Hizume *et al.*, 2001).

In *S. latifolia*, we have used complex DOP-PCR-derived sex chromosome-specific probes to screen a sample BAC library (Lengerova *et al.*, 2004). We have found a sex chromosome-specific, tandemly arrayed sequence that was further characterized by Hobza *et al.* (2006b). The construction of *S. latifolia* Y chromosome-specific library was used to search for sex-linked DNA markers (Hobza *et al.*, 2006a). First, several DNA sequences, either from the sex chromosomes or from the autosomes of *S. latifolia*, were chosen to test the complexity of the DOP-PCR products. The constructed library represents 10,000 clones that cover about 1% of the Y chromosome. On the basis of hybridization of the library with male and female genomic probes, the clone MK17, which revealed to be male specific, was chosen to be characterized in detail to confirm the reliability and feasibility of the selected method. Southern hybridization data confirmed the linkage of MK17 to the Y chromosome.

The microdissected or flow-sorted chromosomes can be used not only for construction and screening of subgenomic DNA libraries but also for direct physical mapping of known genes or markers by PCR with specific primers. The method was originally developed and first used in plants by Macas *et al.* (1993) to map genes coding for vicilin seed storage proteins in the field bean (*Vicia faba*) in a region which includes the centromere and the proximal parts of the short and the long arms of chromosome II. Kejnovsky *et al.* (2001) used the flow-sorted X chromosomes and autosomes as a template for PCR to map a group of *S. latifolia MROS* (male reproductive-organ specific) genes. The use of micro-dissected chromosomes for mapping of genes on the sex chromosomes is a method of choice since genetic mapping is impossible in nonrecombining regions of the Y (W) chromosome. We are currently developing a PCR assay that will help us to map genes not only to the complete sex chromosomes (Fig. 5) but also to particular arms of the submetacentric X chromosome in *S. latifolia.*

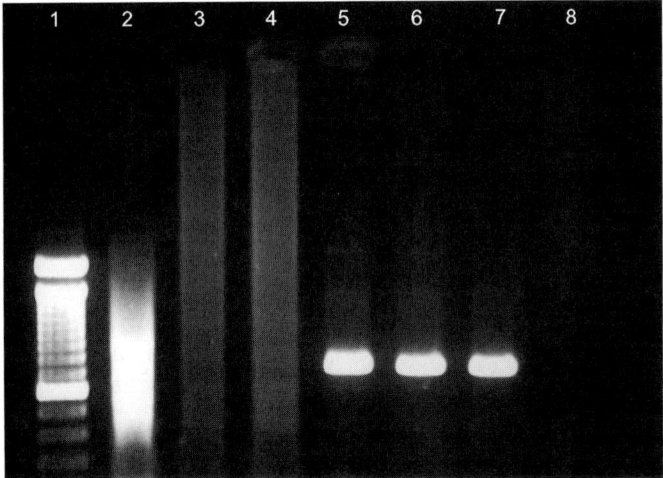

Fig. 5 PCR-amplification of microdissected X chromosomes by different methods (lanes 2–4) and mapping (localization) of *S. latifolia* gene *DD44* on the X chromosome by direct PCR with specific primers using amplified X chromosomes as a probe (lanes 5–8). For amplification of 10 X chromosomes we used DOP-PCR (lane 2), PEP-PCR (lane 3), and I-PEP-PCR (lane 4). The PCR products of amplification step were 100× diluted and 1 μl was used as a template for direct PCR with primers specific for the X-linked gene *DD44*. PCR was carried out with DOP-PCR (lane 5), PEP-PCR (lane 6), and I-PEP-PCR (lane 7) amplified X chromosomes as the template. The amplification using directly the X chromosomes as a template gave no PCR product (lane 8). PCR products were electrophoresed on a 1% agarose gel and stained with ethidium bromide. A 100-bp ladder (Invitrogen; lane 1).

V. Conclusions and Prospects

The advent of techniques in molecular biology is, nowadays, followed by exponential growth of data available on the structure of genomes and epigenomes. Although the efforts to sequence more model plant and animal species bring new insight into the evolution of genomes and chromosomes, we have still gaps in understanding of global chromosomal and subchromosomal changes that lead to the formation of genomes. The dissection of specific region(s) from complex matrix can help us to reduce the amount of analyses needed to unravel a biological context. Laser microdissection has become a method of choice to focus on particular subcellular compartments. Here we have summarized the facilities to analyze the structure and evolution of genomes that result from having the microdissected chromosomal DNA. We focus mainly on the use of techniques of microdissection of plant sex chromosomes and chromosomal arms. We further discuss the methods to amplify dissected DNA, the use of chromosome-specific probes in cytogenetics, and comparative genomics and tools to construct chromosome-specific DNA libraries. It is clear now that laser-assisted microdissection is not only a powerful tool to isolate chromosomes, but specific tissues, cells, and organelles can be also picked up and further analyzed for their DNA, RNA, or protein content (for a review, see Day *et al.*, 2005).

Acknowledgments

We acknowledge funding supports of the Grant Agency of the Czech Republic (521/06/0056) and the Ministry of Education of the Czech Republic (LC06004).

References

Albani, D., Cote, M. J., Armstrong, K. C., Chen, Q., Segal, A., and Robert, L. S. (1993). PCR amplification of microdissected wheat chromosome arms in a simple single tube reaction. *Plant J.* **4,** 899–903.

Arumuganathan, K., Martin, G. B., Telenius, H., Tanksley, S. D., and Earle, E. D. (1994). Chromosome 2-specific DNA clones from flow-sorted chromosomes of tomato. *Mol. Gen. Genet.* **242,** 551–558.

Atanassov, I., Delichere, C., Filatov, D. A., Charlesworth, D., Negrutiu, I., and Moneger, F. (2001). Analysis and evolution of two functional Y-linked loci in plant sex chromosome system. *Mol. Biol. Evol.* **18,** 2162–2168.

Aubele, M., Cummings, M., Walsch, A., Zitzelsberger, H., Nahrig, J., Hofler, H., and Werner, M. (2000). Heterogeneous chromosomal aberrations in intraductal breast lesions adjacent to invasive carcinoma. *Anal. Cell Pathol.* **20,** 17–24.

Blanco, L., Bernad, A., Lazaro, J. M., Martin, G., Garmendia, C., and Salas, M. (1989). Highly efficient DNA synthesis by the phage phi 29 DNA polymerase. Symmetrical mode of DNA replication. *J. Biol. Chem.* **264,** 8935–8940.

Buzek, J., Koutnikova, H., Houben, A., Riha, K., Janousek, B., Siroky, J., Grant, S., and Vyskot, B. (1997). Isolation and characterization of X chromosome-derived DNA sequences from a dioecious plant *Melandrium album. Chromosome Res.* **5,** 57–65.

Chen, Q., and Armstrong, K. C. (1995). Characterization of a library from a single microdissected oat (*Avena sativa* L.) chromosome. *Genome* **38,** 706–714.

Chrenek, P., Boulanger, L., Heyman, Y., Uhrin, P., Laurincik, J., Bulla, J., and Renard, J. P. (2001). Sexing and multiple genotype analysis from a single cell of bovine embryo. *Theriogenology* **55,** 1071–1081.

Day, R. C., Grossniklaus, U., and Macknight, R. C. (2005). Be more specific! Laser-assisted microdissection of plant cells. *Trends Plant Sci.* **10,** 397–406.

Dean, F. B., Hosono, S., Fang, L. H., Wu, X. H., Faruqi, A. F., Bray-Ward, P., Sun, Z. Y., Zong, Q. L., Du, Y. F., Du, J., Driscoll, M., Song, W., *et al.* (2002). Comprehensive human genome amplification using multiple displacement amplification. *Proc. Natl. Acad. Sci. USA* **99,** 5261–5266.

Delichere, C., Veuskens, J., Hernould, M., Baarbacar, N., Mouras, A., Negrutiu, I., and Moneger, F. (1999). *SlY1*, the first active gene cloned from a plant Y chromosome, encodes a WD-repeat protein. *EMBO J.* **18,** 4169–4179.

Dietmaier, W., Hartmann, A., Wallinger, S., Heinmoller, E., Kerner, T., Endl, E., Jauch, K.-W., Hofstadter, F., and Ruschhoff, J. (1999). Multiple mutation analyses in single tumor cells with improved whole genome amplification. *Am. J. Pathol.* **154,** 83–95.

Dolezel, J., Kubalakova, M., Bartos, J., and Macas, J. (2004). Flow cytogenetics and plant genome mapping. *Chromosome Res.* **12,** 77–91.

Dolezel, J., Lysak, M. A., Kubalakova, M., Simkova, H., Macas, J., and Lucretti, S. (2001). Sorting of Plant Chromosomes. *In* "Flow Cytometry, Part B" (Z. Darzynkiewicz, H. A. Crissman, and J. P. Robinson, eds.), 3rd edn., pp. 3–31. Academic Press, San Diego.

Engelen, J. J., Moog, U., Evers, J. L., Dassen, H., Albrechts, J. C., and Hamers, A. J. (2000). Duplication of chromosome region 8p23. 1–> p23. 3: A benign variant? *Am. J. Med. Genet.* **91,** 18–21.

Esteban, J. A., Salas, M., and Blanco, L. (1993). Fidelity of phi 29 DNA polymerase. Comparison between protein-primed initiation and DNA polymerization. *J. Biol. Chem.* **268,** 2719–2726.

Fuchs, J., Houben, A., Brandes, A., and Schubert, I. (1996). Chromosome "painting" in plants—a feasible technique? *Chromosoma* **104,** 315–320.

Fukui, K., Minezawa, M., Kamisugi, Y., Ishikawa, M., Ohmido, N., Yanagisawa, T., Fujishita, M., and Sakai, F. (1992). Microdissection of plant chromosomes by argon-ion laser-beam. *Theor. Appl. Genet.* **84,** 787–791.

Gracia, E., Ray, M. E., Polymeropoulos, M. H., Dehejia, A., Meltzer, P. S., and Trent, J. M. (1997). Isolation of chromosome-specific ESTs by microdissection-mediated cDNA capture. *Genome Res.* **7,** 100–107.

Grimm, D. R., Goldman, T., Holley-Shanks, R., Buoen, L., Mendiola, J., Schook, L. B., Louis, C., Rohrer, G. A., and Lunney, J. K. (1997). Mapping of microsatellite markers developed from a flow-sorted swine chromosome 6 library. *Mamm. Genome* **8,** 193–199.

Hizume, M., Shibata, F., Maruyama, Y., and Kondo, T. (2001). Cloning of DNA sequences localized on proximal fluorescent chromosome bands by microdissection in *Pinus densiflora* Sieb. & Zucc. *Chromosoma* **110,** 345–351.

Hladilova, R., Siroky, J., and Vyskot, B. (1998). A cytospin technique for spreading plant metaphases suitable for immunofluorescence studies. *Biotech. Histochem.* **73,** 150–156.

Hobza, R., Hrusakova, P., Safar, J., Bartos, J., Janousek, B., Zluvova, J., Michu, E., Dolezel, J., and Vyskot, B. (2006a). Isolation of a Y chromosome specific marker closely linked to the region controlling supression of gynoeceum development in *Silene latifolia. Theor. Appl. Genet.* **113,** 280–287.

Hobza, R., Lengerova, M., Cernohorska, H., Rubes, J., and Vyskot, B. (2004). FAST-FISH with laser beam microdissected DOP-PCR probe distinguishes the sex chromosomes of *Silene latifolia. Chromosome Res.* **12,** 245–250.

Hobza, R., Lengerova, M., Svoboda, J., Kubekova, H., Kejnovsky, E., and Vyskot, B. (2006b). An accumulation of tandem DNA repeats on the Y chromosome in *Silene latifolia* during early stages of sex chromosome evolution. *Chromosoma* **115,** 376–382.

Hosono, S., Faruqi, A. F., Dean, F. B., Du, Y. F., Sun, Z. Y., Wu, X. H., Du, J., Kingsmore, S. F., Egholm, M., and Lasken, R. S. (2003). Unbiased whole-genome amplification directly from clinical samples. *Genome Res.* **13,** 954–964.

Houben, A., Field, B. L., and Saunders, V. A. (2001). Microdissection and chromosome painting of plant B chromosomes. *Methods Cell Sci.* **23,** 115–124.

Hozier, J., Graham, R., Westfall, T., Siebert, P., and Davis, L. (1994). Preparative *in situ* hybridization: Selection of chromosome region-specific libraries on mitotic chromosomes. *Genomics* **19,** 441–447.

Jamilena, M., Garrido-Ramos, M., Ruiz Rejon, M., Ruiz Rejon, C., and Parker, J. S. (1995). Characterisation of repeated sequences from microdissected B chromosomes of *Crepis capillaris*. *Chromosoma* **104,** 113–120.

Jung, C., Claussen, U., Horsthemke, B., Fischer, F., and Herrmann, R. G. (1992). DNA library from an individual *Beta patellaris* chromosome conferring nematode resistance obtained by microdissection of meiotic metaphase chromosomes. *Plant Mol. Biol.* **20,** 503–511.

Junker, K., Hindermann, W., Schubert, J., and Schlichter, A. (1999). Differentiation of multifocal renal cell carcinoma by comparative genomic hybridization. *Anticancer Res.* **19,** 1487–1492.

Kejnovsky, E., Vrana, J., Matsunaga, S., Soucek, P., Siroky, J., Dolezel, J., and Vyskot, B. (2001). Localization of male-specifically expressed *MROS* genes of *Silene latifolia* by PCR on flow-sorted sex chromosomes and autosomes. *Genetics* **158,** 1269–1277.

Kubickova, S., Cernohorska, H., Musilova, P., and Rubes, J. (2002). The use of laser microdissection for the preparation of chromosome-specific painting probes in farm animals. *Chromosome Res.* **10,** 571–577.

Lasken, R. S., and Egholm, M. (2003). Whole genome amplification: Abundant supplies of DNA from precious samples or clinical specimens. *Trends Biotechnol.* **21,** 531–535.

Lengerova, M., Kejnovsky, E., Hobza, R., Macas, J., Grant, S. R., and Vyskot, B. (2004). Multicolor FISH mapping of the dioecious model plant, *Silene latifolia*. *Theor. Appl. Genet.* **108,** 1193–1199.

Ludecke, H. J., Senger, G., Claussen, U., and Horsthemke, B. (1989). Cloning defined regions of the human genome by microdissection of banded chromosomes and enzymatic amplification. *Nature* **338,** 348–350.

Macas, J., Gualberti, G., Nouzova, M., Samec, P., Lucretti, S., and Dolezel, J. (1996). Construction of chromosome-specific DNA libraries covering the whole genome of field bean (*Vicia faba* L.). *Chromosome Res.* **4,** 531–539.

Macas, J., Weschke, W., Bumlein, H., Pich, U., Houben, A., Wobus, U., and Schubert, I. (1993). Localization of vicilin genes via polymerase chain reaction on microisolated field bean chromosomes. *Plant J.* **3,** 883–886.

Matsunaga, S., Kawano, S., Michimoto, T., Higashiyama, T., Nakao, S., Sakai, A., and Kuroiwa, T. (1999a). Semi-automatic laser beam microdissection of the Y chromosome and analysis of Y chromosome DNA in a dioecious plant, *Silene latifolia*. *Plant Cell Physiol.* **40,** 60–68.

Matsunaga, S., Schutze, K., Donnison, I. S., Grant, S. R., Kuroiwa, T., and Kawano, S. (1999b). Single pollen typing combined with laser-mediated manipulation. *Plant J.* **20,** 371–378.

Menczel, K., Galiba, G., Nagy, F., and Maliga, P. (1982). Effect of radiation dosage on efficiency of chloroplast transfer by protoplast fusion in *Nicotiana*. *Genetics* **100,** 487–495.

Monajembashi, S., Cremer, C., Cremer, T., Wolfrum, J., and Gruelich, K. O. (1986). Microdissection of human-chromosomes by a laserbeam. *Exp. Cell Res.* **167,** 262–265.

Muller, S., O'Brien, P. C., Ferguson-Smith, M. A., and Wienberg, J. (1997). Reciprocal chromosome painting between human and prosimians (*Eulemur macaco macaco* and *E. fulvus mayottensis*). *Cytogenet. Cell Genet.* **78,** 260–271.

Pirker, C., Raidl, M., Steiner, E., Elbling, L., Holzmann, K., Spiegl-Kreinecker, S., Aubele, M., Grasl-Kraupp, B., Marosi, C., Micksche, M., and Berger, W. (2004). Whole genome amplification for CGH analysis: Linker-adaptor PCR as the method of choice for difficult and limited samples. *Cytometry* **61,** 26–34.

Potier, M. C., Kuo, W. L., Dutriaux, A., Gray, J., and Goedert, M. (1992). Construction and characterization of a yeast artificial chromosome library containing 1.5 equivalents of human chromosome 21. *Genomics* **14,** 481–483.

Reid, T., Schrock, E., Ning, Y., and Wineberg, J. (1998). Chromosome painting: A useful art. *Hum. Mol. Genet.* **7,** 1619–1626.

Safar, J., Bartos, J., Janda, J., Bellec, A., Kubalakova, M., Valarik, M., Pateyron, S., Weiserova, J., Tuskova, R., Cihalikova, J., Simkova, H., Vrana, J., *et al.* (2004). Dissecting large and complex genomes: Flow sorting and BAC cloning of individual chromosomes from bread wheat. *Plant J.* **39,** 960–968.

Sandery, M. J., Forster, J. W., Macadam, S. R., Blunden, R., Jones, R. N., and Brown, D. M. (1991). Isolation of a sequence common to A- and B-chromosomes of rye (*Secale cereale*) by microcloning. *Plant Mol. Biol. Rep.* **9,** 21–30.

Scalenghe, F., Turco, E., Edstrom, J. E., Pirrota, V., and Melli, M. L. (1981). Microdissection and cloning of DNA from a specific region of *Drosophila melanogaster* polytene chromosomes. *Chromosoma* **82,** 205–216.

Scalzi, J. M., and Hozier, J. C. (1998). Comparative genome mapping: Mouse and rat homologies revealed by fluorescence *in situ* hybridization. *Genomics* **47,** 44–51.

Schermelleh, L., Thalhammer, S., Heckl, W., Posl, H., Cremer, T., Schutze, K., and Cremer, M. (1999). Laser microdissection and laser pressure catapulting for the generation of chromosome-specific painting probes. *BioTechniques* **27,** 362–367.

Schmidt, T., and Heslop-Harrison, J. S. (1998). Genomes, genes and junk: The large-scale organization of plant chromosomes. *Trends Plant Sci.* **3,** 195–199.

Schubert, I., Fransz, P. F., Fuchs, J., and de Jong, J. H. (2001). Chromosome painting in plants. *Methods Cell Sci.* **23,** 57–69.

Scutt, C. P., Kamisugi, Y., Sakai, F., and Gilmartin, P. M. (1997). Laser isolation of plant sex chromosomes: Studies on the DNA composition of the X and Y sex chromosomes of *Silene latifolia. Genome* **40,** 705–715.

Shibata, F., Hizume, M., and Kuroki, Y. (1999). Chromosome painting of Y chromosomes and isolation of a Y chromosome-specific repetitive sequence in the dioecious plant *Rumex acetosa. Chromosoma* **108,** 266–270.

Siroky, J., Hodurkova, J., Negrutiu, I., and Vyskot, B. (1999). Functional and structural chromosome analyses in autotetraploid *Silene latifolia. Ann. Bot.* **84,** 633–638.

Siroky, J., Ruffini-Castiglione, M., and Vyskot, B. (1998). DNA methylation patterns of *Melandrium album* chromosomes. *Chromosome Res.* **6,** 441–446.

Stein, N., Ponelies, N., Musket, T., McMullen, M., and Weber, G. (1998). Chromosome microdissection and region-specific libraries from pachytene chromosomes of maize (*Zea mays* L.). *Plant J.* **13,** 281–289.

Telenius, H., Carter, N. P., Bebb, C. E., Nordenskjodl, M., Ponder, B. A. J., and Tunnecliffe, A. (1992). Degenerate oligonucleotide-primed PCR: General amplification of target DNA by a single degenerate primer. *Genomics* **13,** 718–725.

Tsai, Y. H. (1999). Cost-effective one-step PCR amplification of cystic fibrosis delta F508 fragment in a single cell for preimplantation genetic diagnosis. *Prenat. Diagn.* **19,** 1048–1051.

Vega, J. M., Abbo, S., Feldman, M., and Levy, A. A. (1994). Chromosome painting in plants: *In situ* hybridization with a DNA probe from a specific microdissected chromosome arm of common wheat. *Proc. Natl. Acad. Sci. USA* **91,** 12041–12045.

Veuskens, J., Marie, D., Brown, S. C., Jacobs, M., and Negrutiu, I. (1995). Flow sorting of the Y sex chromosome in the dioecious plant *Melandrium album. Cytometry* **21,** 363–373.

Vyskot, B., and Hobza, R. (2004). Gender in plants: Sex chromosomes are emerging from the fog. *Trends Genet.* **20,** 432–438.

Vyskot, B., Siroky, J., Hladilova, R., Belyaev, N. D., and Turner, B. M. (1999). Euchromatic domains in plant chromosomes as revealed by H4 histone acetylation and early DNA replication. *Genome* **42,** 343–350.

Weikard, R., Goldammer, T., Kuhn, C., Barendse, W., and Schwerin, M. (1997). Targeted development of microsatellite markers from the defined region of bovine chromosome 6q21–31. *Mamm. Genome* **8,** 836–840.

Xiao, Y., Slijepcevic, P., Arkesteijn, G., Darroudi, F., and Natarajan, A. T. (1996). Development of DNA libraries specific for Chinese hamster chromosomes 3, 4, 9, 10, X, and Y by DOP-PCR. *Cytogenet. Cell Genet.* **75,** 57–62.

Yang, F., Carter, N. P., Shi, L., and Ferguson-Smith, M. A. (1995). A comparative study of karyotypes of muntjacs by chromosome painting. *Chromosoma* **103,** 642–652.

Zhang, L., Cui, X., Schmitt, K., Hubert, R., Navidi, W., and Arnheim, N. (1992). Whole genome amplification from a single cell: Implications for genetic analysis. *Proc. Natl. Acad. Sci. USA* **89,** 5847–5851.

Zhou, Y., Wang, H., Wie, J., Cui, X., Deng, X., Wang, X., and Chen, Z. (2000). Comparison of two PCR techniques used in amplification of microdissected plant chromosomes from rice and wheat. *BioTechniques* **28,** 766–774.

PART IV

Laser Tweezers/Optical Trapping

CHAPTER 16

Optical Tweezers: Tethers, Wavelengths, and Heat

Michael W. Berns

Departments of Biomedical Engineering, Developmental and Cell Biology, and Surgery
Beckman Laser Institute, University of California, Irvine, California 92612

Department of Bioengineering, Whitaker Institute for Biomedical Engineering
University of California, San Diego, La Jolla, California 92093

This chapter briefly review the four major methods of optical trapping: (1) directly on to single cells or groups of cells, (2) directly on to organelles and structures inside of the cell, (3) on to a bead as a "handle" to apply force, and (4) on to a bead that has been coated with an antigen or antibody that is moved to the cell membrane for the purpose of activation of a chemical response (no force is applied to the cell). In addition, this chapter discusses the issue of optimal wavelength selection for trapping and the potential temperature rise within the trap.

I. Introduction

Optical trapping (also referred to as *optical tweezers* and *laser tweezers*) in biology had its genesis from the work on atom trapping for which Steve Chu (United States), Claude Cohen-Tannoudji (France), and William D. Phillips (United States) were awarded the Nobel Prize for Physics in 1997 (Chu *et al.*, 1985, 1986).

METHODS IN CELL BIOLOGY, VOL. 82

0091-679X/07 $35.00
DOI: 10.1016/S0091-679X(06)82016-9

The use of an optical trap in biology was first described by Ashkin and Dziedric in 1987. They used 120 mW of an argon ion laser (488 and 514 nm) to trap bacteria and viruses (Ashkin and Dziedric, 1987). Because of the potential damage to cells at these wavelengths due to absorption by numerous biological molecules (cytochromes, porphyrins, and so on), Ashkin *et al.* (1987) described an improved laser trap using infrared light from an Nd:YAG laser at 1.06 μm. With the IR laser they were able to (1) observe the reproduction of bacteria and yeast while held in the trap, (2) manipulate organelles inside the plant alga *Spirogyra* and a variety of Protozoa, (3) demonstrate that individual bacteria could be moved with velocities as high as 500 μm *s*–1 and clumps of cells as fast as 1 μm *s*–1, and (4) move individual bacteria from one sample to another. These series of studies resulted in the introduction of a major new technology in biology. A good review of different kinds of studies that use laser tweezers was published in volume 55 in the *Methods in Cell Biology* series by Michael Sheetz of Columbia University (Sheetz, 1998). The opening chapter in that book is an excellent dissertation on the physical and optical principles governing optical traps (Ashkin, 1998). Additional chapters in that book cover a wide range of applications of optical tweezers and though a bit outdated, it is still highly recommended for researchers and students looking for a good introduction to optical tweezers and their application to problems in biology (that volume has been published in softcover, thus making it affordable for students). Another excellent review that covers laser tweezers as well as the combination of laser tweezers with laser scissors was published in 1999 by the coeditor of this volume (Greulich, 1999).

Following Ashkin's introduction of optical traps for manipulation of bacteria, viruses, protozoa, and unicellular algae numerous investigators have applied optical tweezers to study problems in biology (a recent search of the *PubMed* database under *optical traps/tweezers* resulted in over 500 citations). Generally, four different approaches are taken in the use of single beam optical traps in biology. The *first* is to apply the optical trap directly to the entire cell or groups of cells as in Ashkin's original studies on unicellular organisms (Ashkin and Dziedric, 1987; Ashkin *et al.*, 1987), the studies by Tadir and colleagues on human spermatozoa (Nascimento *et al.*, 2006; Tadir *et al.*, 1989, 1990; see also Chapter 21 by Shao *et al.*, this volume), and on the fusion of two mammalian cells (Steubing *et al.*, 1991).

The *second* approach is to apply the optical trap directly to an organelle or structure *inside* the live cell. These kinds of studies have been performed on such diverse structures as the cytoplasmic filaments in plant cells (Ashkin and Dziedric, 1989), microtubule-associated mitochondria in the large amoeba Reticulomyxa (Felgner *et al.*, 1998), chromosomes on the mitotic spindle of vertebrate cells (Berns *et al.*, 1989; Hong *et al.*, 1991; Liang *et al.*, 1993), and the nucleoli in a fungal plant cells (Berns *et al.*, 1992).

The *third* approach is to use optically refractive 1- to 10-μm diameter beads (polystyrene, carboxylate, or latex) as "handles" for the application of the optical force. A good review of the procedures of bead preparation, tweezer calibration, and tethering to the cell membrane for studies on membrane biomechanics see the review by Dai and Sheetz (Dai and Sheetz, 1998). In one series of studies, 1-μm diameter latex beads were placed on the surface of mouse fibroblasts and laser trapping forces were used either to restrain the beads as a mimic of extracellular resistance or to apply a calibrated force to the beads in order to apply an external load to spatially defined sites in order to estimate traction forces exerted by the cell (Choquet *et al.*, 1997). In another cell surface application, beads were bound to integrins in the cell membrane of human umbilical vein endothelial cells (HUVECs) that had been transiently transfected with a double-fluorescent-labeled Src biosensor. By measuring the wave of FRET signal it was possible to detect a series of signal transduction events within the cell in response to the applied force (Wang *et al.*, 2005; see Chapter 18 by Botvinick and Wang, this volume).

In other studies at the single molecule level, 1-μm diameter beads were attached to single 16- to 100-μm length DNA molecules which were subsequently stretched in several different configurations (Perkins *et al.*, 1994). As the DNA relaxed, it closely followed a path similar to its original contour. This and other observations on the relaxation of the DNA molecule provided direct evidence in support of several major assumptions of polymer physics. In subsequent DNA-stretching studies by Steven Block and his colleagues (Wang *et al.*, 1997), a laser tweezer interferometer was built that provided feedback control in response to the amount of force exerted on a DNA molecule that was stuck to a glass slide on one end and tethered to a bead on the other end. As the microscope stage was moved the DNA molecule was stretched and the amount of laser tweezer force exerted on the bead could be modulated in response to the DNA's elasticity. In this way they were able accurately to measure the force extension (F–x) of different lengths of DNA molecules. In another single-molecule study, beads were conjugated to single molecules in order to study molecular motors. In this study, beads were conjugated to the molecule kinesin (Kuo and Sheetz, 1993). It was possible to reversibly "stall" the translocation of single microtubules by single kinesin molecules by applying laser trapping forces to streptavidin-coated latex beads. The stalled microtubules escaped optical trapping forces of 1.9 pN. This study established the use of optical traps to measure force parameters of single macromolecules.

The *fourth* approach in the use of optical tweezers has been to use the trap as a way to translocate beads coated with specific proteins to specific regions of the cell membrane/surface in order to detect/activate specific cell surface receptors. One of the first studies using this approach was to map the sensitivity of immune T cells with respect to the polarity and number of receptors for Ca^{2+}on the cell surface (Wei *et al.*, 1999). Contact with antigen-presenting cells (APCs), such as B lymphocytes, initiates an activation cascade within T lymphocytes, including a rise in cytosolic calcium. In this study, calcium imaging combined with an optical trap enabled the T cell contact requirements and polarity to be investigated at the single-cell level. APCs or anti-CD3 mAb-coated beads were trapped with a laser and placed at different locations along the T cell. The T cell has a distinct morphology when migrating, with a broad migrating edge and a narrow tail (Fig. 1). It was found that T cells were threefold more sensitive to bead-coated APC contact made at the leading edge than with contact made at the tail. Anti-CD3 antibody-coated 6-μm beads induced calcium

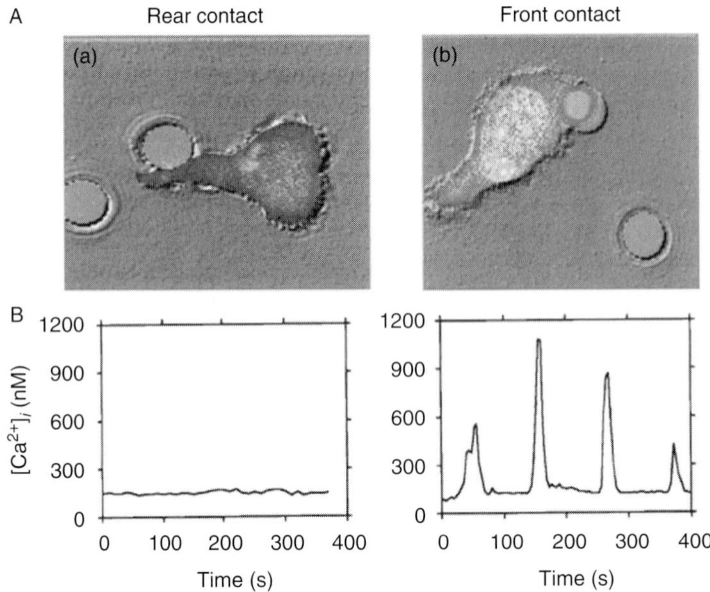

Fig. 1 Optical trapping of antibody-coated bead identifies functional T cell polarity. (A) Bright field and fluorescence images are shown as an overlay. Oregon green and Fura Red dyes were preincubated with the T cells for 90 min. Cells produced an emission shift from red to green when $[Ca^{2+}]$ was elevated. Fluorescein-conjugated antibody-coated 6.2-μm diameter beads were presented to the cell surface using the optical trap. Beads were presented either to the tail (a) or the leading edge (b) of the cell. Time course of the change in $[Ca^{+2}]$ was estimated every 4 s. See original paper, Wei *et al.* (1999), for further details. Reprinted with permission of the U.S. National Academy of Sciences. Copyright (1999) National Academy of Sciences, USA. (See Plate 15 in the color insert section.)

signaling with approximately tenfold higher frequency on contact with the leading edge of the T cell than on contact with the trailing edge. Alterations in antibody density (2–500 per micrometer) and bead size (1- to 6-μm in diameter) were used to determine the spatial requirements and the minimal number of receptors which must be engaged to transmit a positive signal. The authors suggest that activation of no more than 340 T cell receptors (~1% of the total on the cell) is sufficient to initiate Ca^{2+} signaling and that the minimal contact area was $\sim 3\mu^2$. This was one of the first studies using protein-coated beads to activate specific cell-surface receptors and to demonstrate a head–tail polarity of the receptors on the cell surface.

II. Optimum Trapping Wavelength

Arthur Ashkin's first laser trap used the blue-green wavelengths of the argon ion laser (Ashkin and Dziedric, 1987). However, it soon became apparent that for most biological objects those wavelengths were substantially absorbed, therefore making them unsuitable for laser trapping experiments. The first study that examined the range of wavelengths for optical trapping examined the region from 700 to 840 nm using an argon ion-pumped Ti:Sapphire laser (Vorobjev *et al.*, 1993). In this study the double-arms of a single chromosome in metaphase of mitosis were exposed to the laser trap for periods of 0.3 s to 5 min at wavelength intervals of 20 nm for the wavelength ranges 700–740 nm and 780–840 nm, and for intervals of 5 nm for the wavelength range 740–780 nm. Laser power in the trapping focal spot was kept at 130 mW for all the wavelengths tested. The criterion for no significant effect was that the cell could undergo a normal mitosis following the laser exposure. Significant effects were either permanent chromosome bridges followed by an abnormal mitosis or mitotic arrest without separation of the chromosomes and cytokinesis. The results revealed that the greatest percentage of abnormalities (100%) occurred at 760 nm and the lowest percentage was observed at 820 nm (40%).

Though the previous study defined an optimal wavelength region for the tunable Ti:Sapphire laser, it did not extend far enough into the near-IR and IR regions to be widely useful to investigators who were using either the 1.06-μm Nd:YAG laser or longer wavelengths from the Ti:Sapphire laser. A study examining the cloning efficiency of cells exposed to trapping wavelengths from 700 nm to 1.06 μm provided the needed data (Liang *et al.*, 1996). In this study individual cells were exposed to focal spot powers of 88 and 176 mW for periods of 3, 5, and 10 min and then the ability of the single cell to grow into an isolated clone was determined (Fig. 2). In this study, a total of 2471 cells were exposed and followed to determine clonal growth. The results indicated that 950–1000 nm appeared to be the spectral region with the least amount of deleterious effects. Also, there was a reasonably good trapping region around 800–830 nm. This region corresponded to the optimal trapping region of 820 nm described in the previous chromosome irradiation experiment (Vorobjev *et al.*, 1993). Similarly, both studies suggest that 750–760 nm is most deleterious, and trapping in this wavelength region must be avoided. Of particular interest was the apparent drop in cloning efficiency past 1000 nm. This is of particular interest because many laser tweezer experiments use the 1060-nm wavelength Nd:YAG laser. Whereas this would not make a major difference in studies in which latex or other nonabsorbing beads are used, it does pose questions for investigators who are using the Nd:YAG trapping laser for direct exposure to cells or their organelles. Whether this would pose an equivocation of these studies must be carefully considered. Though some thermal effects would be expected with this wavelength, the amount of heating produced by the average powers and irradiances used would probably not be severe enough to affect the experiments (see Section III for discussion of thermal effects in a laser trap). However, one study has shown that there can be good trapping at 700 nm without short-term damage to cells. This study used a 76-MHz Ti:Sapphire laser and had 400 mW in the objective focal point (Berns *et al.*, 1992).

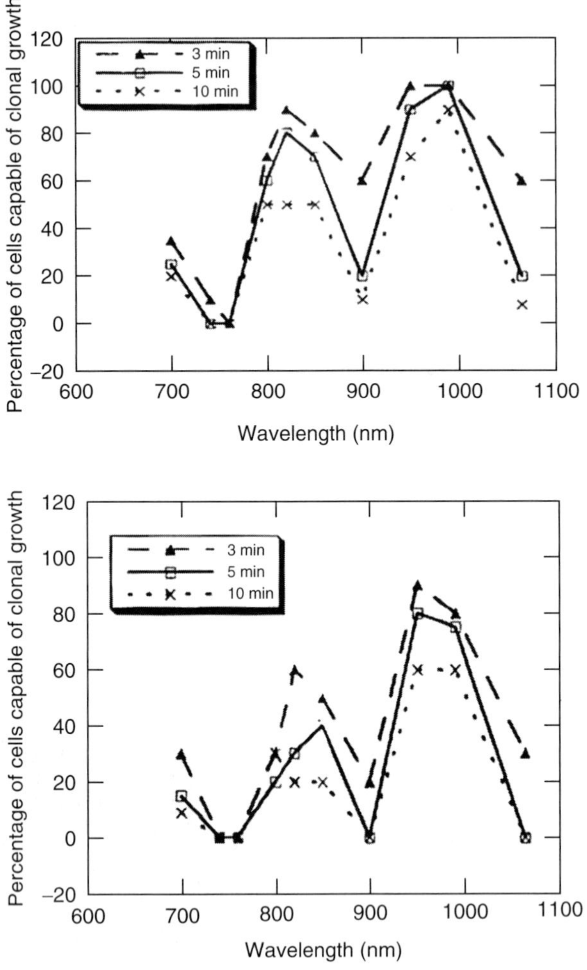

Fig. 2 Clonability of Chinese hamster ovary (CHO) cells following exposure to different trapping wavelengths for different time periods with 88 mW (upper graph) and 176 mW (lower graph) in the laser focal spot. See Liang *et al.* (1996) for details. Courtesy of the Biophysical Society.

III. Thermal Effects in a Laser Trap

The determination of how much heat is generated in an optical trap was first reported in 1994 (Liu *et al.*, 1994). In this study the temperature-dependent fluorescence emission from Laurdan (6-dodecanoyl-2-dimethylamino naphthalene) embedded in the membrane of 10-μm diameter liposomes was used (Fig. 3). This dye decreases in its 450-nm peak emission intensity with temperature rise until the phase transition temperature of the liposome membrane is reached at 42°C at

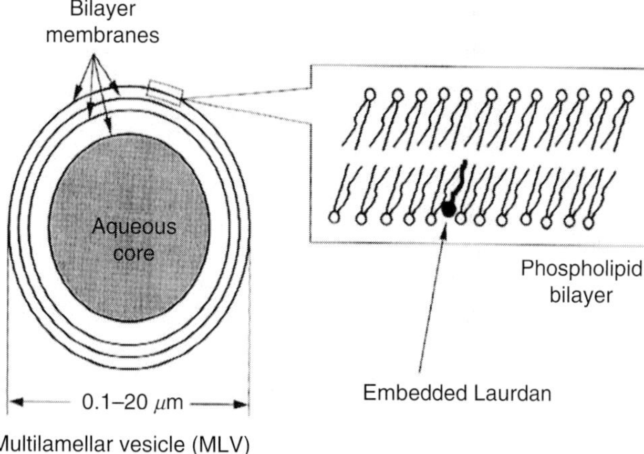

Fig. 3 Structure of multilamellar liposome vesicle, consisting of a phospholipid bilayer. A Laurdan dye probe is embedded within the bilayer membrane and used to sense localized temperature changes [see Figs. 4 and 5 and further details in Liu *et al.* (1995)]. Courtesy of the Biophysical Society.

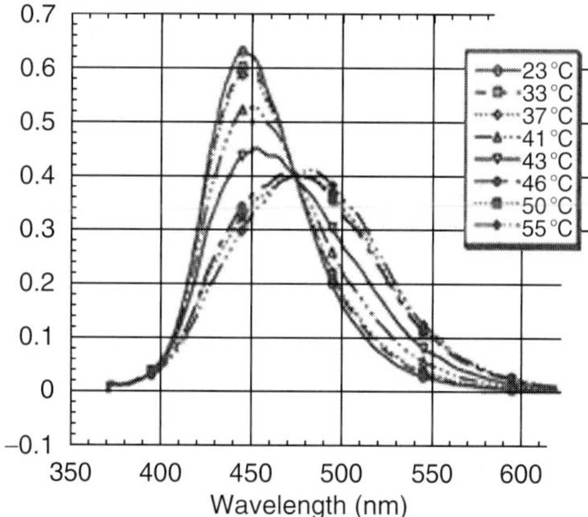

Fig. 4 Temperature-dependent emission spectra from a 10-μm diameter liposome. The emission redshifts by 40 nm, above the liposome transition temperature @ 38–42°C. For further details see Liu *et al.* (1994).

which a 40-nm red Stokes shift is observed. This is followed by an additional linear decease in intensity with temperature rise at the peak emission wavelength of 490 nm (Fig. 4). Using this "fluorescence thermometer" it was possible to measure temperature change for 23–55°C. The result was a straight line fit of 1.1°C/100 mW of laser power in the focal spot (Fig. 5).

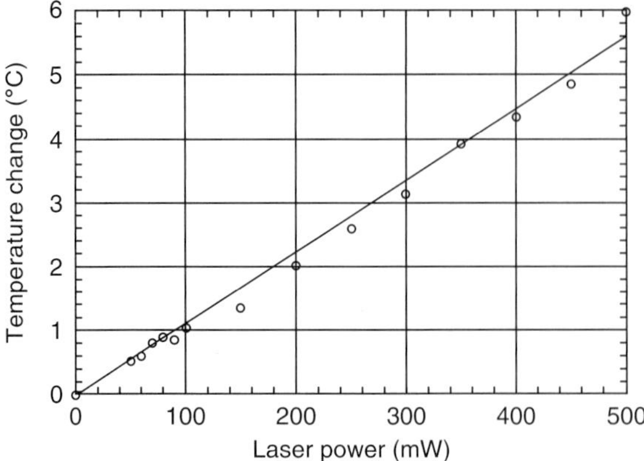

Fig. 5 Experimental data on the temperature changes induced at a given incident laser power. The straight line slope fits a temperature rise of 1.1°C/100 mW laser power in the focal volume. See Liu *et al.*, (1995). Courtesy of the Biophysical Society.

On the basis of the Laurdan-bound liposome "fluorescence thermometer" results, studies were conducted using Chinese hamster ovary (CHO) cells (Liu *et al.*, 1995). Similar to the liposome studies, a decrease and shift in fluorescence was observed with increasing temperature. For the range of laser powers examined, a heating rate of 1.15 ± 0.25°C/100 mW was observed. From these studies it appeared that though there is a temperature rise when a 1.06-μm laser trap is used, the magnitude of the heating, 1.1°C/100 mW, is probably insignificant for most laser trapping studies. However, for studies that use laser trapping powers above 200 mW in the laser focal point, temperature rise could be significant. In addition, the thermal rise when other trapping wavelengths are used may be different.

IV. Conclusions

The purpose of this chapter has been briefly to review the four ways in which laser traps are generally used. This has not been an extensive discussion, but rather a discussion that is intended to provide a basis for an understanding of how laser traps are of general use in biology. Studies have not been discussed that employ multiple trapping beams (Berns *et al.*, 1998; Simmons *et al.*, 1996), rapidly scanning of a single trapping beam to give the appearance of multiple traps (Molloy, 1998), axicon-based traps (see Chapter 21 by Shao *et al.*, this volume), or the use of traps to rotate birefringent crystals (see Chapter 19 by Parkin *et al.* and Chapter 20 by Mohanty and Gupta, this volume). A discussion of the optimal laser trapping wavelengths has been presented because this is often overlooked in designing trapping experiments. It is often assumed that either the 1.06-μm Nd:YAG laser wavelength or another wavelength close to it from other solid state lasers is the best. However, there are traps that employ diode lasers in the near-IR as well as the

Ti:Sapphire laser systems that are good. In general, the wavelength is not that critical when latex or polystyrene beads are used as handles for the trapping beam, unless there are specific absorption characteristics associated with the molecules that are conjugated to the beads that would result in absorption at the trapping wavelengths. Finally, the thermal effects of the trapping beam must be considered. Though it has been shown that the rate of temperature increase is around 1°C/100 mW of laser trapping power in the focal spot, elimination of temperature rise as a mitigating side effect cannot be entirely ruled out, especially when laser powers of 300 mW and above are used. Temperature increases of even a few degrees centigrade conceivably could increase the rate of reactions in the focal volume and may even have a magnified effect on temperature-sensitive biochemical pathways.

References

Ashkin, A. (1998). Forces of a single-beam gradient laser trap on a dielectric sphere in the ray optics regime. *Methods Cell Biol.* **55,** 1–25.

Ashkin, A., and Dziedric, J. M. (1987). Optical trapping and manipulation of viruses and bacteria. *Science* **235,** 1517–1520.

Ashkin, A., and Dziedric, J. M. (1989). Internal cell manipulation using infrared laser traps. *Proc. Natl. Acad. Sci. USA* **86,** 7914–7918.

Ashkin, A., Dziedric, J. M., and Yamana, T. (1987). Optical trapping and manipulation of single cells using infrared laser beams. *Nature (Lond.)* **330,** 769–771.

Berns, M. W., Aist, J. R., Wright, W. H., and Liang, H. (1992). Optical trapping in animal and fungal using a tunable near-infrared titanium-sapphire laser. *Exp. Cell Res.* **198,** 375–378.

Berns, M. W., Tadir, Y., Liang, H., and Tromberg, B. (1998). Laser scissors and tweezers. *Methods Cell Biol.* **55,** 71–98.

Berns, M. W., Wright, W. H., Tromberg, B. J., Profeta, G. A., Andrews, A. A., and Walter, R. J. (1989). Use of a laser-induced optical force trap to study chromosome movement on the mitotic spindle. *Proc. Natl. Acad. Sci. USA* **86,** 4539–4543.

Choquet, D., Felsenfeld, D. P., and Sheetz, M. P. (1997). Extracellular matrix rigidity causes strengthening of integrin-cytoskeleton linkages. *Cell* **88,** 39–48.

Chu, S., Bjorkholm, J. E., Ashkin, A., and Cable, A. (1986). Experimental observation of optically trapped atoms. *Phys. Rev. Lett.* **57,** 314–317.

Chu, S., Holman, L., Bjorkholm, J. E., and Ashkin, A. (1985). Three dimensional viscous confinement and cooling of atoms by resonance radiation pressure. *Phys. Rev. Lett.* **55,** 48–51.

Dai, J., and Sheetz, M. P. (1998). Cell membrane mechanics. *Methods Cell Biol.* **55,** 157–170.

Felgner, H., Grolig, F., Muller, O., and Schliwa, M. (1998). *In vivo* manipulation of internal cell organelles. *Methods Cell Biol.* **55,** 195–202.

Greulich, K. O. (1999). "Micromanipulation by Light in Biology and Medicine," 300 pp. Berlin, Birkhãuser.

Hong, L., Wright, W. H., He, W., and Berns, M. W. (1991). Micromanipulation of mitotic chromosomes in PTK_2 cells using laser induced optical forces ("optical tweezers"). *Exp. Cell Res.* **197,** 21–35.

Kuo, S. C., and Sheetz, M. P. (1993). Force of single kinesis molecules measured with optical tweezers. *Science* **260,** 232–234.

Liang, H., Wright, W. H., Cheng, S., He, W., and Berns, M. W. (1993). Micromanipulation of chromosomes in PTK2 cells using laser microsurgery (optical scalpel) in combination with laser induced optical force (optical tweezers). *Exp. Cell Res.* **204,** 110–120.

Liang, H., Vu, K. T., Krishnan, P., Trang, T. C., Shin, D., Kimel, S., and Berns, M. W. (1996). Wavelength dependence of cell cloning efficiency after optical trapping. *Biophys. J.* **70,** 1529–1533.

Liu, Y., Cheng, D. K., Sonek, G. J., Berns, M. W., Chapman, C. F., and Tromberg, B. J. (1995). Evidence for localized cell heating induced by infrared optical tweezers. *Biophys. J.* **68,** 2137–2144.

Liu, Y., Cheng, D. K., Sonek, G., Berns, M. W., and Tromberg, B. J. (1994). Microfluorometric technique for the determination of localized heating in organic particles. *Appl. Phys Lett.* **65,** 919–921.

Molloy, J. E. (1998). Optical chopsticks: Digital processing of multiple optical traps. *Methods Cell Biol.* **55,** 205–216.

Nascimento, J. L., Botvinick, E. L., Shi, L. Z., Durrant, B., and Berns, M. W. (2006). Analysis of sperm motility using optical tweezers. *J. Biomed. Opt.* **11,** 044001-1 to 044001-8.

Perkins, T. T., Smith, D. E., and Chu, S. (1994). Direct observation of tube-like motion of a single polymer chain. *Science* **264,** 819–822.

Sheetz, M. P. (ed.) (1998). Laser tweezers in cell biology. "Methods in Cell Biology" 228 pp. San Diego, Academic Press.

Simmons, R. M., Finer, J. T., Chu, S., and Spudich, J. A. (1996). Quantitative measurements of force and displacement using an optical trap. *Biophys. J.* **70,** 1813–1822.

Steubing, R. W., Cheng, S., Wright, W. H., Numajiri, Y., and Berns, M. W. (1991). Laser induced cell fusion in combination with optical tweezers: The laser cell fusion trap. *Cytometry* **12,** 505–510.

Tadir, Y., Wright, W. H., Vafa, O., Ord, T., Asch, R. H., and Berns, M. W. (1989). Micromanipulation of sperm by a laser generated optical trap. *Fertil. Steril.* **52,** 870–873.

Tadir, Y., Wright, W. H., Vafa, O., Ord, T., Asch, R. H., and Berns, M. W. (1990). Force generated by human sperm correlated to velocity and determined using a laser generated optical trap. *Fertil. Steril.* **53,** 944–947.

Vorobjev, I. A., Liang, H., Wright, W. H., and Berns, M. W. (1993). Optical trapping for chromosome manipulation: A wavelength dependence of induced chromosome bridges. *Biophys. J.* **64,** 533–538.

Wang, Y., Botvinick, E. B., Zhao, Y., Berns, M. W., Usami, S., Tsien, R. Y., and Chien, S. (2005). Visualizing the mechanical activation of Src. *Nature (Lond.)* **434,** 1040–1045.

Wang, M., Yin, H., Landick, R., Gelles, J., and Block, S. M. (1997). Stretching DNA with optical tweezers. *Biophys. J.* **72,** 1335–1346.

Wei, X., Tromberg, B. J., and Cahalan, M. D. (1999). Mapping the sensitivity of T cells with an optical trap: Polarity and minimal number of receptors for Ca(2+) signaling. *Proc. Natl. Acad. Sci. USA* **96,** 8471–8476.

CHAPTER 17

Cellular and Colloidal Separation Using Optical Forces

Kishan Dholakia,* Michael P. MacDonald,* Pavel Zemánek,† and Tomáš Čižmár†

*SUPA, School of Physics and Astronomy
University of St. Andrews, Fife, KY16 9SS Scotland

†Institute of Scientific Instruments ASCR, v.v.i.
Academy of Sciences of the Czech Republic
61264 Brno, Czech Republic

The separation or sorting of cellular and colloidal particles is currently a central topics of research. In this chapter, we give an overview of the range of optical methods for cell sorting. We begin with an overview of fluorescence and magnetically activated cell sorting. We progress to describing methods at the microfluidic scale level particularly those exploiting optical forces. We distinguish between

0091-679X/07 $35.00
DOI: 10.1016/S0091-679X(06)82017-0

what we term passive and active schemes for sorting. Optical forces pertinent to the sorting schemes are described, notably the gradient force and the optical radiation pressure (or scattering force). We discuss some of the most recent advances. This includes techniques without fluid flow where we have either stationary or moving light patterns to initiate separation. Further methods have shown how using an externally driven flow either counter-propagating against a light field (optical chromatography) or over a periodic light pattern (an optical potential energy landscape) may result in the selection of particles and cells based on physical attributes such as size and refractive index. We contrast these schemes with the field of dielectrophoresis where electric field gradients may separate cells and also briefly mention the upcoming area of light-induced dielectrophoresis which marries the reconfigurability of optical fields with the power of dielectrophoresis.

I. Introduction

The use of the modern optical microscope in unison with laser technology has fueled a revolutionary advance in cellular and molecular biology. A number of microscopic methods have allowed scientists to view how proteins and lipids behave and how their interactions govern the intricate mechanics, maintenance, and function of the intracellular world. Real-time observation and tracking of cellular processes with appropriate fluorescent tagging has yielded a wealth of bioscience (Prasad, 2003).

In parallel with the outstanding advances in microscopy, there have been other complementary advances in the use of laser light at the cellular scale. Most importantly, there has been the exploitation of the forces of light to trap, cut, move, and, more recently, sort biological and colloidal materials. As a key example, optical tweezers (Ashkin and Dziedzic, 1987) allow micron-sized particles and cells to be trapped, moved, and generally manipulated without any physical contact and is discussed in other chapters of this volume. This area of research, more broadly termed optical micromanipulation, is undergoing a growth of activity at the current time with a particular emphasis in the biological and colloidal sciences. In biology, this methodology has revolutionized our understanding of molecular motors. Rather than trapping just one or two objects at a time, there is interest in creating an array of trap sites. Such multiplexed optical trapping may be implemented by the use of holographic, interferometric, and acousto-optic devices (Grier, 2003; Molloy *et al.*, 2003; Neuman and Block, 2004). Such multiple optical traps may create what is termed as an *optical potential energy landscape*. This chapter deals with the concept of sorting of cells and colloidal samples. Particle motion on optical landscapes is central to the newly emerging forms of separation.

At the microscopic scale and nanoscale, many disciplines seek methods for accurately and efficiently separating colloidal, cellular, and other biological particles. Such selection plays a pivotal role in enabling studies in biology and medicine. As an example, at the nanoscale the technique of gel electrophoresis permits sorting of DNA by size. The isolation of specific cell subpopulations is central to the advancement of cell-based therapies for cancer, autoimmune diseases, and genetic disorders. This includes the ability to select stem cells from a population. Stem cell populations are key to important areas in modern medicine—in regenerative medicine, sources of stem cells can be exploited to provide new disease-free tissue. Tumor stem cell populations are key for disease development and for successful therapy. Methods for separating and investigating stem cell populations are thus fundamental to providing a rational basis for improved disease understanding and new forms of therapy.

Excellent macroscopic schemes exist to perform high-throughput, multiparameter cell separation, typically based on the conventional flow cytometer. High yield in such a system is typically realized with inputs in excess of 10^5 cells/s. However, the ability to miniaturize a cell sorting system and move toward a microfluidic basis for this technology, we may gain some advantages. Smaller number of cells may be separated and yet deliver a high yield. Reagent use is dramatically limited in such a small environment and the method may find favor with rare or precious cells, for example primary cells that may not lead to large cell populations. If the device is small and inexpensive, it may offer a disposable, sterile platform for cell separation that would bring the technology of cell separation to a wider number of researchers in the biosciences.

This chapter focuses on some of the emergent methods for cell and colloidal particle sorting at the microscale using minimal volumes of analyte and employing optical forces. Much of this work has been performed in only the last 5 years and it is a young and vibrant area of research. We cannot hope to cover the whole of this field in depth in this chapter, but believe we may give the reader a flavor of the work being performed, the shortcomings and challenges that lie ahead, and a representative list of references. We begin by reviewing established methods for cell sorting using macroscopic apparatus with large throughputs and discuss how they are being miniaturized. We focus on the use of optical forces to separate cells with or without markers attached to them, distinguishing what we term *passive* sorting from *active* sorting. We then review sorting techniques both with and without a microfluidic flow present. We conclude with a short comparison to the method of dielectrophoresis for cell sorting at the microscale.

II. A Brief Review of Fluorescence-Activated Cell Sorting (FACS) and Magnetically Activated Cell Sorting (MACS)

The fluorescence activated cell sorter first appeared in the late 1960s and is based on a flow cytometer. This device records the properties of a single cell as it traverses a laser beam. In doing so, the cell scatters light and this is then recorded on suitably placed detectors. A FACS machine is a special type of flow cytometer: fluorescent markers (fluorescently labeled monoclonal antibodies) allow specific cells to be recognized and subsequently separated. Analysis and separation based on a wide range of parameters may be implemented. A review by the inventors of FACS in 2002

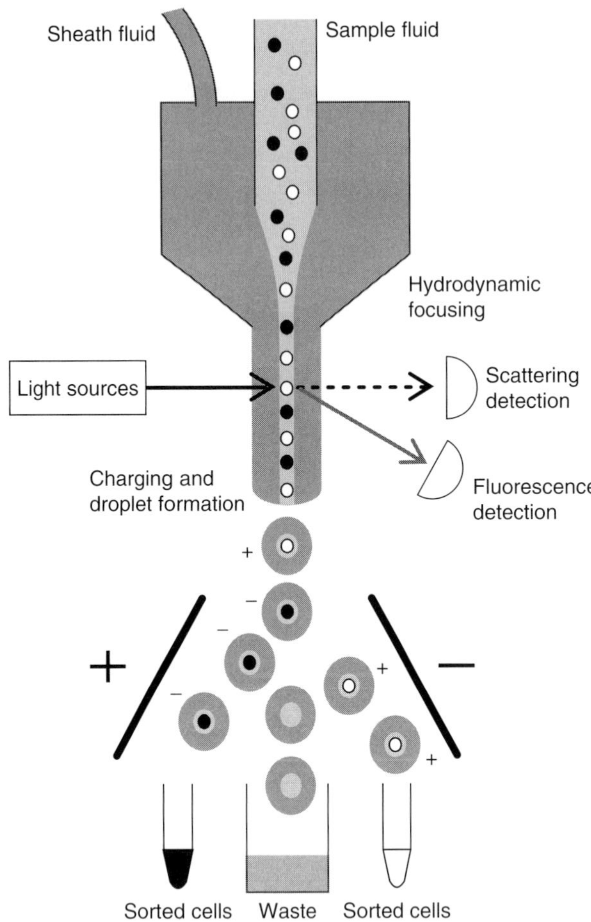

Fig. 1 Diagram of FACS machine. Cells have been fluorescently tagged (shown as black or white), with some cells remaining untagged. On the basis of the fluorescence signal detected, a charge is applied to individual droplets such that they can be deflected into separate chambers.

stated that FACS can record 12 different fluorescent colors and 2 scattering parameters, which could all be used for sorting (Herzenberg *et al.*, 2002). The technique can measure the cell size, volume, or viscosity. DNA or RNA content as well as the presence of surface antigens. FACS has a variety of uses and has been used in the diagnosis of leukemia, lymphoma, immunodeficiencies, and for compatibility in transplants.

Figure 1 shows a schematic of a cell sorting device. Air pressure pushes cells out of a nozzle at high speed: a liquid jet (e.g., saline) is combined with the cells and acts as a sheath flow. An acoustic vibration is coupled to the nozzle's tip, leaving a trail of cyclical imprints onto the liquid's surface. Surface tension pulls at the waists between the imprints, forcing the jet to separate into regularly spaced droplets. The cells, first carried by a cylinder of liquid, are thus distributed among a string of discrete droplets. After the cell leaves the nozzle region, it passes through the waist of one or more tightly focused laser beams which are at appropriate wavelengths to perform scattering or fluorescence excitation of the cell markers. The scattered and fluorescence light from these interactions is collected and analyzed.

If a cell meets one or more criteria set by the user, an electrical charge may be applied to the droplets containing cells of interest as they separate from the main jet. In this manner, droplets with different cell types are directed toward separate collection vials by a static electrical field (Fig. 1). The speed of modern electronics and detection systems means one may easily analyze cells at speeds greater than 10^5 events per second. Cell separation rates of the order of 10,000 per second are standard. We refer the reader to the excellent texts on fluorescence-activated cell sorting for further reading (Givan 2001).

Fluorescence-activated cell sorting (FACS) is very powerful and may discriminate cells within a large parameter space; however, in several instances, one may just wish to separate two sets of cells quickly, and at less expense, than a traditional FACS machine. Here magnetically activated cell sorting (MACS) is worth considering. For immunomagnetic separation, cells are incubated with paramagnetic microbeads that are coated with appropriate antibodies. Subsequently, the cells of interest or the unwanted cells may be sorted by use of one or more magnets in a suitable array. The magnetic method is, to some extent, limited by the constraints one places on the parameters for separation and the obvious need for suitable antigens on the cell surface for the paramagnetic beads to bind to. The method is useful for rapid bulk separations or as a precursor to more elaborate sorting schemes.

In the next sections, we turn our attention to recent work using optical forces to instigate cell and colloidal separations at the microscopic scale. We begin by exploring the optical forces which are relevant to sorting and the role of a microfluidic flow therein.

III. Optical Forces for Cell and Colloidal Sorting: Theoretical Considerations

We distinguish four key regimes for the optically mediated sorting process at the microfluidic scale, which are represented in the phase diagram shown in Fig. 2. Later in this chapter, we will look at experiments with emphasis particularly on phases 1–3.

Region (1): Static fluid, static pattern: light-induced flow and separation of microobjects

The most basic form of separation exploits the differences in the affinity of colloidal and biological particles to a stationary optical potential energy landscape with no fluid flow present. Any affinity differences would manifest themselves as

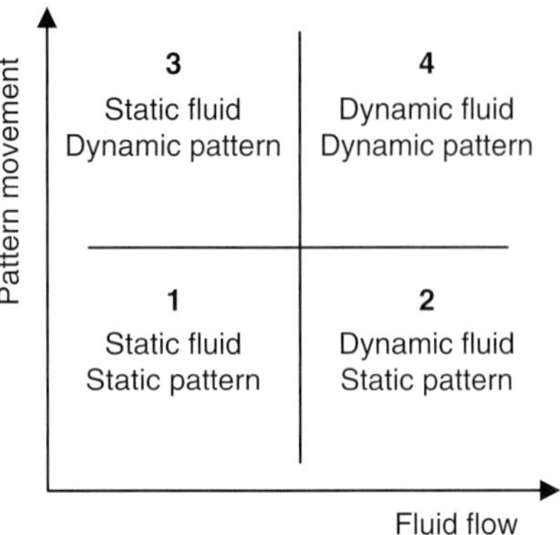

Fig. 2 The dynamic phases available for optical separation. As the analyte can be either static or dynamic, in the presence of a static or a dynamic optical pattern, we have four different phases within which to work.

particle sorting and separation in the absence of flow. Motion of objects can be investigated and this is achieved solely by optical forces. How fast the particle moves over the landscape depends on the energy depth of the potential landscape well (trap). The time a particle needs to overcome a potential barrier via Brownian motion to the particle size via Brownian motion. Directed particle motion was obtained by tilting the periodic optical landscape. Overall, this means that particles of different size, shape, refractive index, and composition move differently across the optical landscape. For example, we shall see that red and white blood cells may be separated by placing them in a tailored circularly symmetric light beam pattern or spherical objects of desired size placed into a three-beam interference field will be transported in opposite directions (Zemánek *et al.*, 2004b).

Region (2): Dynamic fluid, static pattern: microfluidic optical sorting

Flowing particles through an optical potential energy landscape [two-dimensional (2D) or 3D interference pattern (a lattice), holographically produced array, time-shared array] will facilitate the separation of these objects based on their physical characteristics such as size, shape, and refractive index. Flow through an optical lattice can readily lead to sorting as a function of size and/or refractive index-related deflection (MacDonald *et al.*, 2003; Pelton *et al.*, 2004).

Region (3): Static fluid, dynamic (moving) pattern

By creating a time-varying optical landscape (e.g., scanning the beam to create, for example linear interference fringes moving perpendicular to the fringe axis), we can transport particles as if on a mechanical conveyor belt and use the fact that different particle species respond differently. From a physical perspective, it is another way by which we can introduce a tilted or biased optical landscape. Naturally, this sensitivity of a given particle size on the landscape varies as a function of particle size and relative refractive index between the object and the media. The traveling landscape provides a convenient means of pumping and actuation, but the distance the particles are carried will depend on many factors, including the shape and extent of the envelope of the optical pattern itself: as the intensity in the pattern falls off more strongly, interacting particle species will be carried further than those that respond weakly to the optical forces. Since this dynamic phase of operation does not have a fluid flow to remove sorted particles or cells, particles will accumulate at the position for which the optical conveyor belt is no longer strong enough to transport them, leading to a fractionated column of different particle species which may then be extracted.

Region (4): Dynamic fluid, dynamic pattern, microfluidic sorting with enhanced particle separation

The final regime utilizes a motional light pattern within a microfluidic flow. To date, this is a little explored regime even though it offers more degrees of freedom with respect to tuning the sorting process. A particular advantage of this combination is that deflected particles can be moved out of the polydisperse flow more quickly, leading to fewer particle–particle interactions that might otherwise lead to undesirable behavior of the particles (e.g., clustering, jamming, incomplete deflection, and/or separation). This will allow higher efficiencies to be achieved in the sorting process, something that may be key to future implementations of all optical sorting.

As a precursor to looking at the techniques being developed in all four regimes described above, we will now discuss the behavior of particles within an optical trap. Optical forces generate a mechanical effect on atoms, molecules, and particles right up to the size of microscopic colloidal particles and single cells. An optical trap may readily generate piconewton forces in a noninvasive manner on cellular and colloidal particles. A particle in a trap behaves as a highly overdamped harmonic oscillator with a stiffness in the range of 0.05 pN/nm (Molloy *et al.*, 2003; Neuman and Block, 2004). The object may escape from a trap due to thermal activation and the dynamics of motion may change in the presence of flow.

This issue of thermal activation was studied over 60 years ago, Kramers (1940) elucidated the dynamics of particles in a double-well potential which can be approximated by closely spaced optical traps. It was shown that the mean time (Kramers time) τ_K to get the object over an energy barrier of height ΔU can be described by an exponential law of the form $\tau_K = R\exp(\Delta U/k_B T)$, where T is the temperature, k_B the Boltzmann constant, and R is a quantity depending on the potential curvature at the maximum and minimum. Microscopic colloidal particles or cells in an optical potential represent a powerful means by which to study such activation.

Two components of optical force can be distinguished at the nanoscale. The *gradient force* is dependent upon which may scale linearly with the particle volume and is dependant upon the particle polarisability. This pushes the object to the higher (lower) intensity place if its refractive index is higher (lower) than the surrounding medium. The *scattering force* is proportional to the square of the particle volume and pushes the object in the direction of propagation of the incident light. Therefore, the gradient force spatially localizes the object while the scattering force pushes it away along the beam propagation axis. The object is trapped at equilibrium, (a position of zero total force) and any small object deviation from this position results in restoring force proportional to displacement. Since the bigger object is strongly pushed out of the trapping region of the gradient force, trapping can be reached only in tightly spatially localized field intensity maxima. In contrast, the optical force has a weak dependency on the object size, if the trapped object has dimensions one order larger than the wavelength.

Straightforward single beam trapping is a possible method of optical sorting, if we select cells from a microfluidic environment and divert them into the channel of interest. The use of extended optical potential energy landscapes (periodic light patterns) is also of interest. Such periodic patterns may trap cells and colloid in the bright parts of the light field, if they are smaller than the pattern period and of higher refractive index than the surrounding medium. However, if the objects become larger than the pattern period or pitch they could settle with their center in the dark parts of the field. Intermediate-sized objects may not even sense the presence of the periodic field pattern and stay largely unaffected by the optical landscape. This, in turn, implies that sorting of objects is based on their size and polarisability. We concentrate here on particles in extended optical potential energy landscapes.

In one dimension, particle behavior in an optical-standing wave was theoretically studied by Zemanek and colleagues for nanoparticles (Zemánek *et al.*, 1998) and microparticles showing a size-dependent effect (Lekner, 2005; Siler *et al.*, 2006; Zemánek, 2002, 2003). They showed that the sensitivity of a spherical object of radius a to the periodic light pattern of period Λ can be described analytically for a weakly polarized particle (with its refractive index close to that of the surrounding medium) using the force acting on it perpendicular to the fringes:

$$F(x) \approx F_0 \left[\sin\left(\frac{2\pi a}{\Lambda}\right) - \cos\left(\frac{2\pi a}{\Lambda}\right)\frac{2\pi a}{\Lambda} \right] \sin\left[\frac{2\pi(x - x_0)}{\Lambda}\right] \quad (1)$$

where x_0 is the position of the closest peak to the beginning of axis x. The first square bracket (size term—see Fig. 3) has oscillatory behavior passing through zero at $a/\Lambda \approx 0, 0.715, 1.230, 1.735, \ldots$ and having maxima and minima separated by $\Lambda/2$ starting at $a_{\max} = \Lambda/2$. Therefore, if the size term is positive (e.g., particle radius is smaller than 0.715 Λ), the sphere center is localized at $x = x_0 + M\Lambda$, where $M = 0, \pm 1, \pm 2, \ldots$ This position corresponds to the intensity maximum of the fringe. In contrast, if the size term is negative, then the equilibrium position is shifted by $\Lambda/2$ and the sphere center is positioned at the intensity minimum of the fringe. If the size term is close to zero, no matter where the sphere is localized in the periodic pattern, the force equals to zero and the sphere does not feel the periodic landscape. The above-mentioned particle sensitivity to the periodic pattern (that may on

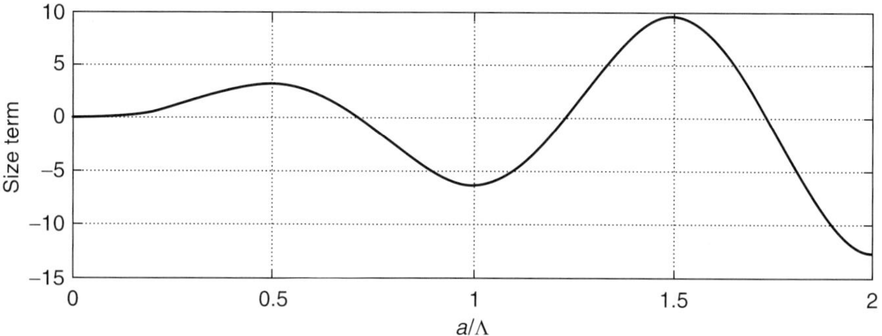

Fig. 3 Plot showing how the size term $\sin(2\pi a/\Lambda) - \cos(2\pi a/\Lambda)2\pi a/\Lambda$ in Eq. (1) depends on the size of the sphere. This dependence is exactly valid only if the refractive index of the object is close to that of the surrounding medium. But the qualitative behavior has more general validity.

occasion be more complex than described here) underpins all four types of optical sorting in periodic light patterns (Fig. 2).

Sorting of nanoparticles according to their optical properties was studied theoretically in a 1D pattern obtained using a three-beam configuration (Zemanek, 2004a). The same field configuration was used later to analyze static sorting of microspheres according to their size or optical properties (Zemanek, 2004b). The particle separation was based on the opposite light-induced particle flow if the particle center settles in a fringe maximum or minimum. The size effect also plays an important role if the sorting is based on thermally activated jumps over a barrier between neighboring equilibrium positions. The lower the barriers, the more frequently a particle will jump out of a given trapping region. This facilitates more rapid motion along the tilted periodic landscape (Reimann, 2002; Tatarkova *et al.*, 2003). In separate studies, nanoparticle sorting using a plasmon resonance excitation has been studied theoretically (Zelenina *et al.*, 2006).

Turning now to sorting in the presence of flow, Pelton *et al.* (2004) presented a theoretical model that took into account the Brownian motion of objects smaller than the trapping wavelength in 2D periodic potentials. Here one needs to examine the effect of the optical forces competing with the Stokes forces (viscous drag) as the external driving term. For such small objects, they approximated the optical forces acting on them and expressed the analytical solution for deflection angles in separable 2D landscapes. The particle motion may be described by a Langevin equation and particles may be locked-into certain directions that are not correlated with the direction (orientation) taken by the fluid flow. They explored the sensitivity of various landscapes to parameters such as size and theoretically showed that, in general, extended periodic landscapes would yield excellent sorting resolution and exponential size selectivity. Alternative theoretical studies including general 2D potentials and the influence of thermal noise have been presented by Lacasta and colleagues (Gleeson *et al.*, 2006; Lacasta *et al.*, 2005). They used the Langevin equation

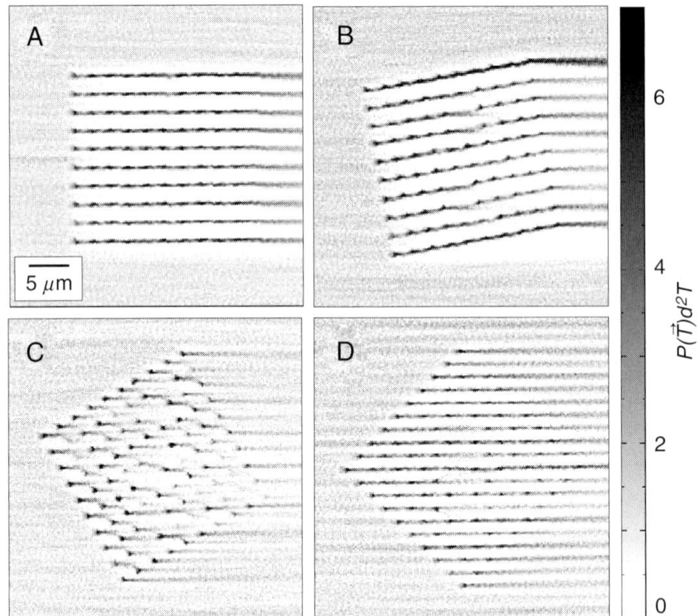

Fig. 4 Kinetic lock-in of colloid flowing through a static optical potential landscape. Depending on the orientation of the landscape to the flow different behavior is seen. Relative probability that a sphere will pass through a point in the field of view, when the direction of the trap lattice is oriented at (A–D). In all figures, the external flow is from left to right. As the behavior of particles with different size or refractive index will diverge, this and similar phenomena were later used to obtain sorting of colloid and cells. Reprinted figure with permission from Korda, P. T., Taylor, M. B., and Grier, D. G. (2002). *Phys. Rev. Lett.* **89,** 128–301, copyright (2002) by the American Physical Society.

to study particle behavior in these potentials and they especially looked for particle trajectories, if they were driven over the periodic structure under different angles and various force amplitudes (Lacasta *et al.*, 2005). This driving force may be exerted, for example, by fluid flow (viscous drag) as previously mentioned. They observed a terrace phenomenon in the dependence of the absolute velocity angle on the direction of the force. This phenomenon has been observed experimentally by Korda *et al.* (2002) (Fig. 4) and termed kinetically locked-in states. Gleeson *et al.* (2006) presented an analytical approach to this phenomena based on the over-damped Fokker–Planck equation. They derived a first-order approximation to the average velocity vector $\mathbf{v}$ in the form

$$\mathbf{v} = \mathbf{F} + \frac{1}{(2\pi)^4} \int d\mathbf{K} \frac{k^2 \mathbf{K}}{\bar{T}k^2 - \mathbf{k} \cdot \mathbf{F}} \hat{Q}(\mathbf{k}), \tag{2}$$

where $\mathbf{F}$ is the vector of the uniform driving force, $\bar{T}$ is the dimensionless temperature, and $\hat{Q}(\mathbf{k})$ is defined by $\hat{Q}(\mathbf{k}) = \hat{V}(\mathbf{k})\hat{V}(-\mathbf{k})$, where $\hat{V}$ is the Fourier transform of the potential $V(\mathbf{x})$ (periodic or even random) and $k^2 = \mathbf{k} \cdot \mathbf{k}$. The validity of this approximation is fairly good, especially for higher temperatures and forces.

IV. Overview of Experimental Optical Force-Based Sorting

Techniques which use optical forces to sort cells can be placed into one of two groups. We call the first group of techniques *active* sorting, where an external decision based on probing a particle passing a detection region is part of the sorting mechanism. We have already seen examples of this in our brief discussions of FACS and MACS. The second group may be termed *passive* sorting, where the selection and separation of cells occur without the need for any external decision-making process, but rather are based on the differing affinity of a given object to the light field in the presence of an external drive (the fluid flow). Both of these groups allow for the attachment of markers which give enhanced selectivity, but it is predominantly the active group that relies on these markers while the passive techniques are usually promoted as needing no such markers or tags.

It is also instructive to consider some of the fluid dynamics that will be pertinent to microfluidic cell sorting: at the microfluidic scale, fluid flow is what we term laminar: we are typically in the low Reynolds number regime. The Reynolds number is the ratio of inertial to viscous forces: thus, we are ignoring inertia to a large extent and are reliant solely on viscous forces: turbulence is not present and thus for tasks such as mixing, deflection, and sorting we have to rely, in the absence of optical or other forces, solely on diffusion. Thus, we may have two fluid streams running parallel to one another with little or no mixing taking place. In terms of sorting, as this means for such a geometry, our optical forces may deflect particles of choice readily from one flow to another (Squires and Quake, 2005).

A. Active Sorting with Optical Forces

In active sorting techniques, most of the work has been done in introducing FACS (Galbraith *et al.*, 1999; Givan, 2001) techniques into the microfluidic regime (Fu *et al.*, 1999, 2002; Kruger *et al.*, 2002; Wang *et al.*, 2005). In this instance, the mechanisms for cell identification are very similar to that of macroscopic FACS machines and the concept of hydrodynamic focusing is maintained. The main difference arises in the different methods for cell deflection. In microfluidic FACS, deflection of cells is done in fluid using optical forces (Fig. 5). It is worth noting here that the continued need for the sheath fluid comes largely from the active nature of the technique where a single-file flow of cells is required so that one cell at a time can be analyzed and then subsequently deflected.

Microfluidic versions of FACS machines have been implemented (Applegate *et al.*, 2006; Fu *et al.*, 1999). At the microfluidic scale level, we need to consider the fact that we are likely to have much lower throughput, are aiming to try and separate cells from microliter samples, and indeed deal with the interesting fluid mechanics (Squires

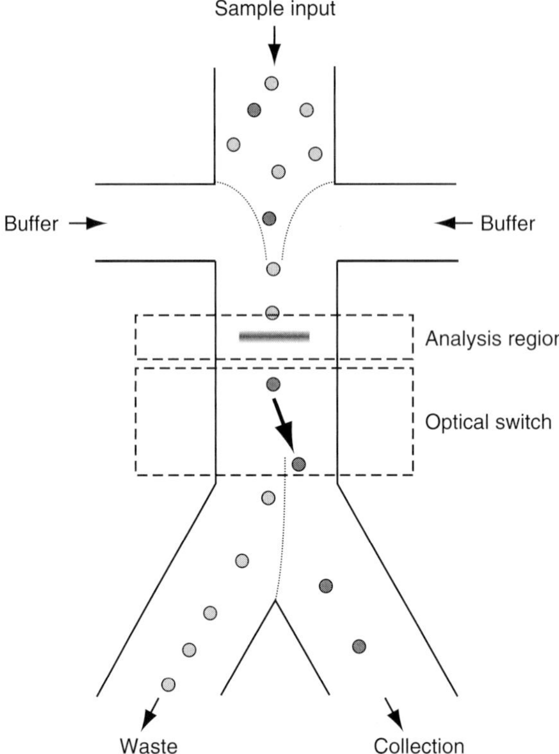

Fig. 5 Schematic representation of a micro-FACS system as demonstrated by Genoptix. Similar to macroscopic FACS, the sample fluid stream is focused by a sheath flow, this flow of single-file particles is then analyzed for fluorescence and scattering, a decision made and based on this decision particles will be deflected into one of two output streams. Reprinted by permission from Macmillan Publishers Ltd.: Wang, M. M., Tu, E., Raymond, D. E., Yang, J. M., Zhang, H. C., Hagen, N., Dees, B., Mercer, E. M., Forster, A. H., Kariv, I., Marchand, P. J., and Butler, W. F. (2005). Microfluidic sorting of mammalian cells by optical force switching. *Nature Biotechnol.*, **23,** 83–87, copyright (2005).

and Quake, 2005). To adapt a FACS machine into a viable microfluidic technology, one must ensure that we retain a good throughput, purity, and recovery of cells that have not been unduly stressed by the sorting process. The use of an optical force switch to divert out selected cells of interest potentially offers a good route to enable a micro-FACS system to be created. The work by Wang *et al.* (2005) showed a new microfluidic implementation of sorting using such an optical switch. The work builds upon the studies of Buican *et al.* (1987), who showed guiding of various types of cells in a gently focused light field and introduced the notion of deflecting out cells of interest into a reservoir of choice.

Cells were hydrodynamically focused into a linear flow through a detection and subsequent deflection region. The Wang cell sorter used two lasers: one at 488 nm to excite green fluorescent protein (GFP) fluorescence in cells. The presence of the desired cell type is indicated by fluorescence recorded onto a photomultiplier.

A 1070-nm fiber laser, further downstream, then acts as an optical-switch deflecting cells of choice into a collection reservoir. To ensure the cells remained viable, tests of the membrane were performed with trypan blue along with an examination of two indicator genes for heat and cell shock.

Other active sorting techniques include simple cell identification via video microscopy with subsequent cell rearrangement using multibeam optical tweezers and techniques more similar to microfluidic FACS where cell identification is done in a microfluidic chamber before particle deflection or placement into separate laminar flows (Buican *et al.*, 1987; Ericsson *et al.*, 2000; Grover *et al.*, 2000, 2001; Oakey *et al.*, 2002; Rodrigo *et al.*, 2002).

The discrimination of FACS methods is very high and is achieved by the sensitivity of the fluorescence detection and a sufficiently strong cell deflection mechanism. The sorting speed is mainly influenced by the speed of the detection unit and response of the control electronics.

B. Passive Sorting with Optical Forces

It is possible to use passive cell sorting techniques in conjunction with markers, such as dielectric tags, where a functionalized dielectric sphere is attached to a specific cell population. However, the trend is towards developing techniques where the need for this pretreatment step is not required. This tag-free approach allows for separation such that the sorted analyte contains cells in their untouched state, with no need to subsequently remove the tags, or to develop the tags in the first place (often the most time consuming and costly part of the cell sorting process).

When sorting is achieved without attaching tags, selectivity is obtained via the intrinsic properties of the cells. These properties are sensed most often as a size difference, but also as a shape and/or refractive index difference. As a result, such tag-free sorting is limited to sorting analyte that contains cell populations which are relatively homogenous (such as blood) rather than to sorting populations that are subject to marked size variations as part of the cells life cycle (e.g., cancer cells).

One form of passive optical force-based sorting takes place where some form of kinetic lock-in of particles occurs, whereby the sorting is obtained via a size- and shape-dependent polarizability in competition with a size- and shape-dependent viscous drag (Korda *et al.*, 2002). This can be achieved either in an array of optical traps or, in a more generalized approach, in a tailored optically induced potential landscape (Ladavac *et al.*, 2004; MacDonald *et al.*, 2003, 2004; Pelton *et al.*, 2004).

It is also possible to obtain passive sorting without the need for flow. This can be done, for example, in a tilted washboard-like optical potential by size-sensitive optical radiation pressure (Paterson *et al.*, 2005; Zemanek *et al.*, 2004b) or by adding some form of size-selective optical-moving pattern (conveyor belt) to perform separation (Cheong *et al.*, 2006; Cizmar *et al.*, 2006; Ricardez-Vargas *et al.*, 2006).

One of the original concepts in passive optical sorting is the idea of optical chromatography as originally proposed by Imasaka (Imasaka *et al.*, 1995; Kaneta *et al.*, 1997) and subsequently refined by Hart and co-workers as well as others (Hart and Terray, 2003; Hart *et al.*, 2004, 2006; Imasaka *et al.*, 1995;

Kaneta *et al.*, 1997; Zhao *et al.*, 2006). In this approach, the balance between the viscous drag experienced by particles in fluid flow is balanced by the radiation pressure of a counter-propagating laser beam. Figure 9 shows an example of optical chromatography being used to separate two different bacterial spores.

Another technique similar in concept to optical chromatography is known as optophoresis. Optophoresis uses a balance between viscous drag and the force exerted on particles by a rapidly scanning laser beam (Zhang *et al.*, 2004). Many more techniques are in development or have been proposed such as those based around flashing optical potentials (Libal *et al.*, 2006; Smith *et al.*, 2007), the particle–light–particle interaction usually referred to as optical binding (Grzegorczyk *et al.*, 2006), optical waveguides (Grujic *et al.*, 2005), optically induced thermophoresis (Shirasaki *et al.*, 2006), or optically induced dielectrophoresis (Chiou *et al.*, 2005).

As to whether passive or active sorting is the most appropriate is dependent on the cells that are being investigated. In general, passive sorting is more flexible as it easily allows for sorting both with and without tags and is simple to combine with a microfluidic system, but this technique is somewhat unproven in the cellular regime. Active sorting, however, has a more established record: the body of work already done in the flow cytometry community means that active techniques like FACS bring with them a lot of valuable experience and established protocols.

In the next two sections, we explore experimental implementations of sorting using optical fields: we focus on methods that are primarily passive but that may be implemented with an active element (akin to fluorescent-activated cell sorting methods), if desired.

V. Flow-Free Optical Methods

In this section, we will explore the use of optical fields to separate objects in the absence of any fluid flow. Without any flow particles are usually trapped but if we make the trap potential shallow we create a metastable state such that these particles may in fact escape due to thermal activation from the optical potential well. This may in turn be exploited to separate particles and cells. In this section we look at experiments that have used static optical energy landscapes for sorting small volumes of analyte. As indicated on the earlier diagram (Fig. 2), we will thus be looking at regions 1 and 3.

The Bessel light field has proven to be an interesting form of optical landscape for micromanipulation in recent years and is a good example of flow-free optical sorting. The Bessel beam is a solution of the Helmholtz equation that exhibits the property of propagation invariance: that is the intensity at a given plane is exactly the same for an idealised version of this beam as at any other plane (McGloin and Dholakia, 2005). This means it is in some sense "diffraction free." This has been exploited for long range optical guiding of microparticles. The zeroth order of such a beam consists of a central bright spot surrounded by a series of concentric rings. The central spot and all of the rings each contain equal amounts of power. Interestingly, if one places colloidal and cellular particles into the outer rings of the light field, one sees that the periodicity of a light pattern makes a profound difference with regard to the equilibrium position for a given object. This is coherent with the above-described size effect. To employ this effect, the landscape was tilted by slight modification of the incident beam so that this tilt biased the motion of the objects to the center of the "nondiffracting" Bessel mode. The objects that pass through the rings have a dependency on the height of the barrier that they have to overcome. A standard Gaussian beam optical tweezer offers a harmonic potential well, from which thermal hopping to adjacent wells can be exploited if two or more focused beams are close to each other. As an example, in a dual tweezers system, the dynamics of particle jumps between them has been studied (Simon and Libchaber, 1992) and they proved that these jumps can be boosted if the depth of both potential wells is modulated with a period two times longer than the Kramers time (Babic *et al.*, 2004). This can also be applied to extended optical potential landscapes such as the Bessel beam, which also have intriguing features relating to equilibria positions for trapped objects that depend on the physical parameters. Notably, objects that are large compared to the optical corrugation respond to the extended envelope of the optical field, while smaller particles respond to the individual rings within this optical landscape.

The Bessel beam was used for blood sorting. Equal concentrations of both mononuclear cells and erythrocytes were suspended in an appropriate culture medium supplemented with fetal calf serum and mixed together in a sample chamber. The Bessel beam used had a 5.0-μm core size and a laser power ranging up to 800 mW. At low powers (up to 300 mW), the majority of cells were transported slowly toward the central core of the Bessel beam, where they are finally trapped by forming a vertical stack at the top of the sample chamber. At higher powers (>400 mW), the biconcave-shaped erythrocytes move toward the central core but before reaching the center, align vertically in the outer rings of the Bessel beam. Once reoriented in this manner, the erythrocytes are locked into the specific ring and are guided upward within that ring. In contrast, the spherically shaped lymphocytes move directly toward the central Bessel core, where they form a vertical stack along the central maximum as described earlier, responding rather to the overlying potential and not being locked within any ring of the Bessel beam.

As the white cells are collected into the center of the beam, they experienced an upward propulsion from the optical radiation pressure from the center of the Bessel beam which thus separated the lymphoctyes and erythrocytes (Fig. 6). A judiciously placed capillary was able to extract out the lymphoctyes from the sample. The addition of colloidal beads as markers may enhance this type of sorting. Streptavidin-coated silica microspheres of $\sim$5 μm in diameter were attached to a T cell subpopulation of mononuclear cells via a mouse CD2 primary antibody and a secondary, biotinylated, antimouse antibody attachment. Attaching silica microspheres aims to enhance this passive method by selection using the beads. They are targeted to a specific subpopulation of cells via antibody–antigen binding. The cells with microspheres attached reacted to the optical landscape more strongly than cells without any attached beads due to the higher refractive index mismatch and

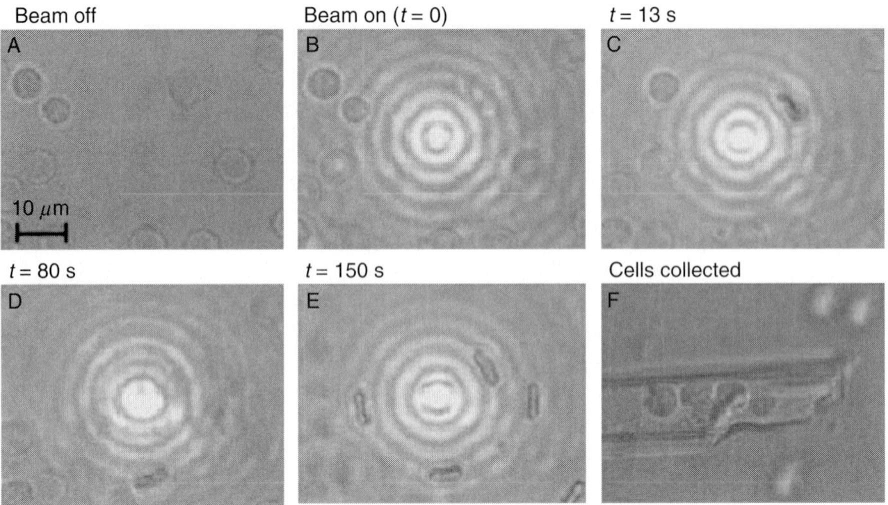

Fig. 6 Sorting of white and red blood cells in a Bessel beam. Red blood cells flip and align to the rings of the Bessel beam becoming trapped in the outer rings. White blood cells move along the gradient of the envelope of the Bessel function such that they are dragged into the center of the Bessel beam and guided away down its central core.

accompanying optical forces. A separation of these T cells from the ensemble of unlabeled cells was achieved in this manner.

A different type of static sorting without the fluid flow that is not based on the thermally activated jumps over a barrier combines light-induced particle flow in opposite directions, if they settle at appropriate parts of a set of bright fringes (Zemanek, 2004b). This type of sorting is especially useful for objects comparable or bigger than the fringe-spacing employed. Experimental separation of 2- and 5-μ m, or 5- and 7-μm polystyrene beads was demonstrated. The sorting speed is proportional to the used laser power as this parameter dictates localization of particles within the fringe and increases the transport velocity along the fringe by light-induced particle flow. The stochastic thermal motion is less dominant in these cases too, and one attains higher sorting precision and discrimination (Fig. 7).

A productive way to enhance optical separation without flow would be to invoke some sort of motional light pattern. In the absence of any flow, this should enhance the throughput of the sorting method while retaining much of the simplicity of the technique (region 2 in Fig. 2). Naturally, the motional speed of the pattern is the critical parameter to control because it establishes the balance of forces such that objects will follow the pattern and will jump over an inter-fringe barrier.

Within this remit, a vibrating fringe pattern has been used to move and separate colloidal particles by Ricardez-Vargas *et al.* (2006). They used a Mach Zender interferometer type arrangement to generate a sinusoidal fringe pattern that they

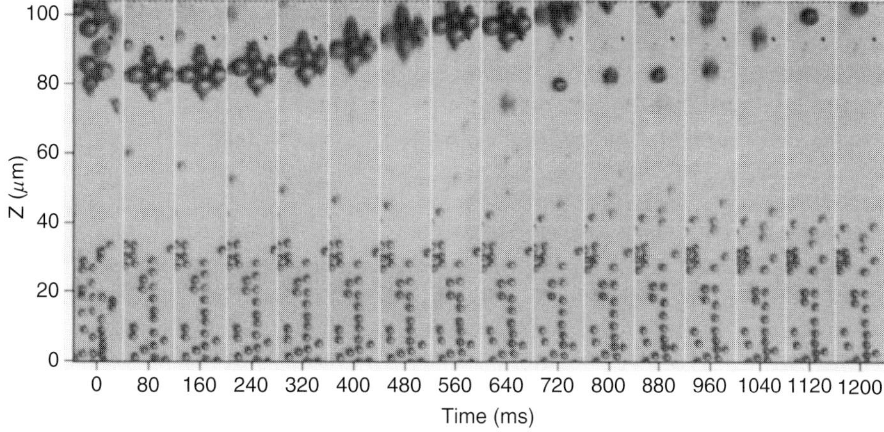

Fig. 7 Example of sorting of 2- and 5-μm objects in a three-beam configuration. The monosize collection of smaller spheres is assembled at the lower part of the figure following the direction of the interference fringes. Bigger spheres are pushed upward.

then projected into the sample plane of the trap. One of the mirrors in the interferometer vibrated using a piezoelectric mount and caused the fringe pattern to oscillate in a saw tooth-like manner in one direction. The size effect was employed to separate 2- and 5-μm polystyrene particles. Moreover, they were also able to separate particles of the same size (5 μm) made of polystyrene and silica, thus showing sorting based on refractive index variations. The sorting speed is dictated by the balance between the optical force pushing the objects forward and the Stokes force resisting this movement, so the higher laser power and shorter fringe distance increases the sorting speed. The discrimination is influenced by the system's ability to move one particle size—to create deep enough groove to suppress the thermally activated jumps between fringes—and keep the other sizes unaffected. Again with increasing laser power and shorter fringe distance, smaller differences in particle sizes can be separated.

In recent work, Cizmar *et al.* (2006) have demonstrated selection and motion of submicron-sized particles near a surface in a moving periodic light pattern created by counter-propagating and interfering beams near a prism surface. This type of optical geometry has proved interesting for large-scale arrangement of microparticles and large area coverage. An added attribute is the fact that no high numerical aperture optics are used and one may freely access and image the particles from above the prism. The motion of the moving standing wave is based on the time variable phase shift between the interfering waves and may be implemented in a number of ways: by an axially movable mirror or introducing a small frequency shift between both beams (e.g. using an angular Doppler shift) with this difference yielding the speed of the pattern.

They obtained sorting so that the pattern moved in one direction while the whole landscape was tilted in the other direction by a higher incident power coming from

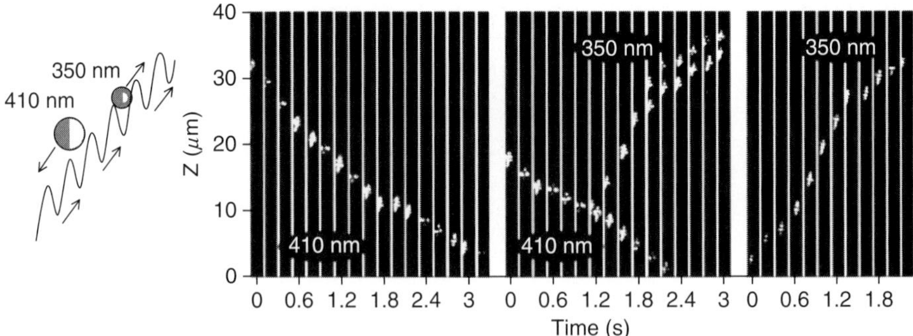

Fig. 8 Principle (left) and examples (middle, right) of sorting of submicrometer-sized colloids of different sizes differing by just 60 nm. Polystyrene beads were used of diameters 350 and 410 nm, the smaller ones are delivered by the moving pattern like in a conveyor belt in the positive direction of the z-axis, the larger beads are insensitive to the periodic modulation of the potential and therefore they fall down due to the tilt of the periodic potential.

one of the incident beams on the prism. Thus, an object confined within the periodic structure followed the pattern motion, while a larger object responds to the envelope of the pattern and thus falls down along the potential landscape. The sorting speed and discrimination follow the same rules as above (Fig. 8).

A more advanced method by which to generate 2D or 3D landscapes that are dynamically reconfigurable is to use the spatial light modulators (SLMs) (Curtis *et al.*, 2002; Eriksen *et al.*, 2002). They work as a phase grating (hologram) where each pixel can be independently addressed so that it has different optical thickness or converting amplitude to phase using the generalized phase contrast method (Rodrigo *et al.*, 2005). This tool can generate motional periodic light patterns with a direct application to the flow-free partile motion termed optical peristalsis (Koss and Grier, 2003). Due to the dynamic properties, the SLM also offers sorting based on optical ratchets. To obtain a ratchet, a sort of asymmetry has to be present in the system—either asymmetric but periodic potentials or a symmetry breaking time sequence of symmetric potentials (Reimann, 2002). Lee and Grier (2005) observed particle flux reversal in a symmetry breaking time sequence of optical traps where different particles experience different potentials and possess different diffusion coefficients. This arrangement can also be used for thermally activated sorting.

Time-dependent optical potential energy landscapes can also lead to sorting via flashing of the laser source rather than spatial scanning of the landscape (Libal *et al.*, 2006; Smith *et al.*, 2007). The frequency and phase of this motion is locked to the oscillation of the optical signal used to produce the landscape, leading to a ratchet effect that gives spatial separation between strongly and weakly interacting particles. It is also possible to achieve the same ratchet-like behavior with a DC flow added orthogonal to the AC motion. This incarnation leads to two spatially separated flows of particles.

VI. Optical Methods with Flow

One of the most promising methods to emerge in recent years has been the combination of microfluidic flow and optical forces where the flow may drive a batch of particles or cells over the potential energy landscape (regions 2 and 4 in Fig. 2). We begin by looking at the method of optical chromatography which fits into this scheme.

A. Optical Chromatography

Chromatography is a separation technique that takes advantage of the differences in partitioning behavior between a mobile phase and a stationary phase to separate the components in a given mixture and is a commonly known technique in biology and chemistry. Optical chromatography is one of the earliest forms of optical sorting with a fluid flow and was first demonstrated over a decade ago (Imasaka *et al.*, 1995). In this method, one uses a weakly focused laser beam in a counter-propagating fluid flow. As the particles flow along the channel, they experience a radiation pressure force from scattering of the light field and are pushed toward the focal region of the beam where they attain the highest velocities. As one might expect, this scattering force that creates the guiding differs depending on the size and refractive index of the particle in question. In the presence of the fluid flow, the radiation pressure (guiding) force pushes the particles against the fluid flow and an equilibrium arises between the competing optical and fluid forces creating regions where particles are held. The distance beyond the focal region of the guiding laser for a given particle is known as its retention distance Z. The guiding force is related to the size of the particle and its refractive index—as stated earlier, this latter physical property is linked to its inherent chemical or biological composition. The method may be used as a powerful analytical tool and the aim is to use it for separation of a diverse range of biological materials such as blood cells, bacteria, yeast cells, pollen even bacterial warfare agents (e.g., *Bacillus anthracis*). If successful, this method could indeed be used for portable biological warfare detectors. Researchers in the United States are now employing this technique in a microfluidic environment (Fig. 9).

As stated, the separation in this method occurs due to the balance between optical forces with Stokes forces from the fluid. The optical radiation pressure force may be given by:

$$F_{\text{rad\_pressure}} = \frac{2n_1 P}{c}\left(\frac{a}{\omega}\right)^2 Q^* \tag{3}$$

where P is the laser power, n_1 is the refractive index of the medium, ω is the beam radius, c is the speed of light, a is the sphere radius, and Q^* is a conversion factor (Hart and Terray, 2003; Hart *et al.*, 2004) that denotes the efficiency of radiation pressure transfer from the light to the trapped object. The retention distance Z is

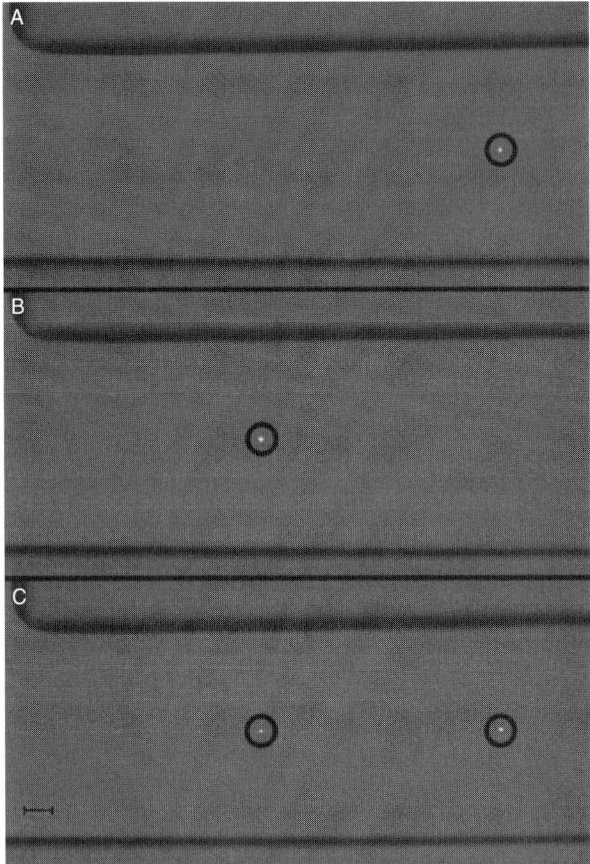

Fig. 9 An example of optical chromatography demonstrated by Hart and colleagues at NRL. Here the interplay between radiation pressure and fluid drag has led to different equilibrium positions for two different spores. Images of (A) *B. anthracis* and (B) *B. thuringiensis* spores optically retained individually, and (C) optically retained simultaneously. The liquid flow was from right to left and the laser was propagating from left to right. Bright spots are due to laser light scatter from the spore (black rings used to highlight position). The laser focal point was positioned in the center of the main channel, 206 μm to the right of the inlet channel edge, seen in the upper left corner of each image. The scale bar represents 100 μm. Reprinted by permission from American Chemical Society: Hart, S. J., Terray, A., Leski, T. A., Arnold, J., and Stroud, R. (2006). Discovery of a significant optical chromatographic difference between spores of *Bacillus anthracis* and its close relative, *Bacillus thuringiensis*. *Anal. Chem.* **78,** 3221–3225, copyright (2006).

determined by equating this formula, given in (3), to the Stokes drag in the fluid. Bacterial and fungal spores as well as various cell types have been separated using optical chromatography. Mulberry pollen and a larger ragweed pollen were held with a difference in retention distances of over 2 mm. The work in this field is on-going and hopes to exploit more subtle differences in biological samples, for example yielding different retention distances for different bacterial strains.

B. Separation Using Microfluidic Flow Over Periodic Optical Energy Landscapes

In contrast to optical chromatography which relies on radiation pressure equating with Stokes drag, other forms of sorting rely on the interplay between gradient forces and viscous drag in a fluid flow. As described earlier, it is possible to achieve kinetically locked-in states when colloidal matter flows across an angled optical potential energy landscape (Korda *et al.*, 2002). These states can lead to spatial separation of different particle species as the flow velocity at which a kinetically locked-in state collapses (where the particles are no longer deflected) is lower for weakly interacting species as compared to those with a large interaction strength (Ladavac *et al.*, 2004; MacDonald *et al.*, 2003, 2004; Pelton *et al.*, 2004).

The key to this technique is to make the path of least resistance across the landscape different for different particle species. Figure 10 illustrates the sorting mechanism. When a particle approaches the optical landscape, it can do one of four things: become trapped in a local intensity maximum, be deflected by local intensity maxima, responds to the envelope function of the landscape, or go through essentially unimpeded. For sorting, one requires some particles to be deflected by hopping or channeling between local trapping sites along a diagonal of the potential landscape, while others are only weakly deflected or experience no net deflection at all. This will clearly lead to spatial separation of colloid and cells according to size, shape, and refractive index. The movement of particles through the landscape can be facilitated by introducing light channels between local trapping sites or going away from the idea of traps entirely and using angled optical guides (e.g., linear interference fringes). Two major experiments in 2003 and 2004 (Ladavac *et al.*, 2004; MacDonald *et al.*, 2003) have shown sorting of colloidal particles in 3D optical lattices and arrays of holographically generated optical traps. Separation angles as high as 45° have been experimentally shown and even low-index particles such as ultrasound contrast agents have been sorted.

One physical incarnation of the microfluidic optical sorting approach consists of a landscape placed within a microfluidic flow chamber. The chamber brings together

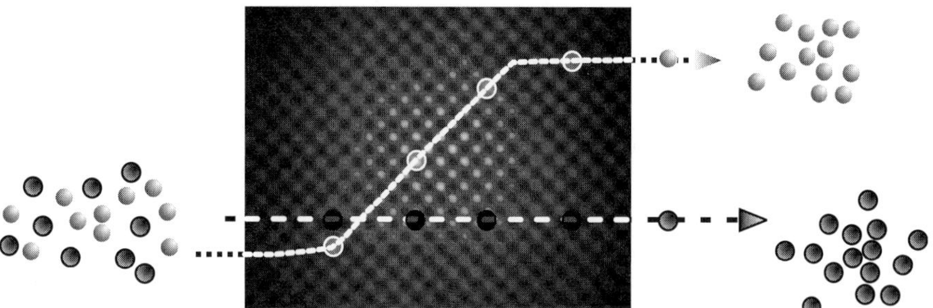

Fig. 10 Sorting in an optical lattice. Two species of particles enter the lattice flowing from left to right. The weakly interacting species (dark) flows straight through while the path of least resistance for the strongly interacting species (light) is at 45° to the flow, leading to physical separation of the two species. This idea can be expanded to more than two species.

two flows and then subsequently reseparates them. One flow contains a polydisperse mixture of particles, the other pure solvent or buffer solution. Strongly interacting species are deflected into the pure flow by the optical landscape and subsequently removed from the mixed flow when the two flows separate.

Although originally proposed for arrays of optical traps, the technique can be extended to the more generalized idea of a potential energy landscape that can be either 2D or 3D. In fact, the sorting behavior depends strongly on the type of landscape used and how it is made. When being produced holographically, near arbitrary 2D landscapes can be produced, but such landscapes are essentially limited to two dimensions unless more complex beam generation algorithms are used. Typically if a landscape is so restricted, only particles in a single plane can be sorted. This is also the case if the landscape is produced by scanning a single laser trap such as with an acousto-optic modulator,where a time-shared array is produced. However, some of the most powerful sorting to date has been shown when using an optical landscape defined using a scanned beam (Milne *et al.*, 2007), where as many as 4 different particle species have been separated simultaneously. To produce a truly 3D landscape, the simplest method is multibeam interference (although interference can only produce symmetric patterns) (MacDonald *et al.*, 2003). 3D landscapes do not require that particles be well confined with a single plane of flow but can instead sort throughout a volume. This has practical benefits even when particles are all denser than their carrier medium, as it removes the crucial alignment of particle flow with optical field that may be required with 2D landscapes.

By allowing for continuous particle throughput and spatially separated flows of sorted particles, this approach gives many advantages over other optical force-based separation techniques. In the particle size range of cells, sorting with this technique is deterministic such that very high purity and efficiency can be achieved and because particles can be flowed through the landscape in parallel (rather than in single file like micro-FACS), it is easily scaleable as long as there is sufficient laser power available.

When sorting cells or other colloidal objects by flowing them over an optical potential landscape, the particle density plays an important role in the behavior of the particles. Importantly, at higher densities, the streams of strongly interacting particles that are being channeled at an angle to the flow can mechanically deflect weakly interacting species. This effect will clearly lead to an error in the output of the sorter. One approach to reducing this problem without having to go to lower input density of particles is to introduce a motional scan in the optical landscape, which corresponds to region 4 in Fig. 2. This scanning effect leads to strongly interacting species being removed more quickly from the sorting area, increasing the nearest neighbor spacing, reducing the frequency of particle–particle interactions, and reducing the number of weakly interacting particles that are deflected (Smith *et al.*, 2007). This regime is least studied from an experimental viewpoint, though there are some notable theoretical predictions in this area (Reichhardt and Reichhardt, 2004).

VII. Dielectrophoresis for Microfluidic Sorting

In parallel with the controlled motion of particles using optical fields, gradients for particle and cell motion may be generated not optically but electrically. The dielectrophoretic (DEP) force is generated by the interaction of an applied electric AC field and the induced electric dipoles in neutral particles (Pohl, 1978). If the applied field is uniform, the two Coulomb forces on the charges on both sides of the particle are equal and opposite and cancel each other out. If the field is non-uniform, however, the Coulomb forces on either side of the particle will not be equal. The resulting force differential leads to a motion of the particle and is referred to as the DEP force. An analytic expression of this force shows that the DEP force is proportional to the gradient of the AC field E squared (Jones, 1995),

$$F_{\text{dep}} = 2\pi a^3 \varepsilon_{\text{m}} \text{Re}[K^*(\omega)] \nabla (E^2) \tag{4}$$

with a the particle radius, ε_{m} the permittivity of the surrounding medium, and $K^*(\omega)$ the Clausius–Mosotti (CM) factor, which depends on the polarizability of the particle and the medium as well as the frequency of the applied AC field; Re $[K^*(\omega)]$ has a value between 1 and $-1/2$. A positive CM factor means that particles are attracted toward higher fields and vice versa.

Standard dielectrophoresis has been used to sort cells based on their differing intrinsic DEP response (Becker *et al.*, 1995; Cheng *et al.*, 1998; Gascoyne and Vykoukal, 2002), but this has potential problems. Just as with optical forces, cell phenotypes or different target cells may show near equivalent intrinsic DEP responses making it difficult to sort them. However, recent work has shown how this may be circumvented (Hu *et al.*, 2005). Cells were harvested and mixed with a biotinylated monoclonal antibody. The cells were incubated with streptavidin-coated polymer beads which attached themselves to these cells. Mixtures thus contained both cells with beads attached and cells without beads attached. The rare target cells in this study were *Escherichia coli* cells with a specific surface peptide antigen that is recognized by a monoclonal antibody.

To create the appropriate electric fields for dielectrophoresis, a quadrupole electrode device was microfabricated using electron beam lithography of gold and titanium onto a glass substrate. The forces on the *E. coli* cells varied dramatically depending on whether they were labeled or not labeled. Tagged cells had forces of ~368 pN, whereas unlabeled cells had on ~57 pN exerted on them. The microfluidic chamber and these forces thus permitted one to readily deflect the labeled cells but not the unlabeled ones. A buffer flow was introduced to the center of the microfluidic chamber and though it parallels the notion of the sheath flow in FACS, in this case it was present for slightly different reason: FACS systems employ the sheath flow to surround the cell flow and to reduce shear stress therein, whereas in this DEP scheme the buffer flow was in the center of the flow stream and essentially once the electrodes were powered, selected cells were readily deflected into this central region and went onto the collection region.

A major issue with microfabricated FACS described earlier is the relatively low throughput: in contrast, the rather high forces combined with the relatively fast flow rates in this DEP experiment, flows of ~3 mm/s near the electrodes and a flow rate of 300 ml/h. This meant that for a single round trip for a given batch of cells, a 200-fold enrichment was observed and a throughput of 10,000 cells per second was recorded, both very high compared to typically what one might expect at this small size scale. It will be interesting to see how well this active cell sorting method works for other cell types and whether such prelabeling (marking) of cells proves to be the method of choice. We note in the passive optical schemes that such marking (making them active) has yet to be fully explored—though the work using the Bessel beam has shown some interesting data.

Dielectrophoresis was realized using *optical* control of the AC electric field (Chiou *et al.*, 2005), as shown in Fig. 11. As a result, trapping in this geometry was demonstrated with up to a 10^5-fold lower power requirement than in conventional optical tweezers requiring only 10 nW/μm^2. This may help realize many optical manipulation effects hitherto inaccessible with conventional optical micromanipulation. In this section we describe this recent development and the data achieved to date.

The fluid is sandwiched between two indium tin oxide (ITO)-coated glass carriers, across which an AC electric field is applied. The amorphous silicon (a-Si) layer acts as a photoconductor that exhibits high resistivity in the absence of illumination. With the laser beam on, electron–hole pairs are generated in the a-Si and the field now drops across the liquid. The a-Si layer, typically 1-μm thick, is coated with a thin film (20 nm) of silicon dioxide to prevent electrolysis. The areas of the cell where DEP forces are at play is therefore entirely controlled by the illuminating beam. The strength of the force, however, is controlled by the applied AC field, and especially its gradient [Eq. (2)]. The gradient depends on the thickness and nature of the a-Si layer (e.g., the diffusion length of the photogenerated electron–hole pairs) which is a key element in determining the smallest spot size, and therefore the resolution, that can be achieved. Calculations show that the diffusion length may readily be less than 500 nm and thus in essence it is the ability of focusing the light to a small spot size that is the key criteria here. Such a cell is made by using ITO-coated glass substrates, sputtering the silicon layer and the thin SiO_2 onto the top surface and mounting them together using, for example, polystyrene spheres as spacers. Since the microfluidic functionality is derived entirely from the pattern projected onto the surface, no further lithographic patterning is necessary. Figure 11 shows some data for this type of sorting.

VIII. Conclusions

There is little doubt that techniques to sort, enrich or isolate small cell populations as well as colloidal samples will be of interest to biologists, material scientists, and chemists alike. Bulk sorting methods exist and are certainly an established

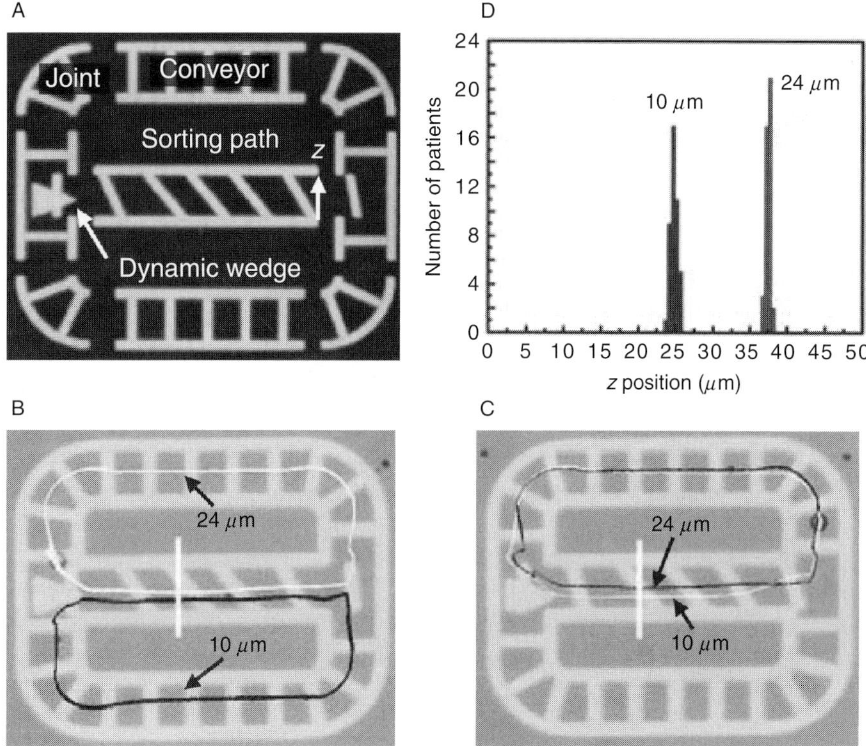

Fig. 11 An example of an integrated virtual optical machine. (A) Integration of virtual components, including an optical sorter path, conveyors, joints, and a wedge. The motion of different components is synchronized. (B and C) Two polystyrene particles with sizes of 10 and 24 μm pass through the sorter path and are fractionated in the z direction owing to the asymmetrical optical patterns. The particle trajectories can be switched at the end of the sorter path by the optical wedge. (D) Optical sorting repeatability test. The white and black loops in B and C represent the particle traces after 43 cycles. The trace broadening at the white bar has a standard deviation of 0.5 μm for the 10-μm bead and 0.15 μm for the 24-μm bead. Reprinted by permission from Macmillan Publishers Ltd.: Chiou, P. Y., Ohta, A. T., and Wu, M. C. (2005). Massively parallel manipulation of single cells and microparticles using optical images *Nature* **436,** 370–372, copyright (2005).

technology: these include fluorescence activated cell-sorting and techniques using magnetic beads attached to cell populations. However, there is the open question as to how we may perform sorting for rare cell types or in situations where we have very small number of cells in the first instance and indeed attain a reasonable purity and throughput for the microfluidic system in light of the challenging dimensions and physics of such laminar flow and low Reynolds number. In this respect, a portable cell sorting methodology based on optics would be highly desirable: there are no doubt prospects for such sorting and some pilot studies have shown promise particularly when it comes to lymphocytes and erythrocytes but all optical microfluidic sorting in a passive scheme remains largely unproven for a wide variety of cells. Both optical methods and DEP schemes with added surface markers (e.g., attached spheres) are certainly viable ways by which to separate cells in a microfluidic environment. The recently developed hybrid method of light-induced dielectrophoresis may possibly offer the best of both worlds, the reconfigurability of light, yet at vastly reduced power levels, and the power of DEP forces over a large area. It is certainly a dynamic and challenging time in the field of cellular and colloidal separation.

Acknowledgments

We thank colleagues in all of our groups for useful discussions on the topics of separation and sorting and Dr. Frank Gunn-Moore for reading the chapter. We acknowledge support from the UK Engineering and Physical Sciences Research Council, Scottish Higher Education Funding Council, and the ATOM-3D network (contract number 508952) funded under the NEST Program of the European Commission framework 6 program, PZ acknowledges the support of the Centre of Modern Optics (LC06007) under the Ministry of Education, Youth, and Sports of the Czech Republic and the ISI Institutional research plan (AV0Z20650511).

References

Applegate, R. W., Squier, J., Vestad, T., Oakey, J., Marr, D. W. M., Bado, P., Dugan, M. A., and Said, A. A. (2006). Microfluidic sorting system based on optical waveguide integration and diode laser bar trapping. *Lab. Chip* **6**(3), 422–426.

Ashkin, A., and Dziedzic, J. M. (1987). Optical trapping and manipulation of viruses and bacteria. *Science* **235**(4795), 1517–1520.

Babic, D., Schmitt, C., Poberaj, I., and Bechinger, C. (2004). Stochastic resonance in colloidal systems. *Europhys. Lett.* **67**(2), 158–164.

Becker, F. F., Wang, X. B., Huang, Y., Pethig, R., Vykoukal, J., and Gascoyne, P. R. C. (1995). Separation of human breast-cancer cells from blood by differential dielectric affinity. *Proc. Natl. Acad. Sci. USA* **92**(3), 860–864.

Buican, T. N., Smyth, M. J., Crissman, H. A., Salzman, G. C., Stewart, C. C., and Martin, J. C. (1987). Automated single-cell manipulation and sorting by light trapping. *Appl. Opt.* **26**(24), 5311–5316.

Cheng, J., Sheldon, E. L., Wu, L., Heller, M. J., and O'Connell, J. P. (1998). Isolation of cultured cervical carcinoma cells mixed with peripheral blood cells on a bioelectronic chip. *Anal. Chem.* **70**(11), 2321–2326.

Cheong, F. C., Sow, C. H., Wee, A. T. S., Shao, P., Bettiol, A. A., van Kan, J. A., and Watt, F. (2006). Optical travelator: Transport and dynamic sorting of colloidal microspheres with an asymmetrical line optical tweezers. *Appl. Phys. B-Lasers Opt.* **83**(1), 121–125.

Chiou, P. Y., Ohta, A. T., and Wu, M. C. (2005). Massively parallel manipulation of single cells and microparticles using optical images. *Nature* **436**(7049), 370–372.

Cizmar, T., Siler, M., Sery, M., Zemanek, P., Garcés-Chávez, V., and Dholakia, K. (2006). Optical sorting and detection of sub-micron objects in a motional standing wave. *Phys. Rev. B* **74,** 035105.

Curtis, J. E., Koss, B. A., and Grier, D. G. (2002). Dynamic holographic optical tweezers. *Opt. Commun.* **207**(1–6), 169–175.

Ericsson, M., Hanstorp, D., Hagberg, P., Enger, J., and Nystrom, T. (2000). Sorting out bacterial viability with optical tweezers. *J. Bacteriol.* **182**(19), 5551–5555.

Eriksen, R. L., Daria, V. R., and Gluckstad, J. (2002). Fully dynamic multiple-beam optical tweezers. *Opt. Express* **10**(14), 597–602.

Fu, A. Y., Chou, H. P., Spence, C., Arnold, F. H., and Quake, S. R. (2002). An integrated microfabricated cell sorter. *Anal. Chem.* **74**(11), 2451–2457.

Fu, A. Y., Spence, C., Scherer, A., Arnold, F. H., and Quake, S. R. (1999). A microfabricated fluorescence-activated cell sorter. *Nature Biotechnol.* **17**(11), 1109–1111.

Galbraith, D. W., Anderson, M. T., and Herzenberg, L. A. (1999). Flow cytometric analysis and FACS sorting of cells based on GFP accumulation. *In* "Methods in Cell Biology," Vol. 58, pp. 315–341. Academic Press Ltd., London.

Gascoyne, P. R. C., and Vykoukal, J. (2002). Particle separation by dielectrophoresis. *Electrophoresis* **23**(13), 1973–1983.

Givan, A. L. (2001). "Flow Cytometry: First Principles." Wiley-Liss, USA.

Gleeson, J. P., Sancho, J. M., Lacasta, A. M., and Lindenberg, K. (2006). Analytical approach to sorting in periodic and random potentials. *Phys. Rev. E* **73**(4), Art. No. 041102 Part 1.

Grier, D. G. (2003). A revolution in optical manipulation. *Nature* **424**(6950), 810–816.

Grover, S. C., Gauthier, R. C., and Skirtach, A. G. (2000). Analysis of the behaviour of erythrocytes in an optical trapping system. *Opt. Express* **7**(13), 533–539.

Grover, S. C., Skirtach, A. G., Gauthier, R. C., and Grover, C. P. (2001). Automated single-cell sorting system based on optical trapping. *J. Biomed. Opt.* **6**(1), 14–22.

Grujic, K., Helleso, O. G., Hole, J. P., and Wilkinson, J. S. (2005). Sorting of polystyrene microspheres using a Y-branched optical waveguide. *Opt. Express* **13**(1), 1–7.

Grzegorczyk, T. M., Kemp, B. A., and Kong, J. A. (2006). Stable optical trapping based on optical binding forces. *Phys. Rev. Lett.* **96**(11), Art. No. 113903.

Hart, S. J., and Terray, A. V. (2003). Refractive-index-driven separation of colloidal polymer particles using optical chromatography. *Appl. Phys. Lett.* **83**(25), 5316–5318.

Hart, S. J., Terray, A., Kuhn, K. L., Arnold, J., and Leski, T. A. (2004). Optical chromatography of biological particles. *Am. Lab.* **36,** 13–17.

Hart, S. J., Terray, A., Leski, T. A., Arnold, J., and Stroud, R. (2006). Discovery of a significant optical chromatographic difference between spores of *Bacillus anthracis* and its close relative, *Bacillus thuringiensis*. *Anal. Chem.* **78**(9), 3221–3225.

Herzenberg, L. A., Parks, D., Sahaf, B., Perez, O., Roederer, M., and Herzenberg, L. A. (2002). The history and future of the fluorescence activated cell sorter and flow cytometry: A view from Stanford. *Clin. Chem.* **48**(10), 1819–1827.

Hu, X. Y., Bessette, P. H., Qian, J. R., Meinhart, C. D., Daugherty, P. S., and Soh, H. T. (2005). Marker-specific sorting of rare cells using dielectrophoresis. *Proc. Natl. Acad. Sci. USA* **102**(44), 15757–15761.

Imasaka, T., Kawabata, Y., Kaneta, T., and Ishidzu, Y. (1995). Optical chromatography. *Anal. Chem.* **67**(11), 1763–1765.

Jones, T. B. (1995). Electromechanics of Particles. Cambridge University Press ISBN-13: 978-0521431965.

Kaneta, T., Ishidzu, Y., Mishima, N., and Imasaka, T. (1997). Theory of optical chromatography. *Anal. Chem.* **69**(14), 2701–2710.

Korda, P. T., Taylor, M. B., and Grier, D. G. (2002). Kinetically locked-in colloidal transport in an array of optical tweezers. *Phys. Rev. Lett.* **89**(12), 128–301.

Koss, B. A., and Grier, D. G. (2003). Optical peristalsis. *Appl. Phys. Lett.* **82**(22), 3985–3987.

Kramers, H. A. (1940). Brownian motion in the field of force and diffusion model of chemical reactions. *Physica* **7**(4), 284–304.

Kruger, J., Singh, K., O'Neill, A., Jackson, C., Morrison, A., and O'Brien, P. (2002). Development of a microfluidic device for fluorescence activated cell sorting. *J. Micromech. Microeng.* **12**(4), 486–494.

Lacasta, A. M., Sancho, J. M., Romero, A. H., and Lindenberg, K. (2005). Sorting on periodic surfaces. *Phys. Rev. Lett.* **94**(16), 188–902.

Ladavac, K., Kasza, K., and Grier, D. G. (2004). Sorting mesoscopic objects with periodic potential landscapes: Optical fractionation. *Phys. Rev. E* **70**(1), Art. No. 010901 Part 1.

Lee, S. H., and Grier, D. G. (2005). Flux reversal in a two-state symmetric optical thermal ratchet. *Phys. Rev. E* **71**(6), Art. No. 060102 Part 1.

Lekner, J. (2005). Force on a scatterer in counter-propagating coherent beams. *J. Opt. A: Pure Appl. Opt.* **7,** 238–248.

Libal, A., Reichhardt, C., Janko, B., and Reichhardt, C. J. O. (2006). Dynamics, rectification, and fractionation for colloids on flashing substrates. *Phys. Rev. Lett.* **96**(18), Art. No. 188301.

MacDonald, M. P., Neale, S., Paterson, L., Riches, A., Dholakia, K., and Spalding, G. C. (2004). Cell cytometry with a light touch: Sorting microscopic matter with an optical lattice. *J. Biol. Regulat. Homeostat. Agents* **18**(2), 200–205.

MacDonald, M. P., Spalding, G. C., and Dholakia, K. (2003). Microfluidic sorting in an optical lattice. *Nature* **426**(6965), 421–424.

McGloin, D., and Dholakia, K. (2005). Bessel beams: Diffraction in a new light. *Contemp. Phys.* **46**(1), 15–28.

Milne, G., Rhodes, D., MacDonald, M., and Dholakia, K. (2007). "Fractionation of polydisperse colloid with acousto-optically generated potential energy landscapes." *Opt. Lett.* **32,** 1144–1146.

Molloy, J. E., Dholakia, K., and Padgett, M. J. (2003). Preface: Optical tweezers in a new light. *J. Mod. Opt.* **50**(10), 1501–1507.

Neuman, K. C., and Block, S. M. (2004). Optical trapping. *Rev. Sci. Instrum.* **75**(9), 2787–2809.

Oakey, J., Allely, J., and Marr, D. W. M. (2002). Laminar-flow-based separations at the microscale. *Biotechnol. Progress* **18**(6), 1439–1442.

Paterson, L., Papagiakoumou, E., Milne, G., Garcés-Chávez, V., Tatarkova, S. A., Sibbett, W., Gunn-Moore, F. J., Bryant, P. E., Riches, A. C., and Dholakia, K. (2005). Light-induced cell separation in a tailored optical landscape. *Appl. Phys. Lett.* **87**, Art. No. 123901.

Pelton, M., Ladavac, K., and Grier, D. G. (2004). Transport and fractionation in periodic potential-energy landscapes. *Phys. Rev. E* **70**(3), Art. No. 031108 Part 1.

Pohl, H. A. (1978). "Dielectrophoresis: The Behaviour of Neutral Matter in Nonuniform Electric Fields." Cambridge University Press, Cambridge.

Prasad, P. N. (2003). "Introduction to Biophotonics." John Wiley and Sons Inc., Hoboken, New Jersey.

Reichhardt, C., and Reichhardt, C. J. O. (2004). Directional locking effects and dynamics for particles driven through a colloidal lattice. *Phys. Rev. E* **69**(4), Art. No. 041405 Part 1.

Reimann, P. (2002). Brownian motors: Noisy transport far from equilibrium. *Phys. Rep.* **361,** 57–265.

Ricardez-Vargas, I., Rodriguez-Montero, P., Ramos-Garcia, R., and Volke-Sepulveda, K. (2006). Modulated optical sieve for sorting of polydisperse microparticles. *Appl. Phys. Lett.* **88**(12), Art. No. 121116.

Rodrigo, P. J., Daria, V. R., and Gluckstad, J. (2005). Dynamically reconfigurable optical lattices. *Opt. Express* **13**(5), 1384–1394.

Rodrigo, P. J., Eriksen, R. L., Daria, V. R., and Gluckstad, J. (2002). Interactive light-driven and parallel manipulation of inhomogeneous particles. *Opt. Express* **10**(26), 1550–1556.

Shirasaki, Y., Tanaka, J., Makazu, H., Tashiro, K., Shoji, S., Tsukita, S., and Funatsu, T. (2006). On-chip cell sorting system using laser-induced heating of a thermoreversible gelation polymer to control flow. *Anal. Chem.* **78**(3), 695–701.

Siler, M., Cizmar, T., Sery, M., and Zemanek, P. (2006). Optical forces generated by evanescent standing waves and their usage for sub-micron particle delivery. *Appl. Phys. B-Lasers Opt.* **84**(1–2), 157–165.

Simon, A., and Libchaber, A. (1992). Escape and Synchronization of a Brownian Particle. *Phys. Rev. Lett.* **68**(23), 3375–3378.

Smith, R. L., Spalding, G. C., Dholakia, K., and MacDonald, M. P. (2007). Colloidal sorting in dynamic optical lattices. *J. Opt. A: Pure. Appl. Opt.*, special issue on optical manipulation. (in press).

Squires, T. M., and Quake, S. R. (2005). Microfluidics: Fluid physics at the nanoliter scale. *Rev. Mod. Phys.* **77**(3), 977–1026.

Tatarkova, S. A., Sibbett, W., and Dholakia, K. (2003). Brownian particle in an optical potential of the washboard type. *Phys. Rev. Lett.* **91**(3), Art. No. 038101.

Wang, M. M., Tu, E., Raymond, D. E., Yang, J. M., Zhang, H. C., Hagen, N., Dees, B., Mercer, E. M., Forster, A. H., Kariv, I., Marchand, P. J., and Butler, W. F. (2005). Microfluidic sorting of mammalian cells by optical force switching. *Nature Biotechnol.* **23**(1), 83–87.

Zelenina, A. S., Quidant, R., Badenes, G., and Nieto-Vesperinas, M. (2006). Tunable optical sorting and manipulation of nanoparticles via plasmon excitation. *Opt. Lett.* **31**(13), 2054–2056.

Zemánek, P., Jonas, A., Sramek, L., and Liska, M. (1998). Optical trapping of Rayleigh particles using a Gaussian standing wave. *Opt. Commun.* **151,** 273–285.

Zemánek, P., Jonas, A., Jakl, P., Jezek, J., Sery, M., and Liška, M. (2003). Theoretical comparison of optical traps created by standing wave and single beam. *Opt. Commun.* **220,** 401–412.

Zemánek, P., Jonas, A., and Liska, M. (2002). Simplified description of optical forces acting on a nanoparticle in the Gaussian standing wave. *J. Opt. Soc. Am. A.* **19,** 1025–1034.

Zemánek, P., Karasek, V., and Sasso, A. (2004a). Optical forces acting on Rayleigh particle placed into interference field. *Opt. Commun.* **240**(4–6), 401–415.

Zemánek, P., Karasek, V., and Sery, M. (2004b). Behaviour of colloidal microparticles in a planar 3-beam interference field. *Proc. SPIE* **5514,** 15–26.

Zhang, H. C., Tu, E., Hagen, N. D., Schnabel, C. A., Paliotti, M. J., Hoo, W. S., Nguyen, P. M., Kohrumel, J. R., Butler, W. F., Chachisvillis, M., and Marchand, P. J. (2004). Time-of-flight optophoresis analysis of live whole cells in microfluidic channels. *Biomed. Microdevices* **6**(1), 11–21.

Zhao, B. S., Koo, Y. M., and Chung, D. S. (2006). Separations based on the mechanical forces of light. *Anal. Chim. Acta* **556**(1), 97–103.

CHAPTER 18

Laser Tweezers in the Study of Mechanobiology in Live Cells

Elliot L. Botvinick* and Yingxiao Wang†

*Department of Biomedical Engineering, Beckman Laser Institute
University of California, Irvine, California 92612

†Department of Bioengineering, Molecular and Integrative Physiology
Neuroscience Program, Beckman Institute for Advanced Science and
Technology, University of Illinois, Urbana-Champaign, Illinois 61801

The study of how cells respond to mechanical stimuli has recently leaped into the forefront of cell science with recent advances in molecular probes facilitating real-time measurements of cell signaling. In this chapter, we will detail the development of a "real-time" molecular probe designed to report the current fractional activated state of Src kinase by changing its spectral output in accordance to local Src states. Src kinase is widely understood to be a key player in the transduction of mechanical stimuli transduced through cell adhesions and focal complexes. To study the local and long-range Src response to localized stresses, an experimental protocol was developed whereby ligand-coated microspheres were adhered to the cell surface and pulled laterally by laser tweezers. This chapter contains a practical

0091-679X/07 $35.00
DOI: 10.1016/S0091-679X(06)82018-2

discussion of system design considerations and force calibration. Image processing, background subtraction, and the construction of an unbiased ratio image are discussed. Methods of analyzing the distribution of activated Src molecules are detailed with examples of cells with varying degrees of mechanostimulation.

I. Introduction

It has become clear that cells can sense their surrounding environment, both chemically and mechanically (Chien *et al.*, 2005). It is, however, not clear how cells perceive the external physical cues and transmit them into intracellular biochemical signals, that is, mechanotransduction. Recent evidence indicates that mechanotransduction is a complex process and many molecules are involved. In particular, integrins, PECAM-1, ion channels, and receptor tyrosine kinases have been implicated to be the mechanosensing elements and trigger intracellular signaling (Chen *et al.*, 1999; Kaufman *et al.*, 2004; Kernan and Zuker, 1995; Wang *et al.*, 1993). Cytoskeletal components, for example actin filaments and microtubules, have been shown to play crucial roles in mediating and transferring subcellular local signals to the whole cell (Ingber, 2003a,b). It appears that different signaling pathways do not function in isolation. Instead, they interact with each other and coordinately regulate cellular functions in response to mechanical stimuli. For example, the activation of VEGF Receptor 2 (Flk-1) by shear stress is dependent on integrins (Wang *et al.*, 2002). Despite these findings, the molecular hierarchy in regulating cellular functions in live cells on mechanical stimulation remains elusive. This chapter describes the design and utilization of Src kinase molecular biosensor to study the real-time spatial response of Src to mechanical stimuli on single cells.

II. Rational

Progress in developing novel fluorescent probes, for example organic dyes and quantum dots, has been enormous in recent years. A large fraction of new probes have been applied not only in basic biological studies but also in molecular imaging, disease diagnostics, and therapeutics. Mechanobiology, as a rapidly developing field, inevitably encompasses the integration and application of these probes. We will briefly describe current research on this aspect not utilizing förster resonance energy transfer (FRET, more commonly referred to as fluorescence resonance energy transfer). In Section III, we will discuss the development of probes using FRET.

A. Fluorescent Probes for Measuring Extracellular Mechanical Stress

Yu-Li Wang has been pioneering the field of measuring the force a cell can exert on the extracellular environment. The protocol he developed has been adopted and used by many other groups. In general, fluorescent beads labeled with either

FITC or Rhodamine are mixed with polyacrylamide. After the polyacrylamide and bead mixture gels, the beads form an irregular grid embedded in the gel. When the cells are cultured on the bead-embedded gel, the force generated by the cells causes the deformation of the gel and displacement of the beads. Since the gel is composed of polyacrylamide, which has a linear stress/strain relationship, the stiffness of the gel can be conveniently controlled by varying the percentage of polyacrylamide in the solution. In a typical experiment, cells are seeded on the gel and as they spread, the positions of the beads change due to stresses the cells exert through their adhesions to the gel. The displaced positions of beads are recorded by a camera. The cells are then trypsinized and rinsed away, and the beads again move as the gel returns to its stress-free state. The bead positions are recorded again. The local displacements of beads before and after the removal of the cells are calculated, which allow the calculation of the stress distribution exerted by the cells on the gel with the help of sophisticated computing algorithm and software (Beningo *et al.*, 2001; Pelham and Wang, 1997).

B. Fluorescent Probes for Imaging Intracellular Mechanical Properties

Other fluorescent probes have been employed to study the intracellular mechanics and molecular dynamics. Green fluorescent protein (GFP) and its derivatives, for example yellow fluorescent protein (YFP), have been fused to targeting molecules such that these molecules can lead the fluorescent proteins (FPs) to specific subcellular locations. By monitoring the position changes of these localized fluorescent markers, the intracellular deformation can be mapped out and the *in situ* mechanical properties can be assessed. For example, a YFP-conjugated mitochondria construct has been used to identify the positions of mitochondria. RGF-coated magnetic beads are then allowed to bind to cell membrane receptors, integrins, and adhere on cell surface. Since integrin engagement can induce the coupling between integrins and cytoskeleton, the beads bind tightly to the cell body. The beads are then mechanically perturbed by a three-dimensional magnetic-twisting device (Hu *et al.*, 2004). This allows the intracellular mechanical properties to be characterized. This method has revealed an anisotropic mechanical property of cells. In other applications, GFP was fused to vinculin or paxillin. On the local mechanical perturbation of the cell edge using a micropipette, local assembly of focal adhesion complexes was observed, which was shown to be dependent on mDia, but not ROCK pathways (Riveline *et al.*, 2001).

III. Methods

A. Designing Custom Probes to Measure Src Kinase Activity

1. Overview of FRET

FRET occurs between two fluorophores if they are in sufficient proximity and if the emission spectrum of the donor fluorophore sufficiently overlaps the excitation spectrum of the acceptor. Any change of the distance and/or relative orientation

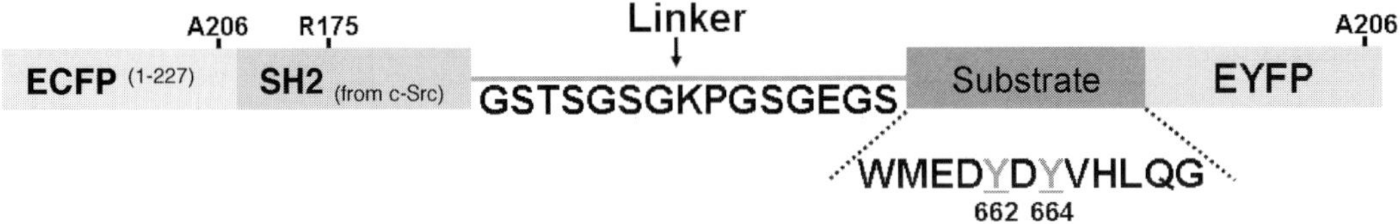

Fig. 1 A schematic cartoon of the Src biosensor composition. (See Plate 16 in the color insert section.)

between the two fluorophores may affect the efficiency of FRET and therefore the ratio of acceptor to donor emission (Ting *et al.*, 2001). Previous studies have shown that fusion proteins with interacting peptide partners sandwiched between two different FPs are capable of monitoring various cellular events in live cells with high spatial and temporal resolution (Miyawaki *et al.*, 1997; Tsien, 1998; Zhang *et al.*, 2001).

We have developed a FRET-based biosensor capable of detecting Src kinase activation. As shown in Fig. 1, this genetically encoded biosensor consists of an N-terminal ECFP, an SH2 domain derived from Src kinase, a flexible linker, a substrate peptide specific to Src phosphorylation, and a C-terminal EYFP. When the Src kinase is in its rest state, the ECFP and EYFP are positioned in proximity as a consequence of (1) the tendency of the wild-type ECFP and EYFP to form antiparallel dimmers, (2) the flexible linker, and (3) the juxtaposition of N- and C-terminals of the SH2 domain. Therefore, strong FRET can occur and the excitation of ECYP at 433 nm leads to the emission of EYFP at 527 nm. When Src kinase is activated it phosphorylates the designed substrate peptide, which displays a high affinity to and binds the bottom pocket of the SH2 domain. This action will lead to the displacement of the EYFP from the ECFP and will decrease the FRET efficiency between these two FPs. The excitation of ECFP at 433 nm then results in the emission from ECFP at 476 nm, as shown in Fig. 2. Hence, the emission ratio of ECFP/EYFP with the excitation of ECFP should serve as a good indicator of the status of Src activation.

2. Characterization of the Src Biosensor *In Vitro* and *In Vivo*

We first examined whether the Src biosensor is sensitive and specific to Src kinase. The chimeric Src biosensor proteins were expressed as N-terminal His_6 tag fusions in *E. coli* and purified by nickel chelate chromatography as described previously (Wang *et al.*, 2005). The purified biosensor protein was subjected to an *in vitro* Src kinase assay, in which the fluorescence spectra of the biosensor were monitored by a fluorescence plate reader (TECAN Safire; emission spectra scan at excitation wavelength of 433 nm) to monitor changes in FRET before and after the addition of kinases. The addition of Src kinase and ATP to the Src biosensor caused a 25% loss of FRET by the biosensor, whereas the addition of other kinases, including Yes, FAK, EGFR, Abl, Jak2, or ERK1, caused only very minor FRET change (<2%). Fyn, another member of Src family kinases, caused

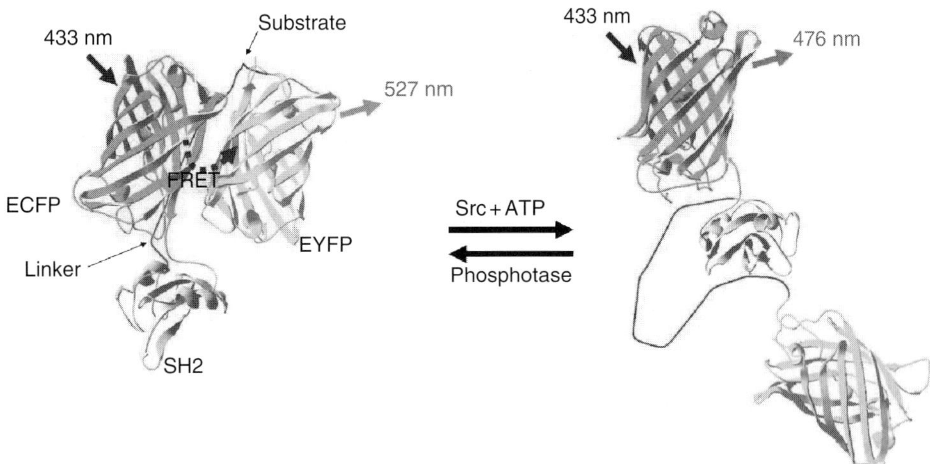

Fig. 2 A schematic cartoon of the activation mechanism of the Src biosensor. (See Plate 17 in the color insert section.)

a moderate 10% FRET change of the biosensor. All these results indicate that the Src biosensor is capable of specifically reporting Src activation *in vitro*. We further subcloned the Src biosensor into a PCDNA3.1 vector for mammalian cell expression. The Src biosensor was transfected into HeLa cells, which were subjected to 50-ng/ml EGF to activate Src kinase (Thomas and Brugge, 1997). EGF induced a 25–35% emission ratio change. The EGF-induced FRET response in HeLa cells was reversed by PP1, a selective inhibitor of Src family tyrosine kinases. When HeLa cells were pretreated with PP1 for 1 h, EGF could no longer induce a significant FRET response, suggesting the specificity of the Src biosensor toward Src in mammalian cells.

3. Monomerization of the Src Biosensor

Because the original Src biosensor was irreversible on activation, we reasoned that the tendency of CFP and YFP to form weak antiparallel dimers may be to blame. We modified the original Src biosensor to produce a monomerized Src biosensor by introducing A206K mutations in both ECFP and citrine (EYFP). In HeLa cells, the monomerized biosensor undergoes a dramatic FRET change on Src activation triggered by EGF. The response of the monomerized Src biosensor is also reversible by EGF washout, hence decreasing the Src activity (Fig. 3).

4. Membrane-Targeting Src Biosensor to Plasma Membrane

It has been shown that Src translocation to the plasma membrane is a prerequisite of Src activation (Thomas and Brugge, 1997). To increase the local concentration of the biosensor and position it close to activated Src, we have genetically modified the

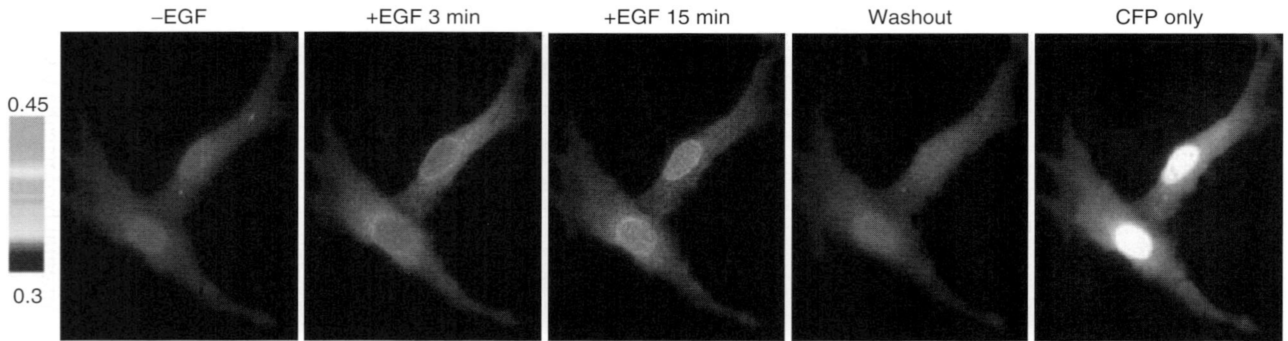

Fig. 3 The monomerized Src biosensor is sensitive to EGF application and reversible on EGF removal. Hela cells were transfected with the monomerized Src biosensor and stimulated with EGF (50 ng/ml) for various time periods as indicated. The cells were subsequently washed with serum-free medium for 15 min (washout). The scale bar on the left represents the CFP/YFP emission ratio, with cold color indicating low Src activity and hot color indicating high levels of Src activation. The representative emission ratio images are shown and the CFP-only image is shown in black and white on the far right. (See Plate 18 in the color insert section.)

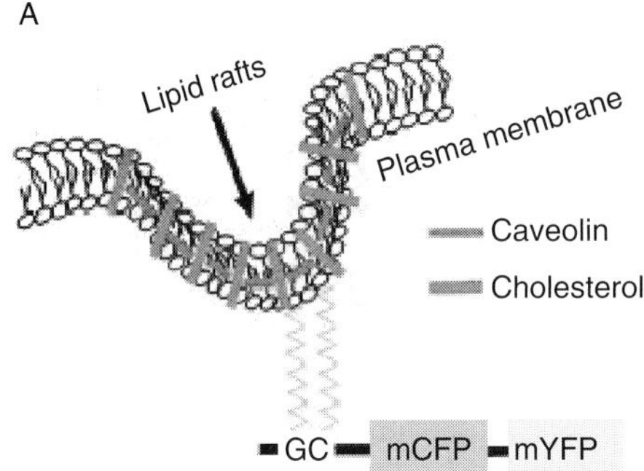

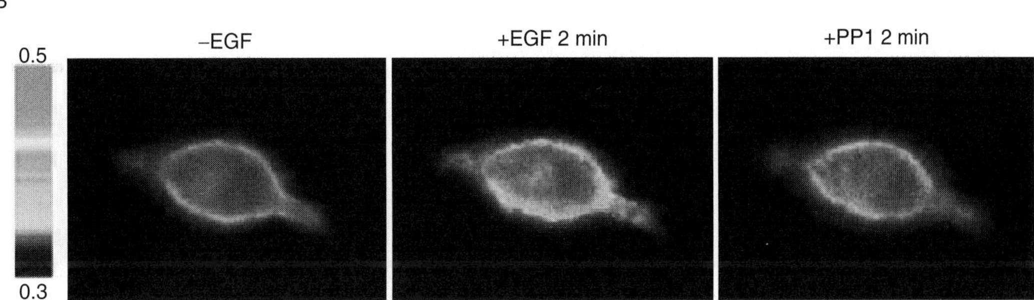

Fig. 4 (A) A schematic cartoon of the modified Src biosensor targeted to the plasma membrane. (B) Hela cells transfected with the membrane-targeted Src biosensor were stimulated with EGF (50 ng/ml) and subsequently incubated with PP1 for the time periods as indicated. The images represent the CFP/YFP emission. The scale bar on the left represents the CFP/YFP emission ratio, with cold color indicating low Src activity and hot color indicating high levels of Src activation. (See Plate 19 in the color insert section.)

Src biosensor. A small peptide derived from Lyn kinase was fused to the front of the Src biosensor so that the biosensor can be tethered to the lipid raft domains in the plasma membrane (Fig. 4A). The FRET response of this membrane-targeted Src biosensor can be induced by Src activation on EGF stimulation and inhibited by PP1, a specific inhibitor of Src (Fig. 4). These results indicate that the membrane-targeted Src biosensor is a sensitive and specific indicator of Src activity.

B. Investigation of Mechanotransduction Utilizing FRET Technology and Laser Tweezers

1. Cell Culture Protocol

One important fundamental biological question we want to address is how cells perceive mechanical signals and transmit them into biochemical signals, that is, mechanotransduction. Since we have developed tools capable of visualizing intracellular biochemical activities and perturbing cells mechanically at subcellular levels, it is natural for us to integrate these two technologies together and investigate the molecular mechanism of mechanotransduction. Since the human umbilical vein endothelial cells (HUVECs) that we are interested in are very difficult to transfect using the lipofectamine method, we first generated a retrovirus version of the Src biosensor (Retromax). The retrovirus carrying the Src biosensor was produced by 293 cells and used to infect HUVECs. The infected cells were serum starved for 24 h before FN-coated beads were seeded onto the cells by the laser tweezers system. Adhesion between FN units and their membrane receptor integrins will induce the adhesion complex to mechanically couple to the cell cytoskeleton. After 20 min of incubation, a calibrated laser tweezers force will be imposed on the bead. The FN-integrin-cytoskeleton coupling then can transmit the force into the cell. FRET responses of the membrane-bound Src biosensor are then monitored and recorded by the fluorescence microscopy system described below.

2. Combining Fluorescence Imaging with Laser Tweezers

Laser tweezers is a method in which the momentum flux of light incident on a particle is transferred, in part, to that particle. When properly constructed, laser tweezers have a local minimum of potential energy within the particle. In this case, the particle can be stably "held" by the focused laser. In our experiments, spherical particles, or microspheres (also referred to as beads in this chapter), are trapped above the focal plane of a high numerical aperture (NA) microscope objective lens within an inverted microscope. To gain a better understanding of the principles of optical tweezers, we highly recommend either Ashkin's paper (Ashkin, 1992) or the Chapter 6 by Nieminen *et al.*, this volume.

The method described below combines both laser trapping of microspheres and fluorescence imaging of a FRET biosensor. Emphasis will be placed on combining the trapping laser and fluorescence excitation light into the microscope, calibrating and applying forces to a microscope attached to the cell surface, and acquiring and calculating the FRET image.

The four manufacturers of research-quality inverted microscopes, have recently modified their platforms or developed add-ons to ease the combination of laser tweezers and nonlaser-based fluorescence excitation. Our laboratory has modified three Carl Zeiss microscopes Nikon (Nikon, Melville, NY), Olympus (Olympus, America, Center Valley, PA), Leica (Leica Microsystems GmbH, Wetzlar, Germany), and Zeiss (Carl Zeiss MicroImaging, Inc., Thornwood, NY), to combined arc lamp excitation and laser tweezers, each in their own way. The three systems have been named RoboLase I, RoboLase II, and RoboLase III. The microscopes implement infinity-corrected optics thereby creating an "infinity space" between the back of the objective and the tube lens of the microscope. Light originating from in-focus objects exits the back of the objective lens and travels through the infinity space as a set of diverging plane waves. The plane waves are refocused by the tube lens to form an image at the intermediate image plane where imaging devices or light detectors are typically mounted. The way to think about combining the light paths is to first consider the fluorescence filter cube placed in the infinity space. The primary function of the filter cube is to house a dichroic filter. The dichroic filter is responsible for directing incoming fluorescence excitation wavelengths and for transmitting outgoing fluorescence emission to/from the objective lens. The dichroic filter reflects light exiting the epi-illumination lens system housed at the back of the microscope stand up through the back aperture of the objective lens. In a correctly aligned Köhler-illuminated epifluorescence system, the arc lamp is imaged (after reflection from the dichroic filter) onto the back focal plane of the objective lens, thus illuminating the specimen plane with a reasonably uniform distribution of plane waves. Spherical waves of fluorescence emission are subsequently collected by the objective lens to exit its back aperture as a diverging set of plane waves incident on the dichroic filter. The filter cube may also hold an exciter filter with a designed passband of excitation wavelengths and an emitter with a designed passband of emission wavelengths. The emitter has sharp cutoff to remove any excitation light which may be scattered or partially back-reflected within the system.

To add laser tweezers to the microscope, the laser light must either pass up through the dichroic filter or reflect upward from the dichroic filter after passing through the epifluorescence light path. In either case, if the laser propagates as a plane wave through the cube (and the infinity space), it will be focused in the specimen plan. It should be noted that laser tweezers typically use wavelengths beyond the design spectrum of the microscope objective and will focus deeper than the specimen plan. This is typically advantageous as discussed below. Microscope manufactures currently design special objective lenses with long enough chromatic correction to cover the laser wavelength. If the filter cube housing the dichroic filter also houses an emitter and an exciter then they must be custom designed to pass the laser where appropriate. Similarly, the dichroic must be custom designed to pass/reflect both the visible and laser wavelengths.

3. The RoboLase Systems

RoboLase I is diagrammed in Fig. 5. It is built on an older model of Zeiss microscope (Zeiss Axiovert 100) with either a 40× (phase III, NA 1.3) or a 63× (phase III, NA1.4) oil immersion objective. RoboLase I makes special use of two Zeiss dual video adapters. The dual video adapters come standard with a 50/50 beam splitter mounted in a removable filter cube designed to hold a rectangular beam splitter (or filter) and two round filters. The filters and beams splitters can be replaced with custom-designed filters. RoboLase I uses an Nd:YVO_4 continuous wave 1064-nm wavelength laser (Spectra Physics, Model BL-106C, Mountain View, CA) for laser tweezers. The lower dual video adapter (dual video adapter I) houses a filter cube (filter cube I) containing only a long-pass dichroic filter that transmits laser light entering the left-hand port of the dual video adapter and reflects visible light coming back from the objective into the upper dual video adapter (dual video adapter II). In Fig. 5, the incoming laser light (dashed line) is shown entering the dual video adapter, while fluorescence and/or red-filtered white

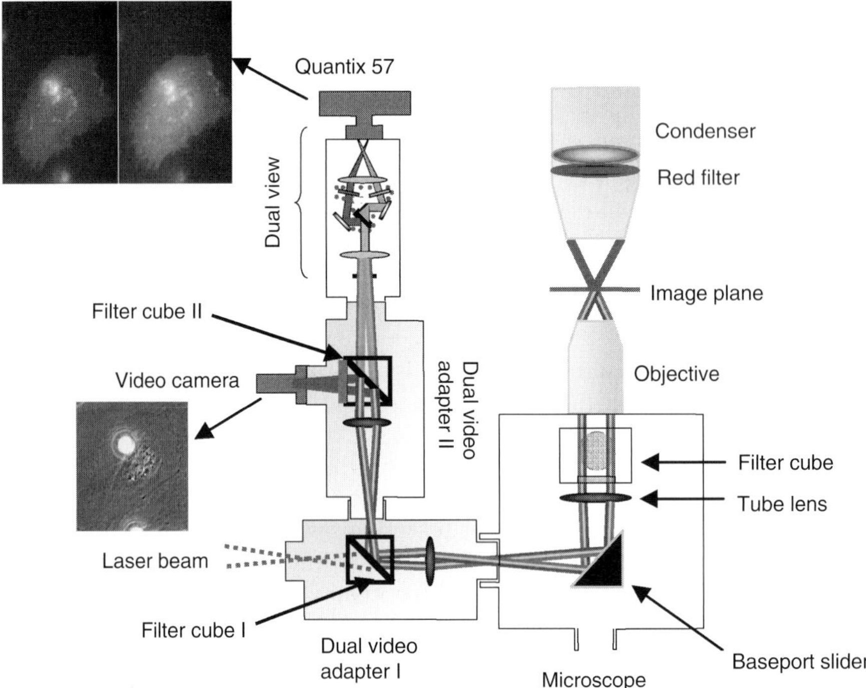

Fig. 5 Diagram of RoboLase I microscope components. An inverted microscope and two dual video adapters create a laser-scissors/FRET-imaging system. Laser light (dashed line) enters dual video adapter I and light collected by the objective lens is reflected by filter cube I toward dual video adapter II and the dual view system. A red filter is placed before the microscope condenser to separate phase-contrast imaging from shorter wavelength fluorescence emissions. Filter cube II reflects phase-contrast light toward the video camera to continuously display phase images of the microsphere and the cell (sample image shown). CFP and YFP light passes into the dual view system which splits the emissions and images two copies of the cell (CFP and YFP) simultaneously on the Quantix 57 CCD camera (sample image shown). (See Plate 20 in the color insert section.)

light (thick rainbow lines) are shown coming from the microscope objective and reflecting upward from filter cube I. In order to separate CFP and YFP fluorescence from wide-field nonfluorescence light (phase contrast), a red band-pass filter (HQ 675/50 M, Chroma Technology Corporation, Rockingham, VT) is placed above the microscope condenser. Visible light entering dual video adapter II is separated by a dichroic filter (short-pass, 650-nm cuton wavelength, Chroma Technology Corporation) housed in filter cube II. The filter cube also houses a second red band-pass filter (HQ 675/50 M, Chroma Technology Corporation). A phase-contrast image is focused onto the CCD camera (Sony, Model XC-75, New York City, NY) as shown in Fig. 5. The shorter wavelength CFP and YFP emissions pass through filter cube II and enter the dual view system (Optical Insights, Tucson, AZ). The dual view uses a 505-nm cuton wavelength long-pass dichroic filter (505 dcxr,

Chroma Technology Corporation) to separate the blue/green light as shown in Fig. 5. Each light path is further filtered by CFP/YFP emission band-pass filters centered at 470 nm (HQ 470/30, Chroma Technology Corporation) and 535 nm (HQ 535/30, Chroma Technology Corporation), respectively. When designing the band-pass filters, take care that they also block the laser wavelength to remove any scattered or reflected laser light in the system. The CFP and YFP representations of the cell are imaged side by side on a single CCD chip (Quantix 57, Photometrics, Tucson, AZ). During experiments, phase-contrast images are used to position the bead with respect to the laser focus by moving the microscope stage. Repeatability is 1 μm at best and limits the accuracy at which force can be controlled.

RoboLase I allows CFP/YFP imaging and laser trapping using standard over the counter equipment and no modifications of the microscope stand. However, it suffers from poor laser light efficiency. Laser light passes through the dual video adapter's lens system, the microscope tube lens, and the microscope objective; none of which are optimized for 1064-nm light. There is only 30% efficiency from outside the microscope to the back of the objective and 30% efficiency through the objective. Taken together, the system efficiency is at best 10%. The optics within our microscope are not antireflection coated for 1064-nm light and significant back reflection from internal lenses of the microscope and the dual video adapter leads to poor net transmission through the system.

External optics of RoboLase I are shown in Fig. 6. Two dielectric mirrors tuned to 1064 nm reflect and orient the beam parallel to the optical table and along the

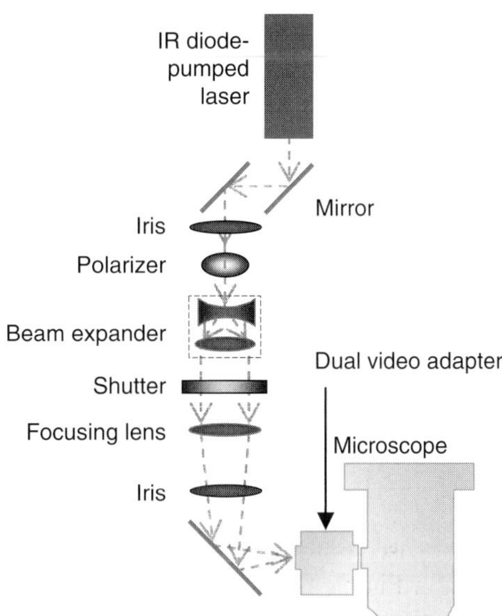

Fig. 6 External laser optics of the RoboLase I system.

optical axis of the microscope. Two antireflection-coated lenses expand the beam (plano-concave lens, $f = -25.5$ mm at $\lambda = 1064$ nm and plano-convex lens, $f = 76.2$ mm at $\lambda = 1064$ nm) in order to fill the back aperture of the microscope objective. A third antireflection-coated lens (biconvex lens, $f = 200$ mm) focuses the beam just beyond the intermediate image plane outside dual video adapter I. This lens can be moved axially in order to shift the focal position of the laser deeper into the cell culture chamber without moving the microscope objective. In this way, the axial-trapping position of a bead can be changed without changing the focal plane of the image. The focal displacement is also necessary to trap a microsphere. If the laser focus is too near the cover glass (i.e., less than the bead radius away), the trap will compress the microsphere as it attempts to pull it down through the glass and the microsphere will be ejected.

The RoboLase II system increases laser transmission efficiency by directing the laser along the epifluorescence excitation light path from the back of the microscope, as shown in Fig. 7 and previously published by Botvinick and Berns (2005). The trapping laser light source is an Ytterbium continuous wave fiber laser with a 5-mm collimator providing randomly polarized TEM00 mode 1064-nm laser output with 10-W maximum power (IPG Photonics Corporation, Oxford, MA). The epi-illumination optics were removed from a Zeiss Axiovert 200M motorized

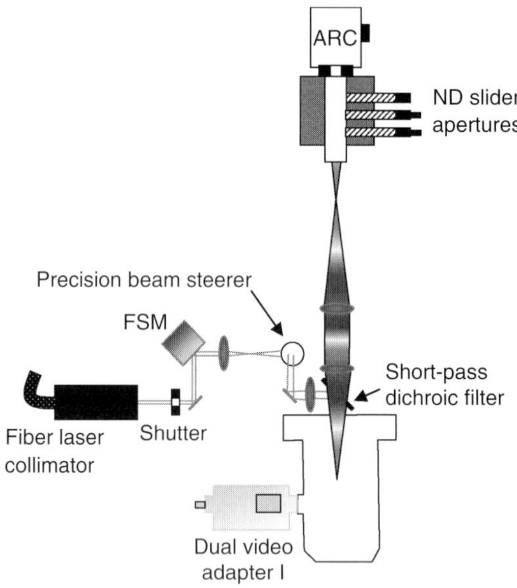

Fig. 7 External optics of the RoboLase II system. The epifluorescence light train was removed and recoupled into the microscope via two achromatic doublets. A dichroic filter combines fluorescence excitation light with laser light behind the microscope stand. The back focal plane of the objective is mapped to the surface of a fast-scanning mirror (FSM) for x, y beam steering via two antireflection-coated lenses.

microscope. The arc lamp and epioptics were positioned distal to the microscope and recoupled via two 400-mm positive achromatic doublets (Newport Corporation, Irvine, CA) antireflection coated for the visible spectrum. A short-pass dichroic mirror with a 900-nm cuton wavelength was placed just behind the microscope stand to reflect laser light into the microscope while passing light from the arc lamp. In this setup, the laser neither passes through a tube lens, the epifluorescence lens system, or any lens not antireflection coated for the laser wavelength. The laser trap is steered within the specimen plane by a motorized two-axis mirror mount placed conjugate to the back focal plane of the objective (FSM-300-01 Fast Scanning Mirror, ER.1 coating, Newport Corporation). The fast scanning mirror has a 3-μrad repeatability along both axis, corresponding to a 0.001-V change at its controller's analog input. Analog signals are created by a 12-bit data acquisition board (PXI 6711, National Instruments, Austin, TX) with 0.005-V resolution in the range -10 to 10 V. With a 63$\times$ (phase III, NA 1.4) objective and the scan and tube lenses, a 3-μrad mirror deflection corresponds to 2.6-nm displacement in the specimen plane. Therefore, RoboLase II has a 13-nm pointing resolution. The laser transmission efficiency is 75% from the laser collimator to the back aperture of the objective and 30% through the objective.

RoboLase II employs a similar strategy to separate phase contrast from CYF/YFP emission as found in the dual video adapter II of RoboLase I. The fluorescent filter cube housed in the reflector turret of the microscope stand is custom designed to pass both arc lamp and laser light through the exciter and to reflect them from the dichroic filter. CFP/YFP emissions are then separated with a dual view-imaging device with identical optics to that of RoboLase I. During experiments, the laser is steered near to the edge of the bead and can be steered throughout the experiment without moving the microscope stage.

In RoboLase III, we mix laser tweezers with multiphoton excitation. In this system, a custom-built filter cube is mounted in the reflector turret so that it is rotated 90° counterclockwise (top view) about the optical axis of the objective lens. The reflection turret is modified so that light enters the left-hand side of the microscope (top view) and is reflected upward into the microscope objective. With the cube in place, light from the epiflorescence port would hit the edge of the mirror and not be reflected in to the objective lens. Instead, RoboLase III uses a TrimScope (TauTech, Columbia, MD) which multiplexes the femtosecond-pulsed laser (MaiTai Broadband, 710- to 990-nm tuning, 100-fs pulse width; Newport Corporation) into multiple beams forming a linear array of diffraction-limited spots in the specimen plane. The TrimScope scans the array to produce multiphoton images. The TrimScope is equipped with a short-pass dichroic filter mounted in the custom filter cube to reflect laser light up into the objective, while passing returning fluorescence emission down toward the camera port. We mounted a short-pass dichroic with cuton wavelength 1000 nm (DCPS 1000, Chroma Technology Corporation) in between the tube lens and scan lens of the TrimScope in order to mix 1064-nm trapping laser with the broad range of femtosecond wavelengths. The efficiency of this system is 50% from the laser head to the back aperture of the objective and about 30% through the objective. But efficiency from outside the microscope to the back aperture is 85% and is limited by the dichroic in the filter cube.

4. Calibrating the Laser Tweezers

In mechanobiology experiments, it is often necessary to maximize force generation on beads adhered to a cell. Large forces are generated by steering the trap focus near to the edge of the microsphere, which is held more-or-less in place by its connections to the cell. According to Ashkin (1992), the maximum transverse force is generated by focusing the laser trap just shy of the microsphere's surface (98% of the radius) in a transverse plane through the center of the microsphere. Near the surface of the microsphere the bead to beam displacement is not linearly related to restoring force and implementation of a quad detector with a linear spring model is not appropriate. However, a quad detector can be used to measure bead to beam displacement with a nonlinear calibration. Microspheres can be visualized by the CCD camera in the red light path (Fig. 5) of the RoboLase systems. Using the Airy disk radius as an approximation of microscope resolution, the spatial sampling frequency and thus accuracy of positioning of the bead image on the CCD can be calculated. With the addition of a secondary zoom lens (not shown), the bead can be sufficiently over sampled to determine the bead position with accuracy greater than the Airy disk radius. Sheetz *et al.* have reported measuring the position of micron-size beads with 5- to 10-nm resolution (Felsenfeld *et al.*, 1999; Gelles *et al.*, 1988).

The force calibration is best performed on a microsphere trapped above a cover glass with no cells in culture. Moving the stage along a square wave trajectory with sufficient amplitude allows the bead to reach force equilibrium before it changes directions. The force equilibrium is reached when viscous forces equate to light forces. The bead to beam displacement should be measured for a range of laser powers or stage velocities so that the observed displacements range from zero to near the bead radius. Since the relationship between the optical force and the laser power is linear, the calibration can be done at a lower laser power and thus lower stage speeds, and then scaled later to higher laser powers. It is not recommended to rely on a mathematical fit to extrapolate the force–displacement relationship. Care should be taken to measure the relationship for the full range of displacements about the operating point (i.e., the microsphere edge). Viscous force can be calculated from stage velocities using the stokes flow approximation

$$F_{\text{viscous}} = 6\pi\eta r v_{\text{stage}}$$

where η is the viscosity of the surrounding fluid (e.g., water at room temperature), r is the microsphere radius, and v_{stage} is the microscope stage velocity. It is not recommended to steer the beam during calibration. The calibration can be repeated for different beam locations in the field of view to determine the spatial distribution

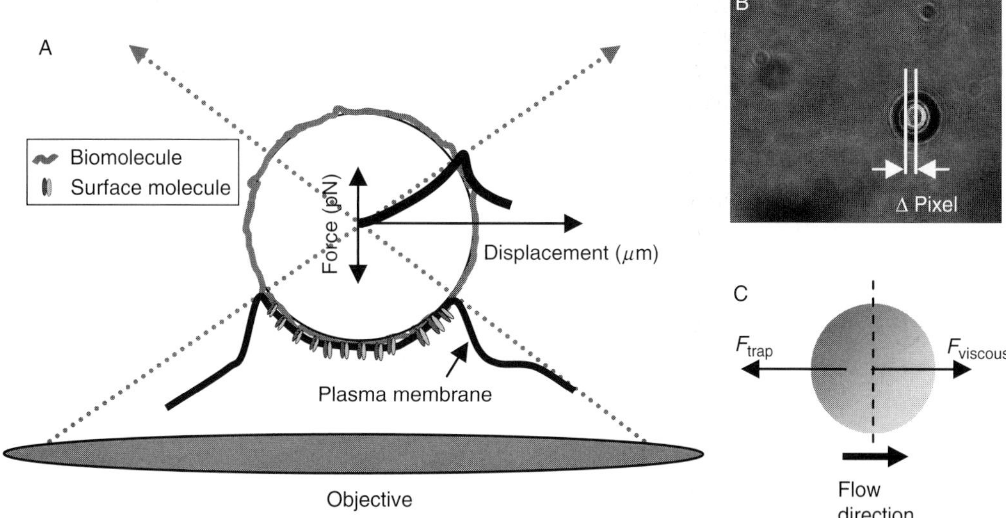

Fig. 8 Diagram of laser-tweezers calibration. (A) During experiments, a microsphere coated with a specific biomolecule is rigidly attached to the cell through the cell's complementary surface molecules. The trapping laser is focused into the microsphere to produce forces. As shown, the focus is aligned with the microsphere center and no force is exerted (ignoring scatter forces). As the focus is steered toward the bead surface, the laser light exerts increasing force on the microsphere as demonstrated by the superimposed graph where "displacement" is the center to focus distance and the "force" acts transversely to pull the center toward the focus. (B) a microsphere freely suspended in water is trapped as the stage moves at constant velocity to the right. The two vertical white lines indicate the bead's center and the focus' horizontal coordinate. The displacement Δx is indicated. (C) Illustration of microsphere in a rightward flow field with viscous and laser trap forces in balance. When the microsphere is in force equilibrium, the viscous force is equal in magnitude to the laser trap force. A calibration is performed by recorded Δx for a set of stage velocities and thus laser forces. During bead adhesion experiments, as Δx is known, the force can be determined.

of trap strength. The trap strength can change throughout the field if the beam is not telescopically steered, and in general will change due to imperfections in system optics. Figure 8 demonstrates the calibration process.

5. Image Processing

Once images have been acquired they must be processed in a manner appropriate to what is being quantified. For example, if the experiment seeks a binary answer as would be the case for a live/dead assay with propidium iodide, image processing is straight forward and most canned algorithms would be appropriate. If the experiment seeks to map out fluorescence intensity, or to determine fine structures within the cell, or near the edge of the cell itself, care must be taken in assigning a value to each pixel. For the case of imaging the Src reporter with one of our RoboLase systems, the following procedures are applied: extraction of the CFP and YFP subimages from the raw image, pixel registration between the two subimages, extraction of the cell from the background, subtraction of background signal, and calculation of the ratio image.

As shown in Fig. 5, the dual view system images both the CFP subimage and the YFP subimage onto the CCD simultaneously. The extraction of the two subimages can be as trivial as dividing the raw image in two, if care is taken while aligning the dual view system. Alternatively, two regions of interest of identical dimension can be extracted around the two subimages. RoboLase II uses an Orca AG (C4742–80–12AG, Hamamatsu, Bridgewater, NJ) CCD camera with 1344 $\times$ 1024 pixels (6.45 μm^2). The CFP subimage is extracted as the first 1344/2 columns and the YFP subimage begins at the 1344/2 + 1 column and ends at the end of the raw image.

The next step is to test the pixel registration between the two subimages. That is, to make sure each pixel CFP(x_i, y_j) corresponds to pixel YFP(x_i, y_j) for all i, j where x and y are the Cartesian coordinates of each pixel within the subimage. We use the corr.m function in Matlab (The Mathworks, Inc., Natick, MA) to compute the Pearson's linear correlation coefficient between the two images. The correlation is called recursively as the YFP subimage is cropped by a sliding region of interest that includes all but a boarder region about the subimage. The CFP subimage is cropped only once so that each calculated correlation represents a relative shift between the two subimages. We then choose the shift which best correlates the two subimages. The two cropped images, shift included, now constitute the new subimages, which are further analyzed as described below.

Now that the two CFP and YFP subimages have been coregistered, image segmentation must be applied to extract the pixels belonging to the cell from pixels belonging to the background. In this way, we can specify for which pixels the ratio should be calculated, and we can determine the proper background value to be subtracted before the ratio is calculated. It is critical to remove the background bias in order to compare ratio values between different cells. The ratio should only be calculated for pixels in which both the CFP and YFP values are above the noise floor. With the dual view system, neutral density filters can be used on the YFP channel in order to balance the relative brightness between the CFP and YFP channels. We do not implement this strategy in our laboratory as the YFP channel does not typically fill up its electron well capacity within the time of image integration ($\sim$1 s). Since the CFP channel is dimmer, it is segmented and the resulting binary mask is applied to both the CFP and the YFP subimages. If image segmentation were calculated from the YFP subimage, pixels in the YFP subimage would be selected for which the corresponding CFP pixels would be in the noise floor. Ratios calculated for these pixels would be inversely correlated to the YFP brightness (where ratio is CFP/YFP) and would not represent a physiological measurement.

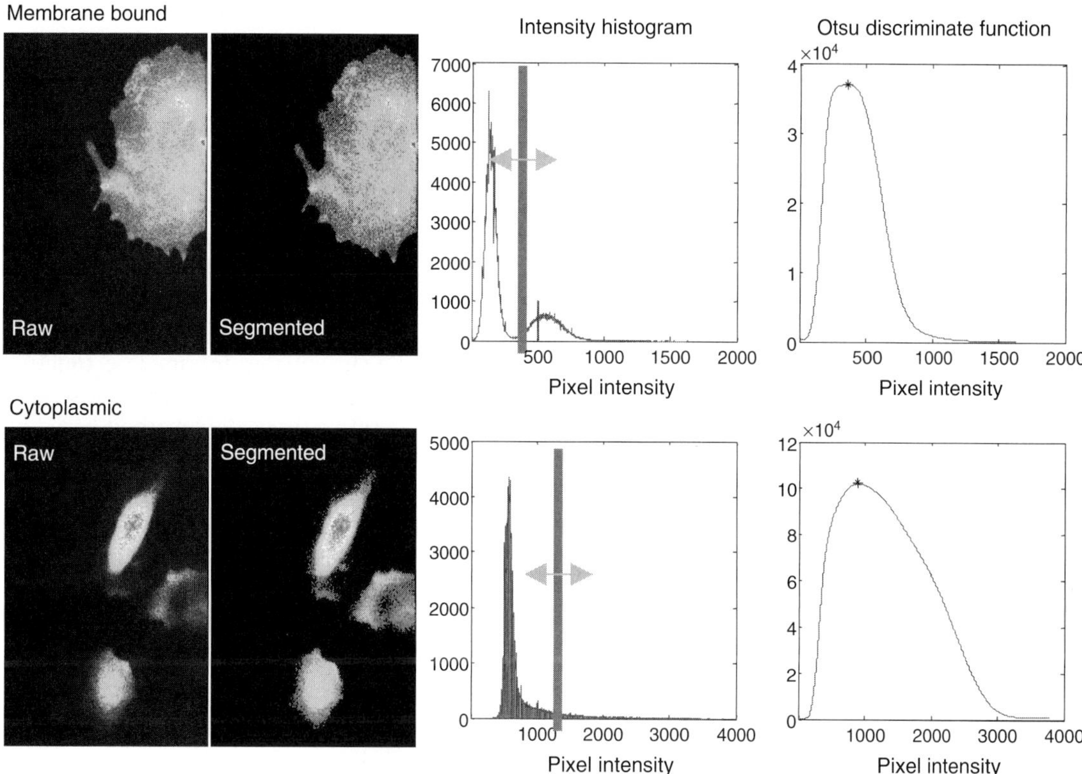

Fig. 9 Two-class thresholding fails on images of cytoplasmically distributed biosensor. (Top row) HUVEC cell labeled with membrane-bound biosensor. A raw image is shown in pseudo-color followed by its segmented image. The image was segmented using the two-class Otsu's method and shown in the next image. The background pixels (black) have been suppressed. The intensity histogram is bimodal and the two-class Otsu's method's discriminate function takes on a maximum value in between the two modes. (Bottom row) Huevec cell labeled with cytoplasmically distributed biosensor. The pseudo-color raw image shows narrow differences between background and cell edge values. The segmented image incorrectly mapped dim pixels at the cell edge and podia to the background. The intensity histogram is not bimodal and choosing the maximum value of the discriminate function does not separate the cell from the background. (See Plate 21 in the color insert section.)

Figure 9 represents two typical cases of image segmentation. The top row contains a raw CFP image, shown in pseudo-color, of a plasma membrane-bound fluorescent biosensor. Fluorescent intensity within the cell is somewhat uniform and application of a two-class Otsu's method threshold correctly removes the cell (shown in pseudo-color) from the background (black). As can be seen by the image's intensity histogram, pixel values have a bimodal distribution. The ubiquitous two-class Otsu's method evaluates a discriminate function for each potential threshold value, as indicated by the sliding gray column in the histogram. Otsu's method chooses the threshold value corresponding to the maximum value of the Otsu's method discriminate function. For the membrane-bound biosensor, the maximum value corresponds to a threshold value which sufficiently separated the background from the cell. The bottom row of the figure shows a raw CFP image of a similar biosensor that is not membrane bound, but distributed within the cytoplasm. This cell is brightest in the nuclear region and very close to background values in the thin podia regions. Application of a two-class Otsu's method mistakenly maps the thinnest region of the podia to the background, and any further analysis on the image would not include those pixels. In the study of mechanobiology, the pixels in the podia are often crucial to the experiment. As can be seen, the intensity distribution is not bimodal so it is not surprising that a two-class system cannot segment the cell well. The maximum value of the discriminate function cannot separate the podia from the background.

Instead, we implement a custom-designed iterative three-class method based on the generalized Otsu's method. Figure 10 demonstrates the method. For each

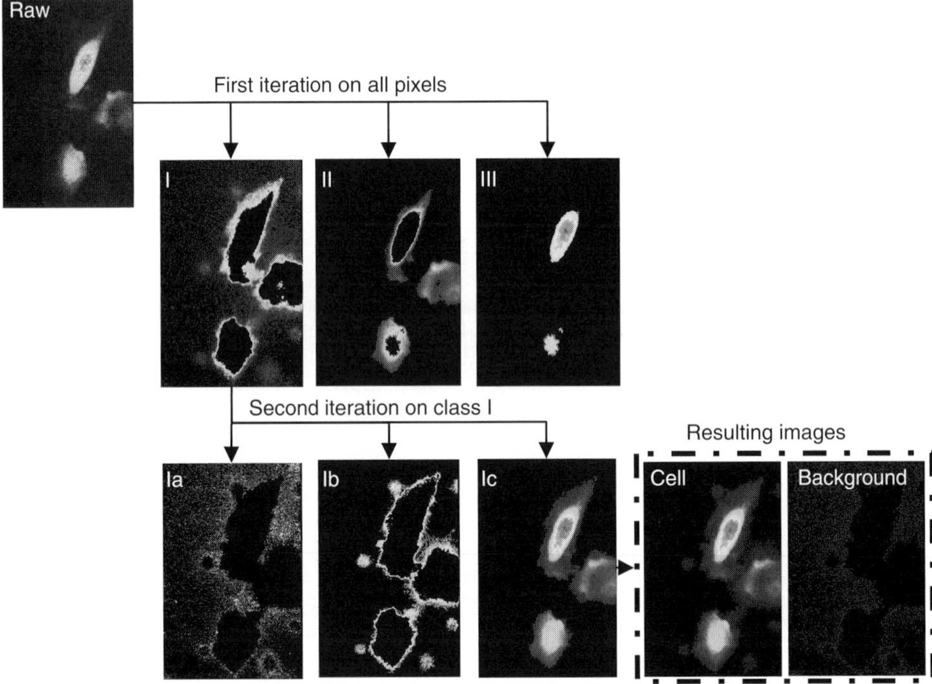

Fig. 10 Segmenting the cytoplasmically distributed biosensor image using an iterative three-class Otsu's method. In the first iteration, three classes are created, the dimmest (class I) containing both background and dim cell pixels. The second iteration acts on class I pixels and creates three new classes, Ia, containing background pixels; Ib, containing dim cell pixels; and Ic, containing the remaining cell's pixels. Two images result representing background and cell (union of Ib and Ic) pixels. (See Plate 22 in the color insert section.)

iteration, the input pixels are separated into three pixel classes. The following Matlab code operates on "input_image", which contains all pixels to be analyzed, and returns two threshold values which create three pixel classes. The code was adapted from Matlab's two-class method and from Liao *et al.* (2001). The algorithm seeks threshold 1 (k_1) and threshold 2 (k_2) within the intensity range $1 < k_1 < k_2 < L$ to create three pixel classes c_1, c_2, and c_3 such that:

$$c_1[1\ldots k_1];\ c_2[k_1+\mathrm{i}\ldots k_2];\ c_3[k_2+1\ldots L]$$

$$\omega_i = \sum_{c_i} P_i \ldots \mu_i = \sum_{c_i} \frac{iP_i}{\omega_i}$$

$$\sigma_b^2 = \omega_1\mu_1^2 + \omega_2\mu_2^2 + \omega_3\mu_3^2$$

where σ_b^2 should be maximized at k_1 and k_2.

```
input_image = double(input_image); % Convert to double precision
max_bin = (max(input_image(:))); % Calculate the max and min values
min_bin = (min(input_image(:)));
[counts,thresholds] = hist(input_image(:),max_bin-min_bin + 1);
L = length(thresholds); % size of intensity value range
P = (zeros(L)); S = (zeros(L)); %initialize P and S

% Calculate first row of P and S
f = counts/sum(counts); % normalize frequencies
P(1,:) = cumsum(f);
S(1,:) = cumsum(f.* (min_bin:max_bin));

%Calculate the rest of P and S
for n = 2:L
  P(n,n:L) = P(1,n:L) – P(1,n – 1);
  S(n,n:L) = S(1,n:L) – S(1,n – 1);

end

% Calculate H
H = S.^2./P;

% calculate initial guess for first threshold
temp = [];
for t1 = 1:L – 2%first threshold
  temp(t1) = H(1,t1) + max(H(t1 + 1,t1 + 1:L));
end
thresh1 = round(mean(find(temp == max(temp))));

% calculate initial guess for second threshold
temp = [];
for t2 = thresh1 + 1:L – 1%second threshold
```

```
    temp(t2) = H(thresh1 + 1,t2) + max(H(t2 + 1,t2 + 1:L));
end
thresh2 = round(mean(find(temp == max(temp))));
% calculate discriminate function between the two initial guesses
for t1 = 1:thresh1
    for t2 = thresh1 + 1:thresh2
        sigma_b_squared(t1,t2) = H(1,t1) + H(t1 + 1,t2) + H(t2 + 1,L);
    end
end

% find max discriminate function value and corresponding thresholds
index = max(max(sigma_b_squared));
[thresh1,thresh2] = find(sigma_b_squared == index);
thresh1 = uint16(mean(thresh1));
thresh2 = uint16(mean(thresh2));

% thresholds reference num_bins and not the absolute pixel value.
% Divide max(I(:)) by num_bins and map thresholds to indices of that array.

thresh1 = round(thresholds(thresh1)); %thresh1 is index pointing to thresholds
    bin centers
thresh2 = round(thresholds(thresh2)); %thresh2 is index pointing to thresholds
    bin centers
```

In Fig. 10, the first iteration of image segmentation operates on the raw image of a cell with a CFP probe distributed in the cytoplasm. The three resulting pixels classes are shown. Class III contains the brightest pixels (nuclear region), and class II contains pixels of the cytoplasmic region surrounding the nuclear region. Class I contains both the background pixels and those of the thin regions of the cell. The second iteration of image segmentation operates on class I pixels to yield three new classes Ia, Ib, and Ic. Ia contains only background pixels, Ib contains the podia and thin edges of the cell, and class Ic contains all remaining cell pixels. Two resulting images are formed, the background pixels (Ia) and the cell pixels (union of Ib and Ic). A binary mask of cell pixels can be constructed from the union of classes Ib and Ic coordinates.

Figure 11 demonstrates the process of transforming a raw CCD image into a ratio image. The raw image contains both the CFP and YFP copies of the cell and is first separated into its two subimages, labeled CFP and YFP in the figure. Pixel registration through image correlation (not shown) is applied and a binary mask (labeled mask) is created by segmenting the raw CFP subimage. CFP and YFP subimages are then masked whereby values are kept if their corresponding mask values are "1" and are set to 0 otherwise. Subimage pixel values corresponding to mask "0" values are averaged as an estimate of the image background intensity. The background estimate is then subtracted from each subimage. The ratio image is calculated by dividing the masked and background-subtracted CFP values by the masked and background-subtracted YFP values:

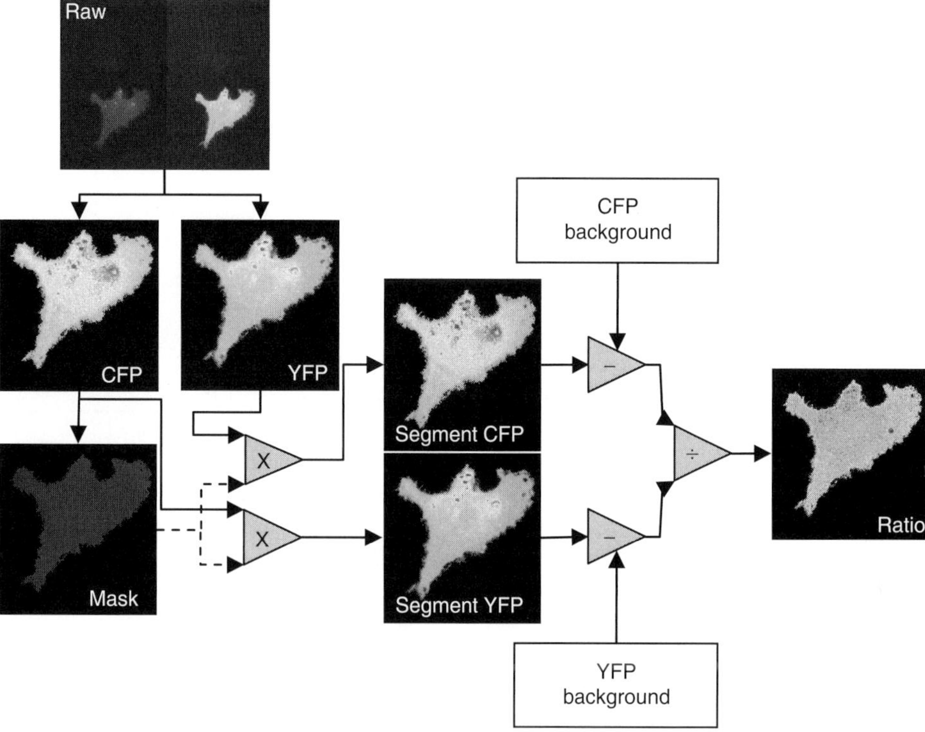

Fig. 11 Demonstration of calculating a ratio image. A raw image contains the CFP and YFP copies of the cell. Two ROIs of identical dimension (not shown) crop the CFP and YFP subimages from the raw image. Two ROIs may have a relative shift to compensate for misalignment of the system (see text). The CFP subimage is segmented to create a binary mask which is applied to both the CFP and the YFP subimages through image multiplication. The CFP- and YFP-segmented images have the mean background value subtracted from each (only subtracting from nonzero pixels) where the mean background is calculated from class Ia pixels (see Fig. 10). The resulting CFP subimage is divided by the resulting YFP image in a pixel-by-pixel manner to form the ratio image. (See Plate 23 in the color insert section.)

$$\text{Ratio}(x_{i,j}, y_{i,j}) = \frac{\text{CFP}(x_{i,j}, y_{i,j})}{\text{YFP}(x_{i.j}, y_{i,j})}$$

6. Quantifying Kinase Activity

Once the ratio image has been calculated, the distribution of ratios within one image and across time can be analyzed to extract physiological measurements from the experiment. Three methods used commonly in our laboratory are evaluation of ratio histograms, analysis of distal pixel activation, and polarity analysis. Figure 12A and B plot the distribution of ratio values for cells expressing the membrane-bound FRET Src reporter. Higher ratio values indicate increased fraction of phosphorylated biosensor within the corresponding cell voxel. Histograms are plotted for two

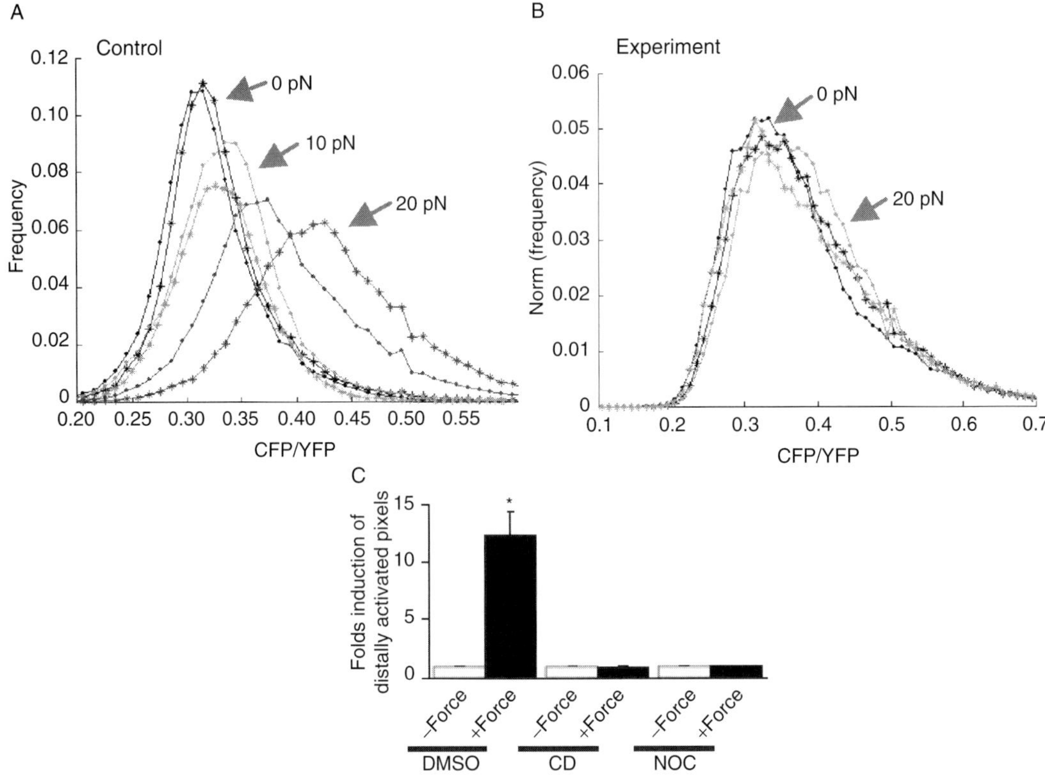

Fig. 12 Analysis of ratio-value frequency distribution and distal activation for HUVECs with membrane-bound Src reporter. (A) Frequency plots of ratio values indicate Src activation within a cell following pulling on an attached microsphere. In a control experiment, the histogram shifts rightward following pulling at 10 pN as more pixels contain a greater fraction of activated Src. Pulling at 20 pN further shifts the histogram and populates the right tail of the distribution indicating a cell-wide trend toward higher ratios and thus great Src activation. (B) Histograms for experimental cell for which long-range mechanotransduction may be compromised. No significant histogram shift occurs following pulling at 20 pN indicating to significant activation of Src throughout the cell. (C) Example of distal pixel activation analysis taken from Wang *et al.* (2005, p. 10). Pixels are counted if they are more than the half the virtual cell radius (see text) from the microsphere center, and greater in ratio value than the 80th percentile ratio from baseline measurements (i.e., before pulling). A low concentration of DMSO was applied to the cell as control and pulling led to a 12-fold increase in activated pixels indicating long-range force activation of Src. Disruption of the cytoskeleton by cytochalasin D or nocodazole suppressed the count indicating that distal locations in the cell did not have an increase in activated Src.

baseline measurements, after pulling on adhered fibronectin-coated 10-μm-diameter microspheres at the indicated force level. The control experiment (Fig. 12A) has two baseline (measurements taken after bead adhesion but before pulling) measurements taken 2 min apart, measurements taken immediately and 5 min after pulling at 10 pN, and measurements taken immediately and 3.5 min after pulling at 20 pN.

As can be seen, the baseline was consistent, while pulling at 10 pN immediately shifted the histogram to the right. This represents an increase in Src signaling throughout the cell as more pixels took on higher ratios as compared to baseline. After 5 min of 10-pN force, the histogram shifted again by a similar distance. On pulling at 20 pN, the histogram shifted to the right again. The shift occurred in both the ratio of peak frequency and in the right-hand tail which was elevated from the floor. The ratio 0.3, which was most frequent at baseline, has been quelled, while the ratio 0.43 which was infrequent at baseline has become most frequent after 3.5 min of pulling at 20 pN. This indicates a significant shift in the Src activation profile in the cell and a global increase in the activated state. Figure 12B shows an experimental cell in which transduction of the force has been compromised. The two baseline distributions taken 3.5 min apart are not different from the distribution measured immediately or 6.5 min after pulling at 20 pN. This type of analysis gives insight into whole cell activation of Src kinase, but does not elucidate the spatial distribution of activated pixels (or ratios) within the cell.

One method of addressing spatial distribution is to examine the distribution of ratios in regions of the cell distal to the attached microsphere. The first step is to define which pixels are distal to the bead. In Wang *et al.* (2005), the boundary separating distal pixels from proximal ones was defined as half the virtual radius of the cell (R)

$$\frac{R}{2}=\frac{1}{2}\sqrt{\frac{A}{\pi}}$$

where A is the area of the cell (or the pixel count of the cell mask). A pixel at coordinate (x, y) is considered distal from the bead center (x_0,y_0) if the Euclidian distance between them is greater than $R/2$ as calculated by:

$$\sqrt{(x-x_0)^2+(y-y_0)^2}>\frac{R}{2}=\frac{1}{2}\sqrt{\frac{A}{\pi}}.$$

The next step is to determine what ratio value within a baseline image of a cell is greater than a percentile of all pixels in that cell (e.g., 80%, 85%, 90%, and so on) at baseline. In subsequent images, the number of pixels that have ratios above that value and are distal to the bead are counted. The fold increase between baseline and pulling can be calculated as a means of measuring the degree to which distal pixels were activated by mechanical transduction through the cell. In Fig. 12C, an 80% threshold was used (Wang *et al.*, 2005). In this figure the control (DMSO) group had a 12-fold induction postpulling, while the two experimental groups (CD, cytochalasin D and NOC, nocodazole) had no significant increase in activated pixels. This method indicates that interruption of the cytoskeleton precludes long-range Src activation as compared to the control.

The spatial distribution can be further examined by measuring Src activation as a function of angular deviation from the direction of pulling. We call it polarity analysis. In Matlab, this operation is easily computed by transforming the original

Cartesian pixel coordinates into polar coordinates. The coordinates of the bead's center must first be subtracted from the coordinates of each pixel, thus moving the origin to the bead center. The (x,y) coordinates can be transformed into polar coordinates (r, θ) by the equations:

$$r = \sqrt{(x - x_0) + (y - y_0)}$$

$$\theta = \tan^{-1}\left(\frac{y}{x}\right)$$

where the four quadrant arctan function "atan2" is called. Positive θ is in the counterclockwise direction, and θ values are adjusted so that 0° corresponds to the direction of pulling. Coordinates are then sorted by increasing θ and the data can be divided into bins, or slices, defined by equally spaced intervals of θ. The distribution of ratio values within each slice can then be analyzed. In Fig. 13, two cells expressing the membrane-bound Src reporter have been analyzed. The top row shows cell 1's ratio images at baseline ($T = 0$ min), after 16.5 min of pulling the bead downward at 10 pN and after 12 more minutes pulling downward at 20 pN. The base of the red arrow and the crosshair indicates the bead center. The cell was divided into 36 slices and the mean and standard deviation of each slice is graphed as a function of angle. Notice that maximum activation occurs at angles 110°–180° which are opposite to the direction of pulling. Cell 2 was subjected to an experimental protocol which may adversely affect long-range mechanotransduction. As can be seen by the ratio images in the bottom row of Fig. 13, the extent of long-range signaling is diminished as compared to cell 1, with little changes away from cell edges. Cell 2 was pulled at twice the force as cell 1, or 20 and 40 pN. The graph of mean and standard deviation of activation as a function of angle shows a slight increase in ratio opposite to the direction of pull after 6.5 min of pulling at 20 pN. After 12 additional minutes pulling at 40 pN, activation is elevated throughout the cell with a maximum still opposite pulling. The maximum activation level of cell 2 after pulling at 40 pN is nearly equal to the minimum activation level of cell 1 after pulling at 20 pN. It should be noted that for both cells the response to each force level was nearly instantaneous, and did not appreciably increase as the force was held at a constant level.

IV. Summary and Future Direction

The integration of genetically encoded FRET biosensors and laser tweezers has been proved to be vital for studying mechanotransduction in live cells with high spatiotemporal resolution. Further development of new FRET-based biosensors and novel implementations of laser tweezers are expected to be the forefront research in this field in the near future. FRET technology can be applied for the biosensor designation of not only tyrosine kinases, but also serine/threonine kinases, proteases, small GTPases, and other signaling molecules. In fact, a FRET-based biosensor

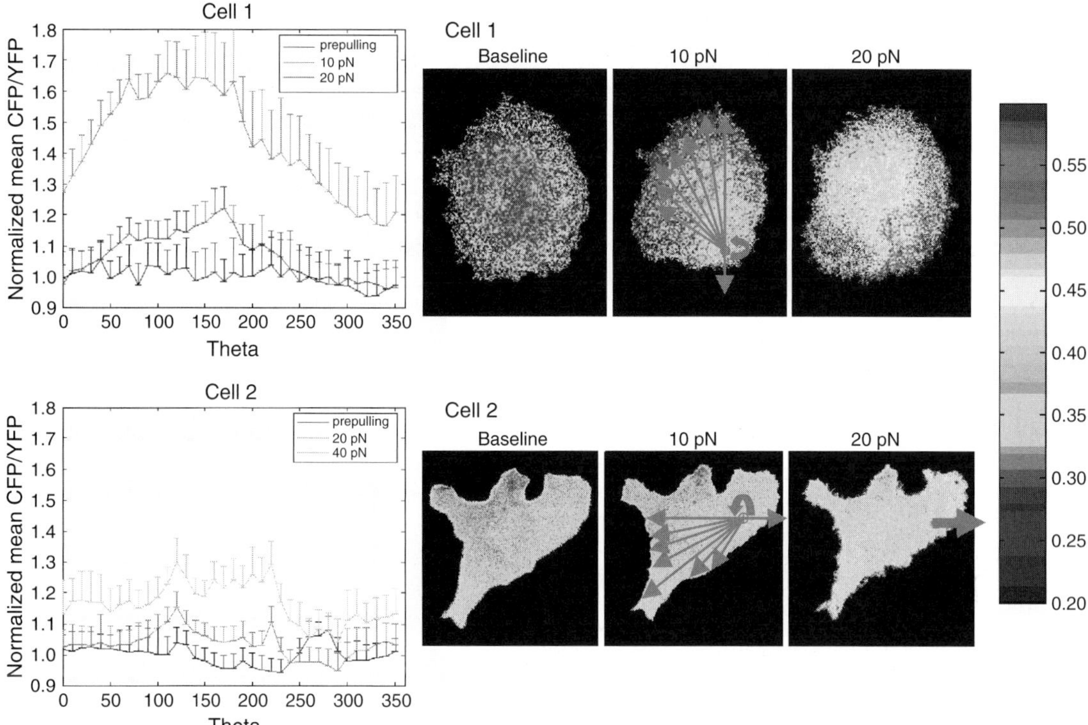

Fig. 13 Polarity analysis of Src activation. Translation of the Cartesian origin to the microsphere center and projection into polar coordinates creates a coordinate system composed of radial distance from the microsphere and angular deviation from the direction of pulling. The cells are sliced into angular bins and mean changes of Src activity can be monitored. (Top row) Cell 1 is a control cell for which an FN-coated bead is attached and pulled downward. Ratio images are shown for baseline and after pulling at 10 and 20 pN. Each cell is divided into slices indicated by red arrows centered about the bead. Graphs of mean and standard deviation as a function of angle demonstrate the polarity in Src activation within the cell. As can be seen, pulling at 10 pN increases activation about 160°, which is opposite the direction of pulling. Pulling at 20 pN substantially increases signaling throughout the cell with peak activation opposite the force. (Bottom row) Cell 2 is a control cell with compromised mechanotransduction. Pulling at 20 pN marginally increases activation, with slightly elevated signal from 110° to 180°. Pulling at 40 pN uniformly shifts the curve vertically with a peak activation value about equal to the minimum activation value observed when pulling cell 1 at half the force. This suggests a compromise in cell 2's long-range and directional sensitivity to force. (See Plate 24 in the color insert section.)

for nonenzymatic membrane receptor molecule integrin has been successfully developed (Kim *et al.*, 2003). The directed evolutionary strategies together with fluorescence-activated cell sorting (FACS) should provide a high-throughput means to develop novel biosensors. Because the excitation and emission wavelengths of CFP and YFP are relatively short and not suitable for *in vivo* imaging, it will be very desirable to develop FP pairs for FRET with longer wavelengths, ideally in the range of dark red or infrared.

The field of optical manipulation is likewise moving in a direction which will open up our understanding of cellular mechanotransduction. In addition to pulling on particles attached to the cell by specific ligand interactions, laser tweezers can be used to apply fluid sheer stress by optically rotating tapped particles (Bishop *et al.*, 2004). In this way, mechanical stresses and shear can be carefully applied to a subregion of the cell through fluid flow. Additionally, the rotating particles can be analyzed to measure the apparent viscosity of the surrounding media. As multiplexing of the trapped beam and advancements in rotating particles (Grier, 2003), both within and outside of cells, become mainstream technologies, the mechanisms of mechanotransduction will unfold from the molecular level up to the function of intact tissues in the normal and disease states.

Acknowledgments

We would like to thank Professor Shu Chien and the Whitikar Institute of Biomedical Engineering for the support, advice, and cooperation on this project. Thanks to Professor Michael Berns for his encouragement, advice, and supervision over the laser tweezers and imaging system, and for the idea that a laser can and should be brought through a microscope. E.B. would like to thank the Beckman Foundation and its Beckman Fellows award for launching his career and allowing him to build the laser tweezers and FRET system. Y.W. would like to thank the Department of Bioengineering and Beckman Institute for Advanced Science and Technology at University of Illinois, Urbana-Champaign, Wallace H. Coulter Foundation for their support.

This project was funded through the Beckman Foundation and through a grant from the US Air Force (AFOSR No. F9620-00-1-0371).

References

Ashkin, A. (1992). Forces of a single-beam gradient laser trap on a dielectric sphere in the ray optics regime. *Biophys. J.* **61**(2), 569–582.

Beningo, K. A., Dembo, M., Kaverina, I., Small, J. V., and Wang, Y. L. (2001). Nascent focal adhesions are responsible for the generation of strong propulsive forces in migrating fibroblasts. *J. Cell Biol.* **153**(4), 881–888.

Bishop, A., Nieminen, T., Heckenberg, N., and Rubinsztein-Dunlop, H. (2004). Optical microrheology using rotating laser-trapped particles. *Phys. Rev. Lett.* **92**(19), 1981041–1981044.

Botvinick, E. L., and Berns, M. W. (2005). Internet-based robotic laser scissors and tweezers microscopy. *Microsc. Res. Tech.* **68**(2), 65–74.

Chen, K. D., Li, Y. S., Kim, M., Li, S., Yuan, S., Chien, S., and Shyy, J. Y. (1999). Mechanotransduction in response to shear stress: Roles of receptor tyrosine kinases, integrins, and Shc. *J. Biol. Chem.* **274**(26), 18393–18400.

Chien, S., Li, S., Shiu, Y. T., and Li, Y. S. (2005). Molecular basis of mechanical modulation of endothelial cell migration. *Front. Biosci.* **10,** 1985–2000.

Felsenfeld, D. P., Schwartzberg, P. L., Venegas, A., Tse, R., and Sheetz, M. P. (1999). Selective regulation of integrin–cytoskeleton interactions by the tyrosine kinase Src. *Nat. Cell Biol.* **1**(4), 200–206.

Gelles, J., Schnapp, B. J., and Sheetz, M. P. (1988). Tracking kinesin-driven movements with nanometre-scale precision. *Nature* **331**(6155), 450–453.

Grier, D. G. (2003). A revolution in optical manipulation. *Nature* **424**(6950), 810–816.

Hu, S., Eberhard, L., Chen, J., Love, J. C., Butler, J. P., Fredberg, J. J., Whitesides, G. M., and Wang, N. (2004). Mechanical anisotropy of adherent cells probed by a three-dimensional magnetic twisting device. *Am. J. Physiol. Cell Physiol.* **287**(5), C1184–C1191.

Ingber, D. E. (2003a). Tensegrity I. Cell structure and hierarchical systems biology. *J. Cell Sci.* **116**(Pt. 7), 1157–1173.

Ingber, D. E. (2003b). Tensegrity II. How structural networks influence cellular information processing networks. *J. Cell Sci.* **116**(Pt. 8), 1397–1408.

Kaufman, D. A., Albelda, S. M., Sun, J., and Davies, P. F. (2004). Role of lateral cell–cell border location and extracellular/transmembrane domains in PECAM/CD31 mechanosensation. *Biochem. Biophys. Res. Commun.* **320**(4), 1076–1081.

Kernan, M., and Zuker, C. (1995). Genetic approaches to mechanosensory transduction. *Curr. Opin. Neurobiol.* **5**(4), 443–448.

Kim, M., Carman, C. V., and Springer, T. A. (2003). Bidirectional transmembrane signaling by cytoplasmic domain separation in integrins. *Science* **301**(5640), 1720–1725.

Liao, P.-S., Chen, T.-S., and Chung, P.-C. (2001). A fast algorithm for multilevel thresholding. *J. Comput. Inf. Sci. Eng.* **17**(5), 713–727.

Miyawaki, A., Llopis, J., Heim, R., McCaffery, J. M., Adams, J. A., Ikura, M., and Tsien, R. Y. (1997). Fluorescent indicators for Ca^{2+} based on green fluorescent proteins and calmodulin. *Nature* **388**(6645), 882–887.

Pelham, R. J., Jr., and Wang, Y. (1997). Cell locomotion and focal adhesions are regulated by substrate flexibility. *Proc. Natl. Acad. Sci. USA* **94**(25), 13661–13665.

Riveline, D., Zamir, E., Balaban, N. Q., Schwarz, U. S., Ishizaki, T., Narumiya, S., Kam, Z., Geiger, B., and Bershadsky, A. D. (2001). Focal contacts as mechanosensors: Externally applied local mechanical force induces growth of focal contacts by an mDia1-dependent and ROCK-independent mechanism. *J. Cell Biol.* **153**(6), 1175–1186.

Thomas, S. M., and Brugge, J. S. (1997). Cellular functions regulated by Src family kinases. *Annu. Rev. Cell Dev. Biol.* **13,** 513–609.

Ting, A. Y., Kain, K. H., Klemke, R. L., and Tsien, R. Y. (2001). Genetically encoded fluorescent reporters of protein tyrosine kinase activities in living cells. *Proc. Natl. Acad. Sci. USA* **98**(26), 15003–15008.

Tsien, R. Y. (1998). The green fluorescent protein. *Annu. Rev. Biochem.* **67,** 509–544.

Wang, N., Butler, J. P., and Ingber, D. E. (1993). Mechanotransduction across the cell surface and through the cytoskeleton. *Science* **260**(5111), 1124–1127.

Wang, Y., Botvinick, E. L., Zhao, Y., Berns, M. W., Usami, S., Tsien, R. Y., and Chien, S. (2005). Visualizing the mechanical activation of Src. *Nature* **434**(7036), 1040–1045.

Wang, Y., Miao, H., Li, S., Chen, K. D., Li, Y. S., Yuan, S., Shyy, J. Y., and Chien, S. (2002). Interplay between integrins and FLK-1 in shear stress-induced signaling. *Am. J. Physiol. Cell Physiol.* **283**(5), C1540–C1547.

Zhang, J., Ma, Y., Taylor, S. S., and Tsien, R. Y. (2001). Genetically encoded reporters of protein kinase A activity reveal impact of substrate tethering. *Proc. Natl. Acad. Sci. USA* **98**(26), 14997–15002.

CHAPTER 19

Optical Torque on Microscopic Objects

Simon Parkin, Gregor Knöner, Wolfgang Singer, Timo A. Nieminen, Norman R. Heckenberg, and Halina Rubinsztein-Dunlop

Centre for Biophotonics and Laser Science
School of Physical Sciences
The University of Queensland
Queensland 4072, Australia

We outline in general the role and potential areas of application for the use of optical torque in optical tweezers. Optically induced torque is always a result of transfer of angular momentum from light to a particle with conservation of momentum as an

0091-679X/07 $35.00
DOI: 10.1016/S0091-679X(06)82019-4

underlying principle. Consequently, rotation can be induced by a beam of light that carries angular momentum (AM) or by a beam that carries no AM but where AM is induced in the beam by the particle. First, we analyze some techniques to exert torque with optical tweezers such as dual beam traps. We also discuss the alignment and rotation which is achieved using laser beams carrying intrinsic AM—either spin or orbital AM, or both. We then discuss the types of particles that can be trapped and rotated in such beams such as absorbing or birefringent particles. We present a systematic study of the alignment of particles with respect to the beam axis and the beam's polarization as a way of inducing optical torque by studying crystals of the protein lysozyme. We present the theory behind quantitative measurements of both spin and orbital momentum transfer. Finally, we discuss the applications of rotation in optically driven micromachines, microrheology, flow field measurements, and microfluidics.

I. Introduction

Maxwell's electromagnetic theory from 1865 implies that light can carry angular momentum (AM) and at the beginning of the twentieth century Poynting (1909) suggested an experiment to show that light could exert a torque on a birefringent plate. The first experimental confirmation of optical AM transfer came 30 years later, when Beth measured a feeble torque on a birefringent plate suspended on a torsional pendulum, due to changes in the polarization of the light (Beth, 1936). While Beth's (1936) experiment was a great experimental challenge, transfer of momentum from a light beam to an object is nowadays routinely exploited in optical traps. In fact, the principle of optical tweezers is based on exchange of linear momentum between light and matter, and enables translational control over a trapped microparticle (Ashkin *et al.*, 1986). In the last decade, schemes to also transfer AM of light to particles have been investigated. The ability to exert optical torques to rotationally manipulate microparticles has since then developed from an interesting curiosity to seeing deployment in practical applications. The ability to rotate objects offers a new degree of control for microobjects since one gains full three-dimensional (3D) control over a trapped object, with important applications in biotechnology and related areas. Rotation could offer the ability to orient biological specimens such that active enzymes attached to beads could be aligned to latch onto one another. In addition, applying optical torques could allow the investigation of new mechanical properties of biomolecules, for example, the torque necessary to unwind the helix of a single DNA molecule. Spinning microparticles can also be used to realize micromachines, like miniature pumps for microfluidic devices. There are several advantages of optically driven micromachines compared to micromachines based on other principles. Optically driven micromachines can work in hostile environments, like liquids. In addition, there are no connecting cables required to power those machines, and optically driven micromachines are at least one order of magnitude smaller than micromachines based on Microelectromechanical Systems (MEMS) or other techniques. Even biological machines that could function within living cells are feasible. And finally, optically driven micromachines are transparent, making them compatible with many microscopical applications and techniques.

Applications tested so far involve the use of rotating particles as micropumps in microfluidic devices (Knöner *et al.*, 2005). Also the application of rotating particles as actuators to apply spatially limited shear-stress fields to cells have been tested (Botvinick *et al.*, 2006).

II. Ways of Exerting Optical Torque

Optically induced torque is always a result of transfer of AM from the beam to the trapped particle, with conservation of AM as the underlying principle. Consequently, rotation can be induced by an incident beam that carries AM or by an incident beam that carries no AM, but where AM is induced to the beam by the particle. Different schemes, depending on the object to rotate, have been realized and tested, and we begin with those where the incoming light carries no AM.

Perhaps the simplest way to rotate an object is to grasp it with a pair of conventional optical tweezers that can be made to revolve around each other in any plane. This will lead to a rotation of the object with them, if the object is elongated enough to give two separate grips. The technique requires some technical effort, but it is an effective method for true 3D alignment of particles (Bingelyte *et al.*, 2003). In fact, it overcomes the restrictions of many other methods which are limited to the alignment or rotation of particles about the beam axis. The simplest method of producing a dual beam trap is to introduce a pair of beam splitters into the tweezers setup, creating two separate, fully steerable beams (Fällman and Axner, 1997). Two beams can also be implemented by rapid scanning of a single beam between two positions (Visscher *et al.*, 2005) or by use of computer-generated holograms to yield two beams simultaneously (Reicherter *et al.*, 1999). To apply a torque, each of the beams must be deflected by the object to produce two oppositely deflected but displaced scattered beams (Fig.1), introducing thereby a twist into the combined light field so that the torque exerted is equal to the rate at which AM is added to the light field.

Elongated particles also align themselves in a trapping beam with a noncircular intensity distribution. The ends of the elongated spot act on the particle in the same way as the separate beams in the dual beam arrangement referred to above, but rotation is limited to the plane perpendicular to the direction of propagation. One way to generate an asymmetric beam is to introduce an asymmetric aperture, for example, a rectangular aperture, clipping each side of the beam (O'Neil and Padgett, 2002). If the aperture is rotated, the particle will follow the orientation of the rotating beam profile. A macroscopic simulation of the situation in optical tweezers both experimentally and theoretically shows that the AM transfer of such an arrangement can be quite significant (Parkin *et al.*, 2004).

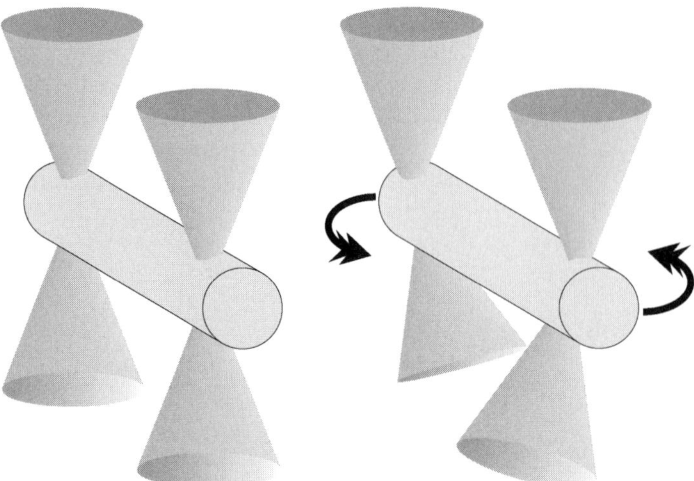

Fig. 1 The orientation of a trapped elongated object can be changed arbitrarily by a dual beam trap, where the individual beams can be moved around each other in any plane. Left: equilibrium trapping position of a rod for stationary beams. Right: changing the beam positions causes a deflection of the beams, which induces orbital AM to the outgoing light field and results in a torque acting on the trapped object.

A similar principle for rotation or alignment is utilized in setups where a dual beam trap has been configured to create an interference pattern. By interfering a beam with an azimuthal structure [e.g., a Laguerre–Gaussian (LG) beam], and another beam with a Gaussian intensity distribution, one can create an interference pattern where the azimuthal angle of the pattern is governed by the phase shift between the two beams. Thus, by changing the path length of one of the beams—which causes the interference pattern to rotate in a controlled fashion about the axis of the spiral pattern—rotation can be induced (Paterson *et al.*, 2001). In its easiest implementation, the interference pattern can be rotated about the axis of the beam by simply tilting a glass plate in one of the beams (MacDonald *et al.*, 2002).

A more elegant way to align or rotate particles, especially if constant particle rotation is the aim, is to make use of a laser beam carrying intrinsic AM.

Light can carry "spin" AM, "orbital" AM, or both. Spin AM is associated with the polarization of the light beam. In a circularly polarized beam, the electric field vector is perpendicular to the direction of the propagation but rotates once in each wavelength. The spin AM *density* is, by definition, independent of the choice of origin of the coordinate system. Linearly polarized light carries no spin AM, circularly polarized either carries $\pm 1\hbar$ per photon, depending on the handedness, and elliptically polarized light can carry an average spin AM between 0 and $\pm 1\hbar$ per photon.

Orbital AM is associated with a circulating structure in the distribution of energy in the beam and is most clearly seen in the LG "doughnut" beams as represented in Fig. 2. Even though the electric field is linearly polarized, the exp ($il\phi$) phase variation turns *the* normally plane wave fronts into helical surfaces, twisting once every l wavelengths.

The AM introduced to the beams in a dual beam trap is largely of this kind. The two forms, spin and orbital, can be present separately or together, in which case they can add or subtract depending on the handedness of the polarization relative

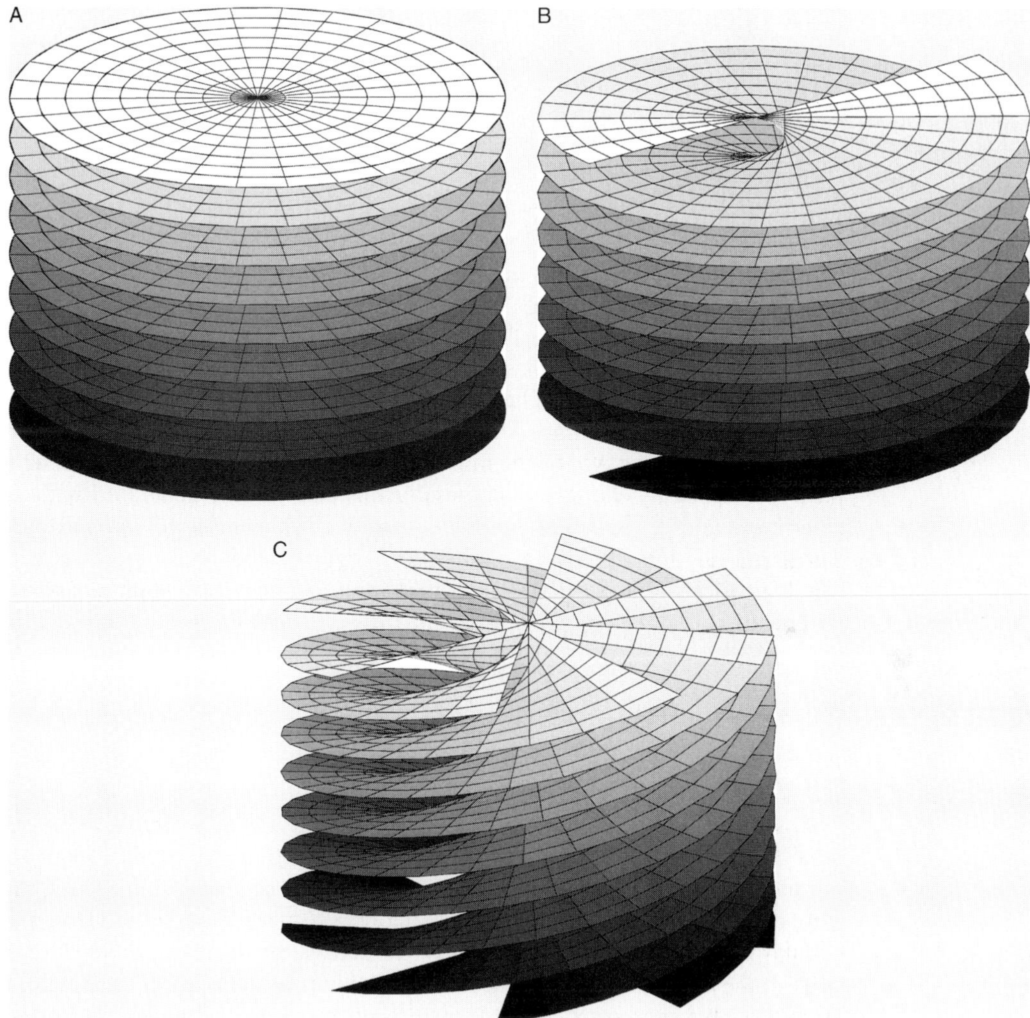

Fig. 2 The phase fronts of a Laguerre–Gauss beam are a function of the azimuthal angle and the topological charge l of the beam. (A) Phase structure for an LG_{00} mode carrying no AM (Gaussian beam). (C) Phase structure for an LG_{02} mode, carrying an AM 2h per photon. (B) LG_{08} mode, carrying 8h per photon.

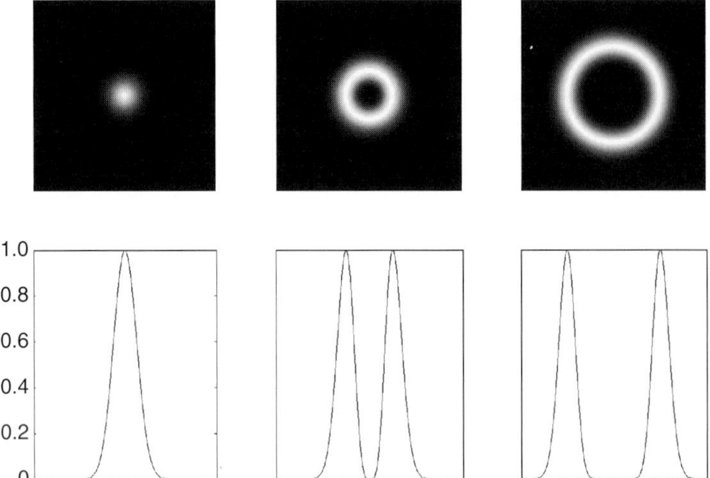

Fig. 3 Intensity distributions of different laser modes. Left: Gaussian beam. Middle: LG_{02}. Right: LG_{08}. The corresponding spatial phase structures of the three beams are shown in Fig. 2.

to the helicity of the beam. The AM such an optical vortex can carry is, at least in principle, any integer l times $\hbar$ per photon, depending on the topological charge. The ostensible advantage of the high AM a doughnut beam can carry is, however, partly cancelled by the fact that these beams can only be generated with limited efficiencies and by comparatively small transfer efficiencies to trapped particles. Furthermore, the diffraction-limited spot sizes in the focus increase with the topological charge l, reducing or even eliminating the 3D trapping capability of such beams (Fig. 3).

III. Types of Objects

A. Absorbing Particles

Absorption is perhaps the simplest transfer mechanism available when the incident beam carries AM. Both spin and orbital AM can be transferred to an absorbing particle. It has been shown in an experiment using absorbing CuO particles trapped in a focused doughnut beam that a change of the polarization from plane to circular caused the rotation frequency to increase or decrease, depending on the sense of the polarization with respect to the helicity of the beam (Friese *et al.*, 1996). However, this method of AM transfer is always accompanied by undesirable heating of the particle and is for that reason not really practical in optical-trapping applications, particularly if high rotation speeds are desirable.

A more elegant approach, that avoids unwanted heating, is to transfer spin or orbital AM to nonabsorbing particles.

B. Birefringent Particles

In order to be able to transfer spin AM to an object, the particle must have the ability to alter the polarization of the incident beam. This is the case if a particle has either an external (shape) or internal (birefringence) anisotropy.

Birefringent objects have two different indices of refraction, depending on the orientation of the electric field vector relative to the optic axis of the crystal. As a consequence, a phase shift between the ordinary and extraordinary components of the beam will be induced, altering the polarization state of the incident light. As long as the particle is oriented in a way where it acts as a phase retarder a torque will act on it.

Uniaxial crystals (such as quartz, calcite, or vaterite, which have all been rotated in optical traps) have one direction where all orientations of linear polarized incident light propagate at the same speed. In other directions, there is an "ordinary wave" which propagates at the same speed in all directions, and an orthogonally polarized "extraordinary wave" the speed of which varies with direction and which is either always greater than (negative uniaxial) or smaller than (positive uniaxial) that of the ordinary wave.

C. Orientation of Birefringent Spheres

Imagine a sphere cut from a birefringent crystal. When it is trapped, such a sphere will tend to align itself in the lowest energy state, which is where the electric field vector experiences the highest permittivity which in turn is associated with the highest average refractive index. For a positive uniaxial crystal (Fig. 4, left), this corresponds to the optic axis of the crystal aligning with the electric field, and hence perpendicular to the propagation direction of the beam. This is most easily seen considering linearly polarized light. With circularly polarized light, one linear component will see a high extraordinary refractive index, n_e, while the other will see only the ordinary refractive index, n_o, but the minimum energy will still correspond to the same direction. This is exactly the ideal orientation for the particle to act as a waveplate. Positive uniaxial materials include quartz and vaterite.

On the other hand, for a negative uniaxial crystal (Fig. 4, right) the ordinary wave is slowest. In this case, for a linearly polarized incident beam the preferred orientation of the optic axis is in the plane of the direction of propagation and the electric field vector. However, for circularly polarized light, the lowest energy configuration will be with the optic axis along the beam propagation direction. In this case, there is no relative retardation and no change in the polarization of the light, hence no torque. Calcite is a negative uniaxial crystal, and it might therefore be expected that it would be difficult to rotate. However, calcite does not adopt a spherical habit, and any asymmetries in the crystal will upset any alignment of the optic axis with the beam (Singer *et al.*, 2006). Thus, although the efficiency may be reduced, in general a torque will act.

The plane of polarization of the incident light, and hence the orientation of the trapped birefringent particle, can easily be changed by the use of a half-wave ($\lambda/2$)

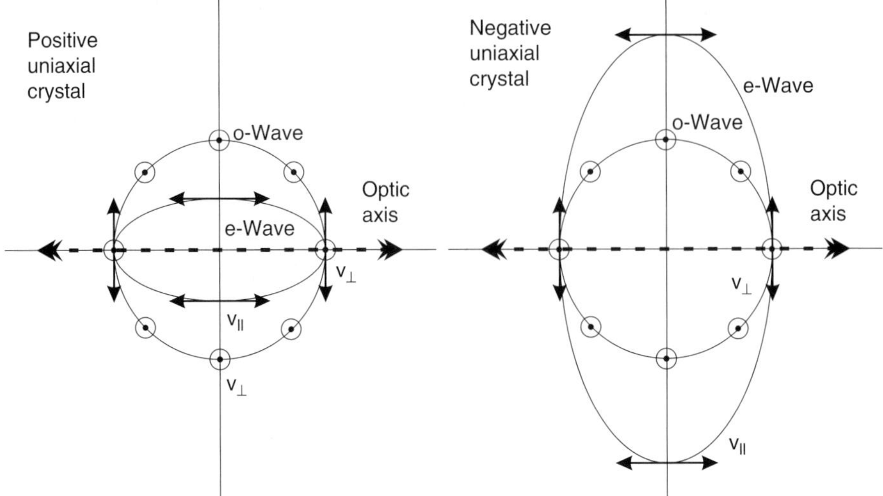

Fig. 4 Polar plots showing propagation velocity of wavelets within uniaxial crystals (Hecht, 1987). Ordinary wave electric field vector (o-wave) is perpendicular to the plane of the page while the extraordinary wave electric field vector (e-wave) oscillates within the plane of the page. For a spherical positive uniaxial crystal, the optic axis aligns with the electric field, which is perpendicular to the trapping beam, to minimize its energy. Therefore, the trapping beam will see a birefringent crystal. For a spherical negative uniaxial crystal, the optic axis aligns perpendicular to the electric field, which is parallel to the trapping beam. In this case, the trapping beam does not see any birefringence in the crystal.

plate in the beam path of the incoming laser beam. Mounting the half-wave plate on a stepper motor enables controlled and fast rotation.

For circularly polarized incident light, the particle cannot follow the optical frequency of the electric field vector, thus a constant change of the polarization state of the light is induced. In this case, the birefringent particle will rotate with a constant rotation rate that is governed by the drag torque of the surrounding fluid. Maximum spin AM will be transferred to a particle if the particle acts as a half-wave ($\lambda/2$) plate, changing the handedness of the incident circularly polarized light (from left to right, or vice versa). This corresponds to a spin AM change of $2\hbar$ per photon, the maximum AM change possible for polarized light. For highly birefringent materials, a phase retardation of $\lambda/2$ is achieved with thicknesses of a few micrometers only. Particles used for rotation experiments, like microscopic calcite (Friese *et al.*, 1998), quartz (La Porta and Wang, 2004) or vaterite crystals (Bishop *et al.*, 2004), are highly birefringent and typically have dimensions of a few micrometers.

Spherical vaterite particles can be grown as shown in Fig. 5 and have the advantage that their drag torque is well defined and can easily be calculated. For that reason they are superior to other particles used for this purpose. Furthermore, the applied torque as a result of spin AM transfer can easily be measured optically (Bishop *et al.*, 2004; Nieminen *et al.*, 2001), as will be explained below. This opens a way for a number of interesting applications where properties of liquids can be studied. On the basis of this technique, a microviscometer has been demonstrated

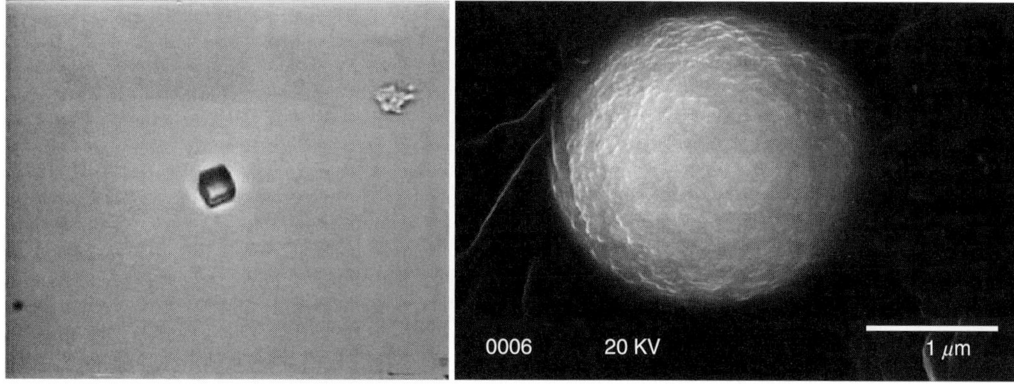

Fig. 5 Left: image of an optically trapped microscopic birefringent crystal (calcite) which can either be spun in circular polarized light or aligned using linearly polarized light. Right: scanning electron microscope (SEM) image of a vaterite crystal. The crystal is almost perfectly spherical and has a quite smooth surface.

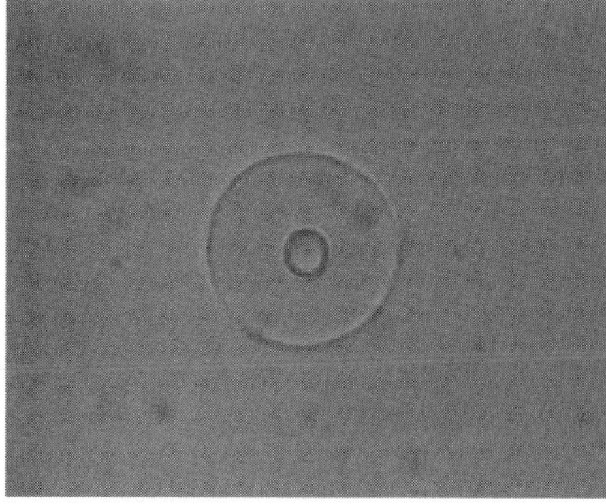

Fig. 6 Optically trapped vaterite crystal inside a hexane-filled lipid-walled vesicle of 16.7 μm in diameter.

using a spherical birefringent crystal as the particle that was trapped and rotated. Measurements of the viscosity of extremely small sample volumes, even inside a microscopic vesicle (Fig. 6), have been demonstrated (Bishop *et al.*, 2004).

The spin AM of an incident laser beam can also be changed by particles that are not axially symmetric. Elongated or flattened particles have different dielectric polarizabilities parallel to their short and long axis, and if comparable to a wavelength in size or smaller, act as birefringent particles giving rise to a polarization change of the transmitted light. This dielectric anisotropy due to the shape of the particle is often referred to as form or shape birefringence. Consequently, form-birefringent particles can also be either aligned in plane polarized or spun in circularly polarized light.

The asymmetry of objects also determines their alignment in the trap with respect to the beam axis. Elongated particles tend to align with their long axis along the

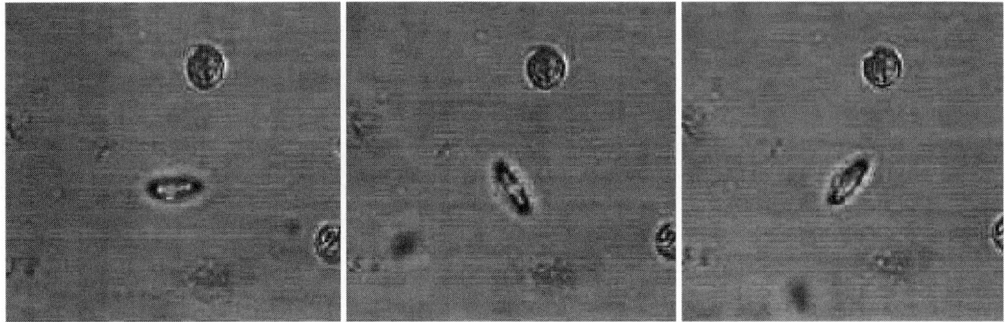

Fig. 7 Alignment of a chloroplast as the plane of polarization of the trapping beam is rotated. The chloroplast (~4-μm long along the long axis) lies in the plane of polarization (Bayoudh *et al.*, 2003).

axis of the trapping beam, because this minimizes the total field energy, analogous to the way a particle is attracted to the point of maximum field strength in normal optical tweezers. Thereby they can lose their asymmetry about the beam axis after being trapped. Consequently, long rod-shaped objects cannot be spun if three-dimensionally trapped. Disk-shaped objects have two long axes and, after aligning one with the beam axis, have a remaining asymmetry about the beam axis, which will align with the plane of polarization of the trapping beam, as shown in Fig. 7.

Elongated birefringent particles orient themselves according to their shape and birefringence with respect to the beam axis.

D. Rotation of Lysozyme Crystals

Alignment of particles with respect to the beam axis was one of the earliest examples of optically induced torque, and even though it is directly relevant to many applications, it has not received a lot of attention (Ashkin *et al.*, 1987). However, in order to be able to transfer AM to a particle, it must retain an asymmetry about the beam axis after being three-dimensionally trapped.

A systematic investigation on the alignment of optically trapped nonspherical birefringent particles by Singer *et al.* (2006) has been performed using crystals of the protein lysozyme. Lysozyme crystals are birefringent, have a well-defined shape, and both their size and aspect ratio can be controlled and changed by parameters of the growing solution, like the pH value and the initial protein concentration. For that reason they are well suited for studies of the dependence of particle alignment in a laser trap on the shape and birefringence. Due to their well-defined morphology they are also suitable for numerical simulations.

Figure 8 shows the growth of a lysozyme crystal in a laser trap. Using such crystals, it was verified that indeed both the shape and the birefringence induce optical torques that tend to align particles with respect to the beam axis. In particular, positive birefringence ($n_o < n_e$) creates a torque that tends to align the optic axis perpendicular to the beam axis, whereas negative birefringence tends

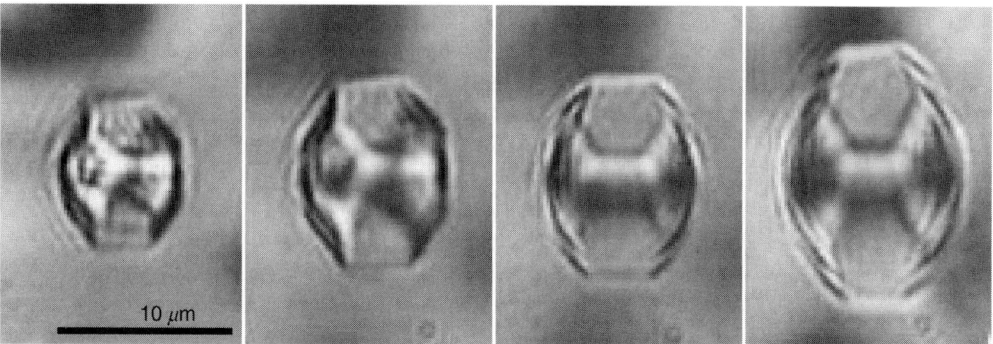

Fig. 8 Growth of a trapped lysozyme crystal. The aspect ratio can be changed during growth, altering the orientation of the trapped crystal.

to align the optic axis normal to the electric field (that can be either parallel or perpendicular to the beam axis). As a consequence, among spherical particles, only positive uniaxial materials can be aligned or spun. For negative birefringent spheres this is not the case. Negative birefringent particles must have a shape anisotropy that prevents unfavorable alignment of the optic axis in order to be able to be spun (Friese *et al.*, 1998).

It has furthermore been shown that the torque acting to align the particle with the beam axis is typically one order of magnitude larger than the torque which aligns the remaining asymmetry with the plane of polarization (Bayoudh *et al.*, 2003).

E. Microrotors

Orbital AM can be transferred to particles that have an asymmetry that alters the orbital AM of the incoming beam. Again, the interaction of light with particles can result in rotation even if the incident light itself does not carry AM. In this case, the particle has to scatter light in a helical manner.

A variety of microparticles have been tested to evaluate their suitability for orbital AM transfer, and their capability to be used in optically driven micromachines. As one can imagine, the shape plays a crucial role for the efficiency of the AM transfer (Nieminen *et al.*, 2004). A promising method to fabricate micro-objects is photopolymerization. In this technique, a laser beam is used to cure a user-defined 3D structure in a photocurable resin. Using two-photon photopolymerization, 3D structures with spatial resolution down to 120 nm can be created (Kawata *et al.*, 2001).

The design for the optimal shape is not trivial. One has to ensure the proper alignment of the particle when three-dimensionally trapped while still having structures with a high transfer efficiency. Galajda and Ormos (2002a,b), who were pioneers in this field, studied the efficiency of a variety of shapes empirically. Shapes included simple helices, "sprinklers," and "conical propellers." They all rotate when trapped with a Gaussian beam, which means they scatter light that is carrying orbital AM. Rotors with blades resembling logarithmic spirals even have the capability to change the direction of rotation, depending on the axial position of the rotor with respect to the focus of the Gaussian beam (Galajda and Ormos, 2002a).

As a focused LG beam has a "doughnut-shaped" intensity profile, with a dark central core, particle orbiting in addition to rotation can be observed. The radius of the circular intensity distribution depends on the topological charge l. Small particles ($r_{\text{particle}} < r_{\text{doughnut}}$) will get trapped in the circumference of the ring-shaped intensity profile rather than in the center of the beam. Those particles will, driven by the AM of the beam, orbit around the beam axis (Gahagan and Swartzlander, 1996; Nieminen *et al.*, 2006).

Among the variety of methods to accomplish AM transfer, transfer of spin AM to a birefringent particle is probably the best suited for actual practical applications for the following reasons. First, spin AM can easily be measured so that the applied optical torque can be determined by purely optical means, making the system well suited for quantitative measurements. Second, the torque can be controlled by changing the polarization state and/or the power of the light. Third, the torque is quite high, typically on the order of $1\hbar$ per photon per second if highly birefringent particles (such as vaterite crystals) are used. Finally, this method can be used with Gaussian beams, ensuring high 3D trapping efficiency.

IV. Quantitative Measurements of Optical Torque

As mentioned above, both spin and orbital AM can be transferred to an object. We have already outlined how the AM of light, be it spin or orbital, can be transferred to an object. We have shown that if the incident beam's AM differs from the transmitted AM a torque will be applied to the optically trapped object.

Here, we describe methods for quantitative measurement of the optical torque resulting from the transfer of AM. This is of particular importance for numerous applications to microscale studies of rotational dynamics.

A. Measuring Spin AM Transfer

As was mentioned before, spin AM arises from light's polarization. If one thinks of a light beam with an arbitrary polarization, for example elliptically polarized, this polarization state can be thought of as the sum of two circularly polarized components: a right circularly polarized component and a left circularly polarized component. (The representation of light as the sum of two linearly polarized components is probably more familiar but is less intuitive for thinking about AM.) The left component consists of photons each with $+\hbar$ of AM and likewise $-\hbar$ for the right component. The relative amplitudes of these two components define the spin AM of the beam that for an arbitrary beam is $\leq +\hbar$ and $\geq -\hbar$.

From the spin AM picture outlined above, a method to optically measure spin AM of a light beam becomes apparent: measure the relative amplitudes of the right and left circularly polarized components. This method is described in detail elsewhere (Nieminen *et al.*, 2001). Here, the key points of the theory are outlined. We start by defining a degree of circular polarization:

$$\sigma_{\mathrm{s}} = \frac{P_{\mathrm{L}} - P_{\mathrm{R}}}{P}$$

where P_{L} and P_{R} are the powers associated with the left and right circularly polarized components respectively, and P is the total power of the beam. The degree of circular polarization allows the total spin AM flux of the beam to be defined:

$$L = \frac{\sigma_{\mathrm{s}} P}{\omega}$$

where ω is the angular optical frequency of the light. However, in order to apply an optical torque to an object, the quantity that we are really interested in is the change in AM flux of the beam as this is equal to the reaction torque on the object which changed the AM of the beam. This reaction torque is given by:

$$\tau = \frac{\Delta \sigma_{\mathrm{s}} P}{\omega}$$

where $\Delta\sigma_{\mathrm{s}}$ is the change in the degree of circular polarization. Therefore, the quantities that need to be measured to evaluate the spin AM transfer are the powers of the two circularly polarized components. As can be seen from the above formula, the laser power at the focus of the trap as well as the wavelength of the laser radiation have to be known.

B. Experimental Design

A schematic of the setup required to measure spin AM transfer is shown in Fig. 9. Like most optical tweezers setups, it is based on a microscope with a high numerical aperture (NA) objective. A high NA condenser is also required to collect as much as possible of the laser light from the optical trap. This is a requirement as the average polarization of the light transmitted through the particle needs to be measured and a significant proportion of the laser power diverges at high angles from the trap. In order to control the rotation or alignment of the trapped particle, a polarizer followed by a quarter-wave plate has to be introduced. This allows the trapping beam to have circular, elliptical, or linear polarization. The final requirement is that the laser light collected by the condenser must be sent to a polarization analyzing system.

The polarization analyzing system consists of three photodetectors, one that measures in a linearly polarized basis and two others that measure in the circularly

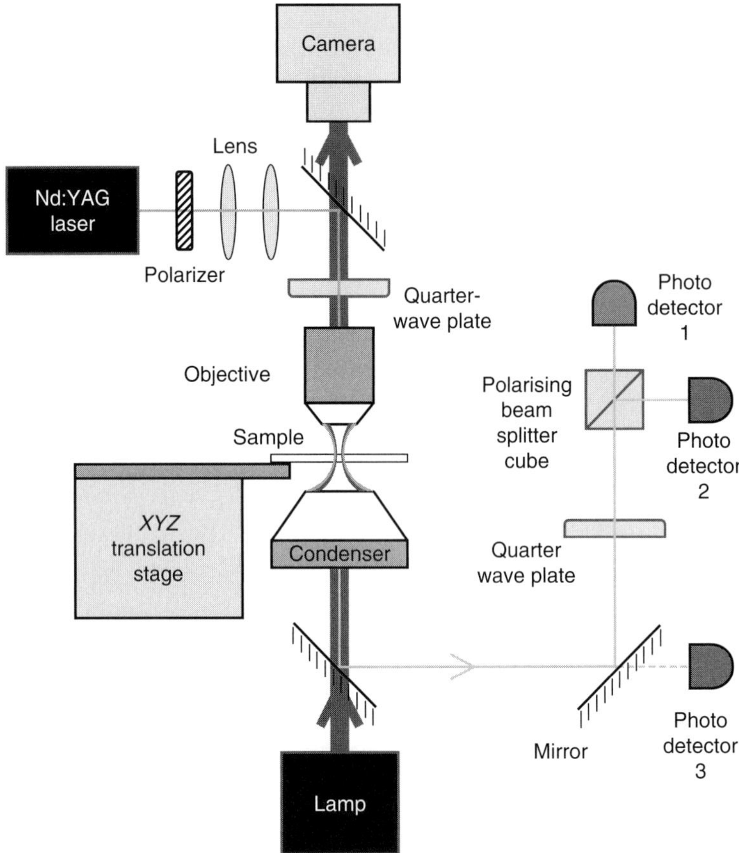

Fig. 9 Experimental setup for spin AM measurements.

polarized basis. The mirror in front of photodetector 3 transmits a very small fraction of linearly polarized light. Therefore, this detector will measure the rotation rate of the trapped birefringent particle. This is because the light transmitted through the particle will, in general, be elliptically polarized, which can be thought of as the combination of a circular and a linear polarization component. The linear component will rotate with the optical axis of the trapped particle so that the variation in the signal measured by photodetector 3 will correspond to the rotating linear component. The twofold optical symmetry of the birefringent particle means that the rotation rate measured by the detector will be twice the rotation rate of the particle. As it is required that photodetectors 1 and 2 measure in the circular basis, a quarter-wave plate is inserted before the polarizing beam splitter cube. Each detector therefore measures an orthogonal circularly polarized component. The degree of circular polarization is found directly from these detectors as they measure the relative amplitudes of the right and left circularly

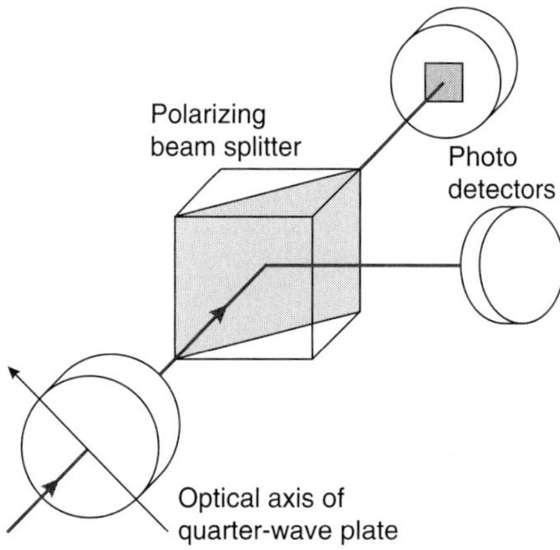

Fig. 10 Required alignment of the quarter-wave plate to the polarizing beam splitter cube for the measurement of transmitted polarization in the circular basis.

polarized components. The alignment required for the quarter-wave plate is shown in Fig. 10. Typical signals measured by the photodetectors are shown in Fig. 11. Applications for this technique are described later in this chapter.

C. Measuring Orbital AM Transfer

There exists an analogy between spin and orbital AM (whorl): spin can be measured by determining the amplitudes of orthogonally polarized components, whereas whorl can be measured by determining the amplitudes of orthogonal LG modes which carry well-defined whorl associated with their azimuthal index. It is possible to measure these modes experimentally, and the technique will be described here. However, it is nontrivial and has not (so far) been implemented to measure torque transfer in optical tweezers.

In keeping with the analogy between spin and orbital AM, we can define a degree of whorl:

$$\sigma_{\mathrm{w}} = \frac{\sum_{l=1}^{\infty} lP_l - \sum_{l=-1}^{-\infty} |l|P_l}{P}$$

where P_l is the power in each LG mode with azimuthal index l and P is the total power given by $P = \sum_{l=-\infty}^{\infty} P_l$. From the degree of whorl, we can find the total orbital AM flux, and the change in this flux gives the torque transfer due to orbital AM.

In order to detect and measure LG modes, computer-generated holograms are used. The interference pattern, or hologram, for a certain LG mode and a plane wave is calculated. An example of such a hologram is shown in Fig. 12. The pattern is written to 35-mm film. Then using a contact print process, it is transferred to a

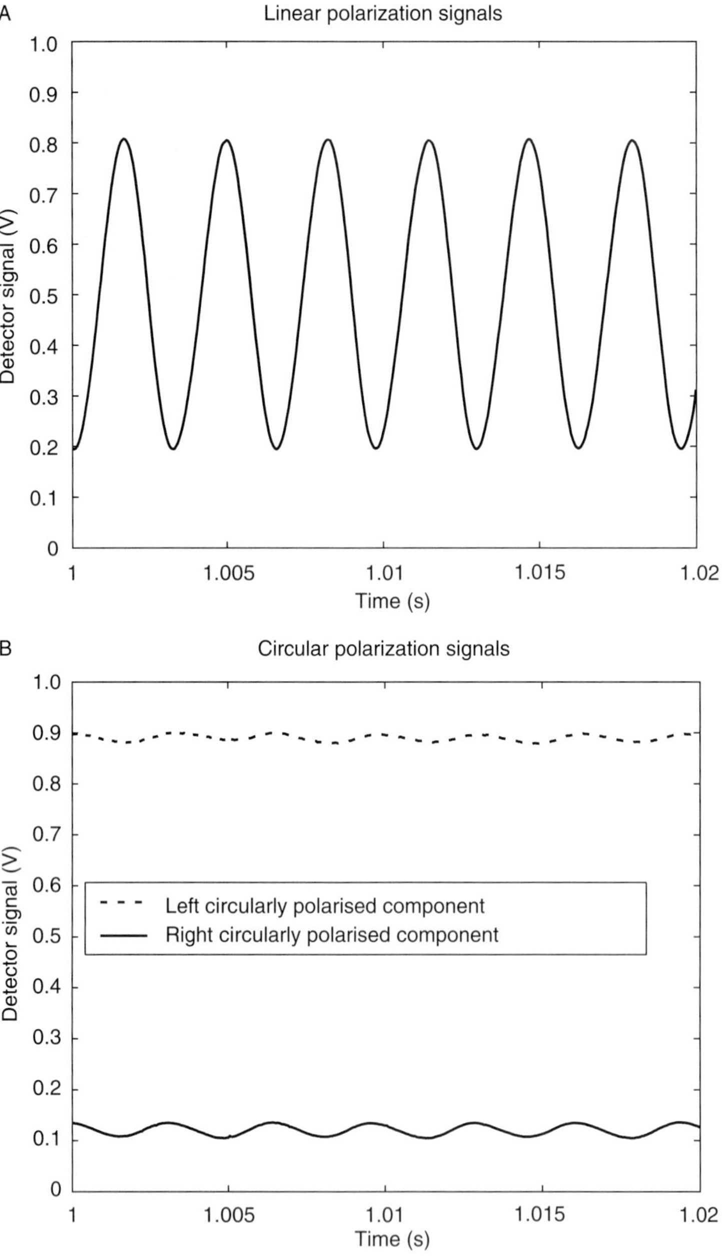

Fig. 11 Signal from photodetector 1 is shown in (A). The rotation rate is half the frequency of this signal. The signals from detectors 2 and 3 are shown in (B).

Fig. 12 An off-axis LG02 phase hologram. The gray scale represents the phase thickness.

holographic plate. A phase hologram is then made by bleaching the emulsion layer of the plate, which creates regions of different phase thickness. A region's phase thickness is determined by its exposure to light during the contact print process. The result is a phase hologram that when illuminated will generate the calculated LG mode (Heckenberg *et al.*, 1992). Such holograms are not 100% efficient and, in fact, generate apart from the desired mode, higher order modes, and transmit some of the original Gaussian mode. Therefore, off-axis holograms are used, which causes the different LG modes to be diffracted to different orders, as depicted in Fig. 13A. These holograms can not only be used to create orbital AM-carrying modes, they can also be used to detect such modes. Figure 13B shows the effect of illuminating the beam with a whorl-carrying mode: a Gaussian mode appears in one of the diffracted orders. The Gaussian can be detected either by coupling into a single mode fiber and detecting the output, or by measuring the intensity at the center of the mode. The fiber detection works as only Gaussian modes propagate down the fiber, so all other modes are filtered out. The central intensity measurement works as only a Gaussian mode will have intensity at the center. All other modes that carry AM must have zero central intensity due to a phase singularity generated by their azimuthal phase structure.

These techniques have been exploited for quantum information applications. Orbital AM states are of interest because of the possibility of multidimensional entanglement. Experiments have shown that orbital AM is conserved during parametric downconversion and the states are entangled (Mair *et al.*, 2001). In some experiments more relevant to optical tweezers, the orbital AM of an arbitrary beam was measured (Parkin *et al.*, 2004). In this work, the torque applied to an elongated phase object by a paraxial laser beam was determined by measuring the power in each of the LG modes present in the forward scattered beam. This modal decomposition method would, in principle, work for any beam. However, the

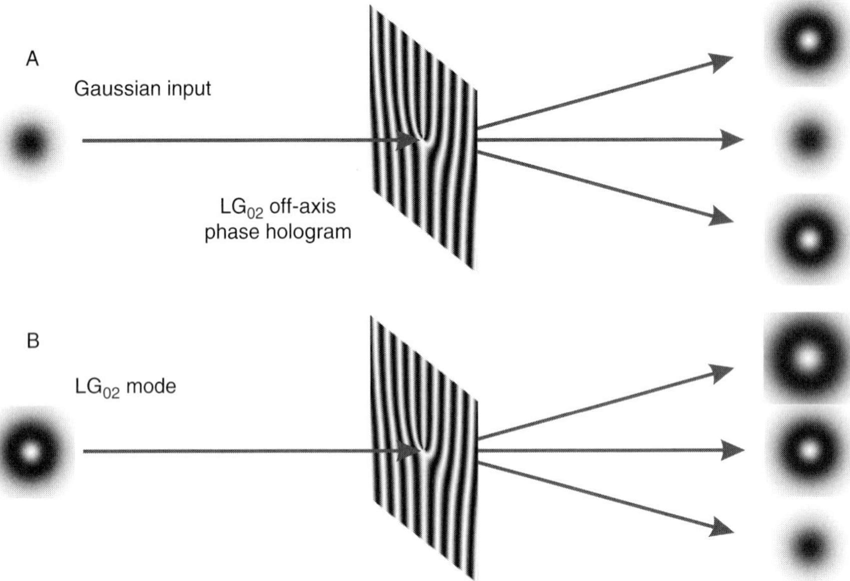

Fig. 13 The output modes from an LG_{02} hologram with a Gaussian input (A), and an LG_{02} input (B).

detection phase hologram would need to be changed a number of times in order to measure the potentially large number of modes that could be present in an arbitrary beam. One possible solution to this problem would be to use a spatial light modulator to generate the detection holograms, which would allow for fast switching between different modes and for a number of modes to be detected at the same time (Gibson *et al.*, 2004).

In order to apply these techniques to optical tweezers, let us consider the effect of an on-axis hologram on an LG_{02} mode. We can see from Fig. 14A that the once Gaussian-free incident beam now contains intensity in the center that signifies a Gaussian in the transmitted beam. The Gaussian detected means the input beam has transferred AM to the phase hologram.[1] The power in the transmitted Gaussian mode tells us how much momentum is transferred. If we consider an asymmetric object (Fig. 14B), we see that it has a similar affect on the transmitted beam, and so we must have transferred AM to the beam. However, just detecting the center of the beam only tells us how much power was coupled from the LG_{02} mode to the Gaussian mode. Using an analyzing hologram (Fig. 14C), coupling to other LG modes can be measured. Incorporating this measurement principle into optical tweezers would result in a setup that looks like Fig. 15.

[1] The reader may note that this is not the case if an equal amount of power is coupled to a higher order LG mode; however, the coupling efficiency tends to be lower for higher order modes.

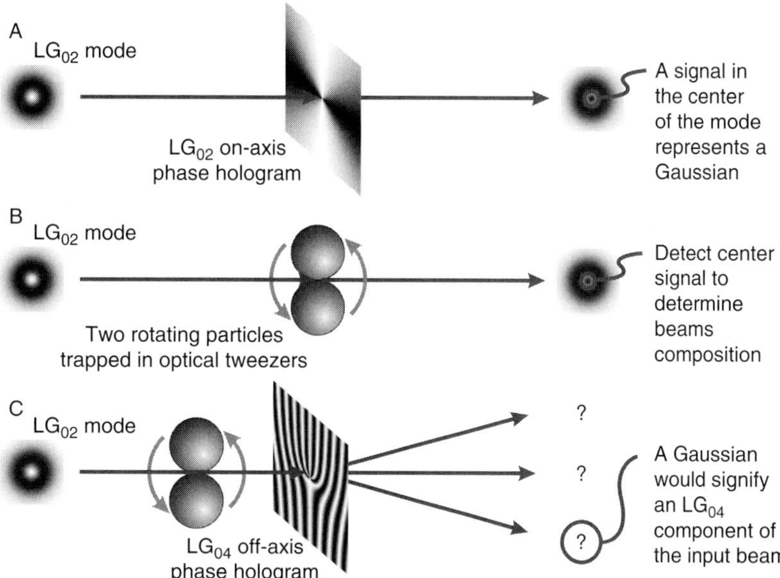

Fig. 14 Output from an LG_{02} on-axis hologram (A), an object composed of two particles which behaves similarly to the on-axis hologram (B), and the output from an analyzing hologram (C), in this case an LG_{04} phase hologram. (See Plate 25 in the color insert section.)

Due to the complexity of this setup, and the difficulty of measuring and quantifying all possible orbital AM-carrying modes in the forward scattered beam, we should consider an alternative technique to measure orbital AM. The spin AM measurement is relatively simple, so we can take advantage of that to measure whorl. The torque acting on two beads in an optical trap has been measured in this way (Parkin *et al.*, 2006). Using an LG_{02} trapping beam the polarization was varied, thus changing the spin component while keeping the whorl component constant. Measuring the change in the rotation rate, while varying the spin component, allowed the whorl component to be determined. Figure 16 shows the relationship between the spin AM transfer and the rotation rate (Parkin *et al.*, 2006). The gradient of the curve depends on the relative magnitudes of the spin and orbital contributions and hence gives the orbital AM transfer.

V. Applications

A. Microrheology

In the same way that optical tweezers have begun to be used to measure the physical and mechanical properties of cells and their membranes, an opportunity exists to explore the properties of more fluid biological media. A number of

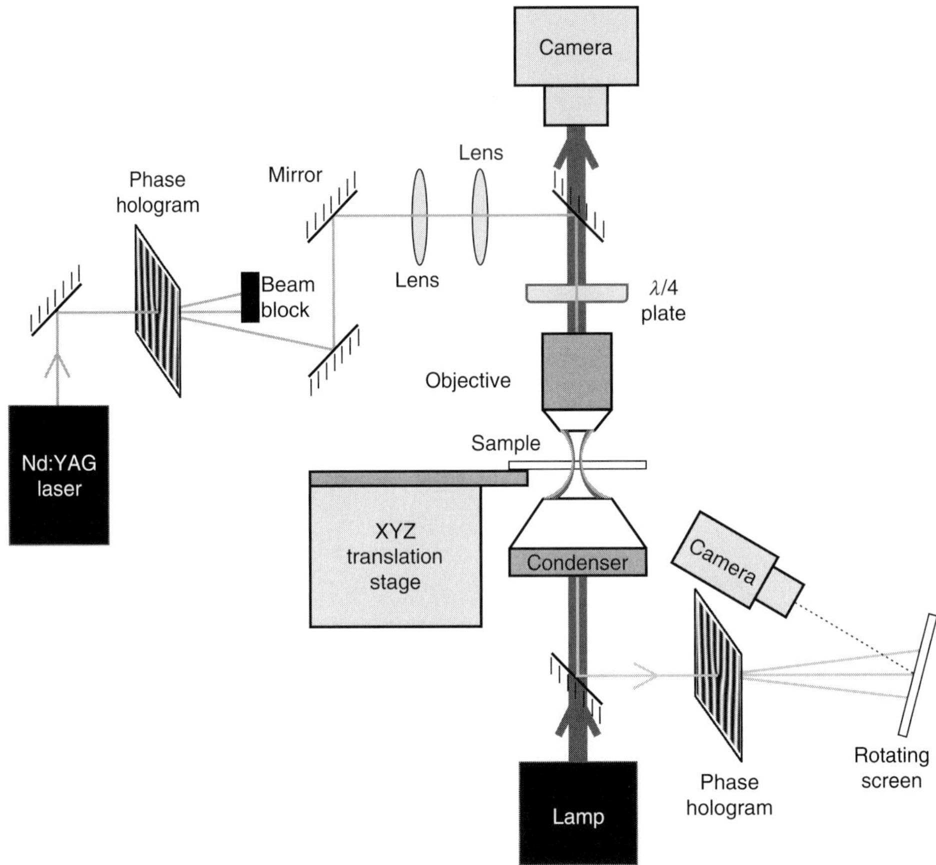

Fig. 15 Figure of the optical tweezers setup, with the output from the microscope being sent to an orbital angular momentum (OAM) analyzing system.

different techniques exist to study the microrheological properties of complex fluids, such as particle tracking, magnetic tweezers, dynamic light scattering, and laser tracking. Waigh (2005) gives an overview of many of these techniques. Here, we will describe a technique to measure viscosity which is based on the spin AM measurement described in the previous section.

We can determine the torque applied to an object trapped in optical tweezers, therefore, in order to measure some property of the surroundings (in this case its viscosity η) we need to have a trapped particle with a simple and well-defined geometry. A sphere is an excellent choice as the viscous drag torque on a sphere in the low Reynolds number limit is given by:

$$\tau_D = 8\pi\eta a^3\Omega$$

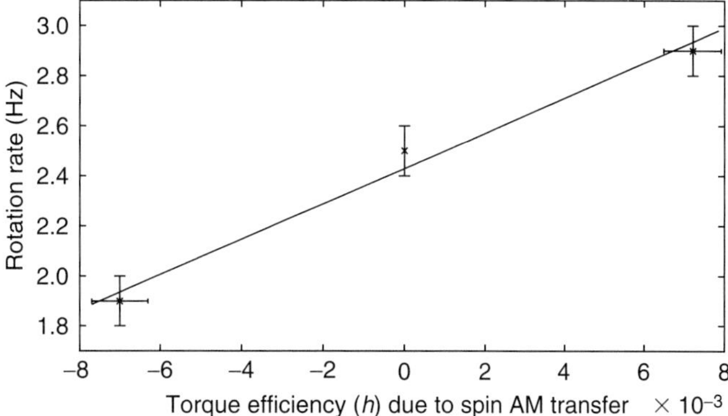

Fig. 16 The rotation rate of two trapped polystyrene spheres as a function of the optically applied torque due to the spin AM of the trapping beam. The torque transfer due to the whorl component was found from the slope.

where a is the radius of the sphere and Ω is its rotation rate. For steady rotation, the optically applied torque is equal to the drag torque, which allows us to measure viscosity:

$$\eta = \frac{\Delta\sigma P}{8\pi a^3 \Omega \omega}$$

As was discussed before, in order to rotate the sphere in optical tweezers, it is required to be birefringent. Fortunately, strongly birefringent spheres exist in the form of vaterite, a calcium carbonate crystal. These crystals form spherical structures in the right growth conditions in the size range of 1–10 μm. A method to grow these crystals is outlined in Bishop *et al.* (2004).

Varying the power of the trapping laser directly varies the torque on the particle and its rotation rate. Figure 17A shows the dependence of the measured viscosity on laser power for methanol. In this case, the viscosity is plotted against power as neither the variation in torque nor rotation rate is responsible for the shape of the dependence. Laser-induced heating is the cause of the effect, due to absorption by the vaterite crystal. Fitting a curve to the data, we can easily determine the room temperature viscosity of the liquid from the intercept, and we also get an idea of the magnitude of the heating. The method also works over a range of particle sizes, as shown in Fig. 17B.

The advantages of this technique over other microrheology techniques are as follows:

Trap stiffness calibration is not required (as is the case in other optical tweezers techniques).

Sample volumes almost as small as the particle itself—picoliters—can be probed.

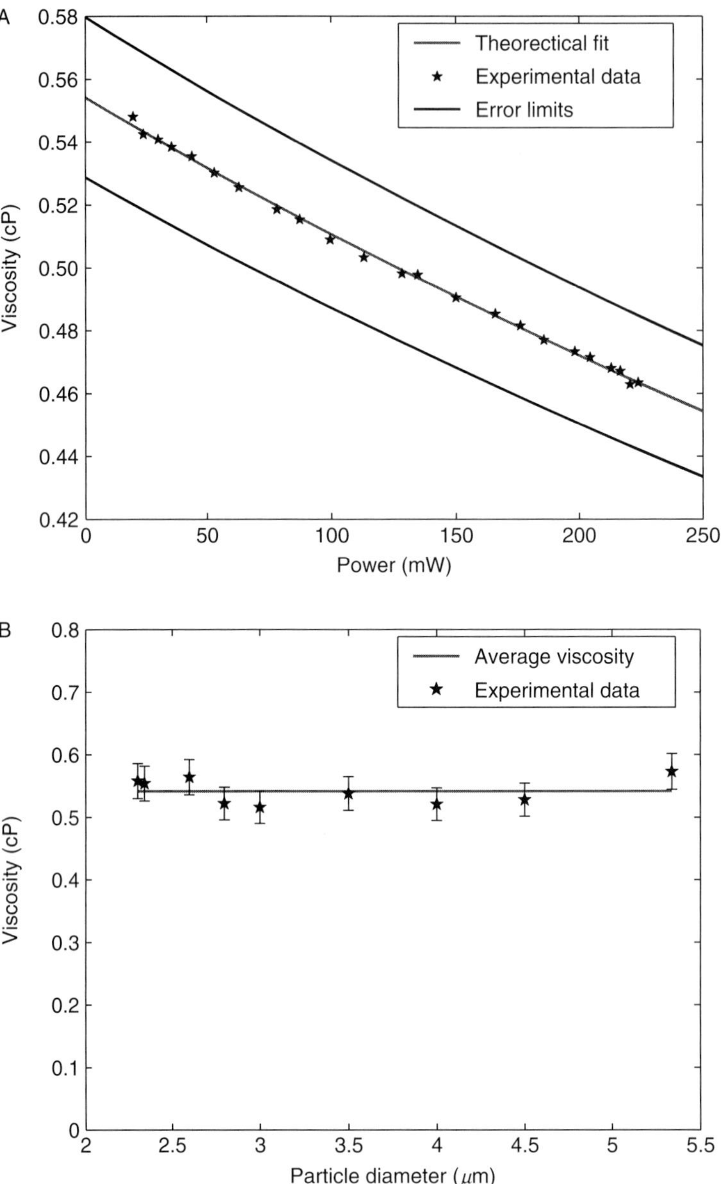

Fig. 17 (A) The viscosity of methanol as a function of trapping laser power. (B) The dependence of the measured viscosity of methanol on the diameter of the vaterite particle used to probe the viscosity.

It is an active technique, so effects such as shear rate dependence can be measured (not possible with particle tracking techniques).

The disadvantages of this technique are as follows:

It requires an accurate measurement of the particle size;

It requires a measurement of the power at the focus;

In complex fluids (non-Newtonian polymer solutions), a depletion layer can form between the probe particle and its surrounding liquid so that the "bulk" viscosity is not measured.

However, some of these difficulties can be overcome. Measurement of the particle diameter is possible by placing the sphere next to a sphere of the same size and measuring the center-to-center distance. This technique allows the diameter to be found within a few percent. The most accurate method to determine the power at the focus is to calibrate it with a liquid with a known viscosity. Then monitoring the laser power outside the microscope is enough to know the power at the focus. The depletion layer that may form is more difficult to account for, but at the same time offers a possibility of studying this phenomenon. Measurements of the flow field around the vaterite sphere, which are described in detail later in the next section, do confirm a depletion layer can exist and also allows the effect to be quantified. However, flow field measurements add unwanted complexity to the system. Another approach would be to change the surface properties of the particle to get better particle–fluid coupling. This is the topic of future research in this field.

B. Flow Field Measurements

Rotation of particles in a liquid environment leads to the generation of fluid flow. The investigation of fluid flow is of great interest for applications of microscopic actuation like micropumps (Ladavac and Grier, 2004) and microstirrers (Lin *et al.*, 2005), where flows with particular direction or flow vortices are required. Apart from visualizing flow direction and speed, measurements of the fluid velocity also elucidate the interaction of the rotating particle with the surrounding fluid. This particle–fluid coupling has far reaching implications for using rotating particles to generate flow or measure physical properties of the system.

To calculate the drag on a particle, a no-slip boundary condition on the particle surface is usually assumed. Yet it has been shown for microscopic and macroscopic systems that this assumption is often not valid. Effects which may occur are surface slip (Bonaccurso *et al.*, 2002), or in polymer solutions a depletion of molecules close to the particle's surface (Chen *et al.*, 2003), as well as shear thinning of the solution at locations of high shear (Krause *et al.*, 2001). Measurement of the flow field can characterize the particle–fluid interaction (Knöner *et al.*, 2005).

Flows in microfluidic devices are often visualized using particle image velocimetry (PIV), where the fluid is seeded with probe particles that are typically fluorescent and of 0.2- to 1-μm diameter (Bown *et al.*, 2005). Particles are tracked by a digital imaging system to obtain fluid velocity values. This technique requires a constant flow of particle suspension through the device and is therefore not applicable to characterize flow created by rotating microscopic objects. Instead, individual probe particles can be trapped and brought to specified positions to probe the flow field.

We have used two methods to probe such flow fields. The passive method is similar to PIV and relies on following the motion of a free particle with the flow (Fig. 18). It was first demonstrated using dual optical tweezers. In this scheme, trap one holds and rotates a vaterite particle that drives the flow, while the second trap brings the probe particle to the desired position, where it is released and follows the fluid flow (Knöner *et al.*, 2005). Fluid velocities are then evaluated by digital microscopy. The technique was extended by using multiple holographic optical tweezers (di Leonardo *et al.*, 2006), yet this is somewhat cumbersome since it requires simultaneous trapping of several probe particles and may furthermore disturb the flow field. The second more active method does not release the probe particle, but holds it in the trap and monitors how much it gets displaced by the drag force applied by the flow field. Motion of the probe particle deflects the beam of a separate detection laser (HeNe) which is registered with a quadrant photodetector (QPD). The signal is directly calibrated against movement with known velocity of a piezo-actuated microscope stage (Fig. 19). Due to the direct calibration, the method does not require knowledge of the fluid viscosity, the trap stiffness, or the probe particle size, and can therefore be used with unknown liquids.

Both methods were used to characterize the flow field in water created by a rotating sphere (Fig. 20). The measured flow field agrees very well with the theoretically expected flow field. The rotating microscopic particle does not experience any slip on its surface, which would result in a drop of measured fluid velocities. These experiments prove the validity of the no-slip boundary condition for microscopic vaterite particles rotating in water.

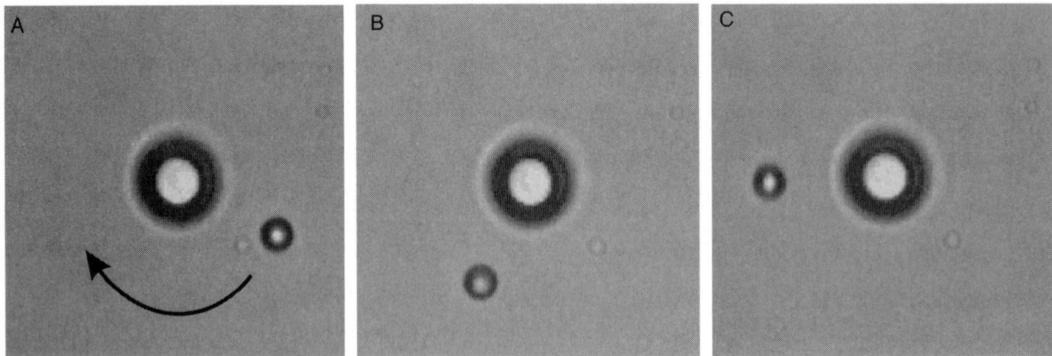

Fig. 18 The 1-μm diameter probe particle and a rotating vaterite sphere are held in dual optical tweezers (A). When the probe particle is released, it follows the streamlines of the flow field generated by the rotating vaterite (B and C).

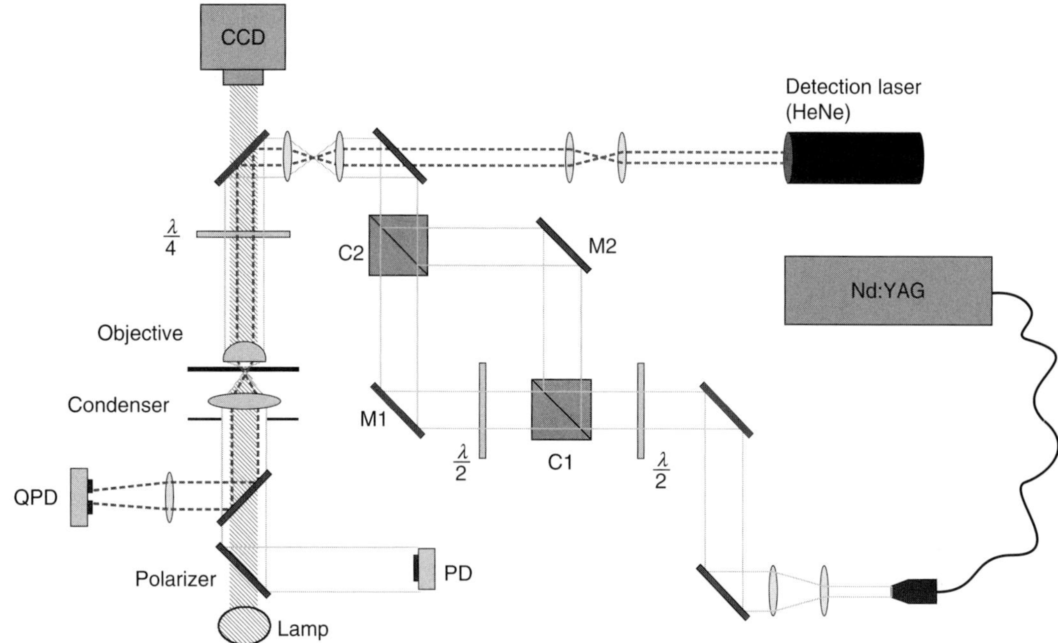

Fig. 19 Dual optical tweezers setup used to measure flow fields generated with optically driven microscopic particles. Two optical traps fully steerable with mirrors M1 and M2 are created by splitting and recombining the trapping beam with beam splitters C1 and C2. A detection laser and a quadrant photodetector (QPD) are used to monitor probe particle displacements.

The use of flow field measurements to quantify non-Newtonian effects in polymer solutions has been demonstrated (Knöner *et al.*, 2005). Flow fields were created in 1.5 g/liter hyaluronic acid (HA) solutions by rotating vaterite spheres. HA is of particular interest since it occurs in the human body in the form of synovial fluid. HA is a linear anionic polysaccharide. The viscoelastic properties of solutions in water strongly depend on the molecular HA concentration (Krause *et al.*, 2001). Vaterite and probe particles were added to a solution of HA (rooster comb, 1.5×10^6 Da average molecular weight) in phosphate buffered saline (PBS).

The velocity profiles obtained from this solution showed a strong deviation from the behavior of Newtonian fluids. Vaterite particles in the size range of 3–4 μm were rotating at 3–5 Hz, considerably slower than in water (~30 Hz). The generated fluid flow around the vaterite was on average 24% lower than expected from the model (Fig. 21). This deviation is significant and much larger than the error in the expected profile (3%). A fit to the points which are more than 7 μm away from the rotating vaterite shows a similar deviation of 25%. At those larger distances, the probe does not disturb the flow, which shows that the deviation is indeed a physical effect and not caused by interference from the probe with the rotating vaterite.

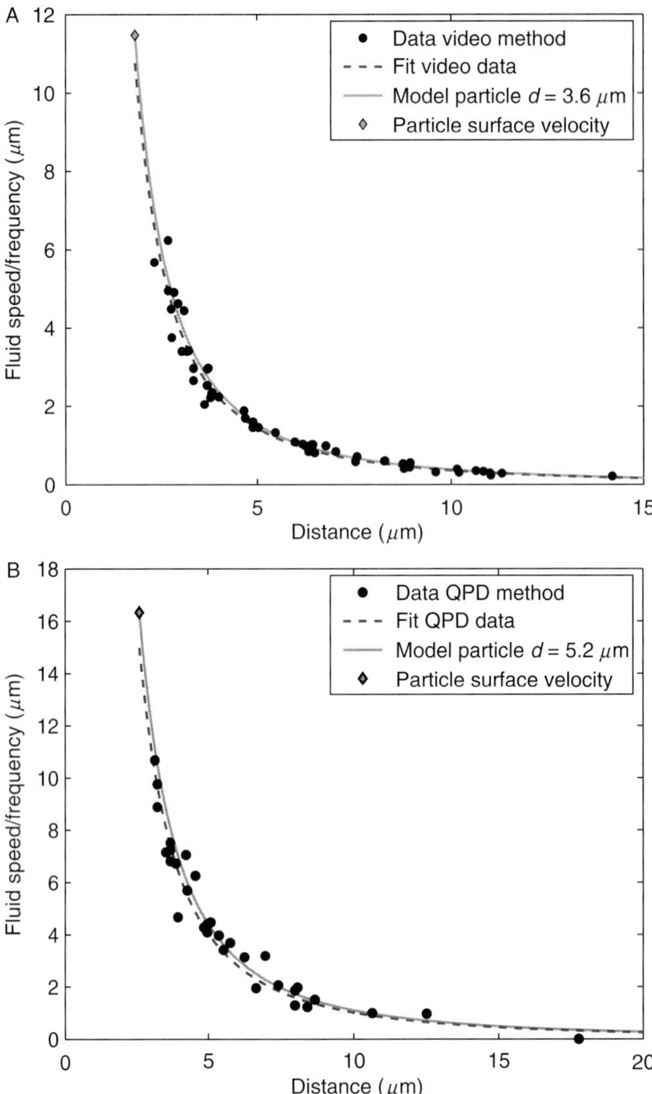

Fig. 20 Fluid velocity profiles in the equatorial plane of a vaterite particle rotating in water. Tracking a freely moving probe with video microscopy (left) and measuring the displacement of a trapped probe with a QPD yield both profiles that agree very well with hydrodynamic theory assuming a no-slip boundary condition.

The reason for the observed deviation is most likely the formation of a polymer depletion layer around the vaterite particle. In such a layer, the concentration of polymer is locally reduced by either electrostatic repulsion between the net charge

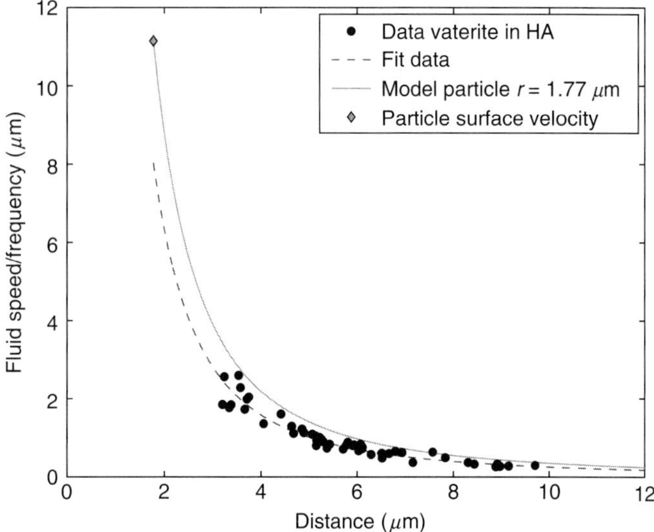

Fig. 21 Fluid velocity profile of a vaterite particle rotating in a solution of 1.5 g/liter HA. The measured profile strongly deviates from the expected profile over the whole range, indicating that the particle–fluid coupling is influenced by the polymer in solution.

on molecule and particle surface or by hard wall interaction, leading to a reduced viscosity in the layer. Shear thinning, which is a reduction in viscosity at places with high shear due to the alignment of polymer molecules with the flow, was not responsible for the deviation since rotation rates and thus generated shear were very low. Flow field measurements could be used to investigate shear thinning by increasing particle rotation rates. The velocity profiles would then show a different slope in regions of high shear, where shear thinning occurs.

Quantification of effects from local inhomogeneities or non-Newtonian behavior is of great importance for applicability of particle-based microrheological techniques, such as the technique described in the previous section. For the case discussed here of depletion layer formation, optical torque measurements were performed to derive the viscosity of the solution. Due to the locally reduced viscosity, vaterite particles rotate faster and the bulk viscosity is thus underestimated. A viscosity of $\eta = 7.9 \pm 0.5$ cP was measured for the 1.5 g/liter HA solution. When the flow field measurements were used to extrapolate the rotation rate at which the particle would rotate without the depletion layer, a viscosity of $\eta = 11 \pm 0.9$ cP was found. This compares very well with the bulk viscosity of $\eta = 12$ cP for the same solution measured with standard rheological techniques (Krause *et al.*, 2001). This illustrates how flow field measurements can be used to characterize biological fluids. These measurements may prove a useful tool for medical diagnosis requiring only tiny samples of body fluids.

C. Progress Toward Pumps

Fabrication of microscopic pumps for the application in microfluidic and lab-on-a-chip devices is the goal of a great research effort. Micropumps are usually classified as displacement or dynamic pumps and employ standard techniques such as moving membranes or rotors (Laser and Santiago, 2004). Yet these conventional micropumps are still relatively large, on the order of 0.1 cm^3 (Zengerle *et al.*, 1995), and not suitable for integration into microfluidic channels that often have a width of only several cell diameters (e.g., 50 μm). Actuation of particles inside these channels with optical tweezers offers new possibilities for pumping of fluids or particle suspensions.

Several geometries have been proposed. Linear actuation of microspheres with arrays of holographic optical tweezers was used as a form of peristaltic pump (Grier and Behrens, 2003; Koss and Grier, 2003). This scheme at present only transports large number of particles and does not allow an effective motion of liquid. An improvement comes from using holographic optical tweezers to create an array of orbital AM-carrying traps (Ladavac and Grier, 2004). Ring-shaped traps are created from high order doughnut modes. Microspheres can be 2D trapped in the ring and propagate around the ring center in the direction given by the charge of the doughnut mode (Fig. 22). Two lines of such traps with opposite charges and thus counter-propagating trapped particles create an effective flow. Although flow is created, several problems with this technique remain. The ring rotation rate of 1.7 Hz and the flow speed of 6 μm/s at 3 W laser power are too low for useful applications. Also, there is no control over the particles that create the flow, they seem to be randomly washed out of the trapping rings and could contaminate samples.

In a promising new approach, we used counter-rotating vaterite particles to create flow. Vaterite particles have the advantage that a single particle can be

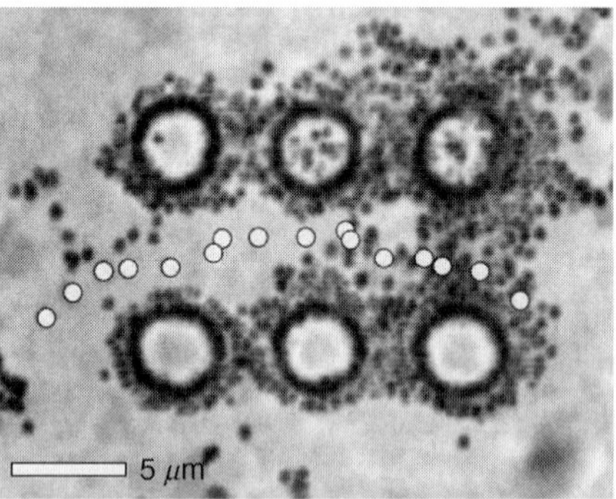

Fig. 22 Two lines of optical traps created with holographic optical tweezers using doughnut modes with opposite charges. High-order doughnut modes carry orbital AM and create ring-shaped traps. Silica particles of 800 nm were 2D trapped and circled the ring centers to create an effective flow. (From Ladavac and Grier, 2004.)

trapped and rotated at high speed without the need for optical vortices or multiple particle arrangements as described above. Furthermore, the flow field measurements discussed in the previous section allow precise modeling of generated flow fields.

Our first experiments to produce a micropump were based on dual beam optical tweezers with opposing circular polarization in each beam to trap two vaterite particles and rotate them in opposite directions (Fig. 23). A fast effective flow is generated between the particles (Fig. 24). At a rotation rate of only 100 Hz, fluid velocities of 200 μm/s are achieved in between the particles. In experiments, vaterite particles of similar size were rotated with frequencies of up to 400 Hz.

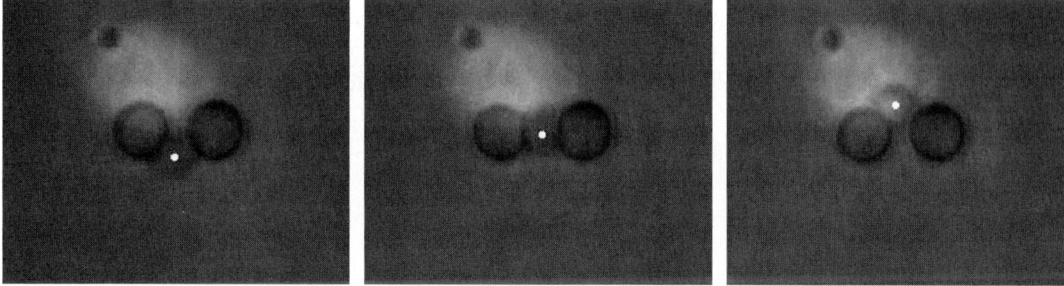

Fig. 23 Two vaterite particles are held in dual beam optical tweezers and counter-rotate, thereby moving a third particle through the space in between them. The ability to rotate these particles at high frequencies allows the generation of high fluid velocities in the space between the particles. (See Plate 26 in the color insert section.)

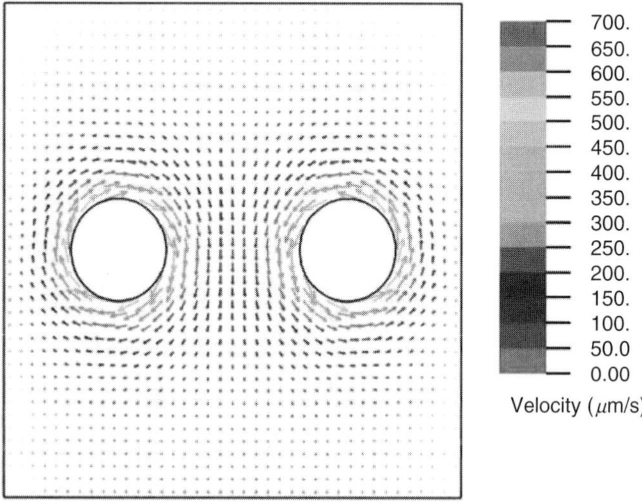

Fig. 24 Fluid flow generated by two spheres rotating in a square microchannel. Rotation rate is 100 Hz, particle diameter 2 μm, center–center distance 4.5 μm. Velocities in the center are 200 μm/s.

This technique was then implemented into a microchannel. The microchannel was produced with two-photon photopolymerization in a rather simple procedure, namely by polymerizing cubes next to one another to build up the microchannel. A vaterite particle was trapped in a simple optical trap, placed in a side pocket and rotated, creating a flow through the channel (Fig. 25). The flow was observed with 1-μm sized polystyrene probe particles.

To effectively move fluids through small microfluidic channels, we propose a scheme where several particles are trapped simultaneously and counter-rotate (Fig. 26). Linear trapping arrays as shown here can be created by using either diffractive elements or acousto-optic modulators, and can then be doubled into two lines with opposite circular polarization by using two beam-splitting cubes as demonstrated in the previous section. Figure 26 shows the computed flow field for particles with a diameter of $d = 2\ \mu$m. Vaterite particles can be grown up to $d = 10\ \mu$m, meaning the scheme can easily be scaled up. Larger particles also create higher fluid velocities. The side chambers in the channel prevent fluid backflow (Fig. 26, right). A similar scheme is possible with straight channels, but the backflow is higher. For the given parameters, a fluid velocity of 135 μm/s is reached in the channel center.

Further improvement in the production of micropumps is possible by using objects other than vaterite particles to create the flow. One very promising route is the fabrication of custom-tailored microparticles, with properties to maximize momentum transfer from the light and flow velocities created. These particles can be assembled using two-photon photopolymerization which will be described in the following two sections.

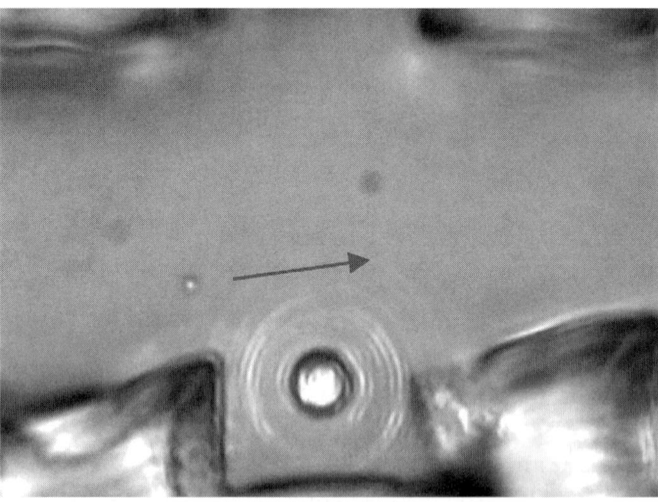

Fig. 25 A vaterite sphere, with a diameter of 5 μm, rotating in a side pocket of a photopolymerized microchannel. A flow is created along the channel and visualized with 1-μm polystyrene probe particles.

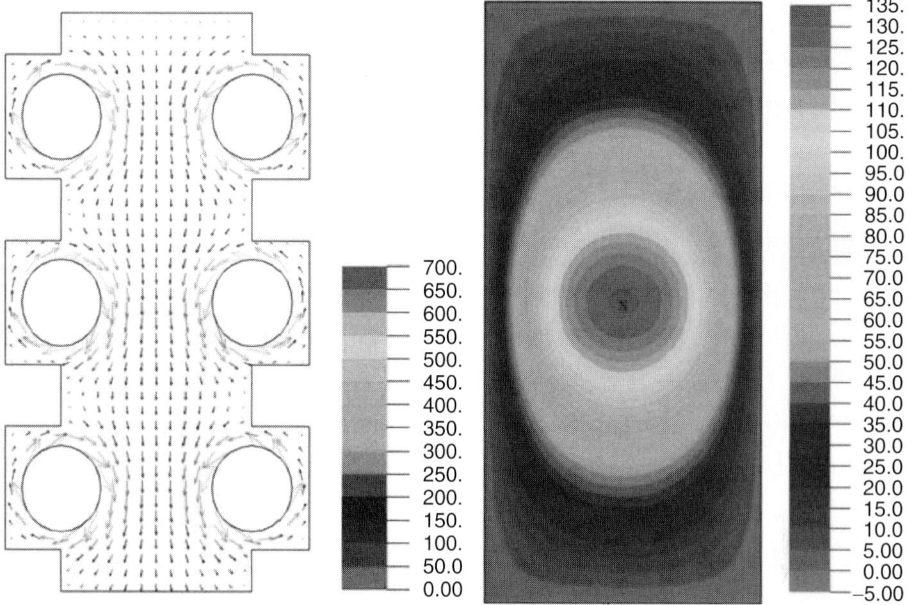

Fig. 26 Proposed design for an effective microfluidic pump. Left: two lines of optical traps are counter-rotating vaterite spheres at a frequency of 100 Hz. Sphere diameter is 2 μm, channel width and height 4.5 and 10 μm, respectively. Right: component of the fluid velocity parallel to the channel, at the output of the channel. Size of the cross-section is 4.5×10 μm. A center velocity of 135 μm/s is achieved, with hardly any backflow. Setting the spheres in a straight microchannel creates higher backflow.

D. Photopolymerization Process for the Production of Micromachines

Objects and processes available to date to apply forces, torques, and flow are limited by the microscopic objects available for micromanipulation. Perhaps the most versatile particles are the previously mentioned vaterite spheres, yet they are prone to degradation due to the long-term unstable crystalline form of vaterite and their noncompatibility with acidic environments. The surface chemistry is not suitable for protein coating, and the maximum torque that can be optically transferred with spin AM is $2\hbar$ per photon. Thus, we are pursuing the development of new complex microscopic tools that do not suffer from these shortcomings and greatly extend the applicability of optical micromanipulation.

Objects with a length scale below 10 μm and complex 3D structure are not easily fabricated. Commercial microscopic mechanical elements are mainly manufactured using lithographic processes, in which a multitude of layers are produced, cured and etched (MEMS). These conventional photolithographic methods are not suitable here as the production effort is dramatically increased for objects with 3D structure, and the resolution is limited. A requirement is that produced structures have to be transparent, so that they can be optically trapped and do not suffer heating by absorption.

A process which fulfills all requirements for fabrication of microscopic structures for optical tweezers is two-photon photopolymerization. The method is suitable to produce complex 3D structures with submicrometer resolution and sizes ranging from 2 to 20 μm. Resin of high optical quality is cured with a two-photon polymerization process induced by femtosecond laser pulses. The process was first used for optical data storage (Strickler and Webb, 1991) and for polymerization of complex 3D structures (Maruo *et al.*, 1997). As mentioned in the introduction to this chapter, Galajda and Ormos (2001) refined the process and were the first to produce microscopic "machines" which could be trapped and rotated in optical tweezers (Galajda and Ormos, 2001). Since then, the technique has been applied to development of a range of microscopic structures. Microscopic elements that are being produced include microoscillator systems (Kawata *et al.*, 2001), mechanical microtweezers, and microsyringes actuated by light (Maruo *et al.*, 2003b), micromechanical cogs on stalks driven by light (Galajda and Ormos, 2002a,b; Maruo *et al.*, 2003a), complex rotors (Galajda and Ormos, 2002a), flat crosses that can be oriented in linearly polarized light (Galajda and Ormos, 2003) and, most recently, integrated optical motors, where a waveguide for the actuating light as well as a suspended microcog are polymerized (Kelemen *et al.*, 2006). Progress in fabrication and handling of those structures promises a range of future biophysical applications.

We fabricate structures for optical rotation by two-photon photopolymerization. UV curing resin NOA63 (Norland Products Inc., Cranbury, NJ) is sandwiched between two coverslips in a sample chamber that is mounted on a piezo-actuated microscope stage. The femtosecond laser, which produces 80-fs pulses of 780-nm wavelength light with a repetition rate of 80 MHz, is focused into the chamber by an NA = 1.3 objective lens. In the focal region, light intensities are extremely high due to the short pulse length and the tight focus (Fig. 27). The resin usually polymerizes only when exposed to light with $\lambda < 400$ nm, but the high irradiance allows two-photon processes, which corresponds to excitation with $\lambda = 780$ nm/2 = 390 nm. Thus, the resin cures. Since the two-photon process depends on the square of the intensity, it occurs only in the focus. Other regions are not cured, resulting in very high specificity and resolution of the process.

To obtain a 3D structure, a template is created—usually a stack of 2D images. The laser focus is stationary and the sample chamber is raster-scanned over a defined grid with the piezo-actuated stage. A shutter controls the exposure of the resin at the specified positions. Since the whole volume has to be raster-scanned, it can be a time-consuming process for big structures. After the photopolymerization process, the unused resin is washed out with acetone, which can then be flushed away with water.

A quantitative study of the resolution of the photopolymerization process was performed with scanning electron microscopy (SEM). Lines with a width of 1–5 pixels and a height of three layers were polymerized (Fig. 28, left). We can verify from this figure the resolution of the system.

The kangaroo was fabricated using an NA = 1.3 oil immersion objective. The results obtained showed that the qualitative comparison between structures fabricated with an NA = 0.65 air objective and an NA = 1.3 oil immersion objective showed that the higher NA yields finer structures. This is expected since the

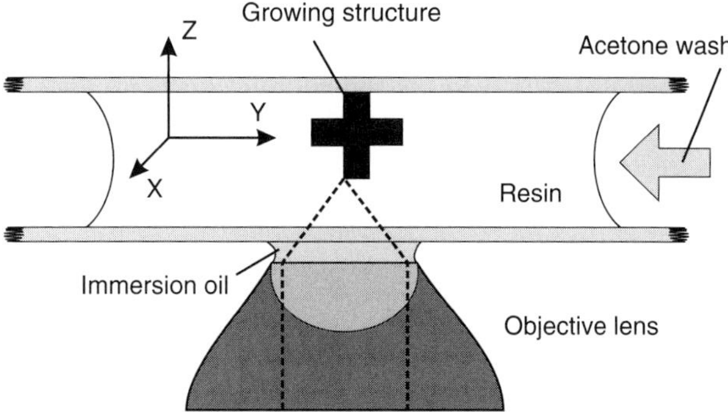

Fig. 27 Scheme for two-photon photopolymerization. The microstructure is grown by exposing the photocurable resin in the sample chamber with the beam from a femtosecond laser, which is focused into the chamber by the high NA objective. The chamber is raster-scanned in 3D to create complex structures.

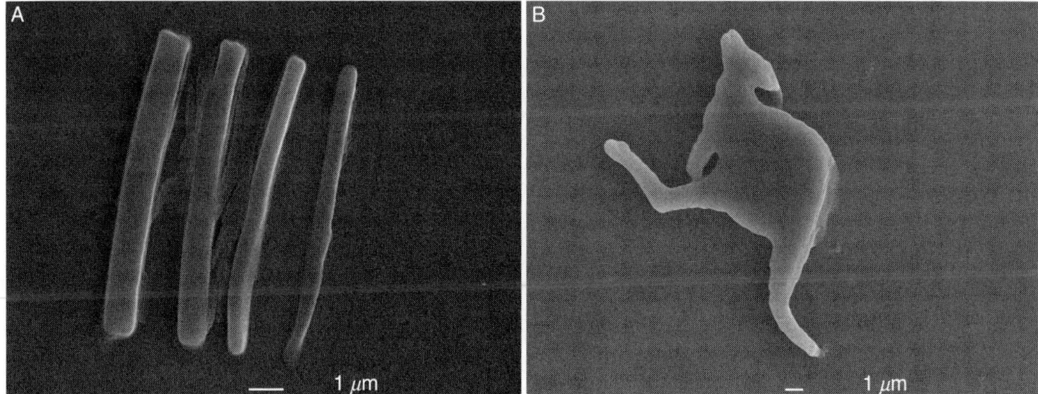

Fig. 28 Photopolymerized grid (left) and kangaroo (right). Single pixels from the raster-scan process with a width of 180 nm are visible (right, tail), demonstrating the high resolution achievable with a two-photon process.

transverse width of the laser intensity distribution in the focal spot is much smaller at higher NA. It also underlines the importance of overfilling the objective back aperture with the laser beam in order to obtain a diffraction-limited spot and high spatial resolution.

E. Micromachine Elements for Efficient Torque Transfer

1. Rotating 3D Cross: High Torque Rotor

There is a need for objects which can be optically 3D trapped and allow an effective transfer of orbital AM. This can be achieved in two ways: either by using an AM-carrying beam and trapping a nonspherical object or by using a Gaussian beam and trapping an object that has a certain degree of helicity. Although the first method will yield higher torques, the technique will be demonstrated by means of the second, which could be more convenient. The object of choice has to fulfill a number of requirements. It needs to have the correct orientation when optically trapped. In this orientation, it needs to present a chiral structure to the trapping beam so that an exchange of orbital AM can take place. One of the simplest, but still very effective, structures to perform this task is a cross that has sideways displaced arms to create chirality and a stalk so that proper alignment in the trap is guaranteed. Such a structure was designed and fabricated using two-photon photopolymerization (Fig. 29).

Following fabrication, the cross was trapped in a linearly polarized Gaussian beam. It began immediately to rotate at a frequency of 14 Hz with 200 mW of laser power. This is substantially faster than what is achieved with form-birefringent structures in circularly polarized light (1 Hz with 1 W of power; Neale *et al.*, 2005) and is even comparable to rotation of birefringent vaterite. Note that the diameter of the cross is 6μm, meaning that the drag is substantially larger than on a 2-μm diameter vaterite particle. Furthermore, one has to remember that this rotation is achieved with a linearly polarized Gaussian beam involving no conversion losses. If additional AM is brought into the system by using LG modes, the rotation rate and applied torque can be substantially increased.

The cross was rotated in biologically compatible medium. Thus, it may be useful for studies on living cells. Applications may include the generation of

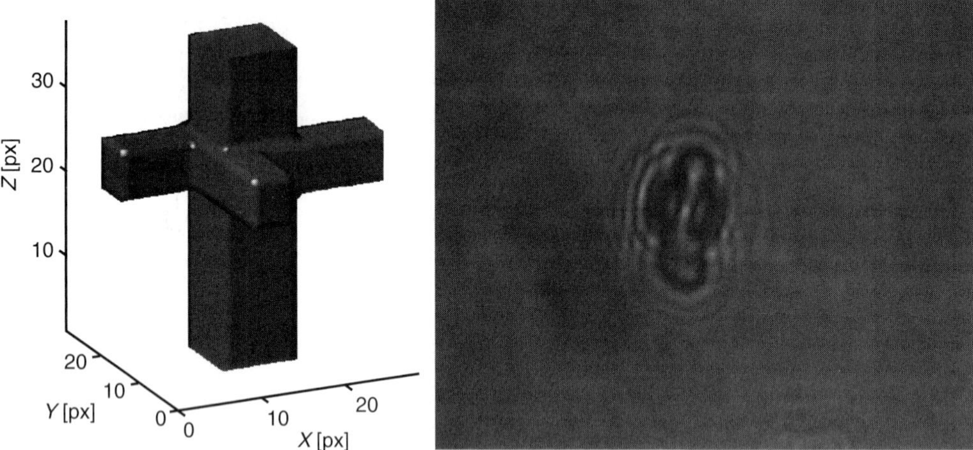

Fig. 29 Left: model for the photopolymerization of a cross. The cross has a stalk for axial alignment in the optical trap, its arms are offset to one side so that it can generate orbital AM from a Gaussian beam, whereby it rotates due to momentum conservation. Right: the actual fabricated cross, lying on its side. The arms are actually thicker than in the model since the focal region in which polymerization occurs is elongated vertically.

controlled high-shear flows, membrane deformation studies, and use as a microdrill. It may also be a useful tool for mixing fluids in a microfluidic device, which is otherwise very difficult due to the laminar nature of microscopic flows. Since high torques can be created with this microrotor, it would also be a candidate for applications in microchannel pumping as described in the above section.

VI. Conclusions

In this chapter, we have discussed how the transfer of orbital AM can be understood and applied. We have discussed both spin and orbital AM transfer and the optical torque created. We have described methods to perform quantitative measurements of these phenomena and discussed some applications that have been made already. These techniques are compatible with the development of the concept of lab-on-a-chip for studies of rotational dynamics on a microscale. They can also be used in the construction of all optically driven micromachines with very versatile geometry.

References

Ashkin, A., Dziedzic, J. M., Bjorkholm, J. E., and Chu, S. (1986). Observation of single beam gradient force optical trap for dielectric particles. *Opt. Lett.* **11,** 288–290.

Ashkin, A., Dziedzic, J. M., and Yamane, T. (1987). Optical trapping and manipulation of single cells using infrared laser beams. *Nature* **330,** 769–771.

Bayoudh, S., Nieminen, T. A., Heckenberg, N. R., and Rubinsztein-Dunlop, H. (2003). Orientation of biological cells using plane-polarized Gaussian beam optical tweezers. *J. Mod. Opt.* **50**(10), 1581–1590.

Beth, R. A. (1936). Mechanical detection and measurement of the angular momentum of light. *Phys. Rev.* **50,** 115–125.

Bingelyte, V., Leach, J., Courtial, J., and Padgett, M. J. (2003). Optically controlled three-dimensional rotation of microscopic objects. *Appl. Phys. Lett.* **82**(5), 829–831.

Bishop, A. I., Nieminen, T. A., Heckenberg, N. R., and Rubinsztein-Dunlop, H. (2004). Optical microrheology using rotating laser-trapped particles. *Phys. Rev. Lett.* **92,** 198104.

Bonaccurso, E., Kappl, M., and Butt, H.-J. (2002). Hydrodynamic force measurements: Boundary slip of water on hydrophilic surfaces and electrokinetic effects. *Phys. Rev. Lett.* **88,** 076103.

Botvinick, E. L., Knöner, G., Rubinsztein-Dunlop, H., and Berns, M. W. (2006). Ultra-localized flow fields applied to the cell surface. *Proc. SPIE* **63,** 26–30.

Bown, M. R., MacInnes, J. M., and Allen, R. W. K. (2005). Micro-PIV simulation and measurement in complex microchannel geometries. *Meas. Sci. Technol.* **16,** 619–626.

Chen, D., Weeks, E. R., Crocker, J. C., Islam, M. F., Verma, R., Gruber, J., Levine, A. J., Lubensky, T. C., and Yodh, A. (2003). Rheological microscopy: Local mechanical properties from microrheology. *Phys. Rev. Lett.* **90,** 108301.

di Leonardo, R., Leach, J., Mushfique, H., Cooper, J. M., Ruocco, G., and Padgett, M. J. (2006). Multipoint holographic optical velocimetry in microfluidic systems. *Phys. Rev. Lett.* **96,** 134502.

Fällman, E., and Axner, O. (1997). Design for fully steerable dual-trap optical tweezers. *Appl. Opt.* **36**(10), 2107–2113.

Friese, M. E. J., Enger, J., Rubinsztein-Dunlop, H., and Heckenberg, N. R. (1996). Optical angular-momentum transfer to trapped absorbing particles. *Phys. Rev. A* **54,** 1593–1596.

Friese, M. E. J., Nieminen, T. A., Heckenberg, N. R., and Rubinsztein-Dunlop, H. (1998). Optical alignment and spinning of laser-trapped microscopic particles. *Nature* **394,** 348–350. Erratum *Nature* **395**, 621.

Gahagan, K. T., and Swartzlander, G. A. (1996). Optical vortex trapping of particles. *Opt. Lett.* **21,** 827–829.

Galajda, P., and Ormos, P. (2001). Complex micromachines produced and driven by light. *Appl. Phys. Lett.* **78,** 249–251.

Galajda, P., and Ormos, P. (2002a). Rotors produced and driven in laser tweezers with reversed direction of rotation. *Appl. Phys. Lett.* **80,** 4653–4655.

Galajda, P., and Ormos, P. (2002b). Rotation of microscopic propellers in laser tweezers. *J. Opt. B: Quantum Semiclass. Opt.* **4,** 78–81.

Galajda, P., and Ormos, P. (2003). Orientation of flat particles in optical tweezers by linearly polarized light. *Opt. Express* **11,** 446–451.

Gibson, G., Courtial, J., and Padgett, M. J. (2004). Free-space information transfer using light beams carrying orbital angular momentum. *Opt. Express* **12,** 5448–5456.

Grier, D. G., and Behrens, S. H. (2003). Optical peristaltic pumping with optical traps US Patent No. 6,639,208 B2.

Hecht, E. (1987). "Optics." Fourth Edition, Addison-Wesley.

Heckenberg, N. R., McDuff, R., Smith, C. P., and White, A. G. (1992). Generation of optical phase singularities by computer-generated holograms. *Opt. Lett.* **17,** 221–223.

Kawata, S., Sun, H.-B., Tanaka, T., and Takada, K. (2001). Finer features for functional microdevices. *Nature* **412,** 697–698.

Kelemen, L., Valkai, S., and Ormos, P. (2006). Integrated optical motor. *Appl. Opt.* **45**(12), 2777–2780.

Knöner, G., Parkin, S., Heckenberg, N. R., and Rubinsztein-Dunlop, H. (2005). Characterization of optically driven fluid stress fields with optical tweezers. *Phys. Rev. E* **72,** 031507.

Koss, B. A., and Grier, D. G. (2003). Optical peristalsis. *Appl. Phys. Lett.* **82**(22), 3985–3987.

Krause, W. E., Bellomo, E. G., and Colby, R. H. (2001). Rheology of sodium hyaluronate under physiological conditions. *Biomacromolecules* **2,** 65–69.

Ladavac, K., and Grier, D. G. (2004). Microoptomechanical pumps assembled and driven by holographic optical vortex arrays. *Opt. Express* **12**(6), 1144–1149.

La Porta, A., and Wang, M. D. (2004). Optical torque wrench: Angular trapping, rotation, and torque detection of quartz microparticles. *Phys. Rev. Lett.* **92,** 190801.

Laser, D. J., and Santiago, J. G. (2004). A review of micropumps. *J. Micromech. Microeng.* **14,** R35–R64.

Lin, C.-H., Tsai, C.-H., and Fu, L.-M. (2005). A rapid three-dimensional vortex micromixer utilizing self-rotation effects under low Reynolds number conditions. *J. Micromech. Microeng.* **15,** 935–943.

MacDonald, M. P., Volke-Sepulveda, K., Paterson, L., Arlt, J., Sibbett, W., and Dholakia, K. (2002). Revolving interference patterns for the rotation of optically trapped particles. *Opt. Commun.* **201,** 21–28.

Mair, S., Vaziri, A., Weihs, G., and Zeilinger, A. (2001). Entanglement of the orbital angular momentum states of photons. *Nature* **412,** 313–316.

Maruo, S., Ikuta, K., and Korogi, H. (2003a). Force-controllable, optically driven micromachines fabricated by single-step two-photon microstereolithography*J. Microelectromech. Sys.* **12,** 533–539.

Maruo, S., Ikuta, K., and Korogi, H. (2003b). Submicron manipulation tools driven by light in a liquid. *Appl. Phys. Lett.* **82,** 133–135.

Maruo, S., Nakamura, O., and Kawata, S. (1997). Three-dimensional microfabrication with two-photon-absorbed photopolymerization. *Opt. Lett.* **22,** 132–134.

Neale, S. L., MacDonald, M. P., Dholakia, K., and Krauss, T. F. (2005). All-optical control of microfluidic components using from birefringence. *Nat. Mater.* **4,** 530–533.

Nieminen, T. A., Heckenberg, N. R., and Rubinsztein-Dunlop, H. (2001). Optical measurement of microscopic torques. *J. Mod. Opt.* **48,** 405–413.

Nieminen, T. A., Heckenberg, N. R., and Rubinsztein-Dunlop, H. (2004). Computational Modelling of optical tweezers. *Proc. SPIE.* **5514,** 514–523.

Nieminen, T. A., Higuet, J., Knoner, G., Loke, V., Parkin, S., Singer, W., Heckenberg, N. R., and Rubinsztein-Dunlop, H. (2006). Optically driven micromachines: Progress and prospects. *Proc. SPIE* **6038,** 237–245.

O'Neil, A. T., and Padgett, M. J. (2002). Rotational control within optical tweezers by use of a rotating aperture. *Opt. Lett.* **27**(9), 743–745.

Parkin, S., Knöner, G., Nieminen, T. A., Heckenberg, N. R., and Rubinsztein-Dunlop, H. (2006). Measurement of the total optical angular momentum transfer in optical tweezers. *Opt. Express* **14**(5), 6963–6970.

Parkin, S. J., Nieminen, T. A., Heckenberg, N. R., and Rubinsztein-Dunlop, H. (2004). Optical measurement of torque exerted on an elongated object by a noncircular laser beam. *Phys. Rev. A* **70,** 023816.

Paterson, L., MacDonald, M. P., Arlt, J., Sibbett, W., Bryant, P. E., and Dholakia, K. (2001). Controlled rotation of optically trapped microscopic particles. *Science* **292,** 912–914.

Poynting, J. H. (1909). The wave motion of a revolving shaft, and a suggestion as to the angular momentum in a beam of circularly polarized light. *Proc. Roy. Soc. London Ser. A* **82,** 560–567.

Reicherter, M., Haist, T., Wagemann, E. U., and Tiziani, H. J. (1999). Optical particle trapping with computer-generated holograms written on a liquid-crystal display. *Opt. Lett.* **24**(9), 608–610.

Singer, W., Nieminen, T. A., Gibson, U. J., Heckenberg, N. R., and Rubinsztein-Dunlop, H. (2006). Orientation of optically trapped nonspherical birefringent particles. *Phys. Rev. E* **73,** 021011.

Strickler, J. H., and Webb, W. W. (1991). Three-dimensional optical data storage in refractive media by two-photon point excitation. *Opt. Lett.* **16**(22), 1780–1782.

Visscher, K., Brakenhoff, G. J., and Krol, J. J. (2005). Micromanipulation by multiple optical traps created by a single fast scanning trap integrated with the bilateral confocal scanning laser microscope. *Cytometry* **14**(2), 105–114.

Waigh, T. A. (2005). Microrheology of complex fluids. *Rep. Prog. Phys.* **68,** 685–742.

Zengerle, R., Ulrich, J., Kluge, S., Richter, M., and Richter, A. (1995). A bidirectional silicon micropump. *Sens. Actuators A* **50,** 81–86.

CHAPTER 20

Optical Micromanipulation Methods for Controlled Rotation, Transportation, and Microinjection of Biological Objects

S. K. Mohanty and P. K. Gupta

Laser Biomedical Applications and Instrumentation Division
Raja Ramanna Centre for Advanced Technology, Indore 452013, India

The use of laser microtools for rotation and controlled transport of microscopic biological objects and for microinjection of exogenous material in cells is discussed. We first provide a brief overview of the laser tweezers-based methods for

0091-679X/07 $35.00
DOI: 10.1016/S0091-679X(06)82020-0

rotation or orientation of microscopic objects. Particular emphasis is placed on the methods that are more suitable for the manipulation of biological objects, and the use of these for two-dimensional (2D) and 3D rotations/orientations of intracellular objects is discussed. We also discuss how a change in the shape of a red blood cell (RBC) suspended in hypertonic buffer leads to its rotation when it is optically tweezed. The potential use of this approach for the diagnosis of malaria is also illustrated. The use of a line tweezers having an asymmetric intensity distribution about the center of its major axis for simultaneous transport of microscopic objects, and the successful use of this approach for induction, enhancement, and guidance of neuronal growth cones is presented next. Finally, we describe laser microbeam-assisted microinjection of impermeable drugs into cells and also briefly discuss possible adverse effects of the laser trap or microbeams on cells.

I. Introduction

The use of lasers for precise manipulation and processing of microscopic objects, such as a single living cell or even objects within a single cell, without compromising the viability of the cell, is finding widespread applications in biological research and technology (Berns *et al.*, 1998; Greulich, 1999). The two important tools for laser-assisted micromanipulation of microscopic objects are the laser optical trap and the laser microbeam. Optical tweezers or laser optical trap (Ashkin and Dziedzic, 1987, 1989; Ashkin *et al.*, 1986, 1987) uses the light of a CW near infrared (NIR) laser for trapping of microscopic objects. A microscopic object illuminated by a laser beam is subjected to three forces. While two of these, scattering and absorption, forces act along the direction of the laser beam, the gradient force, proportional to the gradient of light intensity (as exists in a Gaussian profiled beam), pulls the microscopic objects to the beam axis. Further, in a tightly focused laser beam, the axial gradient force pushes the particle toward the focus, and when this force overcomes the scattering force of the laser beam, the particle is trapped in three dimensions. Following the discovery of optical tweezers in 1986, photonic forces have been used widely to trap and manipulate nonliving as well as living objects, including atoms, molecules, organelles, and cells (Berns *et al.*, 1989; Bustamante *et al.*, 2000; Chu, 1991; Greulich, 1999; Kuo and Sheetz, 1992; Svoboda and Block, 1994). Unlike mechanical microtools, the optical trap is gentle and absolutely sterile and can be used to capture, move, and position single cells or subcellular particles without direct contact or significant damage. Optical tweezers have therefore emerged as an invaluable tool in cell biological and molecular research for a range of applications that includes measurement of forces exerted by molecular motors (Kuo and Sheetz, 1993; Svoboda *et al.*, 1993) or the swimming forces of sperm (Konig *et al.*, 1996; Tadir *et al.*, 1989), studies on polymeric properties of single DNA molecules (Smith *et al.*, 1996), microtubules (Felgner *et al.*, 1996), collagen (Luo and An, 1998), and cells (Bronkhorst *et al.*, 1995; Henon *et al.*, 1999), and mapping protein–DNA interaction (Koch *et al.*, 2002), and so on.

More recently, the use of optical micromanipulation techniques for rotation, transportation, and microinjection of biological materials has received particular attention. The ability of optical tweezers-based methods to orient/rotate microscopic objects can be of immense use in biological research as well as in microfluidics and nanotechnology. Since most of the biological cells or organelles have heterogeneous distribution of receptor molecules over their surfaces, controlled orientation of these will help study interaction between specific regions of these biological objects with other molecules/cells or even an enzyme with a substrate. Besides rotation, development of techniques for controlled transport of microscopic objects is also of considerable current interest for understanding various biophysical processes and for sorting applications. Indeed results suggest that optical tweezers-based transportation methods can be used for biasing intracellular protein movements and thereby control and induce growth of neuronal cells (Mohanty *et al.*, 2005a).

In contrast to CW optical tweezers, the large intensities generated by pulsed laser microbeams at the focal point of the large numerical aperture (NA) microscope objective can be used to cut, perforate, or fuse microscopic objects with submicrometer accuracy (Berns *et al.*, 1981). Laser beams over broad spectral region (UV to NIR) and pulse duration of few nanoseconds down to femtoseconds have been used for microinjection of exogenous materials into single cells. Laser microbeams in combination with optical tweezers provide a powerful tool for creative manipulation and alteration of cells so as to unravel structure–function relationship of several biological objects (Berns, 1998; Berns *et al.*, 1998; Greulich, 1999; Greulich *et al.*, 2000) and has been used for a number of studies in cell and developmental biology such as fertilization research in animal systems (Tadir *et al.*, 1993) and study of chromosome movement (Liang *et al.*, 1993), and so on.

In this chapter, we provide an overview of the use of laser microtools for rotation and controlled transport of microscopic biological objects with emphasis on the work carried out by us in this area. We will also describe laser microbeam-assisted methods for injection of impermeable drugs in cells and discuss possible adverse effects of laser trap or microbeams on cells.

II. Methods for Optically Controlled Rotation of Microscopic Objects

A. Methods for 2D Rotation of Biological Objects

Objects trapped in optical tweezers can be made to rotate by transfer of angular momentum (orbital and/or spin) from the trapping beam. Further, specially shaped objects may be made to rotate like a windmill by conversion of the linear momentum of the trap beam to angular momentum on the object (Galajda and Ormos, 2002). A transfer of angular momentum of the trap beam to the object requires that either the object absorbs a fraction of the trap beam carrying orbital or spin angular momentum (He *et al.*, 1995), or due to its birefringence it changes

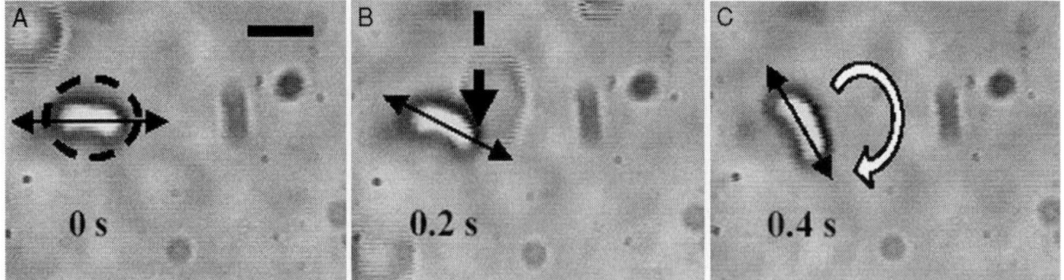

Fig. 1 Time-lapse video images of the orientation of an RBC in a linearly polarized optical tweezers. The RBC was suspended in a buffer having osmolarity of 300 mOsm/kg. (A) The trapped RBC (encircled) tilted in a vertical plane and oriented along the polarization direction of a linearly polarized optical tweezers (~80 mW). (B, C) Orientation of RBC along the direction of viscous flow when subjected to fluid flow in perpendicular direction (shown by dotted arrow) by translation of the stage. All images are in same magnification. Scale bar: 5 μm. (From Mohanty *et al.*, 2006b.)

the ellipticity of the polarization of the elliptically polarized trap beam (Friese *et al.*, 1998). Certain biological objects, such as chloroplasts and discotic RBC, have shape birefringence (Neale *et al.*, 2005) and thus get oriented in plane-polarized optical tweezers. Therefore, rotation of the plane of polarization of the trapping beam, effected by the use of a rotating half-wave plate in the beam path, can be used for rotation of these objects. Figure 1 shows alignment of a discotic RBC along the plane of polarization of the trapping beam (the direction of polarization of the laser beam is shown as double-sided arrow in panel A) and its reorientation by fluid flow (B, C). Because for manipulation of biological objects, absorption of light is not desirable and the objects may not be birefringent (optical or shape), there have been attempts to develop techniques that are not based on these parameters of the object. One interesting approach developed (Paterson *et al.*, 2001) was to generate a rotating interference pattern by varying the path difference between the two interfering Laguerre–Gaussian (LG) and Gaussian beams and thus rotate biological objects trapped in the high-intensity wings of the rotating pattern. However, this approach still has some drawbacks—poor utilization of the trap laser power due to the loss in the generation of the required interference pattern and high susceptibility of the interference pattern to vibrations. Further, it would also be difficult to apply this approach to rotate objects embedded in a turbid medium (e.g., objects inside plant cells) because scattering will degrade the contrast of the interference pattern. We therefore investigated (Dasgupta *et al.*, 2003) the use of an elliptically profiled trapping beam for rotation of biological objects. This approach exploits the fact that in such a trap, an object lacking spherical symmetry orients itself along the major axis of the trap. Therefore, by rotating the elliptical trapping beam, the trapped particle can be rotated around the axis of the laser beam.

For generation of the elliptical trapping beam, Sato *et al.* (1991) proposed the use of a rotating rectangular aperture inside the laser resonator. In addition to the obvious complication of the technique, this approach results in an inefficient and unstable laser operation. Further, the maximum speed of rotation of RBC demonstrated with this approach was 0.1 Hz. A rotating rectangular aperture external to the laser cavity has been also used (O'Neil and Padgett, 2002) to obtain a rotating asymmetric trapping beam pattern. However, this technique also leads to a large loss of the available laser power. For example, the power loss is more than 50% for an aspect ratio of 1:2 and will increase with an increase in the aspect ratio. Further, with both of these approaches, it is difficult to effect a change in elliptical beam profile that is required to trap objects of varying sizes. We have shown that efficient elliptical tweezers can be generated by the use of a cylindrical lens. Rotation of the cylindrical lens leads to a rotation of the elliptical trapping beam and hence of the trapped object (Dasgupta *et al.*, 2003; Yu *et al.*, 2004). This approach not only resulted in better utilization of trapping beam power but also allowed much easier control of the trap beam profile.

A schematic of the experimental arrangement for the generation of elliptical tweezers is shown in Fig. 2A. Varying the focal length of the rotating cylindrical lens (RCL) can change the dimensions of the elliptical beam profile at the object plane. The major axis of elliptical tweezers has to be larger or at least comparable to that of the size of the object(s) to be trapped. However, if very long elliptical tweezers are used for trapping of smaller number of object(s), the laser power will not be utilized efficiently. Further, the objects will have freedom to move along the major axis of the elliptical profile. Since most naturally occurring biological objects do not have spherical symmetry, these can be rotated with rotating elliptical tweezers. The drag force of the medium on the rotating objects determines the limiting speed of rotation. The particle fails to follow the rotating elliptical profile when the drag torque exceeds the torque rotating the trapped object. Therefore, by an increase in the power of the trapping beam the speed of rotation can be increased.

At a power level of 25 mW in an elliptical tweezers (1 μm $\times$ 10 μm) the trapped bacteria could be rotated up to a maximum speed of $\sim$9 Hz beyond which the trapping was not stable. The bacteria trapped at this power level and subjected to rotation at 3 Hz for 10 min were found motile on being released from the trap. This suggests that their viability was not compromised. With the similar trapping configuration, objects inside a plant cell could also be rotated (Dasgupta *et al.*, 2003). However, due to significant scattering by the thick plant cell wall, the light intensity pattern in the cell gets distorted, and intensity is reduced leading to a decrease in the trapping efficiency. Panel B of Fig. 2 shows rotation of a rod-shaped structure (presumably a calcium oxalate crystal) inside an *Elodea densa* plant cell. It is also pertinent to note here that at a given trap beam power, the limiting speed of rotation is determined by the viscosity of its surrounding medium. The particle fails to follow the rotating elliptical profile when the drag torque exceeds the torque rotating the trapped object. Therefore, intracellular rotation of an object might be used to monitor the viscosity in intracellular regions as attempted by Bishop *et al.* (2004) inside a prototype cellular structure. By using line tweezers with dimensions of 1 μm $\times$ 40 μm eight human red and white blood cells along the 40-μm long major

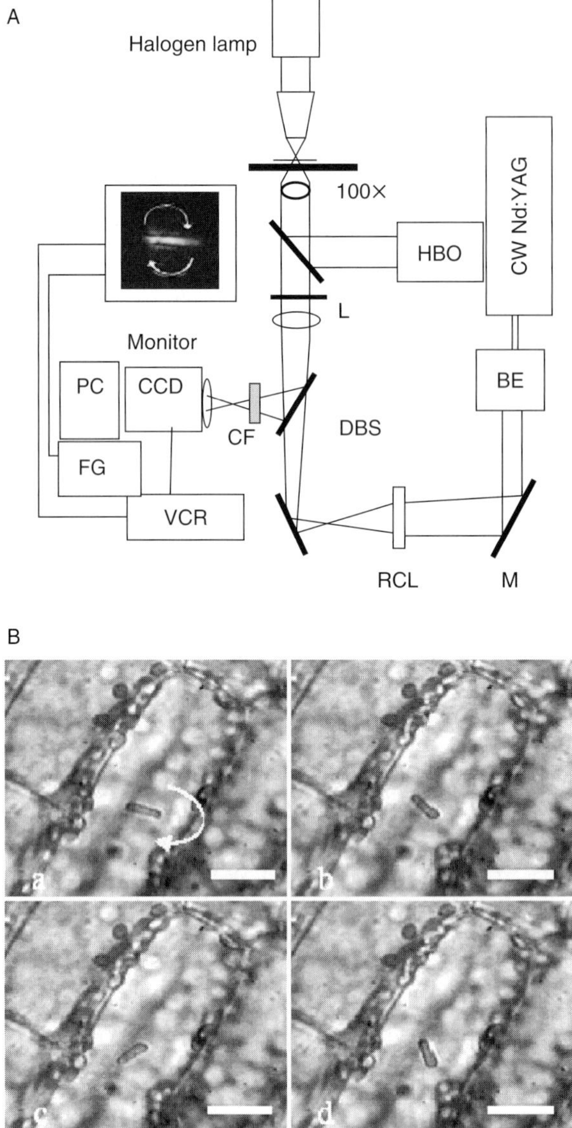

Fig. 2 (A) Schematic of the experimental arrangement. The output of a TEM_{00} mode 1064-nm CW Nd:YAG laser (Solid State Laser Division, RRCAT, India) was expanded using a beam expander (BE), passed through beam-steering and aligning mirror (M) and coupled to a 100× microscope objective (Zeiss Plan Neofluor, Carl Zeiss GmbH, Jena, Germany) through a combination of rotating cylindrical (RCL) and spherical lens (L), to focus it to an elliptical spot in the specimen plane. The dichroic beam splitter (DBS) transmits the CW 1064 nm and reflects the transmitted light/emitted fluorescence to the CCD. An NIR cutoff filter (CF) was placed before the CCD to reject the back-scattered laser light. Bright field and fluorescence images were recorded using a CCD and later on digitized using a frame grabber (FG). (B) Rotation of an intracellular object inside *E. densa* plant cell using the rotating line tweezers. The rod-shaped structure was trapped using 25-mW power and rotated at a speed of 4 Hz. The direction of rotation is shown by arrow (a). Clockwise rotation by angles of 45° (b), 145° (c), and 235° (d). All the images were recorded with the same magnification. Scale bar: 20 μm.

axis of the elliptical profile could be trapped. Because of increased trap dimensions, a larger trap beam power was required to trap and rotate the assembly of cells. At a power level of 40 mW in the trapping plane, the assembly could be rotated at speeds of up to 3 Hz (Fig. 3).

The rotating elliptical tweezers method described here provides a rather simple and efficient approach for controlled rotation of optically trapped microscopic objects varying in size. Compared to previously reported techniques for rotation of microscopic objects, this approach is found to be much simpler, giving better utilization of available laser power and thereby allowing much easier control of the trap beam profile. Demonstration of rotation of intracellular objects in living cells of intact tissue might prove useful in exploring several aspects of developmental biology such as the consequence of reorientation of the chromosomes in the mitotic spindle on cell division.

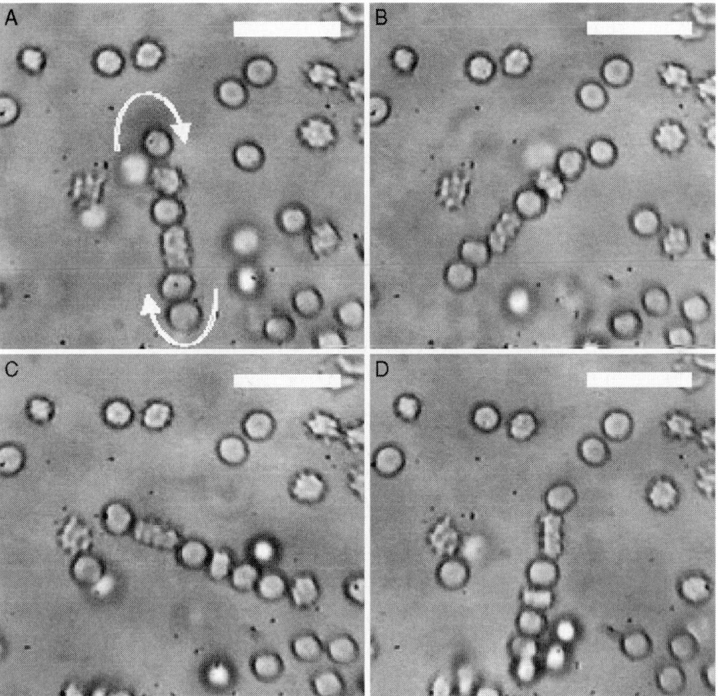

Fig. 3 Rotation of an assembly of human blood cells. The cells were trapped in the line tweezers with 40-mW power at the trapping plane and rotated at 3 Hz. The direction of rotation is shown by arrow (A). Clockwise rotation by an angle of 45° (B), 120° (C), and 190° (D). All the images were recorded with the same magnification. Scale bar: 15 μm.

B. Rotation of Low-Refractive-Index Particles

The methods discussed above are applicable only to objects having index of refraction higher than that of the surrounding medium. Lower refractive index objects are repelled from the regions of high intensity, and therefore trapping of these objects requires specialized trap beam profiles having low-intensity region surrounded with a high-intensity region. Though a rotating Gaussian beam (Sasaki *et al.*, 1992) can be used for trapping low-index objects, this requires complex scanning methods and is not suitable for more complex micromanipulation tasks such as transportation and rotation of a large number of such objects. To circumvent such limitations, various schemes have been proposed, namely optical vortex beams (Gahagan and Swartzlander, 1996, 1998, 1999) and interference pattern (MacDonald *et al.*, 2001), and so on. However, rotation of low-index object(s) using these methods is still difficult to achieve. Use of a high-order Bessel beam offers the advantage that the low-index particle trapped in the dark fringe region can also be rotated by transfer of orbital angular momentum from the Bessel beam (Garcés-Chávez *et al.*, 2002). However, this approach suffers from the drawback of a low throughput (since the laser power is distributed in the multiple rings). In this respect, trapping of low-index particles in dual line tweezers offers a dynamically reconfigurable approach whereby varying the spacing between the two-line tweezers, particles of varying size could be trapped. This configuration, although similar in nature to that using an interference pattern (MacDonald *et al.*, 2001), offers two important advantages. First, in contrast with the interference pattern, here the spacing between the two high-intensity regions can be changed independent of the width of the high-intensity region. Second, here all the laser beam power is distributed in only two high-intensity regions whereas for interference pattern, the laser power is distributed in several high-intensity regions. Therefore, this method can be used for efficient trapping of low-index objects of varying size and can be expected to yield reasonable transverse trapping efficiency even for larger separation between the high-intensity regions required for trapping larger size objects (Mohanty *et al.*, 2007).

The experimental setup used for dual line tweezers was a modified version of the experimental setup shown in Fig. 2A. In order to generate a dual line tweezers, the trap beam was split into two paths using a beam splitter, then combined using another beam splitter, and thereafter coupled to the 100× objective. By carefully tilting the external mirrors, the separation between the two-line tweezers can be varied from tens of micrometer to submicrometer. We used an emulsion of water droplets ($n_{\text{p}} = 1.33$, $\rho_{\text{p}} = 1.00$ g/ml) in acetophenone ($n_0 = 1.53$, $\rho_0 = 1.02$ g/ml) as a model low-index particle system. When shaken vigorously, the solution forms an emulsion of spherical droplets ranging in diameter from $\sim$50 μm down to a few micrometers. For trapping multiple low-index objects, the dimensions of the line traps can be extended up to 40 μm. In this method, the Z-trapping is achieved by confining the particle in the closed intensity void created by the overlap of the two diverging tweezers beam along z-direction. Further, when the particle is placed between the two-line tweezers separated along y-direction (minor axis of the line tweezers), due to the counteracting repulsive force of the two-line tweezers the net force on the particle becomes attractive. For separation larger than the diameter of the particle, the particle located between the two-line tweezers may not overlap with the traps and hence will not experience any restoring force. The restoring force is largest when the separation is close to the size of the object. It is pertinent to emphasize here that the measured transverse trapping efficiency (~0.1 for 15-μm diameter particle trapped with 60 mW in each tweezers, separated by 16 μm) using viscous drag method (Greulich, 1999) is an order of magnitude higher compared to trapping efficiencies (~0.01) reported under similar conditions (particle size 10 μm) for the vortex beam. The large value for transverse trapping efficiency primarily arises due to the fact that for the line tweezers the minor axis of the line focus is much narrower than the width of the high-intensity ring of vortex trap.

Further, simultaneous orientation of the dual line tweezers leads to controlled orientation of the trapped low-index particles (Fig. 4). The sphere assembly trapped between the two-line tweezers (shown as straight arrows) could be rotated continuously in anticlockwise direction (Fig. 4, in panels A–D) by simultaneous rotation of the cylindrical lenses. Both the speed and the direction of the rotation could be easily controlled by a control on the rotation of the cylindrical lenses. At a total

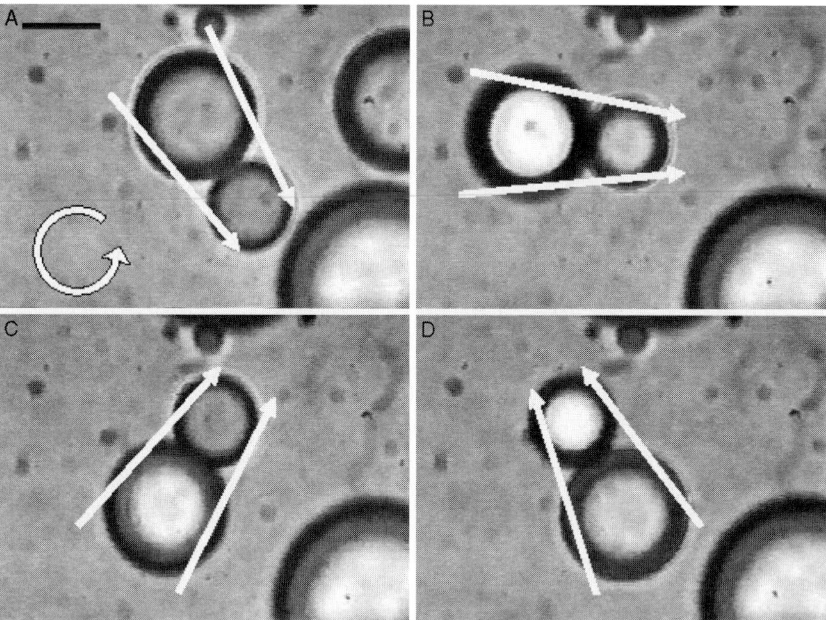

Fig. 4 Time-lapse digitized video images of controlled orientation of low-index spheres using rotating dual line tweezers. The trapped sphere assembly between two separated line tweezers (shown as straight arrows) is being rotated continuously in anticlockwise direction (marked by curved arrows) in panel A–D by simultaneous rotation of the cylindrical lenses. All images are in same magnification. Scale bar: 5 μm.

power of 60 mW in both the beams, rotation speed up to 40 rpm could be achieved (Mohanty *et al.*, 2007). This implies that the optical torque is able to overcome the viscous drag torque ($\tau = 8\pi\eta r^3\omega$, where η is the coefficient of viscosity of acetophenone, r is the radius of the water droplet, and ω is the angular velocity) of ~2.01 ± 0.3 pN μm. This is considerably higher than that achieved by transfer of orbital angular momentum from a high order Bessel beam (Garcés-Chávez *et al.*, 2002), where a particle of similar size (~6 μm) and relative refractive index (1.00/1.33) could be rotated at speeds of only up to ~6 rpm at a trap laser power of 600 mW. This can be attributed partly to the fact that the total power was distributed in the 40 rings of the Bessel beam at the focus.

Controlled manipulation of low-index particles, bubbles in particular, is of considerable interest because of their use for contrast enhancement in ultrasound imaging (De Jong *et al.*, 2002) and enhanced transfection of genes into living cells (Tachibana *et al.*, 1999).

C. Rotation of RBCs in Optical Tweezers

While orientation of cells by optical tweezers-based methods can help in studying their interaction with other cells or substrates, the dynamics of orientation/rotation of specific cells such as RBC in optical tweezers is also of interest as it can provide insights into the changes in their membrane properties. Studies have shown (Mohanty *et al.*, 2004) that RBC, suspended in hypertonic buffer, rotates by itself when optically tweezed. On being trapped, the discotic RBCs first get oriented with their symmetry axes perpendicular to the beam axis. The time-lapse digitized images of change in orientation of an RBC (suspended in a buffer with an osmolarity of 300 mOsm/kg) when subjected to optical tweezing are shown in Fig. 5A. Time required for an RBC to switch from the initial horizontal position to the vertical orientation (estimated from these time-lapse images) was found to depend on both, the trapping power and the osmolarity of the buffer in which the RBC is suspended. At the same trap power level, the time required for an RBC to switch from an initial horizontal position to the vertical orientation was found to decrease with increasing osmolarity.

From the measured rate of change of the orientation angle, the angular velocity (ω) and thus the viscous orientational torque (T_D) can be calculated (Mohanty *et al.*, 2005b) as a function of time using the Eq. $T_D = 6\omega\eta R^3$, where R is the radius of RBC (~3×10^{-6} m) and η being the coefficient of viscosity (~1.005×10^{-3} N.s/m^2). The orientational torque was observed to first increase and then decreases as the plane of the discotic RBC got aligned with the laser beam axis. The peak value of the viscous orientational torque T_D was estimated to be ~1.0 pN μm at 60 mW and ~1.5 pN μm at 85 mW. When the trap beam is switched off, the RBC returns to the original horizontal orientation, within 3–6 s. Even with the trap on, the vertically oriented RBC could be reoriented to the horizontal plane by subjecting the RBC either to a viscous drag (Mohanty *et al.*, 2005b) or by use of additional tweezers. During its reorientation to the horizontal plane the shape of the RBC remains discotic. Since no

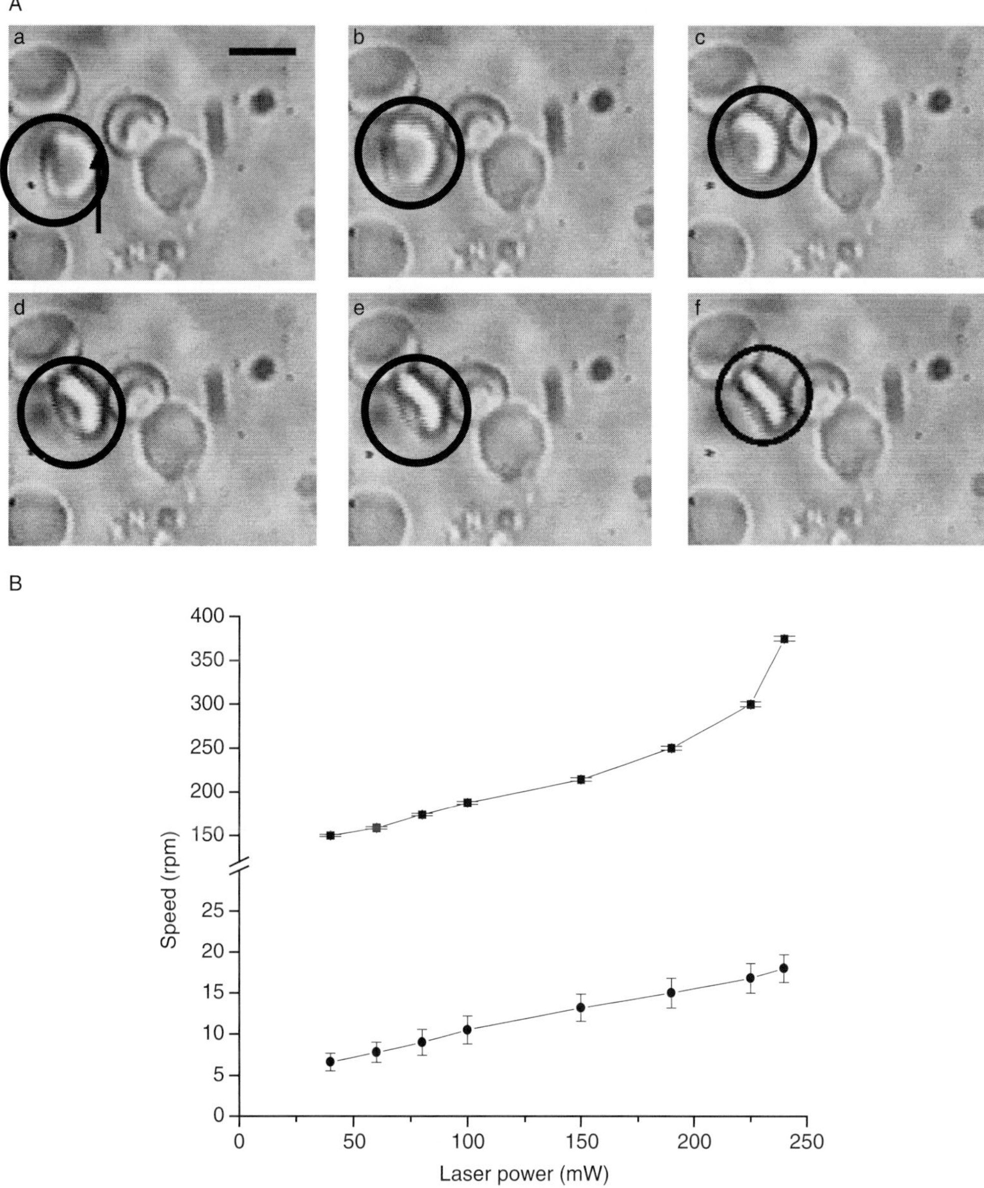

Fig. 5 (A) Time-lapse digitized video images of RBC subjected to optical tweezing with trap beam power of 85 mW. The RBC is shown encircled for clarity and the location of the trap beam is shown by arrow. Time interval between consequent frames was 40 ms. In panel (a) the RBC is in horizontal plane. On being subjected to optical tweezing it gradually orients with its plane along the trap laser beam axis (panels b to f). Scale bar: 5 μm. (B) Dependence of the speed of rotation of RBCs on the power of trap beam. The cells were suspended in hypertonic buffer with osmolarity of 1250 mOsm/kg. Square symbols correspond to measurements made on cells from normal volunteers and circles correspond to measurements made on cells from malaria-infected patients. The values shown represent mean of measurements on 30 cells from each group and error bars indicate standard errors of the mean.

folding of the RBCs to a rod shape as reported by Dharmadhikari *et al.* (2004) was observed under optical-trapping beam, the rotational mechanism suggested by them can be ruled out. The observed edge-on orientation of discotic RBC when optically tweezed in an isotonic buffer is consistent with earlier reports wherein it has been shown that disk-shaped objects get oriented with their symmetry axes perpendicular to the beam axis, since this maximizes the overlap of disk's volume with the region of highest electric field (Cheng *et al.*, 2002; Gauthier, 1997). The shape of the RBC (primarily, the thickness) depends on the osmolarity of the buffer in which it is suspended (Evans and Fung, 1972; Lim *et al.*, 2002). In isotonic buffer (~290 mOsm/kg) it is biconcave in shape; in a hypotonic buffer having an osmolarity of ~150 mOsm/kg, it gets swollen and becomes spherical, and in hypertonic buffer (>800 mOsm/kg) RBC takes a peculiar asymmetric shape.

When RBC in hypertonic buffer is optically trapped, it starts rotating. The rotation of the normal RBC in hypertonic buffer arises due to the torque generated on the cell by transfer of linear momentum from the trapping beam, similar to the previous reports of rotation of specifically structured nonbiological objects (Galajda and Ormos, 2001). However, when the cell suspended in isotonic buffer or in hypotonic buffer is optically trapped, rotation was not observed even up to a trap beam power of 240 mW. This is because in these buffers the shape of RBC is such that a transfer of linear momentum from the trapping beam does not lead to a significant torque on the cell. For a given trap power, the speed of rotation of RBC could be controlled by a change in the osmolarity of the hypertonic buffer. The larger asymmetry of the shape of RBC at higher osmolarity of the suspension leads to higher torque resulting in an increase in rotational speed of RBC. At a trapping power level of 80 mW, RBC was observed to rotate at a speed of ~25 rpm when suspended in buffers having osmolarity ~1000 mOsm/kg, and at the same trap power the rotational speed increased beyond ~200 rpm at an osmolarity of 1250 mOsm/kg. For a given osmolarity, rotational speed also increases with an increase in trap beam power (Fig. 5B).

Several techniques, like micropipette aspiration (Glenister *et al.*, 2002), laminar shear flow (Suwanarusk *et al.*, 2004), and optical tweezers (Suresh *et al.*, 2005) have been used to measure the rigidity of malaria-infected RBC membrane. The results obtained show that the rigidity of malaria-infected RBC membrane is an order of magnitude larger than that for the normal cells. Indeed, enhanced rigidification of RBC is a key feature of the pathophysiology of malaria in which alterations of the spectrin molecular network underneath the phospholipid bilayer occurs due to transport of proteins from the intracellular parasite to the cell membrane. Because of a change in the rigidity of the membrane, the orientational dynamics of RBC infected with malaria parasite is expected to be different from their normal counterparts. Indeed, in contrast to normal RBCs from healthy patients, the malarial parasite infected cells, as confirmed by acridine orange fluorescence staining (Hemvani *et al.*, 1999; Keiser *et al.*, 2002), did not rotate, even at trap power of 240 mW. Interestingly, the nonfluorescent RBCs, in the infected blood sample, were observed to rotate but with significantly lower speeds than the normal RBCs. This can be attributed to the fact that mature parasites release exoantigens that increase the rigidity of the uninfected RBCs in the sample (Nash *et al.*, 1989; Naumann *et al.*, 1991). In Fig. 5B, we show the dependence of the average rotational speed of RBCs on trap beam power (at osmolarity of 1250 mOsm/kg). The average rotational speed of cells from malaria-infected patients can be seen to be an order of magnitude smaller than their normal counterparts. Further, the rotational speed increased much slower with an increase in trap beam power. The large difference in rotational speed at a given trap beam power could be exploited for the detection of malaria-infected cells. An important advantage of this rotational assay is that, even those cells from a malaria-infected blood sample, which do not show acridine orange fluorescence, show a large difference in rotational speed as compared to the normal cells. This would considerably enhance the sensitivity of the present approach as compared to fluorescence staining. Since this approach may be sensitive to circulating antigen-induced stiffness in malaria-infected blood sample, one may detect *Plasmodium falciparum* infection even when the parasites are sequestered in the deep vascular compartment, which is undetectable by fluorescence microscopic examination. Further, in contrast to rapid diagnostics tests (RDTs) based on immunochromatographic methods (Moody, 2002), the parasite densities can be quantified which is needed, for example, to assess parasitological response to chemotherapy.

A 2D array of traps can be used to increase the throughput of this optical tweezers-based rotational assay. This approach for diagnosis of malaria could therefore enhance both sensitivity and the throughput of detection as compared to the current approach based on fluorescence staining. However, the present technique will also be sensitive to other diseases such as thalesimia and leukemia that results in changes in the elasticity of RBC membrane. Therefore, the challenge lies in increasing the specificity of diagnosis. Extensive studies are needed to differentiate among different stages of erythrocytic invasion and also between the four *Plasmodium* species that infect humans, namely *P. falciparum*, *P. vivax*, *P. ovale*, and *P. malariae*.

D. Methods for 3D Orientation of Microscopic Objects

Though the techniques discussed above have been used for rotation of microscopic objects, these are limited to a rotation of objects about the axis of the laser beam. For 3D rotation of trapped object, the most obvious approach is to trap two points of an object by dual optical tweezers and then rotate one point with respect to the other. However, when using Gaussian trap beams even with the highest NA of the objective, it becomes difficult to have two independent trap points closer than few micrometers due to interaction of the two tweezers. Therefore, complete 3D rotation of small microscopic objects is difficult to achieve. Bingelyte *et al.* (2003) created a pair of closely separated optical traps by exploiting the better axial trapping property of LG beams. The pair of closely separated LG optical traps was used to hold different parts of the same object, and the pair of traps could be made to revolve around each other in any plane, rotating the trapped object with them. A maximum rotation rate of 0.2 Hz was achieved.

Another approach (Mohanty and Gupta, 2004) for effecting 3D rotation of a microscopic object, which need not necessarily be trapped, is to apply a tangential force on the object to cause its rotation by focusing a pulsed NIR beam at a point on the periphery of the object. The basic idea used for 3D rotation of an object by a pulsed laser beam is illustrated in Fig. 6A. When a pulsed laser beam is focused at a point on the circumference of an object (assumed spherical for simplicity) it will exert three forces on the object; an axial gradient force (F_{ax} generated due to focusing of the beam along axial direction), a transverse gradient force (F_{tr} generated due to the Gaussian beam profile), and a photokinetic impulse (F_{kin}) due to momentum transfer. For the focusing geometry shown in Fig. 6A, the transverse gradient force would be small, and hence the resultant (F_{res}) of the three forces will act tangentially on the object. The rotation caused by this will be determined by the coefficient of viscosity of the medium and the moment of inertia of the object, and can be controlled via a control on repetition rate of the laser and its pulse energy. A change in the position of the point where the laser beam is focused leads to a change in the axis and the direction of rotation. Further, when the focal plane is above the equatorial plane of the object, the photokinetic impulse and the axial gradient force are in the same direction, leading to higher resultant tangential force. However, when the focal plane is below the equatorial plane of the object, the axial gradient force is in a direction opposite to photokinetic impulse reducing the net tangential force. In configuration where photokinetic impulse and the axial gradient force act in the same direction, an enhancement of photokinetic impulse, say by staining cells with two-photon absorbing dyes, should lead to an increase in the speed of rotation.

For laser microbeam-assisted rotation, the pulsed laser has been used at lower repetition rates to ensure that it does not lead to optical trapping of the object. For simultaneous trapping and orientation of microscopic object using this microbeam method, CW and pulsed lasers need to be combined (Fig. 6B). The use of the technique for intracellular 3D orientation and autofluorescence imaging of motile dividing chloroplasts is shown in Fig. 6C. In order to image division profiles of motile intracellular chloroplasts in *E. densa*, one of the freely streaming chloroplasts was trapped using the CW laser tweezers (Fig. 6C, a). Chloroplasts usually have an ellipsoidal shape, and when trapped they orient with their major axis along the direction of laser beam. Only one of the two sister chloroplasts is visible because the other is in a plane below the plane of the other sister chloroplast. By hitting the top corner edge of the trapped chloroplast with the pulsed laser microbeam, the chloroplast could be reoriented from the vertical plane to the horizontal plane facilitating simultaneous visualization of both the sister chloroplasts (Fig. 6C, b–d, the sister chloroplast that was earlier invisible has been enclosed by a circle for clarity). It may be noted that the chloroplast division in *E. densa* is asymmetric. For example, the size of the two sister chloroplasts (Fig. 6C, d) can be seen to be different. An important advantage of this microbeam-based approach is that, in contrast to other methods based on optical tweezers, here one can use a lower NA microscope objective, which

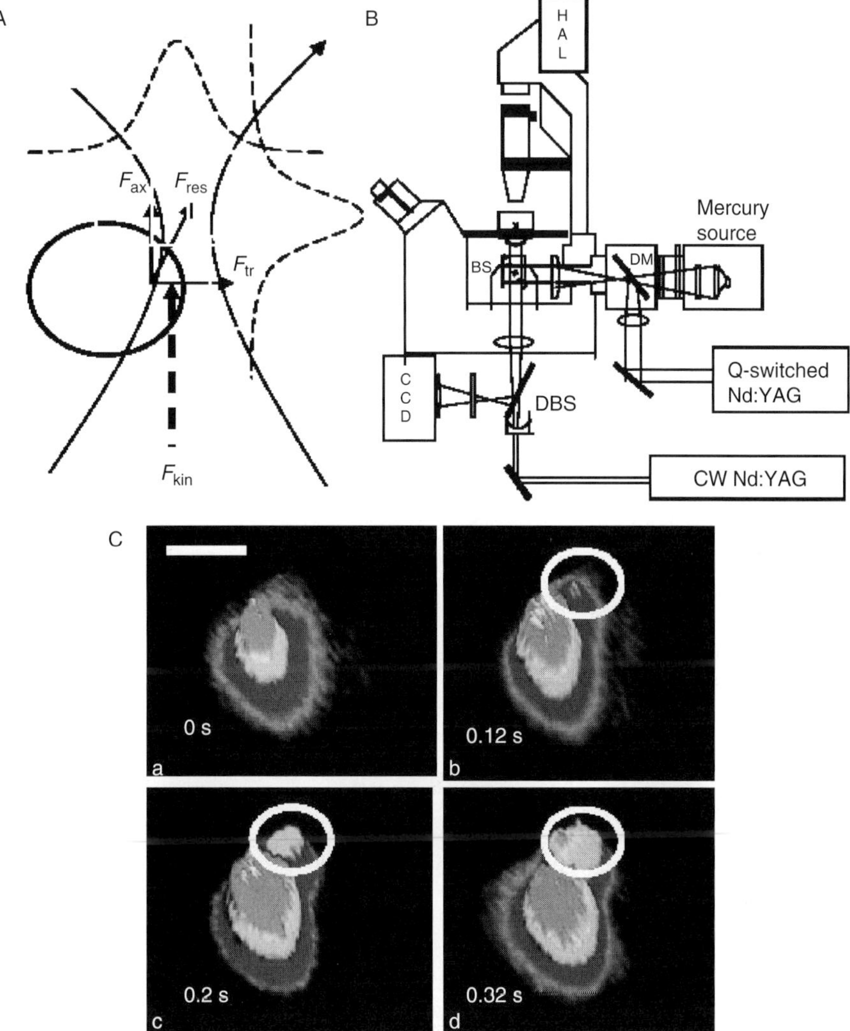

Fig. 6 (A) Forces exerted by a pulsed laser beam focused at a point on the periphery of a spherical object. (B) Schematic of the combined tweezers and microbeam setup. The output of a Q-switched pulsed Nd:YAG laser (pulse energy up to 50 mJ; repetition rate 1–10 Hz, Solid State Laser Division, RRCAT, India) was expanded using beam expander and coupled through the epifluorescence port. The focal position of the pulsed laser beam was varied using an external lens. A dichroic mirror (DM) combines the pulsed 1064-nm beam and the UV-visible light from the excitation source (mercury lamp) transmitted through the excitation filter. The beam splitter (BS) reflects the pulsed beam and transmits the fluorescence emitted from the sample. Another dichroic beam splitter (DBS) is used to transmit the CW trapping beam (1064 nm) and reflect the emitted fluorescence filtered through emission filter to the CCD. (C) Intracellular 3D orientation of motile dividing chloroplasts. (a) Only one of the two sister chloroplasts is visible because the other is in a plane below the plane of the bigger sister chloroplast. (b–d) The chloroplast was reoriented from the vertical plane to the horizontal plane facilitating simultaneous visualization of both the sister chloroplasts. The sister chloroplast that was earlier invisible has been enclosed by a circle for clarity. All images are in same magnification. Scale bar: 5 μm. (See Plate 27 in the color insert section.)

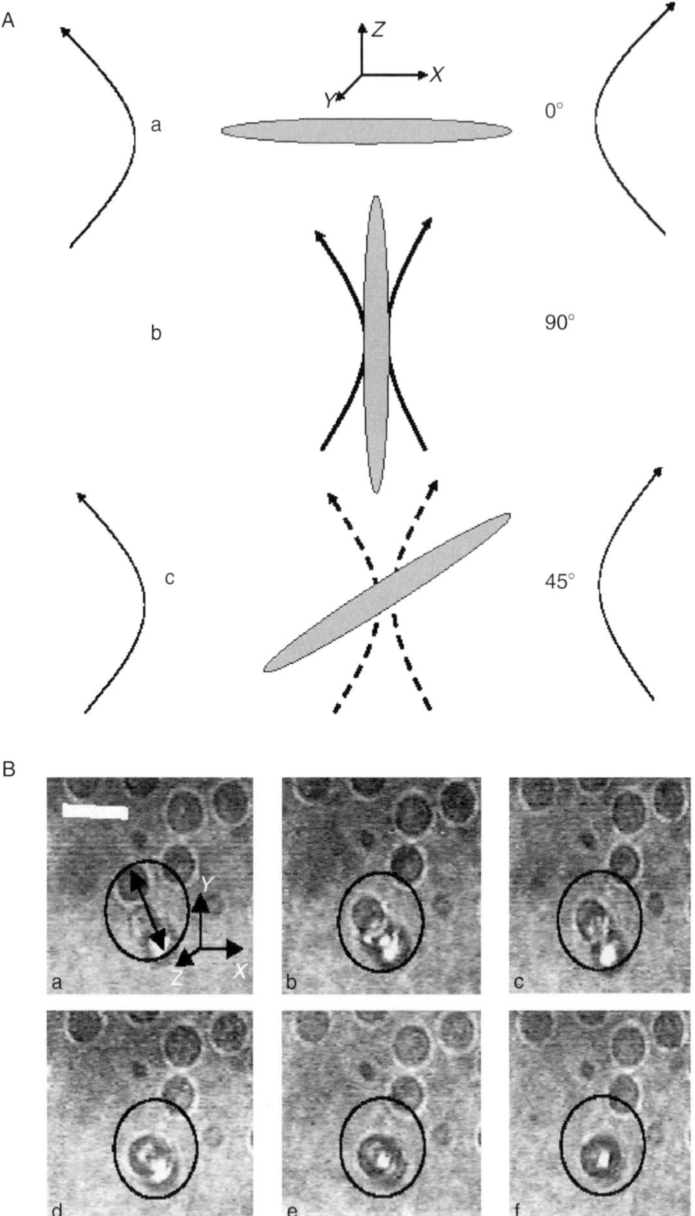

Fig. 7 (A) A schematic to illustrate the principle of 3D orientation of a microscopic object by combined use of elliptical and point tweezers. *xz* Cross-section of the variation of the spot size of trap beam along its direction of propagation for the case of elliptical tweezers is shown in the panel a. In such a tweezers an object lacking spherical symmetry gets trapped along the major axis of the elliptical tweezers (*x*-axis). *xz* Cross-section of the variation of the spot size of the trap beam along its direction of

would permit larger working distance. However, using this pulsed microbeam method, it may be difficult to achieve continuous orientation of a cell with high precision. Further, in some cases, localized damage at the periphery of the rotating microscopic object can occur. Kobayashi *et al.* (2006) have demonstrated the use of a proximal two-beam optical tweezers (both CW) for 3D single-cell rotation by harnessing the light pressure, generated in the optical axis direction. For this purpose, two proximal points of the same object were illuminated from different directions using two beams so that the microscopic object would remain trapped at the center of the two beams, and the light pressure generated could be made to act as rotating torque. In this case, changing the light intensity of the beams could control the rotational velocity.

The above-described methods are limited to the rotation of a single object and also cannot be used for the orientation of an assembly of objects. These limitations can be overcome by a combined use of two (line and point) tweezers. In this approach, the idea is to modulate the intensity of two tweezers holding the same object(s) at two different points. These two tweezers can be either conventional point tweezers or one of them can be line tweezers (Mohanty *et al.*, 2005c). It is well known that in point tweezers, an object lacking spherical symmetry orients with its long axis along the trapping beam since this maximizes its overlap with the high-electric field region of the trapping volume (Gauthier *et al.*, 1999). For the same reason, in elliptical tweezers an object lacking spherical symmetry orients with its long axis along the major axis of the elliptical tweezers. For the case when line and point tweezers overlap, the net electric field will get enhanced in the region of overlap of the two tweezers. The object will therefore orient at an angle with respect to the trap beam direction (z-axis). This is illustrated in Fig. 7A. The angle of orientation can be controlled via a control on the relative intensity of the two trap beams by varying the laser power. The orientation of the elliptical tweezers in the horizontal plane (x, y) and thus the orientation of the object in xy-plane can be controlled by a change in orientation of the cylindrical lens. Therefore, by adjusting the relative powers of the two tweezers, controlled 3D orientation of microscopic object(s) has been achieved (Mohanty *et al.*, 2005c). It is pertinent to note here that for a given trap power, the stiffness of the elliptical tweezers along its major axis will be smaller compared to the transverse stiffness of the point optical tweezers due to a larger dimension for the former. Therefore, larger power levels have to be used in case of elliptical tweezers to make the transverse stiffness for the two tweezers comparable. In order to generate two trapping beams, the laser beam is split into two parts using the beam splitter after it was expanded using a beam expander. A spherical lens was placed in the path of one of the beam to generate

propagation for the case of point optical tweezers is shown in the panel (b). Here, the object gets aligned along the direction of propagation of the trap beam (z-axis). Change in the orientation of the object by the combined use of the elliptical and point tweezers is illustrated in panel (c). (B) Use of an elliptical and point optical tweezers for 3D orientation (panels a–f) of an assembly of three human RBCs. All the images are in same magnification. Scale bar: 10 μm.

point tweezers and a cylindrical lens (which could be rotated about the beam axis) is placed in the path of the other beam to generate the elliptical tweezers. Both the laser beams are again combined using another beam splitter, passed through beam-steering mirrors and coupled to a 100× microscope objective.

Figure 7B shows the digitized images of 3D orientation of an assembly of three human RBCs. The RBCs, placed in a hypotonic suspension, are spherical in shape (diameter ~6 mm). In the line tweezers, the cells assembled along its major axis that is shown by an arrow in Fig. 7B, a. Because of increased trap dimensions (~1 μm × 30 μm) a larger trap beam power (90 mW) is required to trap the assembly of cells. When the point tweezers is positioned at the central RBC (Fig. 7B, b) and the power of the point tweezers is increased, the three cells could be oriented at increasing angles with respect to the *xy*-plane (Fig. 7B, b–d). Further, as expected, the assembly gets oriented along the *z*-axis (Fig. 7B, f) when only the point tweezers (~1 μm × 1 μm, 25 mW) is present. Just by tilting the tweezing laser beam, assembly of microspheres trapped in point tweezers can be tilted to a very small angle ($<5°$) with respect to the *z*-axis (MacDonald *et al.*, 2002). However, with the combined use of both line tweezers and point tweezers and adjusting their relative powers one could orient the assembly of (three) RBCs at angles varying from 0° to 90°. When the elliptical focus was rotated by rotation of cylindrical lens, the assembly of cells could be oriented along the new major axis.

Three-dimensional rotation methods described above can be used for various other applications such as twisting of a single DNA molecule. Such manipulations would enable monitoring of DNA–protein interactions, while the DNA is being rotated or twisted, under full microscopic view. The 3D rotation can also help (Kobayashi *et al.*, 2006) in measuring 2D spectral characteristics from multiple directions, and the 2D data can be converted into 3D spectra by data analysis. This 3D spectral tomographic imaging of single cells can, for example, lead to identification of extremely small amount of a specific protein (contained in the cell nucleus) responsible for the activation of cell division in cancer cells and thus lead to improvement in the early detection and treatment of cancer.

III. Methods for Optically Controlled Transportation of Microscopic Objects

Transport of microscopic objects plays a very important role in several biophysical processes like signaling, locomotion, and so on that are crucial for functioning of a cell (Svitkina *et al.*, 1997). Therefore, the ability to effect controlled transport of intracellular objects in a cell can not only help understand these fundamental processes but also provide a means to manipulate the functionality of living cells. Development of techniques for controlled transport of microscopic objects is also of considerable interest from the point of view of their use in microfluidic devices (Gluckstad, 2004; Terray *et al.*, 2002). There have been several reports on the use of axial light scattering force from a weakly focused or collimated laser beam for transport of particles along the direction of the beam (Buican *et al.*, 1987). If the scattering force dominates the axial gradient force, the particle is propelled along the direction of the beam. The dependence of the scattering force on size and refractive index has also been exploited for the separation of particles in a flowing liquid by having the laser beam propagate in a direction opposite to the direction of liquid flow (Hart and Terray, 2004). The particles become stationary (trapped) when the drag force exerted on the particles by the liquid flow balances the scattering force. However, the scattering force approach cannot be used for transport of particles in a plane transverse to laser beam propagation that is often desirable for manipulation and study of samples under full microscopic view. Transport of trapped objects, in a plane transverse to laser beam propagation, has been carried out either by scanning (Tirlapur *et al.*, 2002; Visscher *et al.*, 1993) the trapping beam or by keeping the trapping beam(s) fixed and moving the stage. Short repetitive sequences of holographic trapping patterns have also been used (Korda *et al.*, 2002) to transport microscopic objects through large arrays of optical tweezers. These methods are limited by the maximum time that could be allowed before the trap beam visits an object to ensure that the object does not diffuse away from the desired direction.

A. Transport of Organelles Within Living Plants Using Optical Tweezers

When the mechanically steerable CW Nd:YAG laser tweezers is placed at a particular locus within the cell of a randomly chosen fully differentiated intact leaf of *E. densa*, spontaneous convergence of the freely moving chloroplasts (Fig. 8A) into the optical trap (arrow marked, 80 mW) is observed (Fig. 8B). Subsequently, after repeated scanning (Fig. 8C) almost all of the trapped chloroplasts could be displaced to one end of the cell (Fig. 8D) without perturbing cytoplasmic streaming in the neighboring cells (Tirlapur *et al.*, 2002). Further, displacement of a single optically trapped chloroplast did not cause any influence on the movement of the nearby chloroplasts. This suggests absence of interconnections (Köhler *et al.*, 1997) between chloroplasts of *E. densa*.

Using the same NIR trapping laser power (80 mW) employed for the manipulation of functional chloroplasts but with the foci of the trapping beam set at different optical depth, we could transport the proplastids (Fig. 8E, relatively smaller in size) distal to the imaging plane (Fig. 8F). In this instance, the transport of these organelles was very rapid and was found in the vicinity of the distal foci of the optical trap away from the proximal imaging plane. Subsequently, when the trapping beam was turned off almost all of the proplastids returned back (within seconds) to the imaging plane (proximal to the trapping foci). It is important to note here that though the proplastids do not have motility like chloroplasts, the out-of-focus trapping configuration allowed interaction of the proplastids with the laser beam and by combination of axial and transverse gradient force, they could be transported in contrast to the action of only transverse gradient force used for the case of transportation of chloroplasts. Such intracellular optical micromanipulation of different organelles in intact living tissue can have potential applications

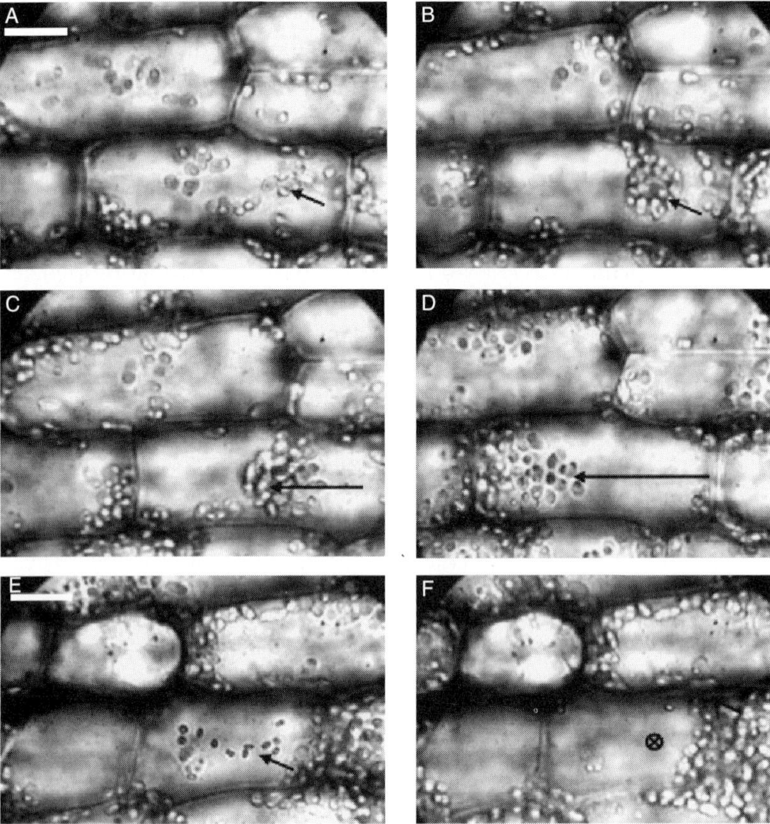

Fig. 8 Time-lapse video images showing intracellular micromanipulation of chloroplasts and proplastids in *E. densa* leaf cell using optical tweezers. The plantlets of *E. densa* were grown in glass water tanks at 27°C for10 h in daylight. The trapping point is marked by arrow. Scale bar: 20 μm.

in cell and developmental biology particularly in studies addressing the issues of organelle inheritance.

B. Method for Transport of Microscopic Objects Using Asymmetric Optical Gradient Force

A major limitation of the scanning optical tweezers approach is that only a handful of objects can be manipulated at a time. This is limited by the maximum time that could be allowed before the trap beam revisits an object to ensure that the object does not diffuse away from the desired direction. We have shown (Mohanty and Gupta, 2005) that a laser trap beam having asymmetric intensity distribution about the center of its long axis can be used for simultaneous transport of a large number of particles in a plane transverse to the laser beam direction. Compared to a symmetric potential well in a conventional laser tweezers where the intensity distribution of the trap beam is symmetric about the beam center, an asymmetry in the intensity distribution of the trap beam leads to an asymmetric potential well. Particles at the steep end of such a potential well will be pulled toward the potential minima, get accelerated, and ejected out along the direction having lower stiffness. It may also be noted that similar behavior is also observed in conventional point optical tweezers. If the focused spot is not symmetric, particles are attracted to the highest intensity point in the trap but are not trapped stably and thrown out of the trap. For realization of the above-discussed approach for transport of objects, we used (Khan *et al.*, 2006; Mohanty and Gupta, 2005) the line optical tweezers setup (Fig. 2A).

A schematic of the mode of the coupling of the laser beam to the microscope objective for generating asymmetric elliptical traps is shown in Fig. 9A. The asymmetry in the intensity distribution of the trap beam (a, panel B) arises due to a tilt of the laser beam (shown by solid arrows) with respect to the axis of the cylindrical lens (shown by the dotted arrows). Line tweezers having asymmetry along the major axis has been used for the transport experiments. The corresponding far-field diffraction pattern is shown in (b) of panel B. It may be noted here that since the laser beam

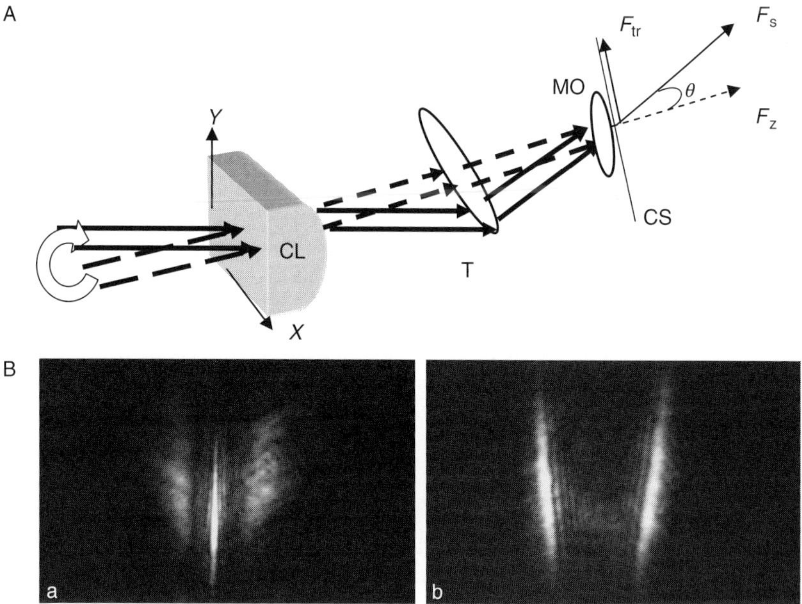

Fig. 9 (A) A schematic of the coupling of the laser beam to the microscope objective for generating an asymmetric elliptical trap; CL, cylindrical lens; T, tube lens inside microscope; MO, microscope objective. The focal length of the cylindrical lens was changed to vary the dimensions of the elliptical beam profile at the object plane. (B) Focused spot having an intensity asymmetry along the major axis (a) and the corresponding far-field diffraction pattern (b).

comes out of the microscope objective at an angle (θ) to the optical axis, a component ($F_{tr} = F_s \sin \theta$) of the scattering force (F_s) exists along the transverse direction also. This force acting along with the asymmetric gradient force helps in transport of the microscopic objects. The total force on the microscopic objects moving along the x-direction (major axis) can be calculated by adding up the asymmetric gradient force with the component of the scattering force. It is pertinent to emphasize here that asymmetric intensity distribution is created only in the direction of the major axis of the elliptical trap beam. In the other two directions, that is direction of propagation of the laser beam and the minor axis of the elliptical trap, the intensity distribution was symmetric resulting in symmetric gradient forces which led to trapping of the objects, as in conventional optical tweezers. Liesfeld *et al.* (2003) have used shadowing of a scanning trap beam to create a similar asymmetric potential trap and used it to transport polystyrene microsphere.

The technique based on asymmetric gradient force allows control on both the direction of transport as well as the speed of transport. The direction of transport can be varied by a rotation of the cylindrical lens to fix the direction of the major axis of the elliptical focus at the desired angle in the transverse plane. The speed with which particles move is determined by the depth and the asymmetry of the potential well which can be controlled by a control on trap beam power and the degree of asymmetry in the intensity profile. In Fig. 10, we show the change effected in the speed of the transport of a single RBC by controlling the trap beam power. The speed of transport could be increased from ~5 μm/s at 38 mW trap beam power to ~100 μm/s at 175 mW.

It is important to note that the asymmetric optical gradient forces, which determine the magnitude of the acceleration and the velocity of projection of

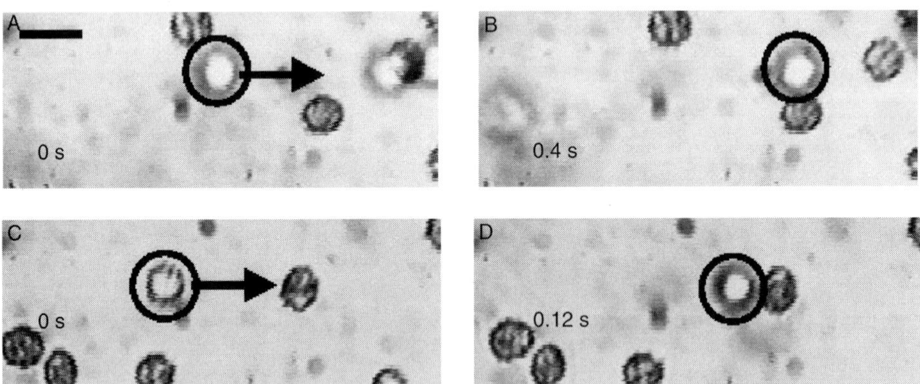

Fig. 10 Bright field digitized time-lapse video images of transport of RBC (shown encircled for clarity) at two trap beam powers. The direction of the arrow points to the direction of transport and the length and the position of the arrow correspond to the length and the position of the major axis of the asymmetric trap beam. For panels A and B, the trap beam power was 38 mW and for panels C and D, the trap beam power was 175 mW. All images are in the same magnification. Scale bar: 10 μm.

particles, depend on the optical and geometrical properties of the particles. The technique can therefore be used for *in situ* sorting of different particles based on difference in their physical properties (either size or optical properties). Such devices by themselves or in combination with other microfluidic devices may find wide range of applications in cell and molecular biology and optically driven micromachines.

C. Use of Asymmetric Line Optical Tweezers for Controlled Guidance of Neuronal Growth

The process of neuronal growth is believed to involve transport of actins in the direction of the growth cone (Mueller, 1999). Ehrlicher *et al.* (2002) have shown that by putting a defocused laser beam spot ahead of the growth cones, the rate of growth of neuronal growth cones can be enhanced. Use of this technique demonstrates the possibility of optical guidance of neuronal growth, which being highly specific may offer important advantages over the existing guidance cues (Ming *et al.*, 2002). However, the major drawback of this method is poor utilization of laser power. Even if the angle of the neuronal growth cone is as large as 10°, it will overlap with less than 1/36th of the area of the circular defocused laser spot. Therefore, only about 1/36th of the total laser power would be used for exerting the desired transverse gradient force.

In order to bias the growth of neuron using asymmetric gradient force, the cell body of neuronal cell (N1E-115) was placed near the high-intensity gradient end of the elliptical focus such that the tip of the growing edge was closer to the lower intensity gradient. The direction of transport was chosen arbitrarily for induction of new growth cones. Compared to a growth rate of 1 ± 1 μm/h observed for unexposed neurons, the lamellipodia extension rate in neuron subjected to asymmetric line tweezers was estimated to be 32 ± 6 μm/h. Irradiation of the neuron with a point tweezers having the same power (120 mW) did not show any enhancement of growth. The asymmetric transverse gradient force could also be used to induce protrusions from neuronal cell body. For short exposures, the optically induced protrusions were found to be transient, lasting a few minutes. However, with prolonged (~20 min) application of asymmetric transverse gradient force, permanent growth cones could be induced (Fig. 11). Figure 11B shows the initiation of extension (encircled) from the cell body after an exposure of 20 min. The protrusion was found to grow with increase in laser exposure time (Fig. 11B–C). The optically induced protrusions showed branching (Fig. 11D) similar to that observed in the natural process of growth. However, the growth rate of these induced cones was found to be lower (15 ± 3 μm/h) than that could be achieved (32 ± 6 μm/h) for the case of natural growth cones.

The growth direction of cones could be changed to almost 90° from the initial direction of growth by controlling the direction of the optical force. Figure 12 shows the use of this control on orientation of a growth cone to bring it into proximity with a desired cone of another neuron. The growth of the cone emerging from the neuron on left (Fig. 12A) was manipulated via the optical force to bring

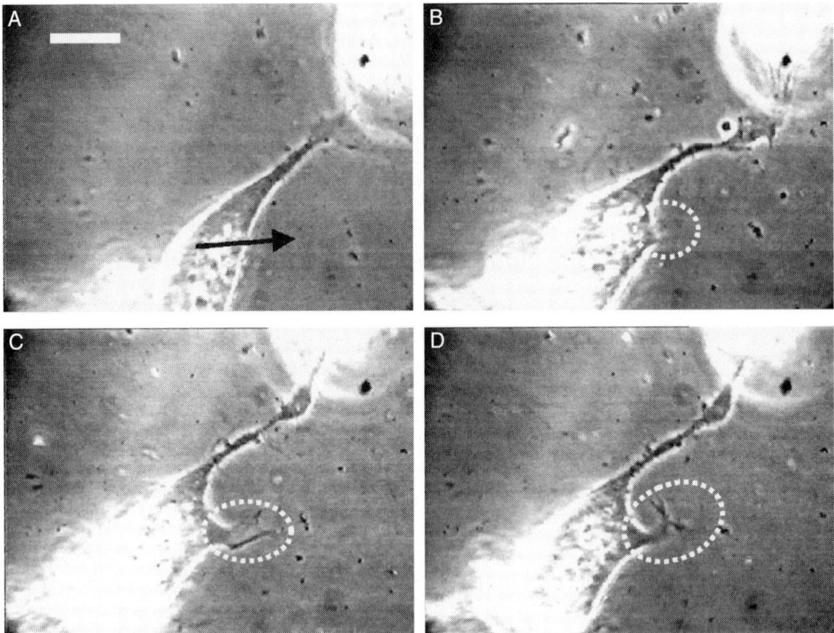

Fig. 11 Time-lapse digitized video images showing optical induction of new growth cone in N1E-115 neuronal cell. A neuronal cell subjected to asymmetric gradient force (1064 nm, 120 mW) in the direction marked by arrow (panel A). Induction of new growth cone (shown encircled) after 20 min (panel B), 25 min (panel C), and 45 min (panel D) of exposure. Natural branching of induced growth cone can be observed after 45 min of exposure (panel D). All images are in same magnification. Scale bar: 20 μm. We used N1E-115 cell line derived from subclone of C-1300 spontaneous mouse neuroblastoma tumor. The cell line was obtained from ATCC, USA, cultured in Dulbeco's minimal essential medium (DMEM, Sigma, USA) containing 10% fetal bovine serum medium (FBS) and antibiotics. Medium was buffered with HEPES (10 mM) and $NaHCO_3$ (1.5 g/liter). Cells were grown on a plane coverslip, which formed the bottom of the microchamber. The cells were incubated at 37 °C in incubator humidified with 5% CO_2 for 18 h. Prior to laser exposure, the medium in which the cells were incubated was replaced with fresh medium.

it in contact with the growth cone of the neuron on right (encircled region in Fig. 12B and C). Such a control may prove useful for development of *in vitro* neuronal circuits and neuroelectronic devices (Kaul *et al.*, 2004). By incorporating fiber-optic probes, such optical methods may be useful for *in vivo* regeneration of severed peripheral nerves in spinal cord injuries (Fawcett, 2002).

The mechanism of this laser-induced neuronal growth is still not clear. Optical force-induced transport of actins and actin-binding proteins in the direction of the growth cone is believed to be responsible for the enhancement of growth as well as the observed induction of new growth cone. However, the role of myosins (Borisy and Svitkina, 2000) in this process cannot be ruled out and would require further studies. It is also pertinent to note here that absorption of trapping light by

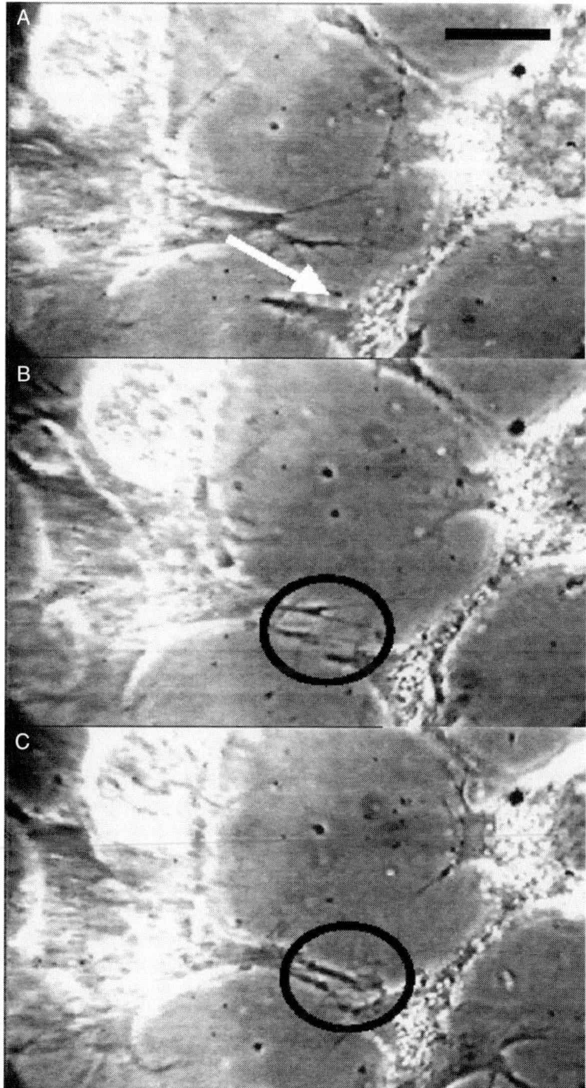

Fig. 12 Use of an asymmetric line optical tweezers for bringing a growth cone of one neuron in proximity to a growth cone of another neuron. The growth of the cone emerging from the neuron on left (panel A) was manipulated via the optical force to bring it in contact with the growth cone of the neuron on right (encircled region in panels B and C). All images are in same magnification. Scale bar: 20 μm.

neurons at 1064 nm may also cause a local rise in temperature (Liu *et al.*, 1995), which in turn might enhance polymerization of actin monomers into filamentous structures and thus lead to growth enhancement (Niranjan *et al.*, 2001). Further, quasi-mechanical strain produced by optical tweezers has also been reported to cause a change in intracellular calcium levels. Since changes in intracellular calcium concentration are also known to regulate (Gomez and Spitzer, 2000) growth cone development and guidance, optical force-induced change in calcium levels in neuronal cell may also be expected to contribute to the observed induction and enhancement of growth cone. Since the symmetric intensity profile line tweezers could not lead to detectable enhancement in growth of neuronal cells even at the highest trap power used (200 mW), the role of temperature rise and force-induced change in calcium levels in the observed laser-induced neuronal growth is believed to be insignificant.

IV. Methods for Laser-Assisted Microinjection into Targeted Cells and Potential Adverse Effect of Focused Laser Beams on Cells

A. Laser-Assisted Microinjection

Microinjection of exogenous genes, fluorochromes, antibodies, and photoactivable compounds into cells is required for a variety of applications in genetics, cell biology, and biotechnology (Greulich, 1999; Kaplan and Somlyo, 1989; Verma and Somia, 1997). Several techniques are being employed for this purpose. These include chemical methods, electroporation, direct microinjection into cells through mechanical means, liposome mediated and recombinant viral vector-mediated transfer. However, all these techniques have some important drawbacks. These techniques are often tedious, require considerable skill, and the efficacy of transfer are highly variable especially with small animal cells. This has motivated exploration of the use of laser-assisted microinjection. This approach offers two important advantages. First, it can be applied to all types of cells and second, it can be used on cells in suspension as well as attached cells.

Lasers in the UV spectral range were the first to be investigated for optoporation (Greulich, 1999; Shirahata *et al.*, 2001; Tao *et al.*, 1987) due to strong absorption by the constituents of the membrane in this spectral region. However, the use of UV light raises concern about the damage to cells or even exogenous DNA being transferred into the cell (Palumbo *et al.*, 1996; Tao *et al.*, 1987). The use of visible laser irradiation at 488 nm for optoporation has also been demonstrated exploiting the fact that the indicator dye, Phenol Red, present in the culture medium has strong absorption at this wavelength. Laser absorption-induced rise in temperature may lead to changes in the membrane permeability (Palumbo *et al.*, 1996; Schneckenburger *et al.*, 2002). However, since several cellular components have significant absorption at this wavelength (488 nm), the possibility of deleterious effects, even at this wavelength, cannot be ruled out. Indeed, reports exist on argon laser-induced cytogenetic damage such as sister chromatid exchange, chromatid and chromosome aberrations (Nakajima *et al.*, 1983). Lasers with wavelengths in the NIR region (700–1100 nm) are finding increasing use (Mohanty *et al.*, 2003; Stevenson *et al.*, 2006; Tirlapur and Konig, 2002) for optoporation because absorption by cellular components in this spectral range is significantly lower. Details of mechanisms of NIR femtosecond laser nanosurgery of cells and tissues are available in the review (Vogel *et al.*, 2005) and in the contribution of Vogel in the present book.

Targeted nanosecond NIR laser-assisted transfer (Mohanty *et al.*, 2003) of impermeable dye (merocyanine 540) into MCF-7 cells in suspension using the setup (Fig. 6B) is shown in Fig. 13. Time-lapse digitized fluorescence images of the cell reveal injection of dye into the irradiated cell through the point of irradiation.

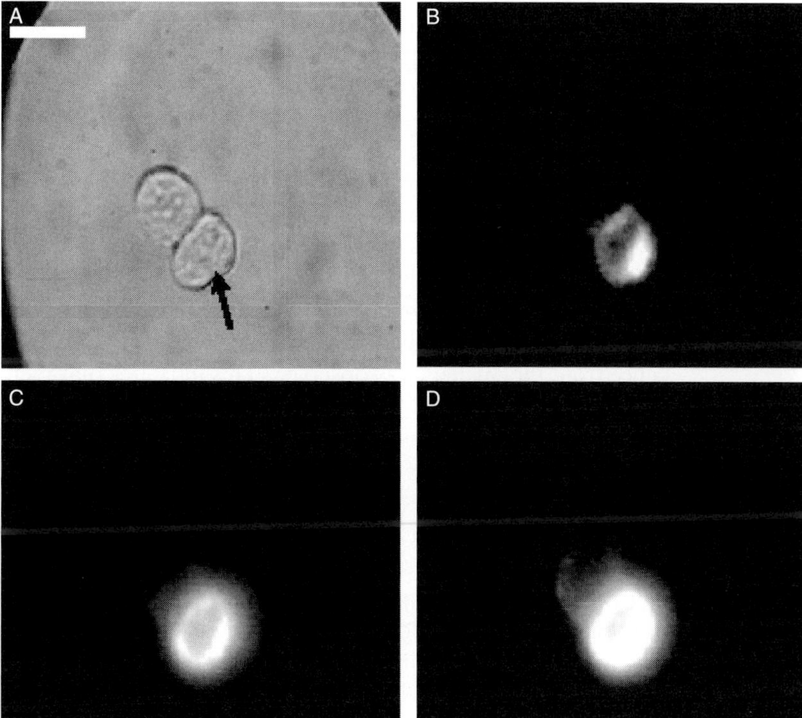

Fig. 13 Laser-assisted transport of merocyanine 540 into MCF-7 cells. The Nd:YAG laser beam (150 μJ per pulse, 17 ns, 10 Hz, energy density $\sim 2.4 \times 10^4$ J/cm^2) was focused at the edge of the membrane of the cell suspended in a medium containing 30-μg merocyanine 540 per ml. The point of laser irradiation is shown by arrow (A). Fluorescence observed from merocyanine 540 injected in the cell after 1 s of laser irradiation (B), after 4 s (C), and after 30 s (D) reveal that dye uptake starts at the point of laser irradiation and entire cell volume is filled up at longer exposure. Merocyanine 540 was excited using 546-nm band-pass filter and 590-nm long pass filer was used to separate fluorescence from excitation light. All the images were recorded with the same magnification. Scale bar: 10 μm. The MCF-7 (human breast adenocarinoma) cells were routinely cultured in Eagle's minimal essential medium (EMEM), supplemented with 10% (v/v) fetal bovine serum, 1-mM sodium pyruvate and antibiotics. The cells were incubated at 37 °C, in 5% (v/v) CO_2 atmosphere. For studying laser-assisted microinjection of dyes, cells grown in monolayer for 48 h were trypsinized, resuspended in EMEM containing 30-μg merocyanine (Sigma) 540 per ml, in presence of serum.

The dye uptake increased with increasing laser exposure (Fig. 13B–D). The dye was observed to preferentially stain the plasma membrane and propagate along the membrane from the point of laser irradiation (Fig. 13B and C). At longer exposure times, the fluorescence from the dye was observed from the entire cell volume with the cell membrane integrity fully preserved (Fig. 13D). Further, even at the highest exposure duration, no dye fluorescence was seen from the other adherent cell. Laser-assisted optoporation could also be used for microinjection of propidium iodide, another impermeable dye, into MCF-7 cells.

Photodynamic treatment of the cells can enhance efficiency of microinjection of dyes into animal cells as reported earlier for mechanical microinjection (Saito *et al.*, 2002). The enhancement is attributed to localized damage resulting from reactive oxygen species (ROS) generated through photodynamic treatment. Because merocyanine 540 has strong two-photon absorption at 1064 nm, injection of impermeable dyes (propidium iodide) in cells prestained by merocyanine 540 was investigated. Time-lapse digitized fluorescence images (Fig. 14) confirm laser dose-dependent injection of propidium iodide into the target cell. The fluorescence from the nucleus of the cell is characteristics of propidium iodide, and the fluorescence observed from the plasma membrane is due to merocyanine 540. It is important to note that in this case, injection of propidium iodide could be achieved at a lower energy densities as compared to cells not stained with merocyanine 540. Further, even 10 min after termination of the irradiation, no significant increase in propidium iodide fluorescence from the nucleus was observed. This indicates that membrane integrity of the microinjected cell remains intact and laser-induced alteration in membrane is transient in nature. The viability of microinjected cells was further confirmed by staining the cells with calcein before introduction of propidium iodide. No efflux of calcein was observed in irradiated cells (Fig. 14D) confirming the viability of propidium iodide injected cells. The nanosecond NIR could also be used (Mohanty *et al.*, 2003) for transfection of a green fluorescent protein (GFP)-plasmid into the MCF-7 cells.

The laser microbeam-based method may be also useful for injecting the photoactivable-caged compounds to study the spatial and temporal release of biological molecule of interest in living cells and also for gene transfection. Since this method is under visual control, it may prove to be useful for introducing photoactivable GFP (Patterson and Schwartz, 2002) and tracking movement of protein in different intracellular locations in real time.

B. Generation of ROS in Cells on Exposure to NIR Laser Tweezers and Microbeam

In order to minimize possible adverse effects to cells being manipulated by a laser beam the deposition of the laser energy in the cell needs to be minimized. Cellular components have strong absorption in the UV and visible regions because of nucleic acids and proteins. Water, the other major constituent of cells, has significant absorption beyond about 1100 nm. In the NIR spectral region (700–1100 nm), only some chromophores of the respiratory chain enzymes, notably cytochrome *c* oxidase, have relatively weak absorption (Karu, 1999). Because of the much

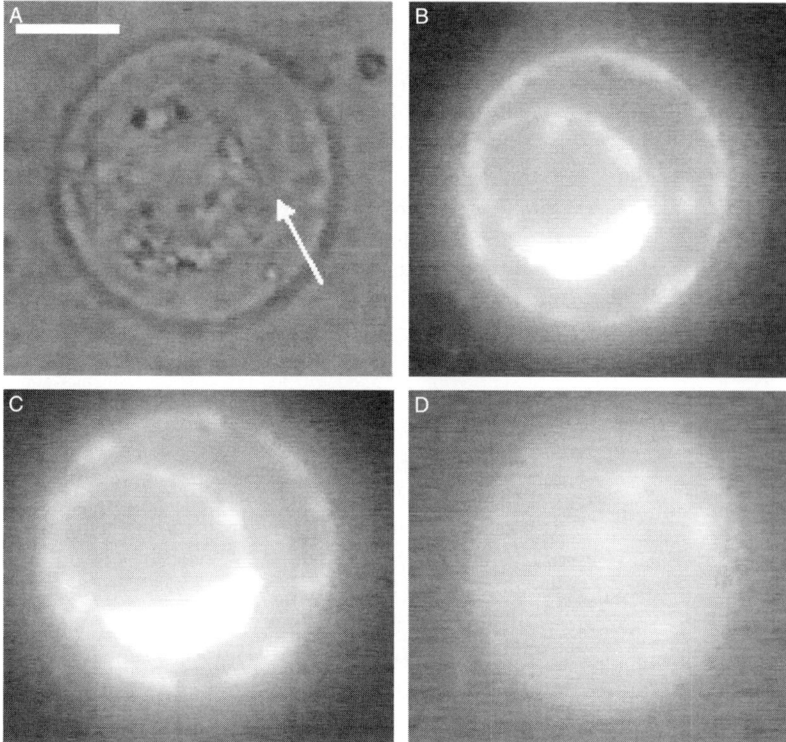

Fig. 14 Laser-assisted transport of propidium iodide (5-μg propidium iodide per ml, Molecular Probes Inc.) into merocyanine 540-stained MCF-7 cell. For experiments on cells with membrane stained by merocyanine 540, the cells were incubated with 7.5-μg merocyanine 540 per ml for 3 min, and unbound dye was removed by centrifugation. The cell pellet was resuspended in a fresh medium containing 5-μg propidium iodide per ml. The point of laser irradiation is shown by arrow (A), accumulation of propidium iodide injected in the nucleus after 1 min of laser irradiation (B), increase in propidium iodide fluorescence from nucleus after 2 min of irradiation (C), and calcein fluorescence after 10 min of laser optoporation (D). The absence of change in calcein fluorescence and morphology indicates that viability of optoporated cell is preserved. Propidium iodide was excited using 546-nm band-pass filter and 590-nm long pass filter was used to separate fluorescence from excitation light. Calcein was excited using 395- to 440-nm band-pass filter and fluorescence emission was monitored using 470-nm long pass emission filter. All the images were recorded with the same magnification. Scale bar: 5 μm. The viability of cells subjected to optoporation was assessed by incubating these with calcein acetomethylester at 10 μM for 10 min followed by resuspension in calcein free medium. Retention of this dye in the irradiated cells determines the viability of cell.

reduced absorption by cells in the NIR spectral range (700–1100 nm), lasers operating in this spectral range are preferred for manipulation of cells with minimal damage. However, even with the use of lasers in this spectral region, the possibility of adverse effects on the cells being manipulated is a matter of concern. Indeed adverse effects, like a decrease in cloning efficiency (Liang *et al.*, 1996) and DNA damage (Mohanty *et al.*, 2002), have been reported in cells exposed to NIR optical-trapping beam. The decrease in cloning efficiency and increased NADH fluorescence observed (Konig *et al.*, 1995) in cells trapped at 760 nm have been attributed to photochemical reactions arising from two-photon absorption in cellular chromophores. While involvement of ROS in the NIR wavelength (800 nm)-induced cellular damage has also been shown (Tirlapur *et al.*, 2001), at wavelengths longer than 800 nm, transient local heating (Liu *et al.*, 1996) of cellular targets is believed to be responsible for cellular damage. Indeed, damage observed at 810 nm was attributed to photothermal process as confirmed by experiments (Leitz *et al.*, 2002) on *Caenorhabditis elegans* carrying heat shock responsive reporter gene. Several studies carried out with 1064-nm laser beam suggest that photothermal processes rather than photochemical effects are responsible for the observed adverse effects.

In order to elucidate the mechanism of laser-induced cell damage at 1064 nm, Mohanty *et al.* (2006a) investigated generation of the ROS in cells exposed to the laser beam. Figure 15 shows the fluorescence image of DCDHF-labeled HeLa cells exposed for various durations with pulsed laser setup (Fig. 6A). DCDHF fluorescence was observed in HeLa cells, after 3 min of laser exposure (160 μJ per pulse, 10 Hz) indicating generation of ROS. A perusal of fluorescence images of the cell, after it was subjected to the laser exposure (Fig. 15A), shows that the intensity around the point of irradiation is much larger than the surrounding region, suggesting that ROS generation was initiated at the laser focal point. With increase in laser exposure, fluorescence was observed from the entire cell (Fig. 15B). Leakage of dye from the cell was observed on further increase in exposure duration. However, DCDHF fluorescence from cells trapped by CW Nd:YAG laser beam (200 mW) was less as compared to pulsed irradiation. While, presence of manitol (quencher for hydroxyl radicals) led to significant reduction of DCDHF fluorescence at lower exposures, sodium azide (singlet oxygen quencher) had little effect on ROS generation. These results suggest the generation of hydroxyl radicals on NIR laser microirradiation.

Since mitochondria are believed to be an important site of ROS generation, alterations in mitochondrial functions following exposure to 1064-nm laser radiation were also studied. The fluorescence images of rhodamine 123-labeled HeLa cells cytoplasm exposed to pulsed laser at average power of 1.6 mW for two different durations are shown in Fig. 15. The fluorescence intensity of laser-exposed cell gradually decreased with exposure time. Further, in presence of manitol no significant change in rhodamine 123 fluorescence was observed (Mohanty *et al.*, 2006a) irrespective of whether cytoplasm or nucleus was irradiated. However, in presence of sodium azide, the rhodamine fluorescence was observed to decrease when cytoplasm was irradiated, and no significant decrease was observed when nucleus was irradiated. This suggests ROS generation occurs preferentially in mitochondria.

It appears more likely that the oxidative stress observed at 1064 nm occurs through localized heating of water. The absorption coefficient of water at 1064 nm is significantly higher ($\sim$0.15 cm^{-1}) than that at 700–850 nm (0.0067–0.042 cm^{-1}). An approximate transient temperature rise of 100°C was estimated in CHO cells due to the pulsed trapping using 1064 nm Nd:YAG laser (40 μJ/pulse, pulse width

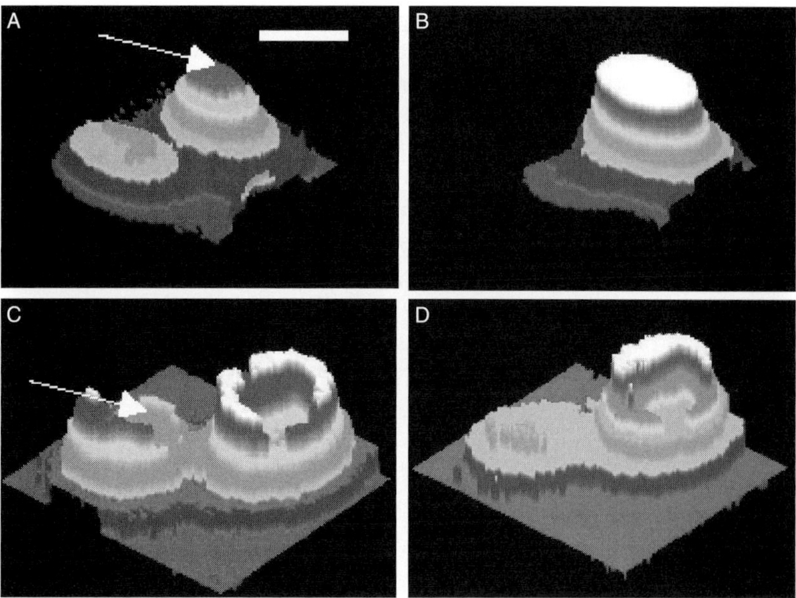

Fig. 15 Digitized time-lapse images of DCDHF (A, B) and rhodamine 123 (C and D)-stained carcinoma of cervix (HeLa) cell exposed to pulsed Nd:YAG laser microbeam. The exposed cell is marked by arrow (A). Fluorescence image of cell after laser exposure of 6 min (A) and 12 min (B). Fluorescence images of cell after exposure of 3 min (C) and 9 min (D). All images are recorded in same magnification. Scale bar: 10 μm. HeLa cells were grown in Eagle's minimal essential medium (EMEM) supplemented with 10% fetal bovine serum and antibiotics. Cultures were maintained at 37 °C in an incubator with 5% CO_2 atmosphere. Cells grown in monolayer for 48 h were trypsinized and after suitably diluting, cell suspension was placed on glass coverslips for irradiation. To study the generation of ROS by laser irradiation, cells were labeled with ROS marker, dihydrofluorescin (DCDHF, 20 μM for 20 min) that produces a green fluorescence when it gets oxidized. Labeled cells were visualized by epifluorescence microscopy (excitation, 450–490 nm; emission, 515–565 nm). In order to identify the type of reactive species involved, cells were also irradiated in presence of a hydroxyl radical quencher (manitol, 20 mM) or singlet oxygen quencher (sodium azide, 1 mM). Estimation of mitochondrial membrane potential was carried out using fluorescent probe, rhodamine 123 (1 μg/ml for 5 min), which enters the mitochondria and the retention of this probe in mitochondria depends on electrochemical and proton gradient. As potential decreases, the marker is progressively excluded from the mitochondria. In order to monitor fluorescence of rhodamine 123, the cells were excited using a 450- to 490-nm band-pass filter and a 515- to 565-nm band-pass filter for emission. (See Plate 28 in the color insert section.)

100 ns; Liu *et al.*, 1996). Transient rise in temperature will be even higher under the experimental condition used in this study (160 μJ/pulse, pulse width 16 ns). For the ~10-μm diameter cell, the transient increase in temperature is expected to get dissipated within a timescale of ~250 μs. However, this timescale is sufficient for ROS generation. At longer exposures and/or higher power levels, higher ROS generation may lead to membrane damage. This is evident from the leakage of DCDHF observed (Mohanty *et al.*, 2006a) at longer exposure of cells to pulsed laser trap/microbeam. The time, at which DCDHF leakage was observed, was also observed

to increase with lowering of the average power of the trap beam. In cells trapped with CW Nd:YAG laser, level of ROS generation was observed to be lower even at high doses (200 mW for 10 min) as compared to pulsed laser (1.2 mW for 10 min). This is consistent with the fact that the temperature rise due to a CW laser beam is expected to be much smaller than that due to a pulsed laser beam. The temperature rise in CHO cells was estimated (Liu *et al.*, 1995) to be 1.15 °C/100 mW of CW laser beam power at 1064 nm. This small temperature rise may not be enough to cause detectable ROS generation in cells exposed for shorter duration and longer exposure of cells to CW trapping beam were required for detectable ROS levels. A wavelength-dependent study on ROS generation in NIR region should prove very useful to understand the relative role of the different chromophores that may have single or two-photon absorption in this range.

V. Summary

Elliptical optical tweezers, generated by placing a cylindrical lens in the path of the trapping beam, have proved to be an efficient method for controlled, continuous rotation of cells or intracellular objects. Further, by the use of dual line tweezers, trapping and rotation of low-refractive-index microscopic particles of varying size can also be achieved. These methods are, however, limited to 2D rotation of objects about the propagation axis of the trap beam. Several methods for controlled 3D rotation and orientation of biological objects have also been discussed, and their use for orienting an optically trapped-dividing chloroplast or an assembly of RBCs has been presented. Rotation of RBC suspended in hypertonic buffer when it is optically tweezed, and the possible use of this approach for the diagnosis of malaria has also been illustrated. The line tweezers having an asymmetric intensity distribution about the center of its major axis can also be used for simultaneous transport of microscopic objects, and the approach has been used successfully for induction, enhancement, and guidance of neuronal growth cones. The use of laser microbeam-assisted microinjection of impermeable drugs into cells and possible adverse effects of the laser trap or microbeams on cells have also been discussed in brief.

Acknowledgments

We thank Mrinalini Sharma and Khageswar Sahu for their comments on the chapter.

References

Ashkin, A., and Dziedzic, J. M. (1987). Optical trapping and manipulation of viruses and bacteria. *Science* **235,** 1517–1520.

Ashkin, A., and Dziedzic, J. M. (1989). Internal cell manipulation using infrared laser traps. *Proc. Natl. Acad. Sci. USA* **86,** 7914–7918.

Ashkin, A., Dziedzic, J. M., Bjorkholm, J. E., and Chu, S. (1986). Observation of a single-beam gradient force optical trap for dielectric particles. *Opt. Lett.* **11,** 288–290.

Ashkin, A., Dziedzic, J. M., and Yamane, T. (1987). Optical trapping and manipulation of single cells using infrared laser beams. *Nature* **330,** 769–771.

Berns, M. W. (1998). Laser scissors and tweezers. *Sci. Am.* **278,** 52–57.

Berns, M. W., Aist, J., Edwards, J., Strahs, K., Girton, J., McNeil, P., Rattner, J. B., Kitzes, M., Hammerwilson, M., Liaw, L. H., Siemens, A., Koonce, M., *et al.* (1981). Laser microsurgery in cell and developmental biology. *Science* **213,** 505–513.

Berns, M. W., Tadir, Y., Liang, H., and Tromberg, B. (1998). Laser scissors and tweezers. *Methods Cell Biol.* **55,** 71–98.

Berns, M. W., Wright, W. H., Tromberg, B. J., Profeta, G. A., Andrews, J. J., and Walter, R. J. (1989). Use of a laser-induced optical force trap to study chromosome movement on the mitotic spindle. *Proc. Natl. Acad. Sci. USA* **86,** 4539–4543.

Bingelyte, V., Leach, J., Courtial, J., and Padgett, M. J. (2003). Optically controlled three-dimensional rotation of microscopic objects. *Appl. Phys. Lett.* **82,** 829–831.

Bishop, A. I., Nieminen, T. A., Heckenberg, N. R., and Rubinsztein-Dunlop, H. (2004). Optical microrheology using rotating laser-trapped particles. *Phys. Rev. Lett.* **92,** 198104, 1–4.

Borisy, G. G., and Svitkina, T. M. (2000). Actin machineryPushing the envelope. *Curr. Opin. Cell Biol.* **12,** 104–112.

Bronkhorst, P. J. H., Streekstra, G. J., Grimbergen, J., Nijhof, E. J., Sixma, J. J., and Brakenhoff, G. J. (1995). A new method to study shape recovery of red blood cells using multiple optical trapping. *Biophys. J.* **69,** 1666–1673.

Buican, T. N., Smyth, M. J., Crissman, H. A., Salzman, G. C., Stewart, C. C., and Martin, J. C. (1987). Automated single-cell manipulation and sorting by light trapping. *Appl. Opt.* **26,** 5311–5316.

Bustamante, C., Smith, S. B., Liphardt, J., and Smith, D. (2000). Single-molecule studies of DNA mechanics. *Curr. Opin. Struct. Biol.* **10,** 279–285.

Cheng, Z., Chaikin, P. M., and Mason, T. G. (2002). Light streak tracking of optically trapped thin microdisks. *Phys. Rev. Lett.* **89,** 108303–108304.

Chu, S. (1991). Laser manipulation of atoms and particles. *Science* **253,** 861–866.

Dasgupta, R., Mohanty, S. K., and Gupta, P. K. (2003). Controlled rotation of biological microscopic objects using optical line tweezers. *Biotechnol. Lett.* **25,** 1625–1628.

De Jong, N., Bouakaz, A., and Frinking, P. (2002). Basic acoustic properties of microbubbles. *Echocardiography* **19,** 229–240.

Dharmadhikari, J. A., Roy, S., Dharmadhikari, A. K., Sharma, S., and Mathur, D. (2004). Naturally occurring, optically driven, cellular rotor. *Appl. Phys. Lett.* **85,** 6048–6050.

Ehrlicher, A., Betz, T., Stuhrmann, B., Koch, D., Milner, V., Raizen, M. G., and Kas, J. (2002). Guiding neuronal growth with light. *Proc. Natl. Acad. Sci. USA* **99,** 16024–16028.

Evans, E., and Fung, Y. C. (1972). Improved measurements of the erythrocyte geometry. *Microvasc. Res.* **4,** 335–347.

Fawcett, J. (2002). Repair of spinal cord injuries: Where are we, where are we going. *Spinal Cord* **40,** 615–623.

Felgner, H., Frank, R., and Schliwa, M. (1996). Flexural rigidity of microtubules measured with the use of optical tweezers. *J. Cell Sci.* **109,** 509–516.

Friese, M. E. J., Nieminen, T. A., Heckenberg, N. R., and Rubinsztein-Dunlop, H. (1998). Optical alignment and spinning of laser-trapped microscopic particles. *Nature* **394,** 348–350.

Gahagan, K. T., and Swartzlander, G. A. (1996). Optical vortex trapping of particles. *Opt. Lett.* **21,** 827–829.

Gahagan, K. T., and Swartzlander, G. A. (1998). Trapping of low-index microparticles in an optical vortex. *J. Opt. Soc. Am. B* **15,** 524–533.

Gahagan, K. T., and Swartzlander, G. A. (1999). Simultaneous trapping of low-index and high-index microparticles observed with an optical-vortex trap. *J. Opt. Soc. Am. B* **16,** 533–537.

Galajda, P., and Ormos, P. (2001). Complex micromachines produced and driven by light. *Appl. Phys. Lett.* **78,** 249–251.

Galajda, P., and Ormos, P. (2002). Rotors produced and driven in laser tweezers with reversed direction of rotation. *Appl. Phys. Lett.* **80,** 4653–4655.

Garcés-Chávez, V., Volke-Sepulveda, K., Chávez-Cerda, S., Sibbett, W., and Dholakia, K. (2002). Transfer of orbital angular momentum to an optically trapped low-index particle. *Phys. Rev. A* **66,** 063402, 1–8.

Gauthier, R. C. (1997). Theoretical investigation of the optical trapping force and torque on cylindrical micro-objects. *J. Opt. Soc. Am. B* **14,** 3323–3333.

Gauthier, R. C., Ashman, M., and Grover, C. P. (1999). Experimental confirmation of the optical-trapping properties of cylindrical objects. *Appl. Opt.* **38,** 4861–4869.

Glenister, F. K., Coppel, R. L., Cowman, A. F., Mohandas, N., and Cooke, B.M (2002). Contribution of parasite proteins to altered mechanical properties of malaria-infected red blood cells. *Blood* **99,** 1060–1063.

Gluckstad, J. (2004). Microfluidics: Sorting particles with light. *Nat. Mater.* **3,** 9–10.

Gomez, T. M., and Spitzer, N. C. (2000). Regulation of growth cone behavior by calcium: New dynamics to earlier perspectives. *J. Neurobiol.* **44,** 174–183.

Greulich, K. O. (1999). "Micromanipulation by Light in Biology and Medicine." Birkhäuser Verlag, Basel, Switzerland.

Greulich, K. O., Pilarczyk, G., Hoffmann, A., Meyer Zu Horste, G., Schafer, B., Uhl, V., and Monajembashi, S. (2000). Micromanipulation by laser microbeam and optical tweezers: From plant cells to single molecules. *J. Microsc.* **198,** 182–187.

Hart, S. J., and Terray, A. V. (2004). Refractive-index-driven separation of colloidal polymer particles using optical chromatography. *Appl. Phys. Lett.* **83,** 5316–5318.

He, H., Friese, M. E. J., Heckenberg, N. R., and Rubinsztein-Dunlop, H. (1995). Direct observation of transfer of angular momentum to absorptive particles from a laser beam with a phase singularity. *Phys. Rev. Lett.* **75,** 826–829.

Hemvani, N., Chitnis, D. S., Dixit, D. S., and Asolkar, M. V. (1999). Acridine orange stained blood wet mounts for fluorescent detection of malaria. *Ind. J. Pathol. Microbiol.* **42,** 125–128.

Henon, S., Lenormand, G., Richert, A., and Gallet, F. (1999). A new determination of the shear modulus of the human erythrocyte membrane using optical tweezers. *Biophys. J.* **76,** 1145–1151.

Kaplan, J. H., and Somlyo, A. P. (1989). Flash photolysis of caged compoundsNew tools for cellular physiology. *Trends Neurosci.* **12,** 54–59.

Karu, T. (1999). Primary and secondary mechanisms of action of visible to near-IR radiation on cells. *J. Photochem. Photobiol. B* **49,** 1–17.

Kaul, R. A., Syed, N. I., and Fromherz, P. (2004). Neuron-semiconductor chip with chemical synapse between identified neurons. *Phys. Rev. Lett.* **92,** 038102, 1–4.

Keiser, J., Utzinger, J., Premji, Z., Yamagata, Y., and Singer, B. H. (2002). Acridine orange for malaria diagnosisIts diagnostic performance, its promotion and implementation in Tanzania, and the implications for malaria control. *Ann. Trop. Med. Parasitol.* **96,** 643–654.

Khan, M., Sood, A. K., Mohanty, S. K., Gupta, P. K., Arabale, G. V., Vijaymohanan, K., and Rao, C. N. R. (2006). Optical trapping and transportation of carbon nanotubes made easy by decorating with palladium. *Opt. Express* **14,** 424–429.

Kobayashi, H., Ishimaru, I., Hyodo, R., Yasokawa, T., Ishizaki, K., Kuriyama, S., Masaki, T., Nakai, S., Takegawa, K., and Tanaka, N. (2006). A precise method for rotating single cells. *Appl. Phys. Lett.* **88,** 131103–131104.

Koch, S. J., Shundrovsky, A., Jantzen, B. C., and Wang, M. D. (2002). Probing protein-DNA interactions by unzipping a single DNA double helix. *Biophys. J.* **83,** 1098–1105.

Köhler, R. H., Cao, J., Zipfel, W. R., Webb, W. W., and Hanson, M. R. (1997). Exchange of protein molecules through connections between higher plant plastids. *Science* **276,** 2039–2042.

Konig, K., Liu, Y., Sonek, G. J., Berns, M. W., and Tromberg, B. J. (1995). Autofluorescence spectroscopy of optically trapped cells during light stress. *Photochem. Photobiol.* **62,** 830–835.

Konig, K., Svaasand, L., Liu, Y., Sonek, G., Patrizio, P., Tadir, Y., Berns, M. W., and Tromberg, B. J. (1996). Determination of motility forces of human spermatozoa using an 800 nm optical trap. *Cell. Mol. Biol.* **42,** 501–509.

Korda, P., Taylor, M., and Grier, D. (2002). Kinetically locked-in collodial transport in an array of optical tweezers. *Phys. Rev. Lett.* **89,** 128301–128304.

Kuo, S. C., and Sheetz, M. P. (1992). Optical tweezers in cell biology. *Trends Cell Biol.* **2,** 116–118.

Kuo, S. C., and Sheetz, M. P. (1993). Force of single kinesin molecules measured with optical tweezers. *Science* **260,** 232–234.

Leitz, G., Fallman, E., Tuck, S., and Axner, O. (2002). Stress response in *Caenorhabditis elegans* caused by optical tweezers: Wavelength, power, and time dependence. *Biophys. J.* **82,** 2224–2231.

Liang, H., Vu, K. T., Krishnan, P., Trang, T. C., Shin, D., Kimel, S., and Berns, M. W. (1996). Wavelength dependence of cell cloning efficiency after optical trapping. *Biophys. J.* **70,** 1529–1533.

Liang, H., Wright, W. H., Cheng, S., He, W., and Berns, M. W. (1993). Micromanipulation of chromosomes in PTK2 cells using laser microsurgery (optical scalpel) in combination with laser-induced optical force (optical tweezers). *Exp. Cell Res.* **204,** 110–120.

Liesfeld, B., Nambiar, R., and Meiners, J. C. (2003). Particle transport in asymmetric scanning-line optical tweezers. *Phys. Rev. E* **68,** 051907, 1–6.

Lim, G. H. W., Wortis, M., and Mukhopadhyay, R. (2002). Stomatocyte–discocyte–echinocyte sequence of the human red blood cellEvidence for the bilayer–couple hypothesis from membrane mechanics. *Proc. Natl. Acad. Sci. USA* **99,** 16766–16769.

Liu, Y., Cheng, D. K., Sonek, G. J., Berns, M. W., Chapman, C. F., and Tromberg, B. J. (1995). Evidence for localized cell heating induced by infrared optical tweezers. *Biophys. J.* **68,** 2137–2144.

Liu, Y., Sonek, G. J., Berns, M. W., and Tromberg, B. J. (1996). Physiological monitoring of optically trapped cellsAssessing the effects of confinement by 1064 nm laser tweezers using microfluorometry. *Biophys. J.* **71,** 2158–2167.

Luo, Z. P., and An, K. N. (1998). Development and validation of a nanometer manipulation and measurement system for biomechanical testing of single macromolecules. *J. Biomech.* **31,** 1075–1079.

MacDonald, M. P., Paterson, L., Sibbett, W., Dholakia, K., and Riches, A. (2001). Trapping and manipulation of low-index particles in a two-dimensional interferometric optical trap. *Opt. Lett.* **26,** 863–865.

MacDonald, M. P., Paterson, L., Volke-Sepulveda, K., Arit, J., Sibbett, W., and Dholakia, K. (2002). Creation and manipulation of three-dimensional optically trapped structures. *Science* **296,** 1101–1103.

Ming, G., Wong, S. T., Henley, J., Yuan, X., Song, H., Spitzer, N. C., and Poo, M. (2002). Adaptation in the chemotactic guidance of nerve growth cones. *Nature* **417,** 411–418.

Mohanty, S. K., and Gupta, P. K. (2004). Laser-assisted three-dimensional rotation of microscopic objects. *Rev. Sci. Instrum.* **75,** 2320–2322.

Mohanty, S. K., and Gupta, P. K. (2005). Transport of microscopic objects using asymmetric transverse optical gradient force. *Appl. Phys. B* **81,** 159–162.

Mohanty, S. K., Rapp, A., Monajembashi, S., Gupta, P. K., and Greulich, K. O. (2002). COMET assay measurement of DNA damage in cells by laser micro-beams and trapping beams with wavelength spanning a range of 308 nm to 1064 nm. *Rad. Res.* **157,** 378–385.

Mohanty, S. K., Sharma, M., and Gupta, P. K. (2003). Laser-assisted microinjection into targeted animal cells. *Biotechnol. Lett.* **25,** 895–899.

Mohanty, S. K., Sharma, M., and Gupta, P. K. (2006a). Generation of ROS in cells on exposure to CW and pulsed near-infrared laser tweezers. *Photochem. Photobiol. Sci.* **5,** 134.

Mohanty, S. K., Mohanty, K. S., and Gupta, P. K. (2006b). "RBCs under optical tweezers as cellular motors and rockers: Microfluidic applications," in Optical Trapping and Optical Micromanipulation III, *Proc. SPIE* **6326,** 62362E.

Mohanty, S. K., Sharma, M., Panicker, M., and Gupta, P. K. (2005a). Controlled induction, enhancement, and guidance of neuronal growth cones by use of line optical tweezers. *Opt. Lett.* **30,** 2596–2598.

Mohanty, S. K., Mohanty, K. S., and Gupta, P. K. (2005b). Dynamics of interaction of RBC with optical tweezers. *Opt. Exp.* **13,** 4745–4751.

Mohanty, S. K., Uppal, A., and Gupta, P. K. (2004). Self-rotation of red blood cells in optical tweezers: Prospects for high throughput malaria diagnosis. *Biotechnol. Lett.* **26,** 971–974.

Mohanty, S. K., Verma, R. S., and Gupta, P. K. (2007). Trapping and controlled rotation of low-refractive-index particles using dual line optical tweezers. *Appl. Phy. B* (in press).

Moody, A. (2002). Rapid diagnostic tests for malaria parasites. *Clin. Microbiol. Rev.* **15,** 66–78.

Mueller, B. K. (1999). Growth cone guidance: First steps towards a deeper understanding. *Annu. Rev. Neurosci.* **22,** 351–388.

Nakajima, M., Fukuda, M., Kuroki, T., and Atsumi, K. (1983). Cytogenetic effects of argon laser irradiation on Chinese hamster cells. *Rad. Res.* **93,** 598–608.

Nash, G. B., O'Brien, E., Gordon-Smith, E. C., and Dormandy, J. A. (1989). Abnormalities in the mechanical properties of red blood cells caused by *Plasmodium falciparum. Blood* **74,** 855–861.

Naumann, K. M., Jones, G. L., Saul, A., and Smith, R. A. (1991). *Plasmodium falciparum* exo-antigen alters erythrocyte membrane deformability. *FEBS Lett.* **292,** 95–97.

Neale, S. L., Macdonald, M. P., Dholakia, K., and Krauss, T. F. (2005). All-optical control of microfuidic components using form birefringence. *Nat. Mater.* **4,** 530–533.

Niranjan, P. S., Forbes, J. G., Greer, S. C., Dudowicz, J., Freed, K. F., and Douglas, J. F. (2001). Thermodynamic regulation of actin polymerization. *J. Chem. Phys.* **114,** 10573–10576.

O'Neil, A. T., and Padgett, M. J. (2002). Rotational control within optical tweezers bye use of a rotating aperture. *Opt. Lett.* **27,** 743–745.

Palumbo, G., Caruso, M., Crescenzi, E., Tecce, M. F., Roberti, G., and Colasanti, A. (1996). Targeted gene transfer in eukaryotic cells by dye assisted laser optoporation. *J. Photochem. Photobiol. B* **36,** 41–46.

Paterson, L., MacDonald, M. P., Arit, J., Sibbett, W., Bryant, P. E., and Dholakia, K. (2001). Controlled rotation of optically trapped microscopic particles. *Science* **292,** 912–914.

Patterson, G. H., and Schwartz, J. (2002). A photoactivable GFP for selective photolabeling of proteins and cells. *Science* **297,** 1873–1877.

Saito, T. K., Muguruma, H., and Mabuchi, K. (2002). Photodynamic assistance increases the efficiency of the process of microinjection in animal cells. *Biotechnol. Lett.* **24,** 309–314.

Sasaki, K., Koshioka, M., Misawa, H., Kitamura, N., and Masuhara, H. (1992). Optical trapping of a metal particle and a water droplet by a scanning laser beam. *Appl. Phys. Lett.* **60,** 807–809.

Sato, S., Ishigure, M., and Inaba, H. (1991). Optical rapping and manipulation of microscopic particles and biological cells using higher-order mode Nd:YAG laser beams. *Electron. Lett.* **27,** 1831–1832.

Schneckenburger, H., Hendinger, A., Sailer, R., Strauss, W. S., and Schmitt, M. J. (2002). Laser-assisted optoporation of single cells. *J. Biomed. Opt.* **7,** 410–416.

Shirahata, Y., Ohkohchi, N., Itagak, H., and Satomi, S. (2001). New technique for gene transfection using laser irradiation. *J. Invest. Med.* **49,** 184–190.

Smith, S. B., Cui, Y., and Bustamante, C. (1996). Overstretching B-DNAThe elastic response of individual double-stranded and single-stranded DNA molecules. *Science* **271,** 795–799.

Stevenson, D., Agate, B., Tsampoula, X., Fischer, P., Brown, C. T. A., Sibbett, W., Riches, A., Gunn-Moore, F., and Dholakia, K. (2006). Femtosecond optical transfection of cells: Viability and efficiency. *Opt. Express* **14,** 7125–7133.

Suresh, S., Spatz, J., Mills, J. P., Micoulet, A., Dao, M., Lim, C. T., Beil, M., and Seuerlein, T. (2005). Connections between single-cell biomechanics and human disease states: Gastrointestinal cancer and malaria. *Acta Biomaterialia* **1,** 15–30.

Suwanarusk, R., Cooke, B. M., Dondorp, A. M., Silamut, K., Sattabongkot, J., White, N. J., and Udomsangpetch, R. (2004). The deformability of red blood cells parasitized by *Plasmodium falciparum* and *P. vivax. J. Infect. Dis.* **189,** 190–194.

Svitkina, T. M., Verkhovsky, A. B., McQuade, K. M., and Borisy, G. G. (1997). Analysis of the actin-myosin II system in fish epidermal keratocytesMechanism of cell body translocation. *J. Cell Biol.* **139,** 397–415.

Svoboda, K., and Block, S. M. (1994). Biological applications of optical forces. *Annu. Rev. Biophys. Biomol. Struct.* **23,** 147–285.

Svoboda, K., Schmidt, C. F., Schnapp, B. J., and Block, S. M. (1993). Direct observation of kinesin stepping by optical trapping interferometry. *Nature* **365,** 721–727.

Tachibana, K., Uchida, T., Yamashita, N., and Tamura, K. (1999). Induction of cell membrane porosity by ultrasound. *Lancet* **353,** 1409.

Tadir, Y., Neev, J., Ho, P., and Berns, M. W. (1993). Lasers for gamete micromanipulations: Basic concepts. *J. Assist. Reprod. Genet.* **10,** 121–125.

Tadir, Y., Wright, W. H., Vafa, O., Ord, T., Asch, R. H., and Berns, M. W. (1989). Micromanipulation of sperm by a laser generated optical trap. *Fertil. Steril.* **52,** 870.

Tao, W., Wilkinson, J., Stanbridge, E., and Berns, M. W. (1987). Direct gene transfer into human cultured cells facilitated by laser micropuncture of the cell membrane. *Proc. Natl. Acad. Sci.* **84,** 4180–4184USA.

Terray, A., Oakey, J., and Marr, D. W. M. (2002). Microfluidic control using colloidal devices. *Science* **296,** 1841–1844.

Tirlapur, U. K., and Konig, K. (2002). Targeted transfection by femtosecond laser. *Nature* **148,** 290–291.

Tirlapur, U. K., Konig, K., Peuckert, C., Krieg, R., and Halbhuber, K. (2001). Femtosecond near-infrared laser pulses elicit generation of reactive oxygen species in mammalian cells leading to apoptosis-like death. *Exp. Cell Res.* **263,** 88–97.

Tirlapur, U. K., Mohanty, S. K., Jain, B., König, K., and Gupta, P. K. (2002). Non-invasive intra-tissue micromanipulation and 3D sorting of intracellular organelles within intact living higher plants with near infrared laser traps. *In* "Proceedings of DAE-BRNS National Laser Symposium," pp. 287–288, 14–16 November 2002, Sree Chitra Tirunal Institute for Medical Science and Technology, Thiruvananthapuram, India.

Verma, I. M., and Somia, N. (1997). Gene therapy-promises, problems and prospects. *Nature* **389,** 239–242.

Visscher, K., Brakenhoff, G. J., and Krol, J. J. (1993). Micromanipulation by multiple optical traps created by a single fast scanning trap integrated with the bilateral confocal scanning laser microscope. *Cytometry* **14,** 105–114.

Vogel, A., Noack, J., Huttmann, G., and Paultauf, G. (2005). Mechanisms of femtosecond laser nanosurgery of cells and tissues. *Appl. Phys. B* **81,** 1015–1047.

Yu, T., Cheong, F. C., and Sow, C. H. (2004). The manipulation and assembly of CuO nanorods with line optical tweezers. *Nanotechnology* **15,** 1732–1736.

CHAPTER 21

Automated Motile Cell Capture and Analysis with Optical Traps

Bing Shao,* Jaclyn M. Nascimento,*
Linda Z. Shi,† and Elliot L. Botvinick‡

*Department of Electrical and Computer Engineering
University of California, San Diego, La Jolla, California 92093

†Department of Bioengineering, University of California
San Diego, La Jolla, California 92093

‡Beckman Laser Institute, University of California, Irvine, California 92612

Laser trapping in the near infrared regime is a noninvasive and microfluidic-compatible biomedical tool. This chapter examines the use of optical trapping as a quantitative measure of sperm motility. The single point gradient trap is used to directly measure the swimming forces of sperm from several different species. These forces could provide useful information about the overall sperm motility and semen quality. The swimming force is measured by trapping sperm and subsequently decreasing laser power until the sperm is capable of escaping the trap. Swimming trajectories were calculated by custom built software, an automatic sperm tracking algorithm called the single sperm tracking algorithm or SSTA. A real-time automated tracking and trapping system, or RATTS, which operates at video rate, was

0091-679X/07 $35.00
DOI: 10.1016/S0091-679X(06)82021-2

developed to perform experiments with minimal human involvement. After the experimenter initially identifies and clicks the computer mouse on the sperm-of-interest, RATTS performs all further tracking and trapping functions without human intervention. Additionally, an annular laser trap which is potentially useful for high-throughput sperm sorting based on motility and chemotaxis was developed. This low power trap offers a more gentle way for studying the effects of laser radiation, optical force, and external obstacles on sperm swimming pattern.

I. Introduction

Optical forces from a single-beam gradient laser trap can be used to confine and manipulate microscopic particles (Ashkin, 1998, 1991). Laser trapping in the near infrared regime is a noninvasive and microfluidic-compatible biomedical tool which has been widely applied for physiological studies of biological cells (Ozkan *et al.*, 2003; Shao *et al.*, 2004) and organelles (Berns, 1998). Since the late 1980s, researchers have been trapping individual sperm using single spot laser tweezers to study laser–sperm interactions and quantify sperm motility by measuring sperm swimming forces (Araujo *et al.*, 1994; Dantas *et al.*, 1995; Konig *et al.*, 1996; Patrizio *et al.*, 2000; Tadir *et al.*, 1989, 1990).

These studies determined that the minimum amount of laser power needed to hold the sperm in the trap (or the threshold escape power) is directly proportional to the sperm's swimming force [$F = Q \times (P/c)$], where F is the swimming force, P is the laser power, c is the speed of light in the medium, and Q is the geometrically determined trapping efficiency parameter (Konig *et al.*, 1996). Sperm swimming force measurements have been used to evaluate sperm viability by characterizing the effects of sperm cryopreservation (Dantas *et al.*, 1995), comparing the motility of epididymal sperm to ejaculated sperm (Araujo *et al.*, 1994), and investigating the medical aspects of sperm activity (Dantas *et al.*, 1995; Patrizio *et al.*, 2000). In addition, a relationship between sperm motility and swimming pattern was found for human and dog sperm (Nascimento *et al.*, 2006; Tadir *et al.*, 1990). They found that as swimming speed increased, so did average escape power.

In fertility physiology studies, the speed of progression (SOP) score is often used as a key parameter in the determination of overall motility score of a semen sample where motility score = (% motile) $\times$ (SOP score of sample)^2 (Olson *et al.*, 2003). The motility score is used to estimate the probability of a successful fertilization. The SOP score takes on discrete values from 1 to 5, 5 qualitatively represents the fastest swimming sperm, and 1 represents the sperm that exhibits the least amount of forward progression (Olson *et al.*, 2003). Since the SOP score is qualitative, it may be subject to variation between individuals. Computer-assisted sperm analysis (CASA) systems, commercially available since the mid-1980s, have been developed with the goal to obtain objective data on sperm motility that can be used in research, human fertility clinics, and animal breeding programs. A detailed review of CASA can be found in Amann and Katz (2004) and Mortimer (1994). Widely used commercial CASA systems include the HTM-IVOS (Hamilton-Thorne Bioscience,

Beverly, MA), the SM-CMA system (Stromberg-Mika, Bad Feilnbach, Germany), and the Hobson Sperm Tracker (Hobson Sperm Tracking Ltd., Sheffield, UK). There are also several noncommercial laboratory CASA systems (Le Pichon and Quero, 1994; Samuels and Van der Horst, 1986; Stephens *et al.*, 1988; Warchol *et al.*, 1996).

The CASA systems differ in their grayscale bit-depth, search region for finding the sperm in the next frame, frame rate, number of consecutive frames analyzed, grayscale thresholding method, image segmentation method to determine the pixel coordinates of the sperm, head versus midpiece tracking, strategies for handling collisions and/or near-miss cases with other bright particles or sperm, and loss of focus during tracking (Shi *et al.*, 2006a). All, however, calculate the percentage of motile sperm and/or velocity parameters such as swimming speed (curvilinear velocity, VCL), average path velocity, and straight line swimming speed. In addition to these swimming characteristic measurements, sperm swimming force could provide useful information about the overall sperm motility and semen quality.

Most sperm trapping experiments involve single sperm studies and require manually driven video analysis in a frame-by-frame manner in order to measure swimming velocities prior to and after trapping. Additionally, stable 3D laser trapping requires a high numerical aperture (NA) oil immersion objective lens, which characteristically has a very shallow depth of field (a few microns). As a result, swimming sperm routinely move in and out of focus as compared to imaging with a low NA air immersion objective lens. It became necessary to develop a more robust automated sperm tracking algorithm for single sperm trapping studies. This chapter describes an automatic sperm tracking algorithm called the single sperm tracking algorithm or SSTA (Shi *et al.*, 2006a) as well as a real-time automated tracking and trapping system or RATTS (Shi *et al.*, 2006b) that operates at video rate. After initially identifying and clicking the computer mouse on the sperm-of-interest, RATTS performs all further tracking and trapping functions without human intervention. This allows a rapid, high-throughput assessment of a sperm population's swimming force and overall quality.

On a fundamental scientific level, sperm swimming force measurements could provide insight on sperm competition and its effect on the evolution of sperm. In multimale–multifemale mating systems, the female of the species will mate with a number of male partners, as opposed to one single male. In such cases, the sperm of the rival males would compete for fertilization within the female's reproductive tract (Dixson, 1998). Sperm competition and sexual selection have been shown to favor a larger number of sperm being produced and ejaculated. Males from multimale–multifemale mating species have evolved larger testis relative to body weight (Dixson, 1998) and individual sperm with larger midpiece volumes (Anderson and Dixson, 2002) compared to those of males from species that mate with only one partner. It is conjectured that perhaps the sperm from multimale–multifemale mating systems with larger midpiece volumes swim with greater velocities and forces. Another theory has been proposed suggesting that some sperm from these multimale–multifemale mating species are "egg-getter" gametes while others are "kamikaze" sperm that inhibit other males' "egg-getter" sperm (Dixson, 1998). These sperm subpopulations would be differentiated by their physiology and behavior. Although there is valid criticism of the kamikaze sperm aspect of the hypothesis (Dixson, 1998), it is feasible that "egg-getter" sperm do exist as a subpopulation in a semen sample and are distinguishable by their swimming force characteristics. In other words, the "egg-getters" are the strongest and thus first to reach and fertilize the egg. It is also feasible that these "egg-getter" sperm are not solely found in males from multimale–multifemale mating species, but are a general phenomenon. It is the purpose of some of the studies presented in this chapter to use optical traps to address these possibilities and gain an understanding of sperm evolution.

In addition to the single spot laser trapping, an annular (ring-shaped) laser trap is developed to assist sperm analysis. When a sperm-of-interest is caught by the single spot laser tweezers, it needs to be held for a time sufficiently long for motility analysis (Fig. 1A–B). However, very frequently, a second sperm swims through the trapping spot and invalidates the measurement (Fig. 1C–D). One potential function of the annular trap is to form a force shield for protecting the sperm being tested from

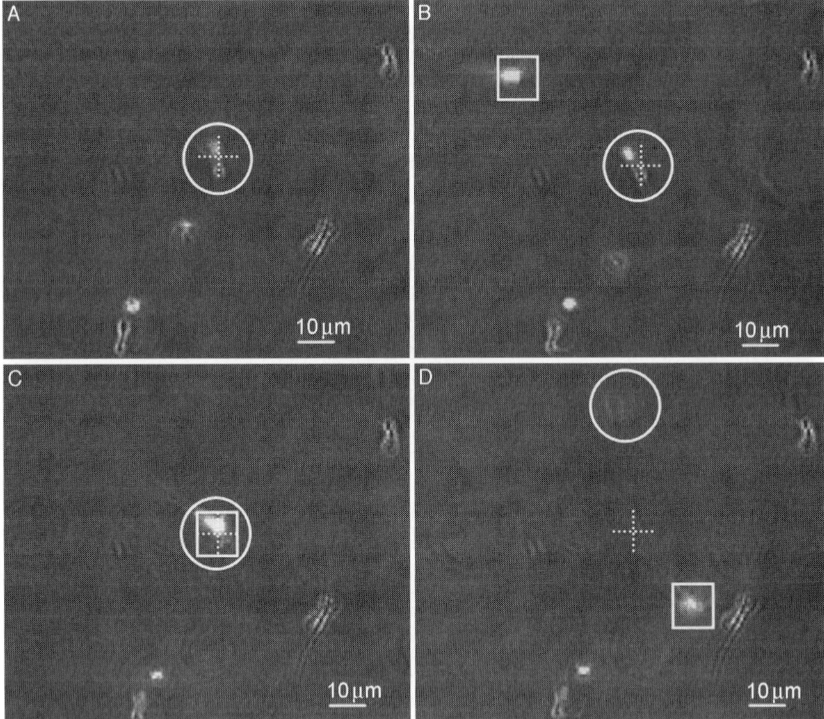

Fig. 1 Interference to single trap sperm analysis introduced by nontargeted sperm. (A) A sperm (inside the white circle) is trapped by a single spot laser trap (indicated by the white dotted cross hair) for analysis. (B) A second sperm (inside the white square) swims toward the trapping spot. (C) The second sperm swims into the trapping spot, and disturbs the analysis of the first trapped sperm. (D) Both sperm swim out of the trap. (Shao *et al.*, 2005).

the interference of other sperm. What is more, the ring-shaped laser trap could potentially be used for parallel sperm sorting based on motility and chemotaxis—a critical feature of sperm in response to the diffusion gradient of chemicals released by the egg and surrounding cells of the cumulus oophorus, which may help to explain infertility and provide new approaches to contraception (Eisenbach and Tur-Kaspa, 1999). Different from single spot laser trap which focuses hundreds of milliwatts on a sperm to immobilize it, in this ring-shaped trap, only tens of milliwatts are applied to a sperm. As a result, the sperm swim along the ring without having to stop. Consequently, the effects of optical force, radiation, and even external obstacles on sperm-swimming pattern and membrane potential (with the aid of the specific fluorescent dye) can be investigated in greater detail.

II. Methods

A. System Hardware and Optical Design

1. Gradient Single Point Trap

The block diagram of the hardware in Fig. 2 and the optical design in Fig. 3 show how the single point gradient trap is generated. The host computer contains an ASUS P5AD2 mother board supporting an Intel Pentium 4 CPU (3.4- GHz, 1-MB L2 onboard cache) with an 802.8-MHz bus and 2-GB RAM. An image

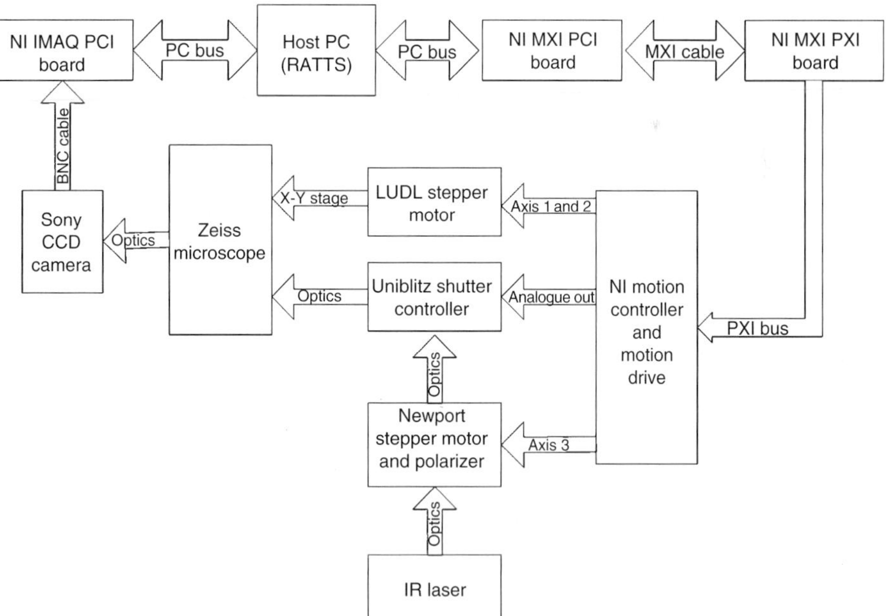

Fig. 2 Block diagram of hardware for gradient single point trap. (Shi *et al.*, 2006b).

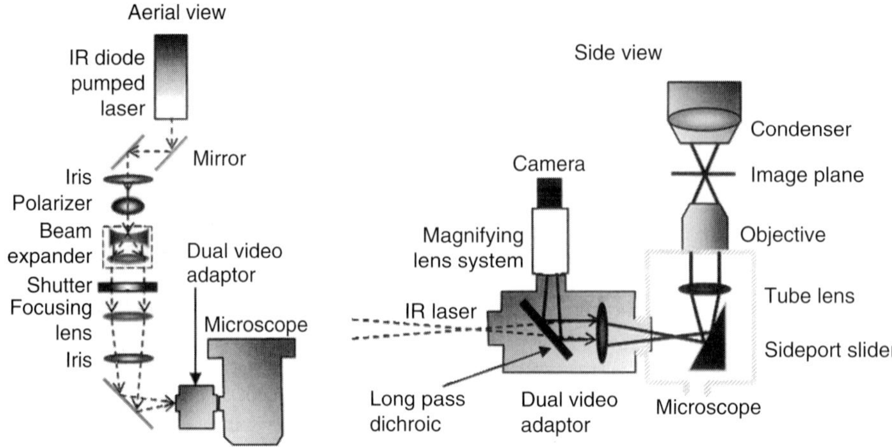

Fig. 3 Optical schematic: layout of microscope path and optical tweezers. (Nascimento *et al.*, 2006).

acquisition board (NI PXI-1407, National Instruments, Houston, TX) is housed in the host computer to digitize analog video signals. A motion controller (7344, National Instruments) is housed in a PXI chassis (NI PXI-1000B, National Instruments) which is controlled from the host PC through a MXI-3 link connected with a fiber-optic cable. An MXI-3 PCI interface card (PCI 8335, National Instruments) in the host computer transmits/receives data to the MXI-3 PXI card in the PXI chassis (PXI 8335, National Instruments) through a bidirectional fiber-optic cable thereby implementing a PCI–PCI bridge.

Laser light from a Nd:YVO_4 continuous wave of 1064-nm wavelength laser (Spectra Physics, Model BL-106C, Mountain View, CA) is reflected from two dielectric mirrors to orient the beam parallel to the table and along the optical axis of the microscope. The beam is expanded by two lenses (plano-concave lens, $f = -25.5$ mm at $\lambda = 1064$ nm, and plano-convex lens, $f = 76.2$ mm at $\lambda = 1064$ nm) in order to fill the microscope objective's back aperture. A mechanical shutter (Uniblitz LS6ZM2, Vincent Associates, Rochester, NY) in the laser path is used to open and close (turn "on" and "off") the optical trap and is controlled by a shutter driver (Uniblitz VMM-D3, Vincent Associates) through two lines of digital input–output from the motion controller. A third lens (biconvex lens, $f = 200$ mm) focuses the beam onto the side port of the dual video adaptor to ensure the beam is collimated at the objective's back aperture (Berns *et al.*, 2006). The light is then coupled into a Zeiss Axiovert S100 microscope and a 40×, phase III, NA 1.3 oil immersion objective (Zeiss, Thornwood, NY) which is also used for imaging (Wang *et al.*, 2005). The dual video adaptor contains a filter cube with a dichroic that allows laser light entering the side port to be transmitted to the microscope while reflecting visible light to the camera attached to the top port for imaging. A filter (Chroma Technology Corporation, Model D535/40M, Rockingham, VT) is placed in the filter cube to block back reflections of IR laser light while allowing visible light to pass.

A four-axis stepper-motor driver (MID-7604, National Instruments) connected to the motion controller drives the x–y stepper motor stage (LUDL Electronic Products Ltd., Hawthorne, NY) of the microscope. Since the laser trap remains stationary near the center of the field of view, trapping a sperm requires the movement of the microscope stage to bring the sperm to the laser trap location. The laser trap location is determined prior to each experiment by trapping 10-μm diameter polystyrene beads suspended in water within a 35-mm diameter glass bottom Petri dish (MatTek Corporation, Ashland, MA). The trap depth within the sample is kept to ~5 μm (approximately one sperm head diameter) above the cover glass. This ensures that the trap geometry is not sensitive to spherical aberrations from the surrounding media.

Laser power in the specimen plane is controlled and attenuated by rotating a Glan laser linear polarizer. A rotary stepper motor mount (PR50PP, Newport Corporation, Irvine, CA) housing the polarizer is controlled by the motion controller and stepper motor driver. The experimenter is able to set the power decay rate (rotation rate of polarizer).

The specimen is imaged at 30 frames per second by an RS-170 standard CCD camera (Model XC-75, Sony, Japan), coupled to a variable zoom lens system (0.33–1.6× magnification) to demagnify the field of view. Video signals from the CCD camera are distributed to a TV monitor, a Camcorder (Sony, Japan) for recording, and the image acquisition board through a video distribution amplifier (not shown) (IN3218HR, Extron Electronics, Anaheim, CA). The images are either analyzed in real-time or recorded via camcorder for future off-line reference and/or analysis. Both scenarios will be described in more detail in Section II.B.

2. Annular (Ring-Shaped) Laser Trap

The experimental setup for annular laser trap system is depicted in Fig. 4. The light beam from a CW Ytterbium fiber laser with 1070-nm wavelength (PYL-20M, IPG Photonics, Oxford, MA) is collimated and expanded via the 3× beam expander (T81-3X, Newport Corporation). For better performance of the trap, a refractive beam shaper (GBS-AR14, Newport Corporation) is used to convert a Gaussian laser beam into a collimated flat top beam. A telescope lens pair shrinks the shaped beam so that the thickness of the light cone input to the objective is equal to the diameter of the back aperture, and the numerical aperture of the trapping beam is maximized (Shao *et al.*, 2005). On Axicon1 (Del Mar Photonics, San Diego, CA), the input beam is divided with respect to the optical axis and bent toward it at an angle $\beta = \arcsin(n\sin\gamma) - \gamma$, where γ is the base angle and n is the refractive index of the axicon. At the back focal plane of the focusing lens (FL), a ring image is formed that is conjugate to the ring focus at the specimen plane. After the tube lens (TL), the laser light is sent into an inverted microscope (Axiovert 200M, Zeiss), and directed into the back aperture of the objective by the dichroic mirror.

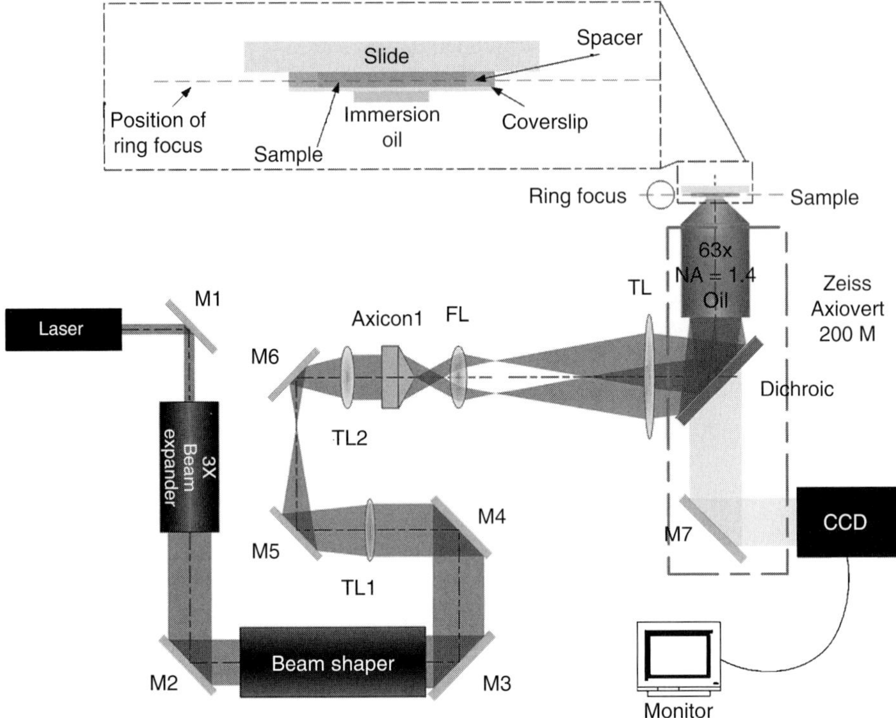

Fig. 4 The schematic diagram of the annular laser trapping system. M1–M7, mirrors; TL1, telescope lens 1; TL2, telescope lens 2; FL, focusing lens; TL, tube lens. (Shao *et al.*, 2007).

B. Offline and Real-Time Sperm Tracking Algorithms

1. Offline SSTA

Most sperm trapping experiments involve single sperm studies and require manually driven video analysis in a frame-by-frame manner in order to measure swimming velocities prior to and after trapping. Additionally, swimming sperm routinely move in and out of focus. As a first attempt to automate tracking prior to and after trapping, video recorded during laser trapping experiments was played back into the HTM-IVOS CASA system (version 12.1, Hamilton Thorne Biosciences, 2004) equipped with an add-on feature for postvideo analysis and access to single sperm data. We found their algorithm failed to accurately report swimming parameters. Three error types are: (1) miscounting a single sperm as multiple sperm in a discontinuous path during transient focus change, (2) miscounting a single sperm as multiple sperm as two sperm paths intersect, and (3) crossover events in which two sperm trajectories are swapped after their paths intersect. These errors are not mutually exclusive. It became necessary to develop a more robust automated sperm tracking algorithm for single sperm trapping studies.

SSTA was first developed in the LabView 7.1 language (National Instrument, Austin, TX) as an off-line analysis tool for the single sperm single point laser trapping and ring trapping studies (Shi *et al.*, 2006a). The microscope stage was controlled by a joystick to locate and place a sperm-of-interest under the laser trap. Sperm was held in the trap with either constant or decaying laser power. Prior to laser trapping, the microscope stage was momentarily halted (3–5 s) to record video footage of the swimming sperm, after which the sperm was positioned in the laser trap. Similar footage was acquired after trapping. The frames of each video sequence were converted off-line into bit map format (BMP) with Adobe Premier Pro 1.5 (Adobe Systems Incorporated, San Jose, CA).

SSTA loads the images saved in the hard drive and displays the first frame of a video sequence. A "sperm-of-interest" (i.e., the tracked sperm) is interactively selected with the computer's mouse. A square region of interest (ROI) is automatically created around the selected pixel. Background intensity was estimated, after calculation of the ROI's intensity histogram, as the sum of the most frequent intensity and the standard deviation. The background intensity was subtracted from each pixel in the ROI and the resulting image was linearly stretched to fill the intensity range [0–255].

SSTA uses custom image segmentation to create a binary image mask from each ROI to identify the sperm-of-interest. Two-class thresholding algorithm was found to be unsuitable for tracking sperm in phase contrast. Tracked sperm may approach bright high-contrast debris, nonmotile sperm, or other motile sperm within the ROI. SSTA calls the LabView function "IMAQ AutoMthreshold.vi," which implements an interactive clustering method seeking four pixel classes within the intensity histogram. Then the LabView function "IMAQ Threshold.vi" is called once for each of the three brightest pixel classes to create a corresponding binary mask. The first (and brightest) class contains the central bright region of a particle, while dimmer pixels lie in the second and third classes. The fourth class contains background pixels. Since the bright particle may be represented in all three classes, it must be "merged" again into a single object. Binary dilation with a 3×3 structuring element is applied to the first class and projected onto the second class, setting corresponding pixel values to zero. The process is repeated on the second class pixels, and projected onto the third class. Resulting masks for the three classes are combined with a binary OR operator to calculate the final mask. The LabView function "IMAQ Particle Analysis.vi" measures the area, centroid (x, y), orientation, the bounding box, and the bounding rectangle diagonal for each particle. An area threshold of 10 μm^2 was applied to remove small debris identified in the final mask.

We found that image segmentation alone could not consistently find the sperm-of-interest in the presence of other high-contrast objects. Speed-check uses the curvilinear swimming speed (VCL) to further filter out interfering objects. SSTA uses the nearest neighbor method (Mortimer, 1994) to associate objects in consecutive image frames. For those two frames, the instantaneous-VCL (IVCL) of an object is calculated as displacement divided by elapsed time. When image segmentation fails to identify the tracked sperm, IVCL will take on an uncharacteristic value as compared to VCL calculated over all previous frames. It was found that over time the VCL asymptotically stabilizes, and variations in IVCL with respect to VCL are bound. Let N_{min} be the minimum number of consecutive frames required to estimate the VCL within a defined range. Let R_{max} define the height of a window about VCL which bounds variations in IVCL. Let VCL_{max} be the species-dependent upper bound on IVCL. SSTA exploits these three physiologically derived bounds for its speed-check feature. That is, the most recently calculated IVCL is checked against the following conditions:

$$\text{IVCL} < R_{max} \times \text{VCL} \quad \text{if } N > N_{min} \tag{1}$$

where N is the current frame number.

As multiple sperm swim freely in the cell chamber, their paths often intersect. Collision-detection is used to detect these events and to engage the algorithm of the next section in order to identify the correct sperm, post intersection. In some cases, two sperm will collide resulting in a change of swimming direction, while in other cases, axially separated sperm pass near each other only appearing to collide. These were termed "real" and "perceived" collisions, respectively (Mortimer, 1994). In the transverse image plane, two sperm may pass very close without colliding (remaining mutually distinguishable). This is termed a "near-miss." It is possible, however, that a single sperm in one frame could be considered the nearest neighbor to both of the two sperm from the previous frame. In such a case, SSTA will regard the "near-miss" as a collision. SSTA employs collision-detection to monitor for these three collisions. From the first frame, that additional particles (including sperm) are detected in the ROI; an additional ROI is created for each particle. The nearest neighbor method is then run in parallel for each particle within its ROI. The size of the ROIs were chosen such that, in accordance with Nyquist sampling theory, two sperm traveling directly toward each other at VCL_{max} would require at least two frames before they could collide, thus avoiding aliasing.

Let N_i be the first frame in which two sperm are detected in the ROI of the tracked sperm. Collision-detection registers a collision if any of the following occurs:

- Two sperm are no longer distinguishable in Frame N_h, where $h > i$, and two distinguishable sperm with different centroid positions (X_j, Y_j) and (X_j', Y_j') are detected in Frame N_j, where $j > h$.
- Two sperm in Frame N_{j-1} share the same nearest neighbor with centroid position (X_j, Y_j) in Frame N_j, where $j > i$, while another sperm in Frame N_j with a different centroid X'_j, is closer to (X_j, Y_j).

Three linearly independent criteria based on measurements of VCL (R_V), net displacement (R_D), and swimming angle (θ) were combined into a single cost function evaluated for each sperm such that the sperm with the greatest cost is identified as the correct sperm. In order to correctly calculate the VCL post collision, ΔN_c was chosen to be twice the minimum frame number N_{min}. Collision-free trajectories were tracked to form distributions of the following two quantities:

$R'_{V_{i,j}}$, the ratio between VCL calculated from frame i to frame $(i + \Delta N_c)$, and VCL calculated from frame 1 to frame i (excluding ΔN_c points at each end) for eachtrajectory j.

$$R'_{V_{i,j}} = \frac{\sum_{h=i}^{i+\Delta N_c} \text{IVCL}_{h,j}/\Delta N_c}{\text{VCL}_{i,j}} \tag{2}$$

$R'_{V_{i,j}}$ measures the deviation of VCL calculated for frames i to $(i + \Delta N_c)$ with respect to VCL calculated on all previous i frames.

$R'_{D_{i,j}}$, the ratio between net sperm displacement observed over ΔN_c frames after and before frame i for trajectory j.

$$R'_{D_{i,j}} = \frac{\sqrt{\left(x_{i+\Delta N_c,j} - x_{i,j}\right)^2 + \left(y_{i+\Delta N_c,j} - y_{i,j}\right)^2}}{\sqrt{\left(x_{i,j} - x_{i-\Delta N_c,j}\right)^2 + \left(y_{i,j} - y_{i-\Delta N_c,j}\right)^2}} \tag{3}$$

$R'_{V_{i,j}}$ or $R'_{D_{i,j}}$ will be unity whenever VCL or net sperm displacement are unaltered by the collision. SSTA transforms $R'_{V_{i,j}}$ and $R'_{D_{i,j}}$ through the Gaussian probability density function

$$f(z) = \mathrm{e}^{-(1/2)(a\times(z-z_0))^2} \tag{4}$$

where z is $R'_{V_{i,j}}$ or $R'_{D_{i,j}}$, z_0 is unity, and the standard deviation $1/\alpha$ serves as a tunable parameter whereby the user can change the "steepness" of the transition from one to zero.

The cost function for swimming angle, θ is chosen to favor small angle changes over large ones. The cost function is defined as:

$$C_{\theta k} = \frac{1 - \cos\theta_k}{2} \tag{5}$$

In Eq. (5), C_θ increases monotonically from zero to one as θ increases from zero to 180°.

The three cost functions for sperm k are multiplied:

$$C_k = C_{Vk} \times C_{\theta k} \times C_{Dk} \tag{6}$$

and the sperm with the highest C_k is selected by SSTA as the correct sperm to continue tracking.

2. Real-Time Automated Tracking and Trapping System

A key feature of RATTS is the ability to track sperm at video rates and to update the microscope stage position to keep a swimming sperm in the field of view. RATTS is also custom coded in LabView to process streaming images, calculate sperm trajectories, and drive the motion hardware. RATTS implements the image segmentation and SSTA. Images are digitized by the image acquisition board and transferred into a continuous buffer from which they are retrieved for image analysis and displayed in the front panel (Graphical User Interface in LabView). Image analysis detects when a tracked sperm has reached a rectangular boundary near the extremity of the camera field of view (about 52% of the field). RATTS moves the microscope stage to position the calculated sperm centroid into the center of the field. Swimming parameters are calculated and saved in a continuously updated data file. Since new images arrive at 30 frames per second, it is necessary to restrain net computation and data writing time to less than 33 ms in order to capture and process each image. RATTS is coded to use the most recent frame in the buffer and consequently if image analysis time is more that 33 ms, an image will be skipped, and the next image will be processed. To benchmark RATTS performance, skipped frames were documented and process times were benchmarked using LabView's timing tools. Sperm were recorded and tracked for extended durations to demonstrate variability in swimming parameters and variation in VCL as a function of track length and integration interval.

The automated trapping feature of RATTS replaces the manual protocol described in the previous section. User input is limited to setting parameters prior to an experiment and selecting, via the computer mouse, a sperm in the field for analysis. The user enters the number of image frames to be captured and stored prior to and after trapping. The user can select the method of laser exposure: (1) laser power is held constant for a fixed duration in the trapping phase of the experiment, (2) laser power is decayed during trapping. Parameters are entered for maximum (or constant) laser power, rate of power decay, and if appropriate, duration of the trap. During the experiment, the user selects a sperm to be analyzed by clicking on its image with the arrow cursor on the front panel of RATTS. The curser coordinate is registered, passed to the tracking algorithm and computation proceeds with no further intervention. Once the specified number of frames has been processed, the stage is moved to place the sperm under the laser trap, and the laser shutter is opened.

In one mode of operation (selected prior to the experiment), the stage update can be performed iteratively a few times before the trap is opened. Since image analysis does not occur during stage movement, errors in the sperm position arise from sperm swimming during that movement and from positioning errors inherent in the rotary encoded stepper motor stage. If the sperm center is not adequately aligned with the laser focus, the scatter force of the trap will push the sperm out of focus, and the sperm will be lost. After the first stage movement, subsequent movements are relatively rapid as the sperm is already near its target position. As mentioned previously, a predictor may also be used to predict the post stage translation position of the sperm.

During the trapping phase of the experiment, RATTS implements an escape detection subroutine to detect the presence of a sperm in the laser trap and to respond if the sperm escapes the trap. The subroutine monitors a small square pixel region (representing ~10 μm per side) centered about the laser trap. Using the SSTA algorithm, the subroutine segments the image within this region and uses size threshold to detect the presence or absence of a sperm. A sperm must remain in the trap for a continuous 15 frames or the subroutine declares a failed trapping attempt.

III. Materials

A. Specimen

Semen samples were collected from several species: domestic dog, Nubian ibex, domestic cat (provided by the Beckman Center for Conservation and Research of Endangered Species, Zoological Society of San Diego), human (provided by IGO Medical Group, San Diego, CA), and rhesus macaque monkey (provided by University of California, Davis, Department of Anatomy, Physiology, and Cell Biology). Semen was either examined fresh or was cryopreserved (stored in liquid N_2, 77 K) for future examination (DiMarzo *et al.*, 1990; Durrant *et al.*, 2000; Harper *et al.*, 1998; Serfini and Marrs, 1986; The Ethics Committee of the American Fertility Society, 1986). Rhesus macaque (*Macaca mulatta*) semen samples collected for fresh analysis were prepared for cooled (5 °C) storage (as opposed to frozen) in the following manner. Four milliliters (mL) of media (0.5 mg of bovine serum albumin (BSA) per 1 mL of Biggers, Whittens, and Whittingham (BWW); Biggers *et al.*, 1971) is added to the monkey semen and rotated for 5 min on a rocker. The coagulum is pulled up onto the side of the tube and is left to sit for 10 min. The top 3.5 mL of sperm supernatant is transferred into a new tube. The sperm motility and concentration is checked. The sample is centrifuged at $300 \times g$ for 10 min then resuspended at 50×10^6 mL^{-1} in EZ-Mixin. The sample is loaded into Whirl-Pak bags ensuring minimal air space then placed in an Equitainer (coolant cans on bottom) and shipped to the lab. The sample arrives at approximately 24 h and is analyzed after warming the sample (37 °C) for 5–10 min.

Frozen semen samples are thawed in a water bath (37 °C) for ~1–2 min (human samples thawed for 10 min). The contents are transferred to an Eppindorf centrifuge tube. For those samples that were preserved in vials (excluding human), the following protocol was used to isolate the sperm. The sample is centrifuged at 2000 rpm for 10 min (centrifuge tip radius is 8.23 cm). The supernatant is removed and the remaining sperm pellet is resuspended in 1 mL of prewarmed media (1 mg of BSA per 1 mL of BWW, osmolarity of 270–300 mmol/kg water, pH of 7.2–7.4; Biggers *et al.*, 1971). Frozen semen samples in straws were not centrifuged. For these, the entire sample was suspended in 1 mL of media (BWW + 1-mg/mL BSA). For the human frozen semen samples, a twice wash protocol was used to isolate the sperm (DiMarzo and Rakoff, 1986; Toffle *et al.*, 1985). Human sperm are suspended in pre-warmed modified human tubal fluid (mHTF) HEPES buffered (osmolarity 272–288 mOsm/kg water, pH of 7.3–7.5) with 5% serum substitute supplement (SSS) filtered through 0.2-μm syringe filter, first five drops were disposed (Irvine Scientific, Santa Ana, CA). Final dilutions of 30,000 sperm per mL of media (either BWW or mHTF) are used in all of the experiments. For single point trap experiments, the specimen is loaded into a Rose chamber and mounted into a microscope stage holder (Liaw and Berns, 1981). Experiments were conducted using an air curtain incubator (NEVTEK, ASI 400 Air Stream Incubator, Burnsville, VA) interfaced with a thermocouple feedback system to achieve 37 °C at the specimen.

For the ring trap experiments, the specimen is loaded into a 120-μm thick chamber with a glass slide as the top and a No. 1 coverslip as the bottom. The sample is kept at room temperature during the experiment.

IV. Results and Discussion

A. Gradient Single Point Trap

RATTS and SSTA algorithms together are used in the gradient single point trap application. The rotation rate of the motorized mount which holds the polarizer is programmed to produce a linear power decay from maximum (100%) to minimum ($\approx$0%) in 10 s. A maximum trapping power of 366 mW in the specimen plane was used in these experiments. Studies were conducted on sperm samples pooled from four dogs to assess the effects of the duration a sperm is exposed to the laser trap as well as the laser trapping power on sperm motility. It was concluded that a maximum duration of 10 s with less than 420-mW laser power posed no significant effect on sperm motility (Nascimento *et al.*, 2006).

A sperm-of-interest is observed for 100 frames (3.33 s) before and after trapping. Once a sperm is trapped, the power decay is initialized within 15 frames (0.5 s). The moment the sperm escapes the trap, the power decay is stopped, and the escape power is recorded by the computer. RATTS calculates the VCL, and exports the (x, y) coordinates of the swimming trajectory for off-line calculation of straight line velocity (VSL), total distance traveled, and ratio of displacement to total distance traveled. The swimming force, or escape power (P_{esc}), and swimming velocity (VCL) distributions are statistically compared between species. P_{esc} is plotted against VCL to determine if there is a relationship between the two parameters for each species. Regressions (robust fitting, bisquare method) are fit to the data and are statistically compared to test the uniqueness of the relationship between swimming speed and swimming force for each species.

Figure 5 plots the average P_{esc} and average VCL for each species. Wilcoxon paired sample tests (Zar, 1984) for equal median (5% significance level) were performed on P_{esc} and VCL distributions of sperm from different species (Table I). The majority of the species distributions of P_{esc} and VCL are statistically different. Specifically, domestic cat, Nubian ibex, and human sperm populations are not similar to any other species. However, rhesus macaque VCL and P_{esc} distributions are equivalent to only that of domestic dog. This suggests that sperm from different species are generally unique in both swimming speed and swimming force.

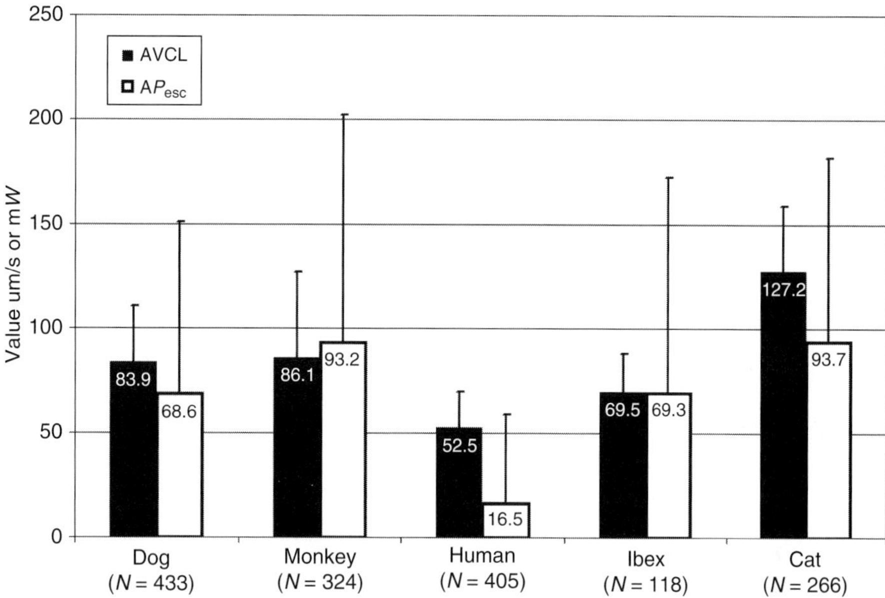

Fig. 5 Average escape power and average swimming velocity for various species: actual average values are listed. Standard deviation shown as error bars. Number of data points (*N*) for each species is listed.

Table I
Comparing VCL and P_{esc} of Different Species[a]

Versus	Domestic dog	Domestic cat	Human	Nubian ibex	Macaque
Domestic dog	X	Unequal VCL medians	Unequal VCL medians	Unequal VCL medians	Equal VCL medians
Domestic cat	Unequal P_{esc} medians	X	Unequal VCL medians	Unequal VCL medians	Unequal VCL medians
Human	Unequal P_{esc} medians	Unequal P_{esc} medians	X	Unequal VCL medians	Unequal VCL medians
Nubian ibex	Unequal P_{esc} medians	Unequal P_{esc} medians	Unequal P_{esc} medians	X	Unequal VCL medians
Macaque	Equal P_{esc} medians	Unequal P_{esc} medians	Unequal P_{esc} medians	Unequal P_{esc} medians	X

[a]The Wilcoxon paired sample test is applied to compare the swimming speed (VCL) and swimming force (P_{esc}) distributions between species. All are tested at 5% significance level (if medians equal, $P < 0.05$, if medians unequal, $P > 0.05$.)

Figures 6–10 plot P_{esc} against VCL for all species analyzed. Figures 6–8 for domestic dog, domestic cat, and human, respectively, show that with increasing swimming speed there is an increase in swimming force. Figures 9 and 10 for the

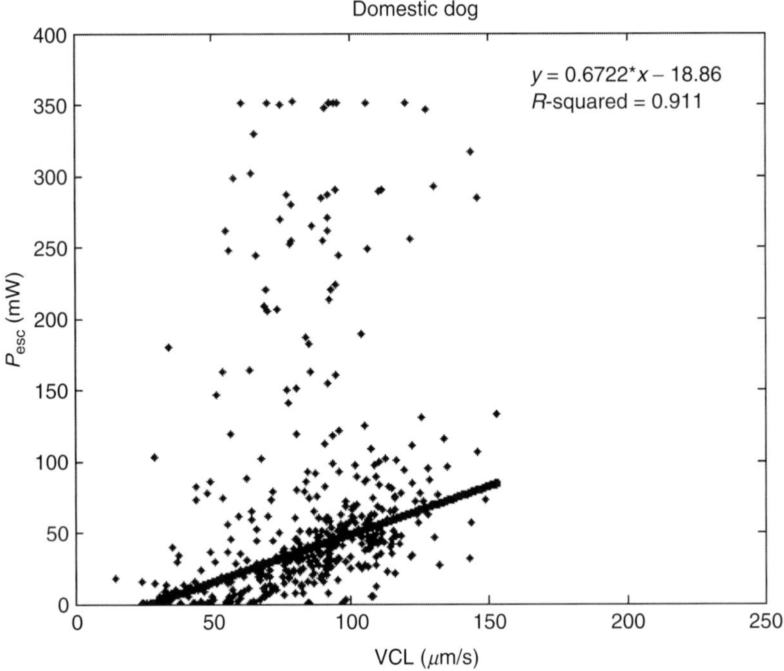

Fig. 6 Domestic dog sperm: P_{esc} (mW) versus VCL (μm/s). A regression is fit to the data set. The inset shows the equation of the regression and the R-squared value.

Nubian ibex and rhesus macaque, respectively, show a random scatter plot, where no obvious relationship between swimming speed and swimming force is identifiable. Regressions using robust fitting (bisquare method) are applied to each data set, whether or not a general trend between swimming force and swimming speed is evident (insets in Figs. 6–10 show the slopes, y-intercepts, and R-squared values). Statistical tests comparing the regression slopes were performed (Zar, 1984; Table II). Interestingly, the majority of the regression slopes are not statistically equal. The regression slope of human sperm is not equivalent to that of any other species. Domestic dog and domestic cat regression slopes are equivalent, as well as those of Nubian ibex and rhesus macaque.

In order to analyze effects of mating type, a direct comparison of the rhesus macaque (an old world monkey with a multimale–multifemale mating system and sperm with relatively larger midpiece volumes) and the human (considered in general to employ a monogamous or polygynous mating system and sperm with relatively smaller midpiece volumes) can be made. Figure 5 shows the dramatic difference between the average VCL and average P_{esc} between the two species, where the macaque P_{esc} and VCL averages are much greater than those of the

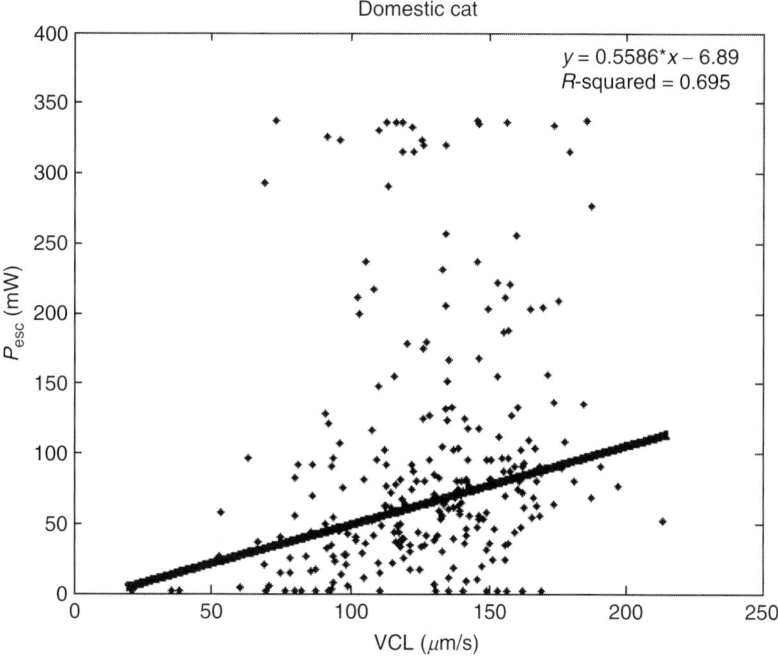

Fig. 7 Domestic cat sperm: P_{esc} (mW) versus VCL (μm/s). A regression is fit to the data set. The inset shows the equation of the regression and the *R*-squared value.

human. This supports the studies discussed earlier that suggest that sperm with larger midpiece volumes swim with stronger forces and faster velocities.

Another key result is the "outlier" sperm that escape the trap at higher laser powers (roughly greater than 300 mW of trapping power) than the majority of the sperm analyzed. In the study done on domestic dog sperm (Nascimento *et al.*, 2006), "outliers" were neither identifiable by their SOP score nor as an independent group using a statistical principal component analysis. The "outlier" sperm are present in semen from every species analyzed and, therefore, are not exclusive to any one particular species. If verified by additional testing, these "outliers" could represent a class of sperm that have unique physiological and behavioral phenotypes. It is possible that the strongest sperm, but not necessarily the fastest, are the sperm that are successful in fertilizing the egg. This would support the theory of "egg-getter" sperm (Dixson, 1998). In this case, as suggested earlier, the theory is not limited to sperm from multimale–multifemale mating species, but rather can be extended to all sperm, regardless of its species mating type. It is also feasible that outlier sperm are in a state of increased energetics (at a higher rate of ATP production and/or consumption) and may respond to physical/chemical barriers in a distinctly different manner than the rest of the sperm in the semen sample.

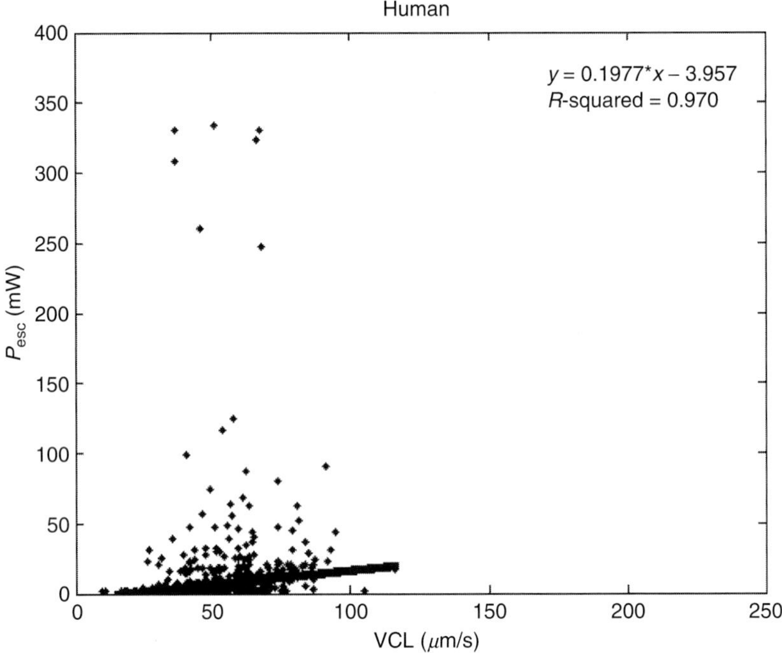

Fig. 8 Human sperm: P_{esc} (mW) versus VCL (μm/s). A regression is fit to the data set. The inset shows the equation of the regression and the R-squared value.

B. Annular (Ring-Shaped) Laser Trap

Five-micrometer diameter silica microspheres (SS06N, Bangs Laboratories Inc., Fishers, IN) are used to evaluate the performance of the annular trap. A microsphere water suspension is put into a 120-μm thick chamber with a glass slide as the top and a No. 1 cover-slip as the bottom. The lateral/axial trapping force is determined by moving the sample stage/objective and measure the escape velocity of the trapped beads. Assuming spherical object symmetry and laminar fluid flow, the fluidic drag on an object is determined from the Navier–Stokes equation. At the escape velocity, v_e, the optical trapping force is equal to the fluidic drag force such that $F_{opt} = F_{drag} = 6\pi\eta r v_e$ (Batchelor, 1991; Smith *et al.*, 1999), where $\eta = 1 \times 10^{-3}$ N S/m^2 for water and r is the radius of the particle.

With an estimated average trapping power of 23 mW per microsphere, a lateral trapping force of 8.14 pN and an axial trapping force is 0.24 pN could be obtained. Figure 11 shows the 3D trapping of 5-μm silica microspheres with 63× oil immersion objective (NA = 1.4, Zeiss, Germany). Figure 11A–B corresponds to two different x-positions of the stage, demonstrating the confinement of particles along the circumference of the ring. Figure 11C–D depicts the lifting of the trapped particles while adjusting the z-position of the microscope objective.

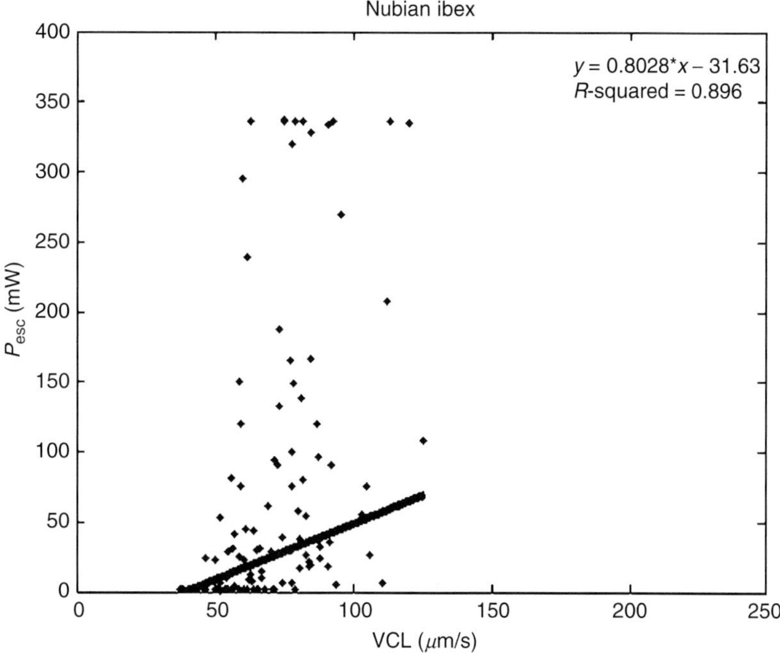

Fig. 9 Nubian ibex sperm: P_{esc} (mW) versus VCL (μm/s). A regression is fit to the data set. The inset shows the equation of the regression and the R-squared value.

For the application of the ring trap on sperm analysis, first, we demonstrate sperm sorting according to their motility parameters. VCL, smooth path velocity (VAP), and amplitude of lateral head displacement (ALH) were measured using custom software that tracks the sperm (Section II.B.1 of this chapter; Shi *et al.*, 2006a). With 25-mW estimated laser power per sperm (total power in the specimen plane divided by the maximal number of sperm the ring could accommodate), sperm could be classified into two groups according to their response to the ring trap. The "fast" group is defined as sperm swimming across the ring with no detectable speed reduction, whereas the "slow" group represents sperm whose swimming pattern are affected by the ring, that is, experiencing a slowing down, a temporary or permanent loss of motility, or a change in swimming trajectory. As could be seen from Fig. 12, the "fast" group has a much higher VCL and VAP than the "slow" group, while their ALH are very similar. The Student's t-test shows a p-value of 1.2×10^{-15} for VCL, 1.2×10^{-16} for VAP, and 0.3 for ALH, which means the "fast" group can be significantly differentiated from the "slow" group with a very high (>99.99%) confidence according to VAP or VCL; however, ALH is a much less reliable way to distinguish the two groups.

Second, the effect of the annular laser trap on sperm motility was studied by making sperm swim along the curvature of the ring trap (Fig. 13). Because of the

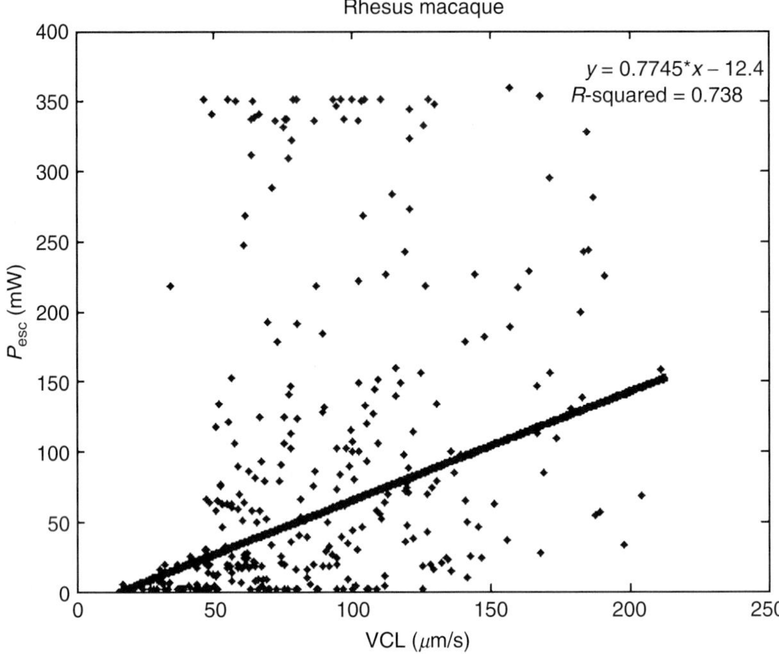

Fig. 10 Rhesus macaque sperm: P_{esc} (mW) versus VCL (μm/s). A regression is fit to the data set. The inset shows the equation of the regression and the R-squared value.

Table II
Comparing Regressions for Different Species[a]

Versus	Domestic cat	Human	Ibex	Macaque
Domestic dog	Equal ($t < 1.96$)	Unequal ($t > 1.96$)	Unequal ($t > 1.96$)	Unequal ($t > 1.96$)
Domestic cat	–	Unequal ($t > 1.96$)	Unequal ($t > 1.96$)	Unequal ($t > 1.96$)
Human	–	–	Unequal ($t > 1.96$)	Unequal ($t > 1.96$)
Nubian ibex	–	–	–	Equal ($t > 1.96$)

[a]Comparison of slopes of regressions fit to the data sets. The t-value of 1.96 was obtained from Zar, 1984, based on an α of 0.05, two tails, and the sum of degrees of freedom for the two species compared.

optical properties of an axicon, the annular laser trap only offers gradient force in the radial direction. Accordingly, trapped particles are free to move in the circumferential direction as long as their tangential movement is not based on their radial action. However, for flagellated cells like sperm, forward movement results from the viscous interactions of the sperm flagellum with the medium (Taylor, 1951), which is coupled with the lateral direction. Therefore, the radial confinement of a sperm swimming along the ring should reduce its motility. To test this hypothesis,

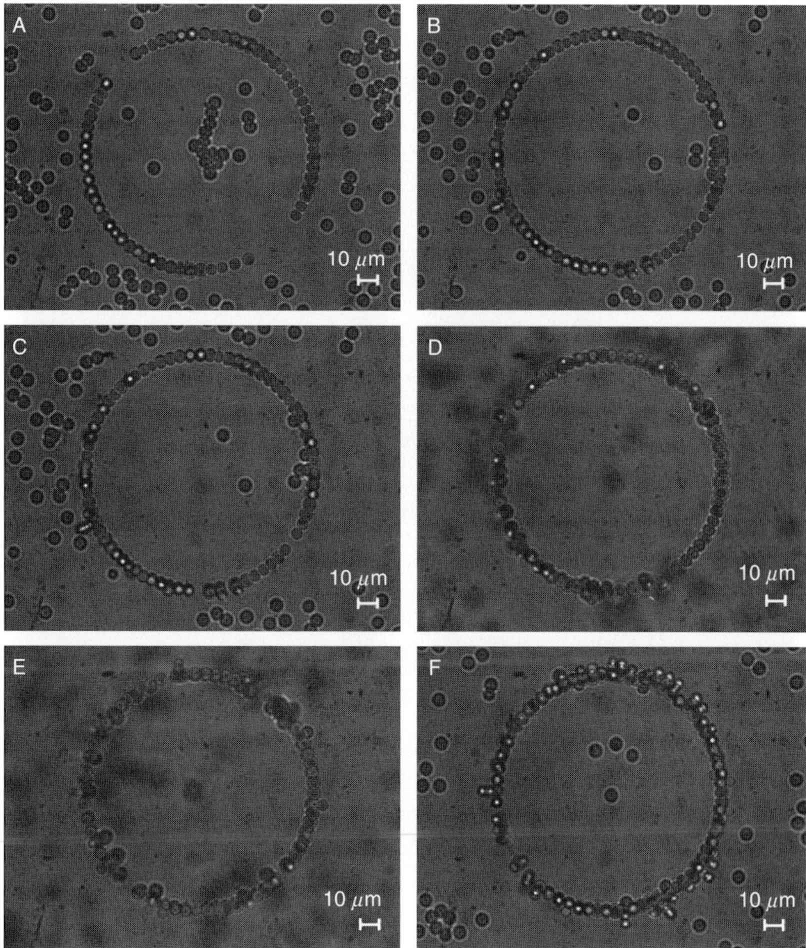

Fig. 11 Sequences of video frames showing (A, B) a ring of trapped particles being moved in the transverse direction, and then (C, D) being lifted 100 μm in the axial direction. (Shao *et al.*, 2006b).

laser power was adjusted so that the swimming pattern of sperm under different amounts of optical force and illumination was examined.

Five types of observations were made, which are identified as "power binary," "power gradient," "fatigue," "load," and "block." In "power binary," the laser is switched on and off so that the swimming parameters are measured for a sperm while it is swimming along the ring and then it is released from the ring when the laser is turned off. "Power gradient" measures the change in sperm swimming pattern as a result of the laser power change. "Fatigue" examines the slowing down of a sperm after it is guided along the ring under a fixed trapping power for an extended period of time. "Load" means a sperm is swimming while pushing another object, such as

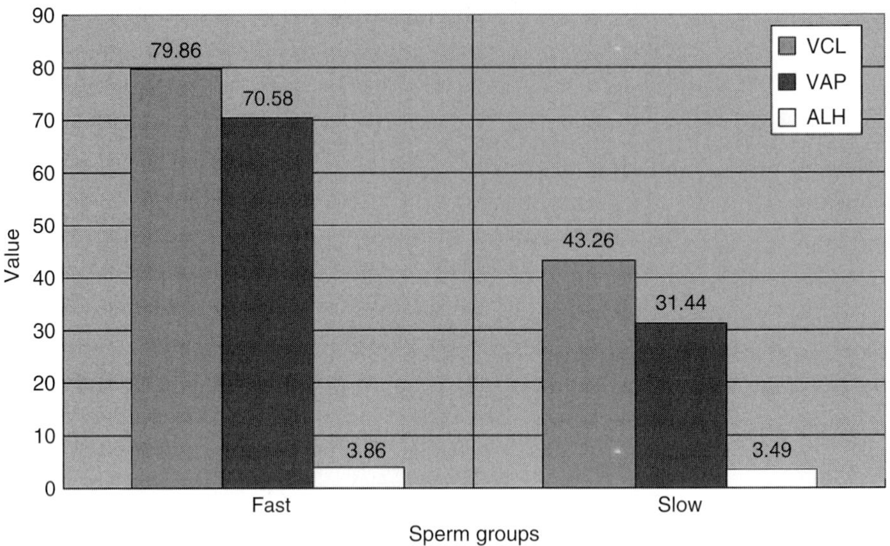

Fig. 12 Parallel sperm sorting with annular laser trap when the estimated average laser power is 25 mW per sperm. Mean value of VCL, VAP, and ALH for "fast" and "slow" sperm from a population of 83 sperm. (Shao *et al.*, 2007).

a dead sperm or cell debris. Finally, "block" studies the behavior of a sperm while its forward movement is interrupted by an obstacle.

A total of 93 sperm were measured and 161 observations were made. For the 64 observations of "power binary," about 80% of the sperm show increased VAP or VCL when they are not swimming along the ring. Only 43% show an increase in ALH. For the 92 occurrences of "power gradient," about 75% of the sperm experience increased VAP or VCL with decreased trapping power, while only 45% show an ALH increase with power decay. Among nine observations of "fatigue," with increased time (typically after 15–20 s along the ring), about 80% have reduced VAP's, 100% have reduced VCL's, and 76% have reduced ALH's. Under the category of "load," 100% of a total of four sperm exhibit a reduction of VAP and VCL, and 75% have a reduction of ALH. For the two sperm "blocked," 100% show a decrease in VAP, VAL, and ALH.

According to the reported linear relationship between laser trapping power and sperm motility (Nascimento *et al.*, 2006; Tadir *et al.*, 1990), it is possible to adjust the average trapping power per sperm so that different thresholds are used for multilevel sorting. When the total input power is limited or fixed, changing the diameter of the annular trap leads to a variation of the trapping power in the different laser spots. This could be achieved by adding two more axicon lenses between the focusing lens and the tube lens, and changing their axial separation (Shao *et al.*, 2006a). As a result, sperm with different motility rates will escape the

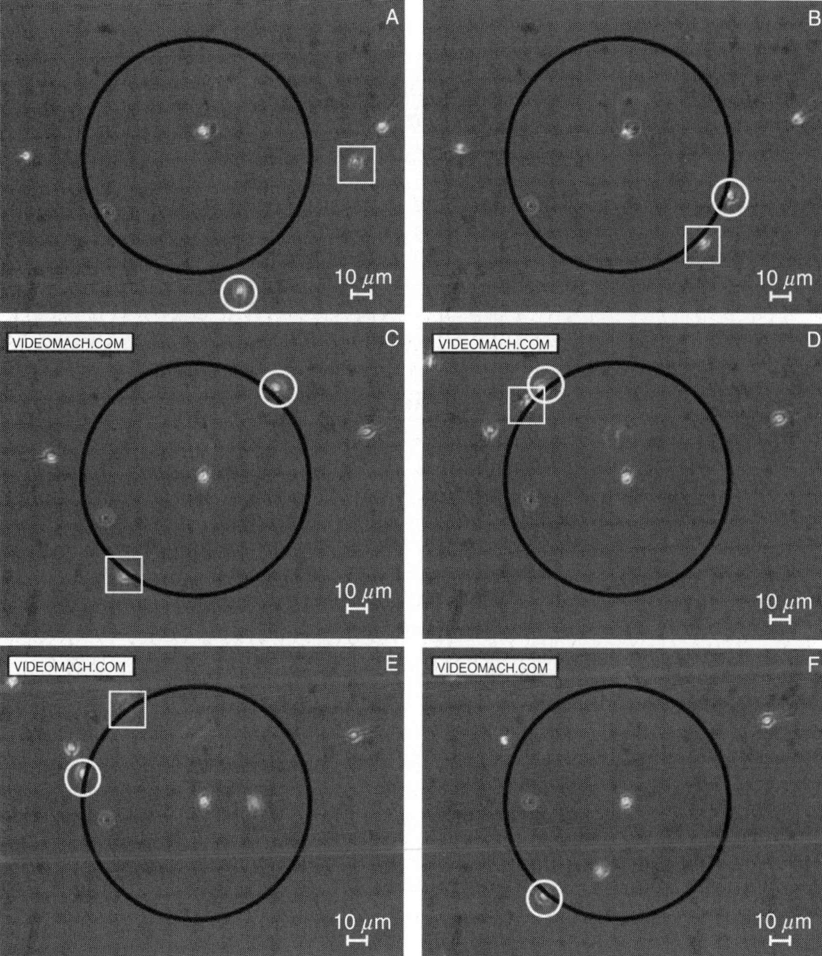

Fig. 13 Guiding two sperm along the ring. (A) Two sperm are freely swimming in opposite directions close to the ring (big black circle). (B) Affected by the optical gradient force from the ring trap, the two sperm start swimming along the curvature of the ring, one clockwise (square), the other counterclockwise (small circle). (C) The two sperm continue swimming in opposite direction. (D) After swimming along the ring for about 180°, the two sperm are about to collide with each other. (E, F) After collision, one sperm (square) is knocked out of the focus, while the other (small circle) keep swimming along the ring. (Shao *et al.*, 2007).

trap at different times. Additionally, since the size of sperm may vary between species, adjusting the size of the ring trap makes it possible to study different species without a complete redesign of the optics. Finally, an adjustable ring can be used to study the diffusion length of an attractant, and monitor sperm swimming behavior under radial confinement (optical gradient force in the radial direction) with various curvatures.

In the current study, the sperm in the chamber swim in random directions. This condition could be changed by introducing a chemo-attractant to the center of the ring trap. Due to chemotaxis, sperm should start approaching the attractant from all the directions. As a result, the ring trap could be used to sort sperm according to their chemotactic response, which would also help in understanding sperm competition and other aspects of fertility.

V. Summary

SSTA has been developed for use in laser trapping studies on sperm motility (Nascimento *et al.*, 2006). The algorithm was developed because the turnkey HTM-IVOS system committed frequent tracking errors as sperm swam out of focus. This is likely due to the high NA objective used in this study, which limits the depth of field to the scale of a sperm head. Consequently, even slight changes in axial position during swimming can significantly change the contrast of a sperm. The HTM-IVOS system is intelligently designed so that the chamber thickness matches the working distance of the objective lens and is therefore not sensitive to slight axial displacements of the sperm. It is reasonable that the algorithm used in that system struggled with the images in the laser trapping application.

To address the focus issue, SSTA uses standard image contrast enhancement and custom four-class thresholding to extract sperm swimming with transient focus quality. SSTA's speed-check feature acts as a check on the thresholding method to find out-of-focus sperm swimming near high-contrast debris. To handle collisions, SSTA tracks all sperm in the ROI prior to and after a collision and uses its post-collision analysis to statistically identify each sperm.

The algorithm and hardware integration presented in RATTS has achieved fast, automated sperm motility analysis with automated laser tweezers force measurements. The automation offered by RATTS has allowed us to investigate the nature of the outlying sperm by removing the inefficient manual method of control and analysis. RATTS allows us to examine not only the swimming speed–escape power relationships but also the effects that the trap may have on sperm motility.

The gradient single point laser trap provides a tool to measure sperm swimming forces. This measurement in conjunction with established swimming characteristic measurements, such as VCL, can help distinguish sperm of different species and provide a baseline for comparison. It also can be used to define subpopulations within a semen sample. Namely, it can identify the sperm that exhibit higher motility characteristics. The potential separation of these sperm from the majority could be of considerable value in both basic and applied infertility research.

The annular laser trap provides the possibility of high-throughput sperm sorting based on motility and chemotaxis. With only tens of milliwatts devoted to each sperm, this new type of laser trap offers a more gentle way for sperm analysis and laser–sperm interaction studies. The strong optical gradient in the radial direction and zero gradient force in the circumferential direction make it possible for sperm to swim along the ring without having to stop. The unique geometrical feature of the "ring" allows a sperm to be confined in the field of view for an extended period of time without having to deal with sharp turns or changes in swimming curvature. As a result, the effect of optical force, laser radiation and external obstacles on sperm swimming pattern and membrane potential (with the aid of the specific fluorescent dye) can now be investigated in more detail. In conclusion, we have described a system that is a prototype of an objective, parallel, and quantitative analytical tool for dynamic studies on cell motility, infertility, and biotropism.

Acknowledgments

This work was supported by funds from the Beckman Laser Institute Inc. Foundation, the Arnold and Mabel Beckman Foundation Beckman Fellows Program award to Elliot L. Botvinick, and a grant from the Air Force Office of Scientific Research (AFOSR #F9620–00–1-0371) to Michael W. Berns. We would like to thank Dr. Barbara Durrant of the Beckman CRES research center of the San Diego Zoological Society for her generosity in providing semen samples of several species, and Professor Stuart Meyers of the University of California, Davis, School of Veterinary Medicine for providing samples of semen from rhesus monkeys. We are particularly grateful to Professors Sadik Esener and Michael Berns for their constant support and advice during the conduct of this work.

References

Amann, R. P., and Katz, D. F. (2004). Reflections on CASA after 25 years. *J. Androl.* **25,** 317–325.

Anderson, M., and Dixson, A. (2002). Motility and the midpiece in primates. *Nature* **416,** 496.

Araujo, E., Jr., Tadir, Y., Patrizio, P., Ord, T., Silber, S., Berns, M. W., and Asch, R. H. (1994). Relative force of human epididymal sperm. *Fertil. Steril.* **62,** 585–590.

Ashkin, A. (1991). The study of cells by optical trapping and manipulation of living cells using infrared laser beams. *ASGSB Bull.* **4,** 133–146.

Ashkin, A. (1998). Forces of a single-beam gradient laser trap on a dielectric sphere in ray optics regime. *Methods Cell Biol.* **55,** 1–27.

Batchelor, G. K. (1991). "An Introduction to Fluid Dynamics." Cambridge University Press, Cambridge.

Berns, M. W. (1998). Laser scissors and tweezers. *Sci. Am.* (*Int. Ed.*) **278,** 52–57.

Berns, M. W., Botvinick, E., Liaw, L., Sun, C.-H., and Shah, J. (2006). Micromanipulation of chromosomes and the mitotic spindle using laser microsurgery (laser scissors) and laser-induced optical forces (laser tweezers). *In* "Cell Biology: A Laboratory Handbook" (J. E. Celis ed.), 3rd ed., Vol. 3, pp. 351–363. Elsevier Press, Burlington.

Biggers, J. D., Whitten, W. D., and Whittingham, D. G. (1971). The culture of mouse embryos *in vitro*. *In* "Methods of Mammalian Embryology" (J. C. Daniel, Jr., ed.), pp. 86–116. Freeman, San Francisco.

Dantas, Z. N., Araujo, E., Jr., Tadir, Y., Berns, M. W., Schell, M. J., and Stone, S. C. (1995). Effect of freezing on the relative escape force of sperm as measured by a laser optical trap. *Fertil. Steril.* **63,** 185–188.

DiMarzo, S. J., and Rakoff, J. S. (1986). Intrauterine insemination with husband's washed sperm. *Fertil. Steril.* **46,** 470–475.

DiMarzo, S. J., Huang, J., Kennedy, J. F., Villanueva, B., Herbert, S. A., and Young, P. E. (1990). Pregnancy rates with fresh versus computer-controlled cryopreserved semen for artificial insemination by donor in a private practice setting. *Am. J. Obstet. Gynecol.* **162,** 1483–1490.

Dixson, A. (1998). "Primate Sexuality: Comparative Studies of the Prosimians, Monkeys, Apes, and Human Beings." Oxford University Press, Oxford.

Durrant, B. S., Harper, D., Amodeo, A., and Anderson, A. (2000). Effects of freeze rate on cryosurvival of domestic dog epididymal sperm. *J. Androl. Suppl.* **21,** 59.

Eisenbach, M., and Tur-Kaspa, I. (1999). Do human eggs attract spermatozoa? *BioEssays* **21,** 203–210.

The Ethics Committee of the American Fertility Society (1986). The new guidelines for the use of semen for donor insemination. *Fertil. Steril.* **6**(Suppl.), 85.

Harper, S. A., Durrant, B. S., Russ, K. D., and Bolamba, D. (1998). Cryopreservation of domestic dog epididymal sperm: A model for the preservation of genetic diversity. *J. Androl. Suppl.* **19,** 50.

Konig, K., Svaasand, L., Liu, Y., Sonek, G., Patrizio, P., Tadir, Y., Berns, M. W., and Tromberg, B. J. (1996). Determination of motility forces of human spermatozoa using an 800 nm optical trap. *Cell. Mol. Biol.* (*Noisy-le-grand*) **42,** 501–509.

Le Pichon, J. P., and Quero, J. C. (1994). Microcomputer-based system for human sperm motility assessment and spermatozoon track analysis. *Med. Biol. Eng. Comput.* **32,** 472–475.

Liaw, L. H., and Berns, M. W. (1981). Electron microscope autoradiography on serial sections of preselected single living cells. *J. Ultrastruct.* **75,** 187–194.

Mortimer, D. (1994). "Practical Laboratory Andrology." Oxford University Press, New York.

Nascimento, J., Botvinick, E., Shi, L. Z., Durrant, B., and Berns, M. W. (2006). Analysis of sperm motility using optical tweezers. *J. Biomed. Opt.* **11**(4), 044001.

Olson, M. A., Yan, H., DeSheng, L., Hemin, Z., and Durrant, B. (2003). Comparison of storage techniques for giant panda sperm. *Zoo. Biol.* **22,** 335–345.

Ozkan, M., Wang, M. M., Ozkan, C., Flynn, R. A., and Esener, S. (2003). Optical manipulation of objects and biological cells in microfluidic devices. *Biomed. Microdevices* **5,** 47–54.

Patrizio, P., Liu, Y., Sonek, G. J., Berns, M. W., and Tadir, Y. (2000). Effect of pentoxifylline on the intrinsic swimming forces of human sperm assessed by optical tweezers. *J. Androl.* **21,** 753–756.

Samuels, J. S., and Van der Horst, G. (1986). Sperm swimming velocity as evaluated by frame lapse videography and computer analysis. *Arch. Androl.* **17,** 151–155.

Serfini, P., and Marrs, R. P. (1986). Computerized staged-freezing technique improves sperm survival and preserves penetration of zona-free hamster ova. *Fertil. Steril.* **45,** 854–858.

Shao, B., Esener, S., Nascimento, J., Berns, M. W., Botvinick, E., and Ozkan, M. (2006b). Size tunable three-dimensional annular laser trap based on axicons. *Opt. Lett.* **31,** 3375–3377.

Shao, B., Nascimento, J., Botvinick, E., Berns, M. W., and Esener, S. (2006a). Dynamically adjustable annular laser trapping based on axicons. *Appl. Opt.* **45**(25), 6421–6428.

Shao, B., Shi, L. Z., Nascimento, J., Botvinick, E., Ozkan, M., Berns, M. W., and Esener, S. (2007). High-throughput sorting and analysis of human sperm with a ring-shaped laser trap. *Biomed. Microdevices*, DOI 10.1007/S10544-006-9041-3.

Shao, B., Vinson, J., Botvinick, E., Esener, S., and Berns, M. W. (2005). Axicon-based annular laser trap for sperm activity and chemotaxis study. *In* "Optical Trapping and Optical Manipulation-II" (K. Dholakia, and G. C. Spalding, eds.), *Proc.* SPIE, **5930,** 1–11.

Shao, B., Zlatanovic, S., and Esener, S. C. (2004). Microscope-integrated micromanipulation based on multiple VCSEL traps. *In* "Optical Trapping and Optical Manipulation" (K. Dholakia, and G. C. Spalding, eds.), *Proc.* SPIE **5514,** pp. 62–72.

Shi, L. Z., Nascimento, J., Berns, M. W., and Botvinick, E. (2006a). Computer-based tracking of single sperm. *J. Biomed. Opt.* **11**(5), 054009.

Shi, L. Z., Nascimento, J., Chandsawangbhuwana1, C., Berns, M. W., and Botvinick, E. (2006b). Real-time automated tracking and trapping system (RATTS). *Micros. Res. Tech.* **69,** 894–902.

Smith, S. P., Bhalotra, S. R., Brody, A. L., Brown, B. L., Boyda, E. K., and Prentiss, M. (1999). Inexpensive optical tweezers for undergraduate laboratories. *Am. J. Phys.* **67,** 26–35.

Stephens, D. T., Hickman, R., and Hoskins, D. D. (1988). Description, validation, and performance characteristics of a new computer-automated sperm motility analysis system. *Biol. Reprod.* **38,** 577–586.

Tadir, Y., Wright, W. H., Vafa, O., Ord, T., Asch, R. H., and Berns, M. W. (1989). Micromanipulation of sperm by a laser generated optical trap. *Fertil. Steril.* **52,** 870–873.

Tadir, Y., Wright, W. H., Vafa, O., Ord, T., Asch, R. H., and Berns, M. W. (1990). Force generated by human sperm correlated to velocity and determined using a laser generated optical trap. *Fertil. Steril.* **53,** 944–947.

Taylor, G. (1951). Analysis of the swimming of microscopic organisms. *Proc. R. Soc. Lond. A.* **209,** 447–461.

Toffle, R. C., Nagel, T. C., Tagatz, G. E., Phansey, S. A., Okagaki, T., and Warrin, C. A. (1985). Intrauterine insemination: The University of Minnesota experience. *Fertil. Steril.* **43,** 743–747.

Wang, Y., Botvinick, E., Zhao, Y., Berns, M. W., Usami, S., Tsien, R. Y., and Chien, S. (2005). Visualizing the mechanical activation of Src. *Nature* **434,** 1040–1045.

Warchol, W., Warchol, J. B., Filipiak, K., Karas, Z., and Jaroszyk, F. (1996). Analysis of spermatozoa movement using a video imaging technique. *Histochem. Cell Biol.* **106,** 521–526.

Zar, J. H. (1984). "Biostatistical Analysis." Prentice Hall, Englewood Cliffs.

CHAPTER 22

Laser Surgery and Optical Trapping in a Laser Scanning Microscope

Jan Scrimgeour, Emma Eriksson, and Mattias Goksör

Department of Physics, Göteborg University
SE-412 96 Göteborg, Sweden

Optical manipulation opens up many new possibilities for experiments in the field of microbiology and is a very powerful tool for investigating cellular structure. In this emerging field imaging retains an important role, and systems that combine advanced imaging techniques with optical manipulation tools, such as laser scalpels or optical tweezers, are an important starting point for researchers. We present a flexible experimental platform that contains both a laser scalpel and optical tweezers, in combination with confocal and multiphoton microscopy. A simple manipulation of the external optics is used to retain the three-dimensional imaging capabilities

0091-679X/07 $35.00
DOI: 10.1016/S0091-679X(06)82022-4

of the microscopes. Two applications of the system are presented. In the first, the laser scalpel is used to initiate diffusion of a fluorescent dye through *Escherichia coli* mutants, which exhibit abnormal cell division, forming filaments, or chains of bacteria. The diffusion assay is used to assess the potential for the exchange of cytoplasmic material between neighboring cells. The second application investigates the binding of endoplasmic reticulum (ER) to chloroplasts in *Pisum sativum* (garden pea). Individual plant protoplasts are ruptured using the laser scalpel, allowing individual chloroplasts to be trapped and manipulated. Strands of the ER which are attached to the chloroplast are identified. The magnitude and nature of the binding between the chloroplast and the ER are investigated.

I. Introduction

The development of microbiology is closely linked with the development of imaging and manipulation techniques which allow the study of cellular structure or cellular responses to particular stimuli. The biggest developments in this field have come alongside advances in laser technology, which facilitated the development of new imaging techniques, such as confocal and multiphoton microscopy. Today a range of direct, sterile, micromanipulation techniques utilizing lasers are also available. Of these new tools the laser scalpel and optical tweezers are the most popular and, in combination with the now well-established imaging techniques, these provide many new avenues for microbiological research.

Laser scanning confocal microscopy (LSCM) made the optical sectioning of live cells and tissues possible, through the exclusion of out-of-focus fluorescence, and allowed the formation of three-dimensional images (Brakenhoff *et al.*, 1979; Sheppard and Wilson, 1980). Multiphoton microscopy has improved on LSCM by limiting the excitation of fluorescence to the image plane of the microscope (Denk *et al.*, 1990; Speiser and Kimel, 1970). In multiphoton processes, excitation of the fluorophore requires the simultaneous absorption of two or more near infrared (NIR) photons, and it is only in the focal volume of the microscope objective that the photon density is high enough to excite a fluorophore efficiently. There is, therefore, a significant reduction in photobleaching away from the focal plane of the microscope in comparison with confocal microscopy. In addition, the localized excitation means that there is no need for the confocal pinhole, leading to more efficient collection of fluorescence from the sample and the achievement of greater penetration depth in tissue samples.

Recently, lasers have been exploited to add yet more flexibility to the microscopy tool kit by using them for micromanipulation of samples and not only imaging. In this area, two of the major applications are the laser scalpel (Berns *et al.*, 1981) and optical tweezers (Ashkin *et al.*, 1986). These contact-free manipulation techniques use strongly focused laser beams to cut or trap biological specimens, allowing sterile manipulation and processing of sensitive samples. Over the past few years, the number of applications for these tools in the biological sciences has exploded.

Laser scalpels are most commonly formed using pulsed lasers with output wavelengths in the ultraviolet (UV) or the NIR. The high energy densities produced in the focus of a high numerical aperture objective lens cause a number of different physical processes to occur, from bond breaking to ablation or plasma formation, depending on the choice of wavelength and properties of the pulse (Vogel and Venugopalan, 2003; Vogel *et al.*, 2005). As the energy density is high only within the focal volume of the microscope objective, interaction between the laser and biological samples can be extremely localized. The scalpel can, therefore, target structures within living biological specimens, such as single nerve cells (Yanik *et al.*, 2004) or subcellular structures such as the cytoskeleton (Colombelli *et al.*, 2005), without damaging the surrounding material.

In contrast, optical tweezers are formed using continuous wave (CW) lasers and although nominally any wavelength can be used, NIR lasers are often the practical choice when working with biological samples. The most common wavelengths used in the field are 830 and 1064 nm as they match local minima in the absorption of water in biological objects and thus heating and optical damage of the specimen is kept to a minimum (Neuman *et al.*, 1999). Optical tweezers have found many applications in biological research from straightforward micromanipulation to high sensitivity force measurements.

We present two biological applications that show how the application of the laser scalpel, in combination with laser tweezers and laser-based imaging, can be used to gain new insight into the structure of biological organisms. In the first application, molecular diffusion through a cell is induced by using the laser scalpel to selectively damage the outer membrane. This technique is used to study the effect of mutations in the transmembrane protein FtsK, which are known to produce aberrations in cell division in *E. coli*. The second application tackles the question of whether there is a physical connection between the chloroplast and the endoplasmic reticulum (ER) in plant protoplasts from *Pisum sativum* (garden pea). These investigations were carried out in combination with confocal and multiphoton microscopy, and we present systems in which these techniques can be combined with the laser scalpel and optical tweezers. We also discuss how to retain the three-dimensional imaging ability of these systems, and the possibility of using the femtosecond laser, available on the multiphoton microscope as a laser scalpel.

II. Optical Manipulation Combined with Confocal/Multiphoton Microscopy

A. Background

The major advantage of using confocal and multiphoton imaging techniques is the ability to prevent the generation or detection of out-of-focus fluorescence. This allows optical sectioning of a sample and production of three-dimensional images based on a stack of sections taken at different heights through the depth of the sample. However, optically trapping a cell prevents the conventional collection of an image stack, which is produced by moving the objective lens relative to the sample. As the trapped object will follow the movement of the objective, the same section of the object is imaged repeatedly. This problem was first addressed by Hoffmann *et al.* (2000) who used an external lens to alter the divergence of the trapping laser entering the microscope. This additional lens was used to compensate for the movement of the trapped object as the objective moved by altering the focal plane of the optical trap. By synchronizing the movements of the microscope objective and the external lens an object can be imaged in three dimensions.

We have demonstrated a simpler route to this goal (Goksör *et al.*, 2004). The microscope objective remains fixed, and thus the confocal system always scans in the same plane with respect to the cover glass. Optical sectioning is then achieved by translating the object through the microscope's focal plane by changing the divergence of the trapping laser. The divergence of the laser is controlled using an external lens. We can therefore freely choose the step size of the axial translation, and there is no need for a complicated algorithm to synchronize the movement of the external lens relative to the movement of the microscope objective. Although this setup does reduce the dynamic range through which three-dimensional images can be acquired, this is of limited importance during single-cell experiments as there is no requirement to image the surrounding environment of the cell.

B. Experimental Setup

Experimental setups combining optical tweezers and a laser scalpel with confocal and multiphoton microscopes are shown in Fig. 1A and B, respectively. The setups are based around an inverted epifluorescence microscope (TE-300, Nikon, Badhoevedorp, The Netherlands), onto which a commercial confocal or multiphoton microscope is mounted via the side port. The additional laser wavelengths required for the laser scalpel and the optical trap are introduced by a custom-made dichroic mirror mounted in the epifluorescence cassette.

The setups are used in conjunction with both a Biorad MRC 1024, Biorad Laboratories, CA confocal scanning unit and a Biorad Radiance 2000 confocal/multiphoton system. These systems make a variety of excitation wavelengths available, allowing us to choose a system tailored to the immediate application. The MRC 1024 offers two lasers: a three-line argon–krypton laser (488, 547, and 647 nm) and a 405-nm laser diode. The Radiance 2000 offers a four line argon laser (457, 476, 488, and 512 nm), a green He–Ne laser (543 nm), and a red laser diode (637 nm) for confocal microscopy, and a femtosecond Ti:Sapphire laser (Mai-Tai, Spectra-Physics, CA) for multiphoton microscopy. This laser is continuously tuneable between 780 and 920 nm with a repetition rate of 80 MHz and a pulse width of 100 fs. Typically, the pulse width is 150–200 fs at the sample due to dispersion within the optical components of the microscope. Peak powers of over 185 kW can be produced corresponding to an average power of 1.5 W at 800 nm.

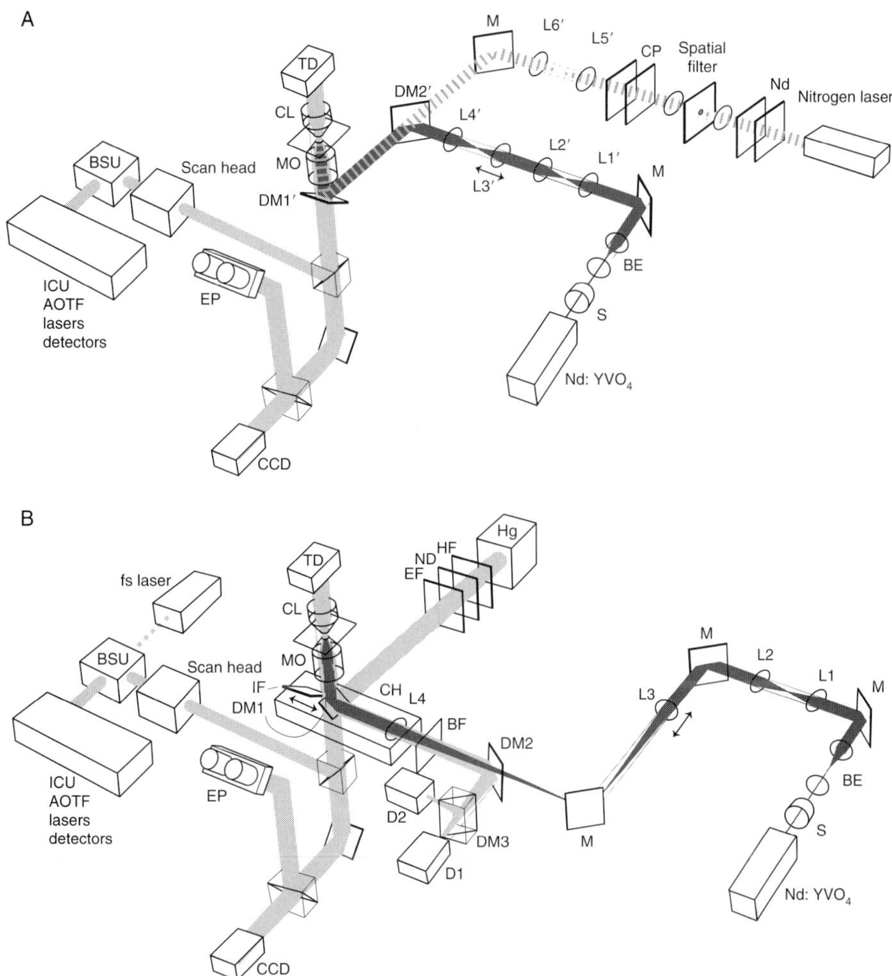

Fig. 1 The experimental setups for the laser scalpel and optical tweezers combined with commercial confocal and multiphoton microscopes. (A) The beam path for the optical microscope with confocal unit is shown in light gray (solid line). This includes the confocal lasers, detectors, and scan head: the ICU, instrument control unit; AOTF, acousto-optic tuneable filter; BSU, beam shaping unit; CH, cassette holder; MO, microscope objective; CL, condenser lens; TD, transmission detector; EP, eyepieces; and a CCD camera for bright field imaging. The beam paths for the laser scalpel and optical trap are shown in light gray (broken line) and dark gray (solid line). The components in these paths are described in detail in the text. (B) The conventional beam path for the multiphoton system is shown in light gray. This is similar to the confocal system but includes a femtosecond laser (fs laser), direct detectors, D1 and D2. Epifluorescence illumination is provided by the components: IF, imaging filter; EF, excitation filter; ND, neutral density filter; HF, heat filter; and Hg, mercury lamp. The beam path for the trapping laser is shown in dark gray and is discussed in the text.

In conjunction with the confocal system (Fig. 1A), the laser scalpel is created using an additional nitrogen laser (337 nm) with a pulse length less than 4 ns and maximum pulse energy of 300 μJ. The required pulse energy is roughly selected using a pair of neutral density filters; these also reduce the power in the beam to a point where it can be safely passed through a spatial filter to improve the mode quality. A pair of crossed polarizers is used to give fine control over the pulse energy entering the microscope. Finally, a pair of lenses, L5′ and L6′, gives control over the divergence of the laser beam and ensures that the laser focus is formed in the image plane of the microscope objective.

The optical trap is formed by a diode-pumped Nd:YVO_4 laser (1064 nm) operating on a Gaussian TEM_{00} mode. The laser is immediately followed by a shutter and a beam expander, and on its way to the microscope the beam is further expanded using lenses L1′ and L2′. L3′ and L4′ are used to control the divergence of the trapping laser. A computer-controlled translation stage allows the lens L3′ to be translated along the optic axis. The outputs from the trapping laser and the scalpel are combined using a dichroic mirror DM2′ before they are reflected along the optical axis of the microscope by DM1′. DM1′ is mounted in the epifluorescence cassette and reflects both the 337- and 1064-nm radiation while it transmits wavelengths between 400 and 900 nm. In some cases, the Nd:YVO_4 laser is replaced by an 830-nm laser diode, and in this case DM1′ is also replaced; where this is the case reference to it is made in the text.

When the Radiance 2000 system is used in multiphoton mode, no additional laser is required to form a laser scalpel. The high power densities produced by the Ti: Sapphire multiphoton source, when it is strongly focused, are used for laser cutting when required during multiphoton experiments (Goksör *et al.*, 2004). The trapping laser is directed into the microscope along the multiphoton collection path which leads to the direct detectors (Fig. 1B). Components L1, L2, L3, and DM1 perform the same functions as L1′, L2′, L3′, and DM1′ in the confocal system, respectively. In this case, L3 is paired with the fixed lens L4 situated in the beam path from the microscope objective to the direct detectors, rather than an external lens. The trapping laser is introduced into the microscope via the arm containing the direct detectors by dichroic mirror DM2.

In both systems, the positions of the laser scalpel and the optical trap are fixed in the center of the microscope objective's field of view. Sample positioning and manipulation are achieved using the microscope's motorized stage to move the sample with respect to this point.

C. Three-Dimensional Imaging

Three-dimensional images are built up by manipulation of the external lens L3 (L3′). This is mounted on a computer-controlled translation stage and has a resolution and repeatability of 2.5 μm. The full range of travel of the translators is 200 mm. The shift in the focus of the optical trap, Δz, in response to a shift in the position of external lens L3 (L3′) Δd, is described by Eq. (1) (Fällman and Axner, 1997), where $f_{\mathrm{objective}}$ and f_{lens} are, respectively, the focal lengths of the microscope objective and the external lens L4. The external translation stage is thus capable of positioning a trapped object with a resolution of 4 nm over a range of 0.32 mm around the focal plane of the microscope objective. The step resolution is, therefore, substantially below the diffraction-limited resolution of the microscope and the spatial resolution of the motorized z-focus of the confocal microscope.

$$\Delta z = \left(\frac{f_{\text{objective}}}{f_{\text{lens}}}\right)^2 \Delta d \tag{1}$$

To demonstrate the three-dimensional imaging in the multiphoton microscope, optically trapped yeast cells with a 4′,6-diamidino-2-phenylindole (DAPI) stained nucleus were observed. When bound to DNA, DAPI has an excitation peak at 358 nm (685 nm for two-photon excitation) and an emission peak at 461 nm (467 nm for 685-nm excitation). The DAPI stain is excited using the femtosecond laser, which is tuned to 810 nm to make use of the high output power available in the near infrared. The resulting fluorescence was detected on one of the direct detectors used for multiphoton imaging. Figure 2 shows a three-dimensional reconstruction of the nucleus of the optically trapped yeast cell produced from a set of optical sections taken with a spacing of 20 nm.

It is important to note that the trapping efficiency is largest when operating near the cover glass. At a distance of 40 μm from the surface of the cover glass, the trapping efficiency is reduced by a factor of 2 due to spherical aberration caused by the refractive index mismatch between the cover glass and the sample. To compensate for this effect, it can be necessary to increase the power of the trapping

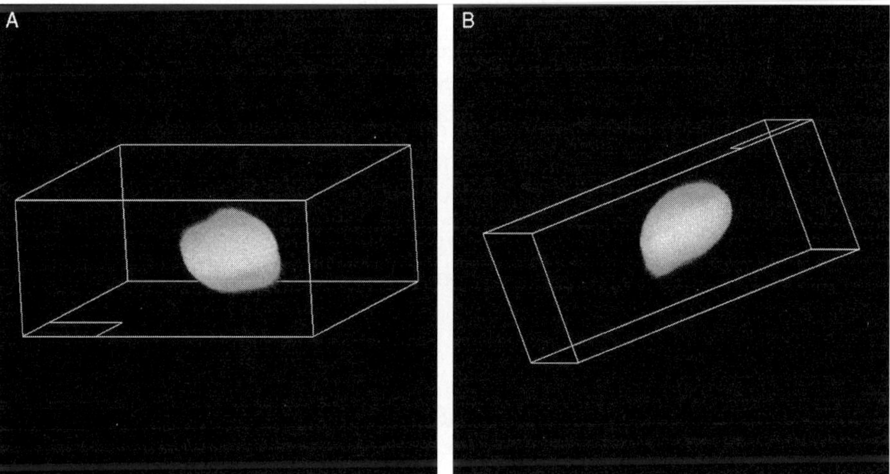

Fig. 2 Three-dimensional reconstructions of the nucleus in an optically trapped yeast cell. The reconstructions were created from a series of optical sections produced by stepping the trapped cell through the microscope's image plane. The separation of the optical sections is 20 nm. The yeast nucleus is shown inside a frame with dimensions $6 \times 6 \times 2.5\ \mu m^3$.

laser. The yeast trapped in this experiment was held using a power of 80 mW at the back focal plane of the objective which corresponds to ~45 mW in the focal plane of the microscope. By providing a suitable choice of wavelength for the trap, this power can be used without damaging the cells. A study of viability in *E. coli* found no change when the cells were exposed to a power 10 times larger than those used in this study (Ericsson *et al.*, 2000).

D. Femtosecond Laser Surgery

In the experimental platforms discussed, optical manipulation is provided by additional lasers which are coupled into the microscope via custom-made components. This approach requires specialist knowledge of optics, particularly when taking laser safety into account. However, in the absence of this knowledge, the femtosecond laser on a multiphoton microscope can usually provide intensities that are high enough to form a laser scalpel (König, 2000; Tirlapur and König, 1999) and may be a good alternative to invest in a dedicated system.

We have demonstrated the potential for this type of system by using the femtosecond laser in our multiphoton microscope to puncture the cell wall of a single spermatozoon (Goksör *et al.*, 2004). To observe this, the spermatozoa were incubated with the fluorescent dyes SYTO 9 and propidium iodide (PI). These dyes fluoresce when bound to DNA; however, only the SYTO 9 can penetrate an intact cell wall. Therefore, a viable spermatozoon will only exhibit fluorescence characteristic of the SYTO 9. When the cell wall is punctured, the PI will enter the cell and, due to its greater binding affinity with the DNA, displace the SYTO 9, producing a change in the characteristic fluorescence observed from the cell.

The results of exposing the spermatozoon to the femtosecond laser scalpel are shown in Fig. 3. The two dyes are excited using the 488-nm argon laser line, and the fluorescence is separated into two channels using dichroic mirrors. Initially, fluorescence is only detected from the SYTO 9, but after the membrane is damaged using the femtosecond laser (1.5-W output at 800 nm), the SYTO 9 signal decreases over time, while PI signal increases.

III. Bacterial Biology: Diffusion Studies in FtsK Cell Division Mutants

A. Background

In *E. coli*, mutations in the transmembrane protein FtsK are known to result in abnormal division of the cells, producing bacterial chains and filaments (Diez *et al.*, 1997, 2000; Liu *et al.*, 1998). The aberrations in cell division present in these mutants are commonly investigated by staining the DNA, the membrane, or the protein of interest to determine the role of the protein in cell separation and the structure of the septum and membrane produced during division. These techniques

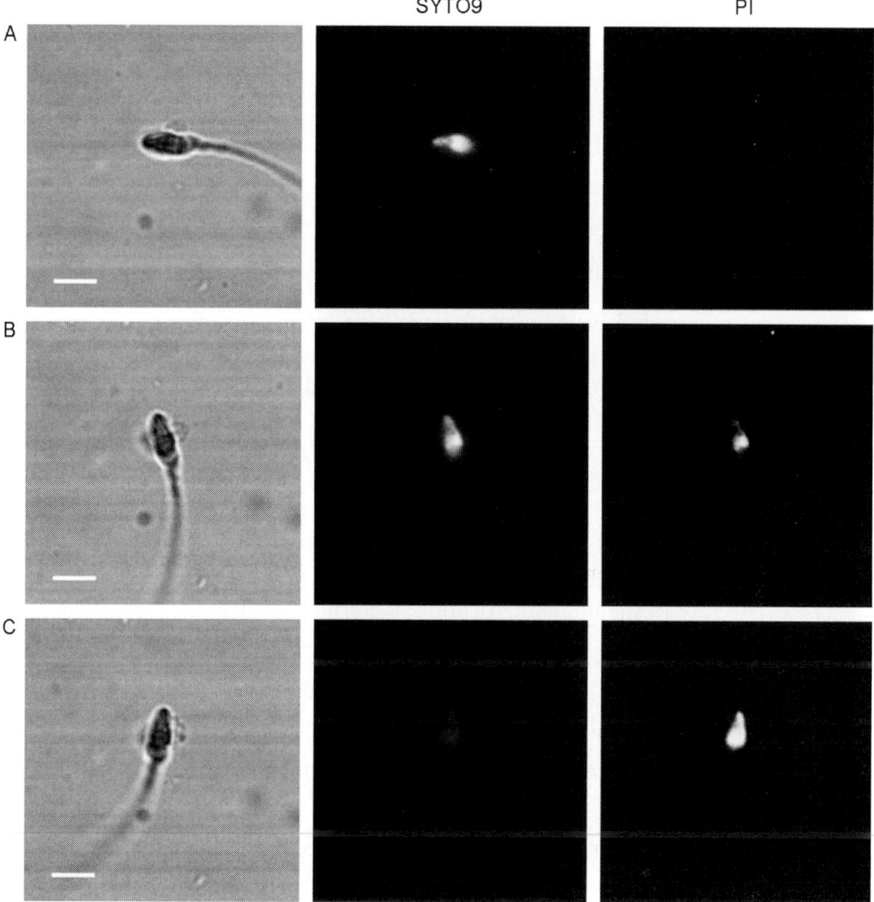

Fig. 3 Membrane cutting in single spermatozoon. Three channels in the confocal microscope were observed, transmission images are shown in the left column, the SYTO 9 fluorescence in the center column, and the propidium iodide (PI) fluorescence in the right hand column. Figures (A–C) show the spermatozoon immediately after exposure with the femtosecond scalpel, and after 4 and 10 min, respectively.

do not, however, always give access to information like membrane permeability or filtering characteristics. This information can only be gained by analyzing vectorial transport, such as molecular diffusion, across the barrier of interest (Goksör *et al.*, 2003).

In this case, the systems of interest were three *ftsK* mutants, one filament-forming mutant, AD579, and two chain-forming mutants, AD10 (*ftsK1::cat*) and AD158 (*ftsK15Δ1264–1329*). Of primary interest is whether there is any exchange of cytoplasmic material through the partially formed septum between adjoining cells in the bacterial chains. The results from the *ftsK* chains are compared with another cell division mutant with mutations in the *ftsI* proteins which form similar chains and are known to have partially formed septa between the cells.

B. Experiment

To investigate intracellular molecular exchange, the fluorescent probe PI was selected. This probe does not penetrate intact cell membranes and emits red fluorescence when bound to nucleic acids. An optical trap, in this case formed by the 830-nm diode laser, was used to arrange the filament- or chain-forming bacteria on a poly-l-lysine-coated cover glass prior to the diffusion analysis. The laser scalpel was used to open the cell pole at the end of the filament or cell chain, and diffusion of the PI through the cells was observed over time in the confocal microscope. The first experiments were performed in the bacterial filament (AD579) and were used to demonstrate that the molecular diffusion could be tracked in the confocal microscope over relatively long periods and distances without problems with photobleaching and focal drift. Results for the *ftsK* filaments (AD579) and the *ftsK* chain (AD158) are shown in Figs. 4 and 5A respectively, which illustrate both the diffusion rate through the system and the distribution of nucleic acids.

In the filament-forming bacteria, the PI diffuses through the complete length of the bacterial filament over the course of ~10 min. However, there are internal zones from which no fluorescence is observed. Since PI binds to all forms of nucleic acids, these regions must be free from DNA and RNA, although the PI can obviously diffuse through them. Although the full nature of these nucleic acid-free zones is not understood, it is possible that a pseudoseptum has been formed by the inner

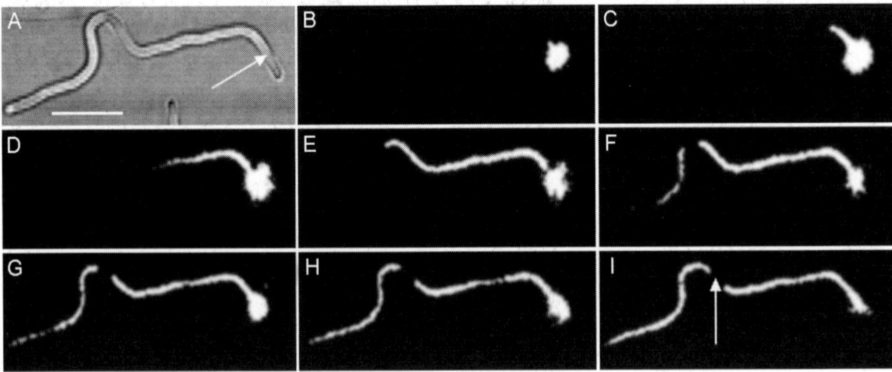

Fig. 4 PI diffusion in an ftsK filament forming mutant (AD579). Diffusion was initiated using laser scalpel to cut a hole in the cell's outer membrane. The exposure site is indicated by the arrow in transmission image in (A). The diffusion of the PI over time was monitored using the confocal microscope. Images (B–I) show the progress of the PI through the cell at 90-s intervals after laser surgery. The arrow in (I) shows the area which is devoid of fluorescence signal, indicating that this compartment does not contain nucleic acids. Scale bar is 5 μm.

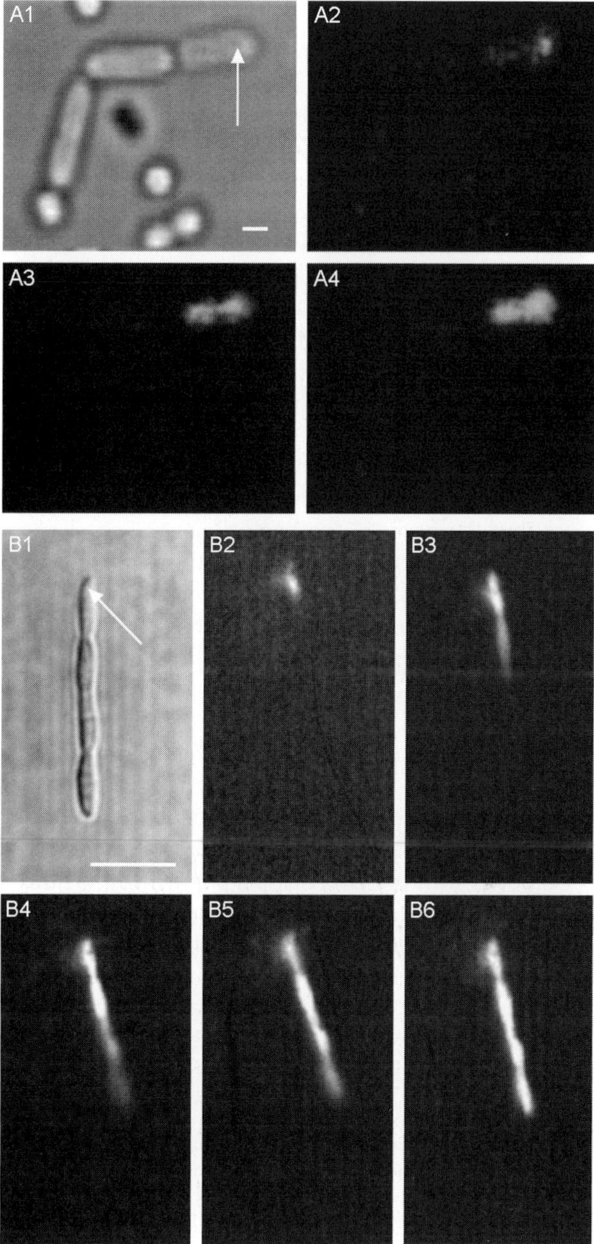

Fig. 5 PI diffusion in (A) a ftsK chain-forming mutant (AD158) and (B) a ftsI chain-forming mutant. Diffusion was initiated by an optimized exposure of the laser scalpel at the sites indicated by the white arrows in transmission images (A1) and (B1), respectively. Diffusion was monitored using the confocal microscope. In the ftsK chain, image (A2) is taken immediately after exposure to the laser scalpel, (A3) was taken 15 s later, while (A4) is taken after 2 h. In the ftsI chain, images (B2–B6) were taken at 60-s intervals, starting immediately after laser surgery. Scale bars are 1 and 5 μm, respectively.

membrane. This picture was confirmed by membrane staining which showed aberrant equatorial membrane structures distributed along the length of the *ftsK* filament (Goksör *et al.*, 2003).

Another important observation from the results of the bacterial filament experiments was a significant outflow of cytoplasmic material around the ablation site. This can be seen as a fluorescent cloud generated in the media outside the cell. There was not only significant ejection of material from the cell immediately after exposure to the laser scalpel but also the aberrant membrane structures were seen to move toward the ablation site, at a speed of ~5 nm/s, as the experiment progressed. While it is not clear how this movement of the cytoplasmic compartment occurs, it is certainly an unwanted effect which should be minimized by further optimization of the power used to create the incision.

In the case of the chain-forming *ftsK* mutants, the power in the laser scalpel was further optimized to create a small discrete incision in the cell membrane and avoid damaging the nearby cell-division zones. This created subdiffraction-limited holes in the bacterial membrane as evidenced by the localization of the fluorescence signal within the body of the bacterium, and the lack of fluorescence generated in the surrounding medium after exposure. This is in contrast to the initial experiment in the filament system where a significant ejection of cytoplasmic material into the surrounding medium was observed. This may place the pore size below 100 nm as one of the major contributors to PI fluorescence in bacterial cells is stable RNA on ribosomes which are about 25 nm in diameter. Under these conditions, it is likely that the scalpel weakened the membrane until one or more pores were created by the breaking of molecular bonds in the membrane.

Diffusion of the PI in the bacterial chain was observed for a period of 2 h. Over this time, no penetration of the probe into the neighboring cells in the chain was observed. Further measurements where other compartments of the chain were exposed with the laser scalpel confirmed that the result was reproduced along the whole chain. This suggests that there is no free exchange of cytoplasmic material between neighboring cell equivalents along the chain and that it is unlikely that the edges of the membrane of the septum have failed to fuse. This can be compared to the case shown in Fig. 5B where a mutation has occurred in the *ftsI* gene. Again a bacterial chain is created with partially constricted septa between adjoining cells. However, in this case the edges of the membrane have clearly failed to fuse, and diffusion of the PI is observed between neighboring cells. This demonstrates that the PI diffusion assay is capable of identifying cells where septa formation is incomplete and further suggests that the chain formation in the *ftsK* mutant is not due to the failure of the septal membrane to fuse.

C. Discussion

This study highlights the use of the laser scalpel to initiate diffusion through a cellular organism. Optimization of the exposure parameters allows very localized, subdiffraction-limited damage to cell membranes allowing an external fluorescent probe, which would normally not cross an intact membrane, to enter the cell. Further, since the UV laser is highly focused, it is only destructive within the focal volume and can therefore be used to target subcellular structures, such as the mitochondria in eukaryotic cells, without affecting the cell wall or other cellular components. The question of whether the mitochondria of some cell types form large connected networks could potentially be addressed using this technique. Laser-induced molecular diffusion therefore provides an additional tool which can be used alongside standard bleaching and diffusion experiments involving green fluorescent protein for analyzing molecular segregation, molecular barriers, vectorial transport phenomena, and molecular filtering processes in a variety of model systems.

IV. Plant Biology: The Binding of Chloroplasts and ER

A. Background

The production of oxygen during photosynthesis is essential for the sustained survival of all oxygen-dependent organisms. In higher plants, the photosynthetic machinery is localized in chloroplasts, which also harbor the synthesis of several important cellular metabolites, like pigments, aromatic amino acids, and fatty acids. The surrounding cell depends on chloroplast metabolism and chloroplasts, in turn, depend on the surrounding cell. One example is the import of precursors from the ER for synthesis of chloroplast membrane lipids (Moreau *et al.*, 1998; Ohlrogge and Browse, 1995).

Confocal microscopy has been used to image the interior of the plant protoplast revealing an ER network with branches that appear to end on the surface of the chloroplasts (Hanson and Köhler, 2001; Staehelin, 1997). This chloroplast–ER association is further hinted at by biochemical analysis of isolated chloroplasts. These studies showed that the lipid synthesis that is commonly attributed to the ER remains is present in highly purified chloroplasts (Kjellberg *et al.*, 2000). However, it is only recently that direct observations of physical connections in the chloroplast–ER association were demonstrated (Andersson *et al.*, 2007).

B. Experiment

To find direct evidence for physical connections between the chloroplast and the ER, protoplasts isolated from leaves of garden pea (*P. sativum*) were investigated using the combined confocal and optical manipulation system, as presented in Fig. 1A. Autofluorescence from the chloroplasts allows them to be identified within the ER network of the cell, which was fluorescently labeled using dihexaoxacarbocyanine ($DiOC_6$; Grabski *et al.*, 1993). One such protoplast is shown in Fig. 6A.

Within this experimental platform, there are several options that can be pursued to determine the relationship between the chloroplast and the ER. One approach

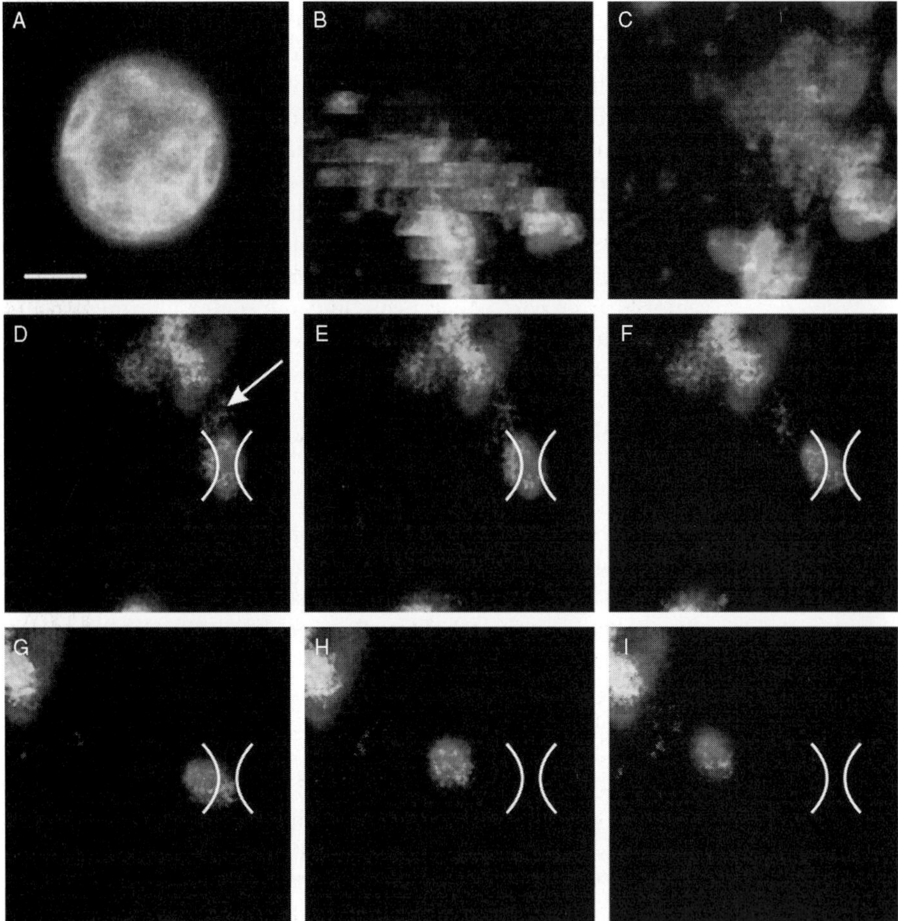

Fig. 6 Manipulation of the chloroplast and endoplasmic reticulum in garden pea (*P. sativum*). (A) An isolated protoplast from *P. sativum*, the chloroplasts (red) is enclosed by a network of endoplasmic reticulum (green). (B) Breakdown of the protoplast on exposure to the laser scalpel. The laser-induced breakdown of the cell occurs within milliseconds of exposure. (C) The contents of the protoplast that have settled onto the cover glass. (D–F) Time series showing the manipulation of a single trapped chloroplast. A chloroplast trapped immediately after breakdown of the protoplast is pulled away from those stuck to the cover glass. A strand of endoplasmic reticulum, indicated by the white arrow in (D), is stretched between trapped chloroplast and those on the cover glass. The position of the optical trap is indicated by the white lines. (G–I) The tension in the endoplasmic reticulum exceeds the maximum force provided by the optical tweezers. The manipulated chloroplast is pulled out of the trap and returns to its original position relative to chloroplasts on the surface of the cover glass. The position of the optical trap is indicated by the white lines. Scale bar is 5 μm. (Data acquired in collaboration with M. X. Andersson.) (See Plate 29 in the color insert section.)

is to manipulate the chloroplast within the protoplast; however, it is difficult to resolve the motion of the chloroplast-associated ER when surrounded by the other contents of the cell. A better approach is to rupture the plasma membrane of a protoplast by exposing it to the laser scalpel. This initiates breakdown of the cell. Its contents are freed and settle onto the cover glass (Fig. 6A–C). Typically, this process requires pulse energy of $\sim$50 μJ from the nitrogen laser, measured prior to the microscope objective. Once the contents of the cell have been released, individual chloroplasts can be trapped and manipulated using the optical tweezers. As the chloroplast is manipulated, a portion of the fluorescently labeled ER is stretched between the trapped chloroplast and its neighbors on the cover glass (Fig. 6D–F). As the ER is stretched, the chloroplast is pulled away from the trap center back toward the chloroplasts stuck to the cover glass. Eventually, the tension in the ER is greater than the force applied by the trap, and the chloroplast is pulled out of the trap and returns to its original position in relation to the other chloroplasts (Fig. 6G–I). This confirms the presence of physical contacts between the chloroplast and the ER.

Isolated chloroplasts, which are often associated with a fluorescent fragment of the ER membrane, can be used to gain insight into the magnitude and nature of this bond. In these chloroplasts, the ER membrane can be manipulated directly, and pulled away from the chloroplast (data not shown). However, this fragment could not be freed from the chloroplast, and would consistently be pulled out of the optical trap and spring back to the chloroplast's surface. This was true even at the maximum output power (5 W) from the Nd:YVO_4 laser, and indicates that the binding force between the chloroplast and the ER is significant.

To place a lower boundary on the magnitude of this interaction, the maximum force applied to the ER–chloroplast system by the optical trap was estimated from an escape velocity calibration (Ghislain *et al.*, 1994). This calibration was performed using the trapped portion of the ER, which was cut from the chloroplast and the rest of the ER using the laser scalpel. The force of the trapped ER was determined by the maximum velocity at which the microscope stage could be moved before the ER escaped from the optical trap. At this point, the force acting on the ER, which is assumed to be approximately spherical, is equal to the drag force applied by the movement of the surrounding fluid and is given by $F = -6\pi a\eta v$, where F is the applied force, a is the radius of the trapped object, η is the viscosity of the fluid, and v is the velocity of the stage. The radius of the trapped ER was estimated from the confocal images, and the maximum optical force applied to the ER fragment was equal to 425 pN.

C. Discussion

The binding of ER to chloroplasts in garden pea (*P. sativum*) was investigated. Protoplasts were ruptured using the laser scalpel, which provides an attractive alternative to chemical isolation techniques as it leaves the protoplast's constituents largely unaffected. Using the optical tweezers, a single chloroplast was pulled free from the ruptured cell, and an attached strand of ER was identified. In isolated chloroplasts, the force exerted on the chloroplast-associated ER by the optical trap was quantified, providing evidence of strong physical connections between these two constituents of the plant cell.

V. Summary

The laser scalpel and optical tweezers are excellent tools for research in microbiology. Laser scalpel and optical tweezers manipulation tools can be combined, in a straightforward manner, with the majority of imaging techniques commonly used in fluorescence microscopy. This includes confocal and multiphoton systems, although some additional control of the optical trap is required to maintain the three-dimensional imaging ability of these systems. We have demonstrated a simple system for retaining this functionality by using a movable external lens to control the divergence of the trapping laser.

The applications where these tools can be applied are diverse, but they often require a straightforward implementation of the techniques in combination with an imaging technique. We have discussed two such applications of our experimental system. In the first, localized damage of cellular membranes can initiate molecular diffusion through a cellular organism giving a vectorial diffusion assay for studying the cellular structure. This technique was used to determine the membrane integrity in *E. coli* cell division mutants by investigating the possible exchange of cytoplasmic material between neighboring cells. Due to the highly localized nature of the damage caused by the laser scalpel, there is the further possibility of applying this type of vectorial diffusion assay to study subcellular compartments such as mitochondria. In the second application, the laser scalpel was used to replace chemical methods for isolating chloroplasts from protoplasts. This allowed the network of ER enclosing the chloroplasts to remain largely intact, so that the optical tweezers could be used to probe the question of whether there were physical connections between the chloroplasts and the ER network. Strong physical connections between these two constituents of the protoplast were identified in *P. sativum*. These experiments provide a good example of what can be achieved with a system that combines both optical manipulations with advanced imaging techniques.

Acknowledgments

The authors wish to thank Professor Thomas Nyström (Department of Cell and Molecular Biology, Göteborg University) and Professor Anna-Stina Sandelius (Department of Plant and Environmental Sciences, Göteborg University) for valuable inputs. We acknowledge funding support from the Swedish Research Council, the SSF-BioX program, and the European Commission 6th framework program through the project ATOM-3D.

References

Andersson, M. X., Goksör, M., and Sandelius, A. S. (2007). Optical manipulation reveals strong attracting forces at membrane contact sites between endoplasmic reticulum and chloro plasts. *J. Biol. Chem.* **282,** 1170–1174.

Ashkin, A., Dziedzic, J. M., Bjorkholm, J. E., and Chu, S. (1986). Observation of a single-beam gradient force optical trap for dielectric particles. *Opt. Lett.* **11,** 288–290.

Berns, M. W., Aist, J., Edwards, A. J., Strahs, K., Girton, J., McNeill, P., Rattner, J. P., Kitzes, M., Hammer-Wilson, M., Laiw, L.-H., Siemens, A., Koonce, M., *et al.* (1981). Laser microsurgery in cell and developmental biology. *Science* **213,** 505–513.

Brakenhoff, G. J., Blom, P., and Barends, P. (1979). Confocal scanning light microscopy with high numerical aperture immersion lenses. *J. Microsc.* **117,** 219–232.

Colombelli, J., Reynaud, E. G., Rietdorf, J., Pepperkok, R., and Stelzer, E. H. K. (2005). *In vivo* selective dynamics quantification in interphase cells induced by pulsed ultraviolet laser nanosurgery. *Traffic* **6,** 1093–1102.

Denk, W., Strickler, J. H., and Webb, W. W. (1990). 2-photon laser scanning fluorescence microscopy. *Science* **248,** 73–76.

Diez, A., Farewell, A., Nannmark, U., and Nyström, T. (1997). A mutation in the ftsK gene of *Escherichia coli* affects cell-cell separation, stationary phase survival, stress adaptation and expression of the gene encoding the stress protein UspA. *J. Bacteriol.* **179,** 5878–5883.

Diez, A., Gustavsson, N., and Nyström, T. (2000). The universal stress protein A of *Escherichia coli* is required for resistance to DNA damaging agents and is regulated by a RecA/FtsK-dependant regulatory pathway. *Mol. Microbiol.* **36,** 1494–1503.

Ericsson, M., Hanstorp, D., Hagberg, P., Enger, J., and Nyström, T. (2000). Sorting out bacterial viability with optical tweezers. *J. Bacteriol.* **182,** 5551–5555.

Fällman, E., and Axner, O. (1997). Design for fully steerable dual-trap optical tweezers. *Appl. Opt.* **36,** 2107–2113.

Ghislain, L. P., Switz, N. A., and Webb, W. W. (1994). Measurement of small forces using an optical trap. *Rev. Sci. Instrum.* **65,** 2762–2768.

Goksör, M., Diez, A., Enger, J., Hanstorp, D., and Nyström, T. (2003). Analysis of molecular diffusion in ftsK cell-division mutants using laser surgery. *EMBO Rep.* **4,** 867–871.

Goksör, M., Enger, J., and Hanstorp, D. (2004). Optical manipulation in combination with multiphoton microscopy for single-cell studies. *Appl. Opt.* **43,** 4831–4837.

Grabski, S., de Feijter, A. W., and Schindler, M. (1993). Endoplasmic reticulum forms a dynamic continuum for lipid diffusion between contiguous soybean root cells. *Plant Cell* **5,** 25–38.

Hanson, M. R., and Köhler, R. H. (2001). GFP imaging: Methodology and application to investigate cellular compartmentation in plants. *J. Exp. Bot.* **52,** 529–539.

Hoffmann, A., Meyer zu Hörste, G., Pilarczyk, G., Monajembashi, S., Uhl, V., and Greulich, K. O. (2000). Optical tweezers for confocal microscopy. *Appl. Phys. B* **71,** 747–753.

Kjellberg, J. M., Trimborn, M., Andersson, M., and Sandelius, A. S. (2000). Acyl-Coa dependent acylation of phospholipids in the chloroplast envelope. *Biochim. Biophys. Acta* **1485,** 100–110.

König, K. (2000). Multiphoton microscopy in life sciences. *J. Microsc.* **200,** 83–104.

Liu, G., Draper, G. C., and Donachie, W. D. (1998). FtsK is a bifunctional protein involved in cell division and chromosome localization in *Escherichia coli*. *Mol. Microbiol.* **29,** 893–903.

Moreau, P., Bessoule, J. J., Mongrand, S., Testet, E., Vincent, P., and Cassagne, C. (1998). Lipid trafficking in plant cells. *Prog. Lipid Res.* **37,** 371–391.

Neuman, K. C., Chadd, E. H., Liou, G. F., Bergman, K., and Block, S. M. (1999). Characterization of photodamage to *Escherichia coli* in optical traps. *Biophys. J.* **77,** 2856–2863.

Ohlrogge, J., and Browse, J. (1995). Lipid biosynthesis. *Plant Cell* **7,** 957–970.

Sheppard, C. J. R., and Wilson, T. (1980). Image formation in confocal scanning microscopes. *Optik* **55,** 331–342.

Speiser, S., and Kimel, S. (1970). On the possibility of observing photochemical reactions induced by multiphoton absorption. *Chem. Phys. Lett.* **7,** 19–22.

Staehelin, L. A. (1997). The plant ER: A dynamic organelle composed of a large number of discrete functional domains. *Plant J.* **11,** 1151–1165.

Tirlapur, U. K., and König, K. (1999). Near-infrared femtosecond laser pulses as a novel non-invasive means for dye-permeation and 3D imaging of localised dye coupling in *Arabidopsis* root meristem. *Plant J.* **20,** 363–370.

Vogel, A., Noack, J., Hüttman, G., and Paltauf, G. (2005). Mechanisms of femtosecond laser nanosurgery of cell and tissues. *Appl. Phys. B* **81,** 1015–1047.

Vogel, A., and Venugopalan, V. (2003). Mechanisms of pulsed laser ablation of biological tissues. *Chem. Rev.* **103,** 577–644.

Yanik, M. F., Cinar, H., Cinar, H. N., Chisholm, A. D., Jin, Y., and Ben-Yakar, A. (2004). Functional regeneration after laser axotomy. *Nature* **432,** 822.

PART V

Laser Catapulting and Capture

CHAPTER 23

Noncontact Laser Microdissection and Catapulting for Pure Sample Capture

K. Schütze,* Yilmaz Niyaz,* M. Stich,* and A. Buchstaller†

*PALM Microlaser Technologies GmbH
Am Neuland 9 + 12, 82347 Bernried, Germany

†Institute of Pathology, Ludwig Maximillians University
Munich, Germany

The understanding of the molecular mechanisms of cellular function, growth, and proliferation is based on the accurate identification, isolation, and finally characterization of a specific single cell or a population of cells and its subsets of biomolecules. For the simultaneous analysis of thousands of molecular parameters within one single experiment as realized by DNA, RNA, and protein microarray technologies, a defined number of homogeneous cells derived from a distinct morphological origin are required. Sample preparation is therefore a very crucial step preceding the functional characterization of specific cell populations.

0091-679X/07 $35.00
DOI: 10.1016/S0091-679X(06)82023-6

Laser microdissection and laser pressure catapulting (LMPC) enables pure and homogeneous sample preparation resulting in an increased specificity of molecular analyses. With LMPC, the force of focused laser light is utilized to excise selected cells or large tissue areas from object slides down to individual single cells and subcellular components like organelles or chromosomes. After microdissection, the sample is directly catapulted into an appropriate collection vial. As this process works entirely without mechanical contact, it enables pure sample retrieval from morphologically defined origin without cross-contamination. LMPC has been successfully applied to isolate and catapult cells from, for example, histological tissue sections, from forensic evidence material, and also from tough plant matter, supporting biomedical research, forensic science, and plant physiology studies. Even delicate living cells like stem cells have been captured for recultivation without affecting their viability or stem cell character, an important feature influencing stem cell research, regenerative medicine, and drug development. The combination of LMPC with microinjection to inject drugs or genetic material into individual cells and to capture them for molecular analyses bears great potential for efficient patient-tailored medication.

I. Introduction

The interaction of laser light with biological matter is utilized for a variety of applications in life sciences and health care. Lasers focused through microscope lenses provide the unique possibility to micromanipulate biological specimen without any mechanical contact and without affecting their viability. Two different microlaser principles enable unique approaches to life science research.

With *optical traps or tweezers*, suspended cells are caught, moved, or positioned in order to segregate individual cells out of a bulk, and samples are stretched or held for rigidity tests or force measurements. With *laser microbeams*, organelles or subcellular structures are ablated to study cell behavior or embryonic development. Furthermore, cell membranes are opened to microinject drugs or genetic material, and individual cells are fused to obtain specific clones (Ashkin *et al.*, 1987, 1990; Berns, 1998; Berns *et al.*, 1981; Clement-Sengewald *et al.*, 2000; Greulich, 1999; Greulich and Weber, 1992; Heel and Dawkins, 2001; Schütze and Clement-Sengewald, 1994; Schütze *et al.*, 1994; Wiegand *et al.*, 1987).

The methods mentioned above are mostly used for basic research in cell and developmental biology. A unique application of the PALM MicroBeam is the solely laser-based capture of single cells or small tissue areas (PALM, Bernried, Germany). This "state-of-the-art" technique represents an important step toward the understanding of molecular mechanisms underlying intracellular processes, cell–cell interaction, cellular malfunction, or the development of disease.

In brief, a pulsed UV-A laser of a high-beam quality is coupled in an inverted research microscope and focused through microscope objectives onto the sample plane creating an energy-rich, micron-sized laser spot which enables noncontact laser micromanipulation and microdissection (Fig. 1).

The patented technology of "laser microdissection and pressure catapulting" (LMPC) is a unique way to capture individual cells or entire tissue areas in a completely contact-free manner without the risk of contamination with unwanted biomatter. LMPC is used to generate pure and homogeneous samples for various molecular analyses spanning the modern fields of functional genomic and proteomic research in animal science and plant biology as well as in forensic investigations of evidence material. Pure specimen preparations enable a more differentiated cell analysis leading to improved tumor diagnosis and optimized treatment. This allows better understanding of pathogen interaction in plants and thus, the design of more resistant crops or provides pure material for genetic fingerprinting for safe approval of a suspect's guilt.

Up to now, LMPC was applied mainly on fixed cell preparations like cytospins, cell smears, or on histological tissue sections, either formalin-fixed and paraffin-embedded or snap frozen. In recent publications, it was proven that also living cells survive the catapulting process and even stem cells continue to proliferate after laser isolation and laser transfer (Chaudhary *et al.*, 2006; Mayer *et al.*, 2002). Furthermore, whole organisms such as the nematode *Caenorhabditis elegans* (Fig. 2A–C) were transferred by LMPC without impacting their viability (list of publications at www.palm-microlaser.com).

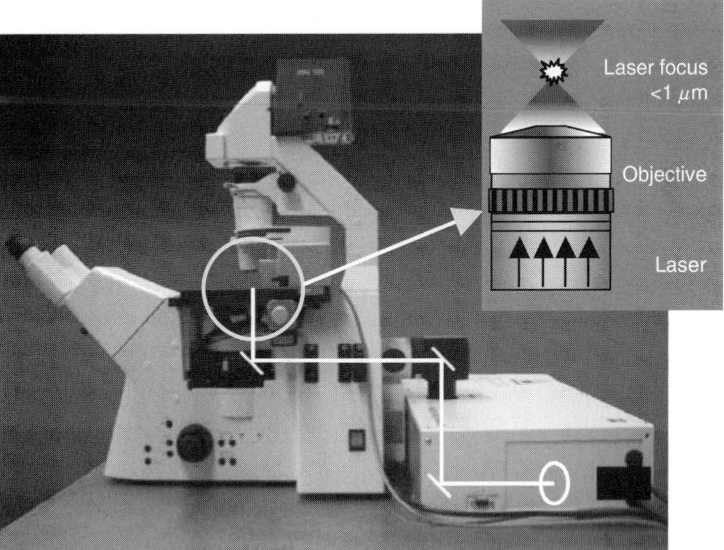

Fig. 1 Schematic setup of a PALM MicroBeam with a UV-A laser coupled via the epifluorescence path into an inverted research microscope and focused through the objective to a spot size of less than 1 μm in diameter.

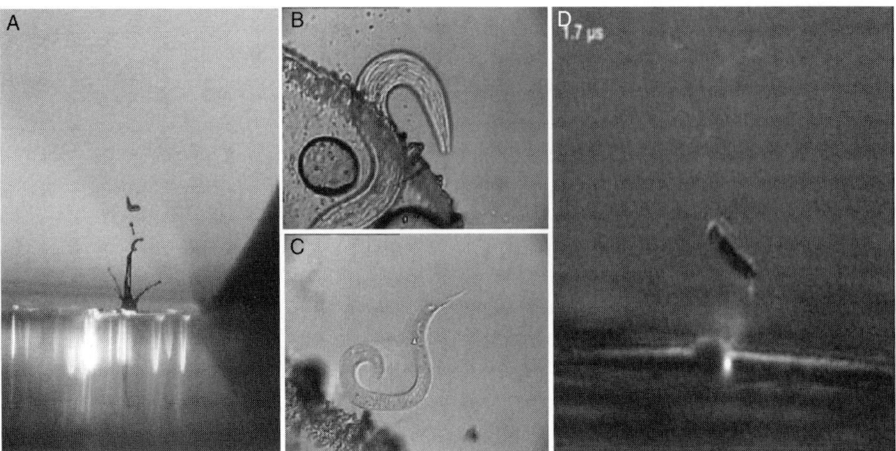

Fig. 2 (A) Catapulting of an entire organism—a nematode *Caenorhabditis elegans* from liquid as visualized with short-time photography (A. Vogel, BMO Lübeck). (B and C) Recovery within the collection cap. (D) Visualization of a microplasma by short-time photography: in the narrow laser focus a spatially restricted microplasma is generated that collapses within fractions of nanoseconds. The gaseous efflux ("plume") is clearly visible in the center. Tumor tissue with a diameter of 40 μm is dissected and catapulted with a single laser pulse. The catapulting process is illuminated by a flashlight ignited 1.7 μs after the laser impulse hit the sample. The ultra-short microplasma exposed the open camera. Transportation speeds of up to 25 μm/s have been measured. (See Plate 30 in the color insert section.)

II. Methods of Noncontact Laser Micromanipulation and Capture

A. Laser Microdissection

Laser microdissection is based on the ablative decomposition of matter utilizing a pulsed UV-A laser that is focused through lenses down to a spot size of less than 1 μm in diameter. At the laser's focal point energy densities are created that enable cutting of the irradiated material.

Laser cutting is based on the physical phenomenon known as "ablative decomposition," which is a photochemical process initiated by the extreme photon density within the narrow laser focal point. This results in an ultrashort, adiabatic expanding microplasma (Fig. 2D), confined to the submicron-sized laser focus. Due to extremely high temperature and pressure, all chemical bonds are disrupted within this narrow region, generating a gaseous efflux that consists of molecular debris and atoms. The microplasma carries a high energy, while the gaseous efflux, also called "plume" (Fig. 2) comprises less energy. The microplasma collapses within a fraction of nanoseconds—a time too short for the transfer of heat into the surroundings. That is why the procedure is called "cold ablation" and can be used for micromanipulation and microdissection of fixed or living cells without the loss of bioinformation or viability.

Nonfocused laser light beyond the laser focus becomes scattered and travels through adjacent areas without any impact on the specimen, as here the photon density is not strong enough to enable ablation. In addition, the wavelengths of the applied UV-A lasers (330–355 nm) are sufficiently far away of the peak absorption of DNA, RNA, or proteins, meaning that unfocused laser radiation traveling through the cells does not interfere with the genome or proteome. Thus, laser microsurgery and microdissection as performed with the PALM MicroBeam represents a reliable and safe method for noncontact cell manipulation as well as precise and pure sample capture (Mohanty *et al.*, 2002; Srinivasan, 1986; Thalhammer *et al.*, 2003; Vogel and Venugupalan, 2003—for further details of the "physics behind" please refer to Chapter 5 by Vogel *et al.*, this volume).

B. Laser Catapulting

During the first microdissection experiments of histological tissue, excised cells were sometimes accidentally propelled out of the tissue section (Schütze *et al.*, 1997). As this effect seemed to be correlated with laser action, it was obvious that the cutting laser can also be used to eject the microdissected samples out of the object plane, that is to catapult them without any mechanical contact directly into an appropriate reaction vial mounted above the objective. This laser phenomenon, named "*L*aser *P*ressure *C*atapulting—LPC," was protected by patent and became well established as unique, highly precise, and entirely noncontact laser capture technology. Like laser cutting, the physics behind LPC seems also to be based on the formation of a microplasma underneath the sample, at the transition between support (i.e., object slide) and specimen. The resulting pressure waves together with the atomic efflux within the "plume" propel the microdissected samples against gravity toward the collection vial, for example a standard microfuge cap. The propulsion of the microdissected area as well as the directive propagation of LPC could be recently visualized using short-time photography and catapulting speeds up to 25 m/s have been measured (Fig. 2A and D) (for further details please refer to Chapter 5 by Vogel *et al.*, this volume).

For homogeneous sample capture, LPC was applied successfully on, for example, cell smears or cytocentrifuged specimens to select for fetal cells enabling noninvasive prenatal analysis (Burgemeister *et al.*, 1999; Hahn and Holzgreve, 2002). Other specimen preparation methods require laser microdissection prior to catapulting. This combined "*L*aser *M*icrodissection and *P*ressure *C*atapulting—LMPC" procedure is the key to pure sample harvesting. In a first step, laser microdissection opens a clear-cut gap around the selected sample to entirely separate it from the surrounding. Depending on the size and nature of the specimen and the applied preparation method, one single or a multitude of laser pulses are required to catapult the samples into the collection vial. If the specimens are prepared on a special, gauzy membrane that serves as a stabilizing backbone, even large dissected areas can be isolated with one single laser pulse keeping their morphology entirely preserved (Fig. 3). From the respective capture vial, the specimens are dissolved and subsequently biomolecules

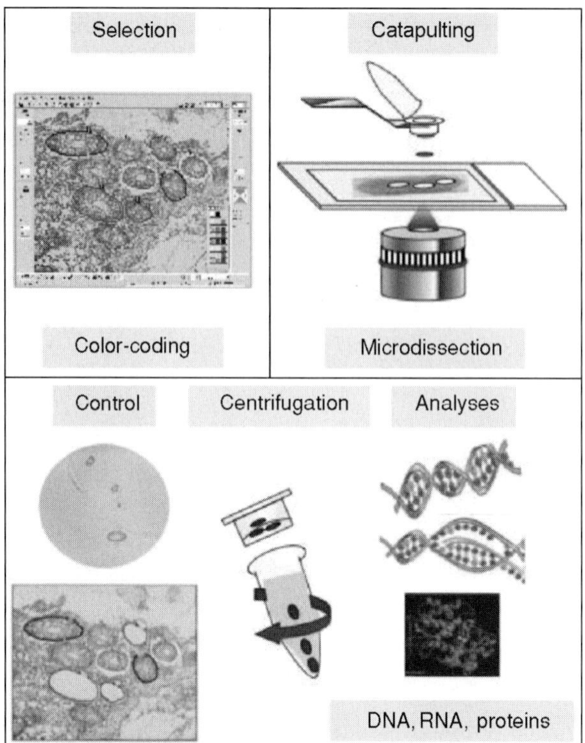

Fig. 3 Workflow of pure sample capture. A histological stained tissue section is mounted onto a thin supporting membrane and placed onto the microscope stage. The specimen of interest are selected on the screen and outlined with different colors using computer graphic tools. The laser precisely cuts along the drawing line and ejects the sample into a cap of a routine microfuge tube, that is filled with buffer or adhesive material. The captured samples are visualized within the cap and the origination tissue section can be used for further extraction. The catapulted specimen is simply centrifuged and ready to be processed for different molecular analyses. (See Plate 31 in the color insert section.)

of interest—that is, DNA, RNA, or proteins—are purified and committed to the corresponding downstream applications such as PCR and microarray hybridization or chromosomal analysis as well as MALDI or SELDI spectroscopy.

The safety of the LMPC procedure for even fragile biomolecules like RNA was already proven in 1998 by the first LMPC experiments on routinely formalin-fixed and paraffin-embedded tissue sections of human-differentiated colon adenocarcinoma (Schütze and Lahr, 1998). Meanwhile, LMPC is described in numerous publications as a well-suited method for the capture of a variety of samples from subcellular organelles up to tissue sections of 3 mm in diameter—solely by the force of focused laser light. In summary, LMPC marks the breakthrough in advanced molecular analyzing methods as it enables the entire noncontact preparation of pure and homogeneous samples (Schütze *et al.*, 2002; Fig. 4).

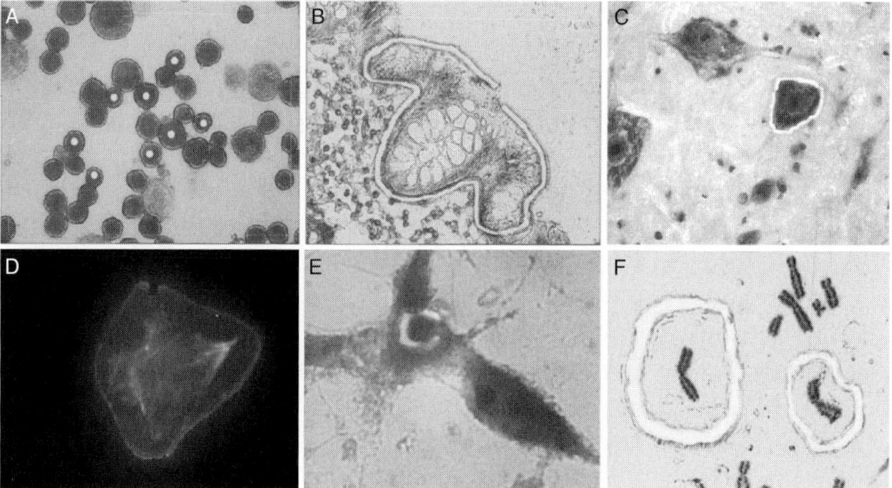

Fig. 4 Various specimen and different preparation techniques suitable for LMPC. (A) Single cells from a cytospin, marked with graphic tools and catapulted. (B) LMPC of a small tissue area from a histological tissue section. (C) LMPC purifies neuron cells from attached glial cells. (D) LMPC performed under fluorescence observation. (E) The so-called "Donut"-capture method prepares subcellular regions, like individual nuclei, and separates them from the remaining cytoplasm. (F) LMPC captures single chromosomes from a metaphase preparation.

C. Automation for Enhanced Throughput Laser Microdissection and Catapulting

The configuration of the latest generation of PALM MicroBeam systems enables complex experimental workflows to be planned and processed automatically. A recently developed robotic unit, the "RoboMover" functions as multipurpose collection device with adapters for versatile capture vials, such as single microfuge caps, a multitude of microfuge tubes, multicap strips, or plates in microtiter format. An especially engineered microscope stage, the "RoboStage" can travel large distances in *x*/*y* directions with high speed and highly accurate relocation of preselected areas. Different customized holders allow collecting samples from single or multiple object slides, LMPC-optimized membrane slides or culture dishes.

The PALM "RoboSoftware," as an intuitive graphical user interface, is facilitating the motion of all components of the fully motorized MicroBeam: the microscope, the "RoboStage," the "RoboMover" capture device as well as the optional fluorescence equipment, which enables LMPC also under simultaneous fluorescence observation. This software includes automated process routines as well as additional functions: a wide palette of computer graphic tools for preselecting the incision path enables the outlining and color-coding of individual areas from up to three object slides concurrently. Selected areas are listed in a protocol, the "Element List," which allows grouping and experimental scheduling and summarizes the outlined samples with their corresponding area measures.

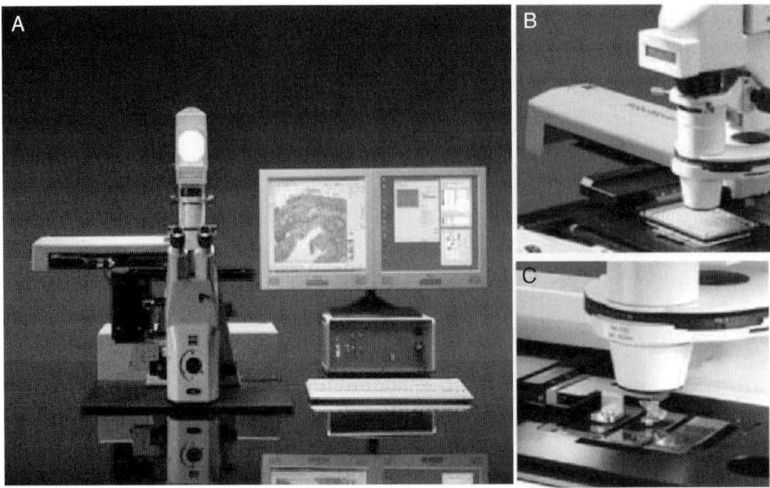

Fig. 5 (A) PALM MicroBeam consists of a motorized research microscope (ZEISS Axiovert 200 M), a UV-A laser with control unit (B) the RoboMover, mounting various collector modules in microtiter format and (C) the RoboStage, holding up to three slides. The combination with image analyzing software enables specimen capture in an enhanced throughput mode.

The high degree of automation can be augmented by image analyzing software modules allowing automated specimen identification through image processing (Niyaz and Sägmüller, 2005). Coupled with any of these software modules, the MicroBeam system is now able to scan, detect, isolate, and finally capture the specimen of interest, for example fluorescently labeled rare cells, metaphase, or FISH-treated cells but also histological stained cells or tissue areas in fully automated manner (Fig. 5).

Meanwhile, the PALM MicroBeam is a well-accepted technology for the generation of pure samples in a fast and efficient work routine. The prevention of contamination achieved by this method is indispensable for reliable molecular analyses in functional genomic and proteomic research (Niyaz *et al.*, 2005).

III. High Versatility in Multiple Fields of Application

In many biological applications, the isolation of intact DNA, RNA or proteins from homogeneous starting material is the most critical step for subsequent analysis. Independent from the kind of analytical approach, the selective acquisition of specific cells or subcellular compounds via LMPC technology leads to consistent results. Furthermore, the starting amount of material can be minimized, as only cells carrying the specific, desired information are harvested. Thus, LMPC supplement studies correlating gene or protein profiles with characteristic cell morphology.

A. LMPC in Genomics and Proteomics

1. Functional Genomics

Modern biomedical research is often interested in the molecular profile of a distinct cell type within a specimen. Such studies rely on the capability of pure specimen preparation, as the cells of interest usually constitute only a small proportion of the specimen and therefore have to be separated from other parts of the tissue. The ability of capturing cells individually has made great impact in the ongoing understanding of cellular physiology and molecular pathology. This is also reflected in the growing number of publications from all fields of life science research. For a more complete overview of publications on microdissection and the respective downstream analyses, refer to the continuously updated homepage of PALM (www.palm-microlaser.com). In this volume, only some of them will be named in the following chapters. Laser microdissection is used for analysis of aberrant genome modifications like loss of heterozygosity (Buerger *et al.*, 2000; Hasse *et al.*, 1999; Huang *et al.*, 2003; Schneider-Stock *et al.*, 2002), chromosomal aberrations (Aubele *et al.*, 2000a; Fassunke *et al.*, 2004a,b; Hartmann *et al.*, 2002; Koene *et al.*, 2004), fluorescence *in situ* hybridization (FISH) (Aarts *et al.*, 2002), comparative genome hybridization (CGH) (Aubele *et al.*, 2000b; Langer *et al.*, 2005; Obermann *et al.*, 2003), and finally for genomic analysis down to single-cell level or even organelles like a single chloroplast (Meimberg *et al.*, 2003) or chromosomes (Fominaya *et al.*, 2005; Thalhammer *et al.*, 2004).

In the field of RNA analysis, the versatility of possible applications is even more pronounced, as histopathological tissue sections are typically composed of a variety of different cell types in a complex three-dimensional architecture. Each of these cell types shows a unique RNA and protein expression pattern depending on neighboring cells within the surrounding tissue. Thus, cDNA microarrays hybridized with microdissected samples usually provide more detailed results with low amount of starting material (Irié *et al.*, 2004; Schlomm *et al.*, 2005; Thelen *et al.*, 2004), enabling transcriptional profiling at its limit using one single microdissected cell (Becker *et al.*, 2002).

2. Functional Proteomics

Identifying proteins and measuring their expression levels can reveal important information regarding a biological process or disease. Particularly in the field of cancer proteomics, the focus is directed on identifying biomarkers of tumorigenesis. It is hoped that these will aid in the detection and diagnosis of cancer as well as in the development of therapies. One obstacle in identifying tissue biomarkers is the complex nature of many samples. The identification of proteins whose expression are specific to certain cells is quite difficult when expression profiles are generated from whole tissue samples. The detection of biomarkers for tumorigenesis may require separate analysis of protein expression in the tumor cells and surrounding tissue. Immunohistochemistry can be used to localize protein expression, but a better approach may be to physically isolate cells, extract proteins, and subject them to analysis. The advantage of laser-assisted microdissection in this kind of application is that cells can be identified by their morphology and marker expression. Using the appropriate protocols, high-quality proteins and nucleic acids are extracted for further analysis (Ball and Hunt, 2004).

The molecular mechanisms involved in most neurodegenerative disorders, such as Parkinson's and Alzheimer's disease, are still unclear. Laser-microdissected samples of single cell types from postmortem brain tissue, for example such as from Parkinson's and Alzheimer's disease, were used for protein expression analysis with MALDI, SELDI, nano-liquid chromatography, or state-of-the-art chip techniques coupled to tandem-mass spectroscopy (Sauber *et al.*, 2004; Schad *et al.*, 2005a; Tannapfel *et al.*, 2003; Wellmann *et al.*, 2002).

These techniques were extended to peptide pattern analysis by MALDI-TOF mass spectrometry. As few as 125 trophoblasts were collected by laser microdissection from frozen tissue sections of human placenta tissue to compare their respective peptide patterns. The extracted peptide mixtures were directly analyzed by MALDI-TOF mass spectrometry. Specific discriminating peptide patterns for trophoblast cells, and surrounding villous stroma cells could be obtained. In future, this method is potentially a suitable instrument for finding specific peptides and identifying proteins that are related to the pathogenesis of trophoblast-linked pregnancy diseases such as preeclampsia (de Groot *et al.*, 2005).

In summary, LMPC is a purification method working under high-resolution microscopic control to isolate the cells of interest for genomic and proteomic analysis keeping track to their origin and function. This enables to correlate cell development with the genomic and proteomic information of cells and to study cellular functioning within the organism. Especially in tumor analysis, this innovative technology has already provided new insights and better understanding of cellular malfunction and tumorigenesis. As a consequence, this will lead to better diagnosis and advanced treatment.

B. LMPC in Botany

Meanwhile laser microdissection is also successfully introduced into many different domains of botany to isolate specific cells in higher plants as they also consist of a multitude of different distinct tissues and cell types, each contributing to the overall performance of the whole organism as reviewed by Kehr (2003). Plant tissue, single plant cells, and their organelles were analyzed for their DNA, RNA, protein, and metabolite profiles. As for example, LMPC was performed to capture individual phloem sectors from vascular-rich flower stalk of *Arabidopsis* to isolate mRNA and probe for presence of transcripts for a phloem potassium channel, sucrose carrier, and proton pump (Ivashikina *et al.*, 2003; Fig. 6A–C).

Metabolite and protein profiling was successfully done on *Arabidopsis thaliana* stems, which have been cryosectioned to preserve cellular structures. Vascular bundles were microdissected, captured, and subsequently analyzed by gas and

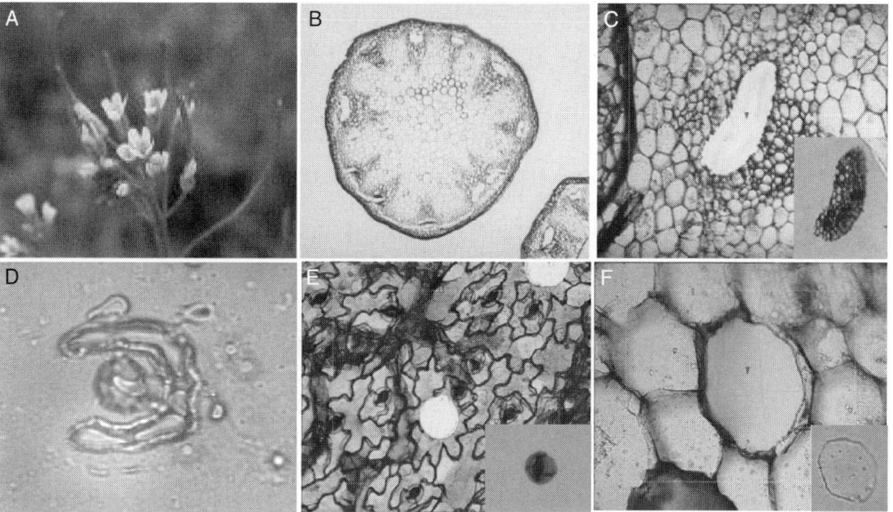

Fig. 6 LMPC in botany. (A) *Arabidopsis* flower. (B) Section through *Arabidopsis* stalk. (C) LMPC of hard phloem—insert: the catapulted sample within the cap features a nicely preserved morphology. (D) LMPC of chloroplasts from *Nicotianum tabacum*. (E) Capture of stomate pores from Toluidine blue stained epidermis cells of *Arabidopsis*. (F) Isolation and capture of pure cytoplasma without cell walls from *Vicia faba* stalk, stained with Toluidine blue. Images courtesy of Professor Hedrich, Botanical Institute, University of Würzburg, Germany.

liquid chromatography time-of-flight mass spectrometry (GC-TOF MS, LC-TOF MS/MS) (Schad *et al.*, 2005a,b).

DNA analysis on hard pollen material became possible after laser microsurgery of pollen grain walls enabling the genotyping of a haploid single pollen (Matsunaga *et al.*, 1999). Further work on DNA with microdissection and chromosomal analysis on plants, for example chromosome painting, linkage analysis, and microcloning, were done by Fominaya *et al.* (2005). Also, differentially labeled X- and Y-chromosomes of a metaphase spreads from fixed mitotic proplasts have successfully been isolated from each other. In combination with a modified FAST-FISH protocol, LMPC represent a quick technique for comparing organization of plant genomes (Hobza *et al.*, 2004).

In the study of heteroplasmy, a single choloroplast out of homogenized tobacco leaf material was isolated using LMPC and its DNA amplified directly by PCR for addressing the question of chloroplast inheritance (Meimberg *et al.*, 2003; Fig. 6D). The isolation and capture of stained as well as unstained single stomatal pores from *Arabidopsis* epidermis or the isolation of cytoplasm without cell wall from *Vicia faba* stalk has successfully been demonstrated (Fig. 6E and F).

The advantage of LMPC in botany compared to currently used techniques, such as limited dilution or needle-based micromanipulation, is the speed and efficiency of the procedure, as it provides the possibility to directly collect and pool into a reaction vial and thus, to prepare high numbers of samples in a short time. LMPC also provides a powerful tool to study the molecular basis of host–pathogen interactions at the cellular or subcellular level. Plant parasitic nematodes cause extensive damage to horticultural and broad acre crops. Thus, it is of economic importance to study the molecular events involved in the induction of specific feeding cells and surrounding gall tissue in host roots following nematode infection (Ramsay *et al.*, 2004). So far, extraction of pure cytoplasmic contents of those cells has been difficult because of their small size but was easy and reliable with LMPC. Thus, this approach will find wide application in cellular and subcellular studies involving other plant–pathogen interactions like the invasion of spores from mycorrhizal fungi (Levy *et al.*, 2003) and much more.

In summary, LMPC enables the differential analysis of individual plant tissues and single plant cells eliminating the averaging effect and allowing the discovery of detailed differences between various cell types. This will greatly contribute to a better understanding and knowledge of plant physiology and will allow to learn more about the pathway of infection preventing horticulture and broad acre crops industry from economic loss. Additional benefits of focused UV-A lasers for "noncontact" micromanipulation in plant biology have been demonstrated in various studies, for example stopping of cytoplasmic motion within lily pollen tubes (Schütze *et al.*, 1989) or the fusion of selected protoplasts (Weber and Greulich, 1992). In addition, laser micromanipulation, for example microperforation for drug ingestion, could significantly contribute to specific crop design and the combination with LMPC to subsequently capture the successfully transfected cells might provide deeper insight into internal molecular processes and pathways.

C. LMPC in Forensics

DNA analysis is an essential tool for modern forensics. To identify individuals, forensic scientists scan certain DNA regions that vary from person to person and use the data to create a DNA profile of that individual. In criminal cases, genetic fingerprinting, generally involves obtaining samples from crime scene evidence like samples of blood, hair, saliva, or semen and from a suspect, extracting his/her DNA, and analyzing it for the presence of a set of specific DNA regions. Usually at the crime scene, evidence is rare and often dispersed in fabrics or soil or mixed with DNA samples from other persons like the victim. Thus, the successful investigation and prosecution of crimes is dependent on the safe capture of pure, uncontaminated cells, and the reliable enrichment of scattered DNA samples, which is exactly the benefit of the "noncontact" LMPC procedure.

In a first forensic approach, the PALM MicroBeam was used to dissect and capture individual cells from evidence material retrieved from postcoital vaginal swabs, sperm swabs, and buccal cell swabs (Seidl *et al.*, 2005). A major obstacle of forensic case work is the scarce amount of DNA, which makes the utilization of routine DNA purification kits difficult. Pure sperm cell capture in combination

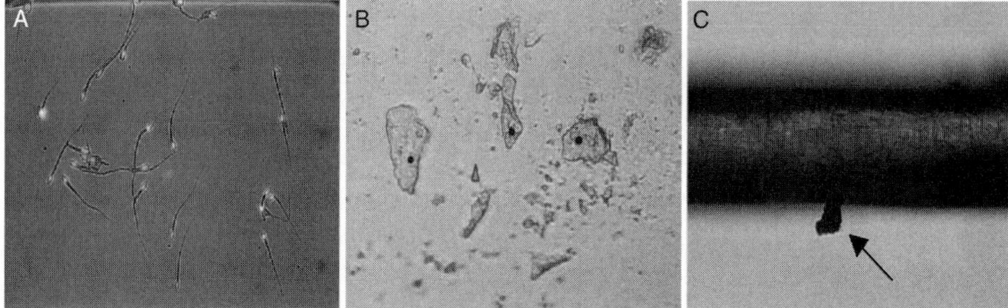

Fig. 7 LMPC in forensics: sampling and enrichment of rare evidence from crime scene using LMPC. Sperm cells from rape case can be catapulted from routine glass slide preparations. (B and C) Epithelial cells could be captured directly from forensic tape or be catapulted from a hair. Subsequent DNA fingerprinting may allow to convict or disburden a suspect and to identify the culprit.

with a customized DNA purification kit successfully provided sufficient DNA amounts for PCR amplification. As the MicroBeam system can also operate with normal glass object slides or even direct from forensic tape a check-up of existing, archived evidence material becomes feasible (Sauer *et al.*, 2004; Fig. 7A–C).

Meanwhile, the LMPC technology is used in State Offices of Criminal Investigations, State Laboratories, Police Departments, and Departments of Justice or Forensics Service Laboratories in several countries around the world and has great potential to be integrated within routine checkups and controls implementing genetic fingerprinting based on pure material.

D. LMPC of Living Cells

The main application of LMPC until now was the precise and pure capture of fixed paraffin embedded or frozen samples for following downstream analysis. Recently, the question arose whether living cells could also be isolated and catapulted from a cell culture dish in the same way and whether they survive the laser treatment as well as the high-speed transportation without harm.

The selective capture of individual living cells would be interesting to study, for example cell growth and development, cell behavior, cell–cell interaction, cell motility, and so on. Furthermore, this technology could become important for clonal expansion or to sort and separate individual cells out of mixtures.

Most of the commercially available cell-sorting systems work with suspended cells (Eisenstein, 2006). To gain pure cell populations, multistep protocols based on cell surface markers are often used, for example fluorescence-activated cell sorting (FACS). Cell loss and stress due to nonphysiological conditions are the main disadvantages of these systems, and often the desired cell populations remain heterogeneous. Sorting of adherent growing cells requires their prior detachment by trypsinization. This may change the cellular parameters because of the applied reagents and the breakdown of the three-dimensional structure. Needles or capillaries mechanically pick cell groups of interest. However, this methods of collection are very time consuming and afflicted with a high risk of contamination. Other systems use "negative cell sorting" by destroying the unwanted cells using laser ablation. These cells are not available to further analysis or propagation. Additionally, the method seems to produce a large amount of dead cells and cellular debris within the culture vial, which might disturb the surviving cells (Koller *et al.*, 2004).

The first experiments using LMPC to capture living cells were performed with bladder carcinoma cells (Mayer *et al.*, 2002). Cells were cultured in a manually assembled, membrane-spanned cell culture chamber. After cell attachment, most of the cell culture medium had to be removed to allow successful catapulting. A small membrane area with ~10–40 cells was outlined, microdissected and, for the transfer into the collection vial, a single laser impulse was placed slightly underneath the membrane. After successful catapulting of the entire membrane-cell stack, the sample was transferred into a fresh culture dish and cell proliferation and growth was observed after a couple of days.

On the basis of these results, the LMPC procedure for the capture of living cells was optimized. The usage of laser systems to micromanipulate living cells requires specialized culture vessels. Routine plastic ware is not suitable, as the laser beams are scattered or absorbed by these materials. Thus, special polyethylene naphtalate (PEN) membrane-spanned culture devices like DuplexDishes and MembraneRings were engineered to facilitate life cell LMPC. In addition, specific cell preparation protocols were developed and standardized for safe capture of living cells and their subsequent recultivation or cloning (Fig. 8A–F). Various tumor cell lines, like HepG2, HeLa, TPC-1, EJ28, have already grown to single colonies after LMPC (Schütze *et al.*, 2003; Stich *et al.*, 2003; von Bally *et al.*, 2006). Even after several rounds of repeated LMPC isolation and recultivation, the cells are still viable and their genome is unaltered as proven with CGH and M-FISH experiments (Langer *et al.*, 2005).

The catapulting of living cells for ongoing cultivation opens a variety of innovative approaches in cell biological research or drug development studies. In addition, LMPC may be combined with experiments concerning cell migration, morphological changes during cell differentiation, and cell behavior, in general. For this purpose, individual cells could be captured and analyzed at defined states or certain time during the experiment to determine the molecular background or to study cellular reactions following drug microinjection or drug treatment.

In summary, the Laser MicroBeam offers the possibility to isolate and retrieve single living cells or a group of cells out of a cell culture dish without harming the selected cells nor interfering with the originating cell culture. Furthermore, due to the precision and contact-free mode of LMPC there is no danger of contamination with unwanted neighboring cells.

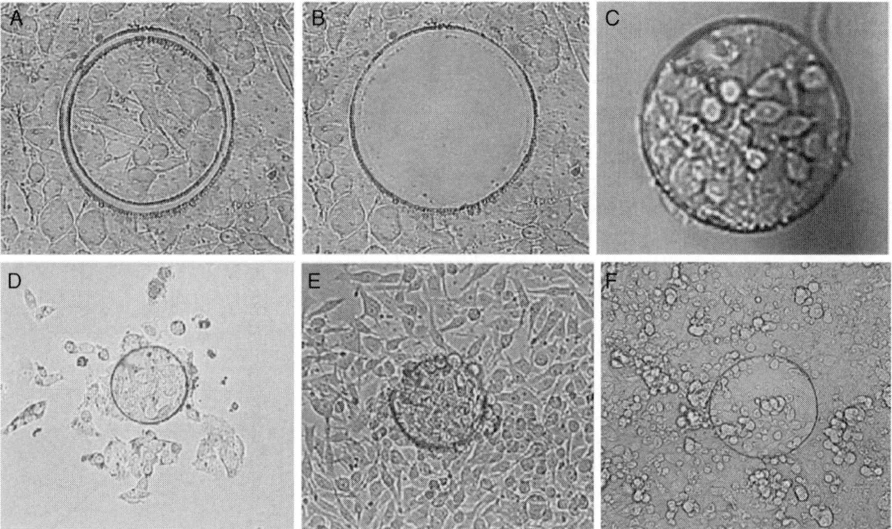

Fig. 8 LMPC of cell areas of bladder carcinoma cells (EJ 28) cultured on a membrane-spanned culture dish. (A) Microdissected area. (B) Cleared area after LMPC. (C) Catapulted cell-membrane stack visualized within a medium-filled capture cap. (D, E) Proliferation of harvested cell areas in a fresh culture dish after 3 and 7 days. (F) Cells migrate into the gap at the originating culture dish demonstrating the safety of LMPC even for those cells adjacent to the laser cut.

E. LMPC of Stem Cells

Stem cells are known as very special cells within the organism. During their lifetime, they keep the ability of self-renewal and differentiation. As they naturally contribute to organ homeostasis, wound healing, and tissue regeneration, replacing dying cells, stem cell research currently stands at the center of regenerative medicine, and tissue engineering. Stem cells can also serve as target cells for drug screening and drug safety tests, and as vehicles for cell-based gene and tumor therapies. Thus, the reliable and safe selection, isolation, manipulation, and recultivation of defined stem cell populations with specific features is essential to efficient tissue regeneration, stem cell therapy, and safe drug discovery. Stem cells are divided in two groups depending on their origin within the body:

- *Embryonic stem (ES) cells* are derived from the inner cell mass of the embryo (blastocyst) and have the ability of dividing indefinitely. Given the appropriate signals, they can differentiate into cells from all three germ layers: into ectoderm, mesoderm, and endoderm. They are called "pluripotent" because they have the ability to generate all the specialized cells that make up the whole organism.
- *Adult stem cells* reside in their stem cell niches in every organ of the adult body such as bone marrow, brain, or liver. They self-renew for the lifetime of the organism and maintain close contact to the differentiated cells in their surroundings. They are called

"multipotent" because they can differentiate into all the specialized cells of the originating tissue.

Tissue and organ transplantation is currently the established therapy used to replace diseased with healthy tissue or organs. It is hampered by a number of problems that are hard to overcome. Often the transplanted organs are rejected by the recipient and life-long immunosuppression causes strong side effects. Donor organs usually are rare and often patients die before an organ becomes available. Transplantation costs are high and usually not affordable to the individual. That is why in the emerging field of regenerative medicine and tissue engineering, embryonic and adult stem cells are used as a renewable source for substitution in diseased or injured organs. To this purpose, stem cells need to be isolated from embryos or adult organisms, then recultivated, differentiated *in vitro*, and administered to the patient.

Unfortunately, adult stem cells represent a very rare and heterogeneous species as, for example only 1 in 50,000 nucleated blood cells is a stem cell. They are usually mingled with various other cell types and are difficult to purify. This fact is reflected in the myriad of available purification protocols, for example selective adhesion to cell culture plastics (Friedenstein *et al.*, 1970; Meirelles Lda and Nardi, 2003), especial designed cocktails of cytokines, and chemicals to selectively kill non-stem cells (Modderman *et al.*, 1994), or complicated multistep protocols using cell surface markers to perform "FACS" or "magnetic-activated cell sorting" (MACS). However, there are many problems regarding those stem cell purification protocols. They are time consuming and cells are subjected to stress due to nonphysiological conditions, and often the desired cell populations remain heterogeneous.

ES cells can be isolated more easily from the blastocyst and can be cultivated and differentiated *in vitro* using specialized protocols. However, during *in vitro* differentiation processes, mixed population of precursor and terminally differentiated cells generally form. This poses a serious problem to the safe therapeutic use of stem cells, as the remaining precursor cells with high proliferative potential can elicit tumor development in the recipient. In most laboratories, selection of those undifferentiated from differentiated cells is currently performed mechanically, using fine needle aspiration. This is a time consuming and tedious work and difficult to control. In other protocols, trypsinization is used to detach cells and embryoid bodies from their culture dish. Trypsinization causes the disintegration of the three-dimensional structure and destroys all the cell–cell interactions essential to ES cell differentiation misguiding the differentiation process.

Thus, LMPC seems to be the method of choice to isolate the desired stem cells in a safe and reliable manner, also allowing a certain degree of automation. In order to prove the safety of laser-based stem cell isolation and its reliability, it had to be demonstrated that the LMPC procedure does interfere neither with the viability nor with the stem cell character of those delicate cells.

1. Identification and Discrimination

The goal of a first set of experiments was to demonstrate that stem cells can adhere and grow on the LMPC facilitating PEN membrane and how a mixture of cells or cells at certain differentiation states could be identified and discriminated as candidates for LMPC.

For this purpose, two differently adherently growing blood- and bone marrow-derived mesenchymal stem cell lines (RM26 and BalbC, respectively), presenting discriminative morphology and stem cell marker expression (Lange *et al.*, 1999), were mixed and plated on UV- or fibronectin-treated MembraneRings. The cells adhered well and their respective morphologies were conserved. While the blood-derived cells have cobblestone-like appearance, an amorph cytoskeleton, and express the membrane stem cell marker CD34 (RM26/CD34+), the bone-marrow-derived cells have a spindle-like morphology, form stress fibers, and lack CD34 expression (BalbC/CD34−) (Fig. 9A–F). Indirect anti-CD34 immunofluorescence

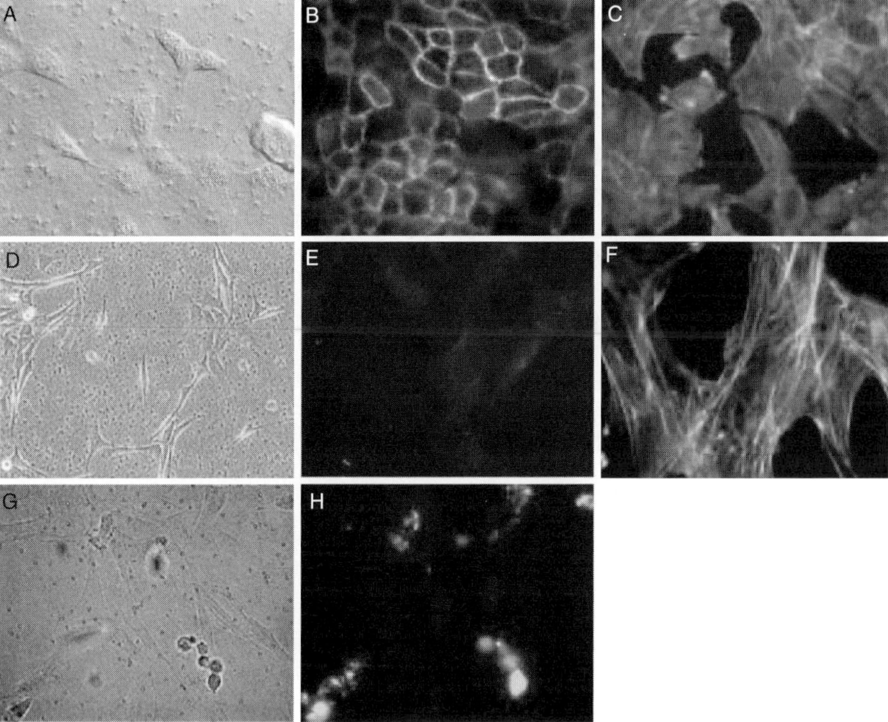

Fig. 9 *Identification* by morphology or *discrimination* by fluorescence marker expression as demonstrated with two different mouse stem cell lines. (A) The cobblestone-like shaped, blood-derived stem cell line RM26 expresses the stem cell marker CD34 (B) and has an amorph cytoskeleton (C). (D) The bone-marrow derived BalbC is spindle-shaped, lacks CD34 expression (E) and generates stress fibers (F). (G) A mixture of the two stem cell lines plated onto a MembraneRing. (H) Indirect anti-CD34 immunofluorescence staining allows clear distinction of the spindle-like BalbC/CD34−, who do not react with anti-CD34, from the fluorescing cobblestone-shaped RM26. (A, D, G) contrast enhancement by a 40× objective with PlasDIC. (See Plate 32 in the color insert section.)

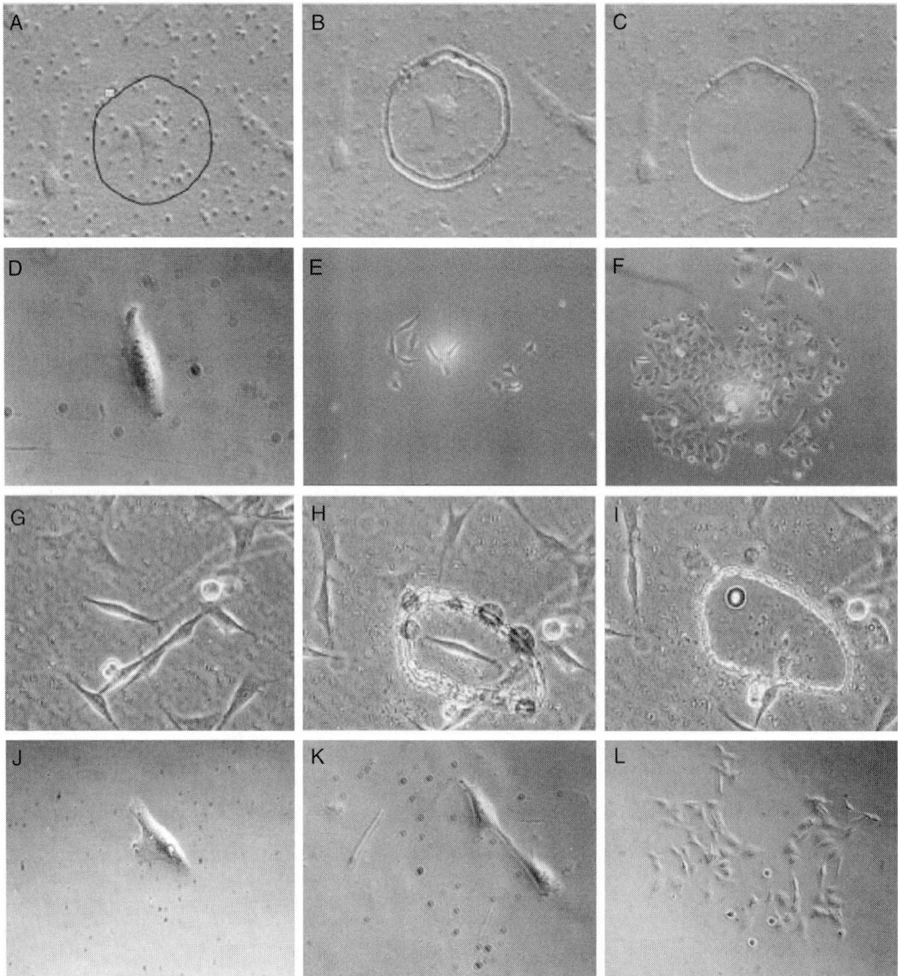

Fig. 10 Clonal expansion. LMPC and recultivation of blood-derived RM26 (A–F) and bone-marrow-derived BalbC (G–L) stem cells, as selected by morphology from a cell mixture (Fig. 9B). (A, G) Prior to LMPC, the culture media was almost completely removed. (B, H) Single cells were microdissected using a 40× objective. (C, I) The cells were catapulted into microfuge caps and (D, J) transferred to 24-well cell culture plates filled with a 1:1 mixture of fresh and conditioned media. (E, K) The cells started to proliferate within 1 day and grew to pure clones. (F) The RM26 stem cells usually grew as clusters of cells, while the BalbC cells had a more dispersed growth (L). This proves the safety of LMPC for stem cell capture and clonal expansion. Contrast enhancement by a 40× objective with PlasDIC.

staining of the living mixed culture grown on the PEN MembraneRing allowed a clear distinction of the two cell types by morphology and marker expression (Fig. 9G and H). Thus, the usage of PEN membrane is suitable for LMPC enabling save discrimination.

2. LMPC for Clonal Expansion

For LMPC cells were either selected according to their morphology only, or according to morphology and CD34 expression. The catapulted cells (Fig. 10) were transferred in a 24-well cell culture plate containing a mixture of fresh and conditioned media. In 90% of the experiments, the catapulted cells adhered well and proliferated quickly. The RM26 stem cells usually grew as clusters of cells, while the BalbC cells had a more dispersed growth. These results clearly demonstrate that LMPC allows the extraction of pure clones from cell mixtures and does not impact stem cell viability.

3. Conservation of Stemness

After 10 days, the originally CD34+ stem cells (blood-derived RM26) still expressed the CD34 stem cell marker as proven by FACS analyses (Fig. 11). Thus, LMPC does not interfere with stem cell character, an important feature for further stem cell research.

4. Human Primary Stem Cells

In a second set of experiments, the experiences with laser-based capture and subsequent cloning of murine stem cell lines were transferred to the even more delicate primary human stem cells, that is bone marrow-derived mononuclear cells (hBMMNC). When plated onto tissue culture plastic, these cells display various morphologies: a smaller, fast-proliferating type that can have one ruffled side and a pseudopodia-like extension on the opposite side and a larger, flat, and more mature type, which proliferates slow or not at all. Also those cells could be selected according to their morphology, they could be purified using LMPC and were successfully recultivated yielding pure clones.

5. Safety and Reliability of LMPC

Both experiments demonstrated that LMPC is well suitable to purify murine stem cell lines as well as delicate human primary stem cells from a mixture of cells according to morphology and marker expression. LMPC does not interfere with stem cell growth and expansion, as the cells remain viable and still express the same marker as the originating cells. Therefore, stem cells can be isolated and captured via LMPC without harming their viability or changing their stem cell characteristics.

6. LMPC of Cell Clusters

Additionally, an important advantage of LMPC for stem cell research is the possibility to catapult large group of cells like embryoid bodies or entire clones, preserving the three-dimensional architecture including all the essential cell–cell interactions involved in ES cell differentiation. Common enzymatic disintegration based on chemicals usually causes a massive change of the stem cells' natural

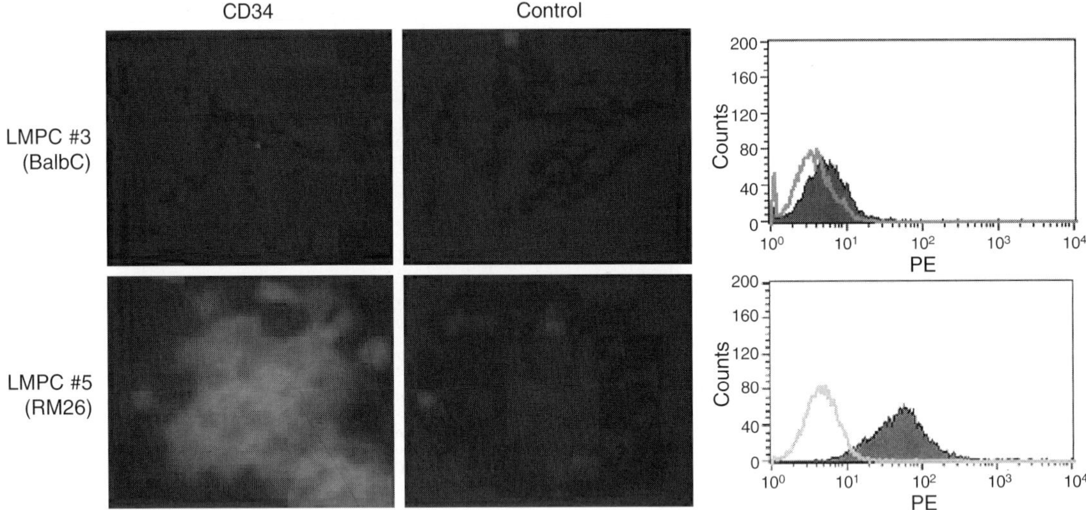

Fig. 11 Conservation of stemness. Ten days after LMPC, the originally RM26/CD34+ stem cell lines still expressed the CD34 stem cell marker as proven by FACS analysis (lower right panel) and immunocytochemistry (lower left panel). This proves that LMPC does not interfere with stem cell character.

conditions and might misguide the differentiation process. With LMPC, the clone is kept intact or, if required could even be cut in part to yield aliquots of desired cell numbers. This might save time and significantly reduces the danger of contamination as compared to the currently performed mechanical needle dissection.

A further promising application of LMPC is the selective isolation of large clones as demonstrated with beating heart myocytes. Cardiomyocytes derived from ES cells enable development of predictive cardiotoxicity models to increase the safety of novel drugs. Heterogeneity of differentiated ES cells limits the development of reliable *in vitro* models for compound screening. Using LMPC, it becomes possible to isolate pure populations of ES-derived cardiomyocytes and subject them to 96-well format for *in vitro* pharmacology assays. The catapulted ES-derived beating foci expressed developmental and functional cardiac markers and functional assays exhibited cardiac-like response to increased extracellular calcium and L-type calcium channel antagonist (Chaudhary *et al.*, 2006). Using LMPC as a "positive cell selection" method remarkably speeded up the selection procedure as compared to negative selection using chemicals. Thus, LMPC has great potential for reliable drug screening and safe drug development providing a novel and efficient tool to pharmaceutical research and industry.

In summary, LMPC of living cells opens new approaches in life science and stem cell research establishing homogenous cell populations from adherent cell and tissue cultures to characterize and clonally expand defined cell types. The possibility to enrich rare cells from mixed cell cultures and to establish pure cell populations provides entirely new approaches to study cell behavior or cell–cell interaction and to analyze the corresponding molecular composition. Especially for stem cell research and regenerative or personalized medicine, LMPC will become an indispensable tool.

IV. Conclusions and Future Challenges

Laser microdissection and catapulting provide pure starting material for the functional analysis of cells and is well established in functional genomics and proteomics research. This fact is confirmed by the increasing number of international publications using laser microdissection as a routine-capturing method.

As the LMPC technology makes its way into life cell research, there will be a further important impact into the emerging field of stem cell research, tissue engineering, and regenerative medicine. Thus, LMPC will become a routine tool for the purification of desired cell groups, for the capture of single cells from cell culture, and even more challenging, from fresh biopsies for recultivation and clonal expansion. Cancer research will benefit due to the provision of pure and homogeneous cell material, which allows a quicker understanding of the underlying molecular processes of cellular malfunction. In combination with drug screening, LMPC will enable the development of more efficient and safer drugs, as well as patient-tailored medical treatment.

The combination of cell migration or wound healing experiments with LMPC will generate a better understanding of cellular processes and associated molecular events that initiate and maintain cell motion and metastasis formation. These experiments could be supported by an innovative digital holographic system that enables imaging and process controlling in a fast and entirely marker-free way. The *D*igital *H*olographic *M*icro-*I*nterferometry (DHMI) technology can monitor cell behavior in all three dimensions and can quantitatively measure changes in volumes or refraction indices over time (von Bally *et al.*, 2006).

LMPC combined with laser microinjection, laser-induced cell fusion, or laser scanning cytometry will confirm itself as key technology in basic research, drug development, or regenerative medicine. The resulting benefits will be the development of better drugs for patient treatment and efficient methods for tissue regeneration.

All the above-mentioned procedures will greatly benefit from extended automation of laser micromanipulation and laser capture technologies, which is an ongoing process within divers research laboratories (Botvinick and Berns, 2005) and R&D departments of corresponding companies. Thus, the PALM MicroBeam with its LMPC system exerts explicit influence on academia, economy, and health care—that is, a real key technology.

Acknowledgments

These studies were partly supported by grants from the German Ministry of Education and Research (BMBF: Mikroso: 13N8257 and PhoNaChi: 13N8466). The authors want to thank Susana Soria-Lopez for excellent technical assistance as well as Sieglinde Hinteregger for outstanding project administration and the colleagues of the PALM application laboratory for experimental studies and technical support.

References

Aarts, W. M., Bende, R. J., Vaandrager, J. W., Kluin, P. M., Langerak, A. W., Pals, S. T., and van Noesel, C. J. M. (2002). *In situ* analysis of the variable heavy chain gene of an IgM/IgG-expressing follicular lymphoma. *Am. J. Pathol.* **160**(3), 883–891.

Ashkin, A., Dziedzic, J. M., and Yamane, T. (1987). Optical trapping and manipulation of single cells using infrared laser beams. *Nature* **330,** 769–771.

Ashkin, A., Schütze, K., Dziedzic, J. M., Euteneuer, U., and Schliwa, M. (1990). Force generation of organelle transport measured *in vivo* by an infrared laser trap. *Nature* **348,** 346–348.

Aubele, M., Cummings, M. C., Walch, A., Zitzelsberger, H., Nahrig, J., Höfler, H., and Werner, M. (2000a). Heterogeneous chromosomal aberrations in intraductal breast lesions adjacent to invasive carcinoma. *Anal. Cell. Pathol.* **20**(1), 17–24.

Aubele, M., Mattis, A., Zitzelsberger, H., Walch, A., Kremer, M., Welzl, G., Höfler, H., and Werner, M. (2000b). Extensive ductal carcinoma *in situ* with small foci of invasive ductal carcinoma: Evidence of genetic resemblance by CGH. *Int. J. Cancer* **85,** 82–86.

Ball, H. J., and Hunt, N. H. (2004). Needle in a haystack: Microdissecting the proteome of a tissue. *Amino Acids* **27,** 1–7.

Becker, A. J., Chen, J., Paus, S., Normann, S., Beck, H., Elger, C. E., Wiestler, O. D., and Blümcke, I. (2002). Transcriptional profiling in human epilepsy: Expression array and single cell real-time qRT-PCR analysis reveal distinct cellular gene regulation. *Neuroreport* **13**(10), 1327–1333.

Berns, M. W. (1998). Laser scissors and tweezers. *Sci. Am.* **278,** 62–67.

Berns, M. W., Aist, J., Edwards, J., Strahs, K., Girton, J., McNeill, P., Rattner, J. B., Kitzes, M., Hammer-Wilson, M., Liaw, L. H., Siemens, A., Koonce, M., *et al.* (1981). Laser microsurgery in cell and developmental biology. *Science* **213,** 505–513.

Botvinick, E. L., and Berns, M. W. (2005). Internet-based robotic laser scissors and tweezers microscopy. *Microsc. Res. Tech.* **68,** 65–74.

Buerger, H., Gebhardt, F., Schmidt, H., Beckmann, A., Hutmacher, K., Simon, R., Lelle, R., Boecker, W., and Brandt, B. (2000). Length and loss of heterozygosity of an intron 1 polymorphic sequence of egfr is related to cytogenetic alterations and epithelial growth factor receptor expression. *Cancer Res.* **60,** 854–857.

Burgemeister, R., Schütze, K., Minderer, S., and Gloning, K. P. (1999). Single fetal cells separated by the Laser MicroBeam technique. *BIOforum Int.* **3,** 119–121.

Chaudhary, K. W., Barrezueta, N. X., Bauchmann, M. B., Milici, A. J., Beckius, G., Stedman, D. B., Hambor, J. E., Blake, W. L., McNeish, J. D., Bahinski, A., and Cezar, G. G. (2006). Embryonic stem cells in predictive cardiotoxicity: Laser capture microscopy enables assay development. *Toxicol. Sci.* **90,** 149–158.

Clement-Sengewald, A., Schütze, K., Sandow, S., Nevinny, C., and Pösl, H. (2000). PALM® Robot-MicroBeam for laser-assisted fertilization, embryo hatching and single-cell prenatal diagnosis. *In* "Photomedicine in Gynecology and Reproduction" (P. Wyss, Y. Tadir, B. J. Tromberg, and U. Haller, eds.), pp. 340–351. Karger, Basel.

de Groot, C. J., Steegers-Theunissen, R. P., Guzel, C., Steegers, E. A., and Luider, T. M. (2005). Peptide patterns of laser dissected human trophoblasts analyzed by matrix-assisted laser desorption/ionisation-time of flight mass spectrometry. *Proteomics* **5,** 597–607.

Eisenstein, M. (2006). Divide and conquer: A key element of performing good cell-biology experiments is starting with exactly the right cells. *Nature* **441,** 1179–1185.

Fassunke, J., Blümcke, I., Lahl, R., Elger, C. E., Schramm, J., Merkelbach-Bruse, S., Mathiak, M., Wiestler, O. D., and Becker, A. J. (2004a). Analysis of chromosomal instability in focal cortical dysplasia of Taylor's balloon cell type. *Acta Neuropathol.* **108,** 129–134.

Fassunke, J., Majores, M., Ullmann, C., Elger, C. E., Schramm, J., Wiestler, O. D., and Becker, A. J. (2004b). *In situ*-RT and immunolaser microdissection for mRNA analysis of individual cells isolated from epilepsy-associated glioneuronal tumors. *Lab. Invest.* **84,** 1520–1525.

Fominaya, A., Linares, C., Loarce, Y., and Ferrer, E. (2005). Microdissection and microcloning of plant chromosomes. *Cytogenet. Genome Res.* **109,** 8–14.

Friedenstein, A. J., Chailakhjan, R. K., and Lalykina, K. S. (1970). The development of fibroblast colonies in monolayer cultures of guinea-pig bone marrow and spleen cells. *Cell Tissue Kinet.* **4,** 393–403.

Greulich, K. O. (1999). "Micromanipulation by Light in Biology and Medicine: The Laser Microbeam and Optical Tweezers." Karl Otto Greulich, Birkhäuser, Basel, Boston, Berlin.

Greulich, K. O., and Weber, G. (1992). The light microscope on its way from an analytical to a preparative tool. *J. Microsc.* **167**(2), 127–151.

Hahn, S., and Holzgreve, W. (2002). Prenatal diagnosis using fetal cells and cell-free fetal DNA in maternal blood: What is currently feasible? *Clin. Obstet. Gynecol.* **45,** 649–656.

Hartmann, A., Schlake, G., Zaak, D., Hungerhuber, E., Hofstetter, A., Hofstaedter, F., and Knuechel, R. (2002). Occurrence of chromosome 9 and p53 alterations in multifocal dysplasia and carcinoma *in situ* of human urinary bladder. *Cancer Res.* **62,** 809–818.

Hasse, U., Tinguely, M., Opplinger Leibundgut, E., Cajot, J. F., Garvin, A. M., Tobler, A., Borisch, B., and Fey, M. F. (1999). Clonal loss of heterozygosity in microdissected Hodgkin and Reed-Sternberg cells. *J. Natl. Cancer Inst.* **91,** 1581–1583.

Heel, K., and Dawkins, H. (2001). Laser microdissection and optical tweezers in research. *Today's Life Sci.* **13,** 42–48.

Hobza, R., Lengerova, M., Cernohorska, H., Rubes, J., and Vyskot, B. (2004). FAST-FISH with laser beam microdissected DOP-PCR probe distinguishes the sex chromosomes of Silene latifolia. *Chromosome Res.* **12,** 245–250.

Huang, Q., Choy, K. W., Cheung, K. F., Lam, D. S. C., Fu, W. L., and Pang, C. P. (2003). Genetic alterations on chromosome 19, 20, 21, 22, and X detected by loss of heterozygotity analysis in retinoblastoma. *Mol. Vis.* **9,** 502–507.

Irié, T., Aida, T., and Tachikawa, T. (2004). Gene expression profiling of oral squamous cell carcinoma using laser microdissection and cDNA microarray. *Med. Electron Microsc.* **37,** 89–96.

Ivashikina, N., Deeken, R., Ache, P., Kranz, E., Pommerrenig, B., Sauer, N., and Hedrich, R. (2003). Isolation of AtSUC2 promoter-GFP-marked companion cells for patch-clamp studies and expression profiling. *Plant J.* **36,** 931–945.

Kehr, J. (2003). Single cell technology. *Curr. Opin. Plant Biol.* **6,** 617–621.

Koene, G. P. J. A., Arts-Hilkes, Y. H. A., van der Ven, K. J. W., Rozemuller, E. H., Slootweg, P. J., de Weger, R. A., and Tilanus, M. G. J. (2004). High level of chromosome 15 aneuploidy in head and neck squamous cell carcinoma lesions identified by FISH analysis: Limited value of β2-microglobulin LOH analysis. *Tissue Antigens* **64,** 452–461.

Koller, M. R., Hanania, E. G., Stevens, J., Eisfeld, T. M., Sasaki, G. C., Fieck, A., and Palsson, B. O. (2004). High-throughput laser-mediated *in situ* cell purification with high purity and yield. *Cytometry A* **61,** 153–161.

Lange, C., Kaltz, C., Thalmeier, K., Kolb, H. J., and Huss, R. (1999). Hematopoietic reconstitution of syngeneic mice with a peripheral blood-derived, monoclonal CD34−, Sca-1+, Thy-1(low), c-kit+ stem cell line. *J. Hematother. Stem Cell Res.* **4,** 335–342.

Langer, S., Geigl, J. B., Gangnus, R., and Speicher, M. R. (2005). Sequential application of interphase-FISH and CGH to single cells. *Lab. Invest.* **85,** 582–592.

Levy, A., Chang, B. J., Abbott, L. K., Kuo, J., Harnett, G., and Inglis, T. J. J. (2003). Invasion of spores of the arbuscular mycorrhizal fungus *Gigaspora decipiens* by Burkholderia spp. *Appl. Environ. Microbiol.* **69,** 6250–6256.

Matsunaga, S., Schütze, K., Donnison, I. S., Grant, S. R., Kuroiwa, T., and Kawano, S. (1999). Single pollen typing combined with laser-mediated manipulation. *Plant J.* **20,** 371–378.

Mayer, A., Stich, M., Brocksch, D., Schütze, K., and Lahr, G. (2002). Going *in vivo* with laser-microdissection. *Methods Enzymol.* **356,** 25–33.

Meimberg, M., Thalhammer, S., Brachmann, A., Müller, B., Eichacker, L. A., Heckl, W. M., and Heubl, G. (2003). Selection of chloroplasts by laser MicroBeam microdissection for single-chloroplast PCR. *BioTechniques* **34,** 1238–1243.

Meirelles Lda, S., and Nardi, N. B. (2003). Murine marrow-derived mesenchymal stem cell: Isolation, *in vitro* expansion, and characterization. *Br. J. Haematol.* **123,** 702–711.

Modderman, W. E., Vrijheid-Lammers, T., Lowik, C. W., and Nijweide, P. J. (1994). Removal of hematopoietic cells and macrophages from mouse bone marrow cultures: Isolation of fibroblast-like stromal cells. *Exp. Hematol.* **22,** 194–201.

Mohanty, S. K., Rapp, A., Monajembashi, S., Gupta, P. K., and Greulich, K. O. (2002). Comet assay measurements of DNA damage in cells by laser microbeams and trapping beams with wavelengths spanning a range of 308 nm to 1064 nm. *Radiat. Res.* **157,** 378–385.

Niyaz, Y., and Sägmüller, B. (2005). Non-contact laser microdissection and pressure catapulting: Automation via object-oriented image processing. *Med. Laser Appl.* **20,** 223–232.

Niyaz, Y., Stich, M., Sägmüller, B., Burgemeister, R., Friedemann, G., Sauer, U., Gangnus, R., and Schütze, K. (2005). Noncontact laser microdissection and pressure catapulting—sample preparation for genomic, transcriptomic, and proteomic analysis. *Microarrays Clin. Diagnos., Methods Mol. Med.* **114,** 1–24.

Obermann, E. C., Junker, K., Stoehr, R., Dietmaier, W., Zaak, D., Schubert, J., Hofstaedter, F., Knuechel, R., and Hartmann, A. (2003). Frequent genetic alterations in flat urothelial hyperplasias and concomitant papillary bladder cancer as detected by CGH, LOH, and FISH analyses. *J. Pathol.* **199,** 50–57.

Ramsay, K., Wang, Z., and Jones, M. G. K. (2004). Using laser capture microdissection to study gene expression in early stages of giant cells induced by root-knot nematodes. *Mol. Plant Pathol.* **5,** 587–592.

Sauber, C., Sägmüller, B., Neumann, M., and Kretschmar, H. A. (2004). Identification of proteins in post-mortem human brain tissue by laser microdissection/pressure catapult and nano-LC/MS/MS. *Agilent Application Note 5989–0895EN.*

Sauer, U., Ehnle, S., and Burgemeister, R. (2004). Laser pressure catapulting(LMPC) of sperm cells and optimisation of DNA isolation. *PALM Application Note*, 1–4.

Schad, M., Lipton, M. S., Giavalisco, P., Smith, R. D., and Kehr, J. (2005a). Evaluation of two-dimensional electrophoresis and liquid chromatography-tandem mass spectrometry for tissue-specific protein profiling of laser-microdissected plant samples. *Electrophoresis* **26,** 2729–2738.

Schad, M., Mungur, R., Fiehn, O., and Kehr, J. (2005b). Metabolic profiling of laser microdissected vascular bundles of *Arabidopsis thaliana. Plant Methods* **1,** 2.

Schlomm, T., Luebke, A. M., Sultmann, H., Hellwinkel, O. J., Sauer, U., Poustka, A., David, K. A., Chun, F. K., Haese, A., Graefen, M., Erbersdobler, A., and Huland, H. (2005). Extraction and processing of high quality RNA from impalpable and macroscopically invisible prostate cancer for microarray gene expression analysis. *Int. J. Oncol.* **27,** 713–720.

Schneider-Stock, R., Boltze, C., Jaeger, V., Stumm, M., Seiler, C., Rys, J., Schütze, K., and Roessner, A. (2002). Significance of loss of heterozygosity of the RB1 gene during tumor progresson in welldifferentiated liposarcomas. *J. Pathol.* **197,** 654–660.

Schütze, K., and Clement-Sengewald, A. (1994). Catch and move—cut or fuse. *Nature* **368,** 667–670.

Schütze, K., and Lahr, G. (1998). Identification of expressed genes by laser-mediated manipulation of single cells. *Nat. Biotechnol.* **16,** 737–742.

Schütze, K., Becker, I., Becker, K. F., Thalhammer, S., Stark, R., Heckl, W. M., Böhm, M., and Pösl, H. (1997). Cut out or poke in—the key to the world of single genes: Laser micromanipulation as a valuable tool on the look-out for the origin of disease. *Genet. Anal.* **14**(1), 1–8.

Schütze, K., Becker, B., Bernsen, M., Björnsen, T., Brocksch, D., Bush, C., Clement-Sengewald, A., van Dijk, M. C. R. F., Friedemann, G., Heckl, W., Lahr, G., Lindahl, P., *et al.* (2002). Part 1: Manual dissection and laser capture microdissection, pp. 307–313 and Part 2: Laser Pressure Catapulting. *In* "*DNA Microarrays, A Molecular Cloning Manual*" (D. Bowtell, and J. Sambrook, eds.), pp. 331–356. CSHL Press, New York.

Schütze, K., Burgemeister, R., Clement-Sengewald, A., Ehnle, S., Friedemann, G., Lahr, G., Sägmüller, B., Stich, M., and Thalhammer, S. (2003). Non-contact live cell laser micromanipulation using PALM MicroLaser systems. PALM Scientific Edition No. 11 ISBN No. 3-9808893-0-0.

Schütze, K., Clement-Segewald, A., and Ashkin, A. (1994). Zona drilling and sperm insertion with combined laser microbeam and optical tweezers. *Fertil. Steril.* **61,** 783–786.

Schütze, K., Reiss, H. D., Becker, H., Monajembashi, S., and Greulich, K. O. (1989). Laser microsurgery on pollen tubes. *Ber Bunsenges Phys Chem (Int. J. Phys. Chem.)* **93,** 249–252.

Seidl, S., Burgemeister, R., Hausmann, R., Betz, P., and Lederer, T. (2005). Contact-free isolation of sperm and epithelial cells by Laser Microdissection and Pressure Catapulting. *Forensic Sci. Med. Pathol.* **V1–2,** 153–158.

Srinivasan, R. (1986). Ablation of polymers and biological tissue by ultraviolet lasers. *Science* **234,** 559–565.

Stich, M., Thalhammer, S., Burgemeister, R., Friedemann, G., Ehnle, S., Lüthy, C., and Schütze, K. (2003). Live cell catapulting and recultivation. *Pathol. Res. Pract.* **199,** 405–409.

Tannapfel, A., Anhalt, K., Häusermann, P., Sommerer, F., Benicke, M., Uhlmann, D., Witzigmann, H., Hauss, J., and Wittekind, C. (2003). Identification of novel proteins associated with hepatocellular carcinomas using protein microarrays. *J. Pathol.* **201,** 238–249.

Thalhammer, S., Lahr, G., Clement-Sengewald, A., Heckl, W. M., Burgemeister, R., and Schütze, K. (2003). Laser microtools in cell biology and molecular medicine. *Laser Phys.* **13,** 681–691.

Thalhammer, S., Langer, S., Speicher, M. R., Heckl, W. M., and Geigl, J. B. (2004). Generation of chromosome painting probes from single chromosomes by laser microdissection and linker-adaptor PCR. *Chromosome Res.* **12,** 337–343.

Thelen, P., Burfeind, P., Grzmil, M., Voigt, S., Ringert, R. H., and Hemmerlein, B. (2004). cDNA microarray analysis with amplified RNA after isolation of intact cellular RNA from neoplastic and non-neoplastic prostate tissue separated by laser microdissections. *Int. J. Oncol.* **24,** 1085–1092.

Vogel, A., and Venugupalan, V. (2003). Mechanism of pulsed laser ablation of biological tissues. *Chem. Rev.* **103,** 577–644.

von Bally, G., Kemper, B., Carl, D., Knoche, S., Kempe, M., Dietrich, C., Stutz, M., Wolleschenksky, R., Schütze, K., Stich, M., Buchstaller, A., Irion, K., *et al.* (2006). New methods for marker-free live cell and tumour analysis. *In* "Biophotonics. Visions for a better Health Care" (J. Popp and M. Strehle, eds.), pp. 301–360. Wiley-VCH, Weinheim.

Weber, G., and Greulich, K. O. (1992). Manipulation of cells, organelles and genomes by laser microbeam and optical trap. *Int. Rev. Cytol.* **133,** 1–41.

Wellmann, A., Wollscheid, V., Lu, H., Ma, Z. L., Albers, P., Schütze, K., Rhode, V., Behrens, P., Dreschers, S., Ko, Y., and Wernert, N. (2002). Analysis of microdissected prostate tissue with ProteinChip arrays—a way to new insights into carcinogenesis and to diagnostic tools. *Int. J. Mol. Med.* **9,** 341–347.

Wiegand, R., Weber, G., Zimmermann (Schütze), K., Monajembashi, S., Wolfrum, J., and Greulich, K. O. (1987). Laser-induced fusion of mammalian cells and plant protoplasts. *J. Cell Sci.* **88,** 145–149.

CHAPTER 24

Laser Capture Microdissection and Laser Pressure Catapulting as Tools to Study Gene Expression in Individual Cells of a Complex Tissue

Benayahu Dafna, Socher Rina, and Shur Irena

Department of Cell and Developmental Biology
Sackler School of Medicine, Tel-Aviv University, Israel

0091-679X/07 $35.00
DOI: 10.1016/S0091-679X(06)82024-8

Laser capture microdissection (LCM) method allows the selection of individual or clustered cells from intact tissues. LCM enables to pick cells from tissues that are difficult to study individually, to sort the anatomical complexity of tissues, and to make the cells available for molecular analyses. This technology provides an opportunity to uncover the molecular control of cellular fate in the natural microenvironment.

It is a difficult task to obtain cells from skeletal tissues, such as cartilage, periost, bone, and muscle, that are structured together and do not exist as individual organs. LCM allows isolation of desired cells from the native tissue environment for the analysis of gene expression. We earlier described the selection of cells from skeletal tissues that were analyzed for expression of transcription factors, receptors for cytokines, nuclear receptors, and functional genes such as alkaline phosphatase and structural proteins. Current results acquired using the LCM technology demonstrate expression of known genes that are in agreement with their reported *in vivo* and *in vitro* function in skeletal cells. The obtained knowledge will provide molecular information in the context of the cell and tissue biology. Such analysis will enable a reliable interpretation of function of known and novel genes expression in the skeletal tissues under various physiological conditions.

I. Introduction

The laser capture microdissection (LCM) method allows the selection of individual or clustered cells from intact tissues. LCM provides the opportunity to uncover the molecular control of cellular fate in their natural microenvironment. The technique allows selecting the cells from tissues that are difficult to study individually and to sort the anatomical complexity of these tissues, making the cells available for broad molecular analyses.

Skeletal tissues develop from mesechymal stem cells (MSCs). The mesenchymal pluripotent cells differentiate to chondrogenic, osteogenic, hematopoietic-supporting, and muscle (skeletal and cardiac) cells. Major contribution of mesenchymal cells is their differentiation into chondrocytes, matrix deposition that proceeds with osteogenesis, and bone matrix mineralization. Skeletal tissues are active throughout life in skeletal growth and remodeling. The cells' differentiation is regulated by multiple factors, including systemic hormones and local regulatory factors affecting proliferation and cell differentiation. Mesenchymal cells produce extracellular matrix (ECM) proteins, which create the skeletal cell niches affecting the cells function. Differentiation capacity is accompanied by a particular pattern of gene expression which reflects the cell maturation stage. Mesenchymal cells were reported for their plasticity. The cells' fate is coordinated by activation or repression of genes implicated in regulation of their lineage commitment. LCM technique enables to categorize the gene expression profile in intact tissues (Benayahu, 2000; Benayahu *et al.*, 2005; Shur *et al.*, 2001, 2005, 2006) and will allow discovering new regulatory pathways. Such achieved knowledge will define the molecular mechanisms on the distinct cells from skeletal tissues that will shed light on the cells function and molecular alterations led to diseases or changes of physiological status such as aging.

II. Rationale

The advantage of using RNA, isolated from specific tissues with subsequent utilization for the molecular analysis of cells, retrieved from the native *in vivo* environment. Even a small amount of RNA allows to demonstrate the gene expression that has been achieved using the LCM technique.

The use of LCM allows overcoming the problem of tissue heterogeneity and the procedure of selecting defined cells to minimize the contamination from neighboring unrelated tissues. It became reality now to extract small number of cells to retrieve RNA from different types of cells from the intact tissue. This will provide reliable interpretation of function for the tissue-specific gene expression. Analysis of tissue-specific gene expression *in vivo* will enable to determine the specific tissue control systems required for their differentiation. In additions, such analysis will enable discovery of tissue-specific genes and will provide detailed description of *in vivo* molecular pathways.

III. Methods

A. Specimen Preparation

Wear gloves, use RNase-free instruments and reagents. Wipe down the instruments with alcohol before use.

- Best results are obtained using freshly prepared cryosections.
- Laser cutting (microdissection) of the samples of frozen sections.
- Laser pressure catapulting (LPC) of the sample and collection into the cap.
- Getting the collected cells from the cap into the tip of the tube.

B. Preparation of Slides

Sections are mounted on regular 1-mm thick sterile glass slides or membrane covered slides. When using membrane-mounted samples, the dissected membrane acts as a backbone for the selected area/cell and can therefore be catapulted with a single laser shot from a remaining "bridge" at the border. The morphological integrity is completely preserved with this procedure. Dissection and catapulting are performed under 40 $\times$ objective.

C. Preparation of Frozen Sections

Prepare your sections onto sterile slide.

Slides can be fixed or deep frozen at $-80\,^{\circ}$C.

It is recommended to thaw–frozen sections on slides in 1-ml PBS containing 40 units of ribonuclease inhibitor prior to the fixation of the slides.

D. Fixation

After mounting the sections on sterile glass slide, there are many possibilities to fix the sections. For RNA isolation, ethanol fixation is recommended. This is performed after 20 s of air drying and dipping the mounted sections for 90 s into 70% ethanol followed by dipping in RNase-free water. Perform your preferred staining (Section III.F).

E. Removing the Freeze-Supporting Substance

In case when OCT or another tissue-freezing medium is used, it is important to get rid of the media on the slide before laser microdissection because these liquids will interfere with laser efficiency. Removal of the medium is done very easily by gently washing the slide for 1 min in water. If the sections are stained, the supporting substance is removed "automatically" in the aqueous staining solution or the diluted ethanol. Frozen sections should always be allowed to dry for 60 min at room temperature before use for LCM.

F. Histological Staining Methods

Sections may be used for LCM as unstained if areas of interest are recognizable based on morphology or can be stained for better visualization.

Hematoxylin/eosin (H&E) is a routine histological staining which results with nuclei stained in blue and cytoplasm in pink/red.

H&E staining of cryosections is suitable also for RNA preparation.

Slides are treated carefully with solutions prepared with RNase-free water.

The whole procedure must be quick to minimize the activity of RNases. Slides are immersed for 3 min in Mayer's hematoxylin solution, 3 min rinsing in RNase-free water, and so on.

Quick-increasing ethanol series of 70, 96, and 100% ethanol are used, and slides are subsequently let to air dry for 60 min.

G. Specimen Isolation and Collection

1. Laser Cutting (Microdissection) of the Samples

Search for an interesting area on the section; observe the area at the screen. Pipette 5 μl of absolute ethanol onto the depiction (a little distaining of the section may happen, but the drying is much quicker). This stage improved immediately the visualization of the cells after having contact with the ethanol. Now mark the cells or cell area with the software tool. After the rapid ethanol evaporation, catapult the marked cells or cell areas.

H. Laser Pressure Catapulting of the Samples

1. Catapulting into the Cap

Catapult Buffer contains 0.5-μM EDTA pH 8.0, 20-μM Tris pH 8.0; optional: add freshly prepared proteinase K 20 mg/ml 100 μl, use ddH_2O DEPC treated.

The cap needs to contain some solution for "capturing" the capulting samples that will stick to the cap. One option is autoclaved mineral oil (PCR oil) or Catapult Buffer in the inner ring of the cap. Be aware that aqueous solutions will dry out after a while. The catapulted cells or cell areas will stick onto the wet inner surface of the cap and will not fall down after the catapulting procedure. For single cells or very small areas, spinning down from oil may be difficult and therefore aqueous solutions should be preferred.

The more recent approach is to use silicon caps. These are clear adhesive caps without liquid, thus there is neither danger of RNases activity in the absence of water nor evaporation and crystal formation during extended sample harvesting. On the other hand, low energy is needed for catapulting due to short target distance.

When using membrane-mounted samples, the dissected membrane acts as a backbone. Thus, the selected area/cell can therefore be catapulted with a single laser shot from a remaining "bridge" at the border. The morphological integrity is completely preserved with this procedure.

2. Looking into the Cap to Visualize the Catapulted Samples

To control the efficiency of catapulting, it is possible to look into the collection device (e.g., PCR-cap) with the 5×, 10×, 40×, and 63× objectives. By using the software function "go to checkpoint," the lid is moved out of the light path and the cap can be lowered further toward the objectives for looking inside. Normally, most catapulted areas/cells can be found within the small inner ring of the caps.

3. Getting the Collected Cells from the Cap into the Tip of the Tube

After microdissection, in standart caps, the fluid from the cap is spun down in a bench centrifuge (5–10 min, 13,000 rpm) and samples can be stored for the later use. For RNA extraction, first appropriate lysis buffer is added to the PCR tube and after closure mixed inversion. Then the lysate is spun down as above (for future RNA isolation/analysis the tube is stored at −80 °C).

IV. RNA Isolation

A. Some Special Tips for Working with RNA

Working with RNA is more demanding than working with DNA because of the chemical instability of the RNA and the ubiquitous presence of RNAses. DEPC treatment of water and solutions is strongly recommended.

- Designate a special area for RNA work.
- Clean benches with 100% ethanol or a special cleaning solution, for example RNaseZap or similar.
- Always wear gloves. After putting on gloves, do not touch surfaces or equipment to avoid reintroduction of RNase to decontaminated material.
- Use sterile, disposable plastic ware.
- Use filtered pipetter tips.
- Glassware should be baked at 180 °C for 4 h (RNases can maintain activity even after prolonged boiling or autoclaving).
- Purchase reagents that are RNAse-free.
- All solutions should be made with diethylpyrocarbonate-(DEPC) treated H_2O.
- Treat all used material with DEPC.
- For best results use either fresh samples or samples that have been quickly frozen in liquid nitrogen or at −80 °C. (This procedure minimizes degradation of RNA by limiting the activity of endogeneous RNAses.) All required reagents should be kept on ice.
- Store RNA, aliquot in ethanol or RNA buffer, at −80 °C, that is maintain the RNA stable at this temperature for long term. Store prepared slides also at −80 °C.
- RNA is not stable at elevated temperatures, therefore avoid high temperatures (>65 °C) since these affect the integrity of RNA.

To ensure, RNase-free slides are sterile by autoclave.

Sections that are mounted on autoclaved slides that slides maybe treated to reduce the chance of contamination with exogenous RNases with RNase-ZAP followed by two separate washings in DEPC-treated water and drying at 37–55 °C for 30 min up to 2 h.

1. Procedure of Sample Collection of Laser Microdissection and Pressure Catapulting

Use an autoclaved cap/tube, pipet 3–6 μl of Catapult Buffer into the middle of the cap (see below for optional method).

Put the cap/tube into the cap/tube holder.

Perform laser microdissection and LPC of selected cells or cells areas into the cap.

Remove the cap/tube from the cap/tube holder and put it onto a 0.5-ml microfuge tube containing lysis buffer or respectively pipet lysis buffer into the tube with the attached cap.

Mix and lyse the cells by inversion.

Centrifuge the sample at full speed for 3 min.

If not going on directly, store the samples in lysis buffer at −80 °C.

2. Preparation of RNA from Catapulted Samples

Sections may be catapulted and stored at −80 °C until RNA harvesting. If using cyrosections you can now go on straight forward with RNA extraction by using kits for the small-scale RNA purification. Several kits are available for small quantity RNA isolation with laser microdissection and pressure catapulting (LMPC).

Trizol (Sigma-Aldrich, Cat. No. T-9429, St. Louis, MO) for extraction by a phenol precipitation-based method.

PALM® RNA-Extraction Kit (Cat. No. 4600–0100, Germany),

RNeasy Mini Kit (Qiagen, Cat. No. 74104, West Sussex, UK),

Absolutely RNA Nanoprep Kit (Stratagene, Cat. No. 400753, La Jolla,CA),

High Pure RNA Tissue kit (Roche, Cat. No. 2033674, Switzerland),

Purescript RNA isolation kit (Gentra, Distrib. Biozym, Cat. No. 212010, Hessisch Oldendorf, Germany)

ArrayPure™ Nano-scale RNA Purification Kit (Epicentre Biotechnologies, Cat. No. MPS04050, Madison, Wisconsin), or your extraction procedure of choice.

V. RNA Quantification

After the isolation steps, it is possible to quantify the RNA using a nanodrop measuring absorbance at A_{260} at the nanometer range. The assay can measure as low as 1 ng/μl, and it is minimally affected by contamination likely to be found in small nucleic acid preparations.

A method preferred for small amount of RNA measurement applies Pico Lab Chip Kit (Agilent). RNA ladder (Ambion) and unknown samples are analyzed on Agilent 2100 bioanalyzer (Agilent Technologies, Palo Alto, CA). One microliter from each isolated RNA sample is analyzed with RNA 6000 PicoAssay employing RNA Pico LabChips. The resulting electropherograms are used to determine RNA integrity and concentration; final RNA concentration can be measured in the range of 0.2–5 ng/μl.

VI. Reverse Transcription and RT-PCR

RNA reverse transcription to make cDNA includes the first-strand cDNA synthesis with subsequent amplification by PCR using gene-specific primers. It could be performed using several reverse transcriptases.

Sensiscript Reverse Transcriptase (Sensiscript RT Kit, Qiagen)

iScript RNase H^+MMLV reverse transcriptase for sensitive detection using 1 pg to 1 μg of input total RNA (iScript Select cDNA synthesis kit, Bio-Rad)

Reverse-iT™ MAX Blend which included the unique combination of RTases: AMV component allows reverse transcription to be carried out up to 57 °C when secondary structure is problematic, and the MMLV provides maximum yields. RNase inhibitor is included in the blend to prevent degradation from RNAse A (*Reverse*-iT™ MAX Blend kit, ABgene).

A. Choice of Priming Method

Both random hexamers and anchored oligo-dT can be used. Random primers generate the most diverse pool of cDNA, whereas oligo-dT primers anneal only to mRNA poly-A tails and thus will be biased toward the 5′ end.

VII. PCR Analysis

PCR analyses are performed by standard semiquantitative PCR with the end-point product detection or real-time detection kinetic PCR. Standard PCR methods use agarose gels for detection of PCR amplification at the final phase or end-point of the PCR reaction based on size discrimination of amplified cDNA. The real-time PCR system is based on the detection and quantitation of a fluorescent reporter which increases in direct proportion to the amount of PCR product in a reaction. By recording the amount of fluorescence emission at each cycle, it is possible to monitor the PCR reaction during exponential phase where the first significant increase in the amount of PCR product correlates to the initial amount of target template (Livak *et al.*, 1995). Specificity of amplification is assessed with a melt study. Presence of specific amplification products is confirmed by the single peak on the melting curve, indicating an absence of nonspecific amplification and primer–dimer formation.

A. Standard Semiquantitative PCR

The reaction is performed in the thin-walled 0.2-ml reaction tube in the total volume of 10 μl, including: 1 μl of template cDNA; 0.2 μl of gene-specific forward and reverse primers (resuspended to a known concentration with sterile TE); 5 μl of PCR mix ×2 (Sigma); PCR grade double distill water up to 10 μl. There are three major steps in a PCR: denaturation at 94 °C, annealing at 54–60 °C and extension at 72 °C, which are repeated for 30 or 40 cycles. In some cases, the following modifications of the standard PCR procedure can be done: (1) Hot-start PCR: to reduce nonspecific amplification by addition hot-start antibody. (2) "Touch-down" PCR: start at high annealing temperature, and then decrease annealing temperature in steps to reduce nonspecific PCR product.

The following step is the analysis of PCR reaction products by agarose gel electrophoresis of a 5-μl aliquot from the total reaction. The products are visualized by UV transillumination of the ethidium bromide-stained gel and compared to the PCR ladder. The ladder is a mixture of fragments with known size to compare with the PCR fragments of the unknown size. The remained reaction products could be stored at −20 °C until needed. The reaction products can be further purified using a number of procedures including PCR or gel purification systems and sequenced in order to prove their specificity.

B. Materials

Tissue-Tek embedding medium
70% alcohol
Hematoxylene and eosin stain
Catapult Buffer (0.5-μM EDTA pH 8.0, 20-μM Tris pH 8.0),
RNA isolation reagents: RNAeasy total kit (Qiagen)
Trizol (Sigma)
SuperScript reverse transcriptase (Qiagen)
PCR mix (Sigma)
Tris borate EDTA (TBE) buffer
SeaKem® LE Agarose (Cambrex)

C. Results

In the earlier study, we analyzed gene expression in skeletal tissues (Benayahu *et al.*, 2005; Shur *et al.*, 2006) and cultured cells (Shur *et al.*, 2001, 2005). Herein, we show an example of LCM of bone and cartilage cells retrieved from vertebra of newborn mice with subsequent gene expression analysis. Frozen tissues were embedded in optimal-cutting temperature (OCT) Tissue-Tek embedding medium. Serial frozen sections of 6 μm were cut using a cryostat and put onto glass slide. Before laser catapulting, slides were fixed in 70% alcohol for 90 s, air-dried and used for microdissection. The laser cutting (microdissection) follows with LPC of the selected areas was performed for each analyzed tissue at low magnification. For different samples or various tissues, the conditions of laser settings, speed, and energy are adjusted accordingly. Figure 1 demonstrates the morphology of

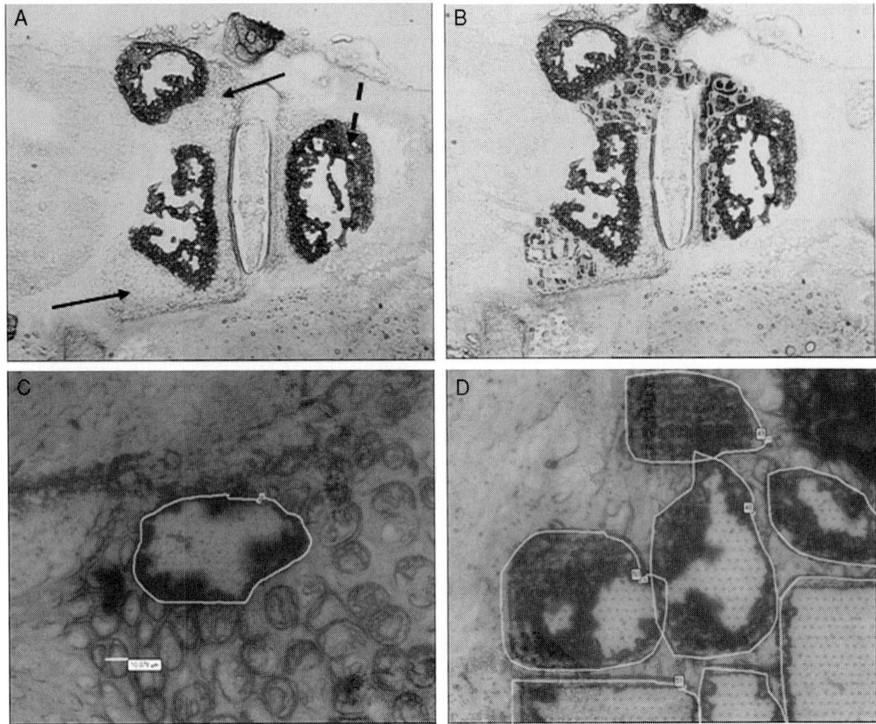

Fig. 1 Morphology of unstained frozen sections before and after capturing by LCM of cartilage and bone. In these sections (Fig. 1A, left panel), you can identify cartilage (arrow) and bone (dash line). Right panel (Fig. 1B) shows the marked areas used for cells capturing by LCM at a low magnification. The lower panels (Fig. 1C and D) demonstrate captured areas at a higher magnification.

unstained frozen section and visualizes tissues before and after capturing by LCM of cartilage and bone. In these sections (Fig. 1A, left panel), one can identify cartilage (arrow) and bone (dash line). Right panel of Fig. 1B shows the marked areas used for cells capturing by LCM. The lower panels of Fig. 1C and D demonstrate captured areas at higher magnification.

Further, the retrieved captured cells were processed for RNA extraction using a phenol precipitation-based technique. The quality of RNA was analyzed using Agilent 2100 bioanalyzer. One microliter from each isolated RNA sample was analyzed with RNA 6000 Pico Assay employing RNA Pico LabChip. The results of an electropherogram (Fig. 2) demonstrated RNA of high quality [RNA integrity numbers 8.1 and 9.5 for bone (2A) and cartilage (2B), respectively]. RNA was taken from each sample and reverse transcribed using Sensiscript Reverse Transcriptase; further PCR amplification of the resulted cDNA was performed with PCR mix and gene-specific primers. The PCR products were separated by electrophoresis in 1% agarose gels in Tris Borate EDTA (TBE) and stained with ethidium

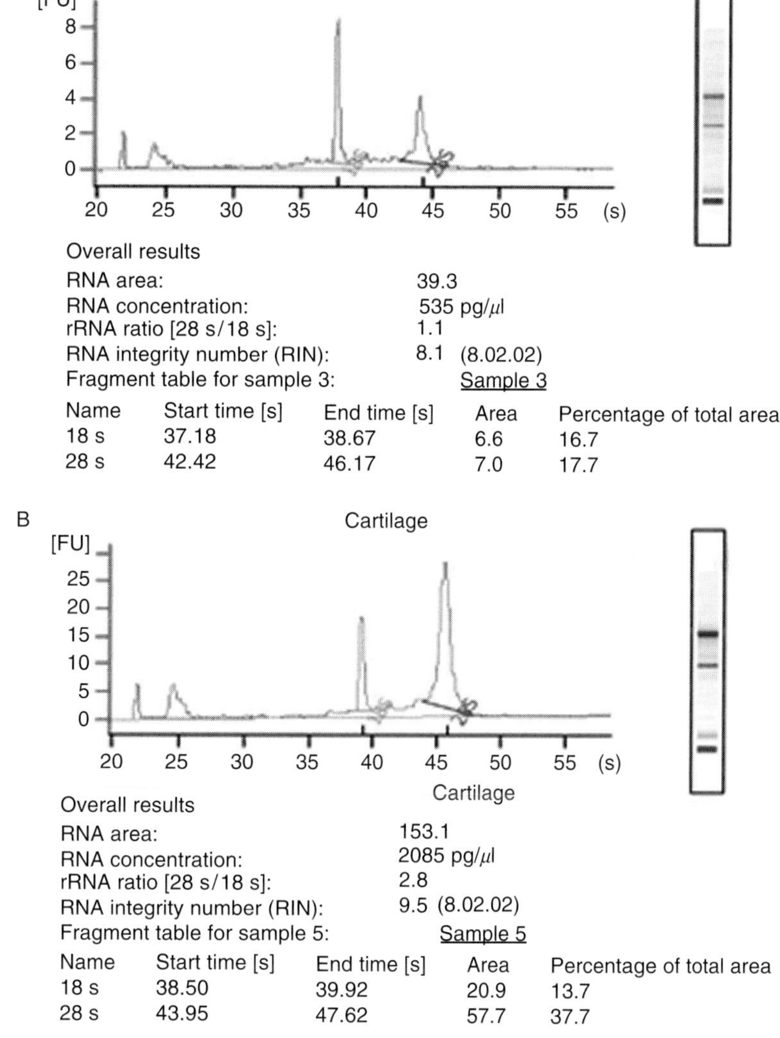

Fig. 2 Analysis of RNA integrity and concentration demonstrated by electropherogram using bioanalyzer. (A) RNA derived from bone tissue and (B) RNA derived from cartilage. Axes *Y* correspond to the fluorescence units (FU), axes *X* to seconds; 18S and 28S rRNA peaks used to calculate the RNA integrity number (RIN) are clearly visible.

bromide for PCR product visualization. G3PDH expression is a positive control, and specific genes amplified were collagen I, collagen II, and osteocalcin (Fig. 3, Table I). Both cartilage and bone were found to express the message for collagen I, while collagen II was expressed in cartilage only, and osteocalcin is expressed by

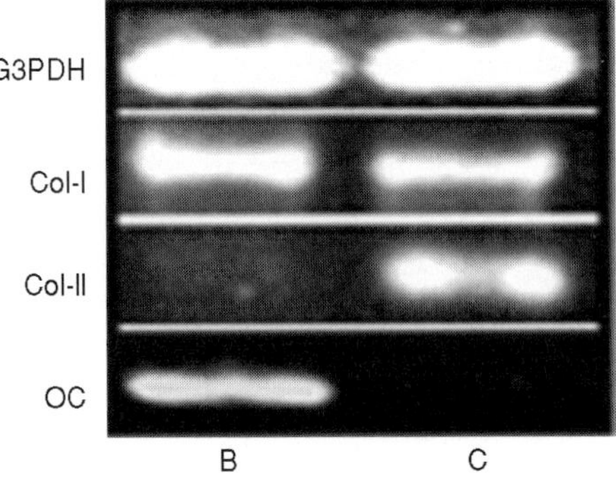

Fig. 3 RT-PCR of RNA idolsted from bone (B) and cartilage (C) tissues were separated by electrophoresis in 1% agarose gels in Tris borate EDTA (TBE) and stained with ethidium bromide. The expression of G3PDH, collagen I (Coll-I), collagen II (Coll-II), and osteocalcin (OC) was analyzed.

Table I
PCR Primers

Primers	Sequence	Expected size (bp)
G3PDH-F	ACCACAGTCCATGCCATCAC	445
G3PDH-R	TCCACCACCCTGTTGCTGTA	
Col I-F	TCTCCACTCTTCTAGTTCCCT	269
Col I-R	TTGGGTCATTTCCACATGC	
Col II-F	CACACTGGTAAGTGGGGCAAGACCG	172
Col II-R	GGATTGTGTTGTTTCAGGGTTCGGG	
OC-F	CATGAGAGCCCTCACA	310
OC-R	AGAGCGACACCCTAGAC	

bone solely. Collagen I is expressed in the bone ECM and also in the fibrocartilage and proliferating cartilage (Fukunaga *et al.*, 2003). Collagen II is a marker for cartilage and osteocalcin for bone (Aubin *et al.*, 1995). Obtained results demonstrate that the quality of isolated RNA is sufficient for further broad molecular analysis for the desired tissues *in vivo* using LCM technique.

VIII. Summary

The LCM approach has been successfully used in many fields and the growing number of publications draw the attention to the methodology (reviewed in Fuller *et al.*, 2003; Lloyd *et al.*, 2005; Shur *et al.*, 2006; Todd *et al.*, 2002). Utilization of the LCM for the molecular analysis of cells from skeletal tissues will enable to study differential pattern of gene expression at the native *in vivo* environment of intact skeletal tissue with reliable interpretation of function for the known genes.

However, variability between samples may be caused either by differences in the tissue samples (different areas isolated from the same section) or by some variances in sample handling in respect to the different kits and still needs the expertise in the multitask technology that combines histology, microscopy work, and dedicated molecular biology.

The concept of LCM will provide results that will pave the way toward high-throughput profiling of tissue-specific gene expression using Gene Chips arrays. This will enable the discovery of tissue-specific genes in the context of tissue function *in vivo*. Detailed description of *in vivo* molecular pathways will make it possible to elaborate on control systems for the repair of genetic or metabolic diseases of skeletal tissues.

Acknowledgments

The study was supported by funding form Ramot at Tel Aviv University and by the Chief Scientist of Ministry of Commerce and consortium "Bereshit" to DB.

References

Aubin, L., Liu, F., Malaval, L., and Gupta, A. K. (1995). Osteoblast and chondroblast differentiation. *Bone* **17,** 775–835.

Benayahu, D. (2000). The hematopoietic microenvironment: The osteogenic compartment of bone marrow: Cell biology and clinical application. *Hematology* **4**(5), 427–435.

Benayahu, D., Akavia, U. D., Socher, R., and Shur, I. (2005). Gene expression in skeletal tissues: Application of laser capture microdissection. *J. Microsc.* **220**(Pt. 1), 1–8.

Fukunaga, T., Yamashiro, T., Oya, S., Takeshita, N., Takigawa, M., and Takano-Yamamoto, T. (2003). Connective tissue growth factor mRNA expression pattern in cartilages is associated with their type I collagen expression. *Bone* **33**(6), 911–918.

Fuller, A. P., Palmer-Toy, D., Erlander, M. G., and Sgroi, D. C. (2003). Laser capture microdissection and advanced molecular analysis of human breast cancer. *J. Mammary Gland Biol. Neoplasia* **8**(3), 335–345.

Livak, K. J., Flood, S. J., Marmaro, J., Giusti, W., and Deetz, K. (1995). Oligonucleotides with fluorescent dyes at opposite ends provide a quenched probe system useful for detecting PCR product and nucleic acid hybridization. *PCR Methods Appl.* **4**(6), 357–362.

Lloyd, R. V., Qian, X., Jin, L., Ruebel, K., Bayliss, J., Zhang, S., and Kobayashi, I. (2005). Analysis of pituitary cells by laser capture microdissection. *Methods Mol. Biol.* **293,** 233–241.

Shur, I., Lokiec, F., Bleiberg, I., and Benayahu, D. (2001). Differential gene expression of cultured human osteoblasts. *J. Cell. Biochem.* **83**(4), 547–553.

Shur, I., Socher, R., and Benayahu, D. (2005). Dexamethasone regulation of cFos mRNA in osteoprogenitors. *J. Cell. Physiol.* **202**(1), 240–245.

Shur, I., Socher, R., and Benayahu, D. (2006). *In vivo* association of CReMM/CHD9 with promoters in osteogenic cells. *J. Cell. Physiol.* **207**(2), 374–378.

Todd, R., Lingen, M. W., and Kuo, W. P. (2002). Gene expression profiling using laser capture microdissection. *Expert Rev. Mol. Diagn.* **2**(5), 497–507.

CHAPTER 25

Laser Capture Microdissection in Comparative Proteomic Analysis of Hepatocellular Carcinoma

Hong-Yang Wang

International Cooperation Laboratory on Signal Transduction
Eastern Hepatobiliary Surgery Institute
The Second Military Medical University
Shanghai 200438, People's Republic of China

0091-679X/07 $35.00
DOI: 10.1016/S0091-679X(06)82025-X

Hepatocellular carcinoma (HCC) is one of the most frequent visceral neoplasia worldwide and is a multifactorial and multistage pathogenesis that finally leads to the deregulation of cell homeostasis. A main problem with the analysis of HCC tissue samples, either at the level of proteins or genes, is the heterogeneous nature of the sample. Laser capture microdissection (LCM) may allow the more ready identification of differences in protein expression of selected cell types or areas of tissue and allows procuring a microscopic region as small as 3–5 μm in diameter. Here we applied the LCM to the isolation of hepatocyte for comparative proteomic analysis of hepatitis B-related HCC and surrounding nontumorous tissues. Proteome alterations were observed using two-dimensional polyacrylamide gel electrophoresis and electrospray ionization tandem mass spectrometry, and alterations in the proteome were examined. LCM was found to eliminate hemoglobin from homogenization of the HCC tissue, demonstrating its capacity of resolving the problem of heterogeneity and contamination in tissue samples. Twenty protein spots were selected and eleven proteins significantly altered in the surrounding nontumorous tissues and HCC tissues. Of the proteins that were selected, peroxiredoxin 2, apolipoprotein A-I precursor, 3-hydroxyacyl-CoA dehydrogenase type II, and 14.5-kDa translational inhibitor protein appear to be novel candidates for useful hepatitis B-related HCC markers. This study indicated LCM is a useful technological method in proteomic study of cancer tissue. The proteins revealed in this experiment can be used in the future for studies pertaining to hepatocarcinogenesis, or as diagnostic markers and therapeutic targets for HCC associated with Hepatitis B virus infection.

I. Introduction

Hepatocellular carcinoma (HCC) is one of the most frequent visceral neoplasia worldwide, with an estimated 564,000 new cases in 2000 (Parkin *et al.*, 2001). It is known that most HCCs develop from chronic inflammatory liver disease due to the *Hepatitis B virus* (HBV) infection, *Hepatitis C virus* (HCV) infection, and exposure to carcinogens such as aflatoxin (Stuver, 1998). Like other cancers, the development of HCC is a multifactorial and multistage pathogenesis that finally leads to the deregulation of cell homeostasis (Harris, 1994, 1996). Although the molecular interactions between hepatocytes and the specific etiologic agents of HCC are being elucidated in the past decade (Brechot, 1998; Diao *et al.*, 2001; Smela *et al.*, 2001), the mechanism of hepatocarcinogenesis is unclear.

The term proteome was first stated to describe the set of proteins encoded by the genome (Wilkins *et al.*, 1996). With the completion of the draft sequence of the human genome (Lander *et al.*, 2001; Venter *et al.*, 2001), there is a great deal of interest in the use of functional genomics, especially gene expression-profiling techniques such as DNA microarrays and proteomics, to identify cancer-associated genes and their protein products. In contrast to the genome, the proteome is dynamic and is in constant flux because of a combination of factors. Proteomic technologies allow for identification of the protein changes caused by the disease process in a relatively accurate manner. At the protein level, distinct changes occur during the transformation of a healthy cell into a neoplastic cell, including altered expression, differential protein modification, changes in specific activity, and aberrant localization, all of which may affect cellular function. Identifying and understanding these changes are the underlying theme in cancer proteomics (Srinivas *et al.*, 2001).

Two-dimensional polyacrylamide gel electrophoresis (2D PAGE) has been the mainstay of electrophoresis technology for a decade and is the most widely used tool for separating proteins. The use of narrow immobilized pH gradients (IPG) for the first dimension increases resolving power and can help to detect low-abundance proteins. Electrospray ionization tandem mass spectrometry (ESI-MS/MS) allows the analysis and identification of very small amounts of protein isolated from the gel (Kovarova *et al.*, 2000; Neubauer *et al.*, 1998). These advances have been combined to make 2D PAGE a more attractive option for the analysis of complex protein mixtures.

However, a main problem with the analysis of tissue samples, either at the level of proteins or genes, is the heterogeneous nature of the sample. Laser capture microdissection (LCM) was used as a promising new approach to collect specific cell types from a tissue sample (Bonner *et al.*, 1997; Craven *et al.*, 2002; Zhou *et al.*, 2002). This allows the selective, relatively rapid microdissection of specific areas of tissue using a low-maintenance system that is easy to operate (Emmert-Buck *et al.*, 1996). Such an approach may allow the more ready identification of differences in protein expression of selected cell types or areas of tissue and allows procuring a microscopic region as small as 3–5 μm in diameter.

LCM is rarely applied to the analysis of HCC, even proteomic characterization is recently being applied to a few studies on human HCC cell lines (Yu *et al.*, 2000) and HCC tissues (Lim *et al.*, 2002). It is possible to use LCM to capture cancer structures with high purity, but the amount of samples collected by LCM is often limited, complicating proteomic analysis by 2D gel-based methods. These limitations can be alleviated by the integration of LCM with the highly sensitive ESI-MS/MS technology. We formerly combined the LCM with isotope-coded affinity tag (ICAT) technology and 2D liquid chromatography to investigate the proteomes of HCC (Li *et al.*, 2004). In this chapter, we present the isolation of cancerous tissues using LCM approach, and the systemic identification of extracted proteins by 2D PAGE and ESI-MS/MS. The presence of peroxiredoxin 2 (Prx II) in the plaques was confirmed by immunohistochemistry and Western blot, Prx II was found to be a novel component of HCC in human specimen. These studies demonstrate a powerful new approach for achieving comprehensive analysis of the proteome of HCC.

II. Rationale

LCM techniques have dramatically increased the ease of isolating specific cells from complex tissues for subsequent molecular analyses. Tissue preparation and microextraction protocols allow LCM microsamples to undergo quantitative proteomic analyses.

III. Methods

A. Preparation of Liver Tissue Sample

All the HCC samples were rinsed in sterile phosphate-buffered saline (PBS) and snap frozen in liquid nitrogen within 30 min. Sections (10 μm) were subsequently cut using a Leica CM 1900 microtome. Each slide was stained by hematoxylin and eosin. Both the hematoxylin and eosin solutions contained complete protease inhibitor cocktail.

B. Laser Capture Microdissection

Sections (10 μm) of frozen liver tissue from each sample above were cut and divided into two groups: one was placed directly in lysis buffer (described below) as the control; the other was placed on the slides for being applied with LCM. The slides were stained as described above. Then the sections were alternatively subjected to laser capture. The sections were captured using a 60-μm diameter laser beam at 30- to 80-mW power with pulse duration of 50 msec and machine gun mode with a laser-firing frequency of 1 shot per 500 msec. Typically, 500 shots were taken per cap. Each shot contained 20 cells and ~10,000 cells were obtained per cap. The obtained samples were placed in lysis buffer immediately.

C. 2D PAGE

Sample lysis buffer was based on the urea/thiourea mix previously described by Rabilloud *et al.* (1997). Samples were centrifuged at 40,000 × *g* for 40 min at 4 °C, the supernatant aliquoted and stored at −80 °C until analysis. Protein concentration was determined with a Bio-Rad protein assay kit. First-dimensional electrophoresis IEF was carried out by IPGphor (Amersham Pharmacia Biotech, Uppsala, Sweden). The 18-cm IPG strips (pH 3–10 nonlinear) and samples were applied overnight using the in-gel rehydration method as previously described (Gorg *et al.*, 2000). Samples containing 100- to 200-μg protein for analytical gels were diluted to 350 μl with rehydration solution, which contained 8-M urea; 2% CHAPS; 100-mM DTT; 0.5% (v/v) pH 3–10 IPG buffer. Proteins were focused initially at 400 V for 20 min, and then the voltage was increased to 8000 V within 3 h, and maintained at 8000 V for 7 h for a total of 60 kVh. After the first-dimensional IEF, IPG gel strips were placed in an equilibration solution (6-M urea, 2% SDS, 30% glycerol, 50-mM Tris–HCl, pH 8.8) containing 1% DTT for 15 min with shaking at 50 rpm on an orbital shaker. The gels were then transferred to the equilibration solution containing 2.5% iodoacetamide (IAA) and shaken for another 15 min. Then they were placed on a 30% gradient polyacrylamide gel slab (185 × 200 × 1.0 mm^3), and the strips were overlayed with 1% agarose. Separation in the second dimension was carried out using Protean II xi electrophoresis equipment and Tris–glycine buffer (25-mM Tris, 192-mM glycine) containing 0.1% SDS. The 2D SDS-PAGE was developed until the bromophenol blue dye marker had reached the bottom of the gel.

D. Silver Staining and Image Analysis

Silver staining was performed similarly to the method described by Shevchenko *et al.* (1996). Briefly, after electrophoresis, the gel slab was fixed in 50% methanol, 5% acetic acid in water for 30 min. It was then washed for 10 min with 50% methanol in water and additionally for 3× 10 min with water. The gel was sensitized by a 1-min incubation in 0.02% sodium thiosulfate, and it was then rinsed with two changes of distilled water for 3 min each. After rinsing, the gel was submerged in chilled 0.01% silver nitrate solution and incubated for 20 min at 4°C. Then the silver nitrate was discarded, and the gel slab was rinsed twice with distilled water for 2 min and then developed in 0.04% formalin (35% formaldehyde in water) in 2% sodium carbonate with intensive shaking. After the desired intensity of staining was achieved, the development was terminated by discarding the reagent, followed by washing of the gel slab with 5% acetic acid. Protein patterns in the gels were recorded as digitalized images using a high-resolution scanner. Gel image matching was done with the ImageMaster 2D Elite software 4.01 (Amersham Pharmacia Biotechnology).

E. Nano-Flow ESI-MS/MS Identification of Differentially Expressed Proteins

Gel spots of interest were excised to 1- to 2-mm^2 plugs and rinsed with 25-mM ammonium bicarbonate/50% acetonitrile. In-gel trypsin digestion and peptide extraction were performed as described (Ying *et al.*, 2003). For analysis of peptide mixture by LC-ESI-MS/MS, lyophilized peptide mixtures were dissolved with 5.5 μl of 0.1% formic acid (FA) in 2% ACN and injected by autosampler onto a 0.3 × 1 mm^2 trapping column (PepMap C18, LC Packings) using a CapLC system. Peptides were directly eluted into a quadrupole time-of-flight mass spectrometer (Q-TOF Micro, Micromass) at 200 nl/min on a C18 column (75 μm × 15 cm, LC Packings) using a 1-h gradient. Fragmentation mass lists were generated by the processing of MS/MS raw data with MassLynx3.5 (Micromass) and used to search against SWISS-PROT protein database via Internet available program Mascot (http://www.matrixscience.com).

F. Immunohistochemical and Western Blot Analyses

Formalin-fixed and paraffin-embedded tissues were sectioned in 5-μm thickness. Deparaffinization and rehydration were performed using xylene and alcohol. The sections were treated with 0.3% hydrogen peroxidase for 3 min and blocking antibody for 30 min. The primary antibody used was human Prx II (rabbit polyclonal). Avidin–biotin complex methodology was used. The chromogen was diaminobenzidine, and counterstaining was performed with hematoxylin. For Western blot analysis, tissues were suspended in lysis buffer containing 20 mmol/liter Tris (pH 7.6), 150-mmol/liter NaCl, 5-mmol/liter ethylenediaminetetraacetic acid, 0.5% NP40, and 1-mmol/liter dithiothreitol. Suspensions were sonicated for ~30 sec and centrifuged at 20,000 × *g* for 15 min. Proteins (20 μg) were loaded onto each lane, size fractionated by sodium dodecyl sulfate/polyacrylamide gel electrophoresis, and transferred to nitrocellulose membrane blocked with PBS/5% skim milk/0.01% Tween 20 for 30 min at room temperature. Primary antibody, Prx II polyAb (1:2000), was diluted in blocking buffer, incubated for 1 h with goatradish peroxidase-conjugated secondary antibody, washed, and developed with ECL-Plus (Amersham Pharmacia Biotech).

IV. Materials

IPG strips of pH 3–10 were purchased from Amersham Pharmacia Biotech (Immobiline DryStrip, 0.5 × 3 × 180 mm^3). BioLyte (pH 3–10) was from Bio-Rad, Hercules, CA. SDS, acrylamide, methylenebisacrylamide, TEMED, ammonium persulfate, DTT, urea, Tris, glycine, glycerol, and CHAPS were purchased from Bio-Rad. Silver nitrate and α-cyno-4-hydroxycinnamic acid were from Sigma (St. Louis, MO). Methanol, ethanol, phosphoric acid, acetic acid, and formaldehyde were purchased from Merck Schuchardt, Hohenbrunn, Germany. Trypsin (sequence grade) was obtained from Promega (Madison, WI). Trifluoroacetic acid (TFA) was from Acros (New Jersey). Ammonium bicarbonate was purchased from Sigma (St. Louis, MO). Acetonitrile (HPLC grade) was purchased from J.T. Baker Co. (Phillipsburg, NJ). FA was obtained from Beijing Chemical Co., Ltd. (Beijing, China). Other reagents were purchased from Sigma or Merck.

V. HCC Samples

Human liver tissue samples were obtained from specimens routinely resected from the 10 HCC patients in the Eastern Hepatobiliary Hospital in 2003. Informed consent was obtained from all patients. The stage of the HCC used in the study was Edmondson's grade III according to the pathological data.

VI. Discussion

A. LCM in Sample Preparation for Comparative Proteomics of HCC

The 2D PAGE technique has been used for over 20 years for protein separation. Even in light of most recent developments in the off-gel approaches, 2D PAGE remains extremely powerful for highlighting such posttranslational modifications. For protein identification/characterization, ESI-MS/MS or MALDI-MS is always followed after 2D PAGE, which has been the primary technique for biomarker discovery in conventional proteomic analysis. This technique is uniquely suited for direct comparison of protein expression and has been used to identify protein that is differently expressed between normal and tumor tissues in various cancers. Proteomic characterization is also being applied to human HCC cell lines (Yu *et al.*, 2000) and HCC tissues (Lim *et al.*, 2002).

Even with all the improvements that could be introduced, 2D gels will probably remain a rather low-throughput approach that requires a relatively large amount of samples. The latter is particularly problematic for clinical samples, as such samples are generally procured in limited amounts. Furthermore, tissue heterogeneity complicates the analysis of clinical samples. So far there are several studies of the proteomics of HCC (Lim *et al.*, 2002; Yu *et al.*, 2000); however the results are rather different. Two possibilities can be proposed to explain these results. First, the causes of HCC are different all over the world, for example HBV infection is the main factor in East Asia but HCV infection and alcohol are the prime factors in western countries. The second possible reason should be the heterogeneity of the samples, which might result in different expression patterns of protein profiles. Various tissue microdissection approaches are beneficial to reduce heterogeneity, but they further reduce the amount of sample available. In particular, the use of LCM, which allows defined cell types to be isolated from tissues, yields amounts of proteins that are difficult to reconcile with the need for greater amounts for 2D gels. Such an approach may allow the more ready identification of differences in protein expression of selected cell types or areas of tissue (Petricoin *et al.*, 2002). Here, we applied the LCM to the proteomic study of HCC tissues. The map for the LCM samples was shown in Fig. 1A and the map for the homogenized control was shown in Fig. 1B. They represented a comprehensive view of the major proteins expressed in human HCC tissue. It showed obviously that there were more spots in the homogenized control (949 spots) sample than in the LCM sample (868 spots). However, some spots in homogenization of the HCC tissue could not be detected in the LCM samples, for example, the hemoglobin, which was shown in Fig. 1C and D. LCM technology has been proposed capable of resolving the problem of heterogeneity and contamination in tissue samples. Our results also confirmed this predominance of LCM technology. Although H&E staining was applied to the LCM samples, there was little effect on the resulting protein profile of the two maps, indicating that the preservation of protein profiles after H&E staining was encouraging and LCM could be a useful method in the HCC tissue proteomic study.

B. Differences in Protein Expression Between Surrounding Nontumorous Tissues and HCC Tissues

By employing 2D PAGE technique, we analyzed the proteome of matched pairs of tumor tissue and surrounding nontumorous tissue from HCC patients, which was shown in Fig. 2. The enlarged area of some differential expression proteins

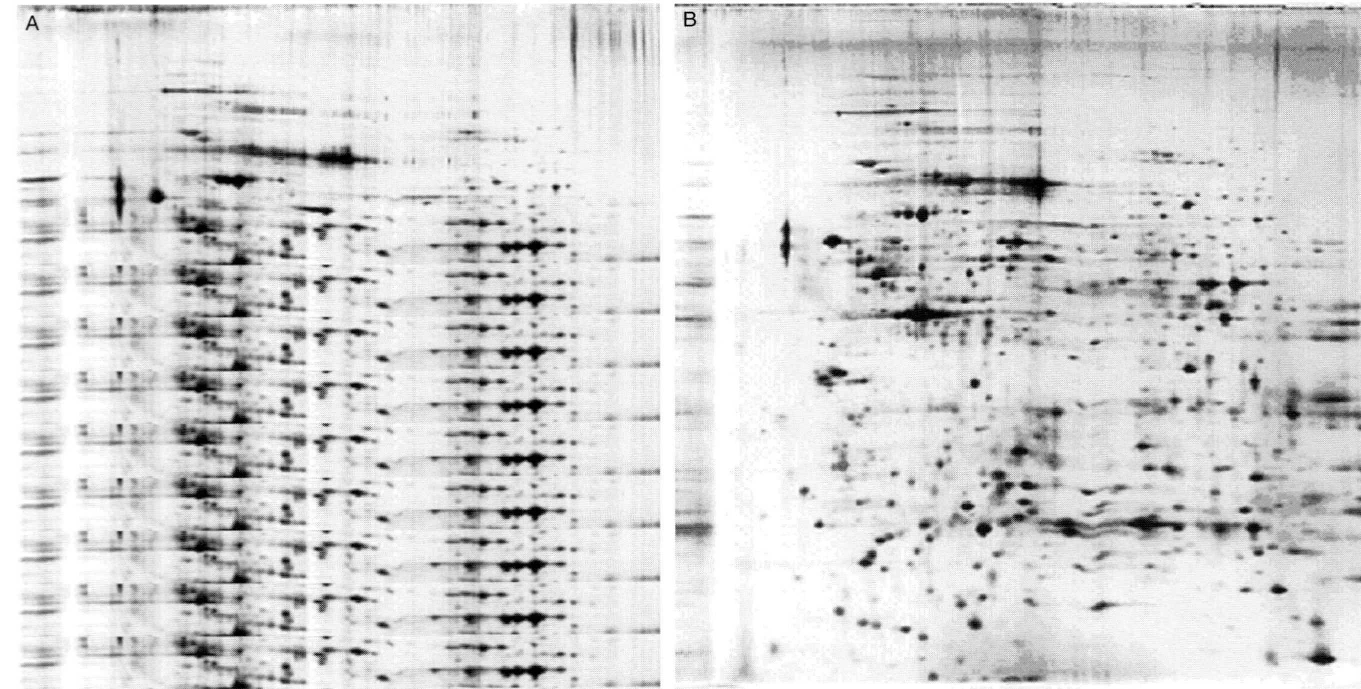

A
B

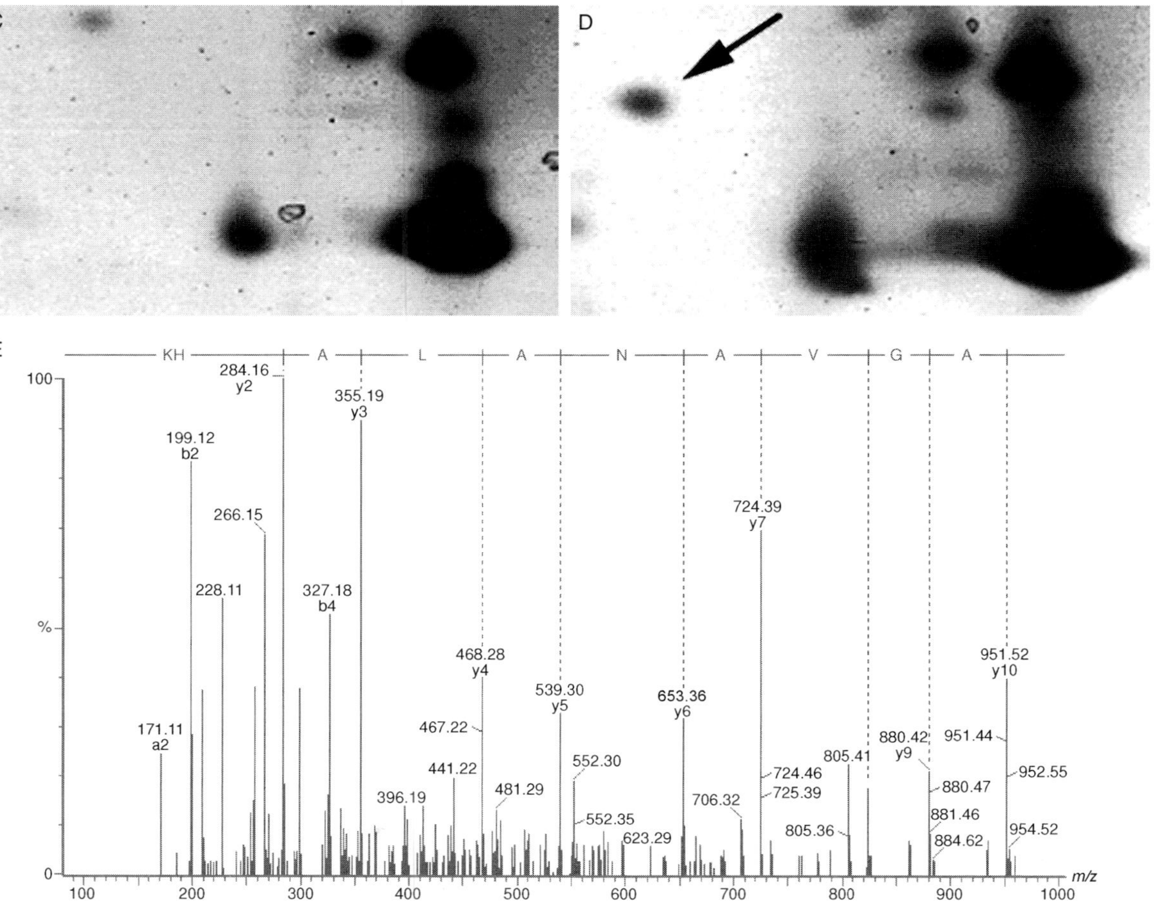

Fig. 1 2-DE map of human liver proteins obtained from liver tissue. 2-DE was performed on an immobilized pH 3–10 strip, followed by the 2D separation on 30% polyacrylamide gels. (A) 2-DE map of HCC tissue after LCM. (B) 2-DE map of homogenized control tissue. It showed that the hemoglobin was removed by LCM from the enlarged maps. (C) Enlarged area of hemoglobin, we could not see the protein from the LCM samples. (D) Enlarged area of hemoglobin, the arrow showed that the protein was in the homogenized control tissue. (E) Nano-flow ESI-MS/MS identification of hemoglobin.

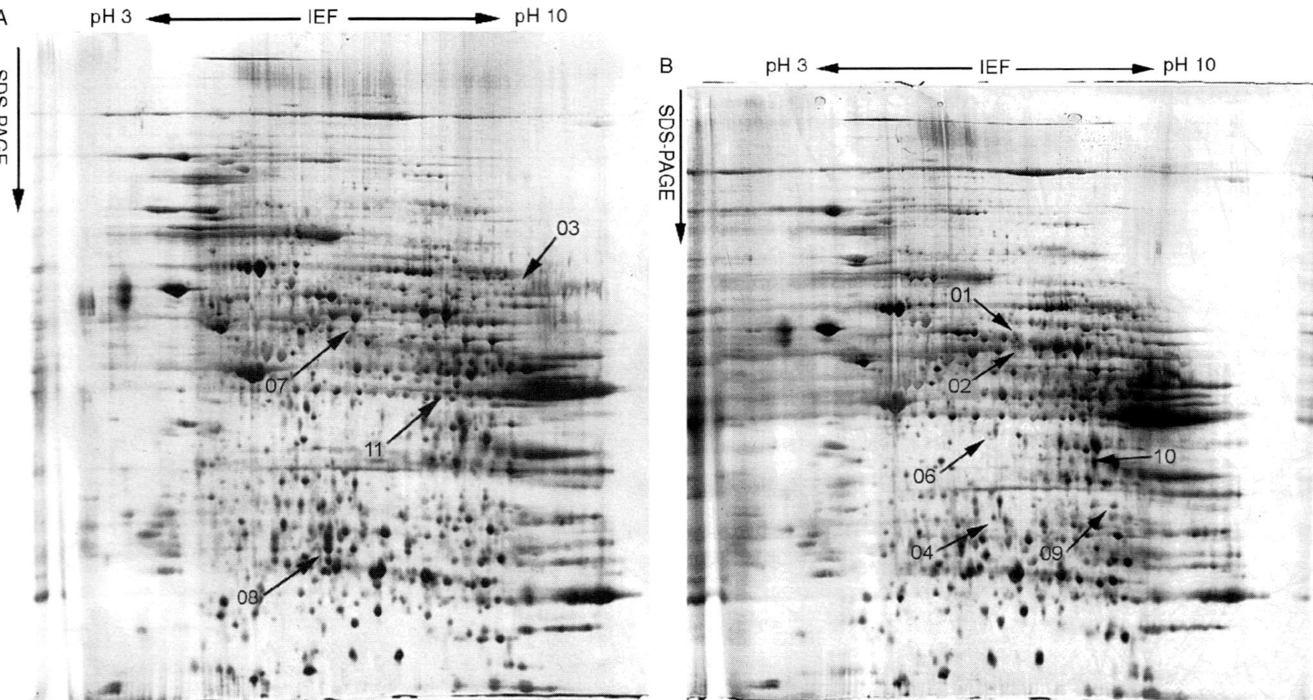

Fig. 2 2-DE maps for (A) HCC tissues after LCM and (B) surrounding nontumorous tissues after LCM. Protein samples were prepared as described in Section II.C , and 100-μg proteins were separated on pH 3–10 nonlinear strips and then on 30% gradient SDS-PAGE. Staining was by silver nitrate.

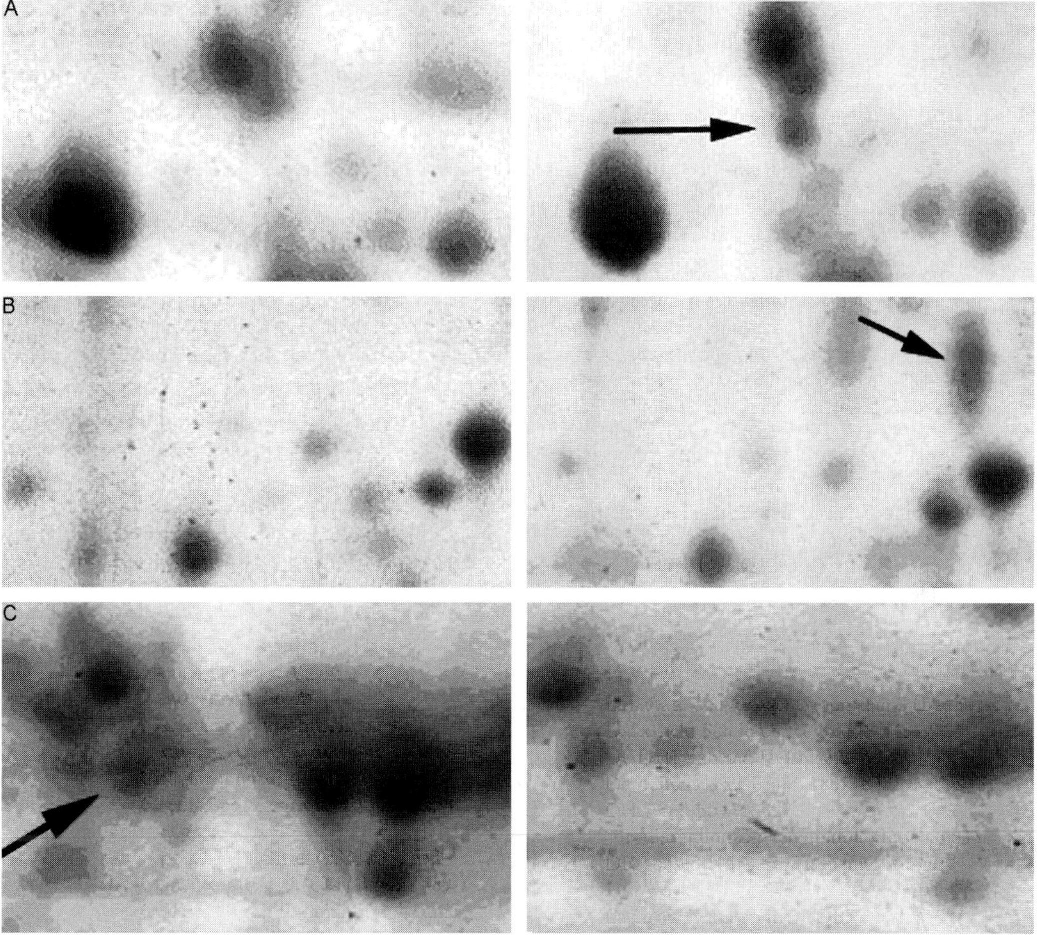

Fig. 3 The types of protein expression patterns obtained from comparisons made between surrounding nontumorous tissue and HCC tissue. Each pattern is visualized by examples of protein spots in 2D gel. (A) The arrow showed that HAD have been upregulated in surrounding nontumorous tissues. (B) The arrow showed Prx II was upregulated in surrounding nontumorous tissues. (C) The arrow showed that PEBP was upregulated in HCC tissues.

was shown in Fig. 3. The analysis was carried out on soluble-fraction proteins with the pH range of 3–10. An image pair consisting of analytical gel images of surrounding nontumorous and cancer tissues of the same patient was comparatively analyzed for each patient. Each sample was examined at least five times. Approximately 100 2D gel images were analyzed. Ratios of normalized spot intensities of cancer to paraneoplastic tissue were calculated, and spots showing threefold difference were selected for each patient. Then, the frequencies of common alteration of the spots were calculated across the patients, and spots showing

80% frequency were submitted to protein identification. One of the 2-DE results was confirmed by Western blot analysis.

C. Differential Proteomic Analysis of Liver Tissues of HCC Patients

The protein spots that altered prominently between surrounding nontumorous tissue and HCC tissue were excised from the gel, subjected to in-gel trypsin digestion, and identified by Nano-flow ESI-MS/MS as described above. From the analysis of the data for gel images, we could select 20 spots. Figure 4A showed the total ion chromatography (TIC) of the Nano-flow ESI-MS/MS analysis of the peptides derived from gel spot 11. Six peptides were matched to the phosphatidylethanolamine-binding protein (PEBP). The MS/MS spectrum of peptide corresponding to 32.36 in TIC was shown in Fig. 4B. The spectrum gave an unambiguous internal peptide sequence of PEBP. Summary of 11 interesting proteins that identified with tandem mass spectrometry was listed in Table I. Of the proteins selected, Prx II, apolipoprotein A-I precursor (Apo-AI), 3-hydroxyacyl-CoA dehydrogenase type II (HAD), and 14.5-kDa translational inhibitor protein (p14.5) (UK114 antigen homologue) appear to be novel candidates for useful HBV-related HCC markers.

D. Immunohistochemical and Western Blot Analyses for Prx II in HCC

To examine whether suppression of Prx II could also be seen in the tissues, immunohistochemical analyses were performed using sections from paraffin-embedded normal liver tissues and HCC tumors using the antibody specific for Prx II. As shown in Fig. 5A, Prx II was not expressed in HCC tissues (left corner) as compared with adjacent nontumorous liver tissues (right corner), when tissue sections containing both normal and HCC lesions were used. This was clearer when each tissue was stained separately, that is, compared with normal tissue (Fig. 5B), Prx II is clearly undetectable (Fig. 5C). To determine the protein expression levels of Prx II, Western blot analysis was also performed using polyclonal antibodies against Prx II (Fig. 6). As expected, Prx II was found to be consistently suppressed in HCC whereas it was shown to be highly expressed in nontumorous tissues.

Prx II is an antioxidant enzyme that reduces H_2O_2 and other reactive oxygen species using thioredoxin as the immediate electron donor (Chae *et al.*, 1994), and its peroxidase activity prevents cells from reactive oxygen species insult (Chae *et al.*, 1999). Prx II is also involved in the cellular-signaling pathways of growth factors and tumor necrosis factor-α, by virtue of its regulation of intracellular H_2O_2 (Kang *et al.*, 1998; Sen, 1998). Prx expression is controversial in certain types of cancer tissues and there is no any other report of the relationship between Prx II and HCC so far. Three types of Prx (I, II, and III) have been shown to be overexpressed in the case of human breast cancer, and it has been suggested that their overexpressions are related to cancer development or/and progression

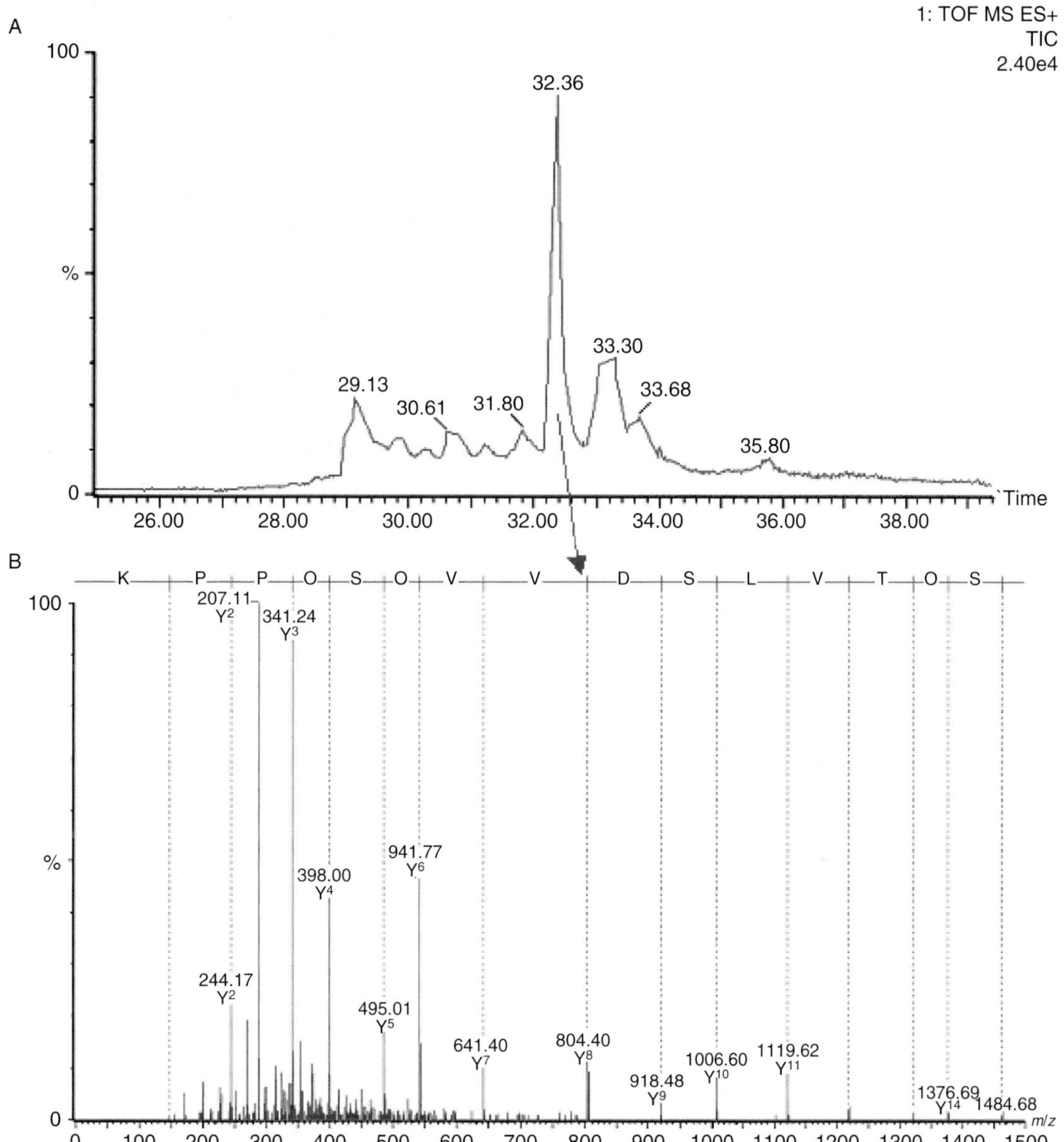

Fig. 4 Nano-flow ESI-MS/MS identification of spot 11. (A) TIC of the Nano-flow ESI-MS/MS analysis of the peptides derived from gel spot 11. (B) MS/MS spectrum of peptide from peak 32.36 in TIC, which corresponds to sequence of GNDISSGTVLSDYVGSGPPK.

Table I
Summary of Proteins Identified with Nano-Flow ESI-MS/MS

Index	Ac	Description	MW/p*I*	Matched peptide	Score
01	P00441	Superoxide dismutase [Cu-Zn] (EC 1.15.1.1)	15795/5.7	HVGDLGNVTADK	30
02	P02766	Transthyretin precursor (prealbumin) (TBPA) (TTR)	15877/5.52	GSPAINVAVHVFR	68
03	P52758	14.5-kDa translational inhibitor protein (p14.5) (UK114 antigen homologue)	14485/8.74	APGAIGPYSQAVLVDR	83
04	P02647	Apolipoprotein A-I precursor (Apo-AI)	30759/5.56	THLAPYSDELR AKPALEDLR ATEHLSTLSEK	34
05	P02647	Apolipoprotein A-I precursor (Apo-AI)	30759/5.56	VQPYLDDFQK THLAPYSDELR DYVSQFEGSALGK LSEDYGVLK	217
06	P32119	Peroxiredoxin 2 (EC 1.11.1.-) (thioredoxin peroxidase 1)	21878/5.66	RLSEDYGVLK QITVNDLPVGR EGGLGPLNIPLLADVTR HVGDLGNVTADK GDGPVQGIINFEQK	196
07	P00441	Superoxide dismutase [Cu-Zn] (EC 1.15.1.1)	15795/5.7	DGVADVSIEDSVISLSGDHC*IIGR GLTEGLHGFHVHEFGDNTAGC*TSAGPH FNPLSR AAQDFFSTC*	186
08	P06132	Uroporphyrinogen decarboxylase (EC 4.1.1.37) (URO-D)	40761/5.77	SPEAC*C*ELTLQPLR GPSFPEPLREEQDLER GSAPPGPVPEGSIR HEVININLK	66
09	P78417	Glutathione transferase omega 1 (EC 2.5.1.18) (GSTO 1-1)	27548/6.23	VPSLVGSFIR EDYAGLKEEFR EDPTVSALLTSEK	231
10	Q99714	3-Hydroxyacyl-CoA dehydrogenase type II (EC 1.1.1.35) (Type II HADH) (endoplasmic reticulum-associated amyloid β-peptide-binding protein)	26923/7.65	GLVAVITGGASGLGLATAER NRPTSISWDGLDSGK LYTLVLTDPDAPSR	66
11	P30086	Phosphatidylethanolamine-binding protein (PEBP) (prostatic-binding protein) (HCNPpp)	20926/7.43	GNDISSGTVLSDYVGSGPPK YVWLVYEQDRPLK C*DEPILSNR LYEQLSGK	304

Proteins that differentially expressed were identified with Q-TOF Micro and the fragmented ion lists were searched with Mascot in SWISS-PROT database. The identified results were further verified by comparing the theoretical MW/p*I* of the proteins with the experimental ones.

C*,These cysteines were modified by iodoacetamide during the equilibrium step of 2-DE.

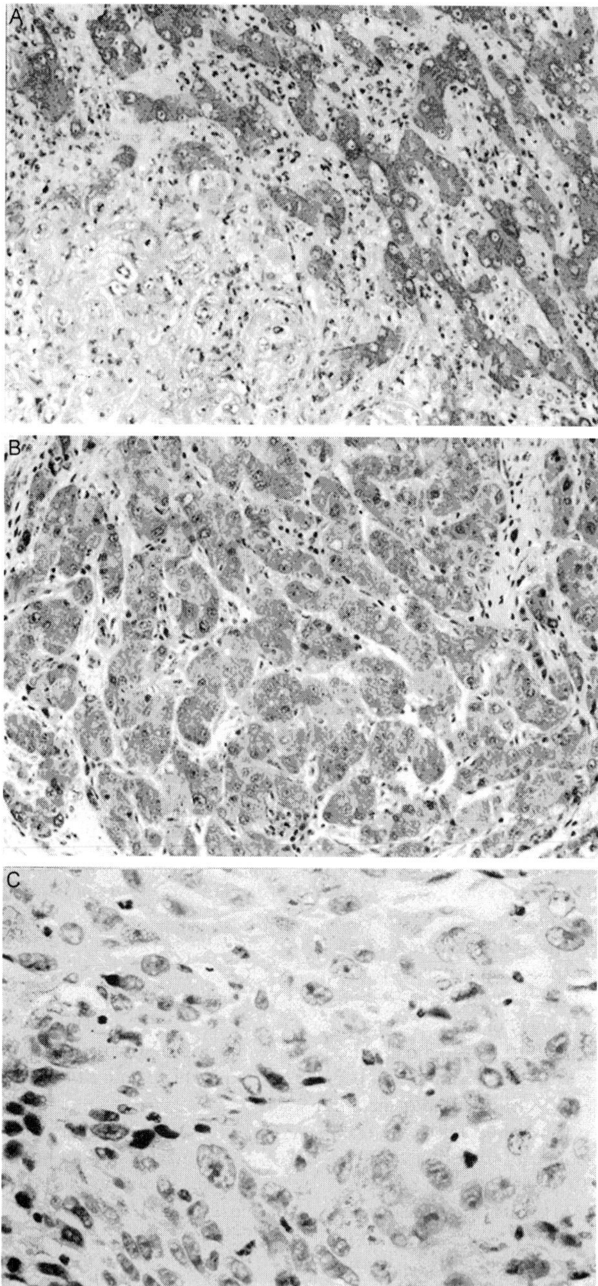

Fig. 5 Immunohistochemical detection of Prx II expression in human normal liver and HCC. (A) Right, normal liver showed strong staining by polyclonal antibodies to Prx II; left, HCC with reduced Prx II expression. (B) Expression of Prx II in normal liver. (C) HCC showed no immunoreactivity for Prx II.

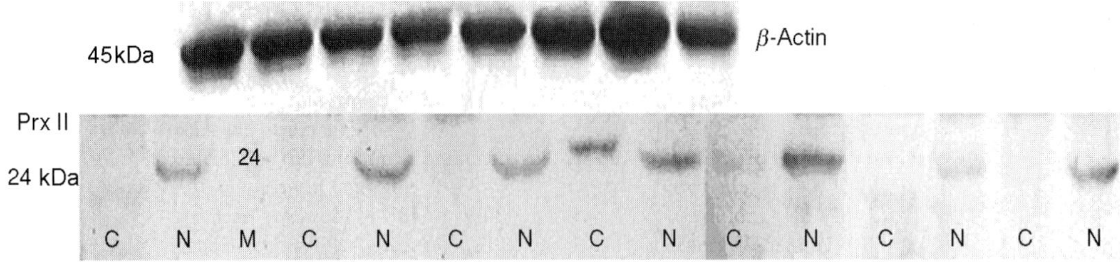

Fig. 6 Protein expression of Prx II in four HCC tissues and the normal tissues by Western blot analysis. Shown here were the 7 representative samples of 17 samples because the remaining samples showed similar or identical patterns of expression. Immunoblotting with a Prx II polyclonal antibody following SDS-PAGE was performed as described in Sections III and IV. We used β-actin as a reference value for comparison between surrounding noncancerous tissue and HCC. N, surrounding noncancerous tissue; C, cancerous tissue; M, marker.

(Noh *et al.*, 2001). The increased expression of Prx I was also detected in lung cancer, thyroid tumors, and oral cancer, and is suggested to constitute a potential tumor marker (Chang *et al.*, 2001; Yanagawa *et al.*, 1999, 2000). However, Neumann *et al.* (2003) found that Prx I functions as a tumor suppressor. Reactive oxygen species are involved in many cellular metabolic and signaling processes and are thought to have a role in disease, particularly in carcinogenesis. Here we showed that Prx II was downregulated in HCC tissues, which suggested that Prx II may function as a tumor suppressor of HCC. Further study is needed to discover the molecular mechanism of Prx II in the hepatocarcinogenesis of HBV infection.

Another interesting finding of the present study was that two enzymes of lipid catabolism were downregulated in cancer tissues: HAD and Apo-AI of which HAD has also been reported by Kim *et al.* (Li *et al.*, 2004). Cancer cells require more cholesterol than normal cells. This requirement may be satisfied by higher hydroxymethylglutaryl-CoA reductase activity. Suto *et al.* (1999) have found that HAD was downregulated in HCC tissues by immunohistochemical analysis. Previous study suggested that fatty acid metabolism might contribute significantly to the HCC (Ockner *et al.*, 1993). HAD catalyzes the third reaction of the fatty acid β-oxidation spiral in eukaryote as a critical housekeeping enzyme (He and Yang, 1996), which indicates a relationship between HAD and HCC. Apo-AI plays an important structural and functional role in lipid transport and metabolism, it serves as an antioxidant enzyme in lipid metabolism. Norton *et al.* (2003) observed that HBV decreased Apo-AI mRNA expression in hepG2 cells, which indicated the association between Apo-AI and HCC. In fact, it has long been recognized that HCC frequently exhibits an accumulation of fatty acids and other lipids in cells (Gibson and Sobin, 1978), which suggests that degradation of fatty acids may be repressed or their synthesis may be enhanced in the lesions. Our result seems to favor the first possibility. Further systematic surveys on lipid metabolism in tumor tissues are required to elucidate the phenotype.

VII. Summary

The proteome differences between HCC and surrounding nontumorous tissue still remain to be elucidated. Therefore, detection and identification of proteins related to HCC, especially those proteins whose expression is different between surrounding nontumorous tissue and liver tumor, might shed light on the molecular mechanism of the carcinogenesis of HCC. In this study, we applied the LCM to the proteomic study of HBV-related HCC and compared the differential expression of proteins in the HCC tissue and the surrounding nontumorous tissue. Twenty protein spots were selected and 11 proteins significantly altered in the surrounding nontumorous tissues and HCC tissues. These proteins may play important roles in tumorigenesis and progression of human HCC, and thus could potentially make useful markers for diagnosis or targets for therapeutic intervention. Further functional analysis is needed for each of these cancer-associated protein candidates.

Acknowledgments

This work was supported by grants of National Development Program for Key Basic Research of China (No. 2001CB510205 and 2001CB510201), National Key Technologies R&D Program of China (No. 2002BA711A02–3 and 2001AA233031), and Key Basic Science Foundation of Shanghai (No. 03DJ14007 and 03dz14024).

References

Bonner, R. F., Emmert-Buck, M., Cole, K., Pohida, T., Chuaqui, R., Goldstein, S., and Liotta, L. A. (1997). Laser capture microdissection: Molecular analysis of tissue. *Science* **278,** 1481–1483.

Brechot, C. (1998). Molecular mechanisms of hepatitis B and C viruses related to liver carcinogenesis. *Hepatogastroenterology* **45**(Suppl. 3), 1189–1196.

Chae, H. Z., Chung, S. J., and Rhee, S. G. (1994). Thioredoxin-dependent peroxide reductase from yeast. *J. Biol. Chem.* **269,** 27670–27678.

Chae, H. Z., Kang, S. W., and Rhee, S. G. (1999). Isoforms of mammalian peroxiredoxin that reduce peroxides in presence of thioredoxin. *Methods Enzymol.* **300,** 219–226.

Chang, J. W., Jeon, H. B., Lee, J. H., Yoo, J. S., Chun, J. S., Kim, J. H., and Yoo, Y. J. (2001). Augmented expression of peroxiredoxin I in lung cancer. *Biochem. Biophys. Res. Commun.* **289,** 507–512.

Craven, R. A., Totty, N., Harnden, P., Selby, P. J., and Banks, R. E. (2002). Laser capture microdissection and two-dimensional polyacrylamide gel electrophoresis: Evaluation of tissue preparation and sample limitations. *Am. J. Pathol.* **160,** 815–822.

Diao, J., Garces, R., and Richardson, C. D. (2001). X protein of hepatitis B virus modulates cytokine and growth factor related signal transduction pathways during the course of viral infections and hepatocarcinogenesis. *Cytokine Growth Factor Rev.* **12,** 189–205.

Emmert-Buck, M. R., Bonner, R. F., Smith, P. D., Chuaqui, R. F., Zhuang, Z., Goldstein, S. R., Weiss, R. A., and Liotta, L. A. (1996). Laser capture microdissection. *Science* **274,** 998–1001.

Gibson, J. B., and Sobin, L. H. (1978). *In* "International Histological Classification of Tumours No. 20" (J. B. Gibson, and L. H. Sobin, eds.), pp. 19–30. WHO, Geneva.

Gorg, A., Obermaier, C., Bouuth, G., Harder, A., Scheibe, B., Wildgruber, R., and Weiss, W. (2000). The current state of two-dimensional electrophoresis with immobilized pH gradients. *Electrophoresis* **21,** 1037–1053.

Harris, C. C. (1994). Solving the viral-chemical puzzle of human liver carcinogenesis. *Cancer Epidemiol. Biomarkers Prev.* **3,** 1–2.

Harris, C. C. (1996). The 1995 Walter Hubert Lecture–molecular epidemiology of human cancer: Insights from the mutational analysis of the p53 tumour-suppressor gene. *Br. J. Cancer* **73,** 261–269.

He, X. Y., and Yang, S. Y. (1996). Histidine-450 is the catalytic residue of L-3-hydroxyacyl coenzyme A dehydrogenase associated with the large alpha-subunit of the multienzyme complex of fatty acid oxidation from *Escherichia coli. Biochemistry* **35**(29), 9625–9630.

Kang, S. W., Chae, H. Z., Seo, M. S., Kim, K., Baines, I. C., and Rhee, S. G. (1998). Mammalian peroxiredoxin isoforms can reduce hydrogen peroxide generated in response to growth factors and tumor necrosis factor-alpha. *J. Biol. Chem.* **273,** 6297–6302.

Kovarova, H., Hajduch, M., Korinkova, G., Halada, P., Krupickova, S., Gouldsworthy, A., Zhelev, N., and Strnad, M. (2000). Proteomics approach in classifying the biochemical basis of the anticancer activity of the new olomoucine-derived synthetic cyclin-dependent kinase inhibitor, bohemine. *Electrophoresis* **21**(17), 3757–3764.

Lander, E. S., Linton, L. M., Birren, B., Nusbaum, C., Zody, M. C., Baldwin, J., Devon, K., Dewar, K., Doyle, M., FitzHugh, W., Funke, R., Gage, D., *et al.* (2001). Initial sequencing and analysis of the human genome. *Nature* **409,** 860–921.

Li, C., Hong, Y., Tan, Y. X., Zhou, H., Ai, J. H., Li, S. J., Zhang, L., Xia, Q. C., Wu, J. R., Wang, H. Y., and Zeng, R. (2004). Accurate qualitative and quantitative proteomic analysis of clinical hepatocellular carcinoma using laser capture microdissection coupled with isotope-coded affinity tag and two-dimensional liquid chromatography mass spectrometry. *Mol. Cell. Proteomics* **3,** 399–409.

Lim, S. O., Park, S. J., Kim, W., Park, S. G., Kim, Y. I., Sohn, T. S., Noh, J. H., and Jung, G. (2002). Proteome analysis of hepatocellular carcinoma. *Biochem. Biophys. Res. Commun.* **291,** 1031–1037.

Neubauer, G., King, A., Rappsilber, J., Calvio, C., Watson, M., Ajuh, P., Sleeman, J., Lamond, A., and Mann, M. (1998). Mass spectrometry and EST-database searching allows characterization of the multi-protein spliceosome complex. *Nat. Genet.* **20,** 46–50.

Neumann, C. A., Krause, D. S., Carman, C. V., Das, S., Dubey, D. P., Abraham, J. L., Bronson, R. T., Fujiwara, Y., Orkin, S. H., and Van Etten, R. A. (2003). Essential role for the peroxiredoxin Prdx1 in erythrocyte antioxidant defence and tumour suppression. *Nature* **424,** 561–565.

Noh, D. Y., Ahn, S. J., Lee, R. A., Kim, S. W., Park, I. A., and Chae, H. Z. (2001). Overexpression of peroxiredoxin in human breast cancer. *Anticancer Res.* **21,** 2085–2090.

Norton, P. A., Gong, Q., Mehta, A. S., Lu, X., and Block, T. M. (2003). Hepatitis B virus-mediated changes of apolipoprotein mRNA abundance in cultured hepatoma cells. *J. Virol.* **77**(9), 5503–5506.

Ockner, R. K., Kaikaus, R. M., and Bass, N. M. (1993). Fatty-acid metabolism and the pathogenesis of hepatocellular carcinoma: Review and hypothesis. *Hepatology* **18**(3), 669–676.

Parkin, D. M., Bray, F. L., and Devesa, S. S. (2001). Cancer burden in the year 2000. The global picture. *Eur. J. Cancer* **37,** S4–S66.

Petricoin, E. F., Zoon, K. C., Kohn, E. C., Barrett, J. C., and Liotta, L. A. (2002). Clinical proteomics: Translating benchside promise into bedside reality. *Nat. Rev. Drug Discov.* **1,** 683–695.

Rabilloud, T., Adessi, C., Giraudel, A., and Lunardi, J. (1997). Improvement of the solubilization of proteins in two-dimensional electrophoresis with immobilized pH gradients. *Electrophoresis* **18,** 307–316.

Sen, C. K. (1998). Redox signaling and the emerging therapeutic potential of thiol antioxidants. *Biochem. Pharmacol.* **55,** 1747–1758.

Shevchenko, A., Wilm, M., Vorm, O., and Mann, M. (1996). Mass spectrometric sequencing of proteins silver-stained polyacrylamide gels. *Anal. Chem.* **68,** 850–858.

Smela, M. E., Currier, S. S., Bailey, E. A., and Essigman, J. M. (2001). The chemistry and biology of aflatoxin B(1): From mutational spectrometry to carcinogenesis. *Carcinogenesis* **22,** 535–545.

Stuver, S. O. (1998). Towards global control of liver cancer? *Semin. Cancer Biol.* **8,** 299–306.

Srinivas, P. R., Kramer, B. S., and Strivastava, S. (2001). Trends in biomarker research for cancer detection. *Lancet Oncol.* **2,** 698–704.

Suto, K., Kajihara-Kano, H., Yokoyama, Y., Hayakari, M., Kimura, J., Kumano, T., Takahata, T., Kudo, H., and Tsuchida, S. (1999). Decreased expression of the peroxisomal bifunctional enzyme and carbonyl reductase in human hepatocellular carcinomas. *J. Cancer Res. Clin. Oncol.* **125**(2), 83–88.

Venter, J. C., Adams, M. D., Myers, E. W., Li, P. W., Mural, R. J., Sutton, G. G., Smith, H. O., Yandell, M., Evans, C. A., Holt, R. A., Gocayne, J. D., Amanatides, P., *et al.* (2001). The sequence of the human genome. *Science* **291,** 1304–1351.

Wilkins, M. R., Pasquali, C., Appel, R. D., Ou, K., Golaz, O., Sanchez, J. C., Yan, J. X., Gooley, A. A., Hughes, G., Humphery-Smith, I., Williams, K. L., Hochstrasser, D. F., *et al.* (1996). From proteins to proteomes: Large scale protein identification by two-dimensional electrophoresis and amino acid analysis. *Biotechnology (NY)* **14,** 61–65.

Yanagawa, T., Ishikawa, T., Ishii, T., Tabuchi, K., Iwasa, S., Bannai, S., Omura, K., Suzuki, H., and Yoshida, H. (1999). Peroxiredoxin I expression in human thyroid tumors. *Cancer Lett.* **145,** 127–132.

Yanagawa, T., Iwasa, S., Ishii, T., Tabuchi, K., Yusa, H., Onizawa, K., Omura, K., Harada, H., Suzuki, H., and Yoshida, H. (2000). Peroxiredoxin I expression in oral cancer: A potential new tumor marker. *Cancer Lett.* **156,** 27–35.

Ying, W., Zhang, K., Qian, X., Xie, L., Wang, J., Xiang, X., Cai, Y., and Wu, D. (2003). Proteome analysis on an early transformed human bronchial epithelial cell line, BEP2D, after alpha-particle irradiation. *Proteomics* **3**(1), 64–72.

Yu, L. R., Zeng, R., Shao, X. X., Wang, N., Xu, Y. H., and Xia, Q. C. (2000). Identification of differentially expressed proteins between human hepatoma and normal liver cell lines by two-dimensional electrophoresis and liquid chromatography-ion trap mass spectrometry. *Electrophoresis* **21,** 3058–3068.

Zhou, G., Li, H., DeCamp, D., Chen, S., Shu, H., Gong, Y., Flaig, M., Gillespie, J. W., Hu, N., Taylor, P. R., Emmert-Buck, M. R., Liotta, L. A., *et al.* (2002). 2D differential in-gel electrophoresis for the identification of esophageal scans cell cancer-specific protein markers. *Mol. Cell. Proteomics* **1,** 117–124.

CHAPTER 26

Rapid Sampling for Single-Cell Analysis by Capillary Electrophoresis

Allison R. Nelson,* Nancy L. Allbritton,*,† and Christopher E. Sims*

*Department of Physiology and Biophysics
School of Health Sciences
University of California
Irvine, California 92697

†Departments of Biomedical Engineering, Chemistry, and
Chemical Engineering and Materials Science
University of California
Irvine, California 92697

Single-cell analyses have found increasing importance in biological investigation. Recent technical advances have made it possible to perform chemical separations of cellular constituents at the level of the individual cell. In this chapter, a laser-based method for the rapid sampling of living cells for chemical analysis using capillary electrophoresis is described. The platform technology consists of ultrasensitive laser-induced fluorescence detection in a capillary mated with a microscope integrated with a pulsed Nd:YAG laser. This platform provides a flexible system for the development of new single-cell biochemical assays of a variety of intracellular enzymes.

0091-679X/07 $35.00
DOI: 10.1016/S0091-679X(06)82026-1

I. Introduction

The study of individual cells has been a driving force in biological investigation for over a century. The analysis of single cells yields rich information that may be obscured by the averaging that occurs when cells are pooled for analysis. Measurements on single cells have enhanced the understanding of cellular biochemical pathways underlying signal transduction and other cell processes. Classic examples of biological phenomenon revealed by single-cell analysis include the temporal patterns of intracellular calcium concentration after receptor activation and the switch-like behavior of signal transduction enzymes in the initiation of embryogenesis (Allbritton and Meyer, 1993; Ferrell and Machleder, 1998). In recent years, modern imaging techniques combined with genetically engineered probes are providing new insights into the molecular mechanisms of cell physiology on a cell-by-cell basis. Moreover, technological progress in separation science has brought to bear the power of chemical separations to the molecular analysis of single cells (Dovichi and Hu, 2003). Chief among these separation technologies is capillary electrophoresis (CE). The separation of the contents of individual cells by electrophoresis has been termed chemical cytometry and CE now has a greater than 20-year history in the analysis and characterization of single cells (Dovichi and Hu, 2003; Kennedy *et al.*, 1989). CE has numerous advantages as an analytical tool for single cells. The dimensions and volumes of the capillary are compatible with those of even small mammalian cells that are tens of micrometers in diameter with volumes of $\sim$1 pl. By virtue of the remarkably high-resolving power of CE and the exquisite sensitivity made possible with laser-induced fluorescence detection, the monitoring of several analytes from one cell can be performed. Accurate quantification of these analytes is also possible. Single-cell analyses first included the amino acid and protein content of neurons and erythrocytes (Hogan and Yeung, 1992; Kennedy *et al.*, 1989; Wallingford and Ewing, 1988). Diverse CE-based analyses have now been accomplished from carbohydrate processing to proteomic studies (Craig *et al.*, 1995; Krylov *et al.*, 1999; Zhang *et al.*, 2000). Our group has had a keen interest in using CE-based techniques for the development of single-cell biochemical assays to monitor the activity of specific enzymes (Lee *et al.*, 1999; Li *et al.*, 2001, 2004; Meredith *et al.*, 2000). Additional applications are continually being reported as the use of CE in cell-based assays expands to probe the intricate biochemistry of single cells.

II. Rationale

Most methods for introducing cellular constituents into the separation channel of the capillary disrupt cell membranes in a detergent-containing buffer to solubilize cell contents. Typically, the cell is exposed to this lysis buffer just prior to loading its contents or after the cell is introduced into the capillary. This lysis method requires several seconds to permeabilize the cell membrane and terminate chemical reactions within the cell. This delay may generate artifacts in the analysis, particularly when studying dynamic cell events. During the period of lysis, chemical reactions may continue or be initiated as the cell's repair mechanisms are activated by the insult. Accurate measurements of enzymatic reactions that can occur on millisecond timescales are particularly challenging (Li *et al.*, 2001). To achieve rapid cell lysis and termination of chemical reactions, we have developed a laser-based method to mechanically shear apart the cell membrane on ultrafast timescales (Rau *et al.*, 2004; Sims *et al.*, 1998). Chemical reactions are terminated by dilution and turbulent mixing of chemical reactants within milliseconds of the lysis event. The cell contents are then rapidly loaded into the capillary with virtually 100% efficiency as electrophoresis is initiated (Sims *et al.*, 1998). To exemplify this approach for rapid cell lysis combined with CE analysis, a typical experiment involving the introduction of a labeled enzyme substrate into a cell followed by identification of substrate and product is described.

III. Methods

A. Fluorescent Reporters

In the current context, a reporter is any molecule, small or large, that can be loaded into a cell and subsequently tracked by CE after it undergoes chemical modifications. The method of detection used most often for these intracellular reporters is fluorescence. A fluorescent label is used due to the extremely low detection limit made possible by laser-induced fluorescence detection in a capillary. With fluorescence, concentrations of the reporter similar to or less than the concentrations of native species may be used, reducing the likelihood of the reporter interfering with cell processes. The reporter may be labeled with a small fluorescent molecule, such as fluorescein, or engineered and expressed within the cell conjugated to a fluorescent protein. The main consideration in the design of a reporter is whether it can be distinguished from cell components after cell lysis and can be electrophoresed efficiently (Fig. 1A). Additionally, the chemical modifications of the reporter that take place within the cell must be such that the modified reporter, or product, can be separated from the original form of the reporter, or substrate. That is, the electrophoretic mobility of the substrate must be altered on conversion to the product. Finally, measures may be necessary to ensure that the reporter is specific for the chemical modifications that one wishes to observe.

Many types of molecules can be used as reporters, including peptides, lipids, and carbohydrates. Some limitations are availability—through commercial sources or custom synthesis—and handling issues, such as stability. For the present discussion, the use of peptide substrates of intracellular enzymes is used to illustrate the potential of the technique. Peptides are synthesized by commercial companies or in-house using solid phase peptide synthesis. The solid support, or resin, bears groups that react with a protected amino acid. The desired sequence is built from the C-terminus to the N-terminus, one amino acid at a time with washing and

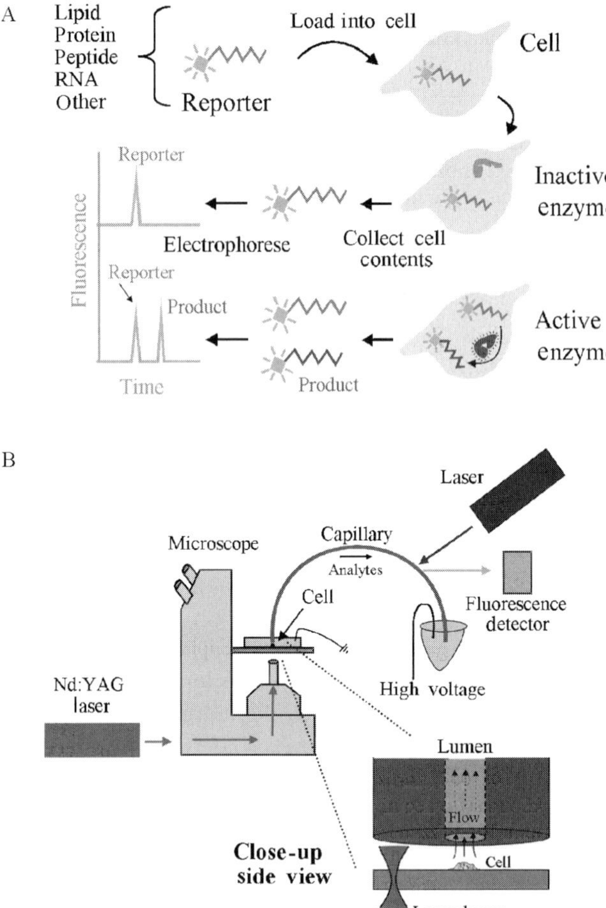

Fig. 1 (A) Schematic of the single-cell biochemical assay. The reporter is loaded into a cell where it undergoes modification via intracellular processes. At the desired time, the cell is rapidly lysed with a laser pulse and the contents are chemically separated and detected using CE with LIF detection. The resulting electropherogram is used to quantify the unmodified and modified forms of the reporter. (B) Schematic of the instrumentation used for cell lysis, CE, and LIF. Close-up view shows the capillary inlet positioned above the cell to be sampled. The laser beam is focused adjacent to the cell so that heating and photodestruction do not occur.

filtration steps between each additional coupling. Modified amino acids such as phosphorylated or myristoylated amino acids may be used if desired. The peptide may be designed so that it possesses multiple functions. For example, a substrate domain can be combined with a protein transduction domain to confer cell uptake (Soughayer *et al.*, 2004). Non-amino acid linkers, such as polyethylene glycol (PEG), may also be incorporated into the peptide to enhance solubility and distance of the substrate domain from other domains within the peptide. After the sequence is completed, the peptide is removed from the resin and deprotected using a TFA cleavage step. Fluorophores, such as fluorescein or Oregon green, are easily added to the peptide's amino-terminus with standard coupling methods or by using a fluorescently labeled amino acid as the N-terminal residue. The peptide is purified using either ion exchange or reverse phase HPLC to remove truncated or deletion peptide byproducts. CE is a high-resolution technique; therefore, the reporter must be pure (>90%) to allow accurate peak identification. The purity of peptides, prepared by HPLC, may be confirmed using mass spectrometry and CE. For long-term storage, the peptide is lyophilized and stored at $-70\,^{\circ}$C. Small portions (~1 mg) may also be dissolved in an appropriate solvent, aliquoted, and stored at $-20\,^{\circ}$C for short-term use. The concentration of this stock solution may be determined by hydrolysis of a small amount of the peptide following the addition of standards, and then the individual amino acids separated, identified, and quantified by HPLC. The digest followed by HPLC analysis also serves as an additional check on the peptide sequence. Sterile, low-retention vials, such as siliconized microcentrifuge tubes, should be used for all aliquots. Even with the use of low-retention tubes, there may be significant nonspecific adsorption of peptide from solutions below ~10 μM. Thus, it is better to make small aliquots (1–5 μl) of a higher concentration, ~1 mM, and then use a fresh aliquot for each day's experiments. To prevent degradation of the peptide and photobleaching of the fluorophore, peptide solutions should be kept on ice in the dark at all times. The stock peptide is diluted in an appropriate buffer depending on the method of cell loading (see below) or in the electrophoretic run buffer to the desired injection concentration for use as a standard. Note that different types of reporters may require special handling. For example, lipids should be stored in glass vials under nitrogen or argon to prevent oxidation. To prevent oxidation or dimerization of peptides containing cysteine residues, oxygen-free solvents should be used.

B. Tissue Culture and Cell Handling

1. Preparation of Cell Chambers

For experiments, cells are placed in disposable cell chambers made using circular #1 glass coverslips and silicone "O"-rings. These are mated using nontoxic glue such as Sylgard 184. Before construction of the chambers, the "O"-rings must be thoroughly cleaned to remove plasticizers and mold-releasing agents which are toxic to cells. The rings are placed in a bottle of soapy water and agitated periodically as they are allowed to soak for 2–3 days. After soaking, the rings are thoroughly rinsed with deionized water, doused in 75% ethanol, and dried. Before attaching the "O"-rings, coverslips are immersed in 75% ethanol to remove any powders or contaminants and allowed to air-dry. Prior to use, Sylgard is mixed with its initiator in a 10:1 ratio. A large quantity of Sylgard with the initiator added may be aliquoted and stored for several months at $-20\,^{\circ}$C for convenience.

The cleaned coverslips are placed on a hot plate at 80 °C. An "O"-ring with its underside coated with Sylgard is centered on each coverslip. The Sylgard is allowed to cure for 5 min. Cell chambers are then placed in a glass Petri dish and autoclaved.

2. Cell Culture and Handling

Standard tissue culture conditions are used that are appropriate for the cell type, and typically require a humidified 5% CO_2, 37 °C incubator. On the day prior to the planned experiments, adherent cells are trypsinized and plated in cell chambers at 20,000–25,000 cells per chamber. This plating density provides 1–2 cells per high-power field-of-view on the day of the experiment. Sparse plating of the cells is desired to enable single cells to be selected and sampled without contamination from surrounding cells. Care must be taken to prevent evaporation from the cell chambers during culture even if the cells are maintained in a humidified incubator. This can be accomplished by placing the glass Petri dish enclosing the cell chambers in a loosely sealed plastic box containing moistened absorbent material such as a paper towel. Immediately prior to use, individual cell chambers are removed from the incubator and placed on the microscope. For nonadherent cells, a dilute suspension of cells is placed into a cell chamber and the cells are allowed to settle. Typically 500 μl of a 50,000 cells/ml suspension is adequate to provide well-spaced cells for the experiment. To aid adhesion of nonadherent cells to the coverslip, cell chambers may be coated with polylysine or other material prior to plating.

3. Cell Loading

Most peptide-based reporters do not traverse the cell membrane. While there are several methods to load membrane impermeant molecules into cells, each has its advantages and disadvantages. Several excellent reviews of this topic have been written to which the reader is referred (Stephens and Pepperkok, 2001; Wadia and Dowdy, 2002). A brief discussion is included here to highlight some of the issues specific to CE-based chemical separations of cellular reporters. Microinjection can be used to introduce a small volume of reporter in solution into the cell. This approach requires the reporter to be soluble at relatively high concentration in a physiological buffer. Depending on the skill of the operator, microinjected cells tend to have a low survival rate which may complicate evaluation of cell physiology. In addition, concentrated reporter leaking from the injection pipette can be nonspecifically adsorbed to the outer surface of the cell and surrounding glass surface. This extracellular reporter can be introduced into the capillary at the time of cell sampling. Since this extraneous reporter is indistinguishable from the intracellular substrate, quantification of the product-to-substrate ratio can be difficult. Pinocytic loading involves induction of membrane vesicles by placing cells in a hyperosmolar buffer followed by intracellular dissolution of these vesicles under hypoosmolar conditions (Okada and Rechsteiner, 1982). Reporter present in the hyperosmolar extracellular fluid is taken up by the cell in these vesicles and released intracellularly. This process enables large numbers of cells to be loaded efficiently, but is stressful and must be optimized for each cell type. Again, nonspecific adsorption of the reporter to extracellular surfaces may complicate subsequent analyses. The reporter may also be introduced by conjugation to a mediator that confers cell uptake, such as a protein transduction domain (Fawell *et al.*, 1994). As an example, a peptide can be synthesized that includes the amino acid sequence "RKKRRQRRR" derived from the HIV-1 TAT protein. This positively charged sequence is readily taken up by cells. Although not fully elucidated, it is thought that uptake is through an endocytic process (Brooks *et al.*, 2005). Peptides, proteins, and other molecules conjugated to this sequence are also efficiently translocated across the cell membrane into the cytoplasm. Although efficient, this approach is not without its drawbacks. The addition of this highly basic peptide can interfere with the kinetic properties of the substrate. This problem can be ameliorated by introducing a linker composed of additional residues between the substrate and translocation domains, or engineering a cleavable linker to enable separation of the two domains after cellular uptake (Soughayer *et al.*, 2004). Peptides containing a myristylated residue have also been used to confer membrane permeability. The hydrophobic nature of this C_{14} moiety may lead to problems with solubility and electrophoresis in aqueous buffers.

C. Microscope Platform

A frequency-doubled (532 nm) Q-switched Nd:YAG laser is aligned into a standard inverted fluorescence microscope (Fig. 1B) as previously described (Sims *et al.*, 1998). When the purpose of the incoming laser pulse is cell lysis, precision of the alignment is not as critical as with applications such as ablation of subcellular structures. Nevertheless, the operator should be experienced in the use of lasers and the basics of alignment into a microscope (Berns *et al.*, 1998). An overview of the process is provided here with attention paid to specific issues related to use of a laser microbeam for the rapid lysis of a cell prior to sampling into a capillary. The laser pulse is aligned to the center of the microscope's field-of-view and is focused at the interface of the coverslip surface and buffer solution. To reduce the risk of eye injury when using a laser input to a microscope, the microscope field is viewed through a CCD camera and monitor rather than the eyepiece. When the goal is to sample a single cell, it is desirable to plate cells sparsely to achieve only one cell per high-power field-of-view. The laser energy is chosen which results in complete ablation of the outer membrane of the target cell without damage to surrounding cells. Due to differences in optics, alignment, and day-to-day variations, specific recommendations are difficult to provide, but a laser pulse energy of 10–100 μJ measured at the back aperture of the objective is typical.

A cell chamber containing the plated cells is fixed to the microscope stage and continuously perfused with a physiological buffer. The targeted cell is positioned $\sim$20 μm from the focus point of the laser pulse and the capillary inlet is positioned 20–30 μm above with the channel centered over the cell (Fig. 2). To sample the cell, a laser pulse is generated which leads to disruption of the

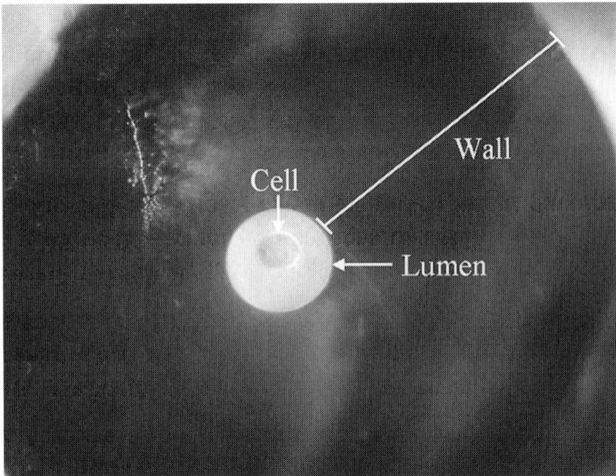

Fig. 2 Shown is a photomicrograph of a capillary positioned ~10 μm directly above a single cell to be sampled (400× magnification). The lumen of the capillary is centered directly above the cell. The capillary wall extends beyond the field of view.

membrane through shear forces created by the resultant cavitation bubble (Rau *et al.*, 2004). By positioning the cell at a distance from the laser's focal point, cell contents are not affected by heat or light generated by the laser pulse. Cell contents are drawn into the capillary and electrophoresed on cell lysis (see above).

D. Capillary Electrophoresis

Details of the system combining CE with a laser-induced fluorescence detection system mated to a microscope have been published previously (Sims *et al.*, 1998). An overview of the capillary preparation and electrophoresis is provided here. A standard polyimide-coated capillary is cut to the desired length to extend from the microscope stage to the detection system. The end of the capillary to be positioned above the cell is referred to as the inlet. The polyimide coating is burned off for 2–3 mm at each end of the capillary which is then prepared for use by flushing with sodium hydroxide (0.1 M) for 12 h followed by deionized, distilled water for 1 h, then dilute hydrochloric acid (0.1 M) for 12 h, and finally water for >1 h. Prior to mounting of the capillary, the inlet is carefully prepared with the aid of a dissecting microscope to achieve a smooth surface perpendicular to the long axis of the capillary. A cleaving stone or capillary-cutting tool (Supelco, Pennsylvania) can be used to aid in this process. Careful preparation of the inlet enables positioning of the capillary channel within ~30 μm of the cell to be sampled to ensure efficient loading of cell contents into the capillary.

A full discussion of methods development for chemical separations by CE is not possible in the space available. The reader is referred to a standard text for further reading regarding wall coatings and buffer additives to optimize separation conditions (Weinberger, 1993). As illustration of the technique to be used, a brief overview using an uncoated capillary is provided. The detection system is placed adjacent to the microscope stage such that the outlet reservoir is at the same height as the stage. This positioning is important to prevent siphon-like fluid flow in the capillary due to mismatches in the fluid levels between the capillary's inlet and outlet. Before the capillary is positioned in the electrophoresis and microscope setup, a detection window is created in the capillary by burning off 5–10 mm of the polyimide coating and cleaning with ethanol. This window is positioned in the detection system with the outlet end of the capillary extending into the high-voltage electrophoresis reservoir that serves as the outlet reservoir. The inlet is mounted to a three-axis micrometer attached to the microscope stage. The capillary channel is then centered in the field-of-view. The system is now ready for cell sampling.

For experiments it is useful to provide a constant flow of heated buffer through the cell chamber to maintain proper temperature and isotonicity of the cells. This perfusion system can be homemade or purchased from a commercial source. Additionally, an objective heater is attached to a 100× oil objective to aid in keeping the cell chamber at the correct temperature. Standard cell culture media contain fluorescent compounds that may interfere with analyte detection, thus a physiological buffer is provided through the flow system. Since cell lysis and loading takes place in this buffer, it is most simple to use the cell chamber as the inlet reservoir for the separation. In this case, the electrophoresis ground electrode is placed in the cell chamber prior to cell sampling. The capillary channel is positioned 10–30 μm above the cell which is then lysed using the Nd:YAG laser. A high voltage is placed across the capillary to initiate electrophoresis simultaneously with the lysis pulse. In some instances, it may be desirable to use a different buffer for electrophoresis than the physiological buffer in the cell chamber. Typical electrophoresis buffers are not compatible with living cells, so that care must be exercised to prevent exposing the cell to the buffer prior to sampling. To accomplish this, a separate inlet reservoir adjacent to the cell chamber and the capillary are filled with the desired separation buffer. The capillary channel is positioned above the cell in the cell chamber as described above. It may be advantageous in this situation to place the capillary outlet reservoir slightly below the level of the microscope stage (1–2 cm) to prevent flow of the cell-incompatible buffer from the capillary onto the cell. After collection of the cell contents into the capillary, electrophoresis is stopped and the inlet is moved to the adjacent reservoir containing the separation buffer. Electrophoresis is once again commenced.

For safety reasons, it is best to have the inlet of the capillary at ground with the high-voltage outlet reservoir shielded within the enclosed detection system. Using a bare capillary with the inlet held at ground, a negative high voltage provides brisk electroosmotic flow in the direction of the outlet helping to load cell contents and speed migration of the analytes to the detection window (Weinberger, 1993). Regardless of the separation conditions, it is important that the capillary be thoroughly washed between runs. Extensive washing will help to rinse cell debris from the capillary to ensure reproducibility between sequential runs. For an uncoated capillary, the wash process requires consecutively flushing the capillary with 10–20 column volumes of NaOH (0.1 M), H_2O, and run buffer between each use. In practice, the capillary is flushed with each solution until 1–2 drops of each has been expelled from the capillary's end.

For single-cell analysis, laser-induced fluorescence is advantageous as it is possible to detect 10^{-20} moles of fluorescently labeled analyte using a fairly simple detection scheme. This detection limit enables quantification of fluorescent analytes present in mammalian cells at nanomolar concentrations (Li *et al.*, 2001; Meredith *et al.*, 2000; Sims *et al.*, 1998). Our laboratory uses a custom-made system as described previously (Sims *et al.*, 1998). Briefly, excitation light of the appropriate wavelength is provided by a laser (e.g., 488 nm for fluorescein). The laser is focused on the capillary lumen while emission is monitored at a 90° angle using a 40× microscope objective to collect and focus the emitted fluorescence signal onto a PMT. For details on the practice of ultrasensitive laser-induced fluorescence detection the reader is referred to one of the excellent reviews on the subject (Johnson and Landers, 2004). The signal from the PMT is amplified and monitored via a computer using standard laboratory software.

E. Data Collection and Analysis

To collect the data and control the hardware, a custom program was written in Testpoint. The program controls the Nd:YAG laser firing, and also initiates electrophoresis by controlling the high-voltage supply. The fluorescence and current are constantly monitored during electrophoresis, and upon run completion, the data is stored in an ASCII format for later manipulation in a graphic software program. The Testpoint program communicates with the hardware through an A/D board. Many graphic software programs are available, such as Origin, which allow for peak identification and peak area calculations.

After the initial setup of the detection and electrophoresis system, it must be tested and adjusted to achieve the detection limits and sensitivity necessary for single-cell experiments. To detect single cells loaded with ~100 nM of the desired reporter, the detection limit of the system should be $\sim 10^{-20}$ moles. The system is tested with a sample of the reporter at known concentrations in the range of 1 μM–10 pM. Solutions of known concentrations should be freshly made from a more concentrated stock solution as discussed previously. The standard is injected for a specified time (5–30 s) at a known height difference (measured between the fluid levels of the inlet and outlet reservoirs, e.g., 1–3 cm). The injected sample is electrophoresed and the peak area is determined. To establish the detection limit, standards are injected at successively lower concentrations until the signal-to-noise of the electrophoretic peak is five times larger than the baseline. To accurately calculate the mass detection limit using a sample of known concentration, the volume of the injection plug must be estimated. The volume loaded depends on several factors: inner diameter of the capillary, capillary length, injection time, and the height difference between the sample and outlet reservoirs. The total volume of an injection plug results from different loading methods that are additive: hydrodynamic loading, spontaneous fluid displacement (i.e., surface tension loading), and diffusion (Fishman *et al.*, 1994a,b; Weinberger, 1993). The mass of the standard in moles is easily determined by multiplying the volume of solution loaded with the concentration of the standard.

F. Example Experiment

First, a capillary is prepared and reporter standards of the substrate and product are electrophoresed multiple times to assure that migration times are consistent, the separation conditions are effective, and the setup is sufficiently aligned with a detection limit appropriate for single-cell experiments. The day before experiments are planned, the cells are plated into chambers and then incubated overnight. On the day of the experiment, the alignment of the fluorescence equipment is checked, the perfusion system is enabled, and the Nd:YAG laser alignment is verified. The cells are loaded with the reporter by the chosen method and allowed to recover in the incubator for an appropriate time. Subsequently, the cell chamber is placed on the microscope stage and perfused for several minutes to maintain physiological temperature and exchange the buffer. An isolated cell is chosen for analysis, and when the capillary has been positioned above the cell, the Nd:YAG laser is fired and electrophoresis is immediately initiated. The trace is analyzed to quantify the amount of substrate and product obtained from the cell (Fig. 3).

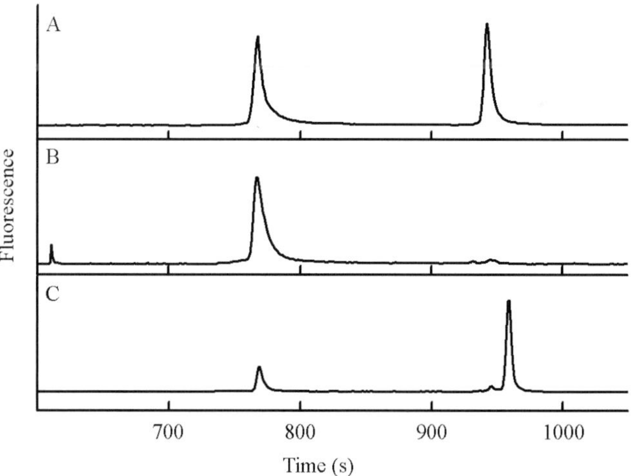

Fig. 3 Example electropherograms from single-cell analyses of protein kinase C (PKC) activation in rat basophilic leukemia cells. Cells were microinjected with a fluorescently labeled peptide substrate for PKC, and then analyzed after exposure to phorbol myristic acid, a pharmacological activator of the kinase. (A) Standards of the nonphosphorylated and phosphorylated peptide in buffer. (B) Electropherogram from an unstimulated cell. (C) Electropherogram from a cell exposed to PMA.

IV. Materials

1. Tissue culture reagents: GIBCO, Grand Island
2. Cell chamber polymer: Sylgard 184; Dow Corning, Midland
3. Physiological buffer: 135-mM NaCl, 5-mM KCl, 1-mM $MgCl_2$, 1-mM $CaCl_2$, 10-mM Hepes, pH 7.4
4. Siliconized microcentrifuge tubes: Fisher (Catalog No. 02-681-311)
5. Silicone "O"-rings: 3/32″ silicone, 11/16″ by 7/8″ McMaster-Carr, Atlanta (Catalog No. 9396K28)
6. Glass coverslips: Fisher (Catalog No. 12-245-102)
7. Fused silica capillaries: 360-μm outer diameter, 30- to 75-μm inner diameter; Polymicro Technologies, Phoenix
8. Video camera: Sony CCD Video Camera, Model XC-77
9. Objective Heater: Bioptechs, Inc., Butler
10. Perfusion system: Biosciences Tools, San Diego
11. Inverted microscope: Model TE300, Nikon, Melville
12. Nd:YAG laser: Surelite I, Continuum, Santa Clara
13. A/D Board: Keithley, Model 3102, Cleveland
14. Testpoint Software: Capital Equipment, Middleboro
15. Origin Software: OriginLab Corporation, Northampton

V. Discussion

A multitude of biochemical assays may be adapted to the single-cell format using the instrumentation described above. The cellular process of interest must have a target, or substrate (found by sequencing, or by the use of a combinatorial library), which can be isolated or synthesized and fluorescently labeled. When this substrate is acted on by a target enzyme, it must yield a product possessing a different electrophoretic mobility so that the two may be efficiently separated. Kinases can be studied since the addition of a phosphate group to the substrate significantly alters the charge-to-mass ratio and therefore, the electrophoretic mobility. Phosphatases, the corollary, may also be targeted. In addition, numerous other assays can be conceived for which reporters can be designed which generate products with unique electrophoretic properties. The approach can be used to measure constitutive enzymatic activity, or cells may be subjected to physiological or pharmacological stimuli or inhibitors followed by analysis. Numerous assays can thus be envisioned to further understanding of individual or multiple enzymatic pathways of the cell.

VI. Summary

We have provided a brief overview of the history of single-cell analysis in conjunction with CE. The technique for laser-based sampling and single-cell analysis developed in our laboratory has been discussed with regards to (1) the selection of a reporter, (2) the culture of cells and the equipment needed to ensure cell viability during experiments, (3) necessary modifications to the microscope platform for laser-based cell lysis, (4) special CE conditions, and (5) data collection and analysis. Additional applications of this technique have also been briefly discussed.

Acknowledgments

This work was supported by NIH grants EB04597 to N. L. A. and CA105514 to C. E. S., N. L. A., and C. E. S. disclose a financial interest in Cell Biosciences, Inc.

References

Allbritton, N. L., and Meyer, T. (1993). Localized calcium spikes and propagating calcium waves. *Cell Calcium* **14,** 691–697.

Berns, M. W., Tadir, Y., Liang, H., and Tromberg, B. (1998). Laser scissors and tweezers. *Methods Cell Biol.* **55,** 71–98.

Brooks, H., Lebleu, B., and Vives, E. (2005). Tat peptide-mediated cellular delivery: Back to basics. *Adv. Drug Deliv. Rev.* **57,** 559–577.

Craig, D., Arriaga, E. A., Banks, P., Zhang, Y., Renborg, A., Palcic, M. M., and Dovichi, N. J. (1995). Fluorescence-based enzymatic assay by capillary electrophoresis laser-induced fluorescence detection for the determination of a few beta-galactosidase molecules. *Anal. Biochem.* **226,** 147–153.

Dovichi, N. J., and Hu, S. (2003). Chemical cytometry. *Curr. Opin. Chem. Biol.* **7,** 603–608.

Fawell, S., Seery, J., Daikh, Y., Moore, C., Chen, L. L., Pepinsky, B., and Barsoum, J. (1994). Tat-mediated delivery of heterologous proteins into cells. *Proc. Natl. Acad. Sci. USA* **91,** 664–668.

Ferrell, J. E., and Machleder, E. M. (1998). The biochemical basis of an all-or-none cell fate switch in *Xenopus* oocytes. *Science* **280,** 895–898.

Fishman, H. A., Amudi, N. M., Lee, T. T., Scheller, R. H., and Zare, R. N. (1994a). Spontaneous injection in microcolumn separations. *Anal. Chem.* **66,** 2318–2329.

Fishman, H. A., Scheller, R. H., and Zare, R. N. (1994b). Microcolumn sample injection by spontaneous fluid displacement. *J. Chromatogr. A* **680,** 99–107.

Hogan, B. L., and Yeung, E. S. (1992). Determination of intracellular species at the level of a single erythrocyte via capillary electrophoresis with direct and indirect fluorescence detection. *Anal. Chem.* **64,** 2841–2845.

Johnson, M. E., and Landers, J. P. (2004). Fundamentals and practice for ultrasensitive laser-induced fluorescence detection in microanalytical systems. *Electrophoresis* **25,** 3513–3527.

Kennedy, R. T., Oates, M. D., Cooper, B. R., Nickerson, B., and Jorgenson, J. W. (1989). Microcolumn separations and the analysis of single cells. *Science* **246,** 57–63.

Krylov, S. N., Zhang, Z. R., Chan, N. W. C., Arriaga, E., Palcic, M. M., and Dovichi, N. J. (1999). Correlating cell cycle with metabolism in single cells: Combination of image and metabolic cytometry. *Cytometry* **37,** 14–20.

Lee, C. L., Linton, J., Soughayer, J. S., Sims, C. E., and Allbritton, N. L. (1999). Localized measurement of kinase activation in oocytes of *Xenopus laevis*. *Nat. Biotechnol.* **17,** 759–762.

Li, H., Sims, C. E., Kaluzova, M., Stanbridge, E. J., and Allbritton, N. L. (2004). A quantitative single-cell assay for protein kinase B reveals important insights into the biochemical behavior of an intracellular substrate peptide. *Biochemistry* **43,** 1599–1608.

Li, H., Sims, C. E., Wu, H. Y., and Allbritton, N. L. (2001). Spatial control of cellular measurements with the laser micropipet. *Anal. Chem.* **73,** 4625–4631.

Meredith, G. D., Sims, C. E., Soughayer, J. S., and Allbritton, N. L. (2000). Measurement of kinase activation in single mammalian cells. *Nat. Biotechnol.* **18,** 309–312.

Okada, C. Y., and Rechsteiner, M. (1982). Introduction of macromolecules into cultured mammalian-cells by osmotic lysis of pinocytic vesicles. *Cell* **29,** 33–41.

Rau, K. R., Guerra, A., Vogel, A., and Venugopalan, V. (2004). Investigation of laser-induced cell lysis using time-resolved imaging. *Appl. Phys. Lett.* **84,** 2940–2942.

Sims, C. E., Meredith, G. D., Krasieva, T. B., Berns, M. W., Tromberg, B. J., and Allbritton, N. L. (1998). Laser-micropipet combination for single-cell analysis. *Anal. Chem.* **70,** 4570–4577.

Soughayer, J. S., Wang, Y., Li, H. A., Cheung, S. H., Rossi, F. M., Stanbridge, E. J., Sims, C. E., and Allbritton, N. L. (2004). Characterization of TAT-mediated transport of detachable kinase substrates. *Biochemistry* **43,** 8528–8540.

Stephens, D. J., and Pepperkok, R. (2001). The many ways to cross the plasma membrane. *Proc. Natl. Acad. Sci. USA* **98,** 4295–4298.

Wadia, J. S., and Dowdy, S. F. (2002). Protein transduction technology. *Curr. Opin. Biotechnol.* **13,** 52–56.

Wallingford, R. A., and Ewing, A. G. (1988). Capillary zone electrophoresis with electrochemical detection in 12.7-Mu-M diameter columns. *Anal. Chem.* **60,** 1972–1975.

Weinberger, R. (1993). "Practical Capillary Electrophoresis." Academic Press, San Diego.

Zhang, Z. R., Krylov, S., Arriaga, E. A., Polakowski, R., and Dovichi, N. J. (2000). One-dimensional protein analysis of an HT29 human colon adenocarcinoma cell. *Anal. Chem.* **72,** 318–322.

PART VI

A Tribute to Sergej Tschachotin

CHAPTER 27

Sergej Stepanovich Tschachotin: Experimental Cytologist and Political Critic (1883–1973)

Karl Otto Greulich,* Alexey Khodjakov,† Annette Vogt,‡ and Michael W. Berns§

*Leibniz Institute of Age Research
D-07745 Jena, Germany

†Division of Molecular Medicine
Wadsworth Center, Albany, New York 12201
Department of Biomedical Sciences, SUNY
Albany, New York 12222
Marine Biological Laboratory
Woods Hole, Massachusetts 02543

‡Max Planck Institute for History of Science
14195 Berlin, Germany

§Departments of Biomedical Engineering
Developmental and Cell Biology, and Surgery
Beckman Laser Institute, University of California
Irvine, California 92612
Department of Bioengineering
Whitaker Institute for Biomedical Engineering
University of California, San Diego, La Jolla, California 92093

Sergej Stepanovich Tschachotin (who also published as Sergej Stepanovich Chachotin, Serge Chakhotin, Serge Tchakhotine in different European languages) was born September 13, 1883 in Constantinople (Istanbul) where his father was vice-consul of the Russian empire in Turkey (Ausstellung "Makroskop", 2006; Vogt, 2007a,b).

0091-679X/07 $35.00
DOI: 10.1016/S0091-679X(06)82027-3

He returned with his family to his mother's home city of Odessa, Russia in 1885 where he went through his early schooling. When he finished his formal schooling, he went to Moscow where he studied medicine from 1901 to 1902 at Moscow University (now Lomonosov Moscow State University). In 1902, Tschachotin joined a student protest against the tsarist regime. All students involved in the uprising were expelled from the University and ordered to return to their native provinces. Because Tschachotin was born in Constantinople, he was deported from the country.

From 1902 until 1912, he lived in western Europe, studying medicine at the universities in Munich, Berlin, and Heidelberg until 1907, and receiving his doctoral degree from Heidelberg University in 1908 with a thesis, "Die Statocyste der Heteropoden." Between 1908 and 1912, Tschachotin conducted scientific research at several institutes in Messina, Villafrance, Triest, and on the island of Helgoland. In Messina he was buried under the rubble during the devastating earthquake of 1908. He had to undergo some very serious surgery and, as the story goes, when he suffered through the surgery he dreamt about tiny little instruments that would allow one to operate on individual cells instead of slashing through organs and tissues. This, plus his earlier encounters with the famous German physicist Wilhelm Conrad Röntgen who told him that biologists are not proper experimentalists because they start their experiments by killing cells and thus abolishing the only unique feature of the object, resulted in his lifelong obsession with microsurgery. In 1912, he started to operate on individual cells with miniature surgical tools that he invented. Many of these tools were quite advanced even by modern standards (Fig. 1). But even more remarkable was the way Tschachotin used these tools. He developed a series of experimental techniques that are still being used in cell physiology. Among these are microperfusion (Fig. 2A), cellular transplantation (Fig. 2B), and instantaneous fixation of individual cells on the microscope stage (Fig. 2C). However, Tschachotin was not satisfied with conventional microsurgery. One of the reasons was that to reach the structure of interest he had to cut through outer parts of the cell, just like surgeons often cut through healthy tissues to gain access to the damaged organ. Tschachotin was searching for a noninvasive surgical approach. He thought that Röntgen's X-rays would serve as an ideal scalpel, but this type of electromagnetic irradiation cannot be focused by glass lenses. So, Tschachotin began to experiment with ultraviolet (UV) light. In 1912, he constructed the first microirradiation apparatus which allowed him to selectively irradiate individual cells or even subcellular organelles. Considering that Tschachotin did not have access to stable UV light sources, the lenses that he used were not corrected for UV, and UV light is invisible, this was a remarkable achievement. Short on technology, Tschachotin used his creativity to overcome problems that would be deemed indissoluble by most researchers. For example, UV light cannot penetrate deep into specimen which seems to preclude its use for irradiation of nuclei in large and relatively opaque cells such as sea urchin eggs. However, Tschachotin was able to overcome this problem by displacing the nucleus in live cell via cell centrifugation (Fig. 3). Further, by combining nucleus displacement and cell flattening he was able to convert sea urchin eggs into an almost ideal object for UV irradiation. This unique creativity was a signature of

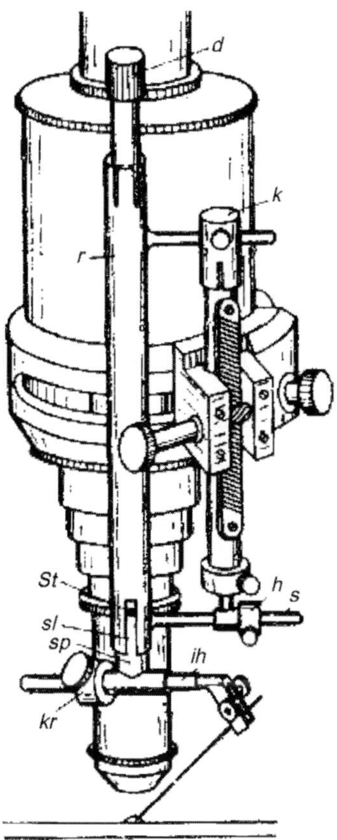

Fig. 1 "Micro-operator" developed by Tschachotin. This versatile instrument was capable of conducting most sophisticated microsurgery on individual cells. A universal instrument holder (*ih*) could hold micropipettes, glass needles, and other similar tools. It was precisely positioned via two independent gear mechanisms (coarse, *K* and fine, *d*). Several independent pivoting collars (*kr, sl,* and *sp*) provided for virtually unlimited angular adjustments so that the cell could be approached from different sides.

Sergej Tschachotin's approach to science (see the publications from 1912 to 1935 in the reference list at the end of this chapter as well as the discussion of his work in Chapter 1 by Berns and Chapter 4 by Quinto-Su and Venugopalan, this volume).

He returned to Russia, and from 1912 to 1917, he was an assistant in the laboratory of the famous Ivan Petrovich Pavlov (1849–1936) in St. Petersburg (which was renamed Petrograd in 1914, Leningrad in 1924, and St. Petersburg again in 1991). Apparently things did not go well in Pavlov's laboratory because Tschachotin was using laboratory equipment to print propaganda leaflets and making nitroglycerin for the revolutionists. Unconfirmed rumor has it that one of Tschachotin's contacts turned out to be an undercover agent and as a result Pavlov's laboratory was searched. Pavlov was so angry that he told Tschachotin to choose either science or politics. Tschachotin chose politics at least for the time being. So he was forced to leave Pavlov's laboratory (Albrecht, 1986, 1987).

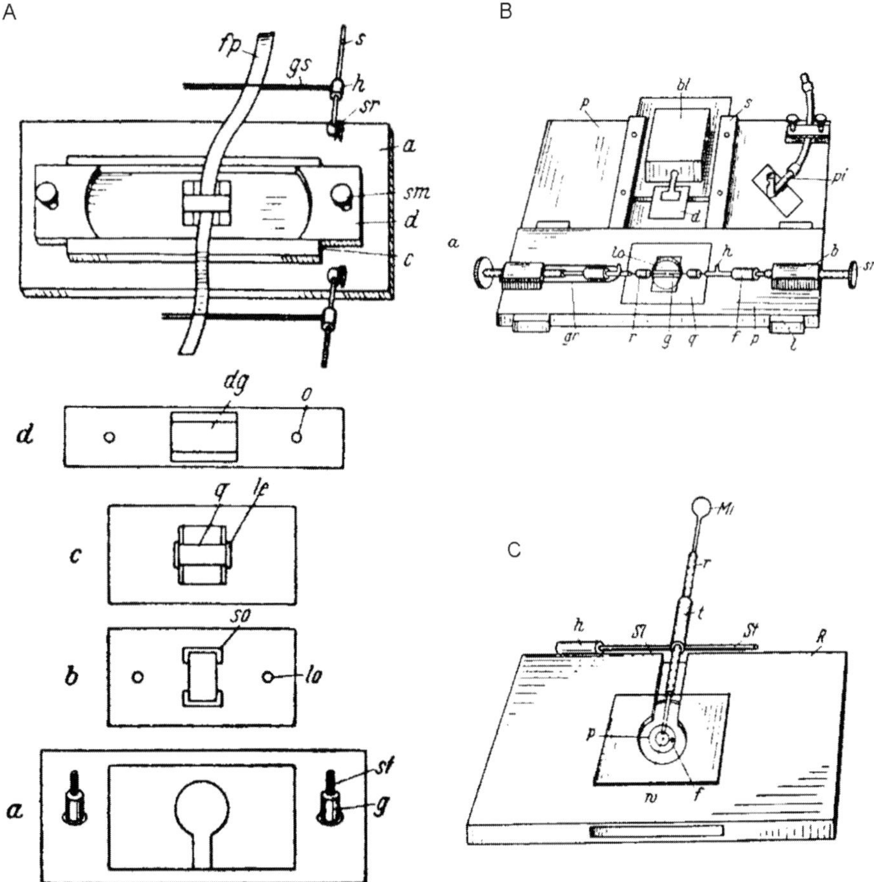

Fig. 2 Examples of instruments and experimental techniques invented by Tschachotin. (A) Microperfusion device for creating a laminar flow around cells on the microscope stage. By sandwiching a series of metal plates (*a–d*) Tschachotin was able to control the geometry of the cultivation chamber. Solutions were perfused by capillary forces via two strips of filter paper (*fp*) on opposite sides of the chamber. (B) Device for cellular transplantation. Using this and similar devices Tschachotin was able to remove nuclei and fragments of cytoplasm from a cell. Further, he used this technique to push two cells against one another, inducing their fusion. During this procedure cells were placed in a thin channel formed between a quartz coverslip (*q*) and base plate. Two high-pitch screws (*sr*) were used to translate pistons (*g*) that pushed the cells together. (C) Microfixator. This elegant device was used to instantly deliver fixatives to the cell. The pipette (preloaded with desired solution, *Mi*) was positioned in the immediate vicinity of the cell and supported by a spring (*Sl* and *St*).

From 1917 until 1927, Tschachotin acted in different capacities as a politician and publisher and was not able to do further scientific research. He was actively involved in the revolution against the tsarist regime in February in 1917. But considered the second, Bolshevik's revolution to be a usurpation, and in 1918 he

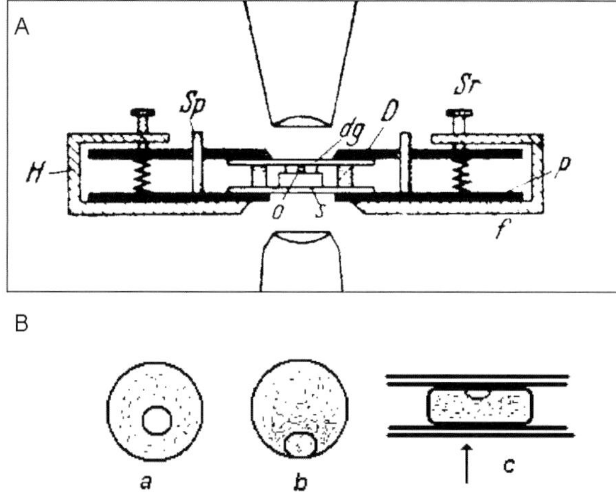

Fig. 3 Tschachotin's approach to irradiation of object that are normally positioned out of the reach of UV microbeams. (A) Cell-flattening device used to make large round cells such as sea urchin eggs amenable to UV microsurgery. (B) A schematic representation of the experiment in which Tschachotin displaced nuclei from the center of the cell to the cortex. Cells were centrifuged so that nuclei (that have higher density than the cytoplasm) moved to the cortex (*a* and *b*). These cells were then additionally flattened (*c*) using device presented in (A). This approach allowed Tschachotin to selectively irradiate just the nucleus or just the cytoplasm.

joined the "White" army which was fighting against the Bolsheviks in the civil war. However, when Tschachotin discovered that some of the White Army generals were considering a separatist peace agreement with Germany, he publicly denounced them and fled the country. It was a difficult decision one that left Tschachotin completely isolated. "White" Russian emigrants never forgave him for leaving the White Army. Yet, for the Bolsheviks he was an enemy who once fought against them.

In 1920, Tschachotin received a professorship at the University in Zagreb, but soon he was denounced as a "red" and lost his position in the following year. Until 1924, he worked for two journals edited by Russian emigrants, "Smena vech" (published in Paris) and "Nakanune" (published in Berlin). Meanwhile, he officially accepted the victory of the Bolsheviks, following which, he met Leonid B. Krassin (1870–1927) in Genoa in 1922 who helped him get a position in the Soviet Embassy in Berlin and to become a Soviet citizen (in 1922). Until 1927, Tschachotin worked for the Soviet Embassy in Berlin, meeting Albert Einstein (1879–1955) among others.

In 1927, he went to Italy to begin his scientific research again. In 1927, Krassin died, Lev D. Trotsky (1879–1940) lost his political influence in Soviet Russia, and Stalin's regime was established. It seems that Tschachotin had well recognized what would happen in the future and cut off his relationship to the Soviets.

From 1927 to 1933, Tschachotin was again a scientist who worked with grants and fellowships in different scientific institutions. In 1927, he started his research at the pharmacological institute of the University in Genoa. Here Tschachotin's research was supported by a grant from the Vatican. This grant was awarded because Tschachotin demonstrated that UV micropuncture induces parthenogenesis in sea urchin and that was considered an explanation for how immaculate conception could occur. Thanks to a grant from the Research Corporation of the United States in 1930, he became a guest scientist in the Kaiser Wilhelm Institute for Medical Research in Heidelberg. The grant was given to him thanks to a reference written by Albert Einstein. In Heidelberg, he wanted to work with Otto Meyerhof (1884–1951), but because the technical equipment was better for his research in the Institute for Physics, headed by Karl Wilhelm Hausser (1887–1933), Tschachotin worked at this institute until 1933.

In spring of 1933, Tschachotin was displaced from the Kaiser Wilhelm Institute because of his political activities against the Nazi's [see Albrecht (1987) about the procedure of displacing him]. This was only too true. Tschachotin, who copublished a booklet on socialist propaganda with Carl (Carlo) Mierendorff (1897–1943) in 1932, was also involved in the disputes of social democrats against the Nazi's in the state of Baden and in Heidelberg. In May 1933, he was arrested but was set free because of his Soviet citizenship. His apartment and his laboratory in the Kaiser Wilhelm Institute were searched by the Gestapo. Sergej Tschachotin had to flee and again became an emigrant. From 1933 to 1934, he lived in Copenhagen where he published his famous booklet against the Nazi's, "Dreipfeil gegen Hakenkreuz" (Fig. 4). After conflicts with Danish social democrats, Tschachotin was exiled again, and he went to France.

From 1934 until 1955, Sergej Tschachotin lived in France, mostly in Paris. He was a freelance scholar, existing partly on grants and fellowships and working in several institutions, among others, the CNRS. In 1939, he published in French as well as in English his most famous book about the psychology of propaganda. During the German occupation in 1941 he was arrested, but was set free after a few months thanks to the appeals of scientists. After his liberation, he conducted cancer research at the CNRS and it is there that, it is very possible, he came into contact with the French hematologist Marcel Bessis, who later with the famous microscopist Normarski, published the first paper on the use of lasers for microbeam irradiation through the microscope in 1962 (see discussion in Chapter 1 by Berns, this volume). Though Tschachotin never used lasers in his research, it is likely that his interactions with the French scientists during 1934–1955 had an influence on the development and use of microbeam irradiation by the French.

From 1945 to 1955, Tschachotin (he published in France as Tchakhotine) conducted scientific research from time to time at the Institute for Physical-Chemical Biology and participated in the antiwar movement "Science-Action-Liberation." In 1955, he went to Italy again and worked in the pharmacological institute in Genoa where he had already been in 1927.

Fig. 4 Cover of Tschachotin's book "Dreipfeil gegen Hakenkreuz," Courtesy: Herbert Zimmermann, Max Planck Institute for Medical Research, Heidelberg, Germany, Tschachotin's host institution during his second stay in Heidelberg. "Dreipfeil gegen Hakenkreuz" means "Three Arrows against the Hakenkreuz." The back cover of the book shows the three arrows which are crossing out the Hakenkreuz. The "Hakenkreuz" refers to the Nazi Swastika symbol. (See Plate 33 in the color insert section.)

In 1958, Tschachotin already 75 years old returned to Russia, now the USSR. From 1958 until his death in 1973, Tschachotin lived in the Soviet Union, first in Leningrad and later in Moscow (Fig. 5). He held senior research positions in different institutes of the Soviet Academy of Sciences. He was honored several times by the Soviet establishment but he was never allowed to travel to foreign countries.

Tschachotin was married five times and had eight sons. According to his wishes he was buried in Corsica where he and his first wife went on honeymoon in 1908.

Below is a reference list of the many scientific publications in the many languages of this "nomadic" scientist. In addition to his scientific publications he also wrote about human oppression and suffering by totalitarian regimes. The drawings presented in Figs. 1–3 are taken form the biography of Tschachotin written by A. Posudin, and published in Kiev, Ukraine in 1995 (Posudin, 1995).

Fig. 5 Sergej Tschachotin (ca. 1960–1970s), from the biography of Sergej Tschachotin published in Russian: Posudin, Yu. I. (1995).

References

Albrecht, R. (1986). Sergej Tschachotin oder "Dreipfeil gegen Hakenkreuz". In: Exilforschung. Jahrbuch, Band 4, Hg. Gesellschaft für Exilforschung. München: edition text + kritik. S.208–228.

Albrecht, R. (1987). "... daß Sie Ihre Tätigkeit einstellen müssen": Die Entlassung Sergej Tschachotins aus dem Heidelberger Kaiser-Wilhelm-Institut 1933. In: Berichte zur Wissenschaftsgeschichte 10, S.105–112.

Ausstellung "Makroskop". (2006). Museum für Photographie, 4.2.–23.4., von Boris Hars-Tschachotin und Hannes Nehls, Berlin.

Benedicanti, A., and Tchakhotine, S. (1956). Azione di alcuni farmaci sugli esseri unicellulari. *Rendiconti dell'Acadcmia Nazionole dei Lincei. Serie VIII* **20,** 19.

Posudin, Yu. I. (1995). "Biophysicist Sergei Tschachotin," p. 92. Natl. Agricult. Univ. Publ., Kiev, Ukraine.

Tschachotin, S. (1904). On transportation of sea urchin gametes to Sankt Petersburg for the purpose of biological experimentation. *Proc. Imperial Acad. Sci. Russia* **7,** 737.

Tschachotin, S. (1907). Ueber die bioelektrischen Stroeme bei Wirbellosen und deren Vergleich mit analogen Erscheinungen bei Wirbeltieren. *Pfl. Arch. Ges. Physiol.* **120,** 565.

Tschachotin, S. (1912). Ueber Strahlenwirkung auf Zellen, speziell auf Krebsgeschwulstzellen und die Frage der chemischen Imitation derselben. *Munch. Medizin. Wochenschr.* **44,** 2379–2381.

Tschachotin, S. (1912). Die mikroskopische Strahlenstichmethode, eine Zelloperationsmethode. *Biol. Zentralbl.* **32,** 623–630.

Tschachotin, S. (1912). Eine Mikrooperationsvorrichtung. *Zs. Wiss. Mikrosk.* **29,** 188.

Tschachotin, S. (1930). Rationelle Organisation von biologischen Instituten. *Abderhalden's Handbuch Der Biolog. Arbeitmethoden* **2,** 2.

Tschachotin, S. (1933). The crisis in scientific research and the way out. *Science* **77,** 426.

Tschachotin, S. (1935). Die Mikrostrahlstichmethode und andere Methoden des zytologischen Mikroexperimentes. *Handb. Biol. Arbeitsmethoden Abderhalden.* **10,** 877–958.

Tschachotin, S. (1939). Experimentelle Kanzerisierung von Embryonalzellen. *Archiv fur Experiment. Zellforschung.* **22,** 422.

Tschachotin, S. (1959). Investigation of localized action of ultraviolet light on cells by the micropuncture method. *Tsitologia (Russ.)* **1,** 614–626.

Tchakhotine, S. (1920). Action localisee des rayons ultraviolets sur le niveau de l'oeuf de l'oursin par radiopuncture microscopique. *C. r. Soc. Biol.* **83,** 1593–1595.

Tchakhotine, S. (1921). Les changements de la permeabilite de l'oursin localisee experimentalement. *C. r. Soc. Biol.* **84,** 464–466.

Tchakhotine, S. (1921). Sur le mecanisme de l'action des rayons ultraviolets sur la cellule. *Ann. Inst. Pasteur.* **35,** 321–325.

Tchakhotine, S. (1929). Attivazione dell'uovo di riccio di mare per mezzo della microraggiopuntura. *Boll. Soc. Ital. Biol. Sper.* **4,** 475–479.

Tchakhotine, S. (1929). La leucocitosi come reazione-indice nel cancro da trapiante del topi e ratti bianchi. *Boll. Soc. Ital. Boil. Sper.* **4,** 470.

Tchakhotine, S. (1929). Studi sulla leucocitosi nel cancro del topi. I.Sulla natura della reazione leucocitaria. *Boll. Soc. Ital. Biol. Sper.* **4,** 571.

Tchakhotine, S. (1935). Recherches physiologiques sur les Protozoaires, faites au myen de la micropuncture ultraviolette. *C. r. Soc. Biol.* **200,** 2217.

Tchakhotine, S. (1935). Floculation localisee des colloides dans la cellule par la micropuncture ultraviolette. *C. r. Soc. Biol.* **120,** 2036–2038.

Tchakhotine, S. (1935). Suppression de la fonction du coeur de la Daphnie par la micropuncture ultraviolette. *C. r. Soc. Biol.* **119,** 1392.

Tchakhotine, S. (1935). L'effet d'arret la fonction de la vacuole pulsatile de la Paramecie par micropuncture ultraviolette. *C. r. Soc. Biol.* **120,** 782–784.

Tchakhotine, S. (1935). Sur le mecanisme de la reaction de la couche superficielle de l'oeuf de la Pholade ala micropuncture ultraviolette. *C. r. Soc. Biol.* **119,** 1394.

Tchakhotine, S. (1935). La partenogenes artificelle de l'oeuf de la Pholade par micropuncture ultraviolette. *C. r. Soc. Biol.* **119,** 1394.

Tchakhotine, S. (1935). La micro-instrumentation pour les recherches de Cytologie experimentale. *Bull. Soc. Franc. Microsc.* **4,** 138–153.

Tchakhotine, S. (1936). Recherches de Cytologie experimentale dans leurs rapports avec le probleme du Cancer. *Le Lutte Contre le Cancer.*

Tchakhotine, S. (1936). La fonction du stigma chez le Flagelle Euglena, etudiee au moyen de la micropuncture ultraviolette. *C. r. Soc. Biol.* **121,** 1165.

Tchakhotine, S. (1936). Les Protozoaires, objets d'experiances en Cytologie experimentale (Recherches faites avec la micropuncture ultraviolette). *Ann. Protisol.* **5,** 1.

Tchakhotine, S. (1936). Sur les reactions du cytoplasme et du vacuome a la micropuncture ultraviolette chez Ascoidea rubescens en fonction du pH du millieu de la permeabilite. *C. r. Soc. Biol.* **121,** 1525.

Tchakhotine, S. (1936). Sur les variations de l'equilibre entre cytoplasme et vacuome chez Ascoidea rubescens, par irradiation au moyen de la micropuncture ultraviolette. *C. r. Soc. Biol.* **121,** 952.

Tchakhotine, S. (1936). Irradiation localisee du myoneme du pedoncule des Vorticelles par micropuncture ultraviolette. *C. r. Soc. Biol.* **121,** 1114.

Tchakhotine, S. (1938). Experiences de micropuncture ultraviolette sur les blastomeres de l'oeuf de Pholades (Ph. Candida et Ph. Crispata). *C. r. Soc. Biol.* **127.**

Tchakhotine, S. (1938). Cancerisation experimentale des elements embryonnaires, obtenue sur des larves d'Oursins. *C. r. Soc. Biol.* **127,** 1195.

Tchakhotine, S. (1938). Parthenogenese experimentale de l′oeuf de la Pholade par micropuncture ultraviolette, aboutissant a une larve vivante. *C. r. Soc. Biol.* **127,** 377.

Tchakhotine, S. (1938). Une technique simple d'entretien des souris au taboratoir. *C. r. Soc. Biol.* **127,** 226.

Tchakhotine, S. (1938). Reactions "conditionnees" par micropuncture ultraviolette dans le comportement d'une cellule isolee (*Paramecium caudatum*). *Arch. Inst. Prophylactique. Paris* **10,** 119.

Tchakhotine, S. (1938). Etudes physiologiques et embryologiques au moyen de la methode des microgouttes. *Trav. Station Zool. Wimereux* **13,** 647–663.

Tchakhotine, S. (1938). Ultramicropuncture ultraviolette, moyen nouveau d'etude de phenomenes colloidaux dans la cellule vivante. *C. r. Soc. Biol.* **127,** 133.

Tchakhotine, S. (1938). Cancer au 3–4-benzopyrene et reaction leucocytairs chez la Souris. *C. r. Soc. Biol.* **127,** 606.

Tchakhotine, S. (1938). Leucocytose et cancers spontanes, greffs et provoques chimiquement par le 3-4-benzopyrene. *C. r. Soc. Biol.* **127,** 606.

Tchakhotine, S. (1938). Heredite du taux leucocytaire du sang chez la Souris. *C. r. Soc. Biol.* **127,** 533.

Tchakhotine, S. (1955). Problemi e metodi di Citologia sperimentale. *Boll. Soc. Ital. Biol. Sperimentale* **31,** 661.

Tchakhotine, S. (1956). Organisation rationnelle de la Recherche Scientifique Paris.

Tschachotin, S. (1933). "Dreipfeil gegen Hakendreuz." Copenhagen, Denmark.

Tschachotin, S. (1939). "Le viol des foules." Gallimard, Paris.

Tschachotin, S. (1940). "The Rape of the Masses: The Psychology of Totalitarian Political Propaganda." George Routledge and Sons, London (Transl from Fr.).

Tschachotin, S. (1946). "Le viol physique des masses—obstacle a une vraie démocratie (brochure)," edn. S.A.L., Paris.

Tschachotin, S. (1952). "Le viol des foules par la propaganda politique," 1st edn. Gallimard, Paris.

Vogt, A. (2007a). Introduction and 21 biographies. In: In memoriam of the displaced scientists from the Kaiser Wilhelm Institute for Medical Research. Heidelberg.

Vogt, A. (2007b). Sergej Tschachotin an Albert Einstein im Dezember 1933 – ein Zeitdokument. (Sergej Tschachotin to Albert Einstein, a letter written in December 1993). In: Dahlemer Archivgespräche, Vol. 12, Berlin (in press).

APPENDIX 1

Practical Aspects of Working with Laser Microbeams and Optical Tweezers: Data and Equations

Karl Otto Greulich

The appendix gives a collection of useful equations and data for construction and use of laser microbeams/scissors and laser/optical tweezers. A number of data and calculations are reproduced from Greulich (1).

A1. The Number of Light Quanta Contained in a Pulse of a Given Energy

For a number of applications, it is useful to have a relationship between the energy of a light quantum and macroscopic energies. For example, it may be interesting, at which power densities photons can start to cooperate and to generate nonresonant two-photon effects. For a general understanding of the interaction of light with matter, such a calculation makes clear that the number of photons involved in an experiment is extremely high, even in attenuated light.

There is no upper limit for the amount of energy, which a beam of light can exchange with matter. However, it was one of the big milestones of science when it was realized that there is a minimum portion (quantum) of energy, which can be exchanged. It is related to the frequency of the light wave

$$E = h\nu$$

where h (the Planck's constant) is 6.6×10^{-27} g cm^2/s.

With $c = \lambda\nu$ one obtains

$$E = hc/\lambda$$

0091-679X/07 $35.00
DOI: 10.1016/S0091-679X(06)82028-5

The product hc is

$$hc = (6.6 \times 10^{-27}\ \text{g cm}^2/\text{s})(3 \times 10^{10}\ \text{cm/s})$$

$$hc = 19.8 \times 10^{-17}\ \text{g cm}^3/\text{s}^2$$

For a quantum of green light (500 nm) the energy can be calculated as

$$E = \frac{19.8 \times 10^{-17}\ \text{g cm}^3/\text{s}^2}{500 \times 10^{-7}\ \text{cm}}$$

$$E = 3.96 \times 10^{-12}\ \text{g cm}^2/\text{s}^2$$

Since

$$10^7\ \text{g cm}^2/\text{s} = 1\ \text{Nm} = 1\ \text{J}$$

the energy of a green light quantum is

$$E = 3.96 \times 10^{-19}\ \text{J}$$

The reciprocal of this gives the number of quanta required to add up to 1 J:

$$2.5 \times 10^{18}\ \text{green light quanta per joule}$$

that is, 2.5 billion billions light quanta per joule. In order to calculate how many quanta of other wavelengths (e.g., 337 nm) are needed to make 1 J, one has just to multiply it with the wavelength ratio (e.g., 337/500). Thus, 1 J corresponds to 1.7×10^{18} quanta of UV light at 337-nm wavelength.

A2. Penetration Depth of Light into Tissue for the Working Wavelengths of Different Lasers

Quite often, only the power density, that is, the power per cm^2 is given. In order to get an impression on how the light of a microbeam acts on a whole volume of material, one needs to know, how deep a laser of a given wavelength penetrates into biological material. The following table gives the depth after which light of an intensity I_0 is attenuated to $I_0/2.71$:

Wavelength (nm) and laser	Penetration depth (μm)
193 (ArF excimer laser)	1
308 (XeCl excimer laser)	50
337 (Nitrogen)	60
550 (Visible laser)	**
1064 (Nd:YAG)	1000
2120 (Ho:YAG)	420
2940 (Er:YAG)	1
10,600 (CO_2)	12

**The penetration depth at visible wavelengths (400–800 nm) depends highly on the specific properties of the material. The values in the infrared are just the inverse absorption coefficients of water. In the ultraviolet, data of Welch *et al.* (2) have been used. See also Furzikov (3).

A3. Heat and Power Density: Use of the Stefan Boltzmann Law

The Stefan Boltzmann law allows to calculate the power density which is emitted by a thermal source. When the power density is known, the temperature can be calculated. For example, the radiation density at sunshine on the Earth's surface can be measured. It is 1345 W/m^2. The Sun's surface is 0.696 million km from the center of the Sun, the Earth's orbit around the Sun at 149.6 million km, that is, at 214 Sun radii. Thus, the power density at the Sun's surface is $214^2 = 45{,}796$ times higher, that is,

$$I = 45{,}796 \times 1345\ \mathrm{W/m^2} = 6.16 \times 10^7\ \mathrm{W/m^2}$$

Using the Stefan Boltzmann law

$$I = \sigma T^4$$

with $\sigma = 5.67 \times 10^{-8}\ \mathrm{W/(m^2K^4)}$ the temperature at the Sun's surface can be calculated as

$$T = \sqrt[4]{\frac{I}{\sigma}} = \sqrt[4]{\frac{6.16 \times 10^7}{5.67 \times 10^{-8}}}$$

$$= \sqrt[4]{1090 \times 10^{12}} = 5.74 \times 10^3\ \mathrm{K} = 5746^\circ\mathrm{C}$$

Correspondingly, the heat generated by complete absorption of a given laser power density can be calculated. For example, a laser pulse of 1-ns duration with an energy of 1 μJ focused to 1 μm (10^{-8} cm^2) corresponds to 10^{11} or 10^{17} W/m^2. Thus,

$$T = \sqrt[4]{\frac{10^{17}}{5.67 \times 10^{-8}}} = \sqrt[4]{18 \times 10^{24}} \approx 2\ \mathrm{Mio\ K}$$

Certainly this is only an upper limit, since even in strong absorbers only a part of the power is absorbed exactly at the surface and heat flow starts immediately. Also, such high temperatures have to be carefully defined. But even an error of a factor 10,000 in the really absorbed energy due to only partial absorption in transparent biological material will result only in a factor of 10 in temperature. In transparent biological material the true value will be considerably smaller.

A4. Time Course of the Dissipation of Heat Generated by a Laser Microbeam

Rationale

As was shown in Section A3, very high temperatures are generated where the laser microbeam hits the target directly. There, any material is ablated. However, for the biologists it is also of interest what happens in the environment of the hit region. The following calculation shows that the heat is dissipated within very short time. Secondary damage in the environment is surprisingly low.

Assumption

An energy E is deposited in a spherical droplet of water (or aqueous solution) with radius r, that is, into a volume

$$V = \frac{4}{3}\pi r^3 = 4.19r^3.$$

When this droplet is heated up to very high temperature by putting an energy E into the volume V, the energy density is

$$D = \frac{E}{4.19r^3}$$

Since a very sharp temperature gradient prevails, the hot spot will expand rapidly with at least the speed of sound, v. The spot radius $r(t)$ at a given time t will be

$$r(t) = r + v \times t$$

The energy per volume, D is also a function of time

$$D(t) = \frac{E}{4.19}(r + v \times t)^3$$

When we define $F(t) = D/D(t)$

$$= \frac{(r + v \times t)^3}{r^3}$$

$$= 1 + \left(\frac{v \times t}{r}\right)^3$$

we can, by simple rearrangement of the last equation calculate the time, after which the energy density is diluted to a given fraction $F(t)$

$$t = \left(\frac{r}{v}\right)\sqrt[3]{F(t) - 1}$$

For example, the time at which the energy density is diluted by a factor of 1,000,000 is then approximately

$$t_{1,000,000} = 10\left(\frac{r}{v}\right)$$

In water, where the velocity of sound is 1 km/s or 1 μm/ns one can write (r/v) in micrometer and will obtain the time t in nanosecond. When r at the beginning of the experiment is 0.3 μm (focusing by a microscope) one obtains

$$t_{1,000,000} = 30 \text{ ns}$$

that is, after 30 ns a temperature of 1 Mio degrees is reduced to an excess temperature of 1°. In these 30 ns, a protein just realizes the environment is "hot" and starts to denature. Due to the short time, however, this process is still reversible.

When the beam diameter is 10 μm (focusing by a lens), the heat is diluted by a factor of 1,000,000 after

$$t_{1,000,000} = 1\ \mu s$$

This time is sufficient for protein denaturation in the environment. This effect occurs in medical laser surgery.

A5. Two- and Three-Photon Absorption

The absorption of light by a single atom or molecule can be described by

$$k = \alpha_1 \times I$$

where k is the rate constant for absorption in photons per second, I is the light intensity in photons per cm^2 and second, and α_1 the molecular absorption cross section in cm^2. The latter is directly correlated to the molar absorption coefficient in the Lambert–Beer law for a macroscopic description of absorption [see Greulich (1), p. 288].

At sufficiently high intensities, an atom or molecule can absorb two, three, or more light quanta simultaneously. The rate constant is then

$$k = \alpha_2 \times I^2$$
$$k = \alpha_3 \times I^3$$

For four-, five-, ... photon processes the equations have to be modified correspondingly. However, when more than a few light quanta cooperate, a macroscopic treatment is adequate and more convenient. The table below gives, for a few selected dye molecules, α_1, α_2, and α_3 at different wavelengths. By using such cross sections one can see that, in order to get a similar number of absorptions per molecule as in one-photon absorption, a 10^4- to 10^5-fold higher intensity is required for the two- and three-photon effect than for one-photon effect.

Molecular Absorption Cross Sections for Dye Molecules

Cross section/		10^{-16} cm^2	10^{-49} cm^4 s	10^{-83} cm^6 s^2
Wavelength		335–362 nm	700 nm	1000 nm
DAPI	345 nm	1.3	0.016	0.25
Dansyl	336 nm	0.17	0.1	0.3
Fura + Ca	335 nm	1.2	1.2	30
Fura	362 nm	1.0	1.1	20
Indo + Ca	340 nm	1.3	0.15	6
Indo-free	345 nm	1.3	0.35	2
APSS	at 800 nm–>		380	
ASPT	at 1064 nm–>		1200	

Data from: Xu *et al.* (5) and (APSS, ASPT) from Cheng *et al.* (1998).

A6. Electric Field Generated by a Laser of a Given Power Density

Light is not only a carrier of energy but also of electric field. Similar to the temperature, the electric field carried by light is related to the intensity I by

$$E^2 = \frac{2\,I}{\varepsilon_0\, c}$$

where $\varepsilon_0 = 8.85 \times 10^{-14}$ A s/V cm is the permittivity constant of the vacuum. An intensity of 1 W/cm^2 corresponds to an electric field of 28 V/cm. Such a field can be produced when a car battery would be connected to two conducting metal plates isolated from each other and separated by 0.5 cm. In turn, a focused laser with 10^{14} W/cm^2 generates an electric field of

$$28 \times 10^7 \text{ V/cm}$$

and this corresponds to the electric field in the electron cloud of many molecules.

A7. Pressure and Acceleration of a Microscopic Object by a 1-W Laser

The force which is exerted by a 1-W laser can be directly calculated

$$F = \frac{W}{c}$$

$$= \frac{1}{3 \times 10^8} \frac{\text{Nm/s}}{\text{m/s}}$$

$$= 3.3 \times 10^{-9}\text{ N} = 3.3\text{ nN}$$

When this radiation is focused, for example, onto a bacterium with a volume of 1 fl or a mass of 1 pg, it will accelerate it over a short distance.
Using 1 N = 10^5 g cm/s^2, one obtains:

$$\text{Acceleration} = \frac{3.3 \times 10^{-9}}{10^{-12}} \frac{\text{N}}{\text{g}} = \frac{3.3 \times 10^{-4}}{10^{-12}} \frac{\text{g cm/s}^2}{\text{g}}$$

$$\approx 3.3 \times 10^8 \text{ cm/s}^2$$

$$= 3.3 \times 10^6 \text{ m/s}^2$$

$$= 330{,}000 \text{ gravitational accelerations}$$

A8. Experimental (Axial and Transversal) Q Values for Different Sphere Sizes and Materials and for Different Focusing

With light pressure as described above, one can balance microscopic objects against gravity, similarly as a ping pong ball can be balanced on a fountain of water. Such an interplay of forces is not exactly what one needs to fix microscopic objects in three dimensions. An additional effect makes life even simpler: A laser beam with higher intensity at the optical axis than in the periphery drives a sphere with suitable properties toward the optical axis. When focused with high aperture, light can pull a microscopic object into the focus even against the action of light pressure due to the action of gradient forces. A particle has to be dielectric in order to feel gradient forces. Metal spheres are not dielectric but plastic spheres and biological cells are! In terms of optics, to be dielectric means not to reflect but to refract light. Therefore gradient forces are, for example, dependent on the refractive indices of the object and of the medium. Additionally, they are dependent on the beam quality of the focused light. Theoretical calculation of such forces is difficult. Thus, one usually quantifies gradient forces semiempirically. A quality factor Q is introduced. The force equation of Section A7 becomes

$$F = Q\left(\frac{W}{c}\right)$$

and

$$P = Q\left(\frac{I}{c}\right)$$

There are several detailed studies available where the value of Q is determined for different experimental situations [Wright *et al.* (4)]. Gradient forces can act axially, that is, against the direction of light propagation and perpendicular to it. Both forces can cooperate to pull an object against light pressure into the focus. For 1-μm silica microbeads, the axial value for Q is of the order of 0.05 (or 5%), the transversal value is approximately 15% of the value for pure light pressure, for which Q is per definition $= 1$ (corresponding to 100%). The values for 20-μm polystyrene microbeads are 0.1 and up to 0.4, that is, 10% and up to 40% of the light pressure value.

The table below summarizes the major results for different sphere sizes, numerical apertures of the microscope objectives, and different polarizations of the lasers. For transversal forces, one has to consider that laser light is polarized. As one can see for the transversal effects from the left part of the table, it makes a difference if the polarization is also transversal or if it is perpendicular to it.

Percentage (100Q) for Different Experiments and Theoretical Predictions (in Brackets)

Type of force	Axial	Axial	Transversal	Transversal	Transversal
Numerical aperture	1.25–1.3	0.8	1.3	1.3	0.8
Polarization			Parallel	Perpendicular	Perpendicular
1-μm silica	0.4–0.6%		15% (16%)	13% (15%)	
5-μm polystyrene	7%		21% (36%)	19% (29%)	
10-μm polystyrene	2.8–8.4%	1.3–1.4%	37% (36%)	29% (29%)	
20-μm polystyrene	10%		41% (36%)	32% (29%)	28% (13%)

Note: Low values, far from cover glass; high values, close to cover glass; in brackets, theoretical values predicted by electromagnetic theory for the 1-mm sphere and by ray optics for the larger ones.

A9. Typical Forces Useful for Optical Trapping

1 dyn = 10^{-5} Newton; pN = piconewton = 10^{-12} Newton; kN = kilonewton.

Classical chemical bonds	
Hydrogen bond	300 pN
Carbon–carbon bond	30,000 pN
Bonds in avidin and streptavidin	
Streptavidin–biotin bond	250 pN
Avidin and biotin bond	160 pN
Avidin and iminobiotin bond	85 pN
Protein–protein interaction	
With a dissociation constant of 10^{-5}	11–22 pN
With dissociation constant of 10^{-7}	32–1400 pN
With dissociation constant of 10^{-9}	74–3300 pN
Antigen–antibody interaction	244 $\pm$ 20 pN
Forces in DNA	
Force to break DNA	480 pN
Elastic structural transition of DNA	65–70 pN
Straightening DNA kinks	6 pN
Maximum force exerted by the mitotic spindle	700 pN
Motility force of a sperm cell	40 pN
Traction force of a locomoting cell	45,000 pN
Optical tweezers (1064 nm), 1 W in object plane	
Pure light pressure (at some distance from focus)	5000 pN
Against light pressure (close to focus)	150–500 pN
Perpendicular to beam propagation	1500 pN
Forces of molecular motors	
Myosin, average force during stroke	2.1 pN
Myosin, peak force during stroke	5.9 pN
Elasticity modules	
Myosin	100 kN/m^2
DNA	1.11×10^{15} kN/m^2

References

1. Greulich, K. O. (1999). (Monography) "Micromanipulation by Light in Biology and Medicine: The Laser Microbeam and Optical Tweezers." Birkhäuser, Basel, Wien, Boston.
2. Welch, A. J., *et al.* (1991). *Photochem. Photobiol.* **53,** 815–823.
3. Furzikov, N. P. (1987). *IEEE J. Quantum Electron* **23**(10), 1751–1755.
4. Wright, W. H., *et al.* (1994). *Appl. Opt.* **33,** 1735–1748.
5. Xu, C., Zipfel, W., Shear, J. B., Williams, R. M., and Webb, W. W. (1996). Multiphoton fluorescence excitation: New spectral windows for biological nonlinear microscopy. *Proc. Natl. Acad. Sci. USA* **93,** 10763–10768.
6. Cheng, P. C., *et al.* (1998). Highly efficient upconverters for multiphoton fluorescence microcopy. *J. Microsc.* **189**(3), 199–212.

INDEX

D

M

N

P

T

VOLUMES IN SERIES

Founding Series Editor
DAVID M. PRESCOTT

Volume 1 (1964)
Methods in Cell Physiology
Edited by David M. Prescott

Volume 2 (1966)
Methods in Cell Physiology
Edited by David M. Prescott

Volume 3 (1968)
Methods in Cell Physiology
Edited by David M. Prescott

Volume 4 (1970)
Methods in Cell Physiology
Edited by David M. Prescott

Volume 5 (1972)
Methods in Cell Physiology
Edited by David M. Prescott

Volume 6 (1973)
Methods in Cell Physiology
Edited by David M. Prescott

Volume 7 (1973)
Methods in Cell Biology
Edited by David M. Prescott

Volume 8 (1974)
Methods in Cell Biology
Edited by David M. Prescott

Volume 9 (1975)
Methods in Cell Biology
Edited by David M. Prescott

Volume 10 (1975)
Methods in Cell Biology
Edited by David M. Prescott

Volume 11 (1975)
Yeast Cells
Edited by David M. Prescott

Volume 12 (1975)
Yeast Cells
Edited by David M. Prescott

Volume 13 (1976)
Methods in Cell Biology
Edited by David M. Prescott

Volume 14 (1976)
Methods in Cell Biology
Edited by David M. Prescott

Volume 15 (1977)
Methods in Cell Biology
Edited by David M. Prescott

Volume 16 (1977)
Chromatin and Chromosomal Protein Research I
Edited by Gary Stein, Janet Stein, and Lewis J. Kleinsmith

Volume 17 (1978)
Chromatin and Chromosomal Protein Research II
Edited by Gary Stein, Janet Stein, and Lewis J. Kleinsmith

Volume 18 (1978)
Chromatin and Chromosomal Protein Research III
Edited by Gary Stein, Janet Stein, and Lewis J. Kleinsmith

Volume 19 (1978)
Chromatin and Chromosomal Protein Research IV
Edited by Gary Stein, Janet Stein, and Lewis J. Kleinsmith

Volume 20 (1978)
Methods in Cell Biology
Edited by David M. Prescott

Advisory Board Chairman
KEITH R. PORTER

Volume 21A (1980)
Normal Human Tissue and Cell Culture, Part A: Respiratory, Cardiovascular, and Integumentary Systems
Edited by Curtis C. Harris, Benjamin F. Trump, and Gary D. Stoner

Volume 21B (1980)
Normal Human Tissue and Cell Culture, Part B: Endocrine, Urogenital, and Gastrointestinal Systems
Edited by Curtis C. Harris, Benjamin F. Trump, and Gray D. Stoner

Volume 22 (1981)
Three-Dimensional Ultrastructure in Biology
Edited by James N. Turner

Volume 23 (1981)
Basic Mechanisms of Cellular Secretion
Edited by Arthur R. Hand and Constance Oliver

Volume 24 (1982)
The Cytoskeleton, Part A: Cytoskeletal Proteins, Isolation and Characterization
Edited by Leslie Wilson

Volume 25 (1982)
The Cytoskeleton, Part B: Biological Systems and *In Vitro* Models
Edited by Leslie Wilson

Volume 26 (1982)
Prenatal Diagnosis: Cell Biological Approaches
Edited by Samuel A. Latt and Gretchen J. Darlington

Series Editor
LESLIE WILSON

Volume 27 (1986)
Echinoderm Gametes and Embryos
Edited by Thomas E. Schroeder

Volume 28 (1987)
***Dictyostelium discoideum:* Molecular Approaches to Cell Biology**
Edited by James A. Spudich

Volume 29 (1989)
Fluorescence Microscopy of Living Cells in Culture, Part A: Fluorescent Analogs, Labeling Cells, and Basic Microscopy
Edited by Yu-Li Wang and D. Lansing Taylor

Volume 30 (1989)
Fluorescence Microscopy of Living Cells in Culture, Part B: Quantitative Fluorescence Microscopy—Imaging and Spectroscopy
Edited by D. Lansing Taylor and Yu-Li Wang

Volume 31 (1989)
Vesicular Transport, Part A
Edited by Alan M. Tartakoff

Volume 32 (1989)
Vesicular Transport, Part B
Edited by Alan M. Tartakoff

Volume 33 (1990)
Flow Cytometry
Edited by Zbigniew Darzynkiewicz and Harry A. Crissman

Volume 34 (1991)
Vectorial Transport of Proteins into and across Membranes
Edited by Alan M. Tartakoff

Selected from Volumes 31, 32, and 34 (1991)
Laboratory Methods for Vesicular and Vectorial Transport
Edited by Alan M. Tartakoff

Volume 35 (1991)
Functional Organization of the Nucleus: A Laboratory Guide
Edited by Barbara A. Hamkalo and Sarah C. R. Elgin

Volume 36 (1991)
***Xenopus laevis:* Practical Uses in Cell and Molecular Biology**
Edited by Brian K. Kay and H. Benjamin Peng

Series Editors
LESLIE WILSON AND PAUL MATSUDAIRA

Volume 37 (1993)
Antibodies in Cell Biology
Edited by David J. Asai

Volume 38 (1993)
Cell Biological Applications of Confocal Microscopy
Edited by Brian Matsumoto

Volume 39 (1993)
Motility Assays for Motor Proteins
Edited by Jonathan M. Scholey

Volume 40 (1994)
A Practical Guide to the Study of Calcium in Living Cells
Edited by Richard Nuccitelli

Volume 41 (1994)
Flow Cytometry, Second Edition, Part A
Edited by Zbigniew Darzynkiewicz, J. Paul Robinson, and Harry A. Crissman

Volume 42 (1994)
Flow Cytometry, Second Edition, Part B
Edited by Zbigniew Darzynkiewicz, J. Paul Robinson, and Harry A. Crissman

Volume 43 (1994)
Protein Expression in Animal Cells
Edited by Michael G. Roth

Volume 44 (1994)
***Drosophila melanogaster:* Practical Uses in Cell and Molecular Biology**
Edited by Lawrence S. B. Goldstein and Eric A. Fyrberg

Volume 45 (1994)
Microbes as Tools for Cell Biology
Edited by David G. Russell

Volume 46 (1995)
Cell Death
Edited by Lawrence M. Schwartz and Barbara A. Osborne

Volume 47 (1995)
Cilia and Flagella
Edited by William Dentler and George Witman

Volume 48 (1995)
***Caenorhabditis elegans:* Modern Biological Analysis of an Organism**
Edited by Henry F. Epstein and Diane C. Shakes

Volume 49 (1995)
Methods in Plant Cell Biology, Part A
Edited by David W. Galbraith, Hans J. Bohnert, and Don P. Bourque

Volume 73 (2003)
Cumulative Index

Volume 74 (2004)
Development of Sea Urchins, Ascidians, and Other Invertebrate Deuterostomes: Experimental Approaches
Edited by Charles A. Ettensohn, Gary M. Wessel, and Gregory A. Wray

Volume 75 (2004)
Cytometry, 4th Edition: New Developments
Edited by Zbigniew Darzynkiewicz, Mario Roederer, and Hans Tanke

Volume 76 (2004)
The Zebrafish: Cellular and Developmental Biology
Edited by H. William Detrich, III, Monte Westerfield, and Leonard I. Zon

Volume 77 (2004)
The Zebrafish: Genetics, Genomics, and Informatics
Edited by William H. Detrich, III, Monte Westerfield, and Leonard I. Zon

Volume 78 (2004)
Intermediate Filament Cytoskeleton
Edited by M. Bishr Omary and Pierre A. Coulombe

Volume 79 (2007)
Cellular Electron Microscopy
Edited by J. Richard McIntosh

Volume 80 (2007)
Mitochondria, 2nd Edition
Edited by Liza A. Pon and Eric A. Schon

Volume 81 (2007)
Digital Microscopy, 3rd Edition
Edited by Greenfield Sluder and David E. Wolf

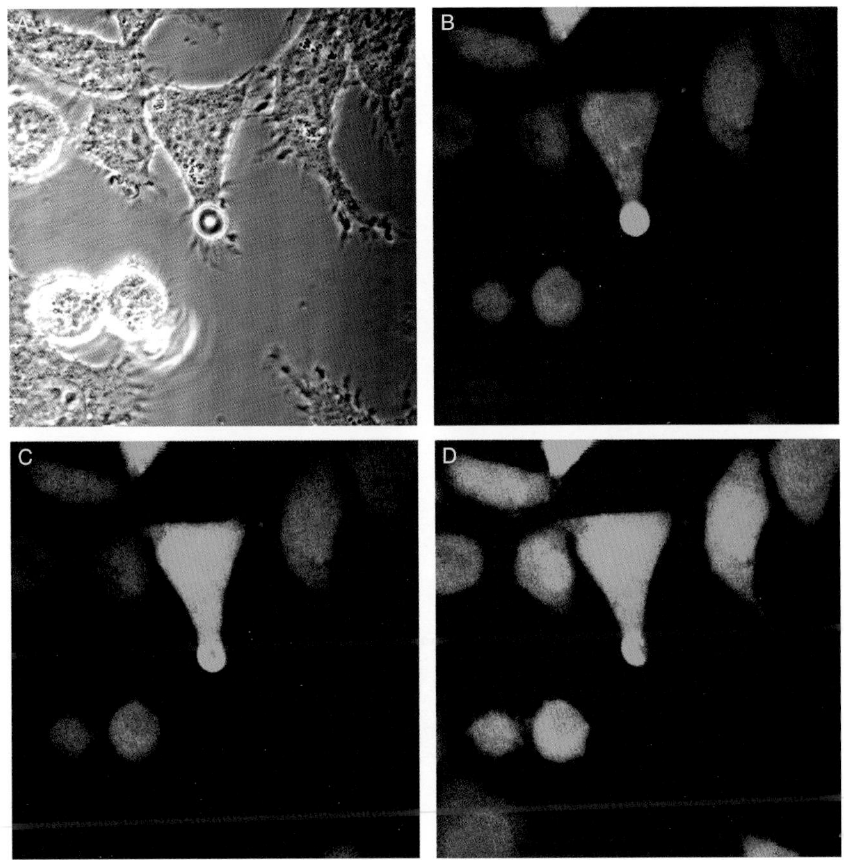

Plate 1 (Figure 2.10)

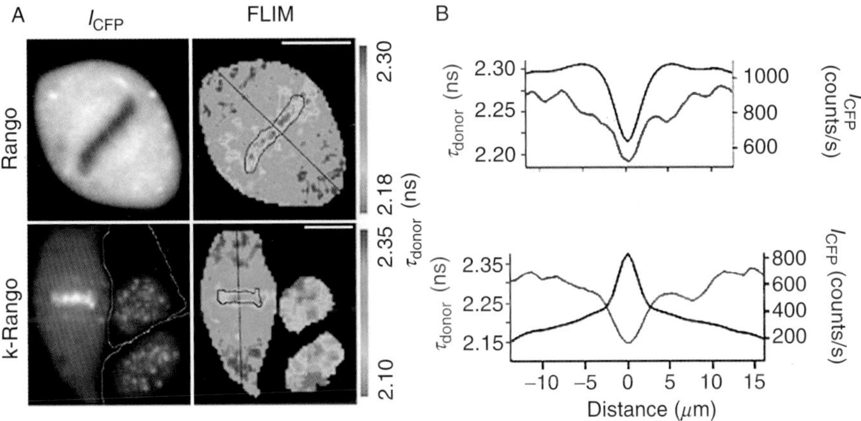

Plate 2 (Figure 3.5)

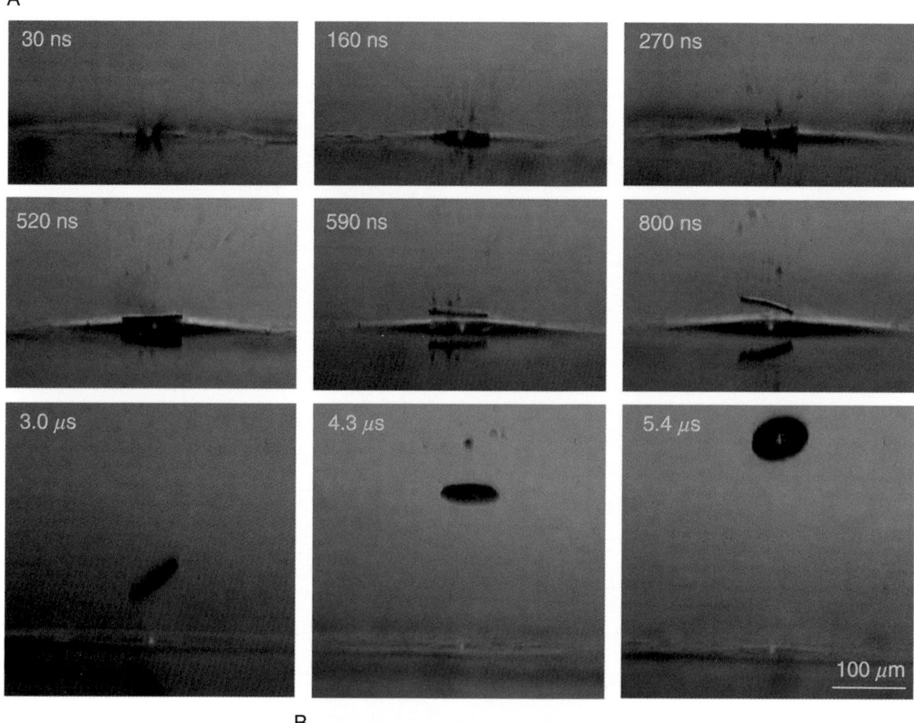

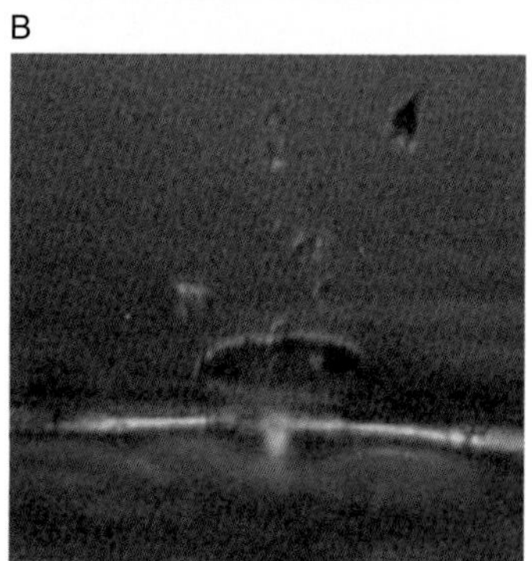

Plate 3 (Figure 5.6)

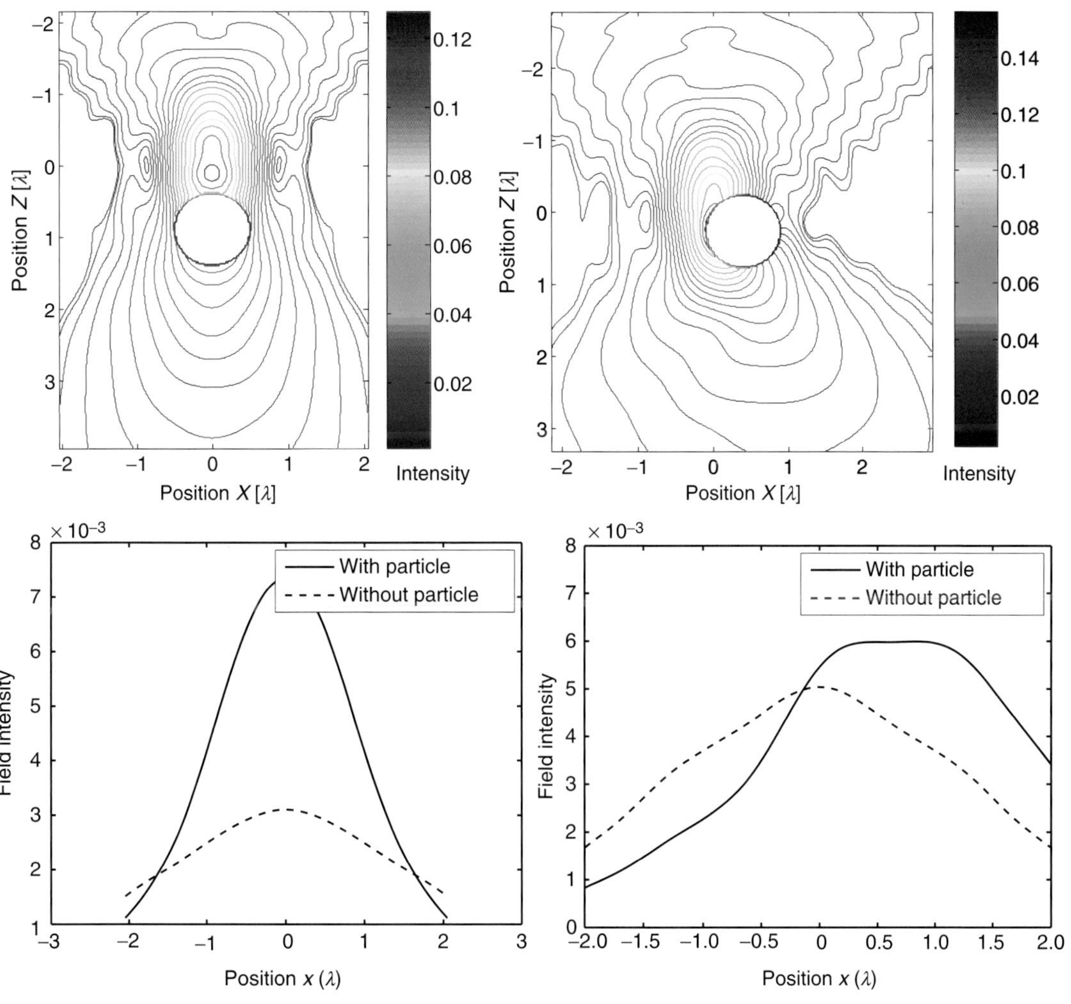

Plate 4 (Figure 6.4)

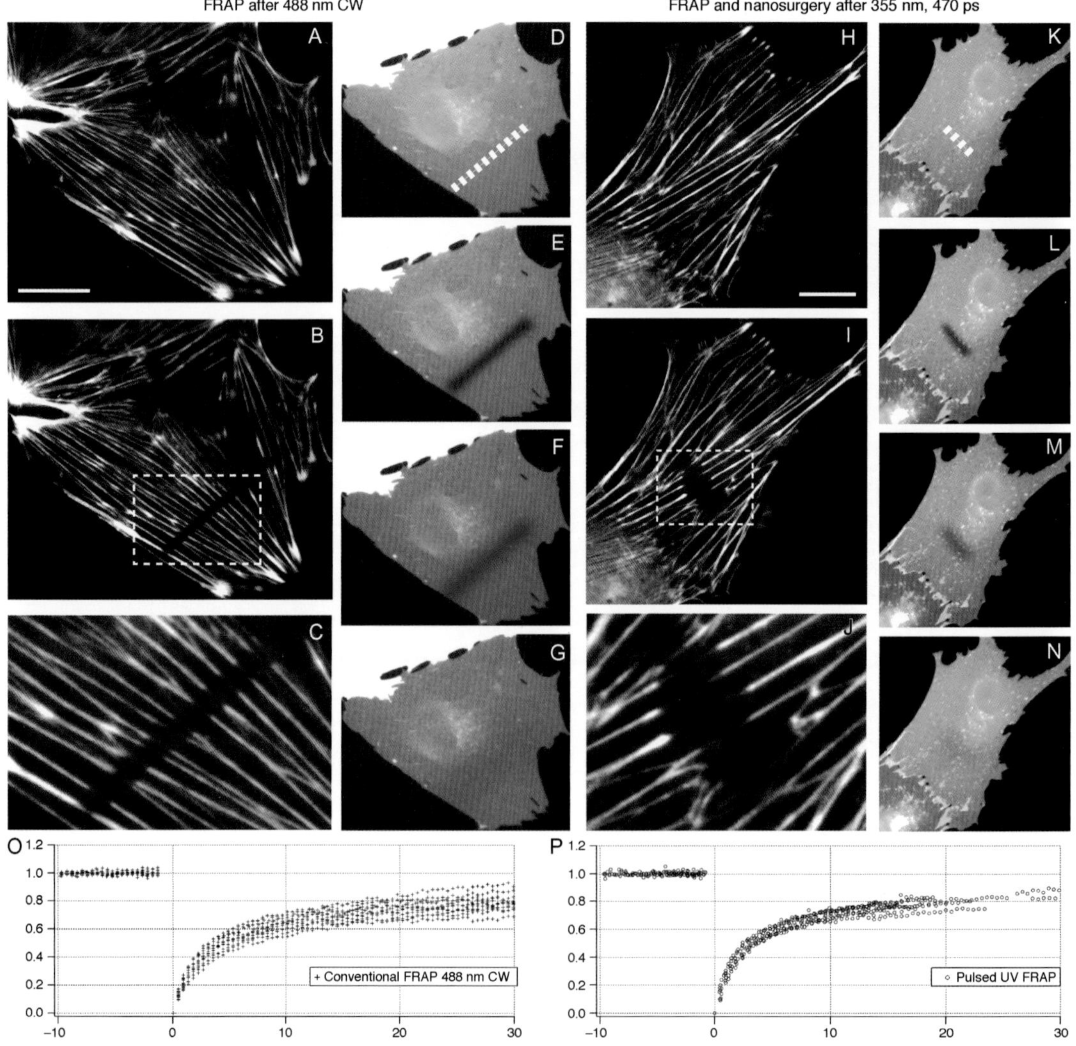

Plate 5 (Figure 8.3)

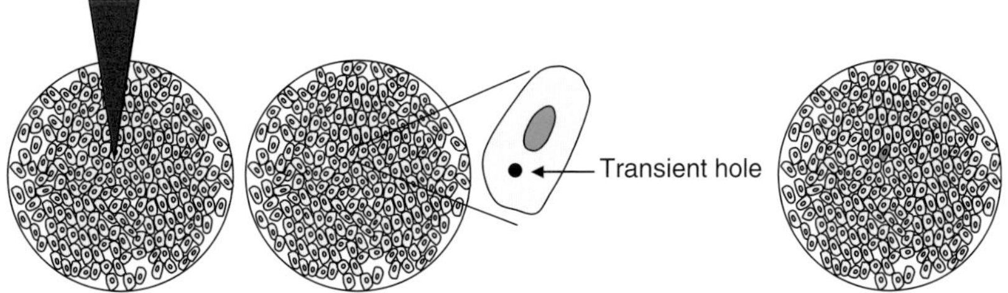

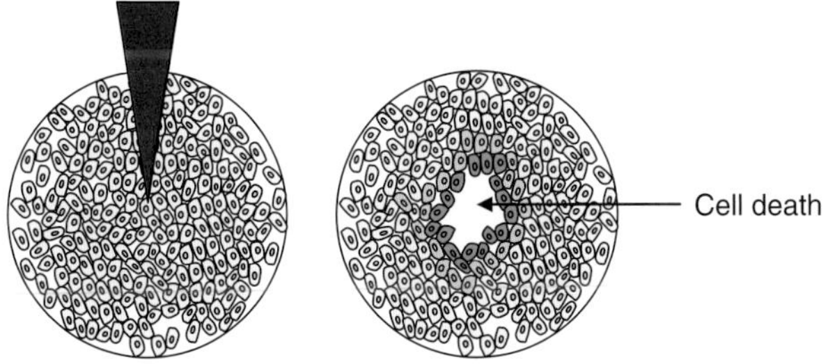

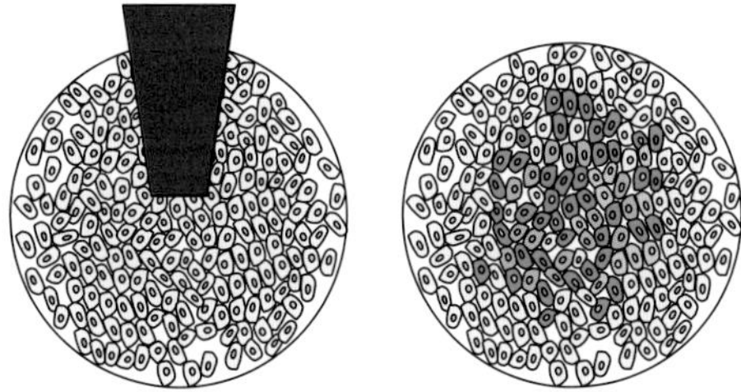

Plate 6 (Figure 10.1)

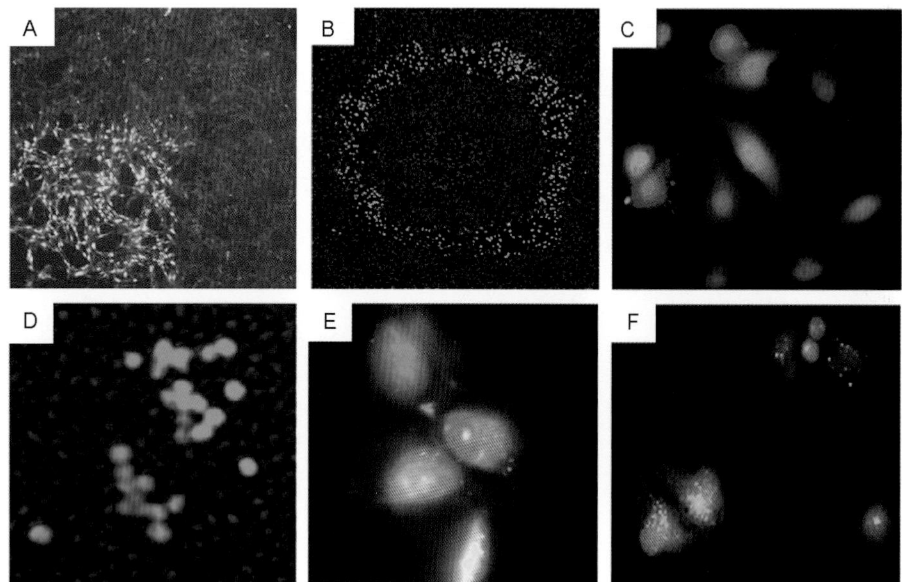

Plate 7 (Figure 10.7)

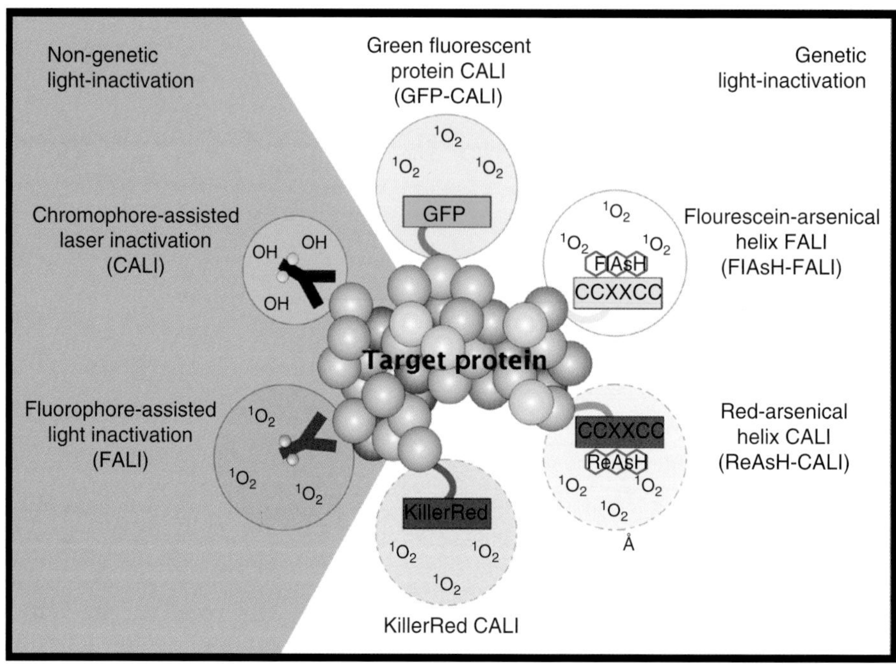

Plate 8 (Figure 11.3)

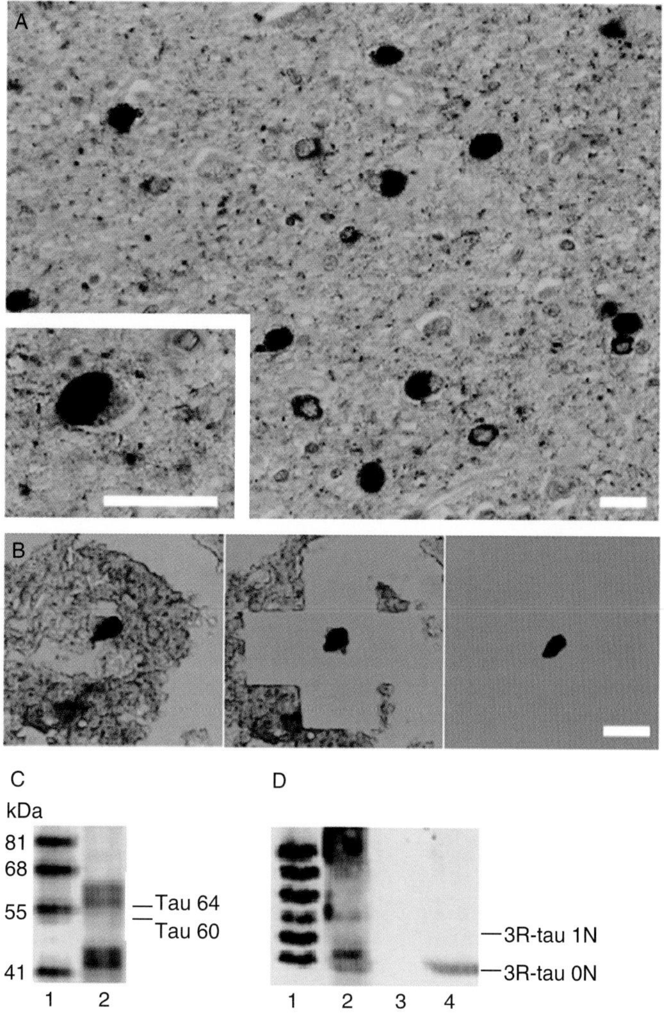

Plate 9 (Figure 12.2)

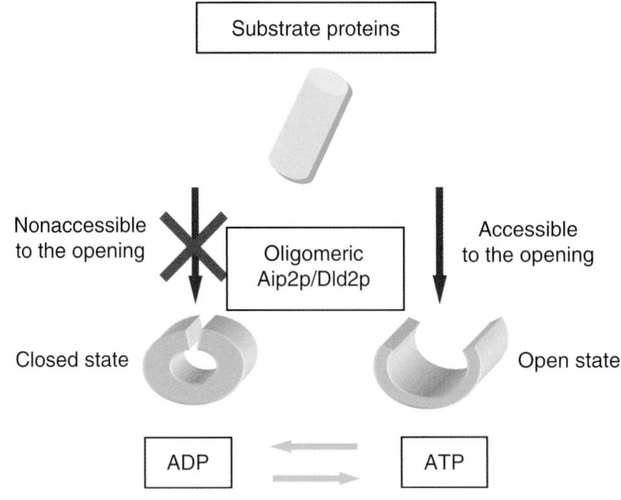

Plate 10 (Figure 12.4)

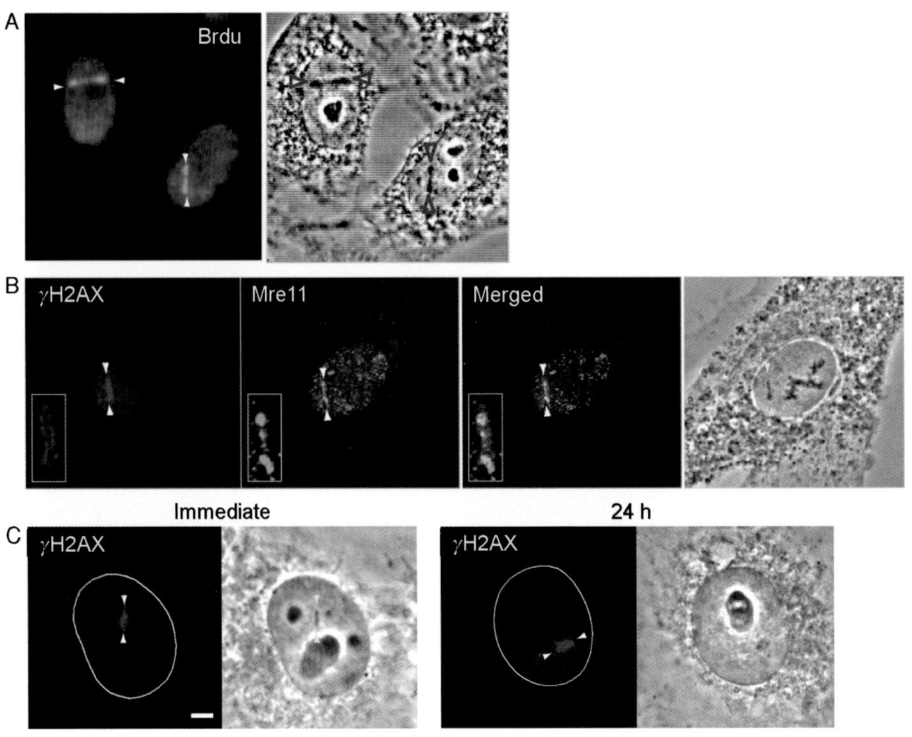

Plate 11 (Figure 13.5)

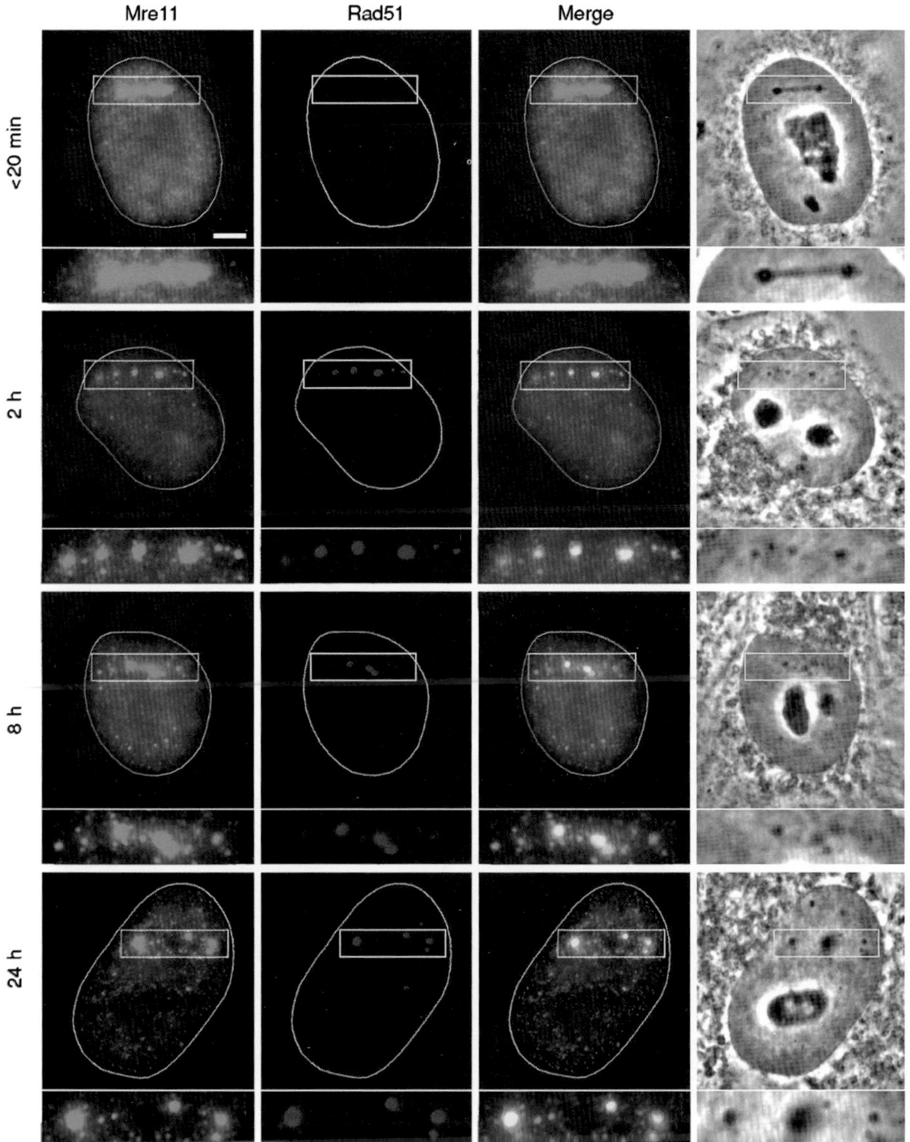

Plate 12 (Figure 13.9)

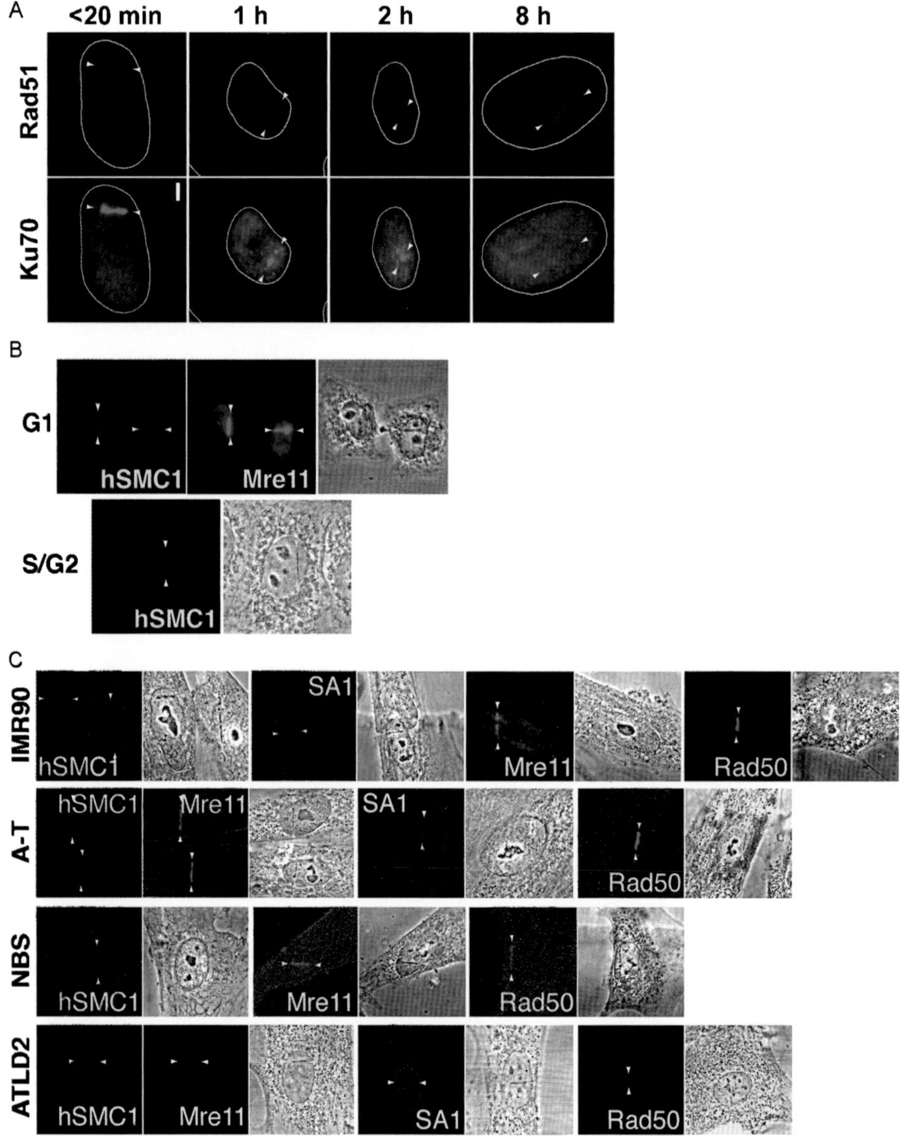

Plate 13 (Figure 13.11)

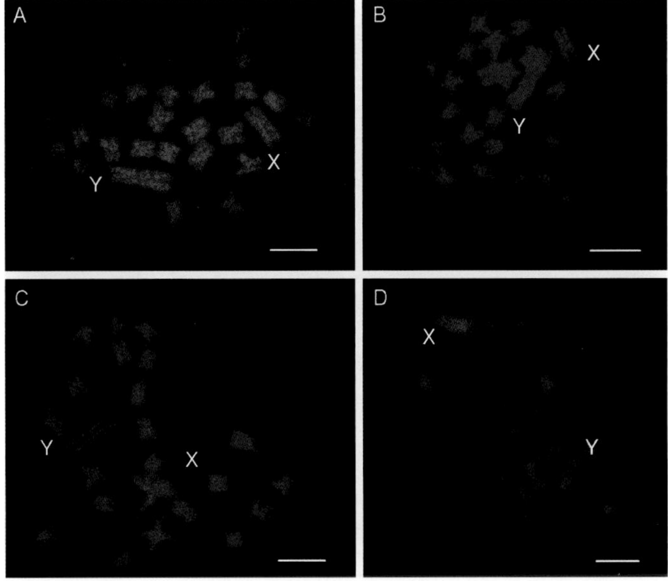

Plate 14 (Figure 15.4)

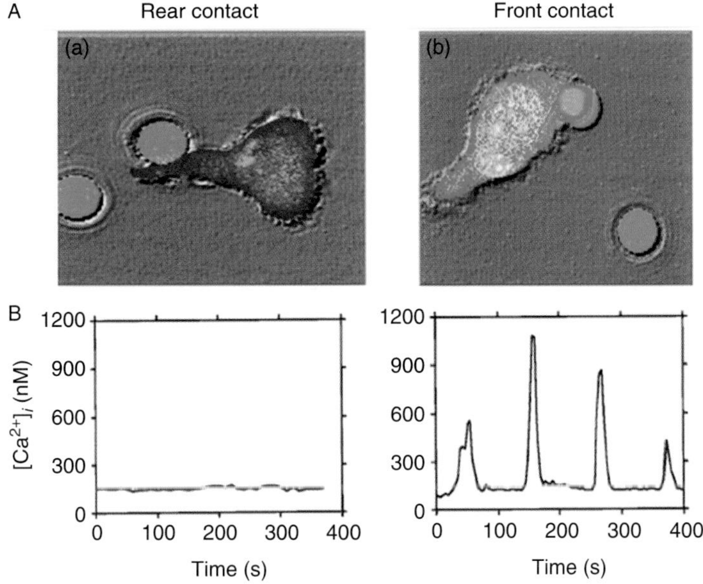

Plate 15 (Figure 16.1)

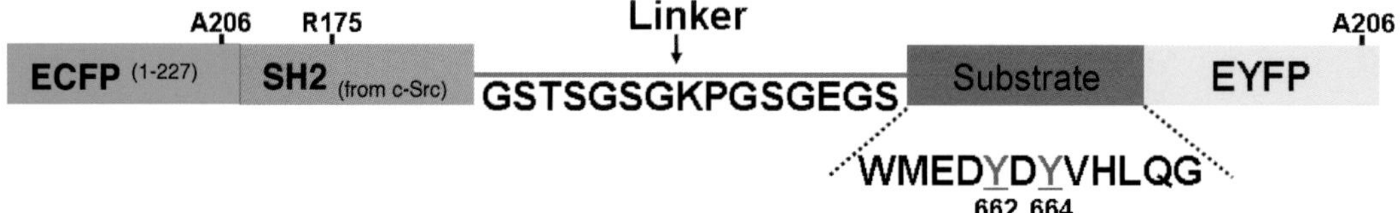

Plate 16 (Figure 18.1)

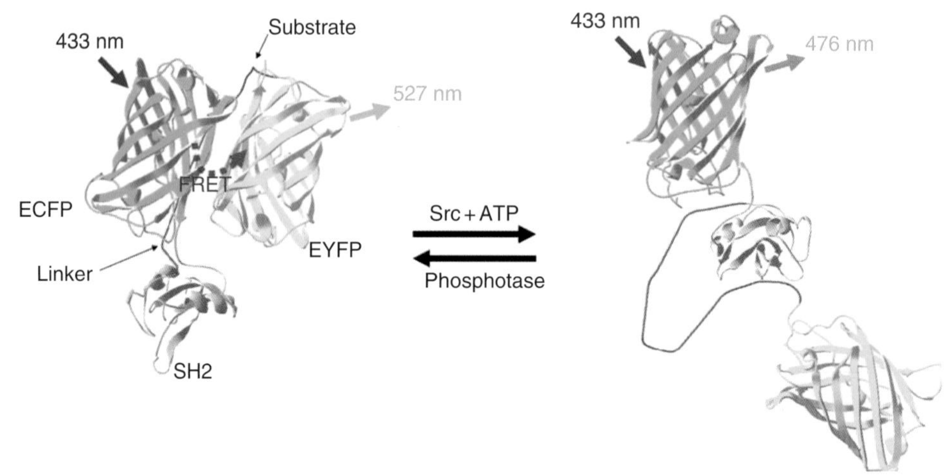

Plate 17 (Figure 18.2)

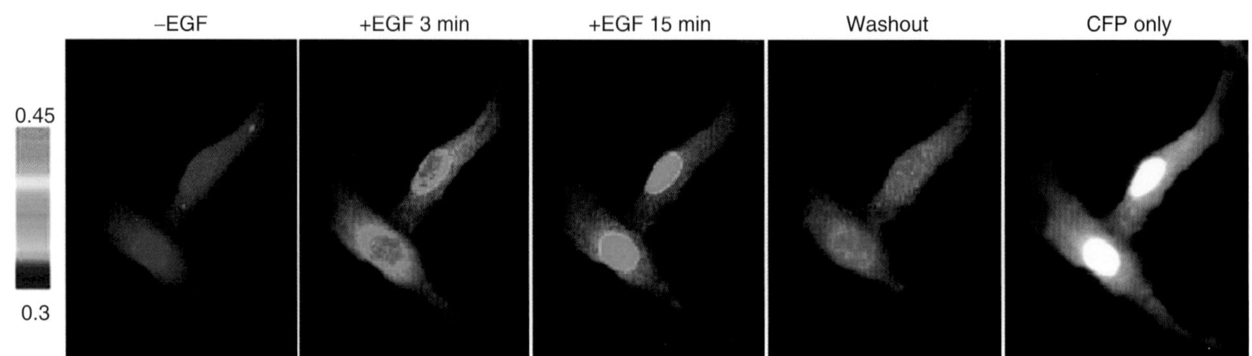

Plate 18 (Figure 18.3)

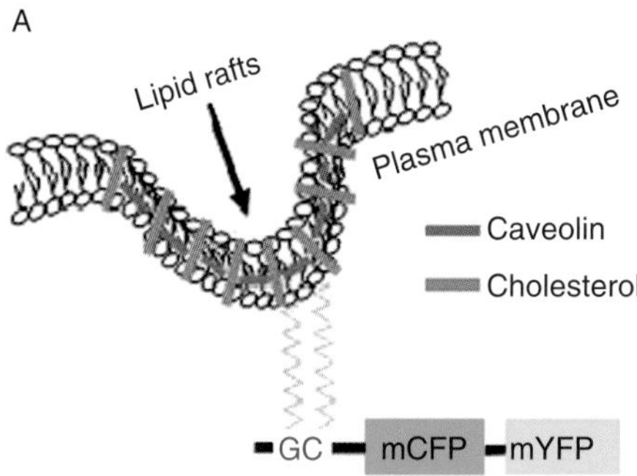

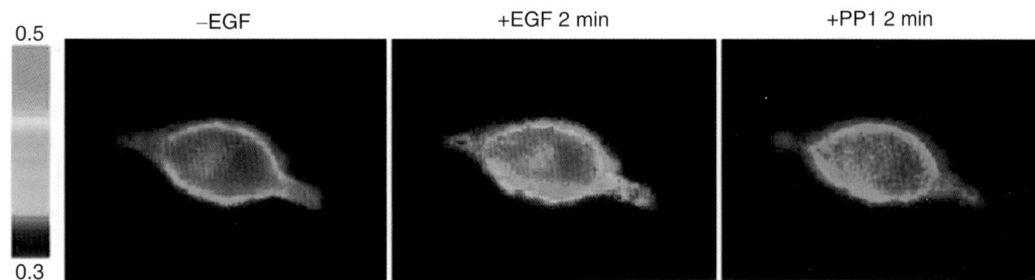

Plate 19 (Figure 18.4)

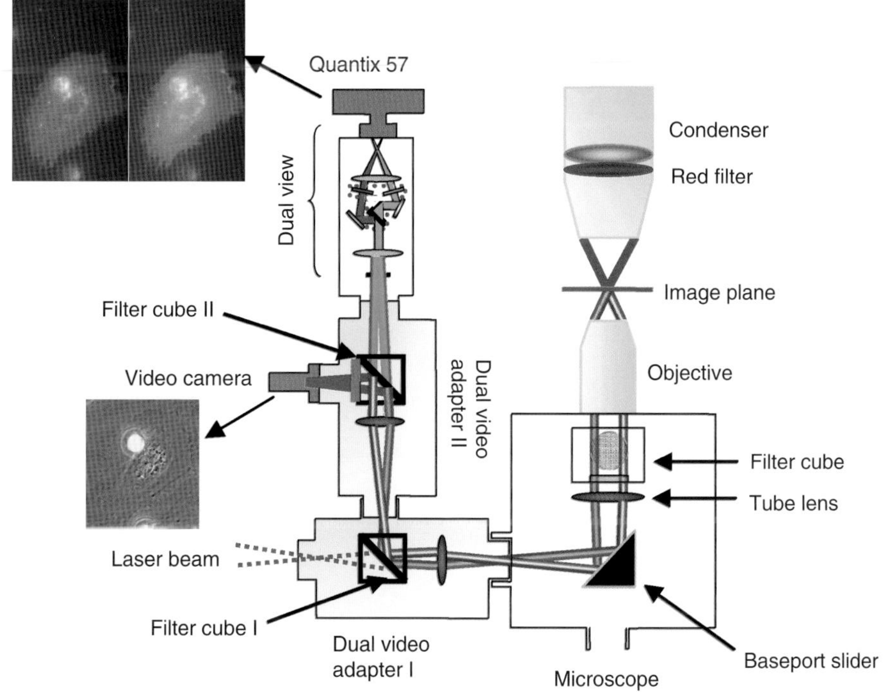

Plate 20 (Figure 18.5)

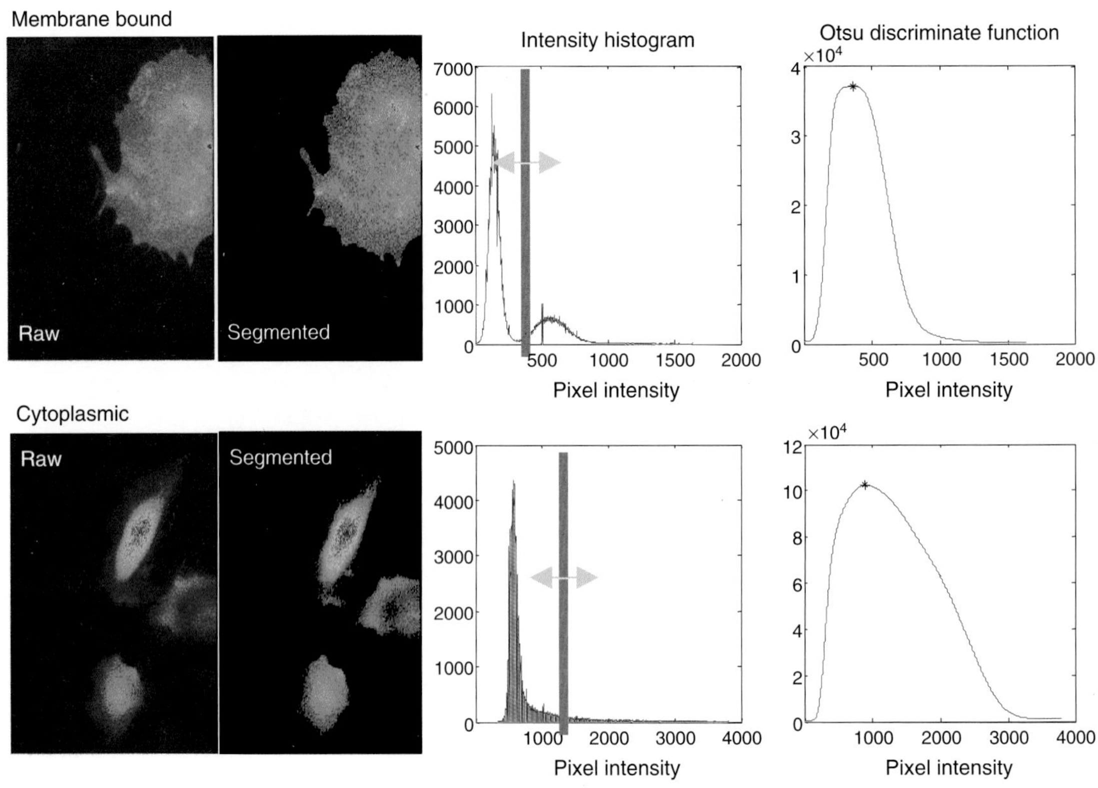

Plate 21 (Figure 18.9)